Radon And Thoron In The Human Environment

Proceedings of the 7th Tohwa University International Symposium

Radon And Thoron In The Human Environment

Fukuoka, Japan 23–25 October 1997

Editors

Akira Katase
Tohwa University, Japan

Michikuni Shimo
National Institute of Radiological Science, Japan

Organized and Sponsored by
Tohwa University and Tohwa Institute for Science

Collaborating Societies
Atomic Energy Society of Japan and The Health Physics Society of Japan

World Scientific
Singapore • New Jersey • London • Hong Kong

Published by

World Scientific Publishing Co. Pte. Ltd.
5 Toh Tuck Link, Singapore 596224
USA office: 27 Warren Street, Suite 401-402, Hackensack, NJ 07601
UK office: 57 Shelton Street, Covent Garden, London WC2H 9HE

British Library Cataloguing-in-Publication Data
A catalogue record for this book is available from the British Library.

RADON AND THORON IN THE HUMAN ENVIRONMENT
Proceedings of the 7th Tohwa University International Symposium

ISBN-13 978-981-02-3443-0
ISBN-10 981-02-3443-0

Preface

The International Symposium on Radon and Thoron In The Human Environment was held on 23 ~ 25 in October, 1997 at Fukuda Minerva Hall of Tohwa University, Fukuoka, Japan.

The plan of this symposium grew in the cooperative atmosphere brewed in a group of radon researchers in western Japan. These people of about 30 members were organized about ten years ago by Dr. S. Okabe on the financial support of Kansai Atomic Energy Association. They had the meetings for the discussion on the member's researches twice a year. These meetings were informal and very stimulating. They made frank and friendly discussions and excited with earnest and sincere debates under the leadership of Dr. Okabe. It was natural that members of the organizing committee for this symposium were selected from these people. Once the plan of this symposium was token at the committee meeting, many invited speakers were proposed. Finally 20 lecturers were decided for invited talks because of the time limit due to the period of the symposium.

After this symposium was announced, many people proposed to present papers. It was a very great pleasure for us to have so many presentations. It was difficult, however, to select the oral presentations from these papers within the limit of available hours. Therefore these papers were decided to be presented at the poster sessions. Considerable hours were arranged for these sessions to be used for full discussions.

The themes of the symposium cover many kinds of research field related to radon and thoron. We think this proceedings give readers the present status of the radon and thoron researches.

We wish to thank all the authors of the submitted papers for their interest in the symposium and for contributing to an exciting and fruitful exchange of ideas. We wish to thank the members of the organizing committee and deeply appreciate the programming committee for reviewing all the presented papers. Last, we would like to thank Dr. Toshinami Fukuda, the president of Tohwa University and Tohwa University for the financial support to this symposium. Thanks also go to the members of the electrical department of Tohwa University and Tohwa Institute for Science for their contributions for the operation of the symposium.

Akira Katase and Michikuni Shimo

Organizing Committee

President:	Toshinami Fukuda	President of Tohwa University
Chairman:	Akira Katase	Tohwa University
Steering:	Takao Iida	Nagoya University
	Kenji Ishibashi	Kyushu University
	Tsuguo Nishikawa	Fukui University
	Shigeru Okabe	Radon Science Research Laboratory
	Michikuni Shimo	National Institute of Radiological Science
	Tadashi Tsujimoto	Energy Research Center Wakasa Bay
Secretariat:	Masakuni Iwahashi	Tohwa University
	Kiyoshi Moriyama	Tohwa University

Programming Committee

Chairman:	Michikuni Shimo	National Institute of Radiological Science
Members:	Kenzo Fujimoto	National Institute of Radiological Science
	Kazunobu Fujitaka	National Institute of Radiological Science
	Kimiko Horiuchi	Otsuma Women's University
	Takao Iida	Nagoya University
	Akira Katase	Tohwa University
	Hiroshi Kojima	Science University of Tokyo
	Tsuguo Nishikawa	Fukui University
	Masanobu Sakanoue	Emeritus Professor of Kanazawa University

９７ラドンシンポジウム　市民生活におけるラドンとトロン
October 23～25, 1997　Fukuda Minerva Hall, Fukuoka, Japan

Table of Contents

A Measurements and calibration

Instruments and methods for measuring radon/thoron and their progeny: calibrations, intercomparisons, and particle size

Andreas C. George, Jason Z. Esposito, and Nancy A. Bredhoff

Radon Testing Corporation of America, Inc.
2 Hayes Street, Elmsford, NY 10523

Abstract

Indoor measurements of the concentrations of radon/thoron and their decay products were intensified worldwide in the last 13 years for the assessment of the radiation dose to workers and the general public. Measurements of particle size have been limited to a small group of universities and government agencies due to their complexity and high cost. A variety and plethora of instruments and methods have been developed using different principles of detection for different measurement applications. Overall, the instruments and methods used today depend on whether radon, thoron or radon and thoron decay products are being measured, the type of radiation being detected, the duration of the measurement, portability, convenience, reliability and cost considerations. The instruments of interest to most of us, use the following detection systems: pulse, current and electret ionization chambers, scintillation detectors with zinc sulfide, alpha particle spectrometers with silicon diodes, surface barrier or diffused junction detectors, registration of nuclear tracks in solid state materials, and gamma ray spectrometry and alpha and beta liquid scintillation spectrometry with NaI(Tl) scintillation crystals or germanium lithium (GeLi) semiconductors and liquid scintillation spectrometry systems. An overview of particle sizing instruments and methods will be presented. The advantages and disadvantages, field applicability and cost of the different instruments will be discussed so the researcher or tester can select the proper one for his or her use.

1. Introduction

The concern and awareness about the health risks from exposure to $^{222}Rn/^{220}Rn$ (radon/thoron) and their decay products was intensified by the discovery of elevated concentrations in several geographical areas of the world This awareness has been very pronounced in the U.S. where very high concentrations of radon were found in more than 10% of the tested areas. Since elevated radon levels were observed throughout the world, local, state, federal government agencies and private individuals initiated national radon surveys using the available short-term or long-term screening measurement techniques. Because of this increased demand, research and development of instrumentation for measuring radon/thoron and their decay products accelerated at a fast pace. Since 1984, more than 11 million measurements were made in the U.S alone, indicating that about 10% of the measurements had radon concentrations higher than 148 Bq m^3 or 4 pCi L^{-1}. About 80% of the residential buildings in the U.S. have not been tested for radon. After recognizing that radon is a significant indoor air pollutant and a radiation problem in the human environment, the European Community (EC) and the Nuclear Energy Agency (NEA) in 1983 established the International Intercalibration and Intercomparison Programme for Radon/Thoron and their Decay Product Measuring Equipment (IIIP). The purpose of the IIIP was to establish traceability and standardization of the methods and instrumentation used around the world. However, in 1989, the NEA stopped its support for the IIIP and the International Atomic Energy Agency (IAEA), established the Coordinated Research Programme (CRP) on Radon in the

Human Environment. In 1992, IAEA, IIIP, EC and the U.S. Environmental Agency (EPA), implemented the International Radon Metrology Programme (IRMP) with the purpose to provide the measurement quality assurance for the CRP. The IAEA, started a project on the assessment and control of radon exposure and between 1989-1994 awarded 14 research contracts, 37 research agreements in Africa, Asia, Australia, the Americas and Europe.

In Europe and in other countries national surveys have been conducted and some hot spots have been found in a few countries. The U.S. appears to be the country with the largest radon problem. Nevertheless, high awareness exists in real estate transactions where only about 16% of the homes are tested for radon. Although, radon decay products are responsible for about 40% of the radiation exposure from natural radiation sources [1] the parent radon supports and controls their airborne concentrations. For this reason, the majority of measurements are made for radon rather than for the decay products and assume an equilibrium factor of 40% between radon and its decay products. The advantages are: radon measurements are simpler more convenient and cost effective. However, there are situations such as in diagnostics, mitigation, research and in risk assessment activities where measurements of radon and thoron decay products are important. Some discussion of instruments and methods for their measurement will be included. The main emphasis of this paper is to describe briefly instruments that are portable and convenient to use to characterize indoor and outdoor environments for their radon/thoron and decay product concentrations, and particle size that provides important information for the assessment of the health risk.

2. Methodology, sampling, and analysis

To obtain complete information on the overall radon problem, several measurements and different instruments and methods are used to measure, identify and quantify radon and thoron sources, airborne and waterborne radon/thoron and their decay product concentrations. With particle size and respiratory deposition they provide the data needed to assess the health risk resulting from inhalation. Ventilation rate measurements are important because of their influence on the airborne concentrations and in the development of control measures necessary to remedy adverse conditions. The necessary measurements for addressing specific needs are:

Measurement of radon/thoron sources

Usually radon in soil may provide some indication how it will impact the indoor airborne concentration Measurement of soil permeability and radon emanation are important in the movement of radon and thoron inside a home. Also, the amount of radon and thoron entering the indoor environment depends on the fraction of diffusible radon, diffusion length and other transport mechanisms. Measurement of ^{226}Ra in building materials for some special situations and radon in water if the supply is an underground source can be very useful. Radon concentration in water >100,000 Bq m^{-3} can be an important contributor to the airborne radon concentration.

Measurement of airborne concentrations of radon/thoron and decay products

Because of the variation of radon/thoron and their decay product concentrations within a building and within a locality, the quality of air and the exposure of the building's occupants should be assessed by measuring the radioisotopic composition of the air they breathe. The instruments and methods used for this purpose will be described in more detail in another part of this paper.

Measurements leading to risk assessment

Although, measurements of the airborne concentrations of radon/thoron and their decay products are important, the amount and site of the deposited radioactivity in the respiratory system are very important. The fractional deposition of radon/thoron decay products depends on particle size, a parameter that is difficult to measure because it requires complex and expensive equipment. However, some facilities mostly universities and government agencies perform size measurements in a limited fashion using equipment such as diffusion batteries and low pressure impactors. Besides particle size measurements some assumptions of the human anatomy and breathing characteristics are made to arrive at the dose to the respiratory tract.

Ventilation rate measurements

The air exchange in a building is very important measurement because of its influence on the airborne concentrations of radon/thoron and their decay products. Although, ventilation measurements are not routinely performed in radon surveys, they are very useful in diagnostics and during mitigation work. There are several techniques for measuring the air exchange and refer the reader to other sources for more information.

Many factors and different protocols must be considered before the appropriate instrument and method is selected to collect the relevant data for the assessment of the airborne radon/thoron and decay product concentrations levels. In single family homes, measurements are made in any room on the first or second floors and in the basement that is often the source of radon and thoron. In multiple dwellings or apartment buildings usually measurements are made in the basement and possibly the first floor only. The choice of an instrument is dependent on availability, funding, trained personnel and on the purpose of the measurement. The three categories of sampling and analysis are:

Grab or prompt (short-term and long-term integrating), continuous grab or prompt sampling

Usually, the test duration is short ranging from instantaneous to 10 minutes providing a quick and accurate result of the concentration at the time of measurement. This type of sampling is used in diagnostic and mitigation activities and in screening of buildings for radon on a small scale searching for "Hot Buildings." Grab sampling is used to cross - check and calibrate other methods and techniques. Grab sampling is not suitable for the assessment of the average indoor radon concentration unless repeated at different times and seasons at considerable cost. Grab sampling is desirable in some real estate transactions if conducted at the beginning and at the termination of a 2-7 day short-term exposure. It requires a trained person to conduct the on-site test using portable equipment and doing the analysis in the field or in a laboratory setting. Instruments that fall into the grab sampling category are scintillation cells, activated carbon collectors and solid state detector monitors or radon and thoron sniffers. Grab sampling for radon/thoron decay products is usually performed with systems that incorporate a filter detector assembly.

Short-term and long-term integrating sampling

Many instruments and methods were developed in the past 15 years to address this need. They are the most popular monitoring devices in which the measurements are carried out over a period of 2-7 days (short-term) and from months to one year (long-term). They are popular because they are passive, very simple, inexpensive and mailable requiring no trained person for handling and for deployment for exposure. In the U.S., they are the instruments of choice for screening purposes. In most situations analysis is done in a laboratory setting. Short-term measuring monitoring devices include activated carbon collectors and electret ionization chambers (2-7 days) and electret ionization chambers and solid state nuclear track detectors for long-term exposure of 3 months to 1 year. There are some passive electronic monitors using solid state detectors. For measuring radon decay products the monitors work on the same principle where air is pumped through a filter detector assembly. The detection is by thermoluminescence (TLD) detectors or by registration of tracks on solid-state nuclear track material or by detecting total alpha activity in silicon chip semiconductor detectors. All these monitoring devices yield a single value representing the average concentration during exposure.

Continuous sampling

In continuous instruments, sampling and analysis occur simultaneously and they provide real-time results. They provide useful information when radon concentrations may change significantly and rapidly. They are very useful when studying the effects of ventilation on indoor radon and thoron levels as well as for the accurate determination of their average concentration levels. They are also used to monitor radon test chambers in which other type of instruments are tested and calibrated. Measurement intervals generally range from a minimum of 15 to 60 minutes with useful results obtainable in 2 to 4 hours of continuous sampling and counting. The disadvantages are the requirement of trained personnel for their operation and for the high cost per measurement. Commercial, continuous instruments are passive and active scintillation cell monitors, pulse ionization chambers and solid-state detector monitors. Active monitors require a pump to pull the test atmosphere continuously through the sensitive volume of the monitor. Continuous decay product instruments, sample the test atmosphere through a filter detector assembly using an air pump. The alpha activity on the filter is integrated over intervals ranging from 30-60 minutes. The detectors are usually solid-state or diffused junction types detecting alpha particles from 2 to 8 MeV.

Measurement of radon sources

The radon potential at best, it can be estimated by making measurements of radon exhalation from building materials, the underlying soil, and from the ^{222}Rn content of the local water supply. The radon exhalation rate can be measured by the rapid accumulation method or by the integrating technique. Rapid measurement of radon flux usually consists of an inverted can sealed against a surface for 1-5 h [2]. Samples of air under the accumulation can, are transferred into scintillation cells for analysis [3]. Integrated measurements of radon flux up to 3 days are made with activated carbon collectors sealed against the surface to be evaluated [4]. Measurement of the ^{226}Ra content of soil and building materials is made utilizing gamma ray spectroscopy. Radon in soil can be measured by transferring the soil gas into a calibrated scintillation cell from accumulated soil gas present in a pipe driven into the soil [3]. Radon concentration in water from underground sources is measured by the more sensitive liquid scintillation counting method [5].

Grab sampling instruments and methods for radon

Scintillation cell: The scintillation cell method is one of the oldest and most widely used for grab or continuous sampling. Scintillation cells vary in size from 0.09-3.0 liters. They have single or double valves and can be filled by evacuation instantaneously or continuously by flowing air through, using an air pump. They are cylindrical, metal, plastic or glass containers coated internally with silver activated zinc sulfide ZnS(Ag) powder [6,7]. The principle of detection is photon counting resulting from the interaction of radon and radon decay product alpha particles with the ZnS(Ag) phosphor. A photomultiplier tube assembly counts the photon events. When used for grab sampling the test air is admitted inside the cell through a filter and the calibration factor follows the theoretical buildup of radon decay products from a pure radon source. The cell can be counted a few minutes after the instantaneous filling and by applying the corresponding calibration factor the concentration of radon can be measured.

Activated carbon collector for radon grab sampling: Usually, in a flow through canister containing 50 g of activated carbon, radon is sampled for 10 min at a flow rate of 10 L min^{-1}. Radon is adsorbed on the carbon bed at 100% collection efficiency. The total volume of air passed through the carbon bed should not exceed 125 liters to avoid breakthrough. The canister is calibrated against a known concentration of radon after gamma counting in a NaI gamma ray spectrometer. Results can be obtained in 3-4 hours usually after counting the radon collector in a laboratory setting. The lower limit of detection (LLD) is 7 Bq m^{-3} when counted a few hours after sampling. A method for measuring both radon and thoron using a similar technique is described by Bigu [8].

The Radometer for grab or sniffing radon and thoron: The Radometer [9], is a battery operated monitor that measures both radon and thoron within 10 min of operation by spectroscopic analysis of the freshly formed charged radon and thoron decay products that deposit directly onto the solid state detector. The test air is admitted into the 6 L sensitive volume by means of a low-power blower. By drying the test air through a desiccant column the sensitivity of the monitor is 0.14 cpm $Bq^{-1}\ m^{-3}$. The instrument is 30 cm in diameter by 35 cm in height and weighs 5.5 kg. It is not commercially available.

The RAD 7 radon monitor: It is very similar to the Radometer and it operates on the principle of electrostatic collection of charged freshly formed decay products of radon and thoron collected onto the surface of a silicon solid-state detector. The sensitive volume of the monitor is 0.6 L with a collection voltage of 2200 V. Dried air is drawn at a rate of 1.0 L min^{-1}. Due to the short half life of ^{218}Po, the instrument recovers in minutes and can be used in a new location. Its dimensions are 24 cm X 19 cm X 29 cm and it weighs 5.5 kg. It has a sensitivity of 0.011 cpm Bq^{-1} m-3. It can be used in the sniffer mode or in the continuous mode depending on the need.

Grab sampling for radon and thoron decay products

This type of measurement is very useful in research and in studies of the behavior of the individual decay product isotopes. All of the available methods are based on the collection of decay products on high efficiency filters and by counting of the alpha radioactivity during or after the end of a short time sample [10]. Detection is either by gross alpha counting or by alpha spectrometry. There are several methods being used to measure the potential alpha energy concentration (PAEC) or working level (WL) only, or the individual concentrations of each decay product and the associated PAEC [10 -14]. The alpha spectrometry method yields more precise measurements [15] but the gross alpha counting method has the advantage of using higher air flow rates and consequently better counting statistics. The detection limit improves with longer sampling time and high flow rate irrespective which method is used. With the advancement in computer technology, sampling and analytical techniques are more-user friendly and suffer less from operator errors.

A commercial instrument designed for grab sampling of radon decay products is the Thomson and Nielsen Instant Radon Progeny Meter (IRPM). With this instrument a quick estimate of PAEC can be accomplished in 5 minutes. This requires an operator to change the filter for more grab samples. However, this instrument can be very useful during mitigation work where results are needed on site. The flow rate through the glass fiber filter and the silicon semiconductor detection assembly is 7.5 L min^{-1}, providing adequate sensitivity with an LLD of 20 nJ m^{-3}. The grab sampling instrument with dimensions of 21 cm X 20 cm X18 cm weighing 5.0 kg costs $2.7K U.S.

Integrating instruments and methods for radon (short-term and long-term)

Most short-term methods for radon are inexpensive and nonelectronic monitors can be used by the general public from simple instructions. They provide quick results and can provide a better estimation of the concentration of radon than a grab sample. Many locations in a large building can be surveyed at the same time.

Activated carbon collectors: For measurements of 2-7 days, the carbon collector has become one of the most popular integrating devices for radon [16 -19]. Activated carbon collectors come in different configurations. The amount of carbon in the collector ranges from 2-3 g when analyzed by the liquid scintillation counting technique to 25-100 g in those that use gamma ray spectrometry for analysis. They may be open-faced or fitted with a diffusion barrier that extends the integration period of the collector up to 7 days. Because humidity in air competes for the active sites of the carbon bed, the diffusion barrier serves to lessen that effect and improve the response of the collector during extreme variations of the concentration of radon. The rate of radon desorption is also reduced substantially. Diffusion barrier collectors provide very accurate average concentrations of radon when exposed in indoor environment. In typical indoor living conditions (temperature 18-25°C and humidity 30-60%), the open-faced collectors perform very well up to 4 days of exposure. At any rate, the effect of humidity can be incorporated into the calibration factor by measuring the net amount of water vapor gained by both the open-faced and the diffusion barrier collectors for different days of exposure. Indoor temperatures from 18-25°C do not affect the adsorption efficiency of the collectors. The gamma ray counting technique is simpler because the exposed collector is counted directly by placing it in contact with the detector. There is no need for sample transfer. The collector can be reused often after regeneration. In the liquid scintillation method the analysis is more tedious because the radon must be extracted from the carbon into the scintillating cocktail with the possibility of degassing some radon. Liquid scintillation counting however, uses a collector with 2-3 g of carbon and lower background. This type of collector is destroyed during the radon extraction process and cannot be recycled. The disadvantage of carbon collectors is the need to analyze them 7-10 days after exposure before the adsorbed radon decays to a point beyond its detection limit. The LLD for collectors analyzed by gamma counting, ranges from 4 to 8 Bq m^{-3} of radon in a measurement interval of 4 days, depending on the type and size. The LLD of collectors using the liquid scintillation counting is 5.0 Bq m^{-3}.

Electret Ionization Chambers: There are two electret ionization chambers in use today; the Rad Elec E-PERM for short-term and long-term exposures and the Radome for long-term exposures only.

Rad Elec E-PERM: The E-PERM [20] consists of 0.2 L, conducting plastic chamber containing a positively charged electret at its base. Air entering the chamber is filtered and the decay of radon generates air ions collected by the charged electret. The electret voltage decreases in proportion to the radon concentration. The initial and final voltage on the electret is measured at the start and at the end of the exposure. The reduction in voltage due to radon exposure is measured with a voltage reader. Thick electrets are more sensitive and they are used for short-term exposures ranging from 2-7 d. Thin electrets being less sensitive are used for long-term exposures ranging from 3 - 6 months. Electrets unlike activated carbon collectors can be read in the field or later at some convenient time. However, as a precaution electrets should be read in the same temperature environment. Because electrets respond to natural radiation background, a correction is necessary to take into account its contribution to the measured value of radon. At high altitudes, the contribution of the background radiation is significant and must be taken into consideration. Quality control (QC) and quality assessment (QA) procedures must be followed rigidly always to obtain reproducibility of results. Also, care must be exercised to avoid discharging the electret accidentally anytime between the initial and the final reading. The sensitivity of the short-term electret is 0.054 V Bq m-3 d^{-1} and that of the long-term is 0.0045 V Bq^{-1} m^{-3} d^{-1}. The LLD of the short-term electret is about 7 Bq m^{-3} in 3 d.

Radome, long-term electret ionization chamber: The Radome is similar to the Rad Elec E-PERM in principle. It consists of 0.068 L, conducting plastic chamber in which a thin specially fabricated Teflon electret is raised slightly from its base to provide a more uniform distribution of the electric field inside the chamber. Like the Rad Elec device, the Radome responds to natural background radiation so a second air tight Radome accompanies the test monitor to measure and compensate for the background contribution. For long exposures the background becomes a very significant contributor to the total voltage reduction. The sensitivity of the Radome is 0.0054 V Bq^{-1} m^{-3} d^{-1} It can accurately measure environmental radon in 90 days. In both the Radome and the Rad Elec, the voltage reader costs about $1.8K U.S.

Solid-state nuclear track detectors: The detection principle is based on the production of alpha particle tracks in solid-state materials such as cellulose nitrate (LR-115) and allyl diglycol carbonate (CR-39) plastic films [21]. After exposure the tracks are made visible by chemical or electrochemical etching usually in NaOH solution. Track density is determined using a microscope by manual scanning and counting or by an automated optical scanning system. Nuclear track detectors require exceptional QC and QA attention from the point of production of the plastic until the time of etching and reading. One of the biggest adverse effects on nuclear track detectors is the presence of electrostatic charges inside the detector chamber. Because of the low sensitivity of the detector, the time of exposure should be long to register an adequate number of tracks above the background of the plastic itself. Nuclear track detectors are not sensitive to beta or gamma radiation. They are simple and easy to handle in the field. They are suitable for measuring the annual average radon exposure. At high altitudes, the range of alpha particles increases and the measurement of radon is overestimated. Detectors using the LR-115 plastic material showed a 40% increase in the radon concentration when exposed at an altitude of 1500 m above sea level [22]. The LLD for a 3 month-long exposure is 5-10 Bq m^{-3} depending on the size of the scanned detector area. There are several manufacturers of nuclear track detectors varying in size and configuration.

Integrating radon and thoron decay product instruments and methods

All integrating instruments operate on the same sampling principle. They are active, by pumping air containing the decay products at a constant flow rate through a filter detector assembly. Measurements are made from periods of 1 d, to 4 wk, or longer. The detector is usually solid-state nuclear track material, surface barrier or diffused junction and alpha scintillation detectors. In the first type, the sensing material is analyzed in a laboratory setting while in the other types the gross alpha counting is carried out *in situ*.

Radon and thoron decay solid-state nuclear track monitor: This battery operated instrument is of interest because it can differentiate between radon and thoron decay products by incorporating collimators and absorbers between the filter and the nuclear track material [23]. Because of its size and light weight, it is being used to monitor the exposure of underground workers. A slightly modified version of this device operates on house electric power for environmental levels of radon and thoron decay products inside buildings. The LLD is 4 nJ m^{-3} for a week-long integrated exposure.

Gross alpha integrating PAEC monitor: This monitor measures the PAEC directly and provides results on site. The first such instrument was developed at EML [24]. Subsequently, a commercial version, the Thomson and Nielsen Radon WL meter very small in size (18 cm X 10 cm X 4.5 cm) and weighing 0.5 kg was developed to address the need for integrated radon PAEC measurements. With a sampling flow rate of 1 L min^{-1} the LLD is 20 nJ m^{-3}. The instrument costs about $2.2K U.S.

Continuous radon instruments

In continuous instruments and methods for measuring radon, sampling and analysis occur simultaneously providing real time measurement results. Actually, these instruments are semi-continuous because they usually integrate the sample for counting intervals ranging from 30 - 60 min. They are passive or active.

femto-TECH Model CRM-550: It is a portable passive monitor (with dimensions of 13.7 cm X 16 cm X 19 cm) weighing 2.3 kg. It is classified as a pulse ionization chamber with an electrometer and open-grid probe with an air ionization probe. Only radon diffuses into the 0.2 L, sensitive volume. The ions from the decay of radon atoms are converted to electrical pulses in an electrometer and then converted to Bq m^{-3}. It provides data usually on hourly basis or the average value during the entire period of exposure. The sensitivity of the instrument is 0.008 cpm Bq^{-1} m^{-3}. The cost of the monitor with a printer is $3.5K U.S.

Atmos 12D radon gas monitor: The Atmos 12D (with dimensions of 22 cm X 50 cm X 37 cm weighing 10 kg), is a portable pulse counting ionization chamber with spectroscopic energy resolution capabilities [25]. The pumped air is filtered and dehumidified before entry into the 0.6 L chamber. To prevent or minimize microphonics, the Atmos 12D is elastically suspended inside a cylindrical chamber that has a volume of 2.0 L. This arrangement protects it from shock and vibration. The sensitivity of the monitor is 0.02 cpm Bq^{-1} m^{-3}. A measurement period of 1 h is required to measure 140 Bq m^{-3} with an accuracy of 10%. The cost of the monitor is $8K U.S.

Alpha Guard MC50 pulsed ion and current ionization chamber: The radon monitor (with dimensions of (40 cm X 17 cm X 15 cm, weighing 4.5 kg), is a portable passive ionization chamber. The sensitive volume is 0.554 L.. It is a very versatile instrument and through the application of three different pulse shaping networks different types of information can be extracted from the chamber signal [26]. It acquires information on low or high concentrations of radon, temperature, humidity and barometric pressure. The sensitivity of the instrument is 0.05 cpm Bq^{-1} m^{-3}. The disadvantage is the high cost at $8K U.S.

Diversified Research Inc. Model Alpha II radon detector: This is a small portable passive monitor (with dimensions of 15.8 cm X 15.8 cm X21.5 cm), weighing 1.6 kg. It uses a solid-state detector. Freshly formed 218Po is detected by impinging on a photodiode. A calibration factor is applied to the alpha count rate that converts the counting rate to Bq m^{-3}. The first two hours of the monitor's operation are used to allow adequate purging of the resulting signal from a previous exposure excluded from the calculated average radon concentration. The instrument's sensitivity is 4.6 counts h^{-1}, with an LLD of 4 Bq m^{-3}. The cost of the instrument with a printer is $1.8K U.S.

Sun Nuclear Model 1023 professional radon gas monitor: The instrument is a small cylindrical portable passive device (with dimensions of 13 cm in diameter and 18 cm high weighing 0.9 kg. It uses a diffused junction photodiode enclosed in 0.02 L chamber. The freshly formed ^{218}Po is collected electrostatically on the detector. Its sensitivity is low being only 0.027 counts h^{-1} Bq-1 m^{-3} and does not discriminate between radon and thoron. Because of its low sensitivity, a measurement point is obtained at a minimum of 4 hr. The instrument costs about $0.8K U.S.

Sun Nuclear Model 1025 radon gas monitor: It is a portable active monitor that can be used both in the grab sampling and continuous mode (with dimensions of 28 cm X 14 cm X 13 cm weighing 2.3 kg. The freshly formed ^{218}Po is collected electrostatically on four diffused junction photodiode detectors inside four chambers. The instrument does not discriminate between ^{222}Rn and ^{220}Rn. Filtered, dried air containing radon is pumped at a flow rate of 1 L min^{-1}. In the continuous mode the sensitivity of the instrument is 1.4 counts h^{-1} Bq^{-1} m^{-3} that is about 3 times better than when operated in the grab sampling mode. The instrument costs $3K U.S.

Continuous scintillation monitors. There are several commercial continuous scintillation cell monitors. The principle of detection is the same by counting photons resulting from the interaction of radon and radon decay product alpha particles with ZnS(Ag) phosphor. The volume of the cells varies from 0.096 - 3.0 L. All scintillation cell monitors are programmable for different sampling, counting and reporting protocols The LLD depends on the size of the cell or sensitive volume and it ranges from 5-37 Bq m^{-3} for a 30 min counting interval. Most scintillation cell monitors are active using an air pump to draw filtered air through one valve of the cell and exhaust from another.

Pylon Model CPRD: This is the only passive monitor using a 0.270 L cell coupled to a photomultiplier tube. Air is admitted by natural diffusion. The monitor costs $4K U.S.
There are many versions of scintillation cell monitors developed in-house by universities, government agencies as well as by the private sector. Some commercial continuous active scintillation monitors are:

Eberline RGM-3: This continuous monitor with the largest scintillation cell (3 L) is the most sensitive monitor among all the others. The air is drawn through the cell at 7 L min^{-1} by an internal air pump. The monitor costs about $4K U.S. The LLD is about 5 Bq m^{-3} Is ideal for monitoring radon calibration facilities.

Pylon AB-5: Uses a 0.277 L cell with a small internal air pump. Integrated counting intervals of 30-60 minutes are used to obtain an average value. The monitor has an LLD of 7 Bq m^{-3} and costs about $4K U.S.

Gemini Research Certifier II GRI-1100 radon and radon progeny monitor: A dual purpose continuous monitor (with dimensions of 32 cm X 25 X 10 cm, weighing 4.8 kg) measures both radon and radon decay products. The flow rate through the 0.096 L cell is 0.25 L min^{-1}. The data are calculated based on 15 min counting intervals allowing for 96 measurements d^{-1}. Because of the small cell and the short counting interval, the LLD is around 10 Bq m^{-3}. The cost of the dual purpose instrument is $8K U.S.

Scintrex RDA-200: Uses a 0.16 L scintillation cell. It is programmable for different sampling intervals. It costs $3K U.S.

Continuous radon and thoron decay product instruments

As with continuous radon instruments radon decay product monitors are semi- continuous because they provide data averaged over integrated periods of 30-60 m. There are no passive instruments in this category. An air pump is used to sample air between the filter and the detector assembly. In some monitors, plateout of radon and thoron decay products occurs when the sampled atmosphere has low particle concentration (<5,000 cm^{-3}). There is no way to correct for the effect of plateout during sampling unless the filter detector assembly is designed as close as possible to represent an open face geometry that may eliminate plateout. A short description of some commercial instruments used mainly in the U.S. and Canada will be presented here, but this is by no means the only ones that are available.

Eberline working level monitor WLM-1A: This portable battery operated instrument (with dimensions of 14.8 cm X 11.8 cm X 20.3 cm weighing 2.6 kg), uses a small internal pump to draw air at flow rates ranging from 0.12-0.18 L min^{-1}. The silicon detector measures the alpha emissions from radon and thoron decay products. The usual operating mode is hourly intervals over periods of 2 d to 1 wk. At the end of sampling, the instrument counts the tail end of the decaying activity and uses it to calculate the contribution of the thoron decay products to the total PAEC. Sampling for a minimum of 2 d is necessary for the measurement of the thoron decay products. The LLD of the WLM-1A is 10 nJ m^{-3}. The cost of the instrument is $3K U.S. and the computer programmer and reader costs an additional $4K U.S. The computer can be used to program several units that will be used to survey a large building at the same time.

Scintrex WLM-30 working level monitor: This instrument (with dimensions of 30 cm X 25 cm X 18 cm, weighing 5.9 kg) is portable with an internal air pump sampling air at a rate of 1 L min^{-1}. The solid-state detector measures all counts from radon and thoron decay products. The data are stored in a 256 multichannel analyzer. Because of the spectrometry capability, the WLM-30 can measure the contribution of thoron decay products. The monitor with an LLD of 20 nJ m^{-3}, costs \$4.5K U.S.

Alpha Nuclear, Alpha PRISM Model 560: This small monitor (with dimensions of 14 cm X 4.9 cm X 6.7 cm weighing 0.41 kg), is used mainly in underground mines for continuous sampling of radon and thoron progeny. It uses a diffusion junction silicon detector. The internal pump draws air at 0.125 L min^{-1}. The LLD is 25 nJ m-^{3} The cost of the instrument with the microcomputer is \$3.8K U.S.

Alpha Nuclear, Alpha Smart 700 or 770A: This large and heavy instrument (with dimensions of 51 cm X 41 cm X20 cm weighing 20 kg), is a laboratory monitor used in a quasi-continuous mode for grab sampling. It is an excellent instrument in research and in monitoring calibration chambers. The monitor and the large and heavy external pump renders the whole system not portable. The pump draws air at 20 L min-1 through the filter assembly. Plateout losses are minimized due to the special design of the filter assembly. Alpha counting is done by two 400 mm2 diffused junction silicon detectors either by gross counting or by alpha spectrometry. For the determination of thoron decay products, the filter is counted by the second detector after the decay products have decayed on the filter. At a minimum, a complete measurement cycle is performed every 45 minutes by advancing automatically the filter onto which the decay products are collected. It can execute 30 measurements of radon decay products and 6 of thoron decay products per day. The LLD is 10 nJ m^{-3} because of the large flow rate. The monitor and the accompanying computer costs \$35K U.S.

Gemini Research, Certifier II GRI-1100 radon and radon progeny monitor: A brief description of this monitor was given in the section on continuous radon monitors. It uses the same sample to make simultaneous measurements of radon and radon decay product concentrations. The filter/detector assembly for radon decay product measurements consist of a membrane filter and a 500 mm^2 solid-state double diffusion silicon detector. The counting efficiency is ~30%. The flow rate through the filter and subsequently through the scintillation cell is 0.25 L min^{-1}. The data are listed every 15 min. The LLD is about 20 nJ m^{-3} due to the small volume of air sampled in 15 min, intervals. Plateout occurs when particle concentration is <5,000 cm^{-3}

Pylon WLx working level monitor: The WLx, (with dimensions of 42 cm X 25 cm X 23 cm weighing 6 kg), can be used for both grab sampling and continuous operation with flexible counting routines. With prolonged counting of a single sample (>250 min), it can measure both radon and thoron decay products. The internal air pump draws air at 0.5 L min^{-1}. It uses a solid-state detector to count the activity on the 25 mm filter. The LLD is 20 nJ m^{-3}. Radon and thoron decay product plateout is about 15% when the particle concentration is <3,000 cm^{-3}.

Instruments and methods for the measurement of thoron

Because of the difficulty of measuring environmental levels of thoron, there has been a lack of commercial instruments for field applications. Routine measurements of thoron are scarce, reserved mostly to some laboratories and universities. There are three known methods that use a combination of detection principles to measure thoron in air and soil gas.

Two-count scintillation method for radon and thoron in soil gas: This method uses the short half life of thoron and its first decay product. Using the instantaneous grab sampling mode, a one minute count is taken in the 2nd minute and another in the 6th minute after filling the cell [27]. In the first count the counting rate from ^{220}Rn and ^{216}Po is substantial. Using mathematics, the calculation of both radon and thoron concentrations can be measured with good accuracy. It is very practical when measuring high concentrations of radon and thoron soil gases.

Delayed-coincidence scintillation cell method: This method uses a scintillation cell in the continuous mode [28, 29]. It uses the ^{216}Po, 145 ms half life. The electronics is set up to record the time of arrival of each pulse. The pulse pairs from ^{220}Rn / ^{216}Po can be resolved from the randomly spaced pulses from the decay of unrelated atoms. This method is adequate for low concentrations of thoron and radon.

Two filter method: This classical method is usually used by research laboratories and in situations where very low concentrations of radon and thoron need to be measured [30-32]. Through a large tube fitted with entrance and exit filters the test atmosphere is drawn at high airflow to maintain a residence time inside the tube below 15 s. In the case of thoron gas, the half life of ^{216}Po is so short that what is collected on the exit filter is ^{212}Pb. The fraction collected can be calculated from diffusion theory and is typically 90%. For thoron gas a high sensitivity can be achieved by increasing the volume of the tube at an airflow of 0.4 m^3 min^{-1}. Levels as low as 0.1 Bq m^{-3} can be measured. For accurate results the entrance filter must have a 100% collection efficiency for the radon and thoron decay products and the exit filter must be devoid of air leaks that bypass the tube. Pushing air through the tube will eliminate inward leakage.

Radometer as a radon and thoron sniffer: This instrument was described briefly in the section on grab sampling techniques for radon where 150 Bq m^{-3} or ^{220}Rn and ^{222}Rn can be measured in 10 min.

Passive integrating monitor for environmental radon and thoron: The monitor consists of two hemispherical diffusion chambers in which two polycarbonate (CR-39) film detector discriminate between the two radioactive gases [33]. Exposures longer than 3 mo. are required for adequate sensitivity.

Thoron decay product instruments and methods

For the measurement of thoron decay products in the presence of radon decay products, the most popular and sensitive method is collecting air on an open face filter for 5 min to 10 h, depending on concentration levels. Counting is done as soon as possible after sampling and continues for several hours to take into consideration the contribution of both radionuclides. The least squares method is used to obtain both concentrations of radon and thoron decay products [13, 34]. High airflows are recommended to improve counting statistics. Measurement of thoron and radon decay products with continuous instruments was described in a previous section using monitors such as the Eberline WLM-1A, the Alpha Nuclear Alpha PRISM, the Scintrex WLM-30 Working level Monitor and the Pylon WLx Working Level Monitor. In the continuous operating mode the monitors must run for a minimum of 24 hours for the accurate measurement of the contribution of thoron decay products to the total PAEC.

3. Instruments and methods for the measurement of the particle size of radon and thoron decay products

The size of the decay products of radon and thoron are important because it affects the amount and location of deposition in the respiratory tract. When first formed, radon and thoron decay products are 0.5-1 nm in diameter. A fraction of them will remain free or unattached (5-10%) and the majority will attach to particles present in the environment. The size of the attached radionuclides range in size from 30-1000 nm in diameter. Particle size classifying instruments are complex and expensive and in need for additional refinement. Measuring the activity size distribution of radon and thoron decay products adds another dimension to the problem. For this reason, activity size distribution measurements are undertaken by a select group of researchers.

There are no commercial instruments to address particle size measurements except what exists at universities and in some government laboratories. Today, there are two basic types of size classifying instruments that measure the activity size distribution of radon and thoron decay products. They are the diffusion battery system and the micro - orifice uniform deposit impactor (MOUDI).

Diffusion battery system: This system consists of two types of batteries, the collimated disc type [35] and the wire screen system [36]. The disc type diffusion battery was first used in underground uranium mines at a flow rate of 1 L min^{-1}. The wire screen battery operating at 25 L min^{-1} was used extensively indoors and outdoors in the New York Metropolitan area. Both battery systems are operated manually and cover the size range from 1-400 nm. Their resolution for sizes above 500 nm is poor. One version of the wire screen battery system became semiautomatic in Germany by Reineking and Posterndorfer and another version was fully automated in 1990 [37]. Both batteries cover the particle range fro m 0.5 to 500 nm.

The MOUDI: This instrument is a more recent development for measuring the size distribution of radon and thoron progeny in the size range from 50 - 5000 nm. Air is sampled at 30 L min^{-1} usually through 8 stages arranged in series. In each stage there is an impaction plate onto which the decay products are collected according to size. Alpha and gamma counting can be used to determine the decay product activities on the impaction plates. Since the MOUDI does not provide information on the ultra fine particles that are <50 nm, a graded screen array (GSA) system consisting of a 60-and and 100-mesh stainless steel screens is used to complement the MOUDI data. By sampling simultaneously through the two different sizing systems a full characterization of the size distribution ranging from 0.5 - 5000 nm is accomplished. The activity penetrating the diffusion batteries and the activity deposited on each of the impaction plates is used to calculate the activity size distribution using the expectation-maximization algorithm [38] refined at the Environmental Measurements Laboratory (EML) in New York [39]. Measurements of the activity size distribution of radon decay products were made with the MOUDI in several locations in a Canadian uranium mine [40].

4. Quality assurance, calibrations, and traceability

The objective of quality assurance (QA) is to assess the quality and confidence placed on the measurement results acquired by an instrument or method. Quality assurance is actually two related activities that of quality control and quality assessment. Every step and procedure taken to make a measurement must be tied into a QA program. That includes the instrument with all its associated components that contribute to a measurement. QA steps for the large variety of instruments and methods used to measure the concentrations of radon, thoron and their decay products should be taken to minimize errors and standardize methods and techniques. Instrument users should abide by and practice the necessary QA procedures for a specific instrument or method in their possession.

Calibration factors for instruments provided by the manufacturers are often inadequate. When an instrument is first acquired, it must be calibrated and occasionally afterwards. The precision and bias of instruments should be evaluated on a regular basis by performing duplicate exposures and by performing blind or spike tests in a well-controlled environment monitored with instruments traceable to a National Institute of Standards or its equivalent. Active instruments and their air moving components such as pumps and flowmeters must be calibrated against standards and in the same conditions of field use. It is easy and economical to do duplicates and blind tests with passive devices but not so practical with the costly and active devices. Calibrations of instruments used to measure radon and thoron decay products must be performed with certified alpha standards with energies >5 MeV and in the exact counting geometry of field use. All active and passive devices should be calibrated under different conditions of exposure to determine the effects of variations in temperature, humidity, static charge, microphonics, background radiation, linearity, dynamic range and for other possible interfering agents.

Intercomparison and Intercalibration exercises: Since 1981, several international intercomparison and intercalibration exercises were conducted for radon, radon decay products and particle size distributions as part of the IIIP and more recently as part of the IRMP. Both programs provided a range of activities with the purpose of getting measurement accuracy, reliability and good quality. Also, these activities helped to promote instrument development and refinement and to train the newcomers in establishing their own radon assessment programs. In the U.S., EPA and DOE provided assistance and training to hundreds of radon testers and mitigators during the first years when radon became a significant indoor air pollutant issue. However, an area that needs attention is intercomparisons for the measurement of soil gas radon and radon exhalation. An international intercomparison for such measurements took place in the New York metropolitan area in June 1995 [3]. The results are encouraging but they suggest that there is need for following up with more intercomparisons and more participants.

5. Conclusions

The instruments and methods used to measure the sources, concentrations of radon, thoron and their decay products and particle size distributions of the decay products have been reviewed for their measurement principle, field application, suitability, sensitivity, accuracy and cost. The main emphasis of this report was to categorize instruments and methods to meet a particular need. At present, the greatest demand is for screening measurements for radon, using simple portable and mailable integrating instruments. For special cases such as diagnostics, research and dose evaluation, measurements of radon, thoron and their decay products with active and passive instruments are recommended as the appropriate monitoring devices. The acquisition of accurate measurements at low concentrations depends on the proper initial calibration of the instrument under conditions of actual use. Of course the operators of active devices must be qualified and be aware of specific QA and QC procedures pertaining to the instrument under their wing. Although, the instruments described here are manufactured by different people, most appear to perform well when used properly and when they are calibrated against known radon, thoron and decay product atmospheres traceable to National Institute Standards or their equivalent.

6. References

[1] National Council on Radiation Protection and Measurements. Evaluation of occupational and environmental exposures to radon and radon daughters in the United States, Bethesda, MD. NCRP Report No. 78, 1984.

[2] Wilkening,, M. H, Measurements of radon flux by the accumulation method. Workshop on Methods for measuring radon in and around uranium mills. E. D. Harward (Editor), Atomic Industrial Forum Inc. 3, 131-138, 1977.

[3] Hutter, A. and E. O. Knutson, An international intercomparison of soil gas radon and radon exhalation measurements, Health Phys., In Press, 1997.

[4] Countess, R. J., Radon flux measurements with a charcoal canister, Health Phys., 31, 455-456, 1976.

[5] Pritchard, H., and T. Gesell, rapid measurement of radon-222 concentrations in water with a commercial liquid scintillation counter, Health Phys., 33, 577-582, 1977.

[6] Lucas, H. F., Improved low-level alpha scintillation counter for radon, Rev. Scient. Instr. 28, 680-683, 1957.

[7] George, A. C., Scintillation flasks for the determination of low level concentrations of radon, In Proceedings of the 9th Midyear Health Phys., Symposium, Denver, CO, 1976.

[8] Bigu, J., Radon-220 determination using activated carbon and a high purity Ge detector, Health Phys., 51, 534-538, 1986.

[9] Negro, V. C., Radometer, a portable field instrument for the rapid measurement of environmental radon and thoron, IEEE Transactions on Nuclear Science , 37, 854-858, 1990.

[10] George. A. C., Measurement of airborne Rn-222 daughters by filter collection and alpha radioactivity detection, In environmental carcinogens, methods of analysis and exposure measurement, International Agency for Research on Cancer, IARC Scientific Publications No. 109, Indoor Air 12, 173-179, 1993.

[11] Kusnetz, H. L., Radon daughters in mine atmospheres. A field method for determining concentrations, Am. Ind. Hyg. Assoc. J. 17,85-88, 1956.

[12] Thomas, J. W., Measurements of radon daughters in air, Health Phys., 23, 783-789, 1972.

[13] Raabe, O. G., and M. E. Wrenn, Analysis of the activity of radon daughter samples by weighted least squares, Health Phys., 17, 593-605, 1969.

[14] Nazaroff, W. W., Optimizing the total alpha three count technique for measuring concentrations of radon progeny in residences, Health Phys., 46, 395-405, 1984.

[15] Martz, D. E,, D.F. Holleman, D. E. McCurdy, and K. J. Schiager, Analysis of atmospheric concentrations of RaA, RaB, and RaC by alpha spectrometry, Health Phys., 17, 131-138, 1969.

[16] George, A. C., Passive integrated measurement of indoor radon using activated carbon, Health Phys., 46, 867-872, 1984.

[17] George, A. C., and T. Weber, An improved passive activated carbon collector for measuring environmental Rn-222 in indoor air, Health Phys., 58, 583-589, 1990.

[18] Cohen, B., and R. Nyson, A diffusion barrier adsorption collector for measuring radon concentrations in indoor air, Health Phys., 50, 457-463, 1986.

[19] Gray, D. J., and S. T. Windham, NAREL standard operating procedures for Rn-222 measurements using diffusion barrier charcoal canisters, U.S. EPA, Office of Radiation Programs, Report No. EPA 520/5-9032, 1990.

[20] Kotrappa, P., J. C. Dempsey, J. R. Hickey, and L. R. Stieff, An electret passive environmental Rn-222 monitor based on ionization measurement, Health Phys., 54, 47-56, 1988.

[21] Alter, H. W. and R. I. Fleischer, Passive integrating radon monitor for environmental monitoring, Health Phys., 40, 693-702, 1981.

[22] Vasudevan, L., and M. E. Mclain, Atmospheric pressure effects on the calibration constant of alpha radon detectors, Health Phys., 66,318-326, 1994.

[23] Rannou, A. and P. Zettwoog, Statistical information on individual dosimetry in underground uranium mines from the measurement of Po-218, Po-214 and Po-212 exposures, Rad. Prot. Dosim., 36, 149-154, 1991.

[24] Latner, N., S. Watnick, and R. T. Graveson, Integrating working level monitor EML type TF-11, New York, U. S. DOE, Report EML-389, 1981.

[25] Baltzer, P., K. Gorsten and A. Backlin, A pulse counting ionization chamber for measuring the radon concentration in air, Nuclear Instr. Methods, Phys. Res. A317, 357-364, 1992.

[26] Genrich, V., Alphaguard multiparameter radon monitor, Frankfurt, Germany, Geritron Instruments, File, EVALSYS2, Doc. Gmbh D-60488, 1990.

[27] Hutter, A., A method for determining soil gas Rn-220 (thoron) concentrations, Health Phys., 68, 835-839, 1995.

[28] Falk, R., H. More, and L. Nyblom, Measuring techniques for environmental levels of Rn-220 in air using flow-though Lucas cell and multiple time analysis of recorded pulse events, Applied Radiat. Isot.43, 111-118, 1992.

[29] Knutson, E. O., A. C. George, P. Shebell and C. V. Gogolak, EML thoron gas measurements, Rad. Prot. Dosim. 56, 263-266, 1994.

[30] Thomas, J. W. and P. C. Le Clare, The two-filter tube method for measuring radon, Health Phys., 18,113-122, 1970.

[31] George, A. C., and E. O. Knutson, Two-filter tube sampling for radon, In EML procedures manual, Environmental Measurements Laboratory, U.S. DOE Report HASL-300 , 28th Ed.

Section 2.2.3.3. New York, NY 1994.
[32] Hutter, A., R. J. Larsen, H. Maring, and J.T. Merril, Radon at Bermuda and Mauna Loa, local and distant sources, J. Radioanal. Nuclear Chem. Articles,193, 309-318, 1995.
[33] Doi, M., S. Kobayashi, and K. Fujimoto, A passive measurement technique for characterization of high-risk houses in Japan due to enhanced levels of indoor radon and thoron concentrations, Rad. Prot. Dosim., 45, 425-430, 1992.
[34] Thiessen, N., Alpha particle spectrometry in radon /thoron program measurements, Health Phys., 67, 632-640, 1994
[35] George, A. C., I. Hinchcliffe , and R. Sladowski, Size distributions of radon daughter particles in uranium mine atmospheres, Am. Ind. Hyg. Assoc. J. 36, 484-490, 1975.
[36] George, A. C., and A. J. Breslin, The distribution of ambient radon and radon daughters in residential buildings in the New Jersey - New York area, Natural Rad. Environment III, CONF-780422, Technical Information Center U. S. DOE, 1980.
[37] Ramamuthi, M., and P. K. Hopke, An automated semi-continuous system for measuring indoor radon progeny activity weighted size distributions, Aerosol Sci. Technol. 14, 82-92, 1991.
[38] Maher, E. F., and N. M. Laird, EM algorithm reconstruction of particle size distributions from diffusion battery data, J. Aerosol Sci. Technol. 16, 557-570, 1985.
[39] Knutson, E. O., Application of the expectation maximization algorithm to the processing of cascade impactor data, the method of log normal components, J. Aerosol Sci. 22, S267-S270, 1991.
[40] Tu, K. W., I. Fisenne, and A. Hutter, Short and long -lived radionuclide particle size measurements in a uranium mine, New York Environmental Measurements Laboratory, U. S. DOE Report EML-588, 1997.

Passive Methods for Environmental Radon and Thoron Monitoring

Takao Iida

Graduate School of Engineering, Nagoya University,
Furo-cho, Chikusa-ku, Nagoya, 464-01, Japan

Abstract

The present paper describes passive radon (^{222}Rn) monitor, some problems with measurements of ^{222}Rn concentrations by passive cup method, a new type of passive ^{222}Rn and thoron (^{220}Rn) cup monitors, and a simple assessment method of dose due to exposure of ^{220}Rn progeny.

A passive cup monitor has been constructed by using a 50 mm radius stainless steel hemisphere. In dwellings with high ^{220}Rn concentrations, it is particularly necessary to measure ^{222}Rn concentration with a monitor having a Rn exchange rate less than 0.1 h^{-1}. On the basis of the ^{222}Rn cup monitor, a new type of passive ^{222}Rn and ^{220}Rn cup monitors was developed. By placing a pair of cup monitors at a 20 cm distance from wall, the concentrations of both gases could be assessed from the α-track densities on the solid state nuclear track detectors (SSNTD). The ^{222}Rn and ^{220}Rn concentrations were surveyed around Nagoya in Japan. The ^{220}Rn concentrations were rather high in the dwellings with soil wall. The results of the surveys have also clarified the relationship between the ^{220}Rn concentrations at a 20 cm distance from wall and the ^{220}Rn progeny concentrations measured in the same dwellings. Then, the annual mean effective dose equivalent due to ^{220}Rn progeny was expected to be 0.67 $mSv \cdot y^{-1}$ in the dwellings with soil wall.

1. Introduction

The effective dose due to inhalation of ^{222}Rn and its daughters in dwellings accounts on average of about one-half of that from all natural sources of radiation [1]. To estimate the effective dose, it is necessary to measure mean ^{222}Rn concentrations in various environments. For the surveys, passive ^{222}Rn monitors that are handy, low-priced and useful for long-term measurements at numerous locations, are suitable. Until now, many types of passive ^{222}Rn monitors have been studied and developed using SSNTD [2, 3, 4], thermoluminescent dosimeters (TLD) [5], activated carbon [6] and photographic film [7] as detectors. Passive ^{222}Rn monitors with SSNTD have been widely used to measure time-average ^{222}Rn concentrations.

In indoor ^{222}Rn surveys in Japan, quite high ^{222}Rn concentrations were sometimes observed by passive methods [8] in Japanese traditional dwellings made of wood and soil walls. High ^{220}Rn progeny concentrations have been often observed in such dwellings [9]. These results suggest that some of the high ^{222}Rn concentrations measured with passive monitors might be influenced by high ^{220}Rn exhalation from soil walls. Consequently, some types of passive integrating cup monitors that could separate indoor ^{222}Rn and ^{220}Rn concentrations have been developed in Japan [10, 11, 12]. Since there are the dwellings with soil walls in all over the world, it is important to evaluate the annual effective dose equivalent dose due to ^{220}Rn and its progeny.

2. Passive ^{222}Rn monitors

2.1 Typical passive ^{222}Rn monitors

A number of passive radon monitors have been developed by different laboratories for monitoring environmental radon. The choice of the method depends on measuring time period and the ^{222}Rn concentration levels. **Figures 1** to **4** shows the typical passive ^{222}Rn monitors. Passive ^{222}Rn monitors with SSNTD can include the following: (1) the bare method [2], (2) the cup method [2, 3], and (3) electrostatic collection method [13]. Moreover, other types of passive monitors with (4) TLD [5], (5) activated charcoal [6], and (6) photographic film [7] have been reported.

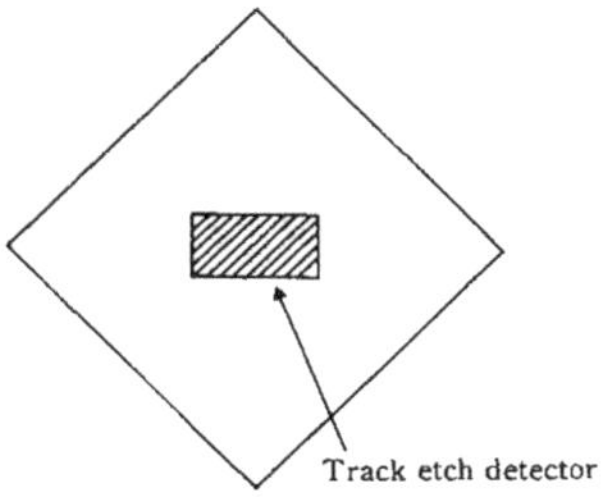

Fig. 1 Passive monitor with a bare SSNTD [2].

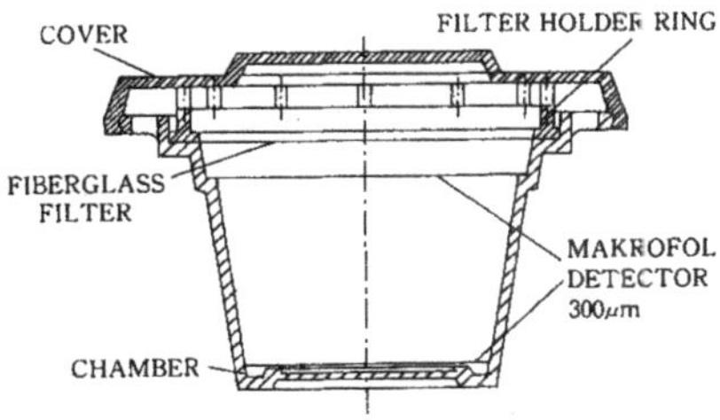

Fig. 2 Typical cup monitor [3].

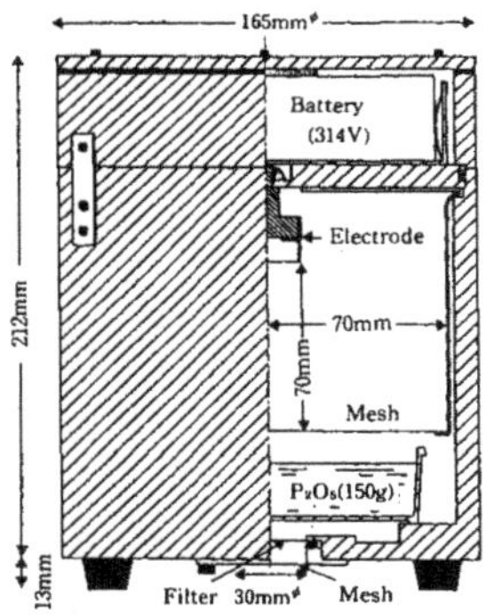

Fig. 3 Electrostatic collection ^{222}Rn monitor [13].

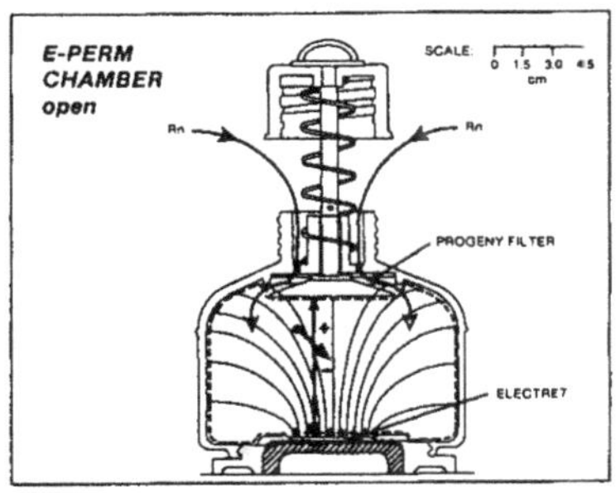

Fig. 4 Electret passive ^{222}Rn monitor [18].

The bare method is very simple as shown in **Fig. 1**, with a bare SSNTD mounted on the wall. However, the accuracy of the bare method using CR-39 (allyl diglycol carbonate) detectors is not good, because bare CR-39 detectors measure not only ^{222}Rn plus ambient daughters, but also daughters plating out on the detectors [14]. Typical cup monitor [3] is shown in **Fig. 2**. The cup method gives a good linear relationship between track density and ^{222}Rn exposure. It has been the main method of indoor ^{222}Rn measurements in Europe and North America [15, 16, 17]. Disadvantages of the cup method are the relatively low sensitivity and the relatively high detection limit. **Figure 3** shows an electrostatic collection ^{222}Rn monitor with SSNTD [13]. The method has high sensitivity so that the method is also suitable for measuring outdoor ^{222}Rn concentration levels. However, this method needs a source of high voltage and a drying agent is not as convenient as the cup method. All passive monitors with TLD use electrostatic collection method [5]. The monitor has the same disadvantages as the electrostatic collection monitor with SSNTD. The detection limit is relatively high, because TLD is irradiated by natural gamma and cosmic radiations during exposure. Activated charcoal method [6] is one of popular passive ^{222}Rn monitor. The exposure time is relatively short from 1 to 7 d since ^{222}Rn collected on the activated charcoal decays by 3.8 d. Collectors are analyzed by gamma counting in

NaI(Tl) systems or liquid scintillation counting techniques. **Figure 4** shows an electret ionization chamber [18]. The E-PERM™ chamber is small, passive integrating device designed for sort- or long-term exposures. The detection limit is relatively high, because the chamber is also irradiated by natural gamma and cosmic radiations. Photographic film with ZnS(Ag) phosphor screen [7] was the first passive ^{222}Rn detector. The passive monitors was used for area and personal ^{222}Rn monitoring in Uranium mine.

2.2 Principle of passive cup method

We have developed several types of passive cup monitors for measuring indoor ^{222}Rn concentrations. It is necessary that the cup wall material is conductive and low α-emitting [19]. A hemispherical shape is suitable for a cup monitor to arrange the incident α-particles on the SSNTD in approximately the same energy. In view of these considerations, the passive ^{222}Rn cup monitor was constructed by mounting a stainless steel hemisphere with a 50 mm radius on an aluminum plate with a 16 mm inner diameter circular window. Offset from the center is a 12 mm diameter circular opening in front of which a membrane filter (Pore size 0.8 um) is positioned. The cross section of the passive monitor is shown in **Fig. 5**.

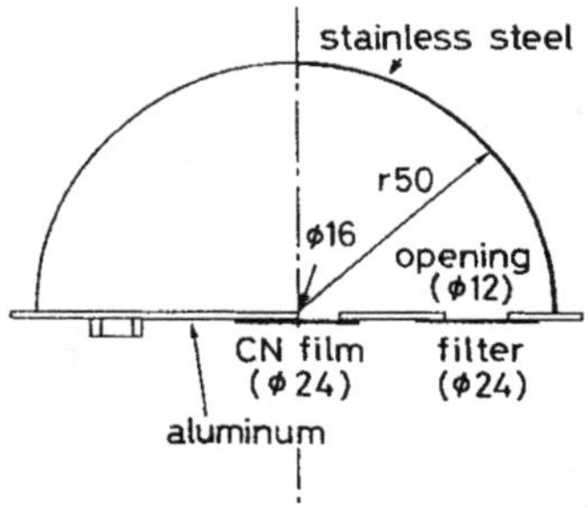

Fig. 5 Cross-section of hemispherical-shaped ^{222}Rn cup monitor.

The principle of the passive cup monitor could be described as follows. The relationship between ^{222}Rn exposure and track density on a SSNTD is given by

$$Q = \frac{N - B \cdot A}{CF \cdot A \cdot T} \tag{2-1}$$

where, Q is the mean ^{222}Rn concentration in $Bq \cdot m^{-3}$, N is the number of tracks counted in field A, B is the background track density in cm^{-2}, A is counting area in cm^2, CF is the calibration factor in tracks $cm^{-2}(Bq \cdot m^{-3} \cdot h)^{-1}$, and T is exposure time in h. Radon-222 atoms in the monitor are exchanged by diffusion through the membrane filter. The ^{222}Rn exchange rate [13] of the monitor is represented by

$$\gamma = \frac{D \cdot A}{\delta \cdot V} \tag{2-2}$$

where γ is the ^{222}Rn exchange rate in h^{-1}, D is the diffusion constant of ^{222}Rn in the filter in $m^2 \cdot h^{-1}$, A and δ are its area in m^2 and thickness in m, and V is the interior volume of the monitor in m^3. The D/δ value of the membrane filter was estimated to be 2.09 $m \cdot h^{-1}$ by Iida et al. [13]. When the window diameter is 12 mm, the ^{222}Rn exchange rate γ is derived to be 0.90 h^{-1} by eq. (2-2). It is considered that the exchange rate for ^{220}Rn is the same as that for ^{222}Rn. When the Rn concentration outside the cup monitor n_{out} is constant, the inside concentration n_{in} is given by

$$n_{in} = \frac{\gamma}{\lambda+\gamma}\,[1 - \exp\{-(\lambda+\gamma)t\}]\,n_{out} \qquad (2\text{-}3)$$

where λ is the Rn decay rate in h^{-1}. Since the decay rates of ^{222}Rn and ^{220}Rn are 0.00755 h^{-1} and 45 h^{-1}, respectively, the ^{222}Rn and ^{220}Rn concentrations inside the monitor approach about 99% and 2.0% of the outside concentrations, after enough time has elapsed. Therefore, one could ignore the influence of ^{220}Rn on the measurement of ^{222}Rn concentrations except when extremely high ^{220}Rn concentrations are encountered.

3. Some problems on the radon measurements by passive cup method

In order to evaluate ^{222}Rn concentrations with the passive cup monitor, we need to consider the following factors with regard to SSNTD: (1) unchanged quality for a long time; (2) reproducibility of the etching process and; (3) low background track density, and moreover, with regard to measurement by cup method; (4) exposure time; (5) effect of environmental factors such as UV light, temperature and humidity; (6) shape and material of cup; (7) α-particles emitted from the inner wall of cup; and (8) the influence of ^{220}Rn.

3.1 Intercomparisons of passive measurements

The track density due to α-particles and the background track density on SSNTDs are believed to follow the Poisson distribution. However, the uncertainty of the cup method was always greater than could be accounted for by Poisson counting statistics, even if the cup was exposed to high ^{222}Rn concentrations under controlled conditions [20, 21]. The sources of error have been considered to be associated with the etching process and track counting.

Intercomparison exercises were performed during the period from January 1989 to September 1991 on passive ^{222}Rn measurements at Critical Assembly Building of the Research Reactor Institute, Kyoto University (KUCA) and Nagoya University. In order to ensure the ^{222}Rn concentrations obtained by passive measurements in Japan, we performed the intercomparisons not only in high ^{222}Rn environment but also in outdoor environment. The exercises were carried out in 4 rounds. Numbers of participating institutions were 4, 6, 12, and 13 for round 1, 2, 3 and 4, respectively. The participants cover almost all Japanese passive ^{222}Rn measurement groups. The results of round 3 and 4 are represented in **Figs. 6** and **7**.

The round 1 intercomparison was carried out from Jan. 16 to 20, 1989 in a room of KUCA of about 470 $Bq\cdot m^{-3}$. The ^{222}Rn exposure evaluated using a flow-through ionization chamber was 35 $kBq\cdot m^{-3}\cdot h$. The mean integrating ^{222}Rn concentration (32.3 ± 2.8) $kBq\cdot m^{-3}\cdot h$ obtained with passive methods gives fairly good agreement with

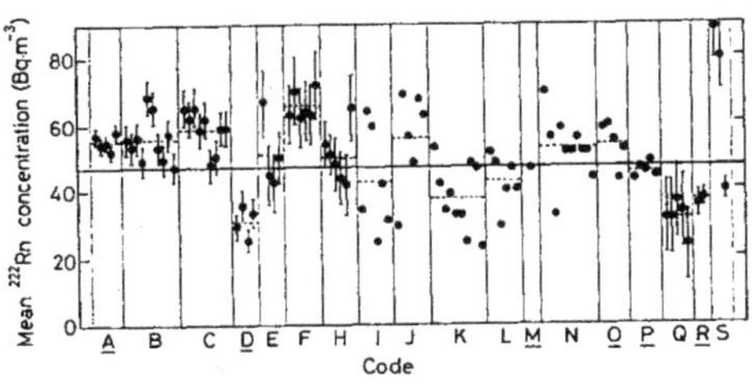

Fig. 6 Result of the round 3 intercomparison of ^{222}Rn from passive monitors, 1 June - 28 July 1990.

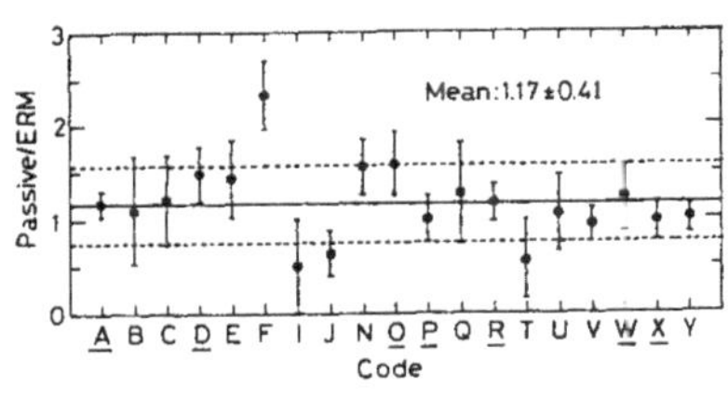

Fig. 7 Result of the round 4 intercomparison.

the ^{222}Rn exposure.

The intercomparison for round 2 was performed from Jan. 20 to Feb. 28, 1989 and for round 3 from Jun. 1 to Jul. 28, 1990 in a preparation room of KUCA of about 50 $Bq\cdot m^{-3}$. The mean integrating ^{222}Rn concentrations obtained with passive monitors were (55.6 ± 12.0) $kBq\cdot m^{-3}\cdot h$ on round 2 and (63.0 ± 12.3) $kBq\cdot m^{-3}\cdot h$ on round 3. These values agree well with the expected ^{222}Rn exposures of 52 $kBq\cdot m^{-3}\cdot h$ and 61 $kBq\cdot m^{-3}\cdot h$, respectively. In **Fig. 6**, monitor E is bare method, monitor B, C, F, H, I, J, K, L, N, Q and S are cup methods and monitors A, D, M, O, P and R are electrostatic collection methods. The standard deviation for each passive method ranges from 5% to 35%. The cup methods show large standard deviation.

The round 4 intercomparison has been carrying out from Jul. 31, 1990 to end of Sep., 1991 on the campus of Nagoya University. The ^{222}Rn concentration has been continuously monitored with an electrostatic radon monitor (ERM) [22] during the same period. The mean outdoor ^{222}Rn concentration at Nagoya is about 6 $Bq\cdot m^{-3}$. Participant has a choice of the exposure time of 2, 3 or 6 months considering the detection limit of his method. The result of ^{222}Rn exposures for 6 months is shown in **Fig. 7**. The vertical axis shows the ratio of the measured values with passive monitor to ERM. The solid line and the broken line represent the average ratio and standard deviation. The code name with underline indicates the method is electrostatic collection. The mean integrating ^{222}Rn concentrations measured with passive monitors agree with the ^{222}Rn exposures. However, the standard deviation is very large because the ^{222}Rn exposure for 6 months is comparable with the detection limit of cup method.

3.2 Influence of ^{220}Rn

Mean ^{222}Rn concentrations in different types of dwellings were measured to test the performance of the ^{222}Rn cup monitor for 3-month intervals from May 1989 to Feb. 1990. The results of the test survey from Nov. 1989 to Feb. 1990 are shown in **Fig. 8**. The locations of the survey were 17 dwellings made of wood and soil wall, 11 dwellings made of concrete, and 3 dwellings made of new building materials (NBM) such as plaster board. One monitor was set in each dwelling except for a concrete laboratory, in which five monitors were set at the same place to study the variations. The code numbers of cup monitors in **Fig. 8** are written down in order of ^{222}Rn concentration according to dwelling types. Average ^{222}Rn concentration levels of two other surveys in summer and autumn were about 2/3 of winter survey shown in **Fig. 8**.

To confirm the reliability of the data obtained with cup monitors, an electrostatic integrating ^{222}Rn monitor (EIRM) [13] was placed in each of three house types. The EIRM used was not affected very much by ^{220}Rn. The mean ^{222}Rn

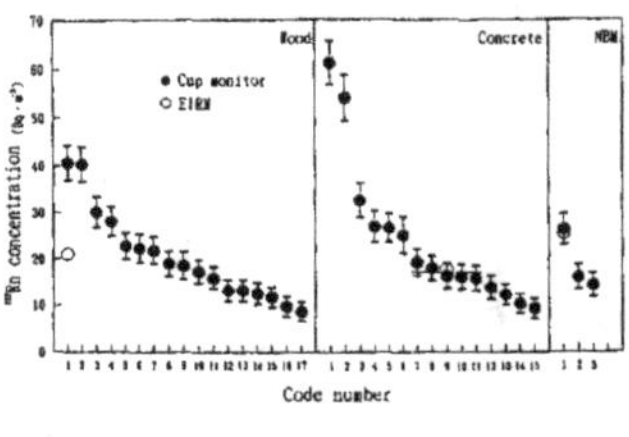

Fig. 8 The results of the test survey from Nov. 1989 to Feb. 1990.

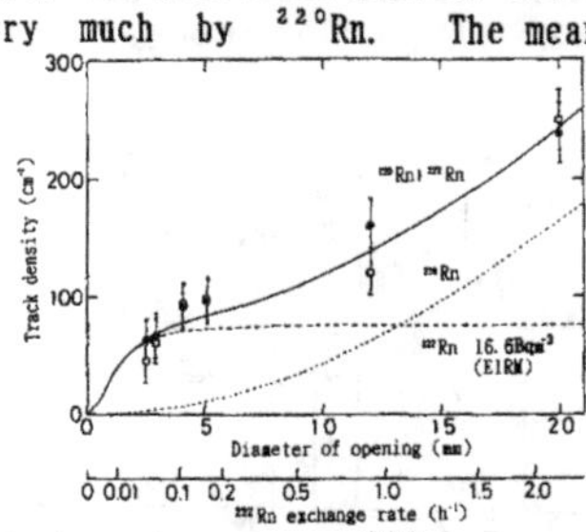

Fig. 9 The influence of ^{220}Rn exhaling from a soil wall on ^{222}Rn concentration measurements. The dashed and dotted lines represent the expected track densities due to ^{222}Rn and ^{220}Rn, respectively. The solid line is the sum.

concentrations measured with five cup monitors of code number 7-11 in the concrete laboratory building and a cup monitor of code number 1 in the NBM house agree well with the EIRM results. However, the mean ^{222}Rn level obtained with a cup monitor of code number 1 in the wooden house was about 2 times that obtained with EIRM. A similar tendency was also seen in both summer and autumn surveys. Very high ^{220}Rn progeny concentrations of 3.0–6.3 Bq·m^{-3} have been observed in the wooden house [11].

These results suggest that, in a wooden house, the measurement of ^{222}Rn concentration with the cup monitor may be affected by only about 2% of ^{220}Rn concentration outside of the monitor.

To verify this supposition, the following measurements were carried out in the wooden house. We made 6 types of cup monitors with different opening diameters of 2.6, 3.0, 4.1, 5.1, 12, and 20 mm by twos. From eq. (2-2), the Rn exchange rates of these monitors are estimated to be 0.042, 0.057, 0.105, 0.163, 0.90, and 2.51 h^{-1}, respectively. The cup monitors and an EIRM were exposed at a distance of 20 cm from the soil wall for one month. The result is illustrated in **Fig. 9**. It is clear that track density depends on the opening diameter. The dashed line represents the expected track density with a cup monitor that was calculated, based on the ^{222}Rn concentration measured with the EIRM. On the other hand, the dotted line shows the track density due to ^{220}Rn that can be calculated from assuming that the difference between the track density measured with cup monitors of opening diameter 20 mm and expected track density resulting from ^{222}Rn concentration is caused by ^{220}Rn. The sum of the dashed and dotted lines is shown by solid line in **Fig. 9**. The observed track densities from cup monitors agree well with the solid line. From the measurement, the mean ^{220}Rn concentration at a distance of 20 cm from the soil wall was estimated to be more than 1000 Bq·m^{-3} by using the Rn exchange rate and a calculated cup monitor calibration factor for ^{220}Rn.

Doi [23] showed also that the ^{222}Rn concentrations measured with the KFK passive ^{222}Rn dosimeters depended on the distance from the surface soil wall. Since the KFK dosimeter detects ^{222}Rn and ^{220}Rn, the higher readings were observed near walls.

4. Passive radon and thoron monitors

Passive ^{220}Rn integrating monitors are suitable for large scale survey of the dwellings with high ^{220}Rn concentration. On the basis of passive ^{222}Rn cup monitor with SSNTD, we have studied and developed a new type of passive integrating ^{222}Rn and ^{220}Rn cup monitors. Indoor ^{222}Rn and ^{220}Rn concentrations can be measured at the same time with the monitors. The cross sections of the passive ^{222}Rn and ^{220}Rn cup monitors are shown in **Fig. 10**. The ^{222}Rn and ^{220}Rn monitors have a 5 mm exchange

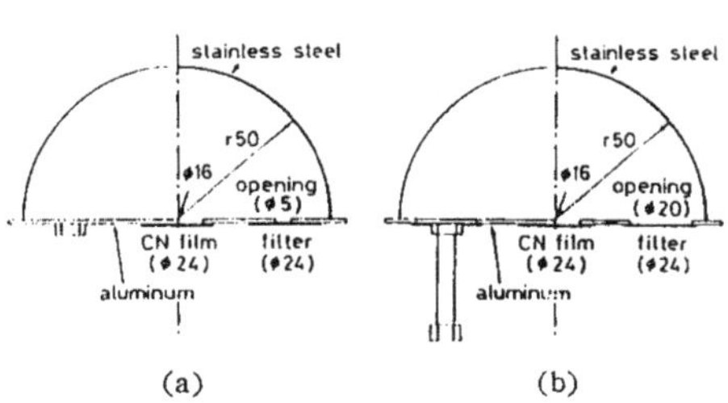

Fig. 10 Passive integrating ^{222}Rn and ^{220}Rn cup monitors (a) for ^{222}Rn measurement and (b) for ^{220}Rn measurement.

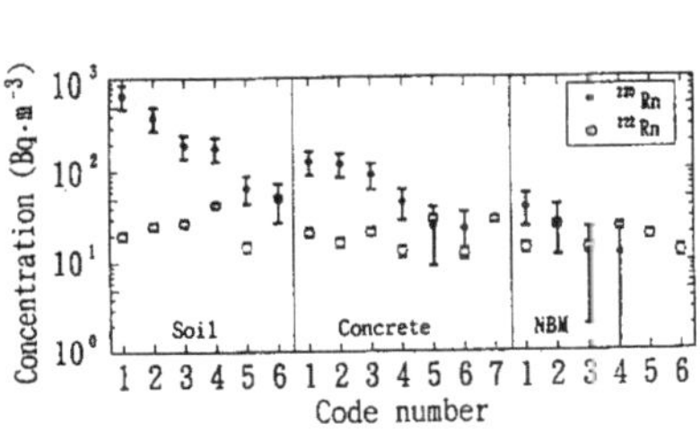

Fig. 11 Result of the survey from 29 Oct. 1991 to 30 Jan. 1992: (●) ^{220}Rn and (○) ^{222}Rn concentrations.

opening and four exchange openings of diameter 20 mm, respectively. Calibration factors of the ^{222}Rn and ^{220}Rn monitors were $(2.15\pm0.19)\times10^{-3}$ and $(5.06\pm0.89)\times10^{-4}$ tracks·cm^{-2}(Bq·m^{-3}·h)$^{-1}$ and agree well with the calculated values. The lower detection limit of the ^{222}Rn and ^{220}Rn monitors were 6.4 and 21 Bq·m^{-3} for an exposure time of 90 d.

Another type of passive ^{220}Rn monitor has been developed and reported by Doi et al. [10] and Miyake et al. [12] recently. Doi's monitor has a structure of two 37.5 mm and 60 mm radius hemispheric chambers with polycarbonate films. The separation of ^{222}Rn and ^{220}Rn concentrations is performed by using two step diffusion method and electrochemical etching. On the other hand, Miyake has performed the discriminative measurement of ^{222}Rn and ^{220}Rn concentrations with an electrostatic passive monitor. From the sizes and shapes of the etch pits on CR-39 detector, the α-particle energy could be determined.

5. Dose assessment method for thoron daughters

5.1 Measurements of indoor ^{222}Rn and ^{220}Rn Concentrations

Small scale surveys of ^{222}Rn and ^{220}Rn concentrations indoors were performed with the cup monitors over three years. An example of the surveys is shown in **Fig. 11.** Mean ^{220}Rn concentrations of the dwellings with soil walls were the highest compared with dwellings made of concrete and NBM.

In some dwellings with soil wall, the ^{220}Rn concentrations have been measured by using six pairs of cup monitors. An example of the results is shown in **Fig. 12.** The ^{220}Rn concentration could be expressed by the following equation [24]:

$$Q(X) = \frac{E_{Tn}}{\sqrt{\lambda_{Tn}\cdot D}} \exp(-\sqrt{\lambda_{Tn}/D})X, \qquad (5\text{-}1)$$

where Q(X) is the ^{220}Rn concentration at a distance of X m from wall in Bq·m^{-3}, E_{Tn} is the ^{220}Rn exhalation rate from wall in Bq·m^{-2}·s^{-1}, λ_{Tn} is the ^{220}Rn decay rate in h^{-1}, and D is the effective diffusion coefficient of ^{220}Rn in air in m^2·, s^{-1}. The E_{Tn} and D calculated by using the least squares method range from 1.6 to 6.4 Bq·m^{-2}·s^{-1} and 1.2 to 7.4×10^{-3} m^2·s^{-1}, respectively.

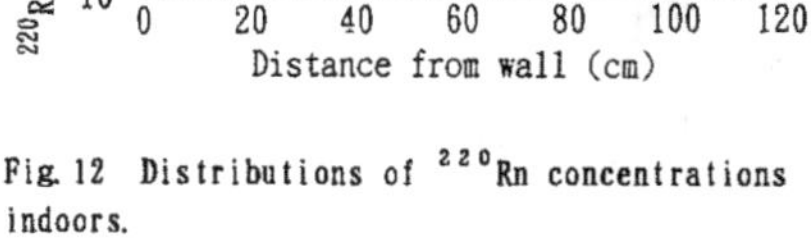

Fig. 12 Distributions of ^{220}Rn concentrations indoors.

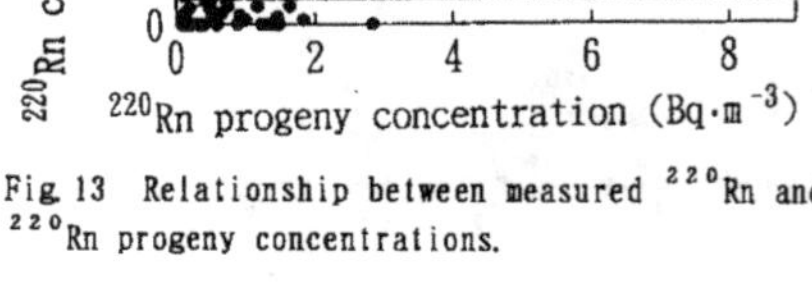

Fig. 13 Relationship between measured ^{220}Rn and ^{220}Rn progeny concentrations.

5.2 Dose assessment method

The effective diffusion coefficient ranges from 1.2 to 7.4×10^{-3} m^2·s^{-1} as described above. If we measured ^{220}Rn concentrations at a distance of 20 cm from wall, Q(0.2), exhalation rate of ^{220}Rn from wall can be calculated to be 0.0099Q(0.2) in ranging from 0.0074Q(0.2) to 0.0124Q(0.2) from Eq. (5-1). Thus, ^{220}Rn exhalation rate could be estimated by uncertainty of only $\pm25\%$ from the ^{220}Rn concentration at

20 cm from wall.

The ^{212}Pb (ThB) concentration in dwelling is expressed as follows [24]:

$$C_B = \frac{\lambda_a \cdot \lambda_B \cdot N_B \cdot S/V + \lambda_v \cdot C_{BO}}{\lambda_B + \lambda_v + \lambda_d}, \qquad (5\text{-}2)$$

where N_B is defined by

$$N_B = E_{Tn}[\lambda_{Tn}(\lambda_B + \lambda_a + \sqrt{\lambda_{Tn}(\lambda_B + \lambda_a)})], \qquad (5\text{-}3)$$

C_B is indoor ^{212}Pb concentration in Bq·m^{-3}, λ_a is the attached coefficient of free nuclei to aerosol particles in s^{-1}, λ_B is the decay rate of ^{212}Pb in s^{-1}, S is the surface area of wall in m^2 from which ^{220}Rn exhales, V is the volume of the room in m^3, λ_v is the ventilation rate of the room air in s^{-1}, C_{BO} is outdoor ^{212}Pb concentration in Bq·m^{-3}, and λ_d is the deposition rate of attached ^{212}Pb atoms in s^{-1}. Here, we assume that the ratio of surface area exhaling ^{220}Rn to volume of the room is about 1/3 and does not vary on dwellings. The values of the another parameters are assumed to be; λ_a = 0.0139 s^{-1} [25], λ_v= 1.39X10^{-4} s^{-1} [26] and λ_d = 5.6X10^{-5} s^{-1} [25]. Moreover, ^{220}Rn progeny concentration is equal to about 1.1 times ^{212}Pb, if ^{212}Pb and ^{212}Bi are in equilibrium. Consequently, ^{220}Rn progeny concentration could be expressed by ^{220}Rn exhalation rate as follows:

$$C_{TP} = 1.3E_{Tn}, \qquad (5\text{-}4)$$

where C_{TP} is ^{220}Rn progeny concentration in Bq·m^{-3}. As E_{Tn}=0.0099Q(0.2) described above,

$$C_{TP} = 0.013Q(0.2). \qquad (5\text{-}5)$$

Equation (5-5) indicates that indoor ^{220}Rn progeny concentration could be estimated from the ^{220}Rn concentration measured with passive cup monitors at 20 cm from wall. In some dwellings, ^{220}Rn progeny concentrations were measured by active methods [9, 27] during the surveys of ^{220}Rn concentrations with cup monitors. **Figure 13** shows the relationship between measured ^{220}Rn and ^{220}Rn progeny concentrations. The correlation coefficient between ^{220}Rn and ^{220}Rn progeny concentrations was 0.79. The relationship between ^{220}Rn progeny and ^{220}Rn concentrations is therefore given by

$$C_{TPm} = 0.015Q(0.2)_m, \qquad (5\text{-}6)$$

where C_{TPm} and $Q(0.2)_m$ are the measured concentrations of ^{220}Rn progeny and ^{220}Rn. The relation of Eq. (5-6) gives fairly good agreement with the theoretical relation in Eq. (5-5). Annual effective dose equivalents due to ^{222}Rn and ^{220}Rn progeny were estimated by using the measured values of small scale surveys. The results classified by building materials are shown in **Table 1**. The effective dose equivalent factors of ^{222}Rn and ^{220}Rn progeny are estimated to be 70 and 280 μSv·y^{-1} per Bq·m^{-3} using a mean breathing rate of 0.8 m^3·h^{-1} indoors and an average occupancy factor of 0.8 indoors [1]. An equilibrium factor between ^{222}Rn and its progeny was assumed to be 0.4 indoors. On the other hand, ^{220}Rn progeny concentration was calculated by

Table 1 Estimation of annual effective dose equivalents due to ^{222}Rn and ^{220}Rn progeny.

	^{222}Rn		^{220}Rn	
Building materials	Concentration (Bq·m^{-3})	H_{eff} (μSv·y^{-1})	Concentration (Bq·m^{-3})	H_{eff} (μSv·y^{-1})
soil	19.1±0.9	530±30	159.7±12.4	670±50
concrete	23.3±1.5	650±40	41.9± 3.3	180±10
NBM	17.5±1.2	490±30	22.9± 3.0	100±10
UNSCEAR 1988[1]		1100		160

Eq. (5-5). The annual effective dose equivalents due to ^{220}Rn progeny in the dwellings with soil wall was calculated to be 0.67 mSv·y^{-1} and may be equal to that due to ^{222}Rn and its progeny.

6. Conclusions

A passive monitor is a simple and handy method for measuring ^{222}Rn concentration. A passive monitor has some unclear factors. In the present study, some problems with the cup method were clarified. It is necessary that the cup wall material is conductive and low α-emitting. The ^{222}Rn concentrations in various living environments differ with the location and season. It is important that the observed track density is not affected by various environmental factors, especially ^{220}Rn. In the dwellings like typical Japanese houses that sometimes have high ^{220}Rn concentrations, care is required to measure ^{222}Rn concentration by the passive cup method. In those dwelling, we should use a cup monitor with a Rn exchange rate less than 0.1 h^{-1}. A pair of passive cup monitors with a different air exchange openings was developed for measuring simultaneously ^{222}Rn and ^{220}Rn concentrations. Indoor ^{220}Rn concentrations were very high in traditional Japanese dwellings with soil walls. The ^{220}Rn concentration decreases exponentially with the distance from wall. The effective diffusion coefficient of ^{220}Rn in dwelling and the exhalation rate of ^{220}Rn from wall were evaluated from the distribution of the ^{220}Rn concentrations. Then, indoor ^{220}Rn progeny concentration could be estimated from the ^{220}Rn concentration at a 20 cm distance from wall. From the results of the surveys, the average annual effective dose equivalent due to ^{220}Rn progeny was expected to be 0.67 mSv·y^{-1} in the traditional Japanese dwellings.

6. References

[1] UNSCEAR, Sources, effects and risks of ionizing radiation, pp.106 (New York, UN), 1988.

[2] Alter, H.W. and R.L.Fleischer, Passive integrating radon monitor for environmental monitoring, Health Phys., **40**, 693-702, 1981.

[3] Urban, M. and E.Piesch, Low level environmental radon dosimetry with a passive track etch detector device, Radiat. Prot. Dosim., **1**, 97-109, 1981.

[4] Bartlett, D.T., P.J.Gilvin, D.W.Dixon, et al., The performance of the NRPB radon personal dosimeter, Radiat. Prot. Dosim., **17**, 139-142, 1986.

[5] George, A.C., A passive environmental radon monitor, HASL 325, pp.25-39 (Springfield, VA: National Technical Information Service), 1977.

[6] George, A.C., Passive integrated measurement of indoor radon using activated carbon, Health Phys., **46**, 867-872, 1984.

[7] Coast-Ribeiro, C., J.Thomas, R.T.Drew, et al., A radon detector suitable for personnel or area monitoring, Health Phys., **17**, 193-198, 1969.

[8] Kobayashi, S., K.Fujimoto, T.Iwasaki, et l., Nation-wide survey of indoor radon concentration in Japan, Proc. 3rd Int. Symp. on Advanced Nucl. Energy Res., Mito, Japan, pp.63-67, 1991.

[9] Guo, Q., M.Shimo, Y.Ikebe, et al., The study of thoron and radon progeny concentration in dwellings in Japan, Radiat. Prot. Dosim., **45**, 357-359, 1992.

[10] Doi, M., S.Kobayashi and K.Fujimoto, A passive measurement technique for characterization of high-risk houses in Japan due to enhanced levels of indoor radon and thoron concentrations, Radiat. Prot. Dosim., **45**, 425-430, 1992.

[11] Guo, Q., T.Iida, K.Okamoto, et al., Measurements of thoron concentration by passive cup method and its application to dose assessment, J. Nucl. Sci. Technol., **32**, 794-803, 1995.

[12] Miyake, H., K. Oda, T. Yamauchi, et al., Discriminative measurement of radon and thoron concentrations with an electrostatic passive monitor, Hoshasen, **19**, 9-18, 1993 (in Japanese).

[13] Iida, T., Y. Ikebe, T. Hattori, et al., An electrostatic integrating ^{222}Rn monitor with cellulose nitrate film for environmental monitoring, Health Phys., **54**, 139-148, 1988.

[14] Abu-Jarad, F., Variation in long-term radon and daughters concentration with position inside a room, Radiat. Prot. Dosim., **3**, 227-231, 1982.

[15] Swedjemark, G.. A. and L. Mjöones, Radon and radon daughter concentrations in Swedish homes, Radiat. Prot. Dosim., **7**, 341-345, 1984.

[16] Fleischer, R. L. and L. G. Tunner, Indoor radon measurements in the New YOrk capital district, Health Phys., **46**, 999-1011, 1984.

[17] Hess, C. T., R. L. Fleischer and L. G. Tunner, Field and laboratory tests of etched track detectors for ^{222}Rn: summer-vs-winter variations and tightness effects in Maine houses, Health Phys., **49**, 65-79, 1985.

[18] Kotrappa, P., Dempsey, J. C., Ramsey, R. W., et al., A practical E-PERM™ (electret environmental radon monitor) system for indoor ^{222}Rn measurement, Health Phys., **58**, 461-467, 1990.

[19] Al-Bataina, B. and J. Janecke, Alpha-particle emission from contaminations in counter materials, Nucl. Instr. Meth., **A255**, 512-517, 1987.

[20] Nelson, R. A., Measurement uncertainties of long-term ^{222}Rn averages at environmental levels using track detectors, Health Phys. **53**, 447-453, 1987.

[21] Pearson, M. D., D. E. Martz, J. L. George, et al., A multiyear quality control study of alpha-track radon monitors, Health Phys. **62**, 87-90, 1992.

[22] Iida, T., An electrostatic radon monitor for continuous measurement of environmental radon, In: Atmospheric radon progeny and environmental radon, Ed. S. Okabe (Atomic Energy Society of Japan, Tokyo) pp. 65-75, 1985 (in Japanese).

[23] Doi, M., K. Fujimoto, S. Kobayashi, et al., Spatial distribution of thoron and radon concentrations in the ambient air of a traditional Japanese wooden house, Health Phys. **66**, 43-49, 1994.

[24] Katase, A., Y. Matsumoto, T. Sakae, et al., Indoor concentrations of ^{220}Rn and its decay products. Health Phys., **54**, 283-286, 1988.

[25] J. Porstendörfer and A. Reineking, Indoor behavior and characteristics of radon progeny, Radiat. Prot. Dosim., **7**, 107-113, 1984.

[26] Y. Matsuo, S. Murakami, N. Miyata, et al., Building and Weather, Asakura, Tokyo, pp. 89, 1986 (in Japanese).

[27] T. Yamasaki, T. Iida, Measurements of thoron progeny using potential alpha-energy monitor in Japan, Health Phys., **68**, 840-845, 1995.

A prospective assessment of ^{210}Po surface collection for estimating ^{222}Rn exposure

Brian Fitzgerald and Philip Karl Hopke

Department of Chemistry, Clarkson University
Potsdam, N.Y. 13699-5810

Abstract

Previous studies have examined the utility of measuring the embedded ^{210}Po in solid materials as an indicator of long term ^{222}Rn exposure. However, in those studies, existing surfaces were employed and problems arose because of the uncertainties in the length of time the surface had been exposed and the concentrations of ^{222}Rn to which it had been exposed. In this prospective studies, pairs of glass surfaces were placed into homes around the United States along with an LR-115-II etched track radon detector for each piece of glass. The results of measurements on a subset of samples are presented. The ^{222}Rn measurements are compared to the ^{210}Po measurements and a relationship is obtained between the two results using a model for the accumulation of activity within an enclosed space. The data is interpreted in terms of the usefulness of the ^{210}Po measurements as a long term ^{222}Rn detector and the value of a new interpretation scheme to make retrospective assessments of the ^{222}Rn exposure within the room.

1. Introduction

In many completed or current case-control radon epidemiological studies of the general population, cumulative radon retrospective exposure assessment is primarily based on contemporary measurements of radon and assuming that past and present levels are equal. Such current measurements cannot take into account the changes that may have taken place over the years in the radon concentrations within these houses [1]. In some studies, the radon measurement may be short term (e.g. less than three months) and most of the epidemiological studies to date have been confined to the houses that the subjects currently occupy. Thus, reliable techniques that can assist in improving the retrospective assessment of exposure to radon and radon decay products are needed.

Measurements using the surface trap technique have been made in which the ^{210}Po activity is trapped by α-recoil implantation, in a thin layer (approximately 100 nm deep) on and beneath the surface of glass [2][3]. A complimentary technique is being developed by other groups that uses porous materials as a volume trap of the long lived radon decay products [4].

The build-up of long lived radon decay products, such as ^{210}Po, in glass and other vitreous surfaces is an experimentally well established phenomenon [5]. The process is driven by α-recoil. The parent radionuclide decays by a-decay and its decay product recoils. Any decay products deposited onto the surface of the glass can therefore recoil into the glass. The activity of the ^{210}Po in the layer on the glass surface is more closely related to the historical mean behavior and characteristics of the airborne decay products than to the radon gas levels. As such, it could be useful in reconstructing past exposure and dose levels than a single contemporary radon gas measurement.

While this present work is specifically directed towards the surface trap method, many questions apply to any method of retrospective exposure assessment. Assuming that the measuring techniques are accurate and well calibrated, a number of questions must be answered.

One question is how representative of the typical conditions within a room or a building is a measurement of ^{210}Po activity on a particular area of a specific piece of glass. The local conditions around a piece of glass such as thermally induced air flow, aerosol gradients, electric fields, etc. can influence the deposition process and will vary from surface to surface within the room.

The quantitative influence on ^{210}Po levels arising from the effects, separately or in concert, of cleaning, accumulation of dirt, grease, ^{210}Po from tobacco smoke and other contaminants on the glass are uncertain. This question has been investigated by Lively and Steck [6]. They found that between 90% and 98% of the ^{218}Po and 50% to 80% of the ^{214}Po was removable by washing. The differences were explained in terms of the thickness of the 'dirt/grease' film on the glass. The thicker film on the glass will lead to more activity being trapped on the surface and so is washable. For ^{210}Po, they measured between 5% and 30% being removable by washing. This variation was again explicable in terms of the amount of 'dirt' on the glass surface. They also detected a deficit in the ^{210}Po adsorbed onto the surface indicating that for activity not embedded into the glass, there is some unknown removal process at work. Cornelis *et al.* [5] indicated that up to 85% of the surface activity may be removable by cleaning.

There is a problem of the reliability of estimation of the age of the glass. It is possible that a piece of glass has been broken and replaced or even moved from other locations at some time in the past. The use of glass from personal mementoes like wedding pictures is generally advantageous as they will usually accompany the subjects through a significant fraction of their adult lives. Because they are often associated with specific events, it may be possible to have more confidence in the age given by their owners.

The above questions highlight the difficulties in using ^{210}Po measurements to make an assessment of the retrospective exposure or dose incurred due to ^{222}Rn and its decay products. However, while exact quantitative assessments are problematic, ^{210}Po measurements can be used as a tool to assess the correctness or otherwise of using contemporary ^{222}Rn measurements as surrogates for historical levels in epidemiological studies.

Previous work has demonstrated both theoretically and experimentally that regardless of the actual indoor exposure conditions within a normally occupied home, an assessment of the radon exposure is adequate for estimating the dose to the lungs [7][8]. If a scheme can be devised to allow investigators to determine whether a contemporary ^{222}Rn measurement is significantly different from the historical average in a dwelling, this result would be useful in epidemiological studies.

If the relationship between ^{210}Po measured in surfaces and ^{222}Rn levels is known, then such surfaces can serve as long term ^{222}Rn detectors; of the order of 2 or more years. As they can be counted with electronic detectors such a pulse ionization chambers (Samuelsson, 1988), they are less prone to saturation than conventional passive ^{222}Rn detectors and could in theory be more environmentally robust and also reusable.

To address some of the questions raised above, this present study was conducted to determine the feasibility of using ^{210}Po in glass measurements as both a contemporary ^{222}Rn detector and an indicator of retrospective exposure or at least changes in exposure.

2. Theoretical Prediction of ^{210}Po Buildup in Glass

^{222}Rn decays into a series of short-lived radio-nuclides followed by a number of long-lived ones. The ^{218}Po, ^{214}Pb, ^{214}Bi and ^{214}Po are all short-lived radio-nuclides. They can undergo a number of process which tend to remove them from the air. Of interest here is the deposition processes, whether they are attached to particles or unattached. The processes are outlined in Figure 1. To predict the partitioning of the radionuclides between the various possible states, a model based on an activity concentration balance is used. To a first approximation, the short-lived radionuclides in a normal room can be modeled with a steady state model. A steady state model assumed that all the rate parameters such as ventilation are constant with time. Also, by assuming that the room has been in a steady state for several hours the short-lived decay products can be assumed to be in radioactive equilibrium. This assumption allows the model to be simplified from a dynamic model involving a first order differential to a simple algebraic equation [9].

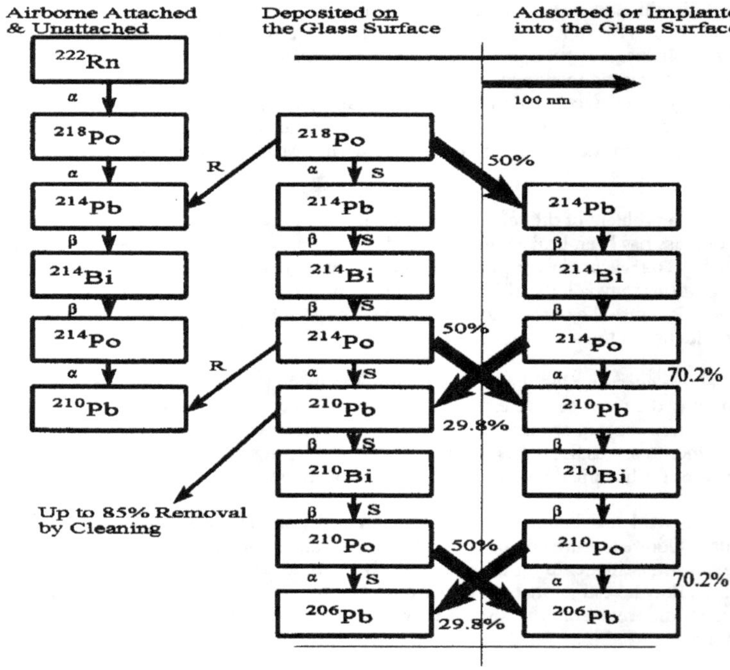

Figure 1. Compartment model for the build-up of ^{210}Po in or on glass surfaces.

The build-up of ^{210}Po in glass has been modeled by Cornelis *et al.* [5]. They have calculated the fraction of the activity that appears on the surface and embedded under the surface (Figure 1). These fractions can be included in the steady state model. The amount of ^{210}Pb and ^{210}Po can be most easily modeled by using the recursive formalism given by Samuelsson [10]. The equations used to generate this model are given by Fitzgerald and McLaughlin [13]. The model was implemented in PowerBasic and allows the calculation of the ^{210}Pb, ^{210}Bi, and ^{210}Po at any time along the life of a piece of glass exposed to a ^{222}Rn atmosphere under the assumption of steady state conditions over a period of time longer than 3 hours. The code allows the equilibrium conditions to be varied in steps to allow the effect of specific parameter changes to be calculated.

3. Contemporary ^{210}Po as a Contemporary ^{222}Rn Indicator

The quantity of ^{210}Po accumulated on a surface is related to the length of the exposure of the surface to the ^{222}Rn atmosphere, the time averaged partitioning of the decay products between airborne and surface deposited and the action of any removal mechanisms such as cleaning. If all rooms were identical in terms of room parameters, then the build-up of ^{210}Po would be identical. Thus, knowing the exposure age of the glass, the ^{210}Po level and the ^{210}Po-^{222}Rn relationship, it would be possible to calculate the 'time-averaged' ^{222}Rn level to which the glass was exposed. This result is only valid if the ^{210}Po was accumulated over the same time period during which the ^{222}Rn measurement was made. It is obvious that not all rooms are the same and thus the relationship between the ^{222}Rn and ^{210}Po will not be precisely the same. Previous work has shown that in domestic dwellings, the exposure conditions tend to be similar as expressed through the equilibrium factor [8] and the build-up of ^{210}Po [3]. It was expected that a relationship between the ^{222}Rn and the ^{210}Po could be established that will be valid for a reasonable range of dwellings.

Table 1. Room parameters used to calculate the theoretical ^{210}Po buildup curves

Constant Parameters	
Attachment Coefficient (b)	4.7×10^{-3} cm^3 hr^{-1}
Unattached Deposition Velocity	5 m hr^{-1}
Attached Deposition Velocity	0.05 m hr^{-1}
Aerosol Recoil Factor	0.83
Variable Parameters	
High Deposition Regime	
Ventilation Rate	0.2 hr^{-1}
Aerosol Concentration	1680 per cm^3
Surface To Volume Ratio	4 m^{-1}
Medium Deposition Regime:	
Ventilation Rate	0.55 hr^{-1}
Aerosol Concentration	16,800 per cm^3
Surface To Volume Ratio	3 m^{-1}
Low Deposition Regime:	
Ventilation Rate	1.55 hr^{-1}
Aerosol Concentration	168,000 per cm^3
Surface To Volume Ratio	2 m^{-1}

Table 1 gives a set of parameters for the steady state room model that allows the calculation of a set of envelopes. These values are regarded as the maximum, minimum and 'base-line/medium' deposition regimes likely to be present in any domestic environment. Therefore, all ^{210}Po measurements should fall within these extremes. Figure 2 shows the results of the calculations using the values in Table 1.

4. ^{210}Po as an Assessment of Historical ^{222}Rn

When making a retrospective assessment of ^{222}Rn exposure, it is assumed that the total ^{210}Po is related to the 'time-average' exposure. While it may be reasonable to deduce an approximate relationship between ^{210}Po accumulated concurrent to the ^{222}Rn measurement, the same cannot be said of the contemporary ^{222}Rn and the ^{210}Po accumulated over a time period longer than the ^{222}Rn measurement. This outcome holds it cannot be positively established that either the historical ^{222}Rn or the historical exposure conditions were constant. If they varied widely, then while the ^{210}Po measured is an indicator of the retrospective conditions to determine 'time-averaged' ^{222}Rn, it would be necessary to have information on the past exposure conditions. However, the contemporary ^{222}Rn and the ^{210}Po measurement can be used to assess whether or not the ^{222}Rn measurement is incompatible with the ^{210}Po measurement and the historical conditions.

If the measured ^{210}Po to ^{222}Rn ratio is plotted against the exposure age of the glass sample, then it should fall within the envelope delimited by the high and low deposition regimes in Figure 2. Figure 2 is a plot of the ^{210}Po production rate per unit ^{222}Rn. Because of the values chosen for the high and low limits, all of the $^{210}Po/^{222}Rn$ values are expected to fall within those limits.

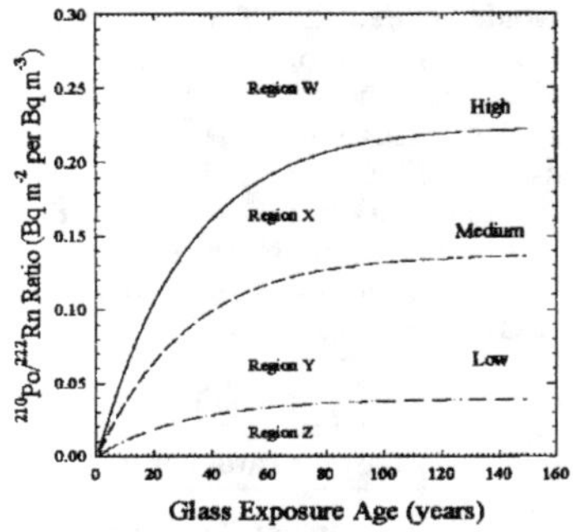

Figure 2. Theoretical Calculations of the accumulation of ^{210}Po on and in a glass surface using the parameters given in table 1.

If the $^{210}Po/^{222}Rn$ ratio falls above the upper limit (region W), then it is reasonable to conclude that the contemporary ^{222}Rn is not representative of the historical levels. The only reason for the $^{210}Po/^{222}Rn$ ratio appearing in region W is if the contemporary ^{222}Rn is lower than the historical average or the ^{210}Po production is higher than though possible. High ^{210}Po production could be due to an external source of additional radioactivity such as cigarette smoke [11]. However, this source of additional radioactivity is unlikely to have a major impact on the observed ^{210}Po surface activity levels. If the value falls below the lower limit into region Z, then while it is likely that the contemporary ^{222}Rn is incompatible with the historical ^{222}Rn levels, as expressed by the ^{210}Po measurement, it is also possible that the ^{210}Po level has been depressed by a removal mechanism, such as cleaning. If the $^{210}Po/^{222}Rn$ ratio falls within the high-low envelope (regions X and Y), it can be concluded that the ^{222}Rn and ^{210}Po are compatible; but only within a wide margin of uncertainty.

The ^{210}Po production envelope as shown in Figure 2 has a factor of 6 difference between the high and low limits. Therefore, a deviation of a factor of up to 6 in the exposure and/or the ^{222}Rn levels is necessary before a data point will fall outside of the envelope. This level of uncertainty is an important limitation on the technique. As plotted, any points falling within the envelope indicate agreement between the ^{222}Rn and the ^{210}Po measurement within a factor of 6.

The interpretation technique outlined in the previous paragraph has the advantage that it responds to the ^{222}Rn levels and to the exposure conditions. This dependency is useful in epidemiological studies as it is the ^{222}Rn and the exposure conditions that determines the inhaled dose. Because of the width of the envelope, this interpretation technique cannot be use to establish if a ^{222}Rn and a ^{210}Po measurement are in agreement; merely if they are not in agreement. However, if the envelope can be narrowed by better definition of various parameters such as the aerosol concentration, then the use of this technique as positive confirmation becomes increasingly valid.

5. Experimental Procedure

The objective of these experiments was to assess if a relationship can be established between contemporary ^{222}Rn measurement and the ^{210}Po accumulated over the same time period. The experimental design was to simultaneous expose a piece of glass and a ^{222}Rn

detector at a large number of sites. The glass and ^{222}Rn detector were placed for the same length of time and in the same place. This concurrence of location eliminates error due to spatial variations in the ^{222}Rn concentration. Efforts were made to ensure that all of the measurements spanned the same length of time.

The glass was purchased in the form of picture frames. Participant were found by contacting the agency in each state with responsibility for that states radon program and requesting information on any measurements indicating ^{222}Rn concentrations greater than 370 Bq m^{-3}. Since many of the measurements were short-term, closed house measurements, it was assumed that many of these values were overestimates. To each participant in the study, two picture frames were dispatched. Accompanying each frame was an LR-115-II detector that was attached to each frame by the participant when the picture was hung. The participants were asked to hang the picture as they normally would. While the majority of the frames were dispatched to homes, 10 were sent to each of 2 different schools. The pictures and the detectors were recalled after 18 months and returned to the laboratory. The glass was removed from the picture frame, wrapped in aluminum foil and stored for 15 months to permit the ingrowth of ^{210}Po. The LR-115-II detectors were returned to the manufacturer, Alpha Spectra Ltd., for processing. A number of blanks, duplicates and spikes were also included for quality control.

To count the accumulated ^{210}Po activity, a flow-through pulse ionization chamber was used. The design is well described by Samuelsson [3]. The basic design is a Frisch grid chamber where the cathode is formed by laying the glass sheet on open top of the chamber, directly above the grid. The chambers used were of two different sizes. The larger chamber was constructed from a plastic outer tube and stainless steel grid and anode. To provide adequate shielding from electronic noise the outside of the chamber was painted with conducting silver paint and grounded through the pre-amplifier. A high voltage of 1800 volts was applied to the anode and 1100 volts to the grid. To minimize the noise, the glass sample was shielded with aluminum foil which was wrapped around the backside of the glass and around the ionization chamber to make contact with the conducting paint. The counting gas was P-10, 10% methane and 90% argon. The chamber was purged for 5 minutes to insure that no oxygen was present. To insure replacement of gas consumed and an over-pressure to prevent oxygen leaking into the chamber, the chamber was operated with a flow rate of 100 cm^3 per minute.

Using standard ^{239}Pu, ^{241}Am and ^{230}Th sources attached to an unexposed glass plate, a reproducible resolution of 50 keV FWHM was attainable. However, when measuring ^{210}Po embedded in the glass sheet, the resolution varied between 100 and 150 keV, and on occasion, some sheets showed resolutions of 200 keV. No clear indication of the mechanism governing the resolution was discernable. However, it is likely to be a combination of the flatness of the glass, the amount of household dust on the glass surface and the efficiency of using the aluminum foil as noise shielding played a role in determining the ultimate resolution achieved.

In the absence of any shielding for the glass, the signal profile from the chamber both showed increased noise and also a slow decrease in the pulse size over time. The signal pulse size decrease was probably due to the buildup of a static charge on the glass plate. Such a charge would tend to decrease the internal electric fields established by the voltages applied to the grid and the anode.

A second chamber design was also used. This chamber was approximately two thirds of the volume of the original model and fabricated from stainless steel. The capacitance was larger and the intrinsic resolution was 100 keV. However, it was more easily grounded.

Each sheet of glass was counted on top of a chamber until approximately 1000 counts were accumulated under the 5.3 MeV peak. Thus, the statistical errors in the counts were approxi mately 3% or less. The results were interpreted using the model outlined above.

6. Experimental Results

The data gathered to date is shown in a series of figures. Each figure is generated from the same complete set of data. Figure 3 shows the ^{222}Rn plotted against the ^{210}Po exposure. This type of plot has been used in previous work [3] to demonstrate the applicability of using ^{210}Po as a retrospective assessment of ^{222}Rn exposure. There appears, from the regression coefficients to be a strong linear relationship, the regression coefficients are given in Table 2. The regression indicates that it should be possible to use the ^{210}Po as an assessment of the total ^{222}Rn exposure.

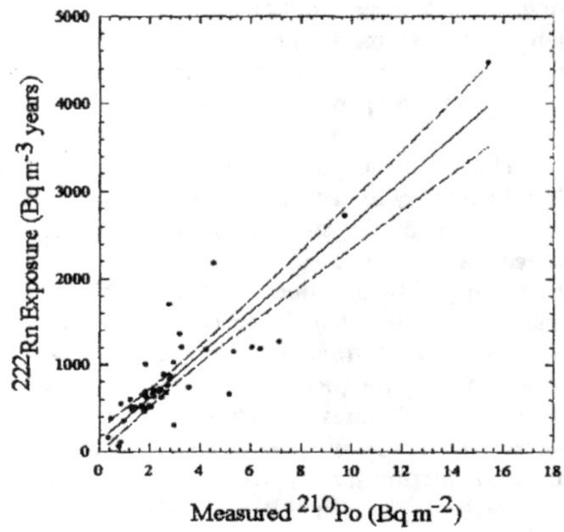

Figure 3. ^{222}Rn exposure as a linear function of ^{210}Po for all data points. The resulting fitted line has a slope of 251 Bq m^{-3} years/Bq m^{-2}, an intercepts of 120 Bq m^{-3} years and an r^2 of 0.81. The dashed lines denote the 95% confidence intervals.

Figure 4 shows the ratio of the measured ^{222}Rn to the value predicted using the ^{210}Po measurement and the linear regression. As is clear from Table 3 and the values shown in the figure, the average ratio is close to unity within one standard deviation. However, individual values can be up to a factor of 6 too high and a factor of 3 too low. This range would indicate that as a single data set, the individual datum point can be incorrect. This may skew studies looking for small effects as it will tend to increase the individual systematic errors.

Table 2. Results of the regression analyses of the various data subsets.

	All Data	High Outliers Removed	High & Low Outliers Removed
y intercept	120	123	53
slope	251	280	284
r^2	0.81	0.94	0.99

Table 3. Ratio between predicted and measured ^{222}Rn for the various subsets.

	All Data	High Outliers Removed	High & Low Outliers Removed
Average ratio	1.1	1.04	1.02
Standard deviation	0.4	0.21	0.17

Figure 5 shows the ratio of the measured ^{210}Po and the ^{222}Rn concentration plotted as a function of the measured exposure age of the glass. Because all the samples were deliberately exposed for the same length of time, all of the data points are clustered around the same region of the x-axis. This type of figure to interpret such data has been used before [12][13]. Using this plot as a data filter before the measurements are employed to estimate retrospective ^{222}Rn concentrations, the reliability of the prediction of individual historical ^{222}Rn values for given individual ^{210}Po measurements can be improved.

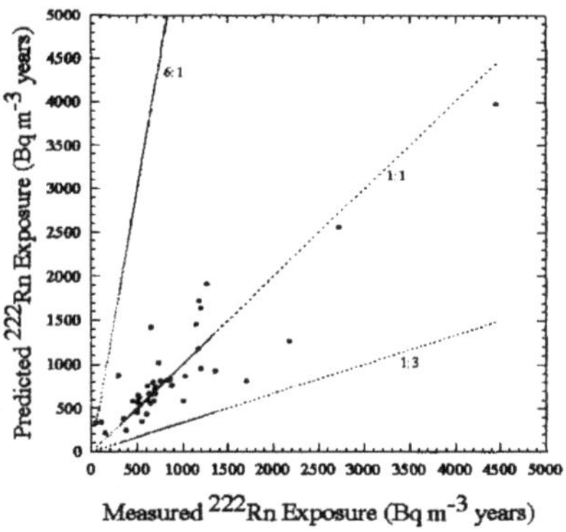

Figure 4. Scatter plot of predicted and measured ^{222}Rn for all data points. The average ratio of the predicted to the measured ^{222}Rn is 1.1 ± 0.4.

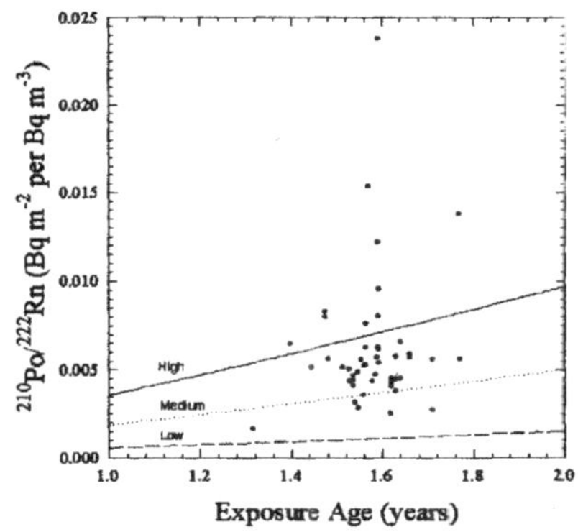

Figure 5. ^{210}Po build-up as a function of glass exposure age, theoretical and measured values for all data points.

7. Analysis of Experimental Results

Using Figure 5 as a data filter the first step is to remove any data points lying above the high extreme. This data is replotted and the regression show in Figure 6 and Table 2. These points correspond to an over production of ^{210}Po relative to the measured ^{222}Rn concentration. Possible explanations for these excess values were previously discussed. Comparison with Figure 3 and the regression data clearly two points of interest. Those points lying significantly below the 95% confidence interval have been removed in Figure 6. The 10 points furthest above the high extreme were removed and these points correspond to the 10 points furthest from the 95% confidence interval. The regression coefficient increased from 0.81 to 0.94. This change has reduced the error in the fit. The ratios of the predicted and measured ^{222}Rn values has been replotted in Figure 5. It can be clearly seen that there is a dramatic reduction in the scatter in the fit. Almost all the data appears to be with a factor of 1.6 on either side of measured value. The average ratio has fallen to 1.02 and the standard deviation has fallen to nearly 50% of the value given in Figure 4. At this point it would appear that using the ^{210}Po generation rate plot to provide physical justification for removing apparent outliers is a valuable aid to accurate reconstruction of the ^{222}Rn exposure.

By the standards that have been set, all of the other points in Figure 5 are valid. As can be seen in Table 1, the low extreme corresponds to an average aerosol concentration of 168,000 particles per cm^3. This value is very unlikely in dwellings and so an argument could be made to reduce this value and so narrow the envelope defined by the high and low extremes. To examine the effects of tightening the room parameters, all of the data points below the medium deposition rate were removed. These data were plotted in Figures 8 and 9. Again the correlation coefficient has been increased.

The 5 points below the medium deposition boundary have been removed. They all appeared above the 95% confidence interval, this mirrors the removal of the points above the high deposition extreme. Also the y axis intercept has been reduced from 123 to 53 Bq m^{-2} years. As can be clearly seen in Table 3 and Figure 9, the predicted data falls with 20% of measured data. The spread of data points has also been reduced and the average ratio between predicted and measured ^{222}Rn exposure is nearly one.

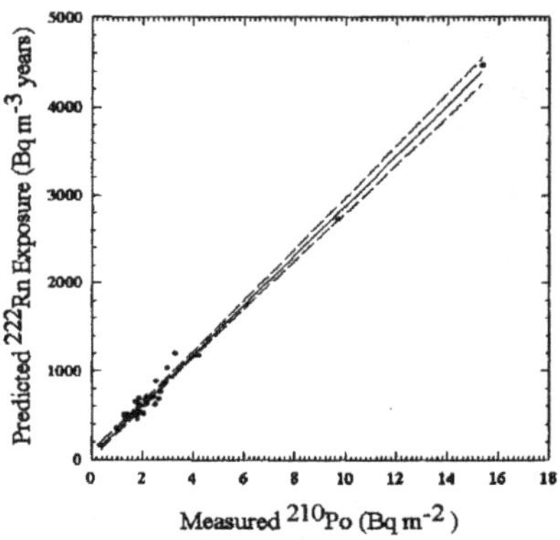

Figure 6. ^{222}Rn exposure as a linear function of ^{210}Po with the "high outliers" removed. The resulting fitted line has a slope of 280 Bq m^{-3} years/Bq m^{-2}, an intercepts of 123 Bq m^{-3} years and an r^2 of 0.94. The dashed lines denote the 95% confidence intervals.

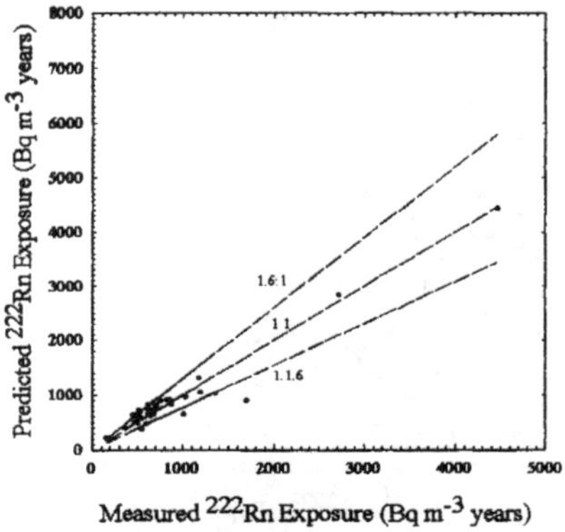

Figure 7. Scatter plot of predicted and measured ^{222}Rn for all data points with the 'high outliers' removed. The average predicted to measured ^{222}Rn ratio is 1.0 ± 0.2.

Overall, it would appear that using the combination of the linear regression analysis and the modeled ^{210}Po generation rate as data filters permits the prediction of the ^{222}Rn exposure within approximately ±60% of the 'true ' value, Figures 6 and 7. If it is possible to narrow the high and low extremes of the envelope, it may be possible to reconstruct the ^{222}Rn exposure to ±20%. The main drawback is that the size of the useful data set is substantially reduced. In this study, the data set was reduced from 47 to 32 points, a reduction of approximately one third.

8. Application of these Techniques

It would appear that it is possible to predict the ^{222}Rn values form the measured ^{210}Po values, but only within a factor of approximately 6. However, using the ^{210}Po to filter the data, it would appear to reduce the error to approximately a factor of 2. As an example of how to utilize this combination of techniques to make a useful determination of the historical ^{222}Rn the following measurements are assumed. Firstly, it is assumed that the relationship between the ^{222}Rn and the ^{210}Po is well established and will hold for the measurement conditions. Measurements are made of the contemporary ^{222}Rn and ^{210}Po accumulation on a piece of glass. In order to be representative of the houses over a long period of time the glass analyzed is the undisturbed wedding picture frame from approximately 30 years ago. If the glass is very old, say 70 plus years then as can be seen in Figure 2, the ^{210}Po may approach equilibrium and so the linear approximation evident in Figure 5 would no longer hold. Most of the data lies in regions X and Y (Figure 2). These data can then be used to calculate retrospective ^{222}Rn exposure with a reasonable expectation that the calculated and the actual unknown retrospective value are within a factor of 2 of each other. However, if a point lies in region W, above the high deposition extreme, this point would be flagged as being an unreliable indicator of retrospective ^{222}Rn concentrations and treated appropriately.

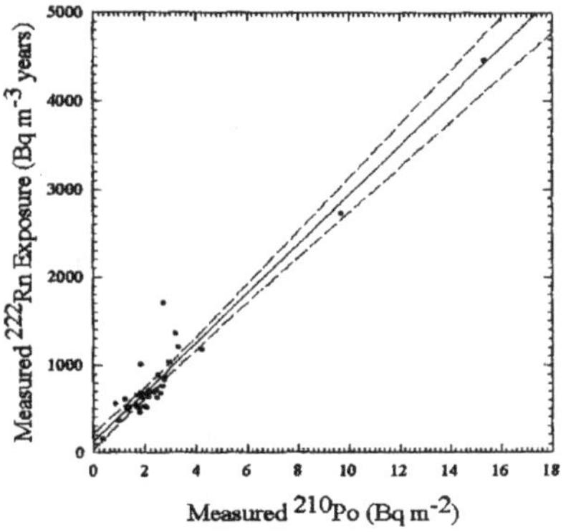

Figure 8. ^{222}Rn exposure as a linear function of ^{210}Po with the high and low outliers removed. The resulting fitted line has a slope of 284 Bq m^{-3} years/Bq m^{-2}, an intercepts of 53 Bq m^{-3} years and an r^2 of 0.99. The dashed lines denote the 95% confidence intervals.

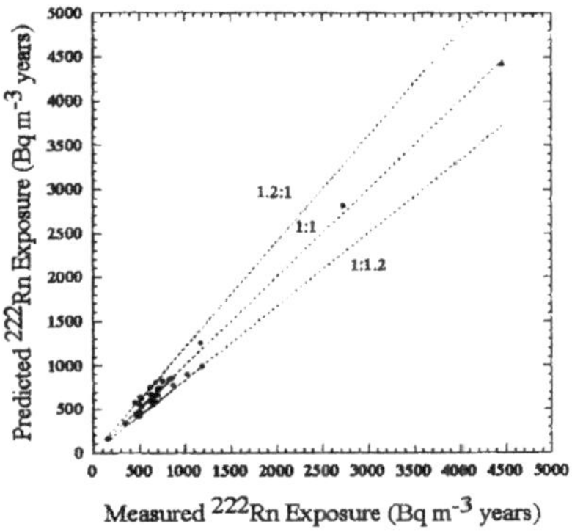

Figure 9. Scatter plot of predicted and measured ^{222}Rn for all data points with the 'high and low outliers' removed. The average predicted to measured ^{222}Rn ratio is 1.0 ± 0.2.

If studies are being performed in a particular area then detailed data maybe available on the likely room parameters in dwellings in those areas. For instance, if an epidemiological study is being performed in the northeastern US and southeastern Canada, then several studies of the aerosol and decay product equilibrium conditions exist; e.g. Hopke *et al.* [8]. These results can be used to redraw the high and low extreme boundaries. This analysis would allow each study to better address the conditions in its geographical area.

9. Conclusions

By using these two interpretative schemes together the reliability of using ^{210}Po as a retrospective predictor of ^{222}Rn exposure is greatly enhanced. It would appear that the ^{210}Po can be used as quantitative measurement of long term contemporary ^{222}Rn concentrations. Using glass samples and with accurate knowledge of their exposure age the ^{210}Po accumulated can be directly converted to a ^{222}Rn concentration and that while the deposition environment will have some impact that most structures are similar enough that the error introduced will be of the order of a factor of two. While glass sheets will not replace detectors such as closed CR-39 they have a number of advantages, they can be already in place in the home, they can be reused and they can be exposed to much higher ^{222}Rn exposures and m much harsher environments . Furthermore, it would appear that by using these two interpretation techniques in tandem that the reliability of individual predictions of retrospective ^{222}Rn exposures can be improved to the point where the ^{222}Rn could be predicted with reasonable confidence to within a factor of 2 of its true value. The next necessary step is to accumulate a larger quantity of similar data in many different environments, such as in various countries and various climates. By doing so it would be hoped to produce a set of relationships between measured ^{210}Po buildup and ^{222}Rn for various dwelling types and locations. The ideal case would be if one single relationship could be established between measured ^{210}Po buildup and ^{222}Rn for all dwelling types.

11. Acknowledgments

The authors of this work wish to acknowledge the financial support of the US Department of Energy under grant DE FG02 90ER61029.

11. References

[1] Steck, D.J., Spatial and temporal indoor radon variations, *Health Phys.*62,351-355, 1992

[2] Lively, R.; Ney, E.P., Surface radioactivity resulting from the deposition of ^{222}Rn daughter products, *Health Phys.* 52, 411-415, 1987.

[3] Samuelsson, C., Retrospective determination of radon in homes. *Nature* 334, 338-340, 1988.

[4] Oberstedt, S.; Vanmarcke, H., Volume traps - A new retrospective radon monitor. *Health Phys.* 70:222-226; 1996.

[5] Cornelis, J.; Landsheere, C.; Poffijn, A.; Vanmarcke, H., Experimental and theoretical study of the fraction of ^{210}Po absorbed in glass. In: Cross, F., ed., *Indoor radon and lung cancer: reality or myth?* Battelle Press, Part 1, 101 -111; 1992.

[6] Lively, R.S.; Steck, D.J., Long-term ^{222}Rn concentrations estimated from ^{210}Po embedded in glass. *Health Phys.* 65, 485-490; 1993.

[7] Vanmarcke, H.; Janssens, A.; Raes, F.; Poffijn, A.; Berkvens, P; Van Dingenen, R., The behavior of radon daughters in the domestic environment : Effect on the effective dose equivalent, In: Hopke, P.K., ed., *Radon and its decay products: Occurrence properties and health effects.* Proc. ACS Symposium, American Chemical Society, Washington D.C., 301- 323; 1987

[8] Hopke, P.K.; Jensen, B.; Li, C.-S.; Montassier, N.; Wasiolek, P.T.; Cavallo, A.J.; Gatsby, K.; Socolow, R.H.; James, A.C., Assessment of the exposure to and dose from ^{222}Rn decay products in normally occupied homes. *Environ. Sci. Technol.* 29:1359 - 1364; 1995.

[9] Knutson, E.O., Modeling indoor concentrations of radon decay products, Radon and its decay products in indoor air. In: Nazaroff, W.; Nero, A., eds. Chapter 5, John Wiley and Sons Inc., New York, 491-504; 1988.

[10] Samuelsson, C., Application of a recursive formula to air sampling of radon daughters. *Nuc. Inst. Meth.* A262: 457-462; 1987.

[11] Martell, E.A., Radioactivity of tobacco trichomes and insoluble cigarette smoke particles. *Nature* 249: 214-217; 1974

[12] Fitzgerald, B., Retrospective radon exposure assessment using alpha track techniques. Ph.D. Thesis; National University of Ireland, University College Dublin; 1994.

[13] Fitzgerald, B.; McLaughlin, J. P., The measurement and assessment of ^{210}Po surface activity on glasses in dwellings, *Environ. Int.* 22, S845-S856, 1997.

Intercomparison of passive radon dosimeters developed by NIRS (Japan) and SSI (Sweden)

Masahiro Doi[1], Rolf Falk[2] and Inger Östergren[2]

[1] Environmental and Toxicological Sciences,
National Institute of Radiological Sciences, Japan

[2] Division of Environmental Monitoring and Dosimetry,
Swedish Radiation Protection Institute, Sweden

Abstract

Solid state nuclear track detectors (SSNTD) are widely applied for the assessment of mean radon concentration over long periods in human environment, because of their simple, reliable specifications.

In Japan, National Institute of Radiological Sciences (NIRS) developed the passive radon-thoron discriminative dosimeter for the nationwide indoor radon and thoron survey project since 1992. The dosimeter contains two polycarbonate disks as SSNTD, which is ready for the low level indoor radon concentration (< 10 Bq m^{-3}) .

In Sweden, Swedish Radiation Protection Institute (Statens strålskyddsinstitut, SSI) developed the passive radon dosimeter for the nationwide survey, which showed the Swedish indoor radon level to be 100 Bq m^{-3} in 1982. As well, the SSI passive radon dosimeter is adopted for the Swedish epidemiological (case control) study on residential radon and lung cancer in 1993.

To compare technical features of above two types of passive radon dosimeters, intercomparison was performed at 24 Swedish houses in Stockholm. Two types of dosimeters were placed side by side in both living room and bed room from March to May (about 90 days) . NIRS and SSI dosimeters showed an average of 48 rooms to be 95.3 ± 11.2 (SEM) and 134.7 ± 14.9 (SEM) Bq m^{-3}, respectively, and their linear regression analysis by least square fitting was Rn (SSI) = 1.31 (±0.03) x NIRS (Rn) + 9.64 (±3.82) with R-square of 0.975. When 30 rooms, of which radon level < 100 Bq m^{-3}, were analyzed, an average radon level shown by NIRS and SSI dosimeters was 45.2 ± 4.6 (SEM) and 70.4 ± 6.1 (SEM) Bq m^{-3}, respectively, and their linear regression was Rn (SSI) = 1.27 (±0.08) x NIRS (Rn) + 13.2 (±3.99) with R-square of 0.905.

Gradient of the linear regression may reflect systematic difference between radon standards in Japan and Sweden. Intercept might be caused by the background uncertainty in either or both dosimeters. This is a first trial for the future cooperative studies between NIRS and SSI.

1. Introduction

Since houses with enhanced radon concentrations are found, the importance of indoor radon has been recognized since 1980s. Now, radon is regarded as the biggest single contributor to natural radiation exposure of the general public [1].

In Sweden, nationwide survey of indoor radon concentration was conducted by Swedish Radiation protection Institute (SSI), and the residential radon level in 506 dwellings was 100 Bq m^{-3} [2, 3] .

The SSI passive radon dosimeter [4] was adopted for the nationwide survey (see Figure 1). As well, it was adopted for the Swedish epidemiological (case control) study on residential radon and lung cancer in 1993 [5].

In Japan, the first national indoor radon survey was commenced in 1985 [6] using KfK passive radon dosimeters, Germany [7] , and its final analysis showed the annual average indoor radon concentration and its standard deviation in 5,717 houses to be 20.8 ± 18.8 Bq m^{-3} , and their median value was 16.0 Bq m^{-3} [8]. In addition, the survey results suggested the high level thoron concentration might be observed mostly in wooden houses of traditional Japanese style with plastered interior walls, and the soil or mud based plaster has been suspected as a source of both thoron and radon because of its thorium and radium content [9]. In Japan, leading studies had noticed indoor thoron and its progeny in the Japanese style wooden houses [10,11] , and their significant contributions to the effective dose of general public have been suggested [12, 13].

To ascertain the contribution of indoor thoron as well as radon, the passive dosimeter which could discriminate between radon and thoron concentrations (NIRS passive radon/thoron discriminative dosimeter, see Figure 1) was developed by National Institute of Radiological Sciences [14, 15] , which is adopted in the second phase nationwide survey of indoor radon and thoron concentrations by Japan Chemical Analysis Center since 1992. Scope of this study is to compare their performances and technical specifications of SSI passive radon dosimeter and NIRS passive radon/thoron discriminative dosimeter directly by in situ measurements in Swedish houses.

2. Material and Methods

The intercomparison between SSI and NIRS passive dosimeters were performed from March to May (approximately 90 days) at 24 Swedish houses in Stockholm region. Two types of dosimeters were placed side by side at both living room and bed room in each house, and the distance from the interior wall to the dosimeters in each site was fixed at 20 cm. Calibration exercises of SSI and NIRS dosimeters were conducted several times in the standard radon exposure chamber in SSI, which has a traceability with standard radon source of National Institute of Standards and Technology, USA [16] .

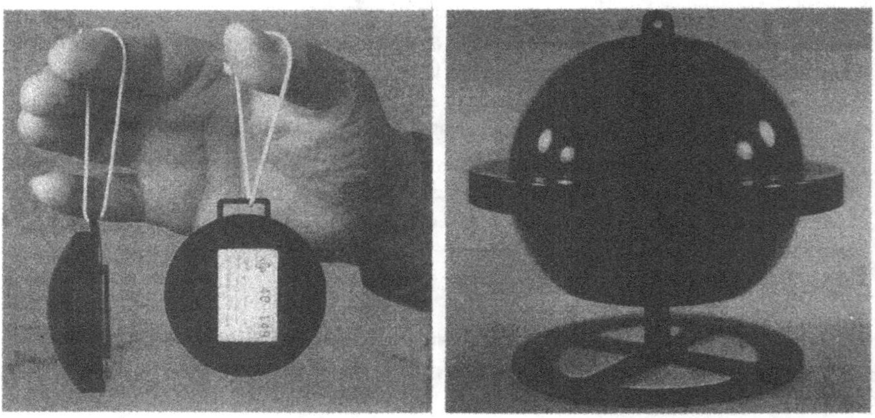

Figure 1 SSI passive radon dosimeter (Left, cited from Reference [3] and NIRS passive radon/thoron discriminative dosimeter (Right, cited from Reference [15])

3. Results

NIRS and SSI dosimeters showed average radon levels of 48 rooms to be 95.3 ± 11.2 (SEM) and 134.7 ± 14.9 (SEM) Bq m^{-3}, respectively, of which linear regression curve by least square fitting was expressed as (see Figure 2) :

$$Rn(SSI) = 1.31\ (\pm 0.03) \times Rn\ (NIRS) + 9.64\ (\pm 3.82) \quad \text{with } R^2 = 0.975 \qquad (3\text{-}1)$$

When 30 rooms, of which radon level < 100 Bq m^{-3}, were analyzed, average radon levels evaluated by NIRS and SSI dosimeters were 45.2 ± 4.6 (SEM) and 70.4 ± 6.1 (SEM) Bq m^{-3}, of which linear regression curve was expressed as (see Figure 3) :

$$Rn(SSI) = 1.27\ (\pm 0.08) \times Rn\ (NIRS) + 13.2\ (\pm 3.99) \quad \text{with } R^2 = 0.905 \qquad (3\text{-}2)$$

When 18 rooms, of which radon level > 100 Bq m^{-3}, were analyzed, average radon levels by NIRS and SSI dosimeters were 178.9 ± 14.6 (SEM) and 241.9 ± 21.4 (SEM) Bq m^{-3}, respectively, of which linear regression analysis was shown as (see Figure 4) :

$$Rn(SSI) = 1.42\ (\pm 0.09) \times Rn\ (NIRS) - 11.8\ (\pm 17.3) \quad \text{with } R^2 = 0.938 \qquad (3\text{-}3)$$

There found no evidence of thoron concentration in the 24 Swedish houses by NIRS radon/thoron discriminative dosimeters. The results of intercalibration exercises between NIRS passive radon-thoron discriminative dosimeter and the Standard ionization chamber in SSI [16] are shown in Table 1.

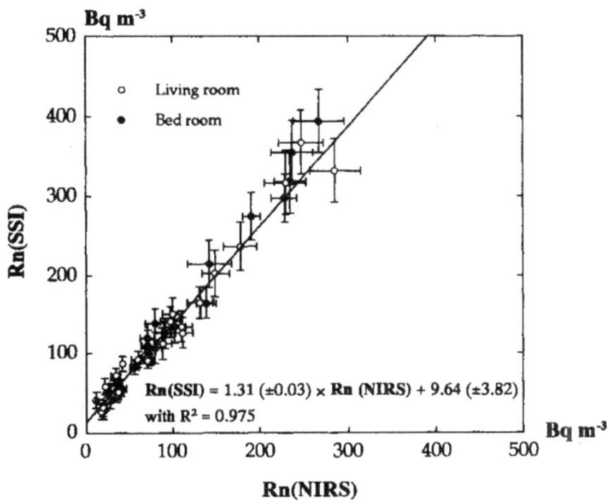

Figure 2 Intercomparison between SSI and NIRS passive radon dosimeters (living rooms and bed rooms in 24 houses)

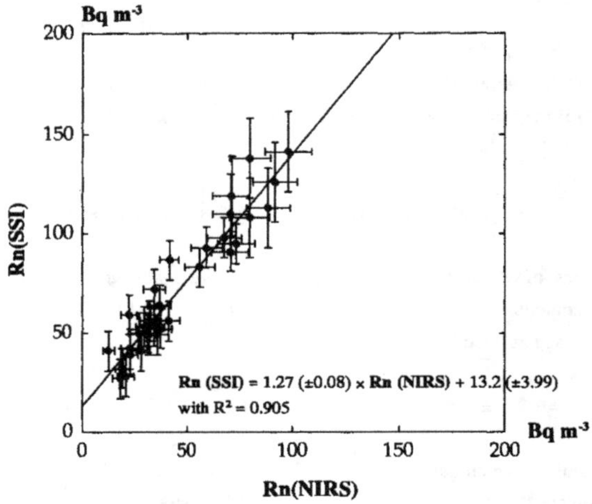

Figure 3 Intercomparison between SSI and NIRS passive radon dosimeters (less than 100 Bq m^{-3} by NIRS dosimeter, 30 rooms)

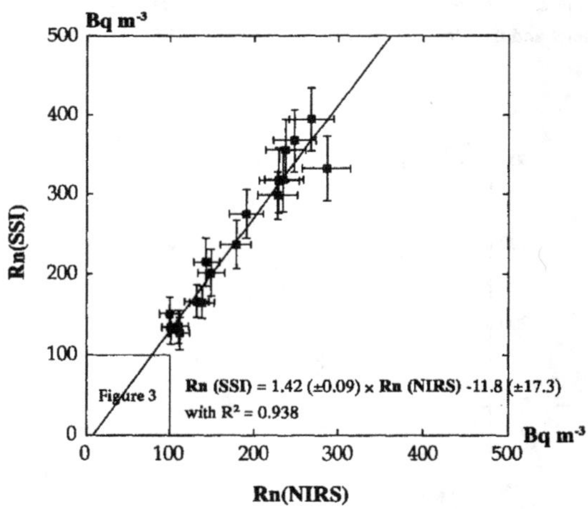

Figure 4 Intercomparison between SSI and NIRS passive radon dosimeters (more than 100 Bq m^{-3} by NIRS dosimeter, 18 rooms)

Table 1 Summary of intercalibration exercises between NIRS passive radon-thoron discriminative dosimeter and the Standard ionization chamber in SSI [16]

		Radon concentration in Bq m-3		Ratio
	Exposure (days)	Ionization Chamber (SSI)	NIRS dosimeter	SSI (IC)/NIRS (passive)
Phase 1	7	2006.3	1545.7 ± 37.7	1.30
Phase 2	4	1568.0	1346.2 ± 50.2	1.17
Phase 3	2	2044.4	1899.2 ± 37.4	1.08

4. Discussion

Gradient of the linear regression, Rn(SSI) / Rn(NIRS) ranged from 1.27 to 1.42, which may reflect systematic difference in radon (or radium) standards or calibration protocol between Japan and Sweden. Table 1 showed the difference of calibration factors given by SSI ionization chamber (Swedish standard) and Japanese domestic standard, ranged from 1.08 to 1.30. Standardization of radon exposure must be established urgently in Japan to certify the international traceability of Japanese radon measurements .

It must be pointed out that NIRS dosimeter uses polycarbonate as SSNTD, which is subjected to electrochemical etching to expand the etch pit diameter from 40 - 200 μm taking advantage of low background track density. Overlapping of etch pits may cause underestimation of high level radon exposures (>200 Bq m^{-3} in 90 days).

Intercepts of the regression curves might be caused by the background uncertainty in either or both dosimeters. But more discussion should be required. This study should be regarded as a first step for the future cooperative and periodical intercalibration and intercomparison exercises between NIRS and SSI.

5. Acknowledgements

This work has been carried out as the cooperative research project between Swedish Radiation Protection Institute (Statens strålskyddsinstitut, SSI) and National Institute of Radiological Sciences (NIRS) in 1995-1996. Thanks to Dr. Fuminori Soga, Director, Division for Planning and Coordination, NIRS for his kind coordination. The deepest gratitude must be expressed to Dr. Yuji Nakamura, Director, Environmental and Toxicological Sciences Research group, NIRS for his invaluable support and encouragement throughout this work.

Much appreciation should be expressed to Dr. Ulf Bäverstam, Huvudenhet för forskning and utveckling, SSI, for his kind coordination throughout this study, and Mr. Hans Mellander and his staffs in SSI for their scientific directions and invaluable support.

The authors would also like to express their hearty appreciation to Dr. Anita Enflo, Forskningssekretariatet, SSI, for her kind arrangement and coordination to realize this cooperative research work.

6. References

[1] UNSCEAR, Sources, Effects and Risks of Ionizing Radiation - UNSCEAR 1988 report, 1988.

[2] Swedjemark, G.A. Radon and its decay products in housing, Ph.D. thesis. Stockholm University, ISBN 91-7146-634-7, Stockholm, 1985.

[3] Clavensjö, B. and Åkerblom, G. The Radon Book, Measures against radon, The Swedish Council for Building Research, ISBN 91-540-5649-7, Ljunglöfs Offset AB, Stockholm, 1994.

[4] Mellander, H. and Enflo, A. Alpha track method used in the Swedish radon epidemiological study, Radiation Protection Dosimetry, 45, 65 -71, 1992.

[5] Pershagen, G. et al. Radon i bostäder och lung cancer, IMM-rapport 2-93, ISSN 1101-2803, 1993.

[6] Kobayashi, S. et al. Nationwide survey of indoor radon concentration in Japan, Proceedings of the third International Symposium on Advanced Nuclear Energy Research, JAERI, Mito Japan, 13-15, March, 1991.

[7] Urban, M. Low level environmental radon dosimetry with a passive track etch detector device, Radiation protection Dosimetry, 1, 97-109, 1981.

[8] Fujimoto, K. et al. Nationwide Indoor Radon Survey in Japan, J. Journal of Health Physics, 32(1), 41-51, 1997.

[9] Asano, K. and Ito, N. Contents of Radioactivity in the several building materials in Japan, Inside document of Building Research Institute, Japan, 1989.

[10] Iida, T.; Sone, K.; Ikebe, Y. Some problems on the measurement of radon concentrations by cup method, Proceedings of the 25th annual meeting of Japan Health Physics Society (in Japanese), Tsukuba Japan, 63 p, 1990.

[11] Yonehara, H.; Aoyama, T. Concentration and distribution of 220Rn and its progeny in a house, Proceedings of the 26th annual meeting of Japan Health Physics Society (in Japanese), Osaka Japan, 16 p, 1991.

[12] Shimo et al. Tn daughters concentration in several indoor and outdoor airs and lung dose due to Rn and Tn daughters-(1), Proceedings of the 24th annual meeting of the Japan Health Physics Society (in Japanese), Nagoya Japan, 25 p, 1989.

[13] Guo, Q. ; Shimo, M.; Ikebe, Y.; Minato, S. The study of thoron and radon progeny concentrations in dwellings in Japan. Radiation Protection Dosimetry. 45: 357-359; 1992.

[14] Doi, M. et al. A passive measurement technique for characterization of high risk houses in Japan due to enhanced levels of indoor radon and thoron concentrations, Radiation Protection Dosimetry, 45, 425-430, 1992.

[15] Doi, M. Development of the passive radon-thoron discriminative dosimeter for indoor and outdoor surveys, Ph.D. thesis, Nagoya University, 1994.

[16] Rolf Falk et al. Calibration of Radon-222 Reference Instrument in Sweden, J. Research, Nationl. Standards and Technol. 95, 115-120, 1990.

Thoron intercalibration procedurers.

Rolf Falk*, Masahiro Doi** and Hans Möre*

*Division of Environmental Monitoring and Dosimetry
Swedish Radiation Protection Institute (SSI)
Stockholm, SWEDEN

**Environmental and Toxicological Sciences,
National Institute of Radiological Sciences(NIRS)
4-9-1, Anagawa, Inage, Chiba 263, JAPAN

ABSTRACT

Solid state nuclear track detectors (SSNTD) are widely applied for measurements of radon concentration in human environment. The SSNTD technique is simple, reliable and suitable for the assessment of mean radon concentration over long periods. A passive Radon-Thoron discriminative dosimeter for indoor and outdoor surveys has been developed at the National Institute of Radiological Sciences(NIRS) and used for large scale national surveys. Due to the short half-life of thoron (55.6 s) and thereby limited time for mixing, large concentration gradients within a dwelling are expected and observed. The calibration of a passive thoron -SSNTD is associated with uncertainties due to, among many other parameters, the difficulties to obtain an atmosphere with known and homogeneous thoron concentration. Due to the short time for the thoron to diffuse in to the sensitive volume of the detector, air movements are suspected to influence the sensitivity.

A walk in thoron-chamber was designed and used for the intercalibration between the NIRS passive Radon-Thoron discriminative dosimeter and the active thoron instrument at the Swedish Radiation Protection Institute(SSI). A homogeneous distribution of the thoron within the chamber was achieved by several inlets to the chamber from the thoron source and by stirring the air in the room with a large-bladed fan. By changing the speed of the fan and using additional fans the air turbulence in the room could be varied. During the intercalibration the thoron concentration was continuously monitored with two flow-through Lucas-cell detectors using the technique of multiple time analysis of pulse events. The radon concentration was monitored with a flow-trough pulse-ionisation detector provided with a delay tube to avoid disturbance from thoron.

The sensitivity of the NIRS passive Radon-Thoron discriminative dosimeter to thoron was found to be affected by the degree of air turbulence. The sensitivity for radon was not affected by air turbulence. The results from simultaneous measurements of thoron showed a systematic difference between the NIRS passive dosimeter and the SSI active instrument. The ratio passive/active detector was closest to one in an atmosphere with high turbulence

1. Introduction

Since a high incidence of houses with enhanced radon concentrations in Europe and America was found subsequently around 1980, public perception of the importance of indoor radon has altered drastically. Now, it is understood that radon is the biggest single contributor to natural radiation exposure of the general public in Sweden. A nation-wide survey of indoor radon concentrations was conducted by the Swedish Radiation Protection Institute, and the average residential radon concentration was found to be 100 Bq/m³ [1]. A large scale indoor radon survey in Japan commenced in 1985 and finished in 1991, which covered some 12,600 rooms of 6,300 houses over 47 districts which was about 0.02% of the total number of homes [2].This survey used KfK passive radon dosimeters developed by Karlsruhe Nuclear

Research Center, Germany [3]. This interim survey found that the geometrical mean indoor radon concentration, in Japan, is 23 Bq/m³, which is about half of the world average [4].
The results of the survey are available for statistical analysis to show the relative risk among inhabitants by their geographical sites. However, survey results suggested that the indoor radon concentration seemed to be higher when the dosimeter was placed near the interior wall, despite the general belief that indoor radon concentration should be uniform. This unexpected phenomenon was mostly observed in wooden houses of traditional Japanese style with plastered interior walls, and the soil or mud based plaster has been suspected as a source of both thoron and radon because of its thorium and radium content [5]. In Japan, some leading studies had noticed the potential risk of indoor thoron and its progeny in the Japanese style wooden houses, and some other studies suggested that the thoron and its progeny would contribute to the general public an effective dose which is comparable to that from radon.
However, the calibration protocol of thoron and its progeny concentration has not yet been studied enough. The scope of this study is to develop the calibration chamber for thoron gas exposure, and to study its technical specifications for different measurement situations.

2. Materials and Methods

2.1 The NIRS passive Radon-Thoron discriminative dosimeter(R-T dosimeter)

The NIRS passive radon-thoron discriminative dosimeter was developed to evaluate low level radon concentrations by discriminating and estimating thoron concentrations differentially. The dosimeter has two opposed hemispherical stainless steel diffusion chambers of 110 mm and 70 mm diameter. A polycarbonate disk (Iupilon® Type FE-2000, 300 mm in thickness, 50 mm in diameter) is installed as SSNTD in each of the two diffusion chambers. Fan shaped air inlets (36.2 cm^2 large) are provided with surfaces covered by glass fibre filters to prevent decay products from entering the detector housing. Since the air exchange rate through these air inlets was 56.8 exchanges h^{-1}, the radon concentration in the 110 mm hemispheric chamber is 99.9% of atmospheric, and the thoron concentration is 55.9% of atmospheric. The 110mm-hemispheric chamber was connected with the 70mm-hemispheric chamber by a pinhole air inlet (1 mm in diameter). Since the air exchange rate through the pinhole air inlet was 0.38 exchanges h^{-1}, 98.0% of radon and 0.47% of thoron concentrations in the 120mm-hemispheric chamber were diffused into the 70mm-hemispheric chamber. After a 2 month exposure in the environment, the polycarbonate disk from the 110mm-hemispheric chamber was subjected to preliminary chemical etching (CE) for 3 h at 30 °C with a blend of 8.0 N KOH (80 % in volume) and C_2H_5OH (20 % in volume) and then was subjected to electrochemical etching (ECE) for 3 h under alternating current of 800V, 2 kHz. In this condition, microscopic alpha tracks with vertical incident energies of 2.5-4.0 MeV, which are mainly attributed to thoron and its progeny, are registered as electrochemical etch pits, of which diameters range from 40 to 300 micron. On the other hand, the polycarbonate disk from the 70mm-hemispheric chamber was subjected to CE for 0.5 h followed by ECE for 3h. In this condition, alpha tracks with vertical incident energies of 0.8-2.7 MeV, which are attributed to radon and its progeny, are registered. Detailed features of the R-T dosimeter is available in [6]Doi et al. (1992).

2.2 The SSI active thoron instrument

The technique used at SSI to measure thoron is based on a delayed coincidence method for alpha-particles from the decay of thoron and its relatively short lived decay product ^{216}Po, (half-life 0.145 s). A multiple time analysis technique of pulse events is applied for the purpose. The Instrument used is a commercial radon monitor (EDA, RD 20), with a 160 ml Lucas-type scintillator cell as detector, working in a flow through mode. Air is sucked through a filter to remove any airborne particles and decay products from thoron and radon before entering the sensitive volume. The flow-rate chosen is about 10 times the volume of the Lucas cell per minute giving a thoron concentration inside the cell practically the same as in the environment, since the half life of thoron is 55.6 s. In the multiple time analysis of pulse events the time intervals between successive pulses are stored in a computer memory as a "time list" for later analysis. The distribution of the time intervals is then used to separate the fraction of alpha-events originating from ^{216}Po from all other alpha-events [7].

The calibration of the instrument is performed with a known air concentration of thoron generated and maintained in a "Bench top" thoron chamber [8].The instrument can also be efficiency calibrated using a pure thoron atmosphere of *unknown* concentration [9].

2.3 The SSI active radon instrument

The instrument used at SSI for continuous measurements of low levels of radon is a commercial radon gas monitor using a pulse-counting ionisation chamber with atmospheric air as filling gas as detector. The instrument (ATMOS 20) has an active volume of 4.9 l giving a sensitivity to the alpha particles from radon good enough to measure 10 Bq/m³ with a statistical uncertainty of 10% within 10 minutes. The instrument is working in flow-through mode and is calibrated against the national radon reference in Sweden [10, 11].The instrument is capable of discriminate between thoron and radon but, for the situation of low radon concentration and high thoron concentration the inlet to the instrument is equipped with a delay-tube corresponding to about 10 half times of thoron to avoid any interference between the two nuclides.

2.4 The thoron chamber

In order to intercalibrate passive and active thoron instruments a simple thoron chamber was designed. The chamber was built like a tent in a basement room at SSI. A simple frame of wood with the size 2.8 * 2.6 *2.6 m³ was covered with polyethylene foils as wall material. The thoron source consisted of a bubbler with natural $Th(NO_3)_4$ in water. Two air pumps were used to continuously extract the toron from the solution and distribute it in the room. Air at a low flow-rate through the bubbler extracted the thoron gas up to the air-volume above the solution. An additional airflow at much higher flow-rate was introduced to this air volume forcing the extracted thoron gas to be distributed by a number of thin tubes to various places in the room at high speed. The outlets of these tubes ended close to the ceiling and the large-bladed fan. With this arrangement a relative homogeneous concentration of thoron was maintained in the thoron chamber.

3. Experiments

Two active thoron instruments were used throughout the series of experiments. Side by side measurements were used for intercalibration of these instruments. Five intercomparison exposures were performed. The first intercomparison was performed with a thoron-concentration of about 100 Bq/m³ and had a duration of 21 days. The other exposures were performed with an thoron-concentration between 2000 and 4000 Bq/m³ during about 3 days. Fig 1 and 2 are examples of recorded radon and thoron concentration during an exposure run.

Fig 1. Measured variation of Tn-concentration during an exposure. Two active detectors used.

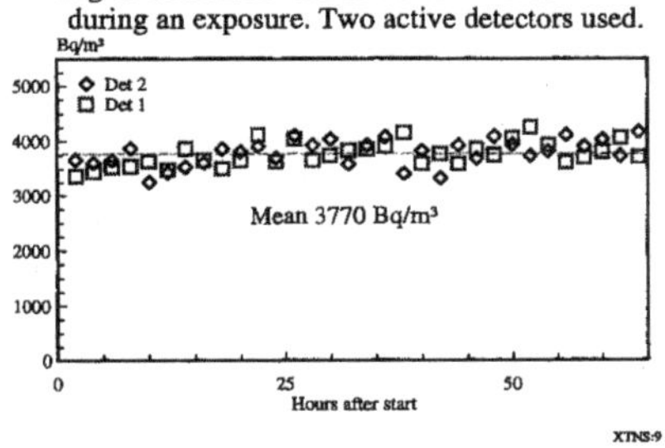

Fig 2. Measured variation of ²²²Rn during an exposure.

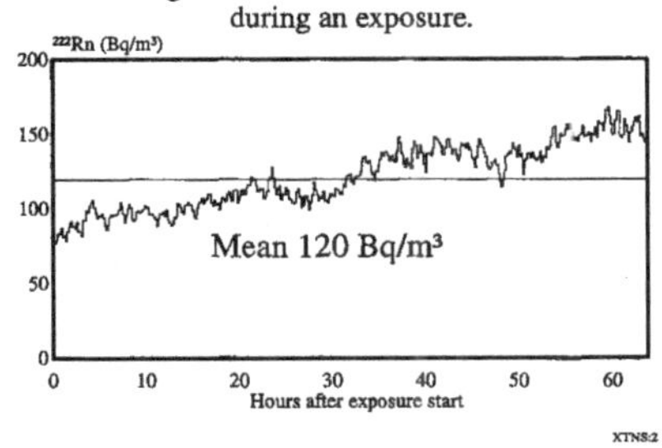

At each run, 12 passive detectors (6 with wind breaker) were exposed in the centre of the room. The active instruments measured the thoron concentration at two locations near the passive detectors.

4. Results

Table 1and 2 and Fig 3 and 4 summarises the results obtained.

Fan Profile	^{222}Rn Bq/m³		
	Active (SSI)	Passive (NIRS)	Ratio passive/active
Very low	170	259	1.52
Low	59	89	1.51
High	113	111	0.98
High	83	91	1.10
Very high	120	153	1.28

Table 1. Results of radon measurements

Fan Profile	^{220}Rn Bq/m³		
	Active (SSI)	Passive (NIRS)	Ratio passive/active
Very low	3580	2069	0.58
Low	2222	1404	0.63
High	77	56	0.73
High	3841	2834	0.74
Very high	3773	3174	0.84

Table 2. Results of thoron measurements

Fig 3. Result from intercomparision measurements of Tn at different degree of air turbulence.

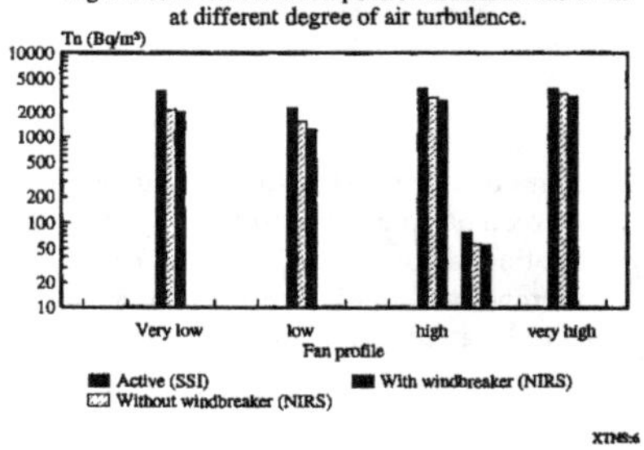

Fig 4. Tn measurements Ratio (Passive/active)

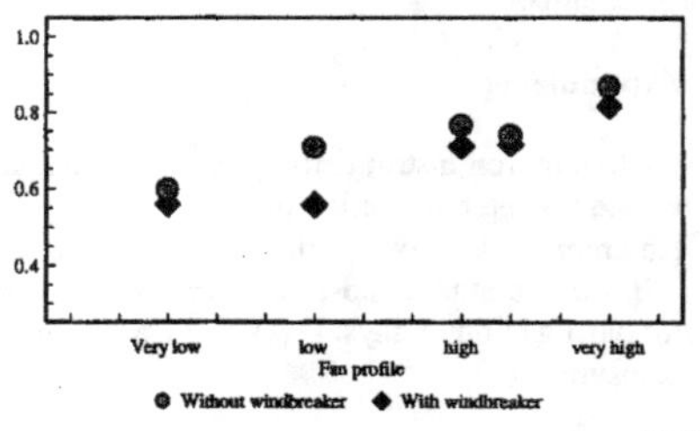

5. Conclusions

The sensitivity of the NIRS passive Radon-Thoron discriminative dosimeter to thoron was found to be affected by the degree of air turbulence, and the effect o the wind breaker is clearly seen. The sensitivity for radon was not affected by air turbulence. The results from simultaneous measurements of thoron showed a systematic difference between the NIRS passive dosimeter and the SSI active instrument. The ratio passive/active detector was closest to one in an atmosphere with high turbulence. The NIRS passive Radon-Thoron discriminative dosimeter showed excellent agreement with the active radon -measurements at low level (about 100 Bq/m³) in the presence of high thoron level (about 3000 Bq/m³). The result of the intercalibration procedures used show the necessity to perform instrument calibration close to the circumstances expected for measurements.

6. References

[1] Swedjemark, G. A. And Mjönes L., Radon and radon daughter concentration in Swedish homes. Rad. Prot. Dosim., 7 no 1-4, 341-345, 1984.

[2] Kobayashi, S. et al. Nationwide survey of indoor radon concentration in Japan, Proceedings of the Third International Symposium on Advanced Nuclear-Energy Research. Japan Atomic Energy Research Institute, Mito Japan, March 13-15, 1991: 63-67.

[3] Urban, M. and Piesch, E., Low level environmental radon dosimetry with a passive track etch detector device. Radiation Protection Dosimetry, 1: 97-109; 1981.

[4] UNSCEAR. Sources and effects of ionizing radiation, New York, United Nations; 1993: 17-18.

[5] Asano, K., Ito, N. Contents of radioactivity in the several building materials in Japan. Inside document of the Building Research Institute, Ministry of Construction, Japan; 1989 (in Japanese)

[6] Doi, M., Kobayashi, S. , Fujimoto, K. A passive measurement technique for characterization of high risk houses in Japan due to enhanced levels of indoor radon and thoron concentrations. Radiation Protection Dosimetry, 45: 425-430; 1992.

[7] Falk, R., Möre, H. and Nyblom, L., Measurement of ^{220}Rn in air using a flow-through Lucas cell and multiple time analysis of recorded pulse events. Rad. Prot. Dosim. No ¼ , 111-113, 1992.

[8]Möre, H., Falk, R. and Nyblom L., A bench-top calibration chamber for ^{220}Rn activity in air. Env. Internat., 22, Suppl 1, 1147-1153, 1996

[9]Falk, R., Möre, H. and Nyblom, L., Measurement techniques for Environmental levels of Radon-220 in air using Flow-through Lucas Cell and Multiple Time Analysis of Recorded Pulse Events. Int.J. Radiat. Appl. Insatrument 43 (½), 111-118, 1992.

[10]Baltzer, P., Görsten, K.G. and Bäcklin, A., A Pulse-counting ionization chamber for measuring the radon concentration in air. Nuclear Instr. Meth., 317(½), 357-364, 1992

[11] Falk, R., Hagberg N., Mjönes L., Möre H., Nyblom L. and Swedjemark G. A. Standards, calibration and quality assurance of ^{222}Rn measurements in Sweden. Nuclear Instruments and Methods in Physics Research A 339, 254-263, 1994.

Control of radon concentrations in a radon chamber

Ching-Jiang Chen, Chi-Chang Liu and Yu-Ming Lin

Radiation Monitoring Center, AEC
823, Cherng Ching Road, 833, Kaohsiung, Taiwan, ROC

Abstract

A walk-in type radon calibration chamber was designed to provide controlled environment for calibration of instrument and evaluation of methodology in Radiation Monitoring Center of Taiwan. However, it takes time to adjust radon concentration in the chamber to meet the requirements of different situations. A mathematical model, based on a set of differential equations, was established to predict radon level in the chamber and to assist the operator to arrange the calibration schedules. The measured results by Lucas cell method were coincided with the theoretical model calculation. The differences between measured data and calculated ones are less than 3% at steady state. Transient state measurements, for example, injection or flush out procedure, caused larger bias that was due to timing lag property of Lucas cell. Radon level can be controlled by regulating injection rate or ventilation rate. Ventilation control method is more practical than injection control.

1. Introduction

The measurement of radon and its daughters concentrations in the environment became an inevitable issue for estimating annual dose from natural environment. To ensure the reliability of measured results, proper calibration is very important. Some of the research centers around the world have established different types of radon calibration chambers to supply stable radon concentration [1-4]. The radon measurement project in Radiation Monitoring Center (RMC) of Taiwan first started in 1980 and has continued till today. Some researchers from other research institutes contacted RMC and requested help in instrument calibration, prompting the design and construction of a radon test chamber in Taiwan. This project started from 1992 and finished in 1993.

Basically, this radon test chamber is a multipurpose one that can be used for both research and calibrations for routine radon measuring instruments. This radon chamber was designed and constructed domestically. For different purpose of measurements, the calibration procedure of instruments would be operated under different radon levels. To meet the requirement, an efficient way to control radon level in the chamber is necessary. The objective of this research is to establish a mathematical model to simulate radon level control to assist the arrangement of calibration schedule. Verification of this model is also discussed by comparing calculated results and measured data.

2. Radon Chamber

The radon test chamber was built in an air-conditioned laboratory to avoid the interference by the environment. The approximate internal dimensions of the chamber are 3.50 m(W)×3.10 m(L) ×2.12 m(H) with a total internal volume of about 23 m^3. An anteroom is provided to enclose the chamber entrance door to prevent radon escape during entrance and exit of personnel. All the chamber walls were built with an outer and inner layers of 1mm thick stainless steel. PU material and wood rods of 50 mm length were filled between the two stainless steel layers for thermal insulation and structural reinforcement. Silicone sealant was applied at steel plate seams to avoid leakage of gas.

Temperature in the chamber is controlled by an air-conditioner and a heater. Temperature can be adjusted between 10~39.9 °C with precision under ±0.1°C. Humidity is controlled by a vibrating wet adder and dehumidifier. Controllable range is between 30%~90% RH with precision under ±1%. The chamber is kept at a slight negative pressure to avoid leakage of radon. All the parameters of temperature, humidity and atmospheric pressure are measured continuously and recorded by a personal computer.

Two radon sources and one thoron source supplied by Pylon Company, Canada were installed near the anteroom. Properties of those sources are available from the manual. Radon concentrations in the chamber can be varied between 250 to 40,000 Bq m^{-3}.Generally it is set to be around 2,000 Bq m^{-3}. A set of 4 fans system is used to ensure radon gas mixed homogeneously in the chamber.

3. Model description

Model of radon chamber operation is demonstrated in **Fig.** 1. Radon gas is injected by a pump. Another pump that is installed on the top exhaust gas to maintain slightly negative pressure in calibration chamber. Therefore the model can be described as below:

$$\begin{bmatrix} \frac{dCs}{dt} \\ \frac{dCc}{dt} \end{bmatrix} = \begin{bmatrix} -(\lambda + \frac{Ri}{Vs}) & \frac{Ri}{Vs} \\ \frac{Ri}{Vc} & -(\lambda + \frac{Ri}{Vc} + \frac{Ro}{Vc}) \end{bmatrix} \begin{bmatrix} Cs \\ Cc \end{bmatrix} + \begin{bmatrix} Csg \\ 0 \end{bmatrix} \tag{1}$$

where

Cs	radon concentration in source chamber (Bq/m^3),
Cc	radon concentration in calibration chamber (Bq/m^3),
λ	decay constant of radon-222 ($t_{1/2}$=3.82 days),
Vs	volume of source (300 ml)
Vc	volume of calibration chamber (23 m^3)
Ri	injection rate (m^3/s)
Ro	exhaustion rate (m^3/s)
Csg	generation rate of radon (Bq/s)

The differential equations can be solved as general form ,

$$Cs(t)=L_1 \cdot exp(a_+ \cdot t)+ L_2 \cdot exp(a_- \cdot t)+Css \tag{2}$$

$$Cc(t)=L_3 \cdot exp(a_+ \cdot t)+ L_4 \cdot exp(a_- \cdot t)+Ccs \tag{3}$$

with coefficients as

$$a_{+,-} = \frac{1}{2}\left[-(2\lambda + \frac{Ri}{Vs} + \frac{Ri}{Vc} + \frac{Ro}{Vc}) \pm \sqrt{(\frac{Ri}{Vs} + \frac{Ri}{Vc} + \frac{Ro}{Vc})^2 + 4\frac{RoRi}{VsVc}} \right] \tag{4}$$

$$L_1 = \frac{a_-(Cs(0) - Css) + (\lambda + \frac{Ri}{Vs})Cs(0) - \frac{Ri}{Vs}Cc(0) - Csg}{(a_- - a_+)} \tag{5}$$

$$L_2 = \frac{a_+(Cs(0) - Css) + (\lambda + \frac{Ri}{Vs})Cs(0) - \frac{Ri}{Vs}Cc(0) - Csg}{(a_+ - a_-)} \tag{6}$$

$$L_3 = \frac{a_-(Cc(0) - Ccs) - \frac{Ri}{Vc}Cs(0) + (\lambda + \frac{Ri}{Vc} + \frac{Ro}{Vc})Cc(0)}{(a_- - a_+)} \tag{7}$$

$$L_4 = \frac{a_+(Cc(0) - Ccs) - \frac{Ri}{Vc}Cs(0) + (\lambda + \frac{Ri}{Vc} + \frac{Ro}{Vc})Cc(0)}{(a_+ - a_-)} \tag{8}$$

$$Css = \frac{Csg(\lambda + \frac{Ri}{Vc} + \frac{Ro}{Vc})}{\lambda^2 + \lambda(\frac{Ri}{Vs} + \frac{Ri}{Vc} + \frac{Ro}{Vc}) + \frac{RoRi}{VsVc}} \tag{9}$$

$$Ccs = \frac{Csg(\frac{Ri}{Vc})}{\lambda^2 + \lambda(\frac{Ri}{Vs} + \frac{Ri}{Vc} + \frac{Ro}{Vc}) + \frac{RoRi}{VsVc}} \tag{10}$$

where

Css steady state radon concentration in source chamber (Bq/m^3),

Ccs steady state radon concentration in calibration chamber (Bq/m^3),

Cs(0) initial radon concentration (t=0) in source chamber (Bq/m^3),

Cc(0) initial radon concentration (t=0) in calibration chamber (Bq/m^3).

The steady state radon concentration *Ccs* in calibration chamber is the level that we need. The $|a_+|$ and $|a_-|$ indicate the time constant from start of injection to the time reach steady state concentration. From equation (10), since the source activity (Csg) is fixed, the only parameters for *Ccs* adjustment are injection rate (*Ri*) and pump out rate (*Ro*).

4. Experiment and discussion

The preliminary test of this model is done by transient test. Radon is injected into the chamber for half hour. Then turned off the pump and close source injection valve. Exhaustion rate is kept at flow rate of 3 LPM. The results is shown in **Fig.** 2. Solid line presents the predicted radon level in the chamber of this test. The square symbols are the measured data by Lucas cell that mounted to the sampling port of the chamber. During the first 15 hours, sampling interval is 10 minute for observing rapid growth behavior. After 15 hours sampling interval is set to be one hour. The measured results of the first 4 hours are lower than the predicted ones because of the time lag property of Lucas cell in continuous flow mode. If a dynamic radon level is utilized for calibration, correction of Lucas cell's timing problem will be necessary.

The steady state test is performed by setting fixed injection and exhaustion rate. The results is shown in **Fig.** 2. The average measured concentration is 2599±55 Bqm^{-3} and the model predicted is 2656 Bqm^{-3} .The bias is about 3% owing to the bias and fluctuation of pumping rate setting.

The results of transient test and stead state test that described above give us confidence to control radon level in the chamber. Divide equation (10) by *Ri/Vc* and obtain a more clear form as

$$Ccs = \frac{Csg}{Vc[(\lambda/Ri + 1/Vs)(\lambda + Ro/Vc) + \lambda/Vc]} \tag{11}$$

An extreme case is discussed that if $\lambda/Ri<<1/Vs$, injection rate can not make any influence to the steady state concentration. That is, it should meet $Ri< Vs\lambda$ to control radon level by injection rate adjustment. For the condition of RMC, injection rate (Ri) should smaller than 0.038 mLmin^{-1} for radon-222 that is not easy to achieve. So, we suggest that keep injection rate fixed just larger than 1 mLmin^{-1} for any level and the radon level control could be achieved by exhaustion rate (Ro)only. Since Vc>>Vs , the equation (11) can be simplified as

$$Ccs(Ro) = \frac{Csg \cdot Vs}{Vc \cdot \lambda + Ro} \qquad (12)$$

This implies that we may adjust the Rn level in the chamber by Ro setting. Equation (12) can also estimate the bias caused by pumping rate setting due to minimum setting interval as 0.2 LPM to be 64 Bq/m^3 that shows bias in **Fig.** 3 is reasonable. The error of model predicted by fluctuation of pumping rate could be derived from equation (12) as,

$$\sigma_{Ccs} = \frac{Csg \cdot Vs}{(Vc \cdot \lambda + Ro)^2}\sigma_{Ro} \qquad (13)$$

The adjustment of different Rn level is shown in Fig. 4. To shorten the time between each level, Ro will be set to zero during increasing concentration procedure till the time of expected maximun concentration predicted by model. On the contrary, the Ro will be set to maximum during decreasing process. Such procedures could be planned before real operation for time saving. The worse control section in the 4 levels is at 1580 Bq/m^3 which is due to the bias of pumping rate setting. However, the bias in the 4 levels are all smaller than 4%.

5. Conclusion

This simplified model is simple, precise enough and easy to use. It is helpful for operator to arrange time schedule in calibration and experimental design. With minor modification, this model can also be applied to the case of other gas calibration system. Although it is suggested to use exhaustion control method for radon-222, for other gas, injection control may be better. If an injection or exhaustion method is for choice before experiment, it could be evaluated by this model too.

6. Reference

[1] Bigu, A Walk-In Radon/Thoron Test Facility, Am. Ind. Hyg. Assoc. J. 45,1984.

[2] Peggie,R.,Solomon,S. B., and Clarke,P., Radon and Radon Daughter Test Chambers For The Radon Reference Facility At The Australian Radiation Laboratory, Commonwealth of Australia,1990.

[3] Wicke, Submitted to the Group of Experts on Radon Dosimetry and Monitoring of the OECD/NEA, The German Radon Calibration And Test Chamber Facility, D-8042, Neuherberg, Germany ,1984.

[4] Sensintaffar,L.,Vancleef,D. J., and Windham,S. T.,U.S. EPA, Description Of A Computer-Based Radon and Radon Decay Product Monitoring System for a Radon Exposure Chamber, National Air and Radiation Environment Laboratory, Montgomery, Alabama, USA, 1991.

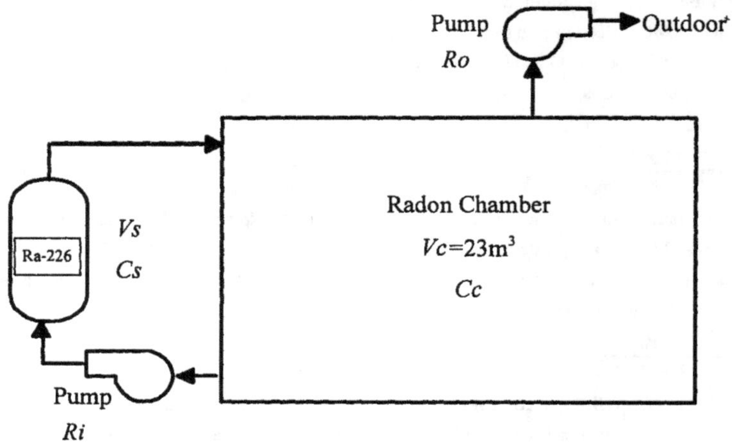

Figure 1. The simplified model of radon calibration chamber.

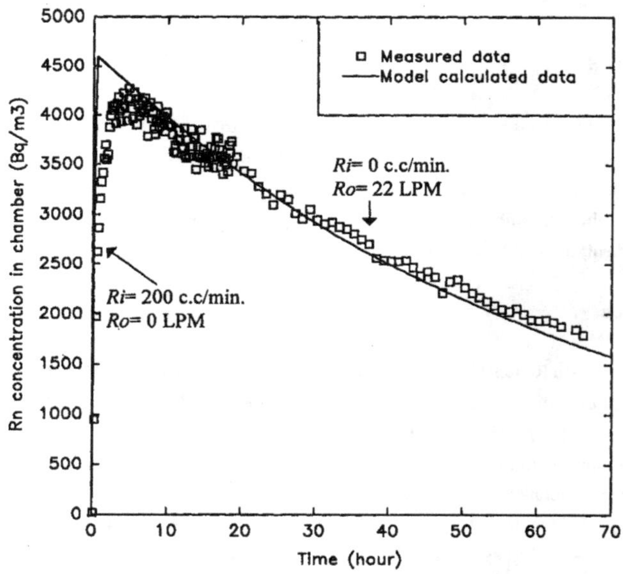

Figure 2. Transient test of the model. After source injection for 0.5 hour, close the injection valve and flush by clean air at a flow rate of 3 LPM.

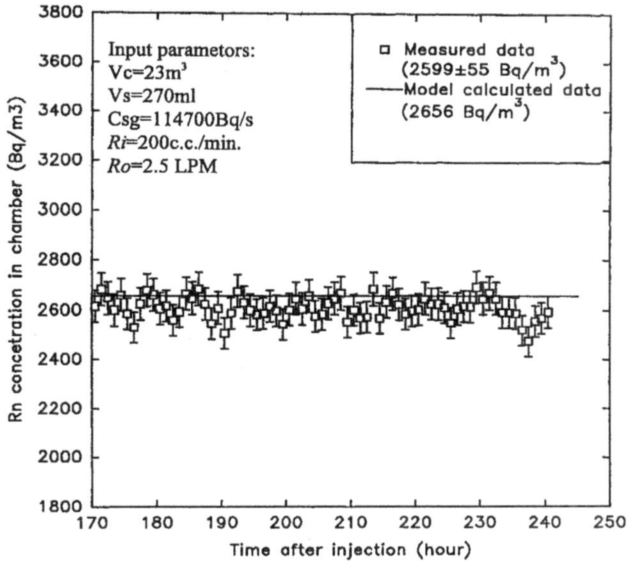

Figure 3. Comparison of radon concentration in the chamber in steady state between calculated and measured results.

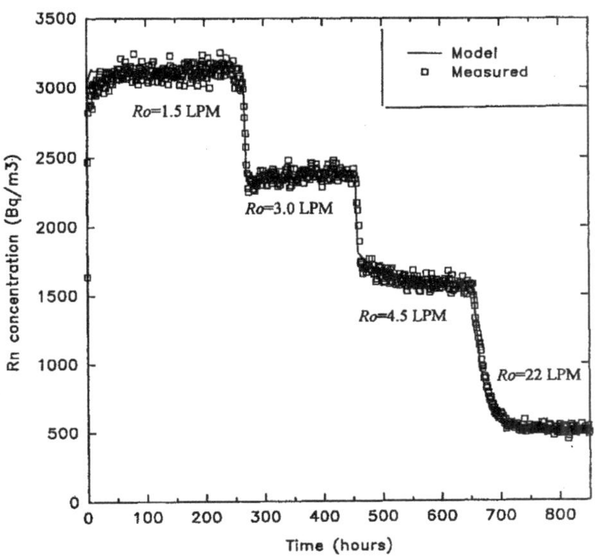

Figure 4. Results of radon level control.

A calibration facility for radon fluxmeter

Li Xianjie, Qiu Shoukang, Zhou Jianliang, Liu Chunkui, Pan Jialin, Yang Mingli

Research Institute of Uranium Mining, Hengyang, Hunan, China

Abstract

Calibration facilities for radon fluxmeter with three kinds of different emanation medium have been developed. The stability of radon flux is 5%, 9% (RSD) respectively. The uniformity of radon flux is 4.5%, 8.5% (RSD) raspectively. These sepcifications fulfill the calibration requirement for radon fluxmeter.

The determination of radon flux of facility takes full account of eliminating the main error source-attenuation effect (including leakage and back diffusion etc.) : not only prevent attenuation and make a relevant correction. Therefore the accuracy of determination is assured.

The calibration, intercomparison of radon flux meter and the quantitively evaluation on the measurement method of radon flux are made to be possible by the successful establishment of this facility.

1. Introduction

Efforts have been made to establish a calibration facility with a stable and uniform radon flux for examining the accuracy of radon fluxmeter, but no successful result has been reported until then. J.N. Hartley et.al.(1983) introduced the application of a calibration facility for calibrating the charcoal canister radon fluxmeter[1] . The facility is made by Bendix Corporation with a thin layer of uniform uranium tailings. It's radon flux is determined on the basis of emanation coefficient, porosity etc. of tailings. The calibration result of charcoal canister, expressing as monitoring efficiency, was frequently much more than 100 % (from 120 % to 160 %), obviously unreasonable. The author had to attribute it to the interference of meteorological factors. R.Colle et.al.(1990) introduced a developing effort for the calibration facility[2]. They took an aqueous solution of radium as the radon source, and a rigid polyethylene sheet as emanation medium. But no stable radon flux had been achieved.

A different approach has been taken by us, and a calibration facility with stable and uniform radon flux as well as a emanation medium simulated to common one is achieved.

2. Facility and operation

2.1 Material

2.1.1 Radon source

Considering the emanation coefficient of common radium-bearing solid material always susceptible to the meteorological variation, a kind of solid radon source with high emanation coefficient was selected. It was made by ourselves. The emanation coefficient of it is (97 ± 4) % , and keeps constant within a wide range of temperature and humidity (-10℃~40℃, 1 %~100 %).

2.1.2 Emanation medium

The basic requirements are: easy to keep the radon flux stable and uniform; simulate the common natural material. The concrete requirements in detail are follows:

a. The radium content is very low, so as to produce radon negligibly. Otherwise, this part of radon emanation will be a variable component.

b. It is not hygroscopic, so as to prevent the effective porosity from change with variation of moisture content.

c. In order to limit the interference on radon flux stability from back diffusion, it is reasonable to choose an emanation medium with effective diffusion coefficient low enough, but the effective diffusion coefficient of emanation medium being simulated is sometimes rather high. Taking account of these two contradictions points, two layers of different materials have been chosen as emanation medium.

d. Soft rubber membrane has been tested as emanation medium. A stable radon flux is achievable at constant temperature; in the temperature range from 18℃ to 31℃, whenever the temperature increase by 3℃, the radon flux will increases by more than 10%. This is due to the significant temperature effect on distribution coefficient.

According to the concrete requirements aforementioned, three kinds of emanation medium has been selected as follows:

a. Polystyrene rigid plastic foam (code name "f")

This is a kind of porous material, but most of the pores are disconnected, So it's effective diffusion coefficient and permeability is moderate. The distribution coefficient of radon between "f" and air has been tested twice, obtaining the value of 0.89 and 0.95 respectively. Considering the uncertainty of test, the distribution coefficient value can be approximately recognized as 1, namely that the mechanism of apportion is negligible, and, different with rubber, no significant effect of temperature variation has been observed. This medium may simulates some building material, moist soil or tailings.

b. Polystyrene rigid plastic foam + gypsum board (code name "g")

Gypsum board is not hygroscopic. The radium content of it is negligible. It's effective diffusion coefficient is higher than polystyrene rigid plastic foam, and may simulate some kind of dry soil or dry mortar.

c. Polystyrene rigid plastic foam + magnetite powder (code name "m")

The magnetite has been washed and baked. Being exposed to the weather, the weight gain is less than 0.1%. It means that the hygroscopicity is negligible. The radon flux from itself is less than 7×10^{-3} $Bq \cdot m^{-2} \cdot s^{-1}$, and negligible for uranium tailings. The grain size of this powder is less than 110 mesh with half of it less than 200 mesh. The effective diffusion is higher than gypsum board. This medium may simulate the dry mill tailings, etc.

2.2 Structure

The calibration facility for radon flux monitor is cylindrical. It is made up of three dismountable parts: "cap", "middle part" and "lower part", as expressed in Fig. 1.

Calibration facilities with three different emanation medium have been made and tested. They are named as L3f, L3fg, L2fm. The alphabet in small letter represents different emanation medium. It, in series, corresponds to: polystyrene rigid plastic foam, polystyrene rigid plastic foam covered with gypsum board, and polystyrene rigid plastic foam covered with magnetite powder. The thickness of the foam is 5cm. The thickness of gypsum board is 2.5cm. The thickness of magnetite powder is 3.7cm or 5.3cm.

2.3 Condition and environment

The radon flux of calibration facility not only depends on it's material and structure, but also depends on the emanation condition that in turn depends on the environmental meteorological parameters as temperature, humidity, air pressure, air flow. Whenever the meteorological parameter has changed, the emanation condition changes correspondingly until a stable emanation condition is reached. It means that the time period is another factor which the unstable emanation state depends on. There are several reports on this point in literature[6][7]. In our experience, when a stable emanation condition has been disturbed by meteorological change, a time period of 1-5 days is needed for a new stable condition being reached.

The permeability of the emanation medium used in this facility is not negligible. For example, when an air pressure difference, e.g. 1 hPa, between two sides of "foam" happens, the pressure difference will disappear in about one minute because of the air permeation through the "foam". In this case, the radon concentration distribution in the emanation medium will change leading up to the change of radon flux. Considering the daily variance of atmospheric pressure is some hPa, and the temperature increase by 1℃ may also cause an air pressure increase in closed space by 3.5 hPa or so, it is important to control the variance of environmental meteorological parameters for the stabilization of radon flux.

2.4 Determination of the radon flux of facility

The calculation formula for radon flux measurement is:

$$J = \frac{V}{S} \cdot \frac{C \cdot \lambda_e}{1 - \exp(-\lambda_e \cdot t)}$$

where V :volume of the vessel, m^3;

C :radon concentration as a function of accumulation time, $Bq \cdot m^{-3}$

λ_e :effective decay constant or attenuation constant, $\sec^{-1}$;

S :area of test surface, m^2;

t :accumulation time, $\sec$.

The λ_e is the summation of every probability of attenuation factors which include radioactive decay, back diffusion, leakage etc. A simple and convenient method for the determination of λ_e has been presented.

3. Test results

The stability of radon flux has been tested for a long period. The environmental temperature, relative humidity and atmosphere pressure as well as other relevant situation has been recorded, which is the value on 9--10 o'clock in every morning. As an example, the test results of radon flux stability on L3f given in Fig.2. The average radon flux of L3f is 1.049 ± 0.092 $Bq \cdot m^{-2} \cdot s^{-1}$ with a RSD of 9%. The average radon flux of L3fg is 1.063 ± 0.053 $Bq \cdot m^{-2} \cdot s^{-1}$ with a RSD of 5%. The average radon flux of L2fm is 0.277 ± 0.032 $Bq \cdot m^{-2} \cdot s^{-1}$ with a RSD of 12%.

The uniformity of radon flux is 9.4%, 8.5% and 4.5%(RSD) respectively.

4. Conclusion

a) The stability and uniformity of radon flux facility fulfill the calibration requirement of radon fluxmeter.

b) The determination of radon flux of facility takes full account of eliminating the main error source-attenuation effect: not only prevent attenuation as much as possible, but also measure the real attenuation and make a relevant correction. Therefore the accuracy of determination is assured.

c) The key to the developing, operating and upgrading of the calibration facility lies in following four aspects: material, construction, condition and environment, determination of radon flux.

5. Reference

[1] J.N.Hartley, et al., 1981 radon barrier field test at Grand Junction uranium mill tailings pile, DOE/UMT-0213, April, 1983.

[2] R.Colle, et al., The NIST primary radon-222 measurement system, J.Res.Natl.Inst.Stand.Technol. 95,155,1990.

[3] K.P.Strong, et al., Effect of moisture content on radon emanation from uranium ore and tailings, Health Phys., 42,1, 27-32,1982.

[4] Wu Guihui, Qiu Shoukang, Emanation coefficient of powdered ore sample as a function of moisture content, Radiation Protection, Vol.4, No.4, 298-304, 1984, (Chinese).

[5] Zhang Zhe, Radon emanation and exhaustion by ventilation, Beijing: Atomic Energy Press, (Chinese).

[6] J.E.Pearson, et al., A soil-gas emanation measurement system used for radon-222, Journal of Applied Meteorology 4, 349-356, 1965.

[7] V.D.Korompay, et al., An instrument for measuring the radon emanation from the floor, roof and walls of underground excavations, Division Report MRP/MRL 79-5(TR). CANMET, Nov. 1978.

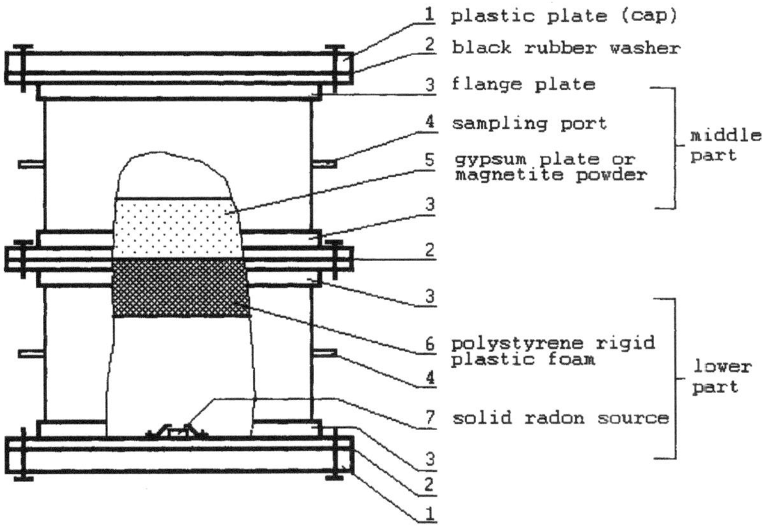

Fig 1.Structure diagram of calibration facility for radon fluxmeter

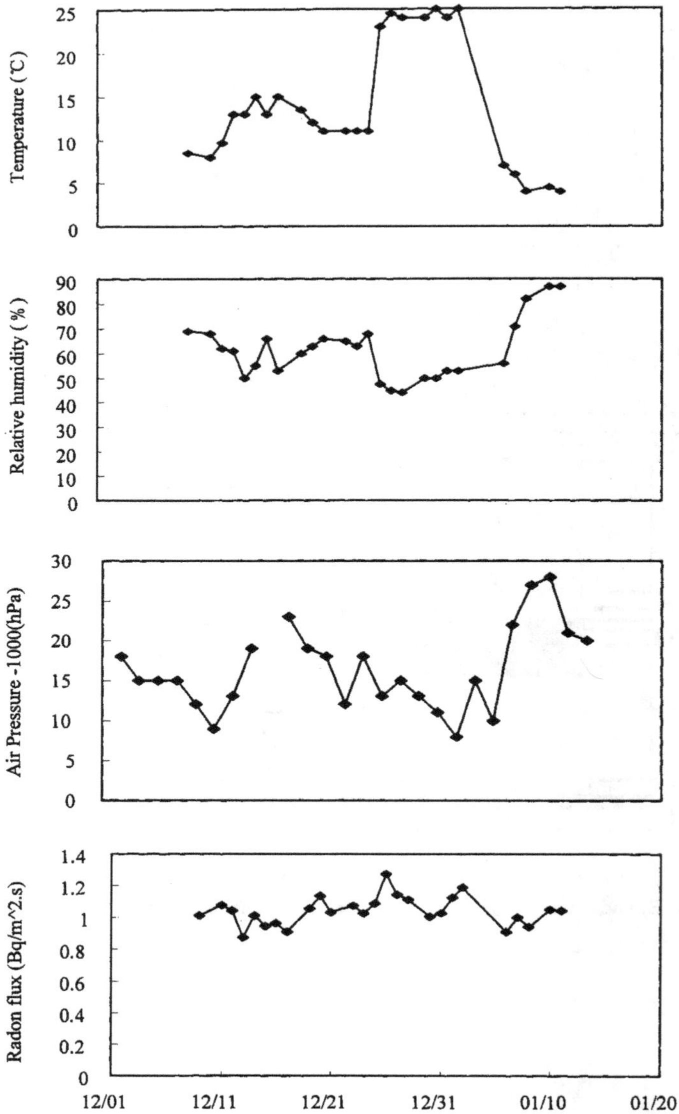

Fig 2. Test results of radon flux stability on L3f

EVALUATION OF FACTORS INFLUENCING MEASUREMENTS OF ENVIRONMENTAL ^{222}Rn CONCENTRATIONS - 1

YOSHIO HOMMA[1], YUKO MURASE[1],, ISAO MURAKAMI[1]
KEIKO HANDA[1] and KIMIKO HORIUCHI[2]

[1]Laboratory for Radiopharmaceutical Chemistry, Kyoritsu College of Pharmacy
1-5-30 Shibakoen, Minato-ku, Tokyo 105, Japan
[2]Department of Environmental Science, Otsuma Women's University
2-7-1 Karakida, Tama-city, Tokyo 206 Japan

Abstract

A commercially available activated charcoal detector for measuring ^{222}Rn concentrations in air was calibrated with known amounts of ^{222}Rn and examined in terms of air luminescence counts, the optimal elusion time, and the interferences from ^{220}Rn and ^{219}Rn. The results indicate that the detector is simple and reliable for conditions normally encountered indoors. The method has been applied to assay indoor ^{222}Rn concentrations in 387 homes in Tokyo and the adjacent four prefectures, which ranged from 0.7 to 140 Bq/m^3 and averaged 22.7 Bq/m^3.

1. Introduction

Investigations at this laboratory of environmental radioactivties were stimulated by reports on activated charcoal detectors for measuring ^{222}Rn concentrations in air [1,2]. Our studies are aimed at finding improved techniques for measuring environmental ^{222}Rn, and at gaining a better understanding of ^{222}Rn exposure problems.

In preceding paper [3], we have modified the integral counting method by extrapolating the integral counting rate curve to the zero detection threshold of the liquid scintillation system rather than to the zero pulse-height.

In this study, first, we have calibrated a commercially available activated charcoal detector with known amounts of ^{222}Rn using a liquid scintillation spectrometer, and second, we have examined the effect of the air luminescence counts on assay of indoor ^{222}Rn concentrations by the method: ^{222}Rn and its α-emitting daughters in the air space above the liquid scintillator produce the air luminescence counts which lead to errors in assay of ^{222}Rn by liquid scintillation counting unless necessary corrections for the counts are made[4,5]. This information will be of value to investigators of environmental ^{222}Rn, since most of the earlier measurements of ^{222}Rn by charcoal detectors with liquid scintillation spectrometers appear to have been more or less subject to uncertainties caused by the air luminescence counts. We have also included a brief discussion of the effects of humidity, temperature variations and the elusion time on the assay of ^{222}Rn by the charcoal detector.

Although the evaluation of the effects described above was the primary concern of this work, we applied the method to the assay of indoor ^{222}Rn concentrations in 387 homes in Tokyo and the adjacent four prefectures.

2. Experimental

Detector cartridges

A commercially available detector cartridge, the Packard PICO-RAD™ ^{222}Rn detector cartridge (Packard Instr. Co., Inc. CT USA), which is henceforth referred to as "detector cartridge", was used in this study. The detector cartridge is made of plastic and its size is identical to that of a standard counting vial for liquid scintillation counting with a tight-fitting lid. A small porous plastic vessel (2.0 cm diameter x 2.5 cm height) in which ca. 1.2 g of activated charcoal and ca. 2.0 g of silica-gel are packed, is located within the detector cartridge at the top.

Standard ^{222}Rn sample and calibration of the detector cartridge

Radon-222 was collected from 0.1 M HCl solution of $^{226}RaCl_2$ with a syringe and added to a standard counting vial which was full of air and closed with a silicon rubber stopper. The ^{222}Rn in the vial was allowed to remain for 3.5 h before measurement. During this time, the daughters ^{218}Po, ^{214}Pb, ^{214}Bi and ^{214}Po came to transient equilibrium with ^{222}Rn. This sample is henceforth referred to as "standard ^{222}Rn sample". The air luminescence counts of the standard ^{222}Rn sample were measured with an Aloka LS spectrometer, Model LSC-3500 (Aloka Co. Ltd. Tokyo, Japan). The counting efficiency for the air luminescence counts was 70.0 % [4].

The detector cartridge and a standard ^{222}Rn sample were put in a plastic vessel, then the silicon rubber stopper of the standard ^{222}Rn sample was uncapped, and the vessel was closed immediately. Thus, the cartridge was exposed to ^{222}Rn of different concentrations for 24 and 48 h within the temperature range of 24-28 and the relative humidity range of 30-40% (Fig.1). The volume of the vessel was in the range of 30-ca.140 L . After the exposure, the cartridge was added 10 ml of Insta-Fluor™ scintillation cocktail and kept at room temperature for an additional 17-18 h. The cartridge was measured with a Packard Tri-Carb LS analyzer, Model 2550 TR/LL using the PICO-RAD™ radon analysis software.

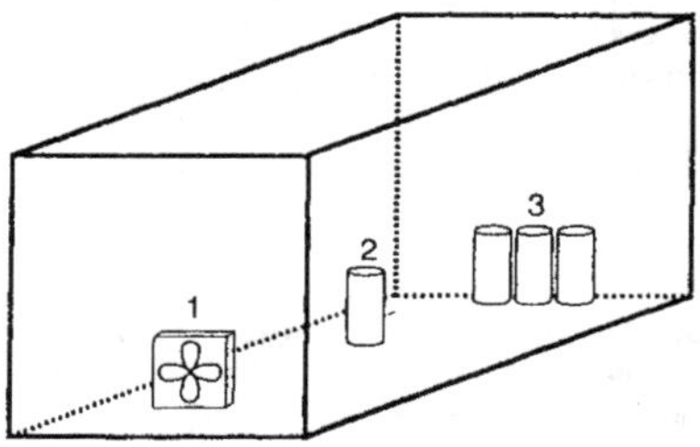

Fig. 1. Exposure of ^{222}Rn to the detector cartridge; 1. Electric fan; 2. the detector cartridge; 3. Standard ^{222}Rn sample.

Measurements of pulse-height spectrum

Pulse-height spectrum of the detector cartridge was measured with the Tri-Carb liquid scintillation spectrometer, Model 2250TR/LL, and was analyzed in order to eliminate the air luminescence counts due to ^{222}Rn, ^{218}Po and ^{214}Po in the air space above the scintillator solution in the detector cartridge.

Determination of elusion time

In order to determine the time required for ^{222}Rn and its daughters to reach equilibrium, eight detector cartridges were exposed to 31.5 Bq/L of ^{222}Rn in a 280-L plastic vessel for 24 h. After the exposure, to each detector cartridge was added 10 mL of Insta-Flour™ scintillation cocktail and measured with the Tri-Carb liquid scintillation spectrometer at various time intervals following scintillation cocktail immersion.

3. Results and Discussion

Calibration

Figure 2 shows the comparison of the average ^{222}Rn concentrations determined by the detector cartridge, A_p, and ^{222}Rn concentration, A_s, which were standardized by the air luminescence method. Agreement between A_s and A_p is satisfactory: The relationship is linear and the value of A_p/A_s is nearly unity. The discrepancy between A_p and A_s is found to be 3.6 % at the very worst. As shown in Table 1, indoor ^{222}Rn concentrations ranged from 0.7 to 140 Bq/m^3, the detector cartridge was proved to give accurate ^{222}Rn concentrations at virtually any exposure level likely to be encountered indoors.

Optimal elution time

As shown in Fig. 3, the count rate of the detector cartridge reach its maximum about 8 h after cocktail immersion, therefore, the optimal elution time is not less than 8 h: In our case, it was usually in the range 8 to 18 h.

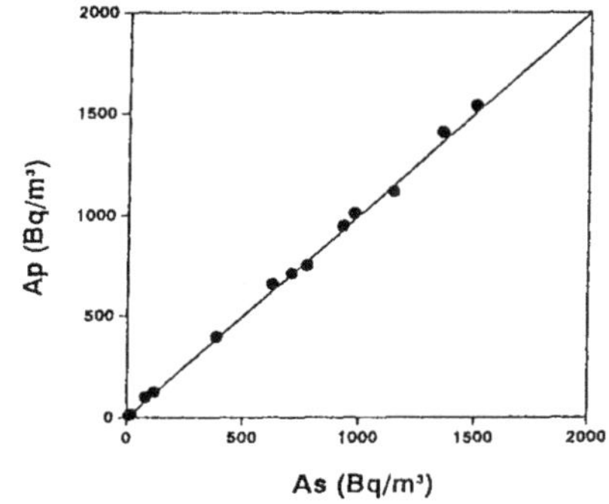

Fig. 2. Radon-222 adsorption by the detector cartridge as a function of ^{222}Rn concentrations. A_p: ^{222}Rn concentration measured by the detector cartridge; A_s: the standardized ^{222}Rn concentration.

Table 1. The distribution of ^{222}Rn activity concentrations in indoor air.

Residents	Number of homes	^{222}Rn Concentration range (Bq / m³)														Average (Bq / m³)
		<10	10~ 20	20~ 30	30~ 40	40~ 50	50~ 60	60~ 70	70~ 80	80~ 90	90~ 100	100~ 110	110~ 120	120~ 130	>130	
Wooden homes	185	40	95	41	7	0	1	0	0	0	0	0	1	0	0	16.6
Concrete homes	188	30	50	38	29	18	7	4	4	2	1	0	1	2	2	29.2
Other homes	14	4	6	2	2	0	0	0	0	0	0	0	0	0	0	16.3
Total	387	74	151	81	38	18	8	4	4	2	1	0	2	2	2	22.7

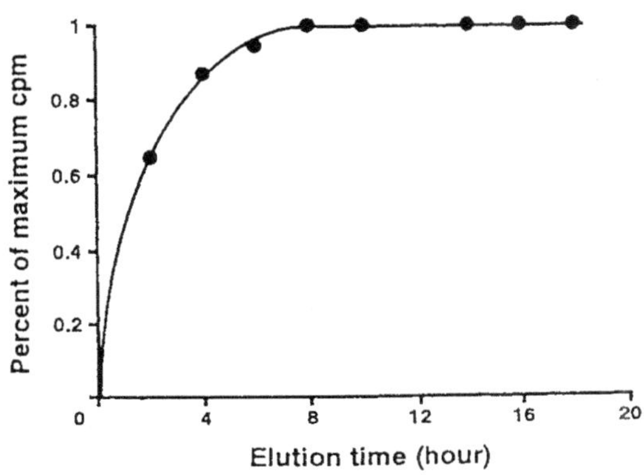

Fig. 3. Determination of an optimal elution time. The count rate of the detector cartridge as a function of the elution time. The count rates have been corrected for decay.

Effect of the air luminescence

Figure 4 shows a representative pulse-height spectrum for the detector cartridge measured with the Packard Tri-Carb 2550TR/LL liquid scintillation spectrometer. Besides the α-peaks of ^{222}Rn, ^{218}Po and ^{214}Po, the air luminescence spectra produced by ^{222}Rn, ^{218}Po and ^{214}Po in the space above the liquid scintillator are also superimposed upon the β-continuum of ^{214}Pb and ^{214}Bi: Since the end-point of the air luminescence spectrum is 18 keV, pulse-heights below 18 keV are due not only to soft β-particles from ^{214}Pb and ^{214}Bi but also to the air luminescence counts. Therefore, the lower window setting of the PICO-RAD™ radon analysis software, 25 keV, is reasonable for the purpose of eliminating the errors due to the air luminescence.

The losses due to the elimination of soft β-particles could be corrected by comparing the measured spectrum in Fig. 4 with that of ^{222}Rn sample which is full of scintillator cocktail; therefore, it does not suffer from possible errors due to the air luminescence counts.

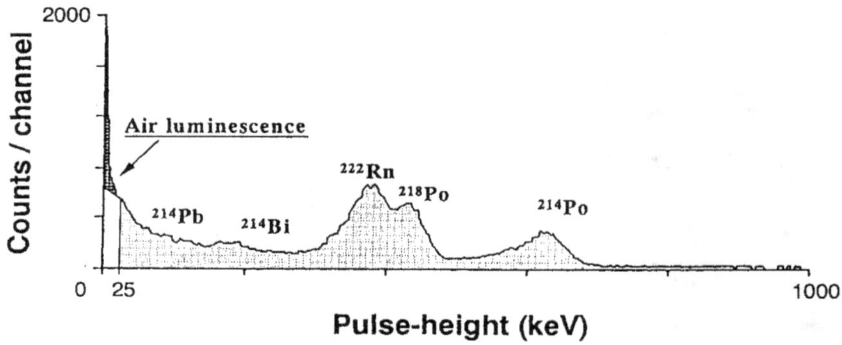

Fig. 4. Pulse-height spectrum of the sample measured with a Packard Tri-Carb liquid scintillation analyzer, Model, 2250TR/LL. Air luminescence due to ^{222}Rn, ^{218}Po and ^{214}Po in the air space above the cocktail are indicated by the shaded area.

Interferences from ^{220}Rn

Radon-220 and its daughter ^{216}Po (0.15 s) will have decayed completely by the time of measurements (usually 12 -22 h after sampling). Moreover, the calculations based on the Bateman equations show that the sum of the activities due to the daughters of ^{218}Po (^{212}Pb, ^{212}Bi, ^{212}Po and ^{208}Tl) at the time of measurements, which is controlled by ^{212}Pb(10.64 h), is about 0.16 % or less of the initial activity of ^{222}Rn. Therefore, it is clear that ^{220}Rn and its daughters would not have a large effect upon the determination of ^{222}Rn, even if the activity of ^{220}Rn is several times larger than that of ^{222}Rn.

A large portion of ^{219}Rn is not allowed to diffuse out of earth and rock into environmental water samples due to its short half-life (3.96 s). Most of the daughters of ^{219}Rn, if present, will have decayed by the time of measurements, and they are seldom present at levels requiring measurement in the environment. Thus, interferences from ^{219}Rn and its daughters can be neglected

Effect of humidity

Experiments which evaluate the effect of humidity on the detector cartridge have been carried out, and the results indicate that adsorption of ^{222}Rn by the detector cartridge is linear for exposures up to 48 h at relative humidity as high as 95 % with temperatures of 30 °C [6]. Therefore, under the ordinary exposure conditions encountered in this study (relative humidity: 31 to 72 %; the duration of exposure: 1 to 2 d; the temperature: 18 to 22 °C), the effect of relative humidity on the adsorption of ^{222}Rn by the detector cartridge would be considered negligible.

4. Assay of indoor ^{222}Rn concentrations

Sampling

Sampling were carried out in Tokyo and the adjacent prefectures-Saitama, Kanagawa, Chiba, and Ibaraki Prefecture - on a total of 387 homes on 18th, 19th, 20th and 21st of November in 1995, and on 9th, 10th, 11th and 12th of December in 1996.

The detector cartridge was exposed for periods of 1 to 2 d at heights 60 to 150 cm above the floor, at temperatures of 18 to 22 °C and relative humidity ranging from 31 to 72 %.

After exposure, the detector cartridge was then sent to our laboratory and 10 mL of a Insta-Flour™ scintillation cocktail (Packard Instr. Co.,Inc. CT USA) was placed into the detector cartridge, which was then capped with a polyethylene cap and shaken for 1 min. The cocktail-charcoal mixture was held for 8 to 18 h to allow chemical and radiological equilibrium to be established. The measurement of the sample detector cartridges has been described above.

Measurements of the detector cartridges

After sampling, the detector cartridges were measured with a Tri-Carb liquid scintillation spectrometer, Model 2250TR/LL (Packard Instr. Co. CT USA), using the PICO-RAD radon analysis software (Packard Instr. Co. CT USA).

Indoor ^{222}Rn concentrations

We have grouped the data into four categories: conventional wooden homes, concrete homes, geographic location of homes, and height of homes and examined the correlations of ^{222}Rn concentrations with those. As shown in Table 1, the average ^{222}Rn concentration for concrete homes is about 1.8 times that for the wooden homes. The results are consistent with the expectations that tight homes will retain more ^{222}Rn than leaky ones: In general, Japanese concrete homes, including an apartment, have tighter structures than conventional wooden homes do. The highest ^{222}Rn concentrations (140.7 Bq/m^3) are found in small, unvented concrete homes. The average value for all homes obtained in this study, 22.7 Bq/m^3, is in essential agreement with previous measurements (23 Bq/m^3)[7].

Since the soil is generally the largest source of ^{222}Rn, it is reasonable that ^{222}Rn concentration would tend to decline with the number of building stories above ground. In this study, however, the height of homes is not highly correlated with ^{222}Rn concentrations. Geographic location of homes seems to be poorly correlated with ^{222}Rn concentrations as well. Larger numbers of samples would be desirable in further studies.

5. Conclusions

The detector cartridge has been proven to give accurate ^{222}Rn concentrations at virtually any exposure level likely to be encountered indoors.

Naturally occurring radon isotopes, ^{220}Rn and ^{219}Rn, and its daughters do not interfere. Moreover, it should be noted that the effect of air luminescence is eliminated by this method. This is especially important because most of earlier measurements of indoor ^{222}Rn by activated charcoal detectors seem to suffer from possible errors due to air luminescence counts.

Acknowledgments- We would like to thank students of our college for their assistance in some of the experimental work. This work was supported in part by the Science Research Promotion Fund of Japan Private School Promotion Foundation.

6. References

[1] Schroeder, M. C., Vanags, U. and Hess, C.T. An activated charcoal-based, liquid scintillation-analyzed airborne Rn detector, Health Phys.,57,43-49, 1989.

[2] George, A. C. and Weber, T. An improved passive activated C collector for measuring environmental ^{222}Rn in indoor air, Health Phys., 58, 583-589,1990.

[3] Homma, Y., Murase, Y. and Handa, K, The detection threshood of a liquid scintillation spectrometer and its application to liquid scintillation, Appl. Radiat. Isot., 45, 341-344, 1994.

[4] Homma, Y., Murase, Y. and Takiue, M. Determination of ^{222}Rn by air luminescence method, J. Radioanal. Nucl. Chem. Lett., 119, 457-465, 1987.

[5] Murase, Y., Homma, Y. and Takiue, M., Effect of air luminescence counts on determination of ^{222}Rn by liquid scintillation counting, Appl. Radiat. Isot., 40, 295-298, 1989.

[6] Shefsky ,S.I., Private communications, 1997

[7] Doi, M., Fujimoto, K. and Kobayashi, S., Electrochemical etching in polycarbonate foil, J. Radiat. Prot. Dosim., 37, 5,1991.

Prototype exposure chamber of radon for animal experiments

Yuji Yamada, Akira Koizumi, Hidenori Yonehara, Michikuni Shimo and Jiro Inaba
National Institute of Radiological Sciences
9-1, Anagawa-4-chome, Inage-ku, Chiba 263, Japan

Abstract

To evaluate a dose conversion factor from the "Working Level of Month" (WLM) of radon to the absorbed dose (mGy), the quality of radon and its progeny was assessed, and exposures controlled for each deposition region were planed as follows: 1) exposure of radon gas to the entire respiratory tract, 2) exposure of "unattached" fractions to the upper respiratory tract, 3) exposure of "attached" fractions onto ultrafine particles to the deep lung, 4) exposure of "attached" fractions onto fine particles to the lower respiratory tract, 5) exposure of "attached" fractions onto coarse particles to the upper respiratory tract. In this preliminary study, a prototype exposure system of radon and its progeny for small rodents was designed. A whole body exposure chamber with a volume of about 0.5 m^3 was used, which it held 20 rats. The aging and mixing chamber separated by the exposure chamber had a volume of about 1 m^3. As career aerosols of radon progeny, carnauba wax aerosols as solid particles, DOS aerosols as liquid particles and NaCl aerosols as hygroscopic particles were selected. These aerosols generated using a vaporization-condensation method and/or an electrical classification method were almost monodisperse with σ_g of <1.2. The monitoring data on biologically related gases showed an importance in the oxygen injection system and the carbon dioxide absorption system.

1. Introduction

Many studies on a dose conversion factor from the "Working Level of Month" (WLM) of radon to the absorbed dose (mGy) have been carried out. Unfortunately, however, it is also true that there has been a not inconsiderable difference in their values. In animal experiments, a difference to the power of 10 was observed e.g., Brooks [1], obtained a value of 0.78 mGy/WLM from the absorbed dose of the fibroblasts in the lungs of a rat, whereas Johnson and Newton [2], obtained a value of 9.8 mGy/WLM from the alveolar macrophages of the same rat.

However, in humans, ICRP Publ.65 [3] determined that epidemiological data gathered on instances of lung cancer in uranium miners showed values of 5 mSv/WLM in workers, whereas the value in the public was 4 mSv/WLM. From the dosimetric approach, a 13.4 mSv/WLM was obtained by Birchall [4], but if there is a slight movement in the aerosol sizes, this value can easily change by as much as either double or a half.

We have to recognize these various figures in dose conversion factor. To reduce the difference, we would like to construct a standard arena for intercomparison. For example, if the attachment coefficient or the career aerosol size are different, even if air with the same amount of radon is inhaled, the deposited activity to each part of the respiratory tract will be different.

Also, even if the activity in each part is uniform, it is highly likely that local sensitivity will be different. Based on the above points, we would like to introduce a deposition controlled exposure in animal studies.

2. Radon exposure system

As shown in Fig. 1, the radon and its progeny exposure system mainly consisted of a radon gas source, an aerosol generator for career aerosols, an aging chamber, an aging and mixing chamber, an exposure chamber and a charcoal bed for radon gas trap. The chambers were made from stainless steel. The internal volume of the exposure chamber was 470 liters, which provided capacity for 20 rats. The test animals were housed in cages. Exposed air laden with radon and its progeny were drawn in via the distribution cone at the top and flowed around the entire chamber in a cyclonic pattern. Soil containing Ra-226 was used as the radon gas source. The radon gas that emanated from the soil always circulated through the aging chamber with a volume of 100 liters. The attached fraction of radon progeny was formed by mixing with career aerosols in the aging and mixing chamber (1,020 liters).

To obtain information on the varied radon concentrations in the aging and mixing chamber over time, a continuous flow Lucas cell (Pylon, model: AB5) was used. For the radon progeny, a radon working level monitor (TracerLab, model: WLM-plus) was used. Its sampling head, with a detector, was placed in the chamber.

By circulating air through the soil, the radon concentration increased from about 10 Bq/ m^3 of the background level to 10^4 Bq/m^3 within half an hour, as shown in Fig. 2. However, it took one day for the concentration stabile. The equilibrium concentration was about one-tenth of the maximum. The radon gas concentration at equilibrium was confirmed to be adjustable up to 1,000 Bq/m^3 when the present radon soil source was used.

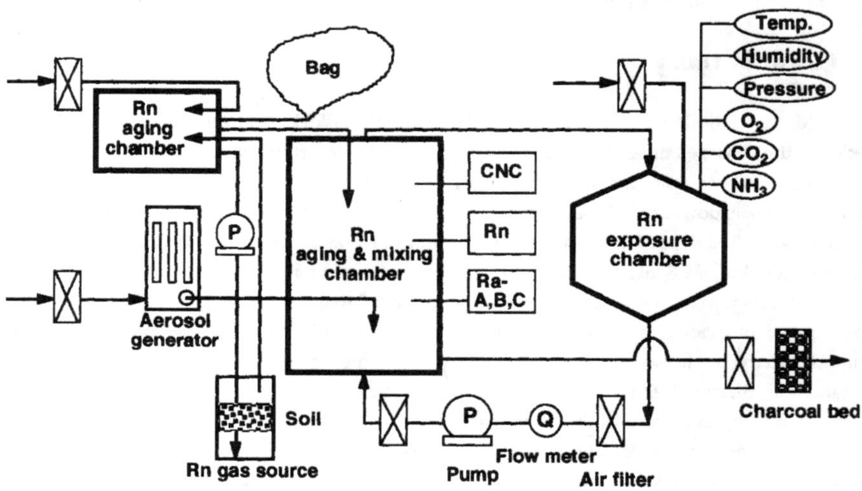

Fig. 1 Radon and its progeny exposure system

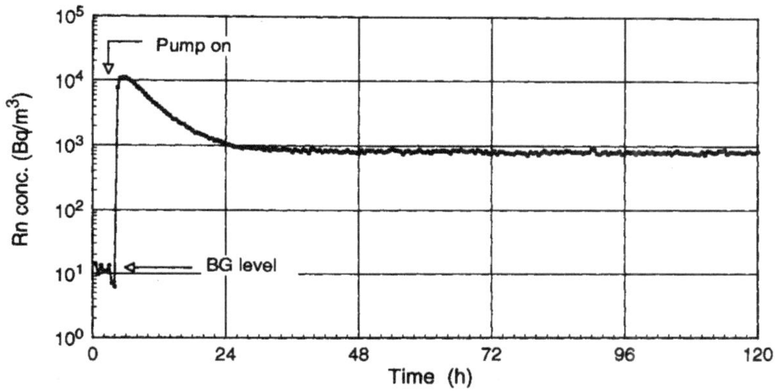

Fig. 2 Stability of radon concentration

3. Career aerosol for the controlled exposures

Inhaled radon gas irradiates the entire respiratory tract. However, its progeny only irradiate a specific region dependant on the size of particles. Exposures controlled for each deposition region were planed in this study, so that career aerosols for the attached fraction of radon progeny were thought much of. The career aerosols were classified into 3 categories based on size : -

1) coarse particles over 1 μm,
2) fine particles between 0.1 to 1 μm
3) ultrafine particles below 0.1 μm.

In addition of these categories, the unattached fraction, consisting of free RaA atoms, was prepared as a 4th category for the controlled exposures. The attached fractions onto the coarse particles and the unattached fractions were expected to be deposited in the nasal cavity and/or upper respiratory tract. The attached fractions onto the fine particles were expected to be mainly deposited in the lower respiratory tract, and those onto the ultrafine particles in the deep lung.

Career aerosols in animal studies must be biologically inert. Three types of career aerosols for the attached fraction of radon progeny were selected : -

1) Carnauba wax as a solid aerosol
2) DOS (dioctyle sebacate) as a liquid aerosol
3) NaCl as a hygroscopic aerosol

These aerosols were generated using the vaporization-condensation method. When monodispersity was required, they were classified by an electrostatic classifier. Size measurements were made with a CNC (condensation nucleus counter, model: TSI-8010) + SDB (screen type diffusion battery) system for less than 0.2 μm and a laser particle counter (PMS, model: HS-LAS) for larger than 0.065 μm.

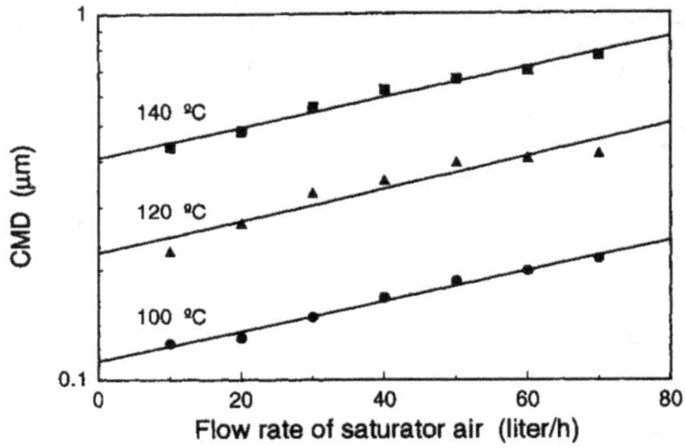

Fig. 3 Dependence of DOS aerosol size on the flow rate of saturator air

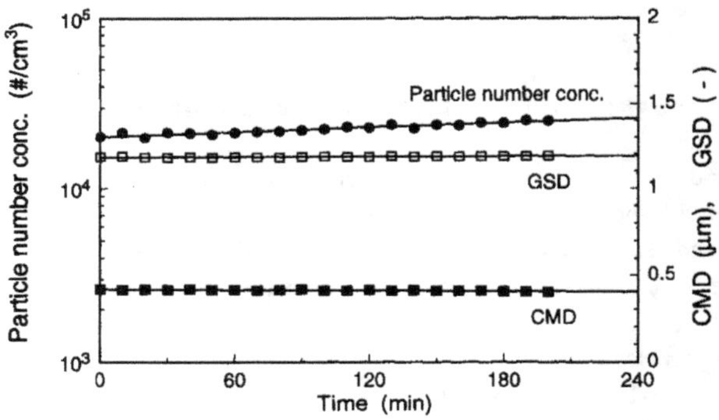

Fig. 4 Stability of DOS aerosol generation

Typical results on career aerosol generation are shown in Fig. 3 and Fig. 4. For the DOS aerosols, the submicron size range was controlled by changing the heating temperature for vaporization and the flow rate of saturator air through the generator. The geometric standard deviations (GSD) of those aerosols were quite small, the size distribution was almost monodisperse. For example, the GSD was only 1.18 when the count median diameter (CMD) was 0.4 μm (Fig. 4). It was confirmed that the aerosol generation was extremely stable for at least 3 hours. The aerosol concentration was over 10^4 particles per cm^3 for all aerosol sizes.

4. Experiments on respiratory gas

Radon exposure experiments were carried out in a closed loop system, so that the respiratory gas control system was essential for long term exposures. To estimate capacities of the oxygen gas supply system and carbon dioxide gas removal system, measurements on these gas concentrations were made for Wister female rats. Each rat was put in a 20 liter volume respiratory gas chamber, and both gas concentrations were monitored using a Galvanic battery sensor and an infrared absorbed analysis method.

Oxygen gas was stably consumed by the rats, and the consumption rate increased in proportion to the body weight, as shown in Fig. 5. The rate per minute, R_{O2} (cm^3/min), is given by

$$R_{O2} = 0.0212\ W_{rat} + 1.74\ , \qquad (1)$$

where W_{rat} is the body weight (g).
And the production rate of carbon dioxide gas, R_{CO2} (cm^3/min), is given by

$$R_{CO2} = 0.0227\ W_{rat} + 0.533\ . \qquad (2)$$

As shown in these equations, R_{O2} is a little higher than R_{CO2} at all times. The same difference had been observed by Nunn [5].

Estimations on the capacities of the oxygen gas supply system and the carbon dioxide gas removal system were done under the following conditions;

Oxygen consumption rate :	3 cm^3/min /100g-weight
Body weight of rat :	300 g
Number of exposed rats :	5 rats/cage x 4 cages/chamber
Chamber volume :	2 m^3 (=2×10^6 cm^3)

Under these conditions, the total consumption of oxygen gas per hour was calculated to be

$$3\ /\ 100 \times 300 \times 5 \times 4 \times 60 = 1 \times 10^4\ (cm^3/h)\ . \qquad (3)$$

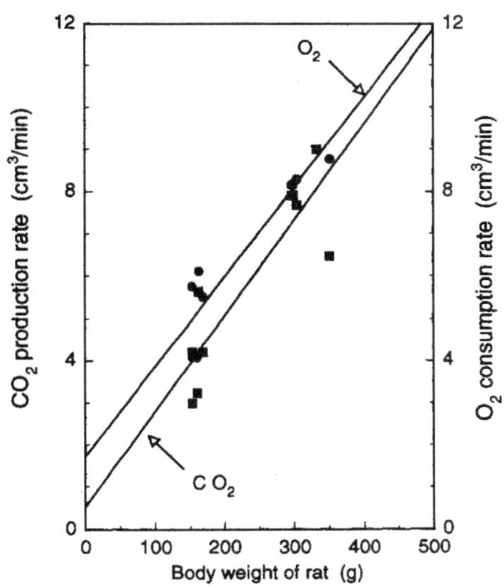

Fig. 5 CO_2 production rates and O_2 consumption rates of rats

Therefore, the oxygen gas concentration in the exposure chamber decreased by 0.5 %/h. If there was no system for oxygen supply, it would take only 5 hours to drop to the permissible limit of 18% for oxygen in air. Therefore, the required capacity for the oxygen gas supply system was estimated as 10^4 cm^3/h (= 10 liter/h), and a carbon dioxide gas removal system with almost the same capacity was also required.

5. Conclusion

A prototype radon exposure chamber was developed. Its basic functions were the same as those of the conventional exposure chambers. A closed loop flow system was employed because of radiological regulations, so that some problems on biologically related gases were caused. It was found that the oxygen injection system and carbon dioxide absorption system were absolutely necessary for a long exposure over 5 hours.

The radon gas concentration was confirmed to be adjustable up to 1,000 Bq/m^3 when the present radon soil source was used. But the concentration was insufficient for a full-scale study on biological effects.

To permit exposures controlled for each deposition region, career aerosols for radon progeny were thought much of, and 4 categories based on particle size were considered, including unattached fractions. Generations and measurements for the aerosols corresponding to different size categories were developed and the validities were experimentally confirmed.

For the next step, a scaled-up study should be carried out. In that case, a horizontal laminar flow chamber should be adapted for a uniform exposure for a long term. A Ra-226 source of over 100 mg should be prepared using a radon gas trap system.

6. References

[1] Brooks, A.L. *et al.*, Effectiveness of radon relative to acute ^{60}Co γ-rays for induction of micronuclei *in vitro* and *in vivo*, Int. J. Radiat. Biol., 66, 801-808, 1994.

[2] Johnson, N.F. and Newton, G.J., Estimation of the dose of radon progeny to the peripheral lung and the effect of exposure to radon progeny on the alveolar macrophage, Radiat. Res., 139, 163-169,1994.

[3] ICRP Publication 65, Protection against Radon-222 at home and at work, 1993.

[4] Birchall, A. and James, A.C., Uncertainty analysis of the effective dose per unit exposure from radon progeny and implications for ICRP risk-weighting factors, Radiat. Prot. Dosimet., 53, 133-140, 1994.

[5] Nunn, J.F., in "Applied respiratory physiology", 3rd Ed., Butterworths, p.72-99 and 379-391, 1987.

Size selective performance of low pressure Andersen sampler and its application to activity size distribution of radon progeny

Keizo Yamasaki

Research Reactor Institute, Kyoto University
Kumatori-cho, Sennan-gun, Osaka 590-04, Japan

Abstract

Size selective collection performance of a low pressure Andersen sampler with some collecting substrates was investigated using 5 kinds monochromatic latex particles labeled with ^{222}Rn progeny. The best performance was obtained in the case of a silicone oil coated stainless steel disk. The sampler was applied to the measurements of the size distribution of the aerosol-attached radon progeny in the nuclear facility of Kyoto University. The size distributions were in fairly well agreement to a log-normal distribution. The geometric mean diameter ranged from 0.20 to 0.27 μm and the geometric standard deviation ranged from 1.7 to 2.4, depending on the kinds of the ambient aerosols. On the other hand, large variances were observed between participants in the domestic intercomparisons on the activity size distribution which were held between Japanese researchers.

1. Introduction

Since differing sized aerosol deposits at different location in the lung, the size distribution of inhaled aerosol-attached radon progeny is important in determining the lung dose [1,2]. Few methods are available on the activity size distribution of radon progeny because of the difficulties encountered in measuring the broad size range of aerosol particles and the low concentration of radon progeny in the environment. The activity size distribution measurements are usually performed with a low pressure cascade impactor or a multichannel wire screen diffusion battery.

A cascade impactor is a very useful instrument for measuring the size distribution of the aerosol particles in various field such as the environmental pollution, the health physics and the atmospheric electricity. A cascade impactor which has been operated at normal pressure was restricted to the minimum size selection of 0.4 μm in diameter. A low pressure cascade impactor is improved to select the minimum size of from 0.02 to 0.06 μm in diameter with operating at the low pressure of few hundreds Pascals. Several researchers apply some types of low pressure cascade impactors (Andersen, Berner, Davis etc.) to measure the activity size distribution of the radon progeny in the environment [1,2,3]. In spite of their careful use, their nonideal behavior is not adequately known [4].

The stage cut of points of cascade impactors are generally obtained from theoretical calculations rather than from experimental calibrations [1]. The usual calculation procedures assume for simplicity that all stages have identical collection characteristics and that a multi jet-multi stage impactor behaves similarly to a single jet-single stage impactor. Effects of factors such as jet Raynolds number, inlet condition, and the nature of collection surface are generally ignored.

Some of the important factors affecting the reliability of impactor data are the wall loss, particle bounce, break up, electrostatic attraction, and the surface nature of collection substrates [4,5]. The complicated natures of these factors are not amenable to theoretical treatments, but by developing the understanding through experimental means, one can minimize their negative effects on impactor performance. This report describes the calibration procedures of size selective characteristics of a low pressure Andersen sampler for some collection substrates using several kinds of aerosol-attached radon progeny with polystyrene latex particles as carriers, and its application to measurements of the activity size distribution of radon progeny in the nuclear facility of Kyoto University.

2. Instrument description

The low pressure Andersen cascade impactor (LP-20-RS, TOKYO, DYLEC Co Ltd.) is a round multi jet impactor which has twelve stages, and samples at a rate of 22.2 ℓ/min with the reduced pressure of -550 mmHg at the last stage. The aerodynamic out off diameters, corresponding to particles collected with the efficiency of 50%, are evaluated with the theoretical calculation to be 9.5, 6.2, 4.2, 2.9, 1.8, 0.95, 0.55, 0.41,0.30, 0.20, 0.13 and 0.056 μm [1]. Particles larger than 0.41 μm are sampled at atmospheric pressure using the first 8 stages, corresponding the Andersen normal type impactor. Four additional stages, operating at a reduced pressure of -75 to -550 mmHg, divide smaller particles. The back filter collects particles smaller than those collected by the last stage. Table 1 lists the design and operational parameters

Table 1. Design and operational parameters of the low pressure Andersen sampler.

Stage number	Diameter of nozzle(cm)	Numbers of nozzles	Pressure (mmHg)	Jet velocity (cms^{-1})	50% cut-off diameter(μm)
0	0.212	98		107	9.5
1	0.121	229		141	6.2
2	0.093	229		238	4.2
3	0.073	229		380	2.9
4	0.054	229		706	1.8
5	0.036	229		1587	0.95
6	0.026	229		3043	0.55
7	0.026	134		5201	0.41
L-1	0.025	110	-75	7603	0.30
L-2	0.025	80	-195	12674	0.20
L-3	0.025	80	-350	17465	0.13
L-4	0.025	110	-550	24799	0.056

Condition (1) Temperature : 23℃ (2) Flow rate : 22.2ℓmin^{-1}

of the Andersen low pressure cascade impactor. The impactor is made of the aluminum, and cylindrically shaped, with a diameter of 10 cm, standing 30 cm high. The impactor is operated by a vacuum pump (OFD–150W,Satoh Vacuum Instruments Co. Ltd.) with a displacement of 150 ℓ/min, with weight of 19kg.

3. Test of size selective characteristics

Measurements of the size distribution of the test radon progeny were carried out by the Andersen low pressure cascade impactor using several kinds of collecting substrates to search their effect on the impactor data. The examined collecting substrates were, clean stainless steel disk, Dow Corning oil-coated stainless steel disk, membrane filter (TM–1, ADVANTEC TOYO Co. Ltd.), teflon binder glass filter (T60A20, Pallflex Products Corp.). Oil-coated disks were obtained by diluting the Dow Corning 200 silicone fluid of 100,000 centistokes viscosity with normal hexane to about 1% solution and coating the solution onto the stainless steel disks.

In those experiments, 5 kinds of polystylen latex particles (Japan Synthetic Rubber Co. Ltd.) which were generated by the atomizer type aerosol generator (KG–02, RION Co. Ltd.) were used as the standard carrier aerosols of radon progeny. Their diameters and standard deviations are listed in Table 2. The size distribution of the generated latex particles were monitored by a laser particle counter (KC–18, RION Co. Ltd.). The latex particles were accumulated at first in a teflon bag (50ℓ) and then transferred to the slightly evacuated test drum (200ℓ) which contained the air with elevated radon concentration (about 1×10^{4}Bqm^{3}) and low aerosol particle concentration (<50cm^{-3}). The test drum was left at least for 4 hours in the drum because the correlation between ^{222}Rn and its progeny concentrations reached to the radioactive equilibrium.

Table 2. Polystyrene latex particles used for the test of size selective characteristics.

Diameter(μm)	S.D.(μm)	C.V.(%)
0.061	0.0038	6.22
0.152	0.0037	2.46
0.309	0.0035	1.13
0.399	0.0060	1.50
0.791	0.0156	2.02

The test radon progeny were sampled on each stage of the impactor for 2 minutes at s flow rate of 22.2 ℓ/min. The alpha particles emitted from sampled radioactivity were sequentially counted with a scintillation counter for 30 seconds after 20 minutes waiting for the decay of ^{218}Po. The number fraction of radioactive particles were found for each stage after decay correction. Since the size distribution was close to a log-normal distribution, logarithmic probability paper was used to plot the integral fraction vs. particle diameter. The smooth line through those data points was used to find the geometrical mean aerodynamic particle diameter, dg, and the geometrical standard deviation, σg.

Figure 1 shows the observed cumulative size distributions of the 0.399 μm monochromatic

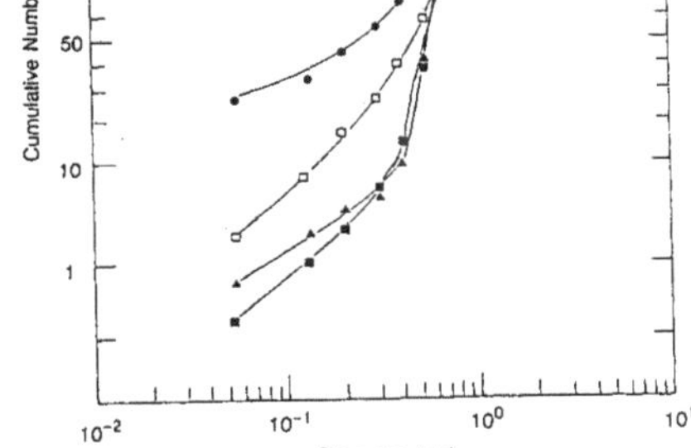

Fig.1 Observed cumulative size distributions on the 0.399μm monochromatics radon progeny for several collecting substrates.

Table 3. Observed size distributions on the 0.399μm monochromatic radon progeny for several collecting substrates.

Collecting substrates	G.M.D. (μm)	G.S.D.
Stainless steel plate	0.25	2.60
Silicone grease coated stainless steel plate	0.53	1.26
Teflon binder filter (T60A20)	0.58	1.46
Membrane filter	0.52	2.16

latex particles labeled with radon progeny for several collecting substrates. The large variation of the observed size distributions may be originated the fact that the latex particle is a solid particle. It is considered that the effects of particle bounce and reentrainment are minimized by the use of liquid aerosols but the results with solid aerosols may differ significantly from those obtained by using liquid aerosols [4,5]. Table 3 shows the experimental values of dg and σg obtained by assuming a log-normal size distribution for several collecting substrates. From these results especially from the values of σg, it was concluded that the silicone oil or grease coated stainless steel disk was a good substrate for the size selective collection of the radon progeny.

On the other hand, it was concluded that the interstage wall loss was negligible small for the measurements of the size distribution of radon progeny from a comparison between the total activity and the sum of the activity collected on each stage.

The theoretically induced size selective characteristics of the low pressure Andersen sampler were experimentally checked with using five kinds of test radon progeny which were produced with the above-mentioned method. The result is shown in Fig.2 and Table 4. From the value of σg, it is considered that this examination has some particle agglomeration problems on the generation of the monochromatic test radon progeny, especially at the part of the smaller particle size. In spite of this disadvantage, it is clear that the experimental size selective characteristics of the sampler differ from theoretical one. Thus, a cascade impactor may need an appropriate calibration procedure including the interstage characteristics for determining the accurate size distribution.

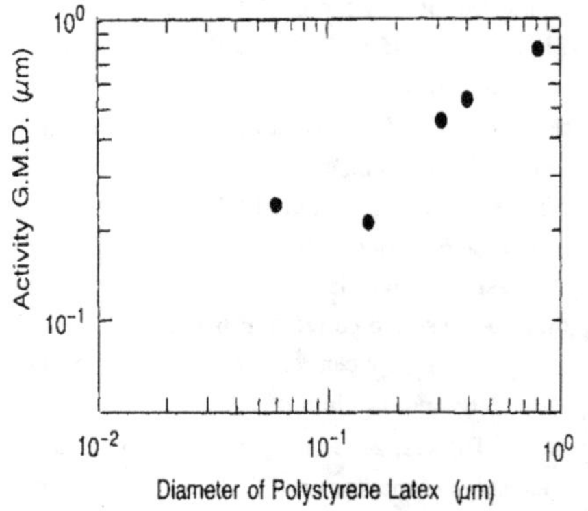

Fig.2 Correlation between the diameter of latex particles and the activity geometric mean diameter obtained with the low pressure Andersen sampler.

Table 4. Summary of the observed activity size distributions for various monochromatic latex particle labeled with ^{222}Rn progeny.

Mean size of latex particle(μm)	G.M.D. (μm)	G.S.D.
0.061	0.24	1.96
0.152	0.21	2.00
0.309	0.46	1.22
0.399	0.53	1.26
0.791	0.79	1.53

4. Intercomparison of activity size distribution

The applicability of the low pressure Andersen sampler to the measurement of the activity size distribution of aerosol-attached ^{222}Rn progeny was examined in the form of the intercomparison experiment. The intercomparison experiment was conducted in a large room of the Kyoto University Critical Assembly Building in the nuclear facility that was airtight and equipped with a re-circulating ventilation system. Participants were Sci. Univ. of Tokyo, PNC-Ningyo, Waseda Univ. Kobe Univ. of Merc. Mar. and Res. React. Inst. of Kyoto Univ. . The first two groups used the wire screen battery method and the others used the low pressure cascade impactor method. ^{222}Rn concentration was maintained in the range from 400 to 600 Bqm^{-3} during experiment. Measurements were carried out under the conditions of low aerosol number concentration (<500 particles cm^{-3}) without artificial aerosol source and high aerosol number concentration (5000~10000 particles cm^{-3}) with additional aerosols generated from a burning incense stick.

Figure 3 shows the typical data of the intercomparison on the activity size distribution of ^{222}Rn progeny which was measured under the condition of high aerosol number concentration. In this figure, participating laboratories are assigned code names A, B, C, D and E to maintain anonymity. Here, code names A, C and D used the low pressure cascade impactor method, and code names B and E used the wire screen diffusion battery method. The activity size distribution of the code name F was calculated using the attachment theory from the aerosol number size distribution which was measured by the laser particle counter. The data at diameter below 0.1 μm of code name C was eliminated owing to the leakage at the lowest stage of the impactor. The data of Fig.3 is summarized in Table 5. The almost all peak diameters except for code name B converged around from 0.2 to 0.3 μm. But the geometrical standard deviations ranged from 1.2 to 2.7, according to the difference of measuring groups and methods. This

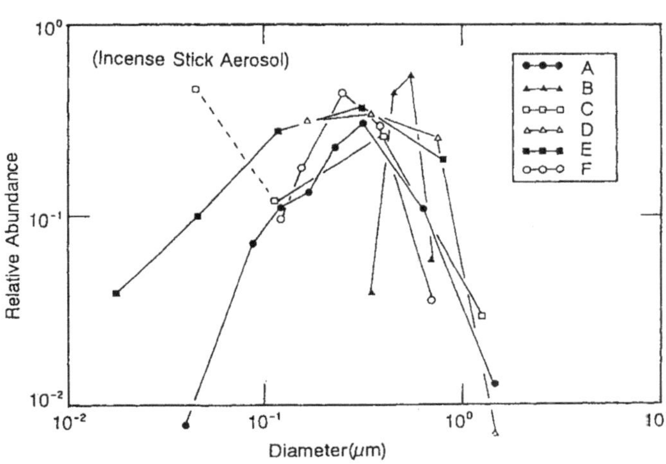

Fig.3 Typical data of intercomparison on activity size distribution of ^{222}Rn progeny (4 May, 1996).

Table 5. Typical data of intercomparison on activity size distribution of ^{222}Rn progeny (4 May,1996).

Code Name	G.M.D.(μm)	G.S.D.	Remarks
A	0.25	1.78	L.P.I.
B	0.48	1.15	S.B.
C	0.19	2.67	L.P.I.
D	0.33	1.85	L.P.I.
E	0.21	2.74	S.B.
F	0.25	1.46	Calculated

variation includes also the uncertainties concerning the definition of aerodynamic and diffusion equivalent (Stokes) diameters. The origin of these differences are unknown at present status but may cause from the mathematical optimization procedure of data acquisition.

5. Conclusion

The size selective performance of the low pressure Andersen sampler was examined experimentally. The best size selective performance was obtained when a silicone oil (or grease) coated stainless steel disk was used for collecting substrate. It was found that the theoretically induced size selective characteristics differed from the experimental characteristics.

The applicability of the low pressure Andersen sampler to the measurement of the activity size distribution of the aerosol-attached radon progeny was discussed in a form of the intercomparison. The size distributions which were measured by the sampler were fairly well agreement to a log-normal distribution. The geometric mean diameter ranged from 0.20 to 0.27 μm and the geometric standard deviation ranged from 1.7 to 2.4, depending on the ambient aerosols. On the other hand, large and serious variances were observed between Japanese participants using the low pressure cascade impactors and the wire screen diffusion battery.

6. References

[1] Yamasaki, K., Okamoto, K. and Tsujimoto, T., Unattached fraction and the size distribution of the radon progeny in air of a nuclear facility, Proc. Asia Cong. Radiat. Prot., 591–594, 1993.

[2] Prostendörfer, J., Properties and behaviour of radon and thoron and their decay products in the air, J. Aerosol Sci., 25, 219–263, 1994.

[3] Tsujimoto, T., Miyake, H., Iida, T., Kimura, Y., Ogawa, T., Morishima, H., Nishikawa, T. and Yamasaki, K., Continuous measurement of size distribution of radon progeny using DRUM Impactor, KURRI Prog. Rep. 1995, 220, 1995.

[4] Rao. A.K. and Whitby, K.T., Non-ideal collection characteristics of inertial impactors − I. Single stage impactors and solid particles, J. Aerosol Sci., 9, 77–86, 1978.

[5] Rao, A.K. and Whitby, K.T., Non-ideal collection characteristics of intertial impactors − II. Cascade impactors, J. Aerosol Sci., 9, 87–100, 1978.

Size distribution of atmospheric radioactive aerosols and its measurement

Yihe Jin, Liya Xu, Guoqiu Fang
Fujian Institute of Radiation Health Protection,
282 Guangrong Road, Fuzhou 350001, China

Michikuni Shimo
National Institute of Radiological Sciences, 4-9-1,
Anazawa, Inage-ku, Chiba 263, Japan

Weihai Zhuo
Graduate School of Engineering, Nagoya University,
Furo-cho, Chikusa-ku, Nagoya 464-01, Japan

Abstract

It is well known that the variation of the size distribution of radioactive aerosols may bring a large fluctuation of lung dose caused by radon and its daughters, and the parameter may vary from one place to another. Sampled by the wire screen-type diffusion battery and calculated by Ikebe analytical method, the size distribution of indoor and outdoor radioactive aerosols at Fuzhou in China were evaluated. Both of the indoor and outdoor size distributions showed a geometric normal distribution, and the arithmetic mean peak diameters for indoor and outdoor air were (0.20 ± 0.09) μ m and (0.27 ± 0.11) μ m, respectively. The result of indoor air was the same as the recommended data by UNSCEAR(1988), while that of outdoor air was slightly higher, but it was similar to that reported by Jacobi in 1980. Furthermore, the correlation between the size distribution and the concentration of radon and/or meteorological factors were also studied, but no significant differences were found.

1. Introduction

Many kinds of aerosols exist in the atmosphere. Some are natural, and others are produced by human activities. Aerosols have been becoming one of important research objects in many fields of sciences. The studies include particle concentration, mass concentration, chemical composition, size distribution, specific radioactivity, optical property, etc. In estimating the lung dose caused by radioactive aerosols, because there is still no perfect method to directly measure by now, the estimation is usually based on the present dose model and the measured radiation energies of aerosols. Due to the variation of model parameters (bronchial morphometry, location of target cells, breathing pattern, particle concentration, size distribution, etc.), the calculated results may fluctuate with the selected parameters. Among them, the activity median diameter (AMD) of the aerosols attached with radon daughters is a sensitive parameter. As reported by NCRP[1], the percent variation in annual dose for a very large median particle size shift from an average of 0.125 μm would range from +96% for the 0.05 μm particle to -20% for the 0.17 μm particle. So more and more attention has been paid to the study of the size distribution of radioactive aerosols.

The diameters of radioactive aerosols range from 0.001 μm to over 100 μm. At present, it is still impossible to measure the whole range by an individual method. From the dosimetric point of view, naturally radioactive aerosols can be regarded as those attached and unattached fractions of radon daughters, their sizes range approximately from 0.001 μm to 0.5 μm. So the wire screen type diffusion battery, which is small in volume and easily controlled, was adopted in this study, and the results of attached fraction of radon daughters at Fuzhou in China were reported in this paper.

2. Materials and Methods

2.1 Principle of Measurement

Whenever uniform radioactive aerosols, with a specific radioactivity Z_0, penetrate through the wire screens, their radioactivity will attenuate to Z due to the diffusion effect. Once the structure of the wire screen (effective volume, wire diameter, screen size, etc.) and environmental conditions (temperature, pressure, etc.) are fixed, the particle penetration ratio ($P=Z/Z_0$) will be the function of airflow rate, particle size (r) and screen number (N)[2]. As the airflow rate can be constant during the sampling period, P is the function of only r and N, i.e. P=R(N, r). For different size particles attached with radon progeny, P can be expressed as the following formula[3]:

$$P(N) = \frac{\int R(N, r)\, f(r)dr}{\int f(r)dr} \qquad (1)$$

where: f(r) is the function of size distribution.

According to the formula (1), once the airflow rate is fixed, the penetration curve can be determined by changing the number of screen. Being mathematically analyzed, the size distribution can be calculated.

2.2 Material and Sampling

Sampling apparatus is shown as Fig. 1. Airflow enters the screen-type diffusion battery

(SDB) which can be installed with 43 chips of wire screens at most from the left to the right.. When more than 43 chips are needed in the diffusion process, other SDBs are sequentially connected. The effective diameter of screen made of stainless steel wire is 40 mm, each screen has 165 meshes, and was adhered to a brass ring(thickness 5 mm, inner diameter 40 mm, outer diameter 60 mm). In order to prevent leakage from the adhered location, a groove was carved on each ring, and an airtight sticky gasket was used to link the screen and the ring. According to the results studied by Sugiyama et.al.[4], the results of size distribution sampled by using different screens with 165~250 meshes were nearly the same, and the results were also similar to others by different types of diffusion battery. So the cheaper one with 165 meshes was used in this study. Having penetrated through the SDB, particles were collected by filters (TOYO, TM-80, 0.8 μ m). The joint between filter and filter holders should also be highly airtighted.

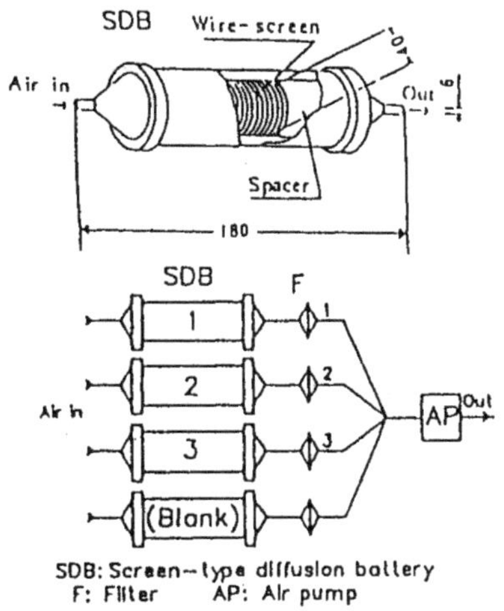

Fig.1 Schematic diagram for measuring the size distribution of radioactive aerosols by screen-type diffusion battery

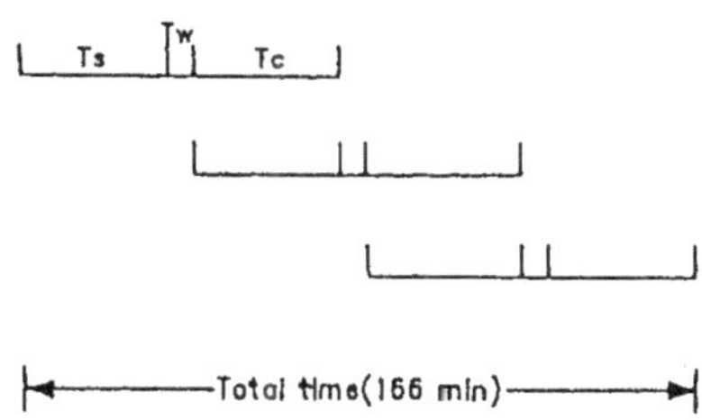

Fig.2 A time table for measurement

T_s: Sampling time (40 min)

T_w: Waiting time (2 min)

T_c: Counting time (40 min)

Both sampling time and airflow rate would play important roles in getting stable and accurate results. They are mainly determined by the detectable threshold and detecting efficiency of the α counter, and the specific activity of radon daughters in air. Having conscientiously done the preparatory experiments, airflow rate for sampling was set at 4 L $\cdot$ min^{-1}, and the sampling time was set for 40 min. In order to keep the flow rate highly consistent and stable during the whole sampling process, flow cells and their regulators were connected to the air-way of each SDB, this is very important to the measuring results.

Various numbers of screens used in each SDB brought out different activities of aerosol particles collected on the filters, and then, the penetration curves could be plotted. To establish a penetration curve, at least 9 data and another blank SDB are needed. That is to say, 10 branches of SDBs and 10 α counters are needed in the meantime. Owing to the limited availability, only 4 branches of SDBs and 4 α counters were used , so sampling for 3 times in each process were required. The process in detail was scheduled as shown in Fig.2. During the 9 different diffusion processes in this study, the numbers of screens used in SDBs were 10, 15, 25, 40, 55, 80, 110, 159, and 269, respectively.

2.3 Method for Data Analysis

Based on the plotted penetration curves, the size distribution of atmospheric radioactive aerosols was calculated by Ikebe analytical method[4]. It was to calculate the size distribution of heterogeneous submicron aerosols by response matrix method. But negative value may exist some time, so a progressively approaching was used to find the approximate solution for f (r). In order to lower the error (less than ± 5%), more than 16-time progressively approaching was required. Typical size distributions were shown as Fig.3.

2.4 Sampling Site

In order to avoid the interference of artificial aerosols, a piece of quiet lawn far away from roads was sited for outdoor sampling, and a laboratory in which few staffs worked was sited for indoor sampling.

2.5 Monitoring Period and Items

After one and a half year of preparation, the regular monitoring were carried out once a week from July 1992 to December 1993. Sampling time were set from 8 am to 12 am. In the case of failure in sampling, another sampling was made in the same week. During the sampling periods, the concentrations of radon and its daughters and meteorological parameters (temperature, atmospheric pressure, humidity, etc.) were also measured and recorded.

3. Results and Discussion

In the living environments, 62 size distribution curves of radioactive aerosols (11 for indoors, 51 for outdoors) were obtained. All of them showed a geometric normal distribution, as shown in Fig.3. The peak diameters of the outdoor air ranged from 0.10 μm to 0.56 μm, their frequency distribution was shown in Fig.4. As shown in Fig.4, the diameters mainly distributed in the range of 0.15~0.40 μm, accounted for over 70% of the whole results. The peak diameter of the indoor mainly gathered around the range of 0.15~0.40 μm, its peak distribution was not analyzed for fewer samples.

The arithmetic mean peak diameters for indoor and outdoor radioactive aerosols were (0.20±0.09) μm and (0.27±0.11) μm, respectively. As shown in Table 1, the calculated result for indoors was the same as the recommended data by UNSCEAR Report(1988)[5]; while the result for outdoor air was slightly higher , but it was similar to that reported by Jacobi, et al.[6].

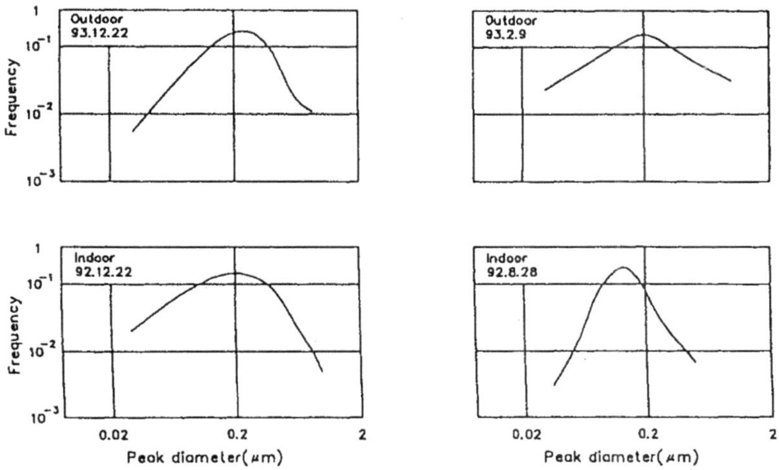

Fig.3 Some examples of size distribution of radioactive aerosols

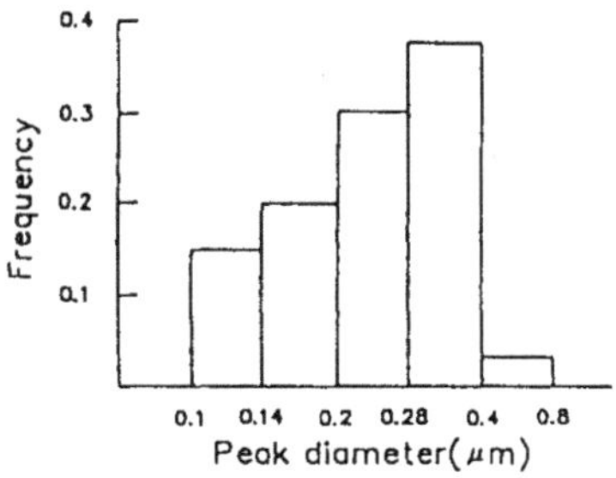

Fig.4 The distribution of peak diameter outdoors

Table 1 Size distribution in other references

Reference	AMD(μ m)		
UNSCEAR(82)	0.2		0.1
UNSCEAR(88)	0.2		0.2
Jacobi-Eist		0.25*	
NCRP No.78		0.05～0.17*	
This study	0.20		0.27

*: Not to be separated for indoor and outdoor air

References about the size distribution of radioactive aerosols in China is scanty, so the recommended data by UNSCEAR(1988) are usually quoted to estimate the lung dose, i.e. the dose conversion coefficients are based on that the AMD are 0.2 μm for both indoor and outdoor aerosols. Based on the results of this study, the lung dose would be over estimated for the contribution of outdoor aerosols at Fuzhou in China if UNSCEAR recommended data were quoted; but for the contribution of indoor air, it was just the same. As about 90% of lung dose is contributed by indoor radon daughters, so it is basically feasible for Fuzhou in China to estimate lung dose by using the recommended data by UNSCEAR (1988). Further studies are still needed to extend this conclusion nationwide.

Furthermore, the correlation between the size distribution and the concentrations of radon and its daughters and/or meteorological factors were also analyzed, no significant correlation was found among the present data.

Acknowledgment—The authors are grateful to the Scientific Research Fund of Fujian Public Health Bureau of P.R.China for financial support.

4. References

[1] NCRP(National Council on Radiation Protection and Measurements). Evaluation of occupational and environmental exposures to radon and radon daughters in the United States. NCRP Report No.78. Bethesda, MD: NCRP; 1984.

[2] Cheng, Y.S.; Yeh, H.C. Theory and calibration of a screen-type diffusion battery. J. Aerosol Sci. 11:313-319; 1980.

[3] Sugiyama. K.; Shimo. M.; Ikebe. Y. Measurement of the size distribution of atmospheric radioactive aerosols by screen-type diffusion battery. Res Lett Atmosph Electr. 6: 67-70; 1986.

[4] Ikebe. Y. Determination of the size distribution of heterogeneous submicron aerosols by response matrix method. Pure Appl. Geophys. 98:197-201; 1972.

[5] UNSCEAR(United Nations Scientific Committee on the Effects of Atomic Radiation) . Source, effects and risks of ionizing radiation. Report to the general assembly with annex. United Nations; New York, 1988.

[6] Jacobi. W. and Eisfeld. K., GSF Report, S-626 (1980)

Electrical Mobility of Radioactive Small Ions(^{218}Po) in the Atmosphere

Sigeru Nakae, Tetuya Yamada, Kazuhiko Miura and Kenji Hara

Department of Physics, Science University of Tokyo
Kagurazaka, Shinjukuku, Tokyo 162, Japan

Abstract

The electrical mobility distributions of radioactive ions are obtained by using the mobility separation chamber and the position sensitive counter. The separation chamber is the condenser of parallel plate type in basically. The face of ion collecting plate is covered with a thin film and serve as the window of proportional counter. The both ends of resistive anode wire of the counter are used as signal output terminals, and the ratios of oneside output pulses to the sum of both ends pulses are proportional to the position of generating alpha pulses. The mobility distributions of radioactive ions can be obtained from the position analysis of a detected events. The resolution of the instrument is about 3 percent in position of alpha decay events. As a results, Bimodal distributions, with peaks of 0.4~0.7 and 1.5~2.7 cm^2/sec.volt., were found on the naturally occurring radioactive ions(^{218}Po) in the room air.

1.Introduction

The physical properties and concentration of the radioactive substances in the air are very important from the viewpoint of atmospheric physics and health physics. The presence of these substances influences the ionization of the atmosphere and the lung dose of humans. Radon in the atmosphere emanates from the soil and rocks, and released from coal-fired power plants. The radioactive decay products of radon are found in the form of small ions in the air. The mobility spectrum of these radioactive ions was determined by the utilization of a Zeleny tube[1]and an Erikson spectrometer[2][3][4], and the alpha-activity of equal length segments of collecting electrode is measured by various devices, such as ZnS scintillators, nuclear emulsions and surface barrier detectors. However, It is difficult to know the properties of ions under the various conditions of atmosphere, because of their time resolution. In this paper, the instrument is developed to determine the electrical mobilities of radioactive small ions continuously and discuss the properties of electrical mobility of ^{218}Po ions in the atmosphere.

2.Experiment

The mobility spectrometer used here is basically the condenser of parallel plate type. The radioactive ions are introduced into the condenser along the surface of one electrode made by Al-plate. The inside face of another plate, on which the ions are deposited, is covered with a thin gold-plated myler film and serve as the window of a proportional counter. The mobility spectrometer was modified to provide a uniform air

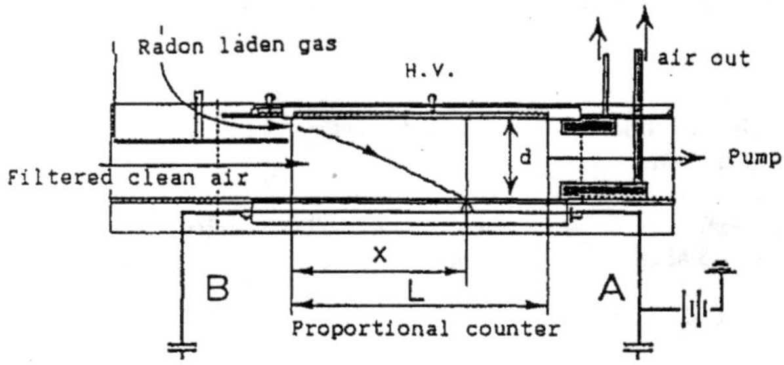

Fig.1. Schematic diagram of the mobility spectrometer of parallel plate type.

flow from two channels at the entrance, one for the filtered radon free air and the other for the radon laden gas. The gas velocities, v, are the same in both channels. This modified spectrometer is schematically shown in Figure 1. The electric potential is also applied between the plates. The vertical electric field, E, is provided through the spectrometer so that the radioactive ions are drawn toward the counter surface and deposited at the distance, x, from the entrance according to their electrical mobilities. The alpha-rays emmited from radioactive ions are detected by the counter through the thin film window. With the known distance, d, between the radon laden gas channel and the detector window plate, the electrical mobility of ions, k, can then be determined from $k = vd / Ex$.

Signal processing

The counter is constructed using RC position encoding. The charge-division method determines the position of the event by calculating the ratio of the charge collected at one end of the resistive anode wire to the total collected charge on both ends. Figure 2 shows a block diagram of the electronics used to read the position of alpha decay events. Each end of resistive anode wire is connected to a charge sensitive amplifier respectively, and is used as signal output terminals. The unipolar outputs of the amplifiers are fed to a sum and invert amplifier. It has two inputs to permit the summation of signals from separate circuits. The summed and inverted signal is furnished through the output terminal, and the inverted signals of individual input pulses are also given at another output terminals. The amplifier has a rise time of < 50 ns, a bandwidth from DC to 7.0 MHz and output impedance is < 0.1 ohm for line drive. The output signal is true only when there is a time coincidence of arrival of the input pulses. Position sensitive detector analyzer at the next stage has two input signals. One of the signals has a peak pulse amplitude proportional to the full energy absorbed from an event in the detector. The other signal has a peak pulse amplitude that is a significant fraction of that for the full-energy pulse, and the fractional ratio indicates the relative point of impingement of the event along the length of the detector. Thus the ratios of oneside output pulses to the sum of both ends pulses are proportional to the position of generating pulses. These pulse ratios can be observed by the use of a multichannel pulse height analyzer.

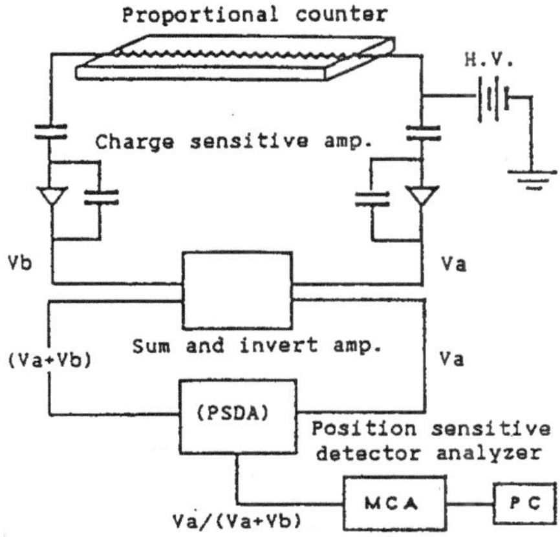

Fig.2. Construction of position sensitive proportional counter.

3.Results and Discussions

The detective device is essentially a rectangular gas flow counter.The cathode of counter is a chrome plated iron mesh covered with thin gold-plated mylar film of 0.85mg/cm^2 thick, which allows alpha particles emitted from the deposited radioactive ions on the surface to penetrate through it. Carbon wire was used as the anode of the counter. Both ends of carbon wire are connected to charge sensitive preamplifiers respectively. Commercial methane of 99 percent purity is circulated through the counter. The values of alpha plateau length and slope of the counter were found to be 250 volts and few percent per 100 volts, respectively. The curve of counting rate versus voltage applied to the counter is shown in Figure 3. The source employed in this measurement was a radium D+E+F source emitting both alpha and beta particles.

The mobility distributions of radioactive ions can be obtained from the position analysis of a detected events. The relation between positions of detected events and observed channels on a pulse height analyzer is shown in Figure 4. In this case, the position of the source is changed in stepwise along the axis of counter. The spatial uncertainty

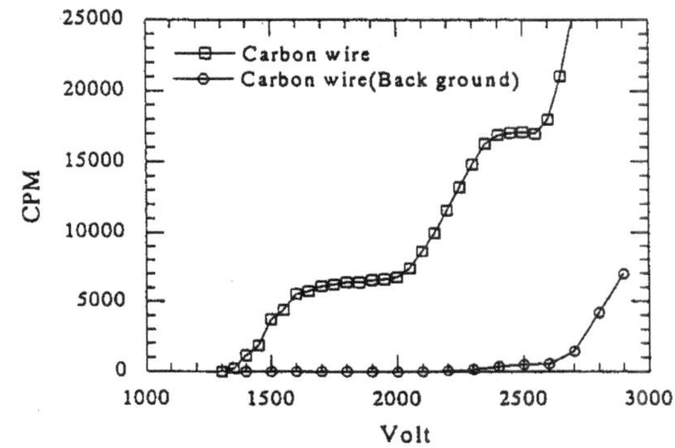

Fig.3. The curve of counting rate versus voltage applied to the position sensitive counter.

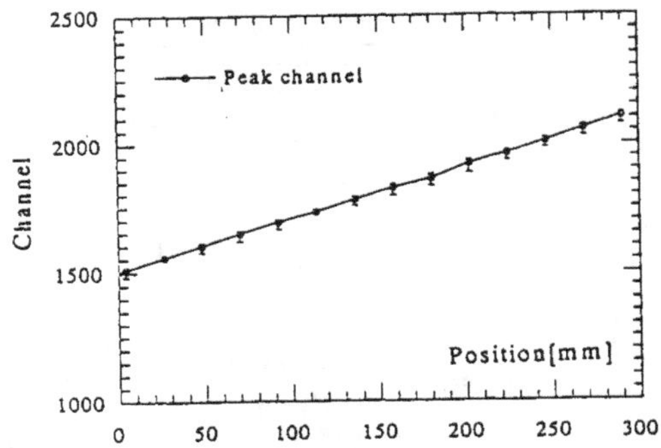

Fig.4. The relation between positions of events and observed channel numbers of a pulse height analyzer.

in position are found as ±4 mm. An alpha particle generally produces an electron-ion track along the trajectory of the particle in the gas flow counter. For a trajectory that is not always perpendicular to the anode of the counter, the concept of an interaction point is not valid, because the interaction is spread over a space interval Δx that has a width $w = d \cos\alpha$, where d is the track length inside the counter and α is the track angle with respect to the anode. For such a particle, the center of charge position is in Δx and depends on the specific ionization of the electron-ion track.

From linear relation between the position of detected events and the observed channels on the pulse height analyzer, the mobility distribution of radioactive ions can be obtained successfully. In this case, the background alpha pulses, caused by radioactive impurities in the counter materials, appeared roughly in uniformity along the axis of counter. Then the signals can be obtained by deducting the background count from the total amount. As a results, bimodal distributions, with peaks of about 0.7 and 2.7 cm^2/sec.volt, are found in winter season on the

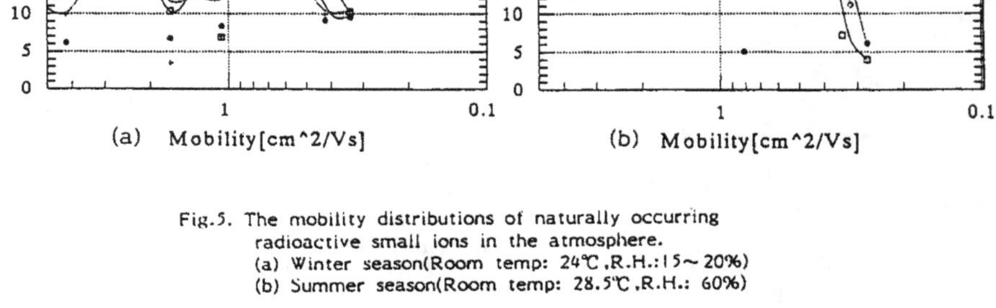

Fig.5. The mobility distributions of naturally occurring radioactive small ions in the atmosphere.
(a) Winter season(Room temp: 24℃, R.H.: 15~20%)
(b) Summer season(Room temp: 28.5℃, R.H.: 60%)

naturally occurring radioactive ions(^{218}Po), and with peaks of 0.4~0.5 and 1.5~2.0 cm^2/sec.volt are found in summer season. Compared with winter season(Temp.24℃,R.H.15~20% on average), they seem to be shift down to the lower mobility side in summer(Temp.28.5℃,R.H.~60%). The record of mobility distributions of radioactive ions is shown in Figure 5.
The reason of bimodal distribution is not clear exactly. However,it seems likely that one peak is due to primary ions and other is secondary ions (namely claster ions), and they are respectively a multiple charged ions and a single charged ones.

4.Conclusion

A new method for obtaining the electrical mobilities of the radioactive ions was developed, such that it is possible to measure the mobility spectrum of radioactive small ions continuously and to clarify the properties of ions under the various conditions of the atmosphere. The apparatus is basically the condenser of parallel plate type, and the electric potential is applied between the plates. The radioactive ions introducing the condenser along one electrode are deposited on the plate surface of opposite polarity, according to their electrical mobilities. The collecting surface is a thin Au-plated myler film and serves as the window of a position sensitive proportional counter using resistance-capacitance position encoding. Then the activity distribution on the collecting surface along the axis of the counter is obtained throughout the sampling period with a multichannel pulse height analyzer. As a result, bimodal mobility distributions are found on the naturally occurring radioactive ions in the room air. In winter season, temperature 24℃, relative humidity 15~20% on the average, it is found with peaks of about 0.7 and 2.7 cm^2/sec.volt. In summer season, temperature 28.5℃, relative humidity 60% on the average, the peaks of 0.4~0.5 and 1.5~2.0cm^2/sec.volt. are found. The values of electrical mobility varies in a given atmospheric condition.

5.References

[1]Blanc,D.,J.Fontan and M.Juan.; On the mobilities of natural ions obtained in filter air. Compte Rendus de l'Academie des Sciences de Paris, 260, 2099-2100,1963.

[2]Bricard,J.,F.Billard,D.Blanc,M.Cabane and J.Fontan.; Detailed structure of the mobility spectrum of the small radioactive ions in the air. Compte Rendus de l'Academie des Sciences de Paris, 263, 761-764, Series B(1966).

[3]Fontan,J.,D.Blanc,M.L.Hubertas and A.M.Mart.; Mesure de la mobilite et der coefficient de diffusion des particules radioactives. Elementary Electrodynamics (S.C.Coronitte and J.Hughes, eds), pp257-267, Gordon and Breach Science Pub.,New York(1969).

[4]Kulju,L.M.,K.D.Chu and P.K.Hopke.; Development of a mobility analyzer for studying the particle producing phenomena related to radon progeny. Radon and Its Decay Products (P.K.Hopke,ed.) pp357-364, American Chemical Society, Washington,D.C.(1987).

Time variation of ^{222}Rn concentration and gamma level in a half-basement room

Takeshi,Iimoto; Hoshio,EguchiI; Toshiso,Kosako and Nobuyuki, Sugiura

Research Center for Nuclear Science and Technology,
The University of Tokyo
2-11-16 Yayoi, Bunkyo-ku, Tokyo 113, Japan

Abstract

Correlation between ^{222}Rn (radon) concentration and gamma level in a half-basement has been discussed. In order to decrease the background count of a whole-body counter (WBC), a ventilation blower of 72 m^3h^{-1} was installed. The device succeeded in a big reduction of radon concentration in the half-basement and then the background of WBC (NaI(Tl)) decreased to the 76.5% of the saturated value. Through a radon saturation test the exhalation rate of radon from concrete wall was estimated as 2.1 $Bqm^{-2}h^{-1}$ by a simple model calculation. In addition, through a ventilation test, the procedure was analyzed by another simple model.

1. Introduction

It is one of the important tasks to decrease the radiation background in order to measure the extremely low level gamma rays. Especially in a basement, a radiation background level is comparatively higher than that on the ground surface as mentioned in the UNSCEAR reports[1][2][3]. The level varies depending on the soil radiation, cosmic rays and ^{222}Rn (radon) / ^{220}Rn (thoron) atoms emanated from soil and concrete. We discussed the correlation between radon series concentrations and gamma level in this present study.

Our whole-body counter (WBC) is placed in a half-basement room of the research center for nuclear science and technology, the university of Tokyo. Before setting up this device in our research laboratory, we had investigated the distribution of the radiation background level in this building in order to determine an adequate position to install the WBC. We especially surveyed the radon concentration because this natural radioisotope might mainly and strongly effect on the measurement of low level gamma rate. **Table 1** shows the distribution of the radon concentrations for three months in the building, measured by the electrostatic collection method which had been developed by IIDA et al[4]. As shown in Table 1, all of these rooms show a high potential radon concentration. Basically the air conditioning devices installed in these rooms are simple air circulation systems which do not connected to the outside. Therefore the radon concentrations could become higher. Unfortunately, we had to place the WBC device in the half-basement room (Room No.1) because of its extremely heavy weight. One side wall of this half-basement room was facing directly to the ground soil. Under this condition, we must make effort on reducing the background of WBC to accomplish a high detection sensitivity.

In order effectively to decrease the gamma background level of WBC due to radon, a ventilation device was introduced in the present research[5][6][7]. The time-variation of the radon concentration following the operation of the device was analyzed by a simple saturation model and a ventilation model. Then the effective exhalation rate of radon from a concrete wall was estimated and verified based on these model calculations.

2. Material and methods

The WBC is located as shown in **Fig.1**. Detectors of the WBC are two scintillation methods, the NaI(Tl) (8" ϕ x4") and the plastic (50x50x15 cm). These detectors were installed in the radiation-shielding-compartment (RSC) made of iron (Show **Fig.2**). The No.1 room is divided into four main rooms; operation room, measurement room, shower room and preparatory room. The WBC system is exactly located in the operation room but the inside air of the RSC is exchanged through the entrance door to that of the measurement room.

Table 1 Radon conc. in the building Bqm^{-3}

Room No.	floor	Jan.	Feb.	Mar.	average
1	half-B	42.5 ± 3.8	47.0 ± 4.2	38.8 ± 3.6	42.8 ± 2.2
2	2nd	18.3 ± 2.5	29.1 ± 3.3	34.1 ± 3.4	27.2 ± 1.8
3	4th	32.5 ± 3.3	57.4 ± 4.7	54.9 ± 4.3	48.3 ± 2.4
4	5th	41.8 ± 3.8	48.6 ± 4.3	67.6 ± 4.8	52.7 ± 2.5

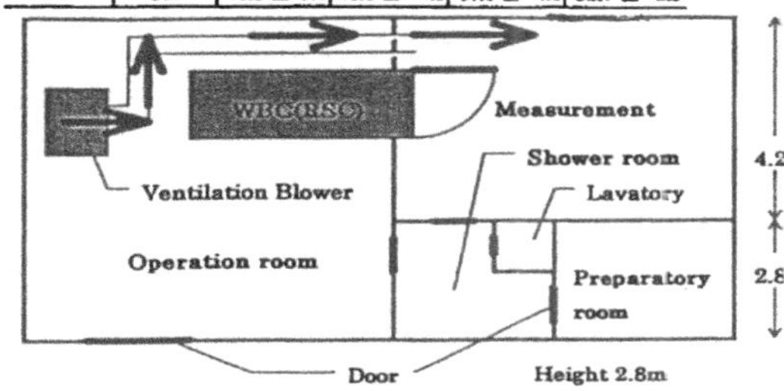

Fig.1 Arrangement of the half-basement room(No.1)

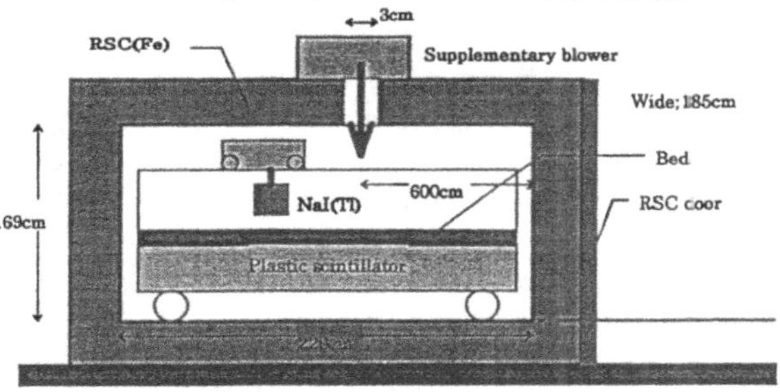

Fig.2 Cross sectional view of WBC (Inside of RSC)

As mentioned above, in order to decrease the background of WBC, a ventilation device was installed. The device is a blower of the maximum flow-rate of 270 m^3h^{-1} which is indicated in the instruction manual. The blower leads a fresh air sampled at 21 m above the ground level to the both insides of the WBC operation room and the measurement room. The fresh air includes 2.0 Bqm^{-3} of radon which is a three-month-average value. In fact, the flow-rate of fresh air leading to the measurement room is estimated as 72m^3h^{-1}. In addition to this, another supplementary blower works effectively to make an air circulation. This is installed on the ceiling of the RSC as shown in Fig.2. This blower assists the air circulation between RSC and measurement room effectively. This combination of the ventilation devices caused the big decrease of radon concentration as shown in the next section.

Radon concentration in the RSC was monitored by the PMT-TEL system (Pylon Co.). This is the continuous electrostatic collection method using ZnS(Ag) scintillation detector. Radon decay product concentration which was expressed by the equilibrium equivalent concentration of radon (EEC) was monitored by the WL meter (Thomson and Nielsen LTD). This is the continuous sampling and gross counting method[8]. Their sampling flow rates are too small (4.0$Lmin^{-1}$ and 1.0$Lmin^{-1}$ respectively) comparing with the volume of RSC and measurement room (63.5 m^3) to effect on the radon concentration and EEC. These devices were operated in the center of the RSC on the bed. They monitored the inside concentrations of RSC at the interval of one hour.

3. Results and discussion

Time variations of radon concentration, EEC and background count of NaI(Tl) detector of the WBC are indicated in **Fig.3.** The NaI(Tl) count indicates the gamma counts whose energies range from 50 keV to 2.0 MeV. The energy analysis ability (FWHM) of the NaI(Tl) system is estimated as 9.0% for ^{137}Cs and 7.1% for ^{40}K. The NaI(Tl) detector do not have a sufficient sensitivity and cannot apparently show the peaks of gamma energies of the background radon series in the natural environment. Radon series emit some gamma rays. These energies and their isotopes are indicated in **Table 2.** Therefore the above range of gamma energies was monitored as the background due to radon in the present study.

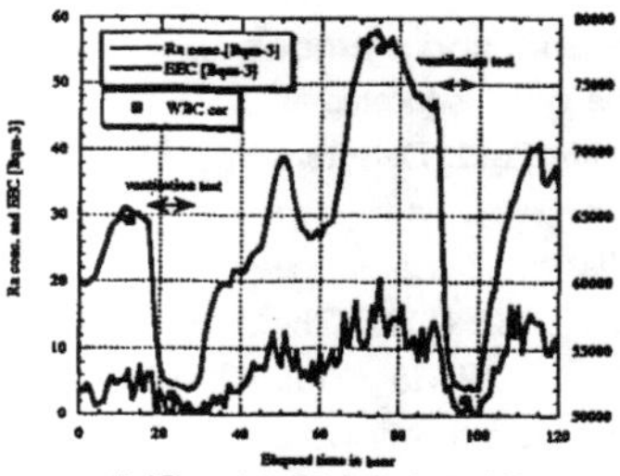

Fig.3 Time variation of Rn/EEC and BG of WBC

Table 2 Gamma energy and radon series

	energy [keV]	ratio [%]		energy [keV]	ratio [%]
^{214}Pb	53.2	1.1	^{212}Pb	239	43.3
	242	7.5		300	3.28
	295	18.5	^{208}Tl	277	6.3
	352	35.8		511	84.5
^{214}Bi	609	44.8		860	12.4
	768	4.8		2614	99.2
	1120	14.8			
	1238	5.9			
	1764	15.4			
	2204	4.9			

The entrance door of RSC was opened between 0 and 113 hours in the elapsed time of Fig.3. The reason is that the door is generally opened to the measurement room during the preparatory period for the practical WBC measurement. During this period, the blowers were operated three times. First is between 16-28 hours, second is between 90-99 hours and the last is between 113-end in the elapsed time. **Table 3** shows the operation situation of the two blowers and the open/close situation of the RSD door.

Table 3 Operation situation

elapsed time in hour	test	Main blower	Supplement	RSC door
0-16	preparation	off	off	open
16-28	ventilation	on	on	open
28-90	saturation	off	off	open
90-99	ventilation	on	on	open
99-113	saturation	off	off	open
113-end	leakage	on	off	close

The first and second tests are standard ventilation tests, which will be discussed at the next in detail. The last one is an RSC leakage test. During the leakage test, the RSD door was closed and the supplementary blower on the RSD ceiling was stopped. Only the main blower was operated. According to the result of the last test, the air shielding ability of the RSD is good as the ventilation rate is estimated as 0.015 h^{-1}. Therefore it shows that once the inside radon concentration of RSD is reduced sufficiently and the RSD door is closed, the main ventilation blowers could be stopped while the practical WBC measurements.

Now we discuss the test between 90-113 hours in the elapsed time. This test can be divided into two categories. One is a ventilation test (90-99 hours) and the other is a saturation test(99-113 hours). At the beginning, the exhalation rate of radon from concrete wall is estimated by the saturation test as follows.

After stopping the ventilation blowers, the radon concentration in the RSD increased. Its source might be the concrete wall and soil around the measurement room. The radon source is here considered as only the wall and its uniformity as a source is assumed for a following calculation. The effective area of the concrete wall is estimated as 87.16 m^2. Then the saturation procedure is indicated as the simple differential formula.

$$dN/dt = Es\,S/V - \lambda N \qquad (1)$$

where,

N = radon concentration in the RSC [Bqm^{-3}]

t = elapsed time [hours]

Es = exhalation rate of radon from concrete wall [$Bqm^{-2}h^{-1}$]

S = area of concrete wall as radon source [m^2] = 87.16

V= effective inside volume of RSC and measurement room [m^3] = 63.50

λ = decay constant [h^{-1}] = 7.552×10^{-3}.

When t=0 (=99 in the Fig.3's elapsed time), then N=4.0 Bqm^{-3} is determined due to the result of the experiment. Based on this condition, the (1) formula was solved. **Fig. 4** shows the saturation curves of radon concentration in the RSC, which are the experimental value and the model calculation value due to (1). The Es of exhalation rate is estimated as 2.1 $Bqm^{-2}h^{-1}$ by the model calculation. Using this result, the ventilation test is discussed at the next.

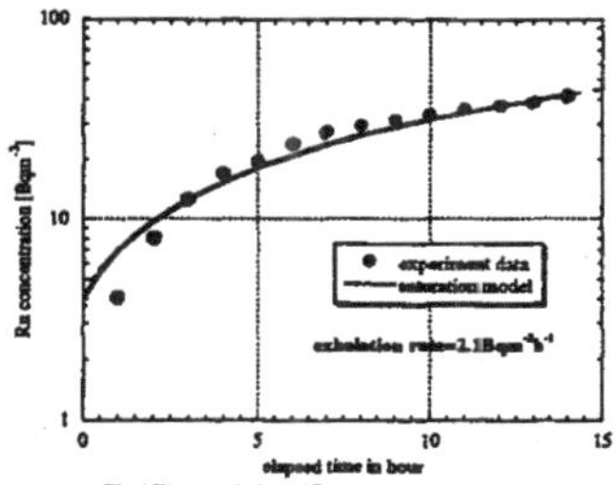

Fig.4 Time variation of Rn conc. during saturation test

Now the ventilation test of 90-99 hours in the elapsed time is also analyzed. The ventilation procedure is indicated as the following simple differential formula.

$$dN/dt = Es\,S/V + N'v/V - (v/V + \lambda)N \qquad (2)$$

where,

N' = outdoor radon concentration introduced to the RSC [Bqm^{-3}] = 2.0

v = ventilation flow speed [m^3h^{-1}] = 72.

When t=0 (=90 in the test's elapsed time), then N=42.17 Bqm^{-3} is determined due to the result of the experiment. Based on this condition, the (2) formula was solved. **Fig. 5** shows the ventilation curves of radon concentration in the RSC, which are the experimental value and the model calculation value due to (2). According to Fig.5, the experimental value seems to delay around 30-40 minutes comparing the model calculation. The delay might be due to the time-lag of the radon monitor. The PMT-TEL measured the two alpha particles, ^{218}Po

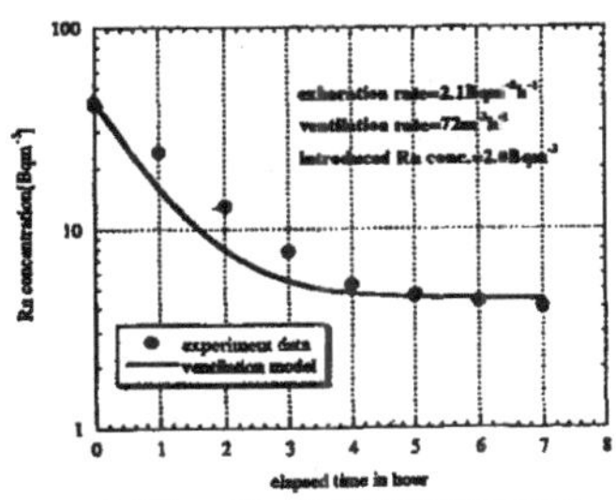

Fig.5 Time variation of Rn conc. during ventilation test

(RaA) and ^{214}Po(RaC′) without doing an energy spectrometry. Therefore this monitor cannot follow the quick and big variations of radon concentration less than one hour due to the half-life of radon decay products. Actually the inside radon concentration was reduced within two hours sufficiently, estimated by the model calculation. Then this present study concluded the blowers succeeded in the sufficient reduction of the inside radon concentration of the RSC in only two hours operation.

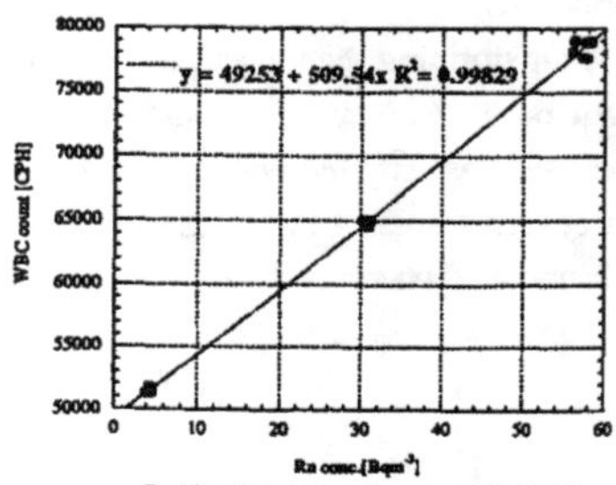

Fig.6 Correlation between Rn conc. and BG of WBC

At last, the variation of the background of WBC is compared with that of radon concentration. As mentioned above, radon monitor has a time-lag, therefore both were compared with each other under the stable condition. The background was actually measured only in some stable periods as indicated in Fig.3. The correlation between the background count of WBC and the radon concentration is shown in **Fig.6**. As shown in Fig.6, the background is directly proportional to radon concentration.

In addition, the correlation between the background count of WBC and EEC is shown in **Fig.7**. However, as shown in Fig.7, the background is not proportional to EEC. The reason might be that the plate-out rate of unattached fraction varies due to the quick air circulation and the inside EEC is not under the stable condition. Therefore EEC could not become the indication value of the background level.

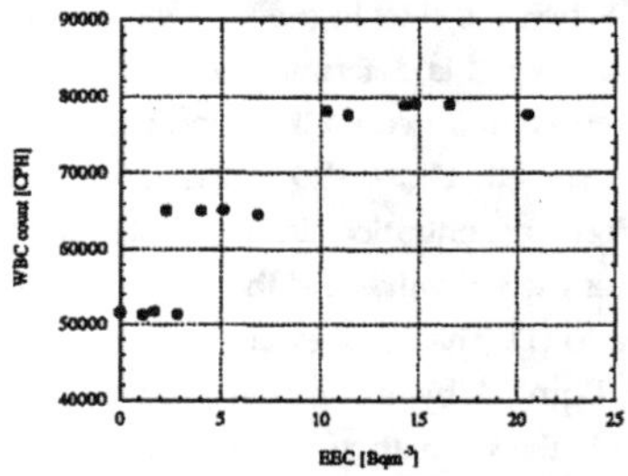

Fig.7 Correlation between EEC and BG of WBC

The approximate formula indicated in Fig.6 shows that the 1 Bqm^{-3} of radon would equal to about 510 counts of background of the WBC. It is difficult to compare this value with the count of a standard gamma souse because the energy range is different each other, but actually this 510 counts of WBC is equal to 2.99 Bq of our ^{137}Cs point source (Detector-source distance is 10 cm). Therefore the decrease of radon concentration due to the ventilation, to the level of 6.8% of the saturated value, leads to the success of an elevation of the sensitivity of WBC. As a result, the background of WBC count decreased to 76.5% for the NaI (Tl) scintillation detector. For a reference, background of the plastic scintillation detector under the bed decreased to 95.9%, which was not in detail mentioned in the present study, however. The cosmic ray might cause the difference of this background reduction effect between two scintillation detectors. In case of the plastic detector, the large sensitive volume for the cosmic rays caused the small reduction effect. The difference should be discussed in the following research.

4. Conclusion

1. In order to decrease a background of WBC, a ventilation device with a supplementary blower was installed. The device leads a fresh air sampled at 21 m above the ground level to the insides of the WBC operation room and the measurement room. The flow-rate is 72m^{3}h^{-1}.
2. Radon concentration in the RSC was monitored by the PMT-TEL system (Pylon Co.), the

continuous electrostatic collection method using ZnS(Ag) scintillation detector. In addition, EEC was monitored by the WL meter (Thomson and Nielsen LTD), the continuous sampling and gross counting method.

3. The air shielding ability of the RSD is good because the ventilation rate is estimated as 0.015 h^{-1}. Therefore it shows that once the inside radon concentration of RSD is reduced sufficiently and the RSD door is closed, the main ventilation blowers could be stopped while the practical WBC measurements.
4. The saturation procedure is analyzed by a simple differential formula. The Es of effective exhalation rate is estimated as 2.1 $Bqm^{-2}h^{-1}$ by this model calculation.
5. The ventilation procedure is analyzed by another simple differential formula. The experimental data seems to delay around 30-40 minutes comparing with the model calculation. The delay is considered to be due to the time-lag of the radon monitor. The blowers succeeded in the sufficient reduction of the inside radon concentration of the RSC in only two-hour-operation. This was estimated by the model calculation.
6. The background of WBC is directly proportional to the inside radon concentration. The 1 Bqm^{-3} of radon equals to about 510 counts of background of the WBC. As a result, the background of WBC for the NaI (Tl) scintillation detector decreased to 76.5% by the presented ventilation technique.
7. The background of WBC is not proportional to EEC. EEC could not become the indication value of the background level.

5. References

[1]United Nations Scientific Committee on the Effects of Atomic Radiation., New York, United Nations, 1993

[2]United Nations Scientific Committee on the Effects of Atomic Radiation., New York, United Nations, 1988

[3]United Nations Scientific Committee on the Effects of Atomic Radiation., New York, United Nations, 1982

[4]Iida T., Ikebe Y., Hattori T. and Yamanishi H., An electrostatic integrating ^{222}Rn monitor with cellulose nitrate film for environmental monitoring, Health Physics, 54, 139-148, 1988

[5]G.A. Roserens and M. Jungck, Radon program in Switzerland - Remedial actions, Radiation Protection Dosimetry, 56, Nos 1-4, 13, 1994

[6]D. Capra, C. Silibello and G. Queirazza, Influence on ventilation rate on indoor radon concentration in a test chamber, Radiation Protection Dosimetry, 56, Nos 1-4, 15-18, 1994

[7]E. Oksanen, Effect of ventilation type on radon concentration at places of work, Radiation Protection Dosimetry, 56, Nos 1-4, 61-63, 1994

[8]Kurosawa R., Iimoto T., Ichiji T. and Tokonami S., Practical interpretative method for indicating values of continuous sampling and gross counting PAEC monitor, Radioisotopes, 45, 14-19, 1996 (Japanese)

New three-count technique for short-lived radon decay products in air

Deyuan Tian Zhizhao Lu*

Laboratory of Industrial Hygiene, Ministry of Public Health, 100088 Beijing

* Suzhou Medical College, 215007 Suzhou, Jiangsu

Abstract

Up to the present, radon and its short-lived decay products in air are usually monitored by means of α detection. But radon progeny, including RaB (^{214}Pb) and RaC (^{214}Bi) which are β and γ emitters, contribute about 90% to the equilibrium equivalent radon concentration (EECRn). Therefore, this paper introduces a new three-count technique by a β detector in the light of radioactive decay law and its boundary conditions during sampling and counting times to solve the Bateman equation. β (even low level β) instruments have been fairly popularized domestically and internationally. It can be used not only as an instrument for radon and its daughters in air, but also as a monitor for β airborne activity in the environment. This new method taps further the latent power of the present instrument and realizes various uses for a unit.

Keywords : radon, three-count technique, equilibrium equivalent radon concentration (EECRn), thoron.

1.Intraduction

It is well known that natural radon, thoron and their short-lived decay products in airborne state may cause serious radiation harm to human health such as inducing lung cancer to those who were exposed to them for a long period. Therefore, detailed studies for how to monitor them were conducted. Today, along with the improvements in science and technology, the monitoring procedures for radon progeny have made an encouraging progress. Tsivoglou et. al. [1] proposed the Tsivoglou method (a three-point method) for monitoring radon daughters in air on the basis of the count-ratemeter in 1950s. Thomas [2] then improved the three-count technique (or modified Tsivoglou method) on the foundation of the scaler in 1970s. In addition, there are some rapid methods for monitoring the radon progeny in air [3], which are all by means of α-particle detection. However, β and γ rays are emitted in the decay process of radon daughters and reports concerning the detection method of them were not seen so far.

2.Principle

A widely used method for measuring radon and its progeny concentration in air involves drawing the air to be measured through a filter in constant flow rate, filtering radon daughters in the air, and then counting the activities on the filter for several specified time intervals after the end of sampling. Radon progeny concentrations in the air are calculated by the countings and coefficients obtained from solving the Bateman equations.

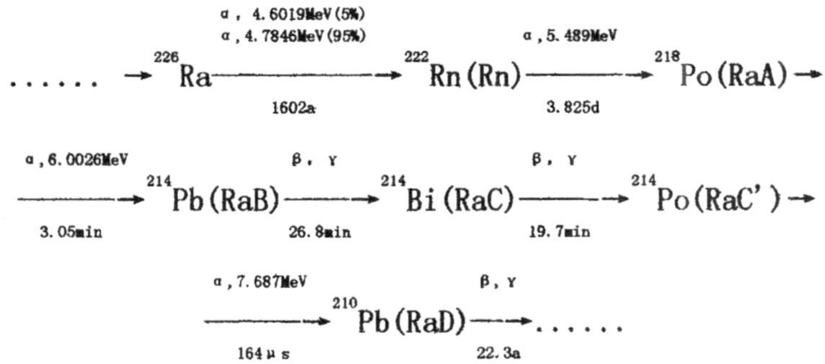

Fig. 1 Main decay diagram of radon and its short-lived decay products

Radon and its short-lived decay products emit α, β and γ rays during the decay process. Their main decay diagram is shown in Fig. 1. Radon (^{222}Rn) decays to form ^{218}Po (RaA), ^{214}Pb (RaB), ^{214}Bi (RaC), ^{214}Po (RaC'), ^{210}Pb (RaD), etc., consecutively. The former four decay products are called radon short-lived daughters because their half-lives are less than 30 min. RaD and the subsequent nuclides have much longer half-lives, for instance 22.3 y for RaD. They are quite stable as compared with the former ones. For this reason, it can be regarded as if RaD and the latter are not to decay, and the multi-progeny decay for radon and its daughters may be simplified to a four-progeny decay. On the other hand, the half-life of nuclide RaC' is very short, only 164 μs. The radioactivity equilibrium between RaC and RaC' appears right away. It can be said that RaC is simultaneously emits one β- and one α-particles and decays immediately to RaD, not through RaC'. Hence, the four-progeny decay is further simplified to a three-progeny decay.

Radon short-lived progeny emit two α-particles in all. For this reason, radon daughter monitoring in air is almost by means of α-particle detection because of its high α-energy and α-pulse signal amplitude, very low background, and the simple structure and easy manufacture of α-detector. On the other hand, the equilibrium equivalent radon concentration (EECRn) is given by [4]:

$$EECRn = 0.105\ C_{RaA} + 0.516\ C_{RaB} + 0.379\ C_{RaC}\quad Bq/m^3 \tag{1}$$

where C_{RaA}, C_{RaB} and C_{RaC} are the concentrations of radon progeny, RaA, RaB and RaC, respectively. If the three radon daughter concentrations in air are known, EECRn can be obtained immediately from Eq. (1), and their dose to the public can be evaluated. The RaC' concentration, $C_{RaC'}$, does not contribute to EECRn as seen from Eq. (1).

Two radon progeny, RaB and RaC, which contribute about 90 % to EECRn are not α-emitter, but are both β and γ emitters. Hence, it is unsuitable to say that α-particle counting for detecting RaA and RaC' is the best choice because radiation doses are caused mainly by α-radiation. On the contrary, β or γ counting is an optimal selection.

Along with the improvement in sciences and technologies, the commercially available high efficiency and large area (well-suited for large area filter from available high volume sampler) β (even low level β) instruments have already been fairly popularized domestically and overseas, such as Model NDY-2, FJ-402, BH-1216, FJ-2600, FJ-2603 and FJ-332 low background counters made in China, and Canberra Model 2401 counting system, etc., made in US. Therefore, it is selected to determine EECRn and the radon daughters in air not by α-detector or expensive α-spectrometer for counting RaA and RaC' which contribute only 10 % to EECRn, but by gross β counting of RaB and RaC with a small contribution of γ radiation. Besides, the efficiency of α-counter is low, only about 10-20 % for the silicon surface barrier detector, but that of the β counter is high, usually about 30-40% for the proportional or G-M counter. Consequently, this paper proposes a new three-count technique for radon and its progeny in air by means of β detection.

Based on the radioactive decay law and relevant boundary conditions during sampling and counting periods, the Bateman equations were solved and got integral gross β counting which includes hidden three unknowns, namely the RaA, RaB and RaC concentrations in air. That is a ternary first-order equation. It is well known that only a set of three simultaneous equations can be solved to get a sole answer when there are three unknowns in it. For this reason, three equations with three observed values are required. Three activity measurements on the filter (i.e., three countings) are to be conducted in the experiment for three different time intervals after the end of sampling. Hence, this method is called the three-count technique. A set of simultaneous linear equations obtained from three countings was set up and solved by the inverse matrix method. The formulas for radon progeny concentration in the air to be measured by the gross β counting method are derived via a IBM computer. According to the UNSCEAR definition [5], the formula of Potential Alpha Energy Concentration (PAEC) is led in. The formula coefficients are a series of complicated exponential operations which are functions of sampling and counting times. A computer program was written to calculate all coefficients rapidly so long as each time parameter inputs are given. These coefficients are simplified constants and do not need to be computed every time for a certain procedure with fixed time parameters.

However, it is possible not to start and / or stop the countings on time in actual situations for a variety of reasons. The data will become invalid at the moment. It is well known that the radon and its daughter concentrations in air change with time and place. Reappeared them is impossible yet even though they are determined again. Therefore, the above-mentioned program is onlined to realize the automatic measurements. Output data from various procedures can be typed out immediately in the light of both relative parameters and measured gross β countings.

3.Results and discussion

To compare conveniently with the Thomas three-count technique [2], the same procedures as his were used, i.e., 5 min sampling time and 2～5, 6～20 and 21～30 min countings after the end of sampling to obtain the three gross β countings, N_1, N_2 and N_3, respectively. If the total β counting efficiency G (including the instrument and filter efficiencies, self-absorption and back-scattering corrections on the filter, etc.), sample flow rate V L/min and background counting rate N_0 cpm (counts per minute) are known, the formulas for the radon progeny concentrations, PAEC and EECRn are:

$$C_{RaA} = (-60.9933N_1 + 29.6053N_2 - 27.9909N_3 + 20.4233N_0) / GV \quad Bq/m^3 \qquad (2)$$

$$C_{RaB} = (+8.50301N_1 - 4.71163N_2 + 5.32176N_3 - 7.44207N_0) / GV \quad Bq/m^3 \qquad (3)$$

$$C_{RaC} = (-3.08556N_1 + 2.90511N_2 - 3.84836N_3 + 3.22043N_0) / GV \quad Bq/m^3 \qquad (4)$$

$$PAEC = (-175.253N_1 + 97.9325N_2 - 90.9556N_3 - 26.6961N_0) / GV \quad 10^{-10} J/m^3 \qquad (5)$$

$$EECRn = (-3.18617N_1 + 1.77840N_2 - 1.65155N_3 - .475117N_0) / GV \quad Bq/m^3 \qquad (6)$$

Generally, the flow rate V, efficiency G and background rate N_0 are determined previously, and their errors can be managed to be reduced as much as possible or neglected. But, statistical counting errors can not be decreased and are dominant in the environmental level. The resultant concentration uncertainties are shown below according to the error resultant rule when only statistics are considered:

$$S_{RaA} = (3720.19N_1 + 876.476N_2 + 783.490N_3 + 268734N_0)^{1/2} / GV \quad Bq/m^3 \qquad (7)$$

$$S_{RaB} = (72.3011N_1 + 22.1994N_2 + 28.3211N_3 + 7295.81N_0)^{1/2} / GV \quad Bq/m^3 \qquad (8)$$

$$S_{RaC} = (9.52067N_1 + 8.43965N_2 + 14.8099N_3 + 2939.46N_0)^{1/2} / GV \quad Bq/m^3 \qquad (9)$$

$$S_{PAEC} = (30713.5N_1 + 9590.78N_2 + 8272.93N_3 + 2826321N_0)^{1/2} / GV \quad 10^{-10} J/m^3 \qquad (10)$$

$$S_{EECRn} = (10.1517N_1 + 3.16270N_2 + 2.72761N_3 + 932.190N_0)^{1/2} / GV \quad Bq/m^3 \qquad (11)$$

where S_{RaA}, S_{RaB}, S_{RaC}, S_{PAEC} and S_{EECRn} are the resultant uncertainties for the RaA, RaB and RaC concentrations , PAEC and EECRn in the air, respectively.

The experiments showed that the results of this new three-count technique agree well with that of the Thomas' method.

For thoron (^{220}Tn) daughters, it can be dealt with in the same way. If the concentrations of thoron and its progeny in the air are to be determined, the decays of two thoron daughters, viz., ^{212}Pb(ThB) and ^{212}Bi (ThC), can be simplified because the other thoron daughter ^{216}Po (ThA) has only a half-life of 0.15 sec. The two-count technique can be used in this case. On the other hand, the five-count technique can be employed to obtain EECRn, EECTn, three (Rn, Tn and total) PAEC, and their five progeny concentrations if radon, thoron and their daughters in the air were measured.

Besides, the airborne sampler and β counter can be used for the monitoring of other airborne β activities, such as those ^{137}Cs, ^{131}I, ^{90}Sr and other β samples suspending in air after nuclear accident. Radon progeny and other β emitters can be distinguished easily on the basis of their different half-lives so as to realize the monitoring. In addition, the β counter can be used as a portable β monitor for other kinds of samples, for example precipitates or wipe tests.

4.Summary

Almost all monitors and dosimeters commercially available now for radon and its progeny in air employ generally the method of detecting α-particle. Their efficiencies are low and detector areas are small normally although the method is good. It means these instruments are well suited for integrated or continuous long time measurements, but not for the rapid and sensitive grab ones at low levels in the environment. This paper develops for the first time a new three-count technique by gross β countings to determine the radon daughters in air. The formula for calculating EECRn is led firstly in so as to realize the combination monitoring of radon and its progeny. This study fully utilizes the high efficiency and large area of the β detector, and is a new pioneering work in radon and its daughter monitoring in the air. The new method taps further the latent power of these β instruments and realizes the various uses for a unit .

5.References

[1] Tsivoglou, E.C., H.E. Ayer and D.A. Holiday, Occurrence of Non-Equilibrium Atmospheric Mixtures of Radon and its Daughters, Nucleonics 11, 40-45, 1953.

[2] Thomas, J.W., Modification of Tsivoglou Method for Radon Daughters in Air, Health Phys., 19, 691-692, 1970.

[3] 田德源，氡子体快速监测，放射卫生，2, 181-182, 1989.

[4] US National Council on Radiation Protection and Measurements, Measurement of Radon and Radon Daughters in Air, NCRP Report № 97, p.129, United States, Bethesda MD, 1988.

[5] UNSCEAR, Ionizing Radiation: Sources and Biological Effects, p.143, 1982.

Development and calibration of a high sensitive radon detector in air

Machiko Nemoto
Faculty of Science, Tokai University
1117, Kitakaname, Hiratsuka-shi, Kanagawa 259-12, Japan

Shigeki Tasaka and Hidemitsu Hori
Faculty of Education, Gifu University
1-1, Yanagido, Gifu-shi, Gifu 501-11, Japan

Kimihiro Okumura, Takaaki Kajita and Yasuo Takeuchi
Kamioka Observatory, ICRR, University of Tokyo
Higashi-mozumi, Kamioka-cho, Yoshiki-gun, Gifu 506-12, Japan

Hiroko Okazawa
Faculty of Science, Niigata University
8050, Ikarashi 2, Niigata-shi, Niigata 950-21, Japan

Abstract

A high sensitive radon detector was developed in order to measure radon concentration for Super-Kamiokande.

Super-Kamiokande is a large water Čerenkov detector located 1000m underground in the Kamioka mine in Gifu Prefecture. It consists of a cylindrical stainless steel tank (39-meter in diameter and 42-meter in height), 50000-ton of purified water, 11146 20-inch PMT, a water and air purification system, electronics systems, and so on. One of the main purposes of Super-Kamiokande is to observe solar neutrinos. The background events in the lower energy region are dominated by beta decay of radon daughters remaining in the water. The radon concentration in the water should be less than about 5mBq/m^3.

As a first step to monitor the radon concentration in water, a high sensitive detector for radon in air was developed. We adopted the electrostatic collection method and used a PIN photodiode to measure the energy of alpha particles from the daughter nuclei of ^{222}Rn. To study the high voltage and absolute humidity dependence of the radon detector, a calibration system was also constructed. As a result of the calibration study, the absolute humidity dependence was clearly observed in the region less than 1.6g/m^3. The calibration factors were 1.8 ± 0.1 (count/day)/(mBq/m^3) at 0.08g/m^3 and 0.75 ± 0.04 (count/day)/(mBq/m^3) at 11g/m^3. As a result, the detection limit at 0.08g/m^3 is 13mBq/m^3/day.

1 Introduction

Kamiokande(KAMIOKA Nucleon Decay Experiment[1]) is an imaging water Čerenkov detector with 4500-ton of purified water. It is located about 1000m underground (2700m water equivalent) in the Mozumi mine[2], which lies in Kamioka town in Gifu Prefecture.

One of the main purposes of Kamiokande was to observe solar neutrinos. It had observed solar neutrinos since December 1986 until February 1995. The energy of solar neutrinos are distributed

[1] Later, it is also called KAMIOKA Neutrino Detection Experiment
[2] stopped the mining operation since October 1994.

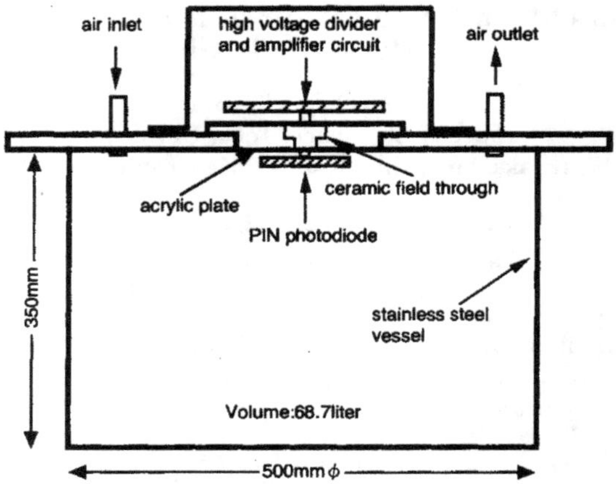

Figure 1: A schematic view of a high sensitive radon detector in air.

from ~0MeV up to ~18MeV. The analysis at Kamiokande used the energy region above 7.0 MeV. The background events in the lower energy region were dominated by beta decay of radon daughters remaining in the water. The radon concentration in the water was $0.5Bq/m^3$ [1].

In April 1996, Super-Kamiokande has started its observation. Super-Kamiokande is a large water Čerenkov detector with 50000-ton of purified water, and makes it possible for us to observe nucleon decay, solar neutrinos, atmospheric neutrinos and so on with high statistic. One of the main purpose of Super-Kamiokande is to lower energy threshold of solar neutrino analysis from 7.0MeV (at Kamiokande) to 5.0MeV, then measure the energy spectrum of solar neutrinos precisely. The current analysis at Super-Kamiokande uses the energy region above 6.5 MeV.

To achieve 5MeV threshold, radon concentration in the water should be less than about 1/100 comparing with Kamiokande. Therefore, the detection limit for radon detector should be less than $5mBq/m^3$, but there was no commercial radon detector which has such a high sensitivity.

As a first step to monitor the radon concentration in water, a high sensitive detector for radon in air is developed. There is a gap between the water surface and the inside of the water tank. In order to keep the dirty mine air out, the gap is filled with purified air (radon-free-air). The high sensitive radon detector which is reported in this paper is used for monitoring radon concentration in the radon-free-air.

2 High sensitive radon detector in air

The principle of the radon detection which is used in this detector is the electrostatic collection of the daughter nuclei of ^{222}Rn, and energy measurement of the alpha decay with a PIN photodiode [2, 3]. Concerning ^{218}Po atoms (one of daughter nuclei of ^{222}Rn), it is reported that more than 90% of ^{218}Po atoms tend to become positively charged[4].

Figure 1 shows a schematic view of the high sensitive radon detector. It consists of a cylindrical SUS vessel, a PIN photodiode, a high voltage divider and amplifier circuit, and a field through. The negative high voltage supplied to the p-layer of the photodiode, and the vessel was grounded. Then, an electric

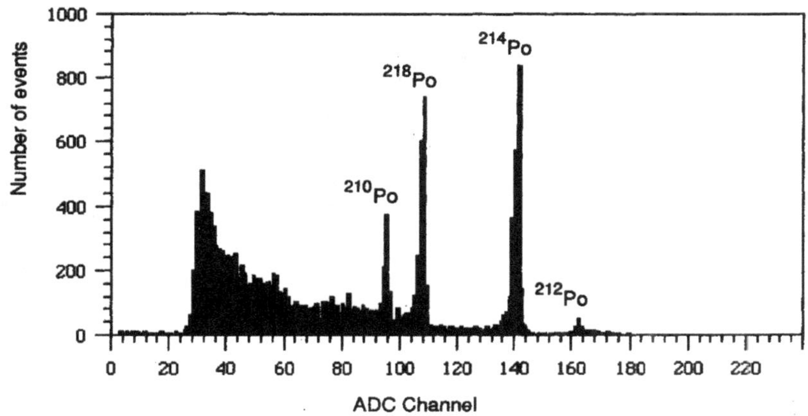

Figure 2: Pulse height spectrum for alpha particles from radon daughter products in the background run during the period between January 1, 1996 and May 1, 1996.

field was produced in the vessel. The daughter nuclei which ionized positively are collected on the surface of the photodiode, then measure the energy of alpha decays.

The dimensions of the SUS vessel are 50cm in diameter and 35cm in height, and the volume is 68.7L. In order to achieve low background level, the inside of the vessel is finished with the electrolytic polishing. The PIN photodiode is isolated from the SUS vessel with an acrylic plate and a ceramic field through.

The PIN photodiode is Hamamatsu Photonics S3204-06 with passivation finish. The glass cover is removed to measure energy of alpha decays. The detection area, sensitive thickness, and capacity are 16mm$\times$16mm, 500μm, and 80pF, respectively. A typical pulse height spectrum is shown in Figure 2. In order to measure radon concentration, the only ^{214}Po peak is used because that energy region has less noise events and no other alpha peak overlapped.

Concerning with the electrostatic collection method, the neutralization effect of ^{218}Po atoms was pointed out [5, 6]. Hence, the humidity dependence of the calibration factor is carefully studied, though ^{214}Po peak is used for the estimation of the radon concentration.

3 Calibration system

Figure 3 shows a schematic diagram of the calibration system. A high sensitive radon detector was connected to a radon source, an ionization chamber, a vapor trap, and a syringe. The radon source consisted of small mine rocks and a 3L glass vessel. The mine rocks were put in the vessel and covered with a membrane filter to keep dusts off. The ionization chamber was used as the standard device to measure the radon concentration in this system. We used 1500cm^3 ionization chamber made by Okura Electric Co. Ltd. The vapor trap consisted of a cooler, a vacuum bottle, and ethanol. We could control the humidity in this system by setting the cooler temperature. The others were a barometer, a dew point meter, a flow meter, and a circulation pump. They were connected with 1/4inch SUS tubes and rigid

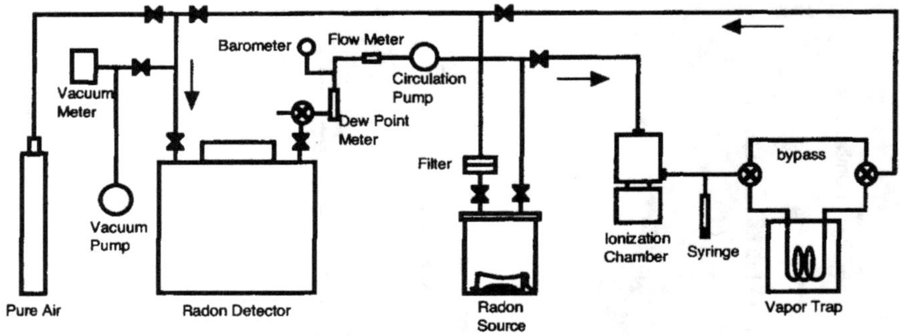

Figure 3: A schematic diagram of the calibration system for the humidity dependence measurement.

nylon tubes. The pure air class A made by Nihon Sanso Co. Ltd. was used as the air in this system. The circulation rate was 3L/min. Before starting calibration, the leak rate was checked, and it was less than 1×10^{-5}cm^3/s.

4 Results

Figure 4 shows the high voltage dependence of the calibration factor for the high sensitive radon detector in air. The absolute humidity were kept at 1.2 ± 0.2g/m^3 during this measurement. The radon source was introduced into the circulation system so as to keep the radon concentration constant. From this measurement, we chose -1500V for the measurement of the humidity dependence.

Figure 5 shows the humidity dependence of the calibration factor for the high sensitive radon detector in air. Three kind of methods to control the absolute humidity were used for this measurement. The syringe was used in Method 1. The cooler and vapor trap were used in Method 2 and 2'. These results agreed well.

From this measurement, the absolute humidity dependence is clearly observed in the region less than about 1.6g/m^3. Above about 2g/m^3, it becomes constant. This behavior is qualitatively consistent with Hopke's report [7].

The calibration factors obtained in this measurement are 1.8 ± 0.1 (count/day)/(mBq/m^3) at 0.08g/m^3 and 0.75 ± 0.04 (count/day)/(mBq/m^3) at 11g/m^3. The detection efficiency of this radon detector at 0.08g/m^3 was estimated at 30%.

The background level of this radon detector was measured by using pure air for 4 months. The result of this background run is 23.0 ± 0.4 count/day. As a result, the detection limit at 0.08g/m^3 is 13mBq/m^3/day if Currie's method is applied [8].

The authors would like to thank Kamiokande and Super-Kamiokande collaborators for their help with this work. They acknowledge the cooperation of the Kamioka Mining and Smelting Company. This work was supported in part by the Japanese Ministry of Education, Science and Culture.

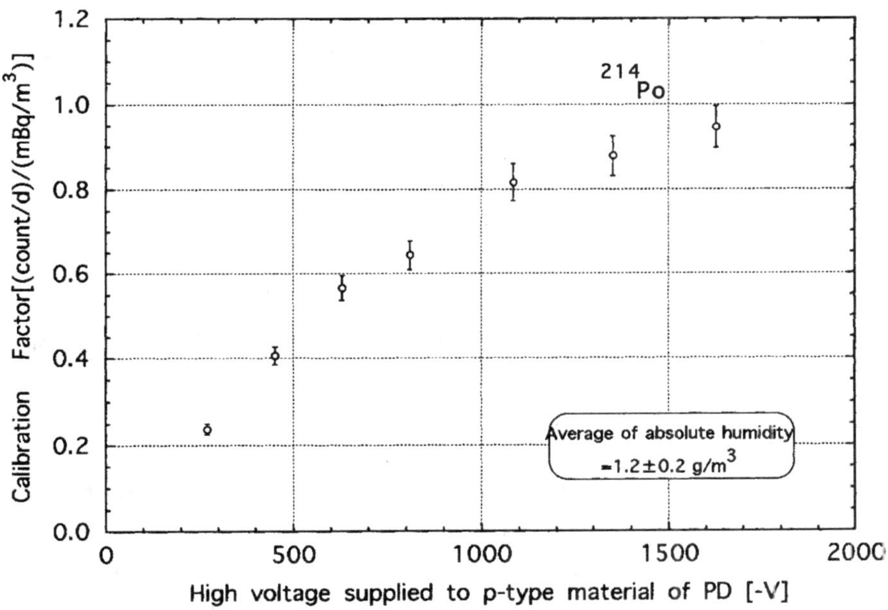

Figure 4: High voltage dependence of the calibration factor for the high sensitive radon detector in air. The absolute humidity are $1.2 \pm 0.2\text{g/m}^3$.

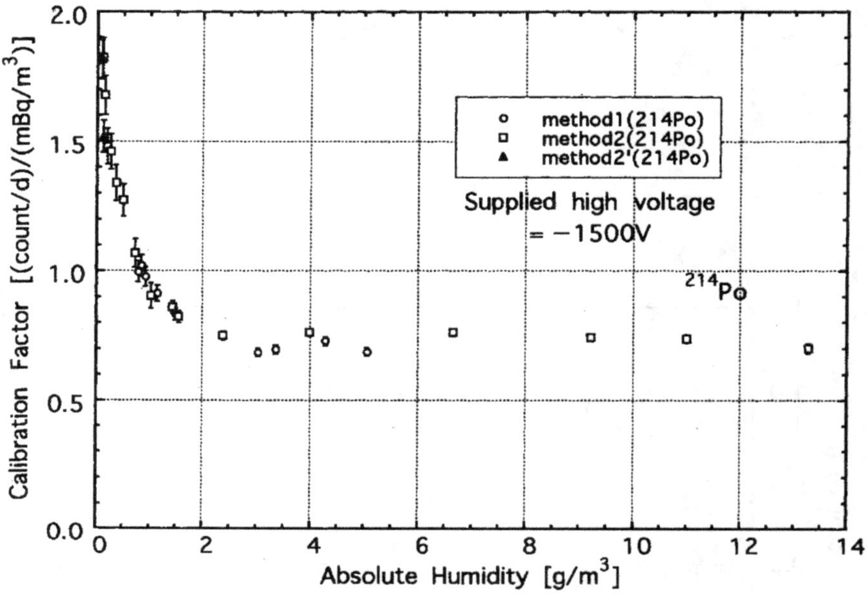

Figure 5: Humidity dependence of the calibration factor for the high sensitive radon detector in air. The supplied high voltages are -1500V.

References

[1] Tasaka,S., Sasaki,Y., Okazawa,H. and Nakagawa.M., Development of an Underwater Radon Detector, Radioisotopes, **43**, 125-133, 1994

[2] Tasaka,S., Development of a new radon detector in the Superkamiokande experiment, ICRR ANNUAL REPORT(APRIL 1994-MARCH1995), ICRR, University of Tokyo, 36-39, 1996

[3] Nemoto,M., Tasaka,S., Hori,H., Okumura,K., Kajita,T. and Takeuchi,Y., Development of High Sensitive Radon Detector with Electrostatic Collection, Radioisotopes, **46**, 710-719, 1997

[4] Kotrappa,P., Dua,S.K., Gupta,P.C. and Mayya,Y.S., Electret, a new tool for measuring concentrations of radon and thoron in air, Health Phys., **41**, 35-46, 1981

[5] Iida,T., Ikebe,Y., Hattori,T., Yamanishi,H., Abe,S., Ochifugi,K. and Yokoyama,S., An electrostatic integration ^{222}Rn monitor with cellulose nitrate film for environmental monitoring, Health Phys., **54**, 139-148, 1988

[6] Howard,A.J., Carroll,S.E. and Stange,W.P., A simple system for radon-in-air concentration determination, Am. J. Phys., **59**, 544-550, 1991

[7] Chu,K.D. and Hopke,P.K., Neutralization for polonium-218, Environ. Sci. Tecnol., **22**, 711-717, 1988

[8] Currie,L.A., Limits for qualitative detection and quantitative determination, Anal. Chem., **40**, 586-593, 1968

Detection efficiency of LR-115 using a new measurement device

Abdelhakim Sabir and Hamid Marah

Laboratoire de Physique Nucléaire Appliquée,
Faculté des Sciences
Ibn Tofail University, Kenitra (Morocco)

Abstract

This paper presents the experimental determination of detection efficiency for the LR-115 detector (expressed as an energetic sensitivity range), as well as the variation of track diameters versus alpha particles energy, and etching time. We used for this purpose a new device consisting of an optical microscope, drawing tube and Graphics Tablet (as digitizer) connected to a computer. This device allows both to count observed tracks , and to determine their diameter, as well as the statistical result processing.

1. Introduction

The passage of alpha particles through solid state nuclear track detectors (SSNTD) creates narrow paths of intense damage wich may be enlarged and revealed by an appropriate treatment. The number of tracks, i.e. the number of particles that have impinged on the surface of detector can be performed by various means. The most tedious one is the direct counting under an optical microscope. Many techniques have been devised to count tracks automatically [1-2]. A semi-automatical device is developped in our laboratory to study radon emanation along border geological faults. We present here the efficiency and angular sensitivity obtained by our technique of the LR-115 nuclear track detector for alpha particles .

2. Methodology

Irradiation

Detector foils were irradiated with collimated alpha particles emitted by an americium source with initial energy of 5.5 MeV. The Energies between 0.5 and 5 MeV were obtained by changing the distance d from the source to LR-115 detector surface in air. Energy E is related to distance d by :

$$d = R(Eo) - R(E) \tag{1}$$

where R(Eo) and R(E) are the alpha ranges in air, corresponding to the energies Eo = 5,5 MeV and E. The range is given by the empirical range-energy equation [3],

$$R(E) = (0.97\,E + 3.26).E^{0.91}) \tag{2}$$

Assuming that the source is thin , geometrical considerations allow to deduce the number of particles which reach on the unit area of detector. Let F represents this number;

$$F(\mathrm{Bq}\,/\,\mathrm{cm}^2) = N\left[1-\left(1+\frac{R^2}{d^2}\right)^{-\frac{1}{2}}\right] \tag{3}$$

where N is the number of incident particles per unit of area of the source, R is the radius of the source. To obtain the same F for all detectors exposed at different distances d_1 and d_2, the irradiation duration must be adapted according the relation (4) ,

$$t_2 = t_1 \frac{1-\left[1+R^2/d_1{}^2\right]^{-1/2}}{1-\left[1+R^2/d_2{}^2\right]^{-1/2}} \tag{4}$$

Etching conditions

After irradiation, the bombarded films were developed in NaOH solution with a molard concentration 2,5 M at three different eching times (80, 100, 120 min). The temperature at which etching takes place was maintened constant to 60 +/- 0,5 °C . Under this conditions the value of the dissolution rate of the non-damaged material is : Vb = 3.4 mm/ .h .

Analysis device

The counting device consists of a binocular microscope fitted out with a bidirectionnel positioning stage, a graphics Tablet (as digitizer) connected to a computer, and a drawing tube (Fig.1) .

Each track visualized under the optical microscope can be counted by acting the mouse of graphics Tablet. The displacement of the mouse on the Tablet is transferred to the microcomputer and recorded in a data file. The measure of track length and diameter needs a preleminary device calibration. To calibrate the device we made1000 measures of an ocular micrometer length. The data are stored in the unit for further calculation of real length and diameter of the track and for statistical analysis.

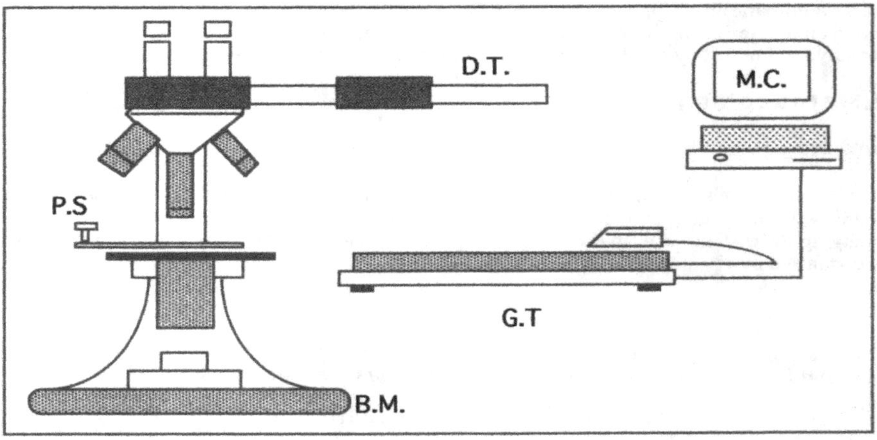

Fig.1: Semi-automatical measurement device: binocular microscope (B.M.) bidirectionnel positioning stage (P.S.), graphics Tablet (G.T.) , drawing tube(D.T.), microcomputer (M.C.)

In order to test our device we determined the tracks diameter on a set of 9 films, irradiated by alpha particles whose energy varies from 1,2 to 5 MeV. Figure 2 presents the comparison of measures to those obtained with an image analyser [4] . Differences between the two series are not statistically significant (Student test) . The mean standard deviation is about 7% for LPNA and 5% for LMN and for the weak diameters it is smaller with the LPNA device than with the LMN image analyser.

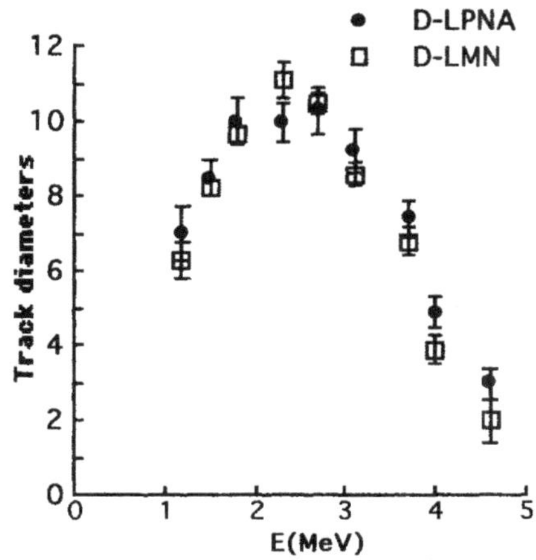

Fig. 2 : Comparison of track diameters with LPNA and LMN devices.

3. Results and discussion

Sensitivity range of LR115

The detection efficiency of alpha particles with different energies and an angle incidence is carefully examined for different eching times of the LR-115 detector to determine the best conditions of utilisation. The detection efficiency is plotted (Fig.3) versus the residual thickness for alpha energies from 0,5 to 5 MeV . The area of sensitivity depends on the eching time, especially for the high energies. The most large sensitivity area ranged between 1.2 and 4.6 MeV for an eching time of 120 mn. This result is in good agreement (Table1) with those obtained by different authors operating in similar conditions [5-6] or using a spark counting technique [7].

Table 1.

LPNA	LMN [5]	[6]	[7]
1.2 to 4,6 MeV	1.5 to 4.5 MeV	1.6 to 4.7 MeV	1.6 to 4.8 MeV

Track Diameters

The variation of track diameters versus alpha energies and for different etching times is shown in Fig.4 . For a given energy the diameter of tracks increases with the etching time. This variation seems to be approximatly linear (Fig.5). For any etching time,the variation D(E) of diameters versus the incident alpha energy has a parabolic form, with a maximum reached for 2,5 MeV. Figure 6 shows that the nature of this variation is similar to that of energy loss in the detector thickness.

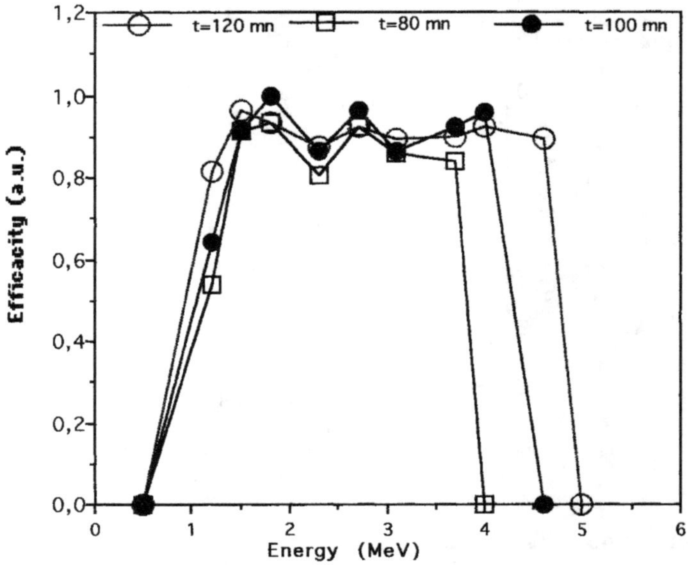

Fig. 3 : LR-115 detector efficiency versus energy at different etching times (t)

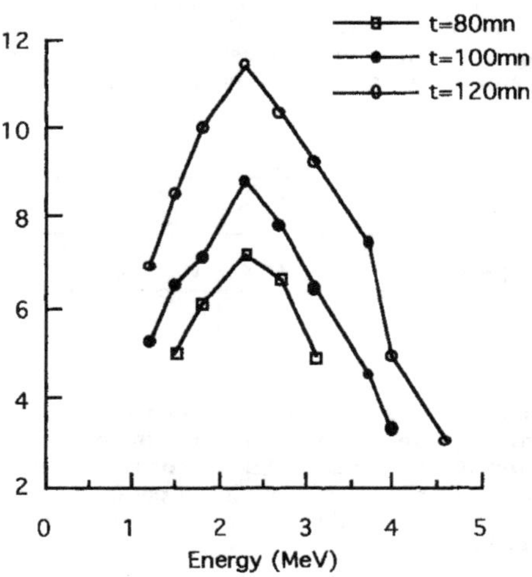

Fig.4 : D(E), Track diameters versus alpha energies at different etching times t .

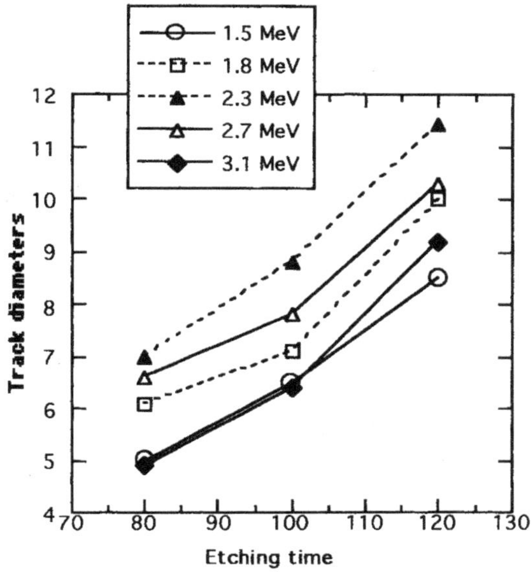

Fig.5: Variation of track diameters versus etching time

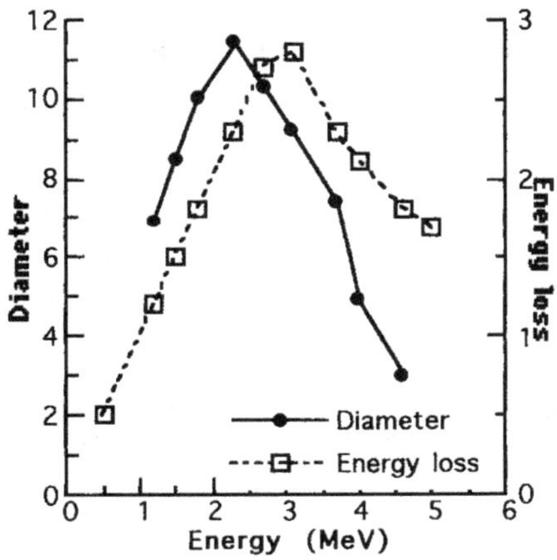

Fig.6 : Energy loss and diameters variation versus incident alpha energy

Critical angle of registration

The critical angles (θ_c) of registration in LR-115 detector were determined under the conditions which gives the best sensitivity of LR-115 detector, i.e an etching time of 120 min and an incident energy ranging from 1,2 to 4,6 MeV. The variation of the critical angle versus incident alpha energy is given in Fig.7. Between 1.2 and 2.5 MeV the values of θc decreases from 30° to 20° . Above 2,5 MeV, θc rapidly increases with an increase in alpha particle energy. A parabola was fitted to the θc values and the following regression function was obtained :

$$\theta_c(E) = 12.089\, E^2 - 62.554\, E + 99.4 \qquad (5)$$

The mean value calculation of this θc (E) regression function gave $\theta c = 31°$.
This result is not far from those obtained under similar etching conditions by other autors: $\theta c = 34°$, 40°, and 36° in references [5], [6] and [7] respectivly.

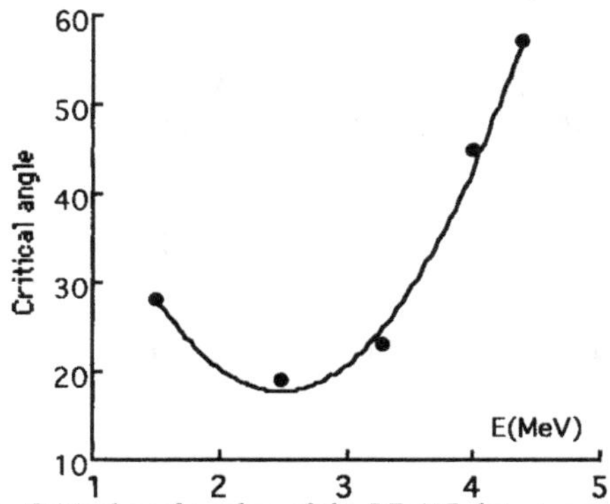

Fig. 7 : Critical angle value of the LR-115 detector versus a particle energies and the fitted regression parabola

4.Conclusion

The experimental study undertaken in the present work has allowed us to determine characteristics of the LR115 for the detection of alpha-particles emitted by the radon and its daughters. The relationships between track diameter, critical angle and alpha particle energy (E) were derived. The empirical detection threshold energies are Emin=1.2 MeV and Emax= 4.6 MeV. The experimental mean value of critical angle is $\theta c = 31°$. The derived fonctions $\theta c(E)$ and D(E) showed a parabolic form like the known empirical curves [8-9]. The new device we used - consisting of an optical microscope, drawing tube and Graphics Tablet (as digitizer) connected to a computer - provides reliable and fast counting of observed tracks , tracks diameter determination, as well as the statistical result processing.

6.References

[1] M. Monnin, Nucl. Instr. and Meth. 173 (1980) 63.
[2] W.G. Cross and L. Tommasino, Radiat. Eff. 5 (1970) 85.
[3] Z. Faj and J. Planinic, Radiat. Prot. Dosim. 35 (1991) 265.
[4] Laboratoire de Microanalyses Nucléaires, Univ. de Franche Comté, Besançon, France
[5] R. Barillon, Franche Comte University Thesis, n°374
[6] A.F. Hafez and M.A. Naim, Nucl. Instr. and Meth, B69 (1992) 373
[7] National Council on Radiation Protection and Measurement.. Report, n°97. Bethesdon. M.D. 5.
[8] J. Kvasnicka, Nucl. Tracks. Radiat. Meas. 11 (1984) 81.
[9] A. Damkjaer, Nucl. Tracks 12 (1986) 295.

B Radon and thoron related to the earthquake

Radon, gas geochemistry, groundwater, and earthquakes

Chi-Yu King

Power reactor and Nuclear Fuel Development Corporation
Tono Geoscience Center
and
Laboratory for Earthquake Chemistry
University of Tokyo
Bunkyo-ku, Tokyo 113, Japan
(On leave from Earthquake Prediction Research, Inc.
381 Hawthorne Ave., Los Altos, CA 94022 USA)

Abstract

Radon monitoring in groundwater, soil air, and atmosphere has been continued in many seismic areas of the world for earthquake-prediction and active-fault studies. Some recent measurements of radon and other geochemical and hydrological parameters have been made for sufficiently long periods, with reliable instruments, and together with measurements of meteorological variables and solid-earth tides. The resultant data are useful in better distinguishing earthquake-related changes from various background noises. Some measurements have been carried out in areas where other geophysical measurements are being made also. Comparative studies of various kinds of geophysical data are helpful in ascertaining the reality of the earthquake-related and fault-related anomalies and in understanding the underlying mechanisms. Spatial anomalies of radon and other terrestrial gasses have been observed for many active faults. Such observations indicate that gas concentrations are very much site dependent, particularly on fault zones where terrestrial fluids may move vertically. Temporal anomalies have been reliably observed before and after some recent earthquakes, including the 1995 Kobe earthquake, and the general pattern of anomaly occurrence remains the same as observed before: They are recorded at only relatively few sensitive sites, which can be at much larger distances than expected from existing earthquake-source models. The sensitivity of a sensitive site is also found to be changeable with time. These results clearly show the inadequacy of the existing dilatancy-fluid diffusion and elastic-dislocation models for earthquake sources to explain earthquake-related geochemical and geophysical changes recorded at large distances. A new paradigm invoking a slowly increasing tectonic-stress field in a broad region of a heterogeneous and faulted crust containing sites along the faults where the materials have been pre-stressed to precarious levels prior to the earthquakes is required.

1. Introduction

Radon in atmosphere, soil air, and groundwater has been studied for the past several decades in many seismically active regions of the world in search of earthquake-related changes that may be useful for earthquake prediction. Earlier findings of radon as well as a number of other geochemical and hydrological parameters have been reviewed in [1-5]. Although anomalous changes were recorded before and after many large and moderate earthquakes, their relation to the corresponding earthquakes has been controversial for several reasons: Similar changes can be caused by changes of other environmental variables such as rainfall, solid-earth tide, and barometric pressure. Such supposedly earthquake-related changes were recorded not only near the earthquakes but also at some so-called sensitive sites hundreds of kilometers away, among many more sites where no such changes were recorded. In this paper, I summarize some recent observations that are much more reliable. The resultant data still show similar features as observed before, thus point out the inadequacy of current conceptual and theoretical models commonly used for explaining the observed earthquake-related changes. Thus I re-iterate a previously proposed paradigm that is capable of providing a reasonable explanation of the observed features involving a low-amplitude and broad-scale tectonic-stress increase, a heterogeneous and faulted crust, and material-property changes not only at the earthquake source but also at the sensitive sites themselves [1, 3].

2. Recent observations

Many recent observations have been made with very reliable instruments over sufficiently long period and with concurrent measurement of meteorological and other environmental parameters, to allow recognition of earthquake-related changes from among background variations. Some measurements are made in areas of intensive studies, where other geophysical parameters such as crustal strain and tilt are also monitored. Comparative studies of high-quality data of various kind recorded simultaneously at many sites together, with proper statistical analysis, are helpful in ascertaining the reality of the recorded anomalies and understanding their underlying mechanisms.

Spatial gas anomalies on active faults. Concentrations of radon and other terrestrial gases have long been observed to be anomalously high along active faults. In recent years many more gas surveys have been conducted, especially in China. In California, King, et. al [6] made some extensive soil-air radon surveys across several different kinds of faults (creeping, locked, freshly broken), and found anomalously high radon values along the faults that showed different characteristic features. They attributed the anomalies to a slow vertical gas flow in or near the fault zones that is capable of changing the subsurface radon-concentration profiles in such a way that an upward flow brings up deeper soil air richer in radon to the detection level [7]. They also measured concentrations of carbon dioxide and oxygen at a selected site and found the results to be consistent with this explanation. Their results indicate that significant radon and other gas anomalies can be generated at a monitoring site locally without the need of long-range radon migration from an earthquake source, as some previous studies had suggested. This mechanism may also explain why sensitive sites have usually been found to be located along active faults, because the mechanically relatively weak fault zones are more compliant to stress changes and have more pore spaces and fluids, which can flow readily when subjecting to a changing tectonic stress. It is also clear that, since radon usually exists in minute amounts, in order to generate a radon anomaly, a flow of some carrier fluid (soil air, groundwater, etc.) is required.

Temporal Radon variations. Soil-air radon concentration has long been known to be affected by changing weather. It varies, for example, with barometric pressure due to the "pumping" effect, lowering pressure tending to draw radon-rich soil air from greater depths to detection level. However, as pointed out by King and Minissale [8], the meteorological variable that affects soil-air radon values the most is significant amount of meteoric water retained in the ground. After analyzing an extensive set of radon data recorded at 60 shallow holes in central California during 1975-85, they found several different patterns of rainfall-related seasonal variations, all of which are attributable to the water-saturation and moisture-retention characteristics of the subsurface soil layers. A wet layer of low gas permeability tends to prevent radon in the radon-rich soil air below the layer from reaching the radon-poor soil air above, thus causing higher measured radon values for detectors situated below. However, if the wet layer is so thick that it reaches below the detector, then the measured radon values tend to be reduced.

Radon in groundwater, especially in deep or confined aquifers, is less perturbed by meteorological variation. The best network of water wells where radon, water temperature, and water level are continuously monitored for the purpose of earthquake-prediction is probably the 12 stations operated by Tokyo University [9]. This network has been in operation for about 20 years and it includes barometric-pressure recorders, with all data telemetered to a computer in the laboratory for quick data analysis. The data recorded by this network permit correction of most environmentally caused changes and thus identification of truly earthquake-related changes. For example, one of the stations (SKE) recorded a significant radon anomaly before the magnitude 7.0 Izu-Oshima-kinkai earthquake in 1978; another station (KSM) recorded radon drops after many large distant earthquakes during periods of relatively high seismic activity in northeast Japan.

Sensitive sites. The earthquake-related radon and water-level data recorded by the above-mentioned Tokyo University network are consistent with the hydrological and geochemical anomaly patterns observed earlier: Earthquake-related changes can be recorded at large distances and only at relatively few sensitive sites [1, 3]. The observed anomalous water-level changes are commonly much larger than, and sometimes in the wrong direction from, the theoretical values calculated on the assumptions that they are proportional to elastic volumetric strain changes and respond to earthquakes in the same way as they do to solid-earth tides [10] and that the earthquakes are modeled as dislocations in a homogeneous isotropic elastic medium.

Another good water-level network is operated by the U. S. Geological Survey in the Parkfield, California, area. Careful analyses of pre- and co-seismic water-level changes recorded by this network also show the inadequacy of using calculated seismic slips alone to explain many of the recorded changes [11, 12].

During the last few years, a nice set of water-level and pressure data has also been recorded at a dense network of wells monitored by Power Reactor and Nuclear Fuel Development Cooperation in Tono area of Gifu prefecture, central Japan. The data clearly indicate the importance of local permeability changes at the individual monitoring sites in producing the various recorded earthquake-related changes [13], as summarized briefly below.

At Tono, water level or pressure has been continuously monitored at more than ten wells in a very small area, within 500 m of the "Tono Mine" , where the largest known uranium deposit in Japan is located. This area is traversed by a east-west trending reverse fault, named Tsukiyosh, with a dip angle of 60-70 degrees to the south, an offset of 30 m, and a thin layer of ordinarily impermeable fault gouge several cm thick. On the south side of the fault is a east-west trending underground gallery 126 m deep. Monitored are nine shallow wells (23-33m deep) and 4 deeper wells (SN1, 90m; SN3, 130 m; TH7, 200m; TH8, 200 m). Most of the wells are located on the north side of the fault where the water pressures are approximately hydrostatic. The TH7 well is, however, located on the south side, where water pressure has been drawn down to about half of the hydrostatic values, probably because of the existence of the water-draining gallery. One of the wells (SN3), located about 400m north of the fault, was previously found to have shown pre-seismic water-level drops and post-seismic increases for three large far-away earthquakes during 1994-95, suggesting that this site might be sensitive to earthquake-related stress changes. Later investigations discovered that water leakage has occurred at about the same times at a 100 m long hole in the gallery that was drilled through the fault to reach the north side. Several attempts were made to stop the leakage, and the hole was mostly sealed in early March 1997. About ten days later, on March 16, 1997, a rare earthquake of magnitude 5.8 occurred about 50 km south of Tono, where the water level/pressure changes recorded at different sides of the fault showed distinctively different features. Recorded at SN1, SN3, and TH8 on the north side were large co- and post-seismic water level /pressure decreases followed by larger increases. But at TH7 on the south side, at about equal distance as TH8 (100 m) from the fault, only a small pressure surge occurred approximately one day after the earthquake, followed by a gradual pressure increase. Also the long-term post-seismic increases at SN1 and SN3 recovered a large seven-month-long water-level drop that had begun in September 1996. More importantly, several pre-earthquake changes were recorded also: A few days before the earthquake, the diurnal variations at SN1, SN3, and TH8 diminished significantly. Beginning about the same time, the recorded water level at a shallow well (KA2) situated about 300 m to the east of TH8 began to drop also. Most of the other shallow wells showed only co- and post-seismic water-level drops.

All the above-mentioned observations in Tono clearly cannot be explained by assuming a co-seismic fault slip at the earthquake source alone; they require local permeability changes near the individual wells for adequate explanation. One possible explanation is that the aquifers tapped by the wells on the north side of the fault had been precariously stressed before the earthquake. Thus a small tectonic-stress increase that caused the earthquake occurrence was sufficient to create also enough interconnected fissures in the local water barriers (especially at the east-west trending and south-dipping fault-gouge zone which can be easily deformed by the north-south tectonic compression in this region), leading to the high sensitivity of SN3 to large distant earthquakes before the March 16 earthquake, the pre-earthquake water leakage into the gallery from the relatively high-pressured north side, the pre-earthquake water-level drop at KA2, and the recorded reduction of diurnal variations at SN1, SN3, and TH8. Additional permeability increase at the water barriers due to the strong seismic shaking may have caused the co- and postseismic water-level/pressure drops at most of the wells on the north side. The fissures were probably healed or sealed soon after the earthquake, leading to the subsequent water-level/pressure increases at the deep wells on the north side. On the other hand, at TH7 on the south side, the water pressure surge one day after the earthquake was probably due to a co-seismically increased water flow from the north side. The subsequent gradual post-earthquake water-pressure increase may have been due to accumulation of water drained from upper aquifers through fissures in horizontal barriers created by seismic-shaking. The above-described complicated patterns of water-level/pressure changes certainly can not be explained solely on the basis of static-stress change generated by fault slip at the earthquake source, for the amplitude of such change is calculated to be only comparable to tidal stresses. To provide a reasonable explanation, we need to assume that the aquifer/barrier system had already been stressed to a precarious level prior to the earthquake, so that a further tectonic stress increase triggered the fissures, even though the increase was probably very small.

This above-described local precariousness may be the reason why a monitoring site is so sensitive to pre-seismic stress changes, while other sites nearby are not. The same hypothesis can also explain the sensitivity variability of a sensitive site, such as what has been observed at the KSM site mentioned previously: The sensitivity was high during periods of heightened seismic activity because the aquifer there was precariously stressed; it was not high after large regional earthquakes, because the local stress decreased to sub-precarious levels after those earthquakes.

Kobe earthquake. Several clear hydrological and geochemical anomalies were recorded in Kobe area before the magnitude 7.2 Kobe earthquake in 1975. They include a radon anomaly at a water well [14], chemical-composition changes in two other wells [15], an increase in groundwater discharge [16], and a radon anomaly in the near-ground atmosphere [17]. In a broader region, some pre-earthquake changes in groundwater temperature, discharge, or color were observed at 4 sites about 100 km to the north [18]. Many post-seismic groundwater changes were also observed; a significant number of such changes do not fit the quadratic pattern of volumetric strain changes calculated for the earthquake on the basis of a elastic dislocation model [18, 19]. At a mineral spring 220 km to the east, the gas flow rate decreased remarkably 3 hours before the earthquake; then it increased, as the gas ratios of helium, nitrogen, and methane to argon increased also after the earthquake [20]. A similarly shaped post-seismic gas-composition change was observed at an auxiliary site too. In contrast, only one of the more than ten strain meters deployed in the source region recorded pre-earthquake changes, while the others, only co-seismic changes [16]. No significant premonitory signals were recorded in the geodetic (GPS) data either.

Other observations of earthquake-related fluid movement at large epicentral distances. Post-seismic water-level and flow changes were observed for some other recent earthquakes also, and they too can be attributed to local permeability changes caused by seismic shaking [18, 21, 22]. Another interesting case is provided by recorded changes of eruption intervals at a periodic geyser north of San Francisco; the intervals showed co-seismic and pre-seismic changes for three large earthquakes 100 to 200 km away [23]. Most surprisingly, the magnitude 7.3 Landers earthquake in 1992 triggered delayed seismicity increases in 14 volcanic and geothermal areas in western United States at distances as far away as 1250 km (17 times as large as calculated source dimensions), but not at other places closer by [24]. This result caused many seismologists to acknowledge the inadequacy of the long-cherished elastic dislocation models in explaining earthquake-related geophysical and geochemical phenomena. Again, earthquake-related fluid movements at sensitive sites are required for explanation [25, 26].

There are some recent reports of observation of earthquake-related thermal-infrared anomalies recorded in satellite images. Such anomalies began days to weeks before the earthquakes and cover broad areas that include or abut the epicenters [27]. A likely mechanism for this phenomenon, if proven true, is earthquake-related changes of near-surface air temperature or composition possibly caused by increased outgassing rate from the earth through newly created fissures in the epicenter regions.

Escaped vs. Dissolved radon. Radon measurements at water wells and springs have so far been made mostly with dissolved radon in the water. Some recent studies suggest that it may be better to monitor escaped radon (radon in gas phase), because it is more responsive to solid-earth tides and shows clearer earthquake-related changes [28].

3. Discussion and Conclusions

Many recent studies of radon as well as other geochemical and hydrological parameters have been improved in measurement quality, in establishing a long-term baseline, in distinguishing earthquake-related signals from environmentally induced background variations, and in multi-disciplinary approach (making simultaneous measurements with other geochemical and geophysical parameters). The resultant data are more reliable, yet they still show features of earlier results: Pre-, co-, and post-seismic changes are recorded at a relatively few sensitive sites up to very large distances from the corresponding earthquake epicenters. Such observations cannot be explained by the existing theoretical models that represent a earthquake by a simple dislocation in a homogeneous, isotropic, porous, elastic half space, and by assuming that dilatancy and fluid diffusion occur only at the earthquake sources. To provide a satisfactory explanation, it is necessary to recognize the heterogeneity of the crust, especially the existence of fault zones, where the tectonic strain is amplified due to their relatively low strength and where crustal fluids (liquids and gasses) are more abundant and more apt to move in response to a broad-scale low-amplitude tectonic stress increase. At a site where the pre-existing stress is infinitesimally close to the fracture stress, a small tectonic stress increase may be sufficient to help reach the fracture-stress level and trigger an earthquake. At other sites, where the pre-existing stresses are close to some precarious levels (i.e., about half of fracture stresses, according to laboratory data [30]), the same tectonic stress increase may be enough to create sufficient number of interconnected fissures to allow increased fluid movements, resulting in various hydrological and geochemical anomalies. Strong co-seismic shaking can further increase gas and water permeability and cause some co-seismic and post-seismic changes at these and other sites. A sensitive site may become insensitive when the local stress is reduced to less-than-precarious level after the occurrence of some major earthquake; thus the time variability of a sensitive site can be understood also.

The above-mentioned non-uniform response of different monitoring sites is not surprising for a realistic earth, which is very heterogeneous. Similar non-uniform responses are well known in earthquake damages, in strong-motion seismic records, and in the occurrence of earthquakes triggered by the 1992 Landers earthquake mentioned in a previous section.

Earthquake-related fluid movement may conceivably be the cause of some other kinds of anomalies, such as electromagnetic. Since the required stress increase is very small, it may not be sufficient to cause detectable deformation in the competent rocks where most strain meters and tilts meters are embedded. Thus such strain measurements, if not modified, are unlikely to catch the expectedly faint premonitory signals. Perhaps this is a good reason why careful studies of radon and other gas-geochemical and hydrological parameters should be continued, despite doubts about their usefulness for earthquake prediction. It is encouraging to see that gas geochemistry has been the theme for many international meetings, including [29] and this symposium.

4. References

[1] King, C.-Y., Gas geochemistry applied to earthquake prediction, J.Geophys. Res., 91, 12,269-12,281, 1986.

[2] Thomas, D., Geochemical precursors to seismic activity, Pure Appl. Geophys., 126, 241-266,1988.

[3] King, C.-Y., Gas-geochemical approaches to earthquake prediction. In: Tommasino, L, Furlan, G., Khan, H. A., and Monnin, M. (eds), Radon monitoring in radioprotection, environmental radioactivity, and earth sciences, 244-274, World Scientific, Singapore, 1990.

[4] Wakita, H., Groundwater observations for earthquake prediction in Japan. In: Proceeding of the International symposium on continental seismicity and earthquake prediction, 494-500, Seismological Press, Beijing, 1984.

[5] Koizumi, N., Review of geochemical research for earthquake prediction, J. JSNDS 41-60, 1997 (in Japanese).

[6] King, C.-Y., B.-S., King, W. C. Evans, and W. Zhang, Spatial radon anomalies on active faults in California, PAGEOPH, 11, 497-510, 1996.

[7] King, C.-Y., Radon emanation on San Andreas fault, Nature, 271, 516-519, 1978.

[8] King, C.-Y. and A. Minissale, Seasonal variability of soil-gas radon concentration in central California, Radiation Measurement, 23, 683-692, 1994.

[9] Wakita, H., Geochemical challenge to earthquake prediction, Proc. Natl. Acad. Sci. USA, 93, 3781-3786, 1996.

[10] Roeloffs, E. A., Hydrologic precursors to earthquakes: a review, Pure Appl. Geophys., 126, 177-209, 1988.

[11] Quilty, E. G. and E. A. Roeloffs, Water-level changes in response to the 20 December 1994 earthquake near Parkfield, California, Bull. Seismol. Soc. Am., 87, 310-317, 1997.

[12] Roeloffs, E. and E. Quilty, Water level and srain changes preceding and following the August 4, 1985 Kettleman Hills, California, earthquake, Pure Appl. Geophys. 149 21-60, 1997.

[13] King, C.-Y., G. Igarashi, M. Ohno, S. Azuma, H. Saito, and H. Wakita, Earthquake -related water-level changes in Tono Area, central Japan, EOS Trans. Amer. Geophys. U., Fall Meeting, 1997 (in Press, manuscript in preparation).

[14] Igarashi, G., S. Saeki, N. Takahata, K. Sumikawa, S. Tasaka, M. Takahashi, and Y. Sano, Groundwater radon anomaly before the Kobe earthquake in Japan, Science, 269, 66-61, 1995.

[15] Tsunogai, U., and H. Wakita, Precursory chemical changes in ground water: Kobe earthquake, Japan, Science, 269, 61-63, 1995.

[16] Fujimori, K., T. Yamamoto, S. Otsuka, and H. Ishii, Observations of crustal movements and discharge at Rokko-Takao station - Change associated with the 1995 Hyogo-ken Nanbu earthquake,

Ann. Disas. Prev. Res. Inst., Kyoto Univ., 38, 287-296, 1995 (in Japanese).

[17] Yasuoka, Y. and M. Shinogi, Variation of atmospheric radon concentration before and after the southern Hyogo prefecture earthquake, Environmental Radiation III, 204-206, October 1995 (in Japanese).

[18] Koizumi, N., Y. Kano, Y. Katagawa, T. Sato, M. Takahashi, S. Nishimura, and R. Nishida, Groundwater anomalies associated with the 1995 Hyogo-ken Nan-bu earthquake, J. Phys. Earth, 44, 373-380, 1996.

[19] King, C.-Y., N. Koizumi, Y. Kitagawa, Hydrogeochemical anomalies and the 1995 Kobe earthquake, Science, 269, 38-39, 1995.

[20] Sugisaki, R., T. Ito, K. Nagamine, I. Kawabe, Gas geochemical changes at mineral springs associated with the 1995 southern Hyogo earthquake (M=7.2), Japan, EPSL, 139, 239-249, 1996.

[21] King, C. Y., D. Basler, T. S. Presser, W. C. Evans, and L. D. White, In search of earthquake-related hydrologic and chemical changes along Hayward fault, Appl. Geochem. 9, 83-91, 1994.

[22] Rojstaczer, S. and W. Wolf, Permeability changes associated with large earthquakes: an example from Loma Prieta, California, Geology, 20, 211-214, 1992.

[23] Silver, P. G. and N. J. Vallette-Silver, Detection of hydrothermal precursors to large northern California earthquakes, Science, 257, 1363-1368, 1992.

[24] Hill, D., P. A. Reasonberg, A. Michael, W. J. Arabaz, G. Beroza, D. Brumbaugh, J. N. Brune, R. Castro, S. Davis, D. dePolo, W. L. Ellsworth, J. Gomberg, S. Harmsen, L. House, S. M. Jackson, M. J. S. Johnston, L. Jones, R. Keller, S. Malone, L. Munguia, S. Nava, J. C. Pechmann, A. Sanford, R. W. Simpson, R. B. Smith, M. Stark, M. Stickney, A. Vidal, S. Walter, V. Wong, and J. Zollweg, Seismicity remotely triggered by the magnitude 7.3 Landers, California, earthquake, Science, 260, 1617-1623.

[25] Linde, A. T., I. S. Sacks, M. J. S. Johnston, D. Hill, R. G. Bilham, Increased pressure from rising bubbles as a mechanism for the remotely triggered seismicity, Nature, 371, 408-410, 1994.

[26] Sturtevant, B., H. Kanamori, E. E. Brodsky, Seismic triggering by rectified diffusion in geothermal systems, J. Geophys. Res., 101, 25269-25282, 1996.

[27] Qiang, Z. J. and C. G. Dian, The thermal infrared anomaly of logical satellite and precursors of impending earthquakes, Proc. 9th Them. Conf. Geol. Remote Sensing, 2, 1,001-1013, Pasadena, CA, 8-11 Feb. 1993.

[28] Zhang, W. and X. Zhang, A contrast study on escaping and dissolved radon's micro-behavior characteristics in groundwater, J. Earthq. Prediction Res., 5, 148-153, 1996.

[29] Dubois, C. (ed.) , Gas geochemistry, Science Review, Northwood, GB, 1995.

[30] King, C.-Y. and G. Luo, Variations of electric resistance and hydrogen and radon emissions of concrete blocks under increasing uniaxial compression, Pure Appl. Geophys., 134, 45-56, 1990.

Radon observation for earthquake prediction

Hiroshi Wakita

Professor Emeritus, University of Tokyo
Bunkyo-ku, Tokyo 113, Japan

Abstract

Systematic observation of groundwater radon for the purpose of earthquake prediction began in Japan in late 1973. Continuous observations are conducted at fixed stations using deep wells and springs. During the observation period, significant precursory changes including the 1978 Izu-Oshima-kinkai (M7.0) earthquake as well as numerous coseismic changes were observed. At the time of the 1995 Kobe (M7.2) earthquake, significant changes in chemical components, including radon dissolved in groundwater, were observed near the epicentral region.

Precursory changes are presumably caused by permeability changes due to micro-fracturing in basement rock or migration of water from different sources during the preparation stage of earthquakes.

Coseismic changes may be caused by seismic shaking and by changes in regional stress. Significant drops of radon concentration in groundwater have been observed after earthquakes at the KSM site. The occurrence of such drops appears to be time-dependent, and possibly reflects changes in the regional stress state of the observation area. The absence of radon drops seems to be correlated with periods of reduced regional seismic activity.

Experience accumulated over the two past decades allows us to reach some conclusions: 1) changes in groundwater radon do occur prior to large earthquakes; 2) some sites are particularly sensitive to earthquake occurrence; and 3) the sensitivity changes over time.

1. Introduction

Since the finding report of precursory radon changes of the 1966 Tashkent earthquake by Professor Sultankhojaev of the Institute of Seismology of Uzbekistan at the International Symposium on Earthquake Prediction which was held in Tashkent in 1973, radon in groundwater has been monitored in many countries in the world, including Japan. About ten years earlier, a possible relation between earthquake and changes in soil gas radon was pointed out by the Japanese scientists [1, 2].

A national earthquake prediction research program in Japan started in the early 1960's. Systematic observations of the radon concentration in groundwater began in late 1973. Continuous monitoring has been conducted at fixed observation sites using deep wells and springs. For earlier years background variations in radon concentration and characteristics of observation sites have been intensively investigated. Anomalous changes are detected through accumulation and analysis of long-term data. Among them, the precursory changes of the 1978 Izu-Oshima-kinkai earthquake (M7.0) were most significant. Also prior to the 1995 Kobe earthquake (M 7.2) a radon anomaly was also observed together with changes in groundwater parameters and in crustal deformation. Besides these precursory changes, coseismic changes, including marked drops of the radon concentration, provide an important clue to finding the relation between changes in regional stress and earthquake occurrence.

2. Observation sites

Figure 1 shows groundwater monitoring sites operated by the University of Tokyo: three in the Tokai region, five in the Izu Peninsula and Izu-Oshima region, one in Kamakura, and three in the eastern part of Fukushima Prefecture (Fig. 1). Groundwater radon as well as other parameters including water level and water temperature are being monitored.

Most of the wells were holes drilled for non-seismogenic studies but some were drilled specifically for our purposes. Suitable wells for earthquake prediction are usually deep or artesian located in an isolated area. As described in a later section, some of the wells are extremely sensitive to earthquake occurrences.

3. Continuous radon monitoring equipment

A radon-monitoring system with a ZnS(Ag) scintillation detector developed by Noguchi and Wakita [3] has been used with several modifications (Fig. 2). The system has the merit of reliable operation for a long-term period and of requiring little maintenance. Groundwater is introduced into the radon detection chamber and simply allowed to overflow into the drainage. In the chamber, part of radon dissolved moves to the air phase above the water phase, and radon is partitioned between the gaseous and water phases according to the distribution coefficient. Alpha activities of radon and its daughters (Po-218 and Po-214) are then detected by a ZnS scintillation surface.

4. Background fluctuations

By examining the long-term variation patterns, the difference between background fluctuations and earthquake-related anomalous changes can be distinguished. Background fluctuations are caused by changes in meteorological conditions, such as, temperature, atmospheric pressure, heavy precipitation, tidal movements and human activities [4, 5]. Since radon emanation in rocks increases with temperature, radon concentration of groundwater is high in the late summer and low in the late winter. The radon concentration is also affected by changes in atmospheric pressure. The lower the pressure, the higher the radon emission. Some of the wells located close to the ocean show fluctuations caused by changes in ocean tides. Pumping activities at adjacent wells may cause noises in radon observation also.

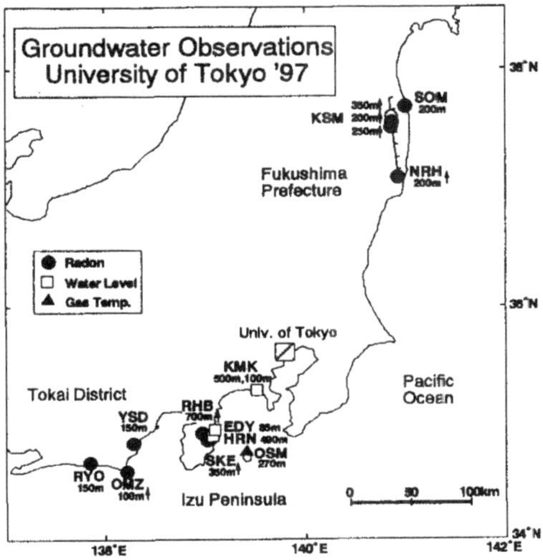

Fig. 1. Distribution of groundwater observation sites operated by University of Tokyo. Arrows indicate artesian wells.

5. Long-term observation data

Figure 3 shows a long-term (18 years) record of radon concentration changes at SKE in the Izu Peninsula, which is at the center of a seismically active region. During this period, the most significant change is the one observed prior to the 1978 Izu-Oshima-kinkai earthquake (M7.0). No other comparable change is visible in the record, which coincides with the absence of other earthquakes of comparative size in the Izu region. As seen in the figure, there are several changes associated with some local seismic and volcanic activities in 1984, 1988 and 1989.

6. Precursory radon changes

6.1. 1978 Izu-Oshima-kinkai earthquake

Figure 4 summarizes the precursory changes of radon and other observations for the 1978 Izu-Oshima-kinkai earthquake [6, 7, 8]. The radon concentration began to decrease in the middle of December 1977, and fluctuated rapidly thereafter. It remained at a lower level for a while and reached the minimum value on January 8, six days before the main shock. Then the concentration began to increase suddenly and returned to the previous higher level two or three days before the earthquake on January 14, 1978 (Fig. 4a). A radon increase caused by seismic vibrations due to the main shock and aftershocks is also noticeable. The pattern and occurrence time of the abnormal radon changes closely resemble those of water temperature (Fig. 4b) and water level (Fig. 4c) in two wells located nearby, and of strain measured by a borehole strainmeter (Fig. 4d) at the tip of Irozaki in the Izu peninsula. It is significant that various kinds of anomalous changes, including groundwater and crustal movement, occurred simultaneously. The similarity between the various anomalies implies that these changes were caused by a single tectonic force in the area before the earthquake.

6.2. 1995 Kobe earthquake

Recently another clear radon precursory change was reported for the 1995 Kobe earthquake (M 7.2) by Dr. Igarashi and others [9] (Fig. 5c). The radon concentration in groundwater at a shallow well located within the aftershock region gradually increased from the initial level measured in October 1994, and

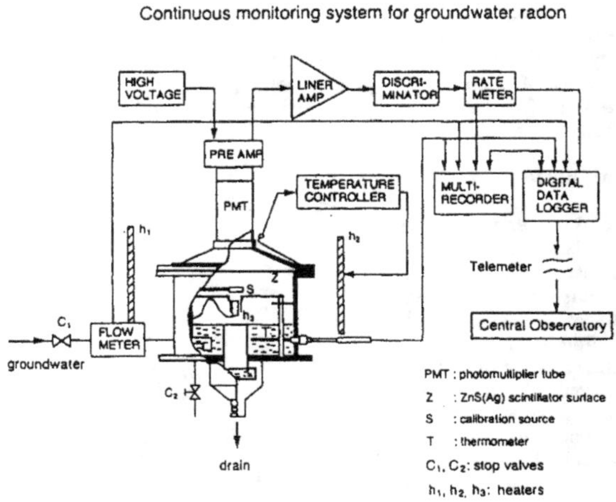

Fig. 2. Continuous monitoring system for groundwater radon.

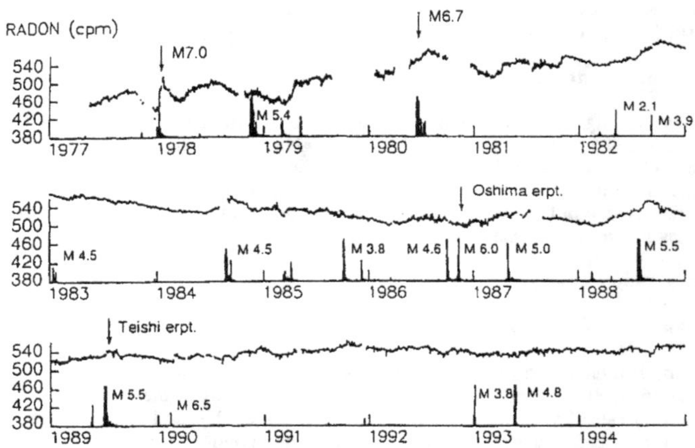

Fig. 3. Long-term variations in the radon concentration observed at SKE site in Izu Peninsula together with relative seismic activity (Wakita, 1996).

reached the highest level on January 8, 1995. It dropped suddenly to the previous background level just before the main shock.

Several other precursory anomalies were also reported for the Kobe earthquake, including changes of chemical components dissolved in groundwater [10, 11]. For example, after August 1994, the chloride ion concentration increased with some fluctuations above a constant level since July 1990, and reached a value of about 10 % in excess just before the earthquake. Groundwater sampled after the earthquake showed even higher values (Fig. 5d). This investigation was made by collecting commercial drinking water in bottles with sampling dates. An idea of extracting a past record of water chemistry was successfully attained by this method.

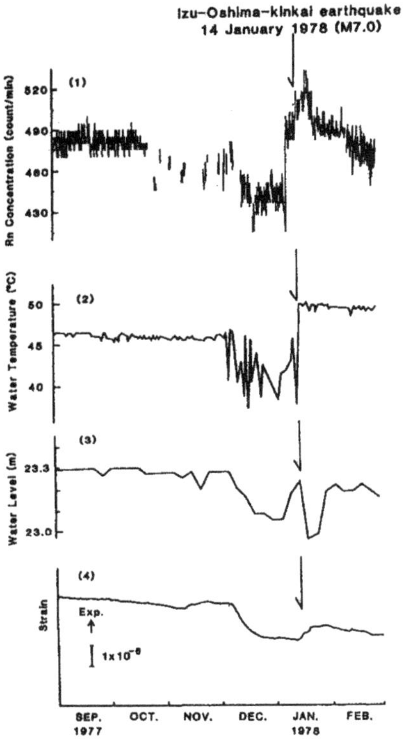

Fig. 4. Precursory changes of the Izu-Oshima-kinkai earthquake (M7.0) on January 14, 1978 [7]. (1) Radon concentration changes in a 350 m well at the SKE (Δ = 25 km). (2) Water temperature in a 500 m well (Δ = 30 km) reported by Geological Survey of Japan. (3) Water level changes in a 500 m well (Δ = 30 km) reported by Geological Survey of Japan. (4) Record of the volumetric strainmeter at Irozaki (Δ = 50 km) measured by Japan Meteorological Agency.

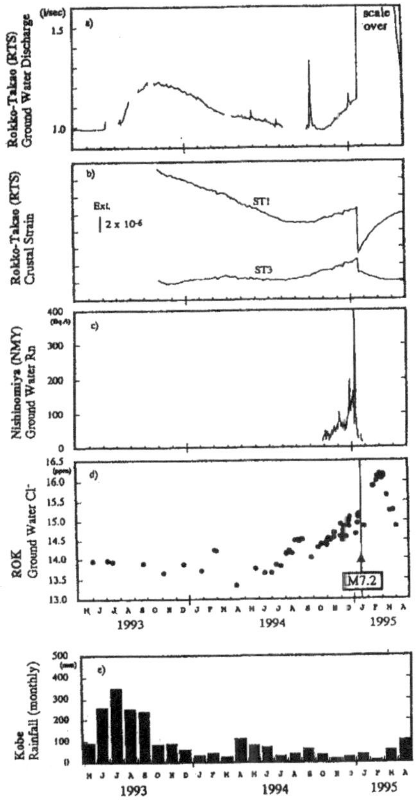

Fig. 5. Summary of precursory changes of the 1995 Kobe earthquake [11].
a) Groundwater discharge rate at Rokko-Takao station. b) Crustal strain changes at Rokko-Takao station. c) ^{222}Rn concetration changes in groundwater at Nishinomiya. d) Chloride ion concentration changes of groundwater at ROK site in Kobe. e) Monthly precipitation at Kobe City.

Scientists at Kyoto University reported an unusual increase in discharge rate of groundwater in a tunnel of the Rokko-Takao Station for crustal movement observation before the main shock. The rate increased by about 7 % starting from early November 1994, and then suddenly increased by 1000 % after the earthquake (Fig. 5a). It is significant that these increases occurred in a season of least precipitation and the total rainfall in 1994 was extraordinarily low [12].

At the same station, some precursory signal was also found in the data of a multi-component borehole strainmeter developed by Professor Ishii of University of Tokyo. As shown in Fig. 5b, a drastic extension in both ST1 (N99°E) and ST3 (N21°W) components started in August 1994 and reached a value of 6.2 and 2.8 μ strain, respectively [12].

A strikingly common feature of the reported anomalies is their beginning time: All of them began several months before the earthquake. The starting points of the anomalies can be interpreted as reflecting the initiation of the preparation stage of the earthquake rupture, although there are not enough constraints for a specific mechanism to explain it [11].

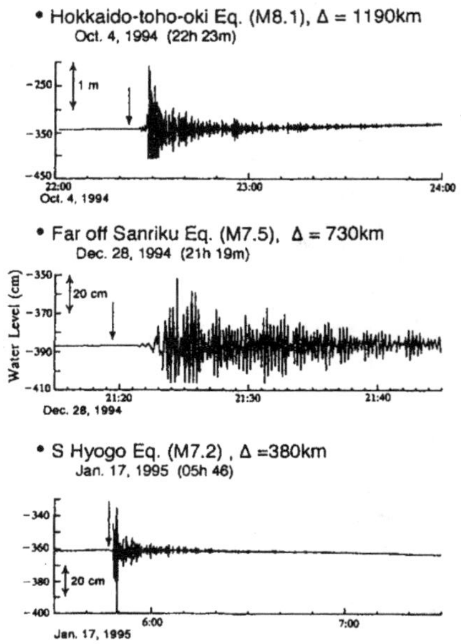

Fig. 6. Water level fluctuations associated with large earthqukes observed at the EDY well in Izu Peninsula. Eq: earthquake [16].

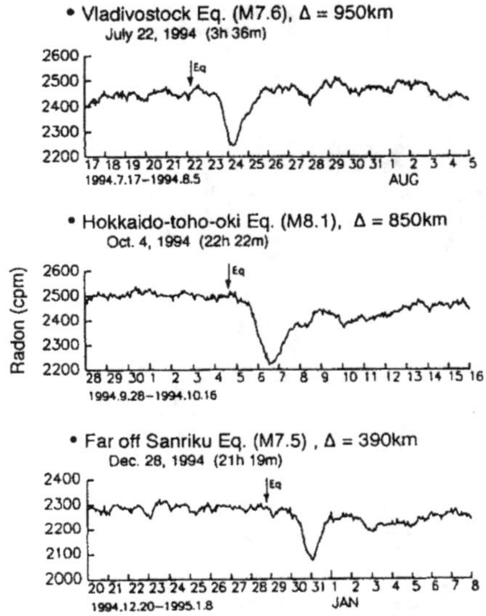

Fig. 7 Coseismic radon drops observed at KSM site. Eq: earthquake [16].

7. "Sensitive wells"

A question has been raised as to why the 1978 radon anomaly was observed at a only single station among others. One answer to this question is the proposition of "sensitive wells". The sensitivity of a sensitive well may also change with time, reflecting changes in regional stress in the area. A good example of sensitive wells is the EDY well in Izu Peninsula, where the water level fluctuates greatly with earthquake occurrence. For example, the water level changes at the time of the 1994 Hokkaido-toho-oki earthquake of magnitude 8.1 were over 2 meters, although the epicentral distance was about 1200 km (Fig. 6). Needless to say, this kind of change is not common to all wells.

Another good example is the KSM site in NE Japan. The radon concentration recorded at the well, drilled on the Futaba fault, decreased after the occurrences of many regional earthquakes [13, 14]. Figure 7 shows some representative cases in 1994, including the Hokkaido-toho-oki earthquake (M8.1). Radon drops were observed for most earthquakes during the period between 1984 and April 1987. Such drops might be interpreted that dissolved radon will be incorporated with newly formed gas bubbles in groundwater by the seismic vibration and trapped in the space of artesian layer, which results lowering the radon concentration in groundwater.

After a series of M6 to 7 earthquakes in the near-by regions, however, the occurrence of radon drops disappeared suddenly. Simultaneously the seismic activity in the northeast Japan calmed down. After a five-year non-responsive period, the sensitivity of the well suddenly recovered in May 1992. Once it recovered, the sensitivity became greater with time such that it reflects earthquakes as small as M4.5. Seismic activity increased also, large shocks including M7.8 (January 15, 1993), M7.8 (July 12, 1993), M8.1 (October 4, 1994), and M7.5 (December 28, 1994) occurred successively. Thus stress accumulation and release may change the physical properties of rocks in the crust, which in turn may change the sensitivity [16]. The disappearance of radon drops occurred again after an M7.5 earthquake of far off Sanriku earthquake on December 28, 1994 (Fig. 8). Thus, the proposition of sensitive wells and their changes in sensitivity with time may be a plausible response to the above-mentioned question. The phenomena of sensitivity change detected by the radon observation may also be recognized in other kinds of observations.

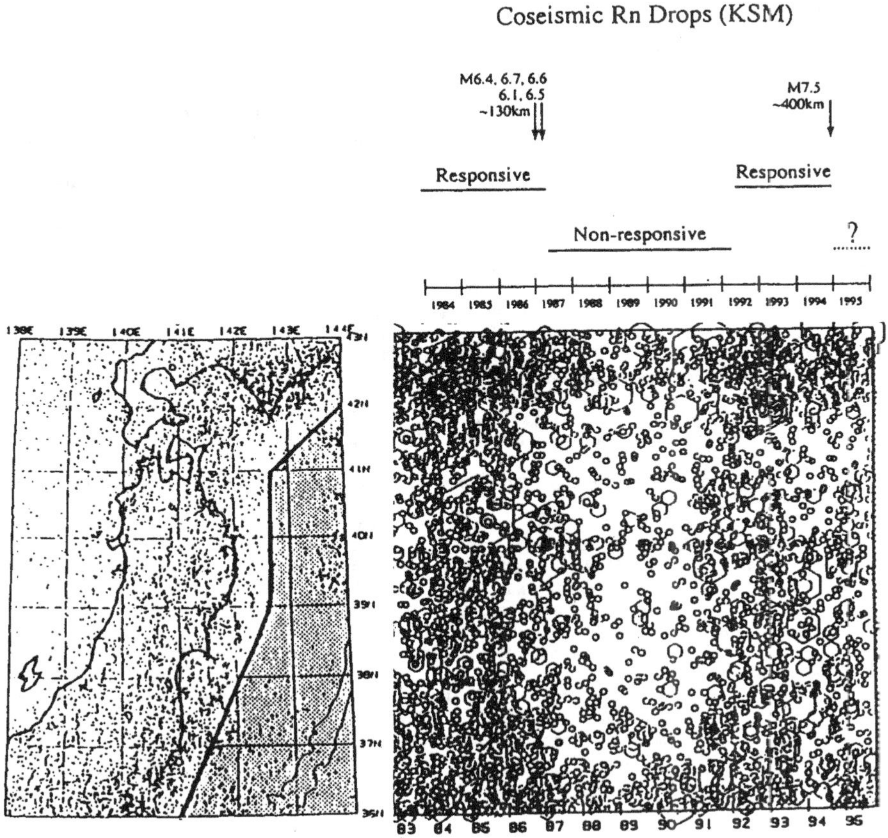

Fig. 8. Appearnace and non-appearance of coseismic radon drops observed at KSM site. Rn: radon. Lower: Deep seismic activity (D> 60 km, M>3) in NE Japan (Tohoku Univ.).

8. Conclusions

Our experience over the past 20 years enables us to draw several conclusions: 1) It is clear that many changes in the radon concentration in groundwater are associated with earthquakes. 2) The earthquake-related changes are not observed at all observation wells but only at a limited number of sensitive wells. 3) Even at sensitive wells, the appearance of earthquake-related signals are likely dependent on the state of stress accumulation in the region. The sensitivity change is interpreted to be caused by changes in physical properties of rocks including the opening and closure of micro cracks.

Detection of precursory phenomena is of primary importance for earthquake prediction. Coseismic signals, similarly, are useful for clarifying the mechanisms of precursory phenomena. The appearance and non-appearance of coseismic radon drops are indicative of the regional stress condition.

Fluids movement in the crust plays a vital role in earthquake occurrence. Besides its original purpose of predicting earthquake, the monitoring of groundwater movement will be useful also to examine other kinds of observations on the ground surface.

The author thanks his colleagues at Laboratory for Earthquake Chemistry, University of Tokyo for help in many aspects and Dr. Chi-Yu King for valuable discussions.

9. References

[1] Hatsuda, Z., Radon content and it's changes in soil air near the ground surface, Mem. Coll. Sci. Univ. Kyoto, Ser. B, 20, 285, 1953.

[2] Okabe, S. Time variation of the atmospheric radon content near the ground surface with relation to some geophysical phenomena, Mem.Coll. Sci. Univ. Kyoto, Ser. A, 28, 99, 1953.

[3] Noguchi, M. and H. Wakita, A method for continuous measurement of radon in groundwater for earth- quake prediction, J. Geophys. Res., 42, 1353-1357, 1977.

[4] Wakita, H., Y. Nakamura, and Y. Sano, Groundwater radon variations reflecting changes in regional stress fields, Earthq. Predict. Res., **3**, 545-557, 1985.

[5] Wakita, H., Y. Nakamura, and Y. Sano, Background fluctuation in groundwater radon observation, J. Phys. Earth, 34, S81-S89, 1986.

[6] Wakita, H., Y. Nakamura, K. Notsu, M. Noguchi, and T. Asada, Radon anomaly: A possible precursor of the 1978 Izu-Oshima-kinkai earthquake, Science, 207, 882-883, 1980.

[7] Wakita H., Precursory changes in groundwater prior to the 1978 Izu-Oshima-kinkai earthquake, Earthquake Prediction, An International Review, D.E. Simpson and P.G. Richard, Eds., Maurice Ewing Series, vol. 4, 527-532, Amer. Geophys. Union, Washington D.C., 1980.

[8] Wakita, H., Y. Nakamura, and Y. Sano, Short-term and intermediate-term geochemical precursors, Pure Appl. Geophys., 126, 267-278, 1988.

[9] Igarashi, G., S. Saeki, N. Takahata, K. Sumikawa, S. Tasaka, Y. Sasaki, M. Takahashi, and Y. Sano, Ground water radon anomaly prior to the 1995 Hyogoken-Nanbu earthquake, Japan, Science, 269, 60-61, 1995.

[10] Tsunogai, U., and H. Wakita, Precursory chemical changes in groundwater: Kobe earthquake, Japan, Science, 269, 61-63, 1995.

[11] Tsunogai, U., and H. Wakita, Anomalous changes in ground water chemistry: Possible precursors of the 1995 Hyogo-ken Nanbu earthquake, Japan, J. Phys. Earth, 44, 381-390, 1996.

[12] Kyoto Univ. and Univ. of Tokyo, Observations of crustal movement and discharge change at Rokko-Takao Station, Rept. Coord. Comm. Earthquake Predict., 54, 695-707, 1995.

[13] Wakita, H., G. Igarashi, Y. Nakamura, and K. Notsu, Coseismic radon changes in groundwater, Geophys. Res. Lett. **16**, 417-420, 1989.

[14] Igarashi, G., and H. Wakita, Groundwater radon anomalies associated with earthquakes,Tectonophys., 180, 237-254, 1990.

[15] Igarashi, G., Y. Tohjima, and H. Wakita, Time-variable response characteristics of groundwater radon to earthquakes, Geophys. Res. Lett., **20**, 1807-1810, 1993.

[16] Wakita, H., Geochemical challenge to earthquake prediction, *Proc. Natl. Acad. Sci. USA*, 93, 3781-3786, 1996.

Geological Fault Exploration and Earthquake Prediction Using Radon

Pao - Shan Weng

Department of Nuclear Science
National Tsing Hua University
Hsinchu, Taiwan 300, R.O.C.

Abstract

Because of the interaction between two postulated tectonic plates where Taiwan is situated, both earthquake and faults are very common in Taiwan. Due to the intensive studies of thermoluminescence dosimeter (TLD) in Taiwan in the past, TLDs were thought as the first choice for radon measurement at the beginning. Later on cellulose nitrate (CN) films had exclusively been used for radon measurements and the technique thus developed was applied to earthquake prediction and fault identification. A spark counter was developed for counting the alpha tracks in the chemically etched CN film. Other radon detectors such as CR-39 chips and electrets were also investigated for the same purpose. Some field experiments are given for illustration.

1. Introduction

Taiwan is located on the boundary between the Eurasian and Philippine Sea plates where active collision and subduction take place very frequently. Such plate tectonic process results in intensive seismicity in this region. Weekly sampling of geothermal water at five hot springs in northern Taiwan to monitor changes in radon content can be traced back to early 1979. Later the samplings were conducted continuously, and finally instead of water radon soil radon was sampled intermittently and this technique was further extended to identify the location of geological fault.

2. Early Efforts

Grab Sampling

The method developed by the Geophysical Laboratory of the University of Southern California (USC), a partner of the Taiwan Earthquake Prediction Research Program implemented by the Institute of Earth Sciences (IES) of Academia Sinica, was used by the IES in the 1980s. Four sampling sites were chosen because they are in or close to the active seismic areas. Water samples were collected in air-tight stainless steel bottles from radon stations located along the east coast and analyzed later in the laboratory.

A few remarkable anomalies stood out in the observed data. A large positive anomaly appeared at one of the sites near the end of 1980 and lasted for several months. During this period three moderate earthquakes (M_L 5.8 - 6.9) occurred in north Taiwan. It was noted that there was a rapid tenfold increase in radon content beginning about 50 days before the M_L 5.8 earthquake. This increase continued for another 100 days until the M_L 6.9 earthquake. After that earthquake the radon content dropped to a new steady level which was about 4 times higher than the level before the anomaly. The M_L 6.9 earthquake was located about 100 km SSE of the site.[1]

In addition to the geothermal water samples mentioned above, radon in CO_2 - rich cold spring water located at Suao (Fig.1) was also collected weekly in duplicate samples. Duplicate groundwater samples were collected weekly in evacuated stainless steel bottles. Radon was stripped from the water sample in the laboratory by helium flushing, followed by condensing at liquid nitrogen temperature. Water vapor and CO_2 were removed by gas filtering through Drierite and Ascarite. The purified radon sample was introduced into a ZnS (Ag) - coated scintillation counting cell and counted with a photomultiplier tube (PMT) after ingrowth of daughter nuclides in about 2 h. This method has a precision of $\pm$ 5%.[2]

Both cold and hot spring samples were measured from July 1980 to December 1983. Seven spike-like radon anomalies (increases of 2 to 3 times the standard deviation above the mean) were observed at 3 out of 4 stations. Following every anomaly except one, an earthquake of M_L above 4.6 occurred within 4 to 51 days, at an epicentral distance of 14 to 45 km, and at a focal depth of less than 10 km.

The distribution of the earthquakes preceded by radon anomalies was skewed in certain directions from the radon stations; the radon stations seemed to be insensitive to earthquakes occurring in the other directions.

At station Chinshan (Fig.1) which is located near a volcanic area, much gas (mainly CO_2) is discharged from the well together with hot water. A very high concentration of radon was detected in the discharged gas; therefore, trapping of gas in water can result in anomalously high radon contents. The radon concentration in water appears to be undersaturated with respect to that in gas. This suggests that hot water is very susceptible to radon loss, and monitoring of radon in gas is more desirable.

Continuous Monitoring

A simple, compact and economic design of continuous radon monitoring (CRM) was used. Its design is shown schematically in Fig.2. The CRM was designed to monitor radon in gas stream. Radon could be stripped off water by artificial pumping system or it was emanated naturally with evolving gas. A cell coated with ZnS (Ag) was used as a scintillator which was coupled with a PMT for the detection of α particles emitted by radon. The activity was counted for 20 min in each hour and recorded as counts per hour (cph) by means of a calculator-printer. Several sets of CRMs have been installed at two hot springs, viz., stations Wulai and Chinshan in northern Taiwan as shown in Fig.1.

At station Wulai, there are a series of hot springs that seep out of submetamorphic argillaceous rocks on a river bank in the hilly area of Wulai. In 1983 two CRMs were installed at the hot springs. Radon was extracted from the water by flushing air through hot water pools with an air pump. The air stream then passed through the radon chamber. Figure 3 shows the scheme of the radon monitoring system. Since the water is rich in bicarbonate, flushing air tended to remove carbon dioxide from the water and raised concentration of carbonate ion which resulted in precipitation of calcium carbonate as shown below:

$$Ca^{2+} + 2HCO_3^- \rightarrow CaCO_3 + CO_2 + H_2O$$

Frequent removal of precipitates was necessary to keep air flowing freely.

At station Chinshan, a well was drilled on a beach underlain by sandstones bordering the Tatun volcanic area. The well discharges hot water at 78 ℃ and CO_2-rich gas. In view of the problems encountered at station Wulai, the CRM system built at station Chinshan was designed in a different manner. Previous study indicates that radon is enriched in gas with respect to water. Therefore, gas was used for radon monitoring instead of water.

The radon station at Chinshan is shown schematically in Fig.4. Discharged fluid from the well head was separated into gas and water. Gas passed through a long copper column for cooling and condensation of steam. The cooled gas was splitted and about 1/10 of the gas passed through the CRM and radon was counted.

At station Suao it is also located in the zone of CO_2-rich springs. The plumbing of the radon monitoring system is shown schematically in Fig.5. An inverted funnel with a rectangular base of 90 cm × 60 cm and a height of 25 cm was placed underwater and half buried in the pebbles at the bottom of the pool. The funnel collected emerging CO_2 bubbles which passed through the plastic tubing and entered the radon monitor.

At station Antung an exploratory geothermal well was drilled in 1978 to a depth of 500 m at the hot spring resort of Antung near Yuli (Fig.1). The hot water (61 ℃) is slightly alkaline (pH=7.9), with dissolved ions consisting mainly of Na^+, Cl^- and SO^{2-}. The discharged fluid contains small amount of gas composed mainly of methane. A gas-water separator was built of bricks and cement on the top of the hot water reservoir in a resort hotel (Fig.6). A peristaltic pump was used to draw the gas through the CRM at a constant rate of 84 mL/min. The pump was turned on 10 min before the beginning of the counting period and turned off after the counting was finished. The magnitude (M_L) 5.9 earthquake that occurred on May 20, 1986 near Hualien (Fig.1) was the strongest earthquake on land. Anomalous radon changes were observed before and after the earthquake at stations Suao and Antung, 65 km to the north and 92 km to the south of the epicenter, respectively. These two stations were in different quadrants of the stress field : Suao was in a compression quadrant and Antung in a dilation quadrant. At the northern station Suao, the radon level increased slowly and steadily by 20% prior to the earthquake and started to decrease afterwards. An abrupt drop of 17% registered 14 days later. These features were recorded by two CRMs connected in series. At the southern station, a spike-like anomalous drop of about 23% was recorded two days before the event. The opposite directions of the anomalies observed at these two stations before the Hualien earthquake may have been a manifestation of the different stress regimes at sites of the two stations prior to the event.[3]

Water samples taken from hot springs, wells and even deep springs for geothermal exploitation were also investigated. Liquid scintillation counter was used to measure α - and β - activities in water samples. An improved measurement method called simple and quick method using toluene as solvent and Butyl-PBD as solute was used as the liquid scintillator. Only 10 mL water sample was needed for each counting. The analysis of counting was performed with the aid of a computer program to solve simultaneous equations.[4] These activities mentioned above were conducted a decade ago.[5]

For the air radon, the sampling method was similar to that developed by Tsivoglou.[6] A 5-min sampling time was followed by 3 counting intervals of gross α emitted from the Whatman No. 41 filter which was counted with a ZnS(Ag) detector.[7]

In the same period back in the 1980s thermoluminescence dosimeters (TLDs) were also used to detect soil radon. The thin TLD $CaSO_4$: Dy-Teflon with thickness of 20 μ m, diam. 6 mm, weight 3.2 $\times 10^{-6}$ kg, surface density 0.113 kgm^{-2} was used. The α-sensitive dosimeter alone will store the energy of α-particles, but are relatively insensitive to the β- and γ-ray background radiation. The determination of radon concentration is made more sensitive if a second α-particle-shielded dosimeter is paired with the α-unshielded dosimeter to measure β- and γ-ray background radiation. The difference gives a more sensitive reading of the α-particle radiation that is present.[8]

A hole excavated in the ground contains the dosimetric housing which consists of an inverted cup to which metallic screen has been stapled or otherwise affixed. To the screen two α-particle sensitive TLDs, one of which is shielded to exclude α-particles but not β- and γ-ray background radiation. A cover secured with rope protects the TLDs.

Though other arrangements for supporting the TLDs may be used, the arrangement mentioned above has the advantage of protecting the TLDs within the cup from mechanical abuse yet exposing them directly to radon from the ground. The screen containing the TLDs should be suspended about 2.6 $\times 10^{-2}$ m away from the bottom of the cup so that radon volume in the cup is utilized on both sides of the TLDs. Soil radon measurement has been one of the major efforts to be used in earthquake prediction and fault exploration in the past which was initiated and carried out by National Tsing Hua University (Fig.1). As mentioned previously, very thin TLDs were used in early days as the α detectors. Nevertheless, the TLDs used were not sensitive enough. It took 1 to 2 weeks to monitor one location. In addition, moist air in the ground poses a potential problem for radon measurement. Therefore, the technique of using cellulose nitrate (CN) films and a spark counter came into our consideration. The CN films are very sensitive to α particles emitted from radon, and the spark counter is quick to count the tracks induced by the α particles and is easily fabricated in a laboratory.

Kodak LR 115 Type 2 is the CN film of choice. It consists of thin film of cellulose nitrate, colored deep red and coated on a 100 μ m thick polyester base. Only the red side is sensitive. The thickness of the sensitive layer is 12 μ m. It must be removed from the base while it is still wet at the conclusion of the washing stage after chemical etching. Naturally occurring α particles emitted by radon have heavy mass, multiple charges and high velocity which make CN films suitable to detect radon. Calibration was performed using a standard radon chamber where known activity of radon is generated from the standard liquid radium source. The CN films were etched in solution consisting of 10^{-3} m^3 pure CO_2 - free water and 0.08 kg NaOH. After the solution was left for 1 h, films were immersed in the solution at a constant temperature of 60 ℃ for 90 min, and then washed for about 1 h. To avoid wrinkles in the etched films, the wet film was placed as flat as possible on paper towels for drying; care must be exercised in handling the wet film to avoid damage.

There are two ways to count the number of α tracks registered. One is to use a microscope for manual counting and the other is to use a spark counter. Both of these two methods were used in this study.

Spark counting is a simple technique for automatic counting and for magnification of etched perforation in thin film track-etch detectors. A schematic drawing of the spark counter developed in our laboratory is shown in Fig.7. The diagram is self-explanatory.[9,10] The dried CN film was placed on the metal electrode and covered with aluminized Mylar with the aluminum side toward the track-etch film. When a high voltage was applied, a spark passed through one of the holes and evaporated a large hole in the thin aluminum electrode. The spark then jumped from one hole to another until it passed through all holes. The number of sparks indicated the number of etched holes. The only source of error displayed is due to counting statistics.

Consistent data were obtained only after trying several and repeatable counting procedures and spark voltage. It was found that at 600 V two sparks for each track would enlarge the etched holes and polarized the film. The counts at 450 V were used to determine the actual signal level.

3. Applications in Earth Science

The results of 7-month monitoring period from October 1983 to April 1984 are shown in Fig.8. The two monitoring stations on the campus of National Tsing Hua University (NTHU) show consistent results. The ordinates are taken as twice the background in order to show the radon peaks more clearly. All the earthquakes presented registered 3 and above on the Richter scale. The radon appeared a few days before the earthquake. Based on the past experiences, when a radon peak appeared, there was a 10% chance that an earthquake would follow. If the radon peak was sufficiently high, there was 30% to 50% chance that an earthquake registering 3 or above on the Richter scale might occur.[11] Very high peaks definitely indicated that a large earthquake would follow.

The NTHU campus is located at Hsinchu (Fig.1) on the west coast of Taiwan. The spike-like daily soil-radon in early 1988 were detected at the NTHU campus monitoring site prior to M=3.9 earthquake occurred on February 28, 1988 as illustrated in Fig.9. Since this earthquake occurred locally at Hsinchu, good radon-seismicity correlation was observed. However, another earthquake occurred at Hualien on the east coast of Taiwan on May 28, 1988 did not show good correlation between radon and seismicity since Hualien is about 100 km away from Hsinchu (Fig.1). The long-term downward trend of radon took about 2 months as indicated in Fig.9 as the dashed profile.[12]

Based on the experiences just discussed, we tried to use similar techniques in uranium ore exploration. Though we failed to find any uranium mines in Taiwan, we explored the faults instead. At the Hsincheng fault and the nearby anticlines shown in Fig.10, radon peaks appeared again just like in the cases of earthquake monitoring. The radon peaks for the fault and anticline are higher than those for surrounding area. The results were confirmed with the resistivity exploration method performed by the Energy and Mining Research/Service Organization of the Industrial Technology Research Institute.

Recently ^{222}Rn concentrations in some Taiwan hot spring spa waters have been monitored using commercially available electret ion chambers (EIC). The use of two EIC units for measuring radon concentrations in the air phase and voltage dependent correction factors for subtracting gamma contributions resulted in improved accuracy.[13] For air radon CR-39 detectors have also been used.[14]

4. Conclusion

The soil radon emanation rate can well serve as an indicator for geological fault exploration and earthquake prediction. More systematic studies are needed to obtain more accurate and reliable data.

Acknowledgments

This is an invited lecture presented at the 7th Tohwa University International Symposium on Radon and Thoron in the Human Environment held at Fukuoka, Japan during October 23-25 ,1997.

5. References

[1] Tsai, Y.B., Teng, T.L., Yeh, Y.H., Yu, S.B., Liu, K.K., and J.H. Wang, Status of earthquake prediction research in Taiwan, R.O.C., Bull. Inst. Earth Sci., Academia Sinica 3, 1-26 ,1983

[2] Liu, K.K., Yui, T.F., Yeh, Y.Y., Tsai, Y.B., and T.L. Teng, Variations of radon content in groundwaters and possible correlation with seismic activities in northern Taiwan, PAGEOPH 122, 231-244 ,1985

[3] Liu, K.K., Yeh, Y.H., and T.L. Teng, Groundwater radon variation associated with the May 20, 1986 Hualien, Taiwan earthquake, Bull. Inst. Earth Sci., Academia Sinica 6, 85-94 ,1986

[4] Cheng, S.S., Measurement of radon activity in water samples with liquid scintillation counter, M.S. Thesis, Institute of Nuclear Science, National Tsing Hua University, 1981.

[5] Lin, Y.M., and C.J. Chen, Monitoring of radon in water of Taiwan, Hoken Butsuri 18, 23-27 ,1983

[6] Tsivoglou, E.C., Ayer, H.E., and D.A. Holady, Occurrence of nonequilibrium atmospheric mixtures of radon and its daughters, Nucleonics 11, 40-45 ,1953

[7] Mei, T.Y.C., Wu, C.F., and P.S. Weng, Monitoring of environment radon-222 in selected areas of Taiwan Province of the Republic of China, Proc. Int. Congr. Int. Radiat. Protect. Asso. Book of

Papers, Vol.III, pp.327-330, Jerusalem, Israel, March 9-14 ,1980

[8] Ho, W., and P.S. Weng, Measurement of radon emanation rate in soil with thermoluminescent dosimeters, Int. J. Appl. Radiat. Isot. 32, 521-522 ,1981

[9] Huang, S.H., Lin, T.H., Su, C.S., and P.S. Weng, Analysis of sparking techniques for counting and magnification of etched holes in cellulose nitrate films, Nucl. Tracks and Radiat. Measure. 11, 115-121 ,1986

[10] Mao, C.H., and P.S. Weng, A miniature spark counter for public communication and education, Nucl. Technol. 77, 92-96 ,1987

[11] Weng, P.S., Cheng, H.H., Hsu, C.C., and K.H. Sun, The applications of nuclear technology in reactor siting, Nucl. Technol. 76, 60-67 ,1987

[12] Weng, P.S., Hsu, C.C., and K.K. Sun, Fault exploration and earthquake prediction at Tsing Hua Open-pool Reactor site using soil radon, Proc. Int. Symp. Res. Reactors, pp.93-68, NTHU, Hsinchu, Taiwan R.O.C. Dec. 6-9 ,1988

[13] Sabol, J., Weng, P.S., and C.H. Mao, Monitoring of ^{222}Rn in Taiwanese hot spring spa water using a modified electret ion chamber method, Health Phys. **68**, 100-104 ,1995

[14] Mao, C.H., and P.S. Weng, Radon concentrations in tunnel, Nucl. Sci. J. 31, 405-410 ,1994

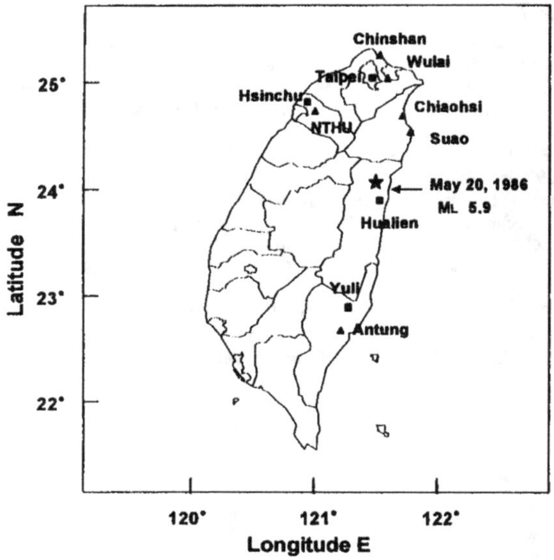

Fig.1 A map of Taiwan showing the epicenter of the May 20, 1986 earthquake (★) and the radon monitoring stations (▲) and cities (■)

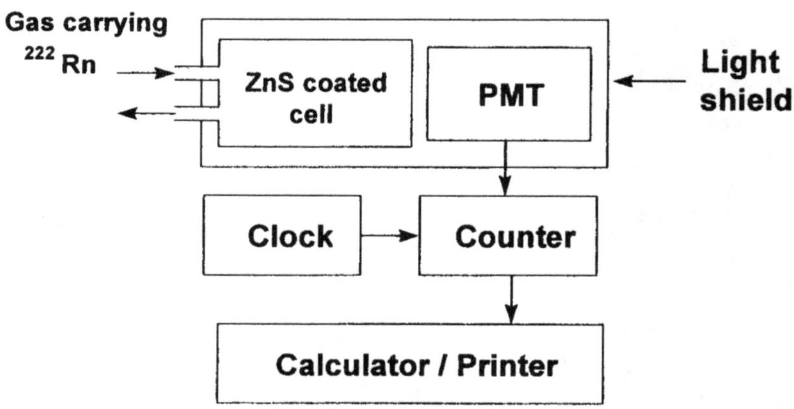

Fig.2 Schematic diagram of the continuous radon monitor

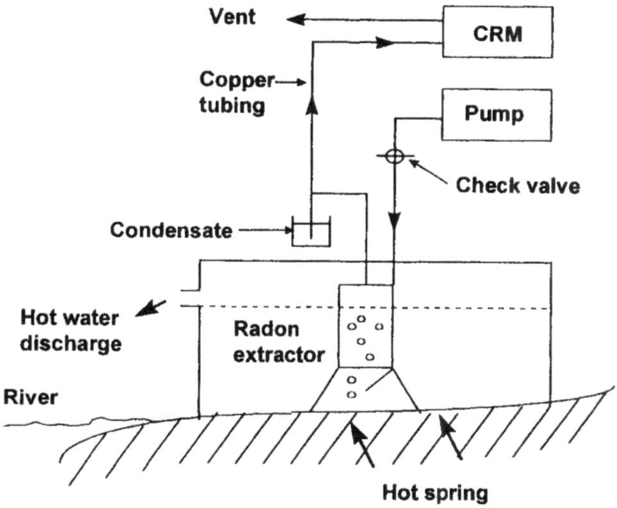

Fig.3 Schematic diagram of continuous radon monitoring system at station Wulai

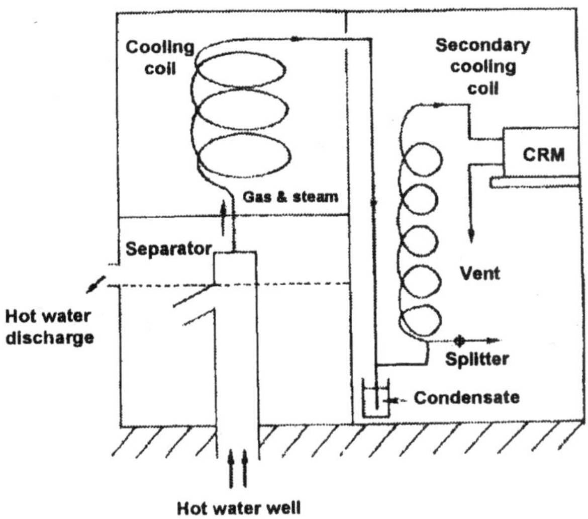

Fig.4 Schematic diagram of continuous radon monitoring system at station Chinshan

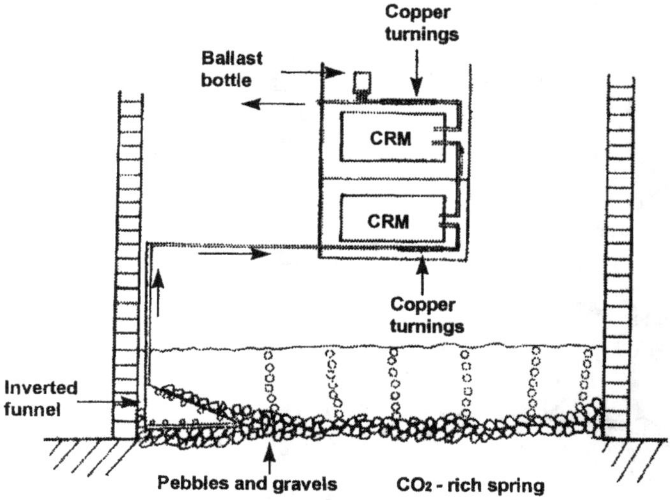

Fig.5 A schematic diagram of the radon monitoring system at station Suao. The pool, shown as widthwise, is one of the several CO_2 - rich springs in Suao

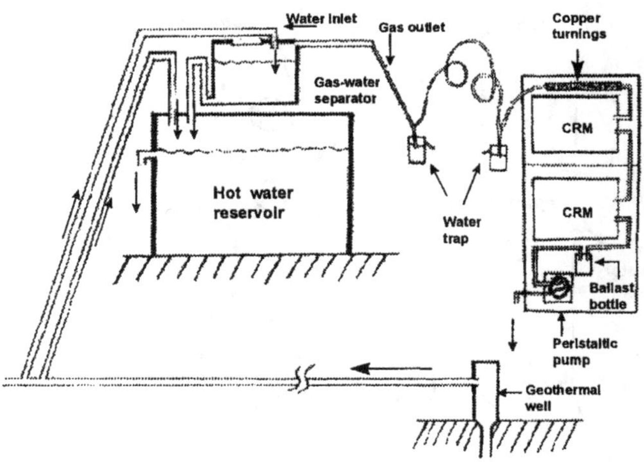

Fig.6 A schematic diagram of the radon monitoring system at Antung

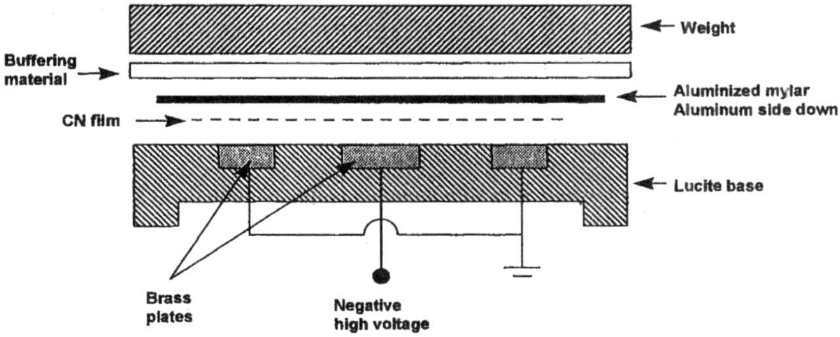

Fig.7 Schematic diagram of a spark counter

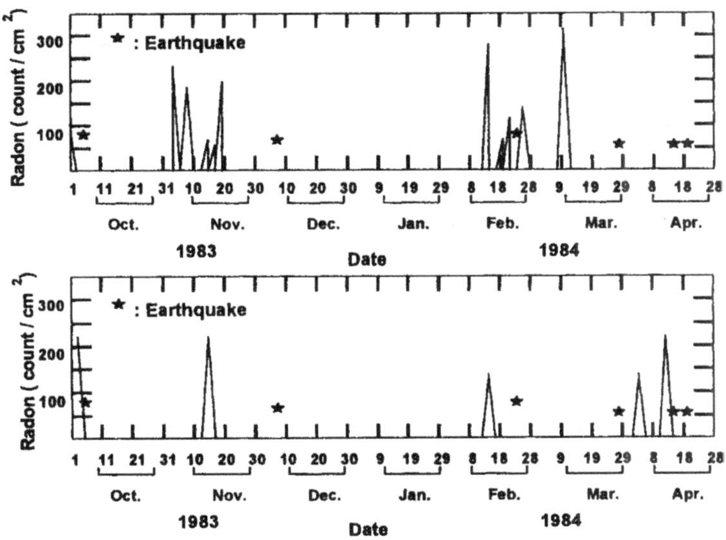

Fig.8 Radon peaks and related earthquakes monitored with CN films at two stations on the campus of National Tsing Hua University

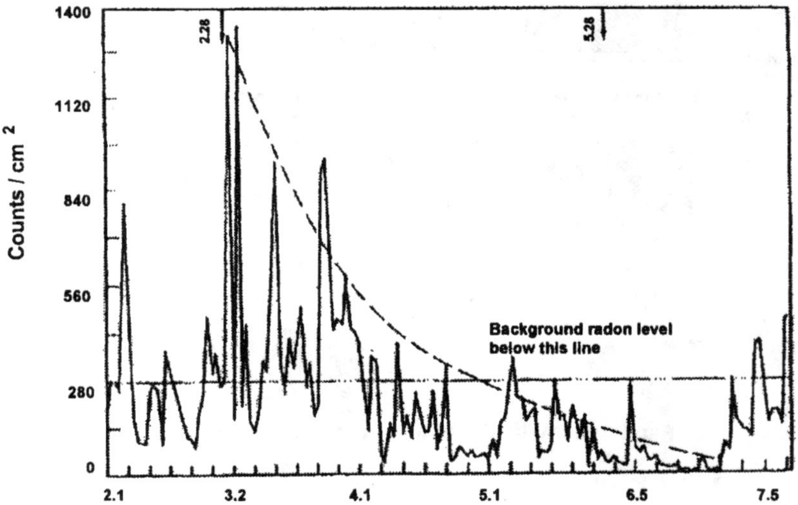

Fig.9 Good radon-seismicity correlation observed at Hsinchu

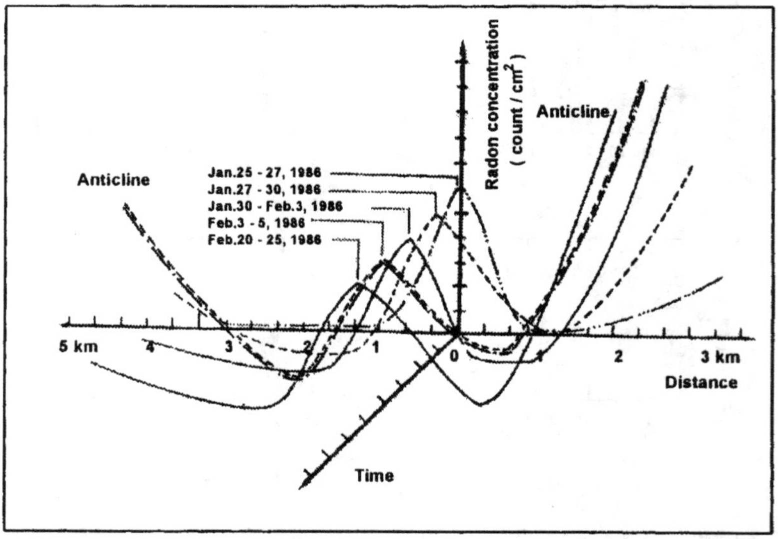

Fig.10 Radon peaks associated with Hsincheng fault and anticline

Research in using radon content variation to predict earthquakes in China

ZHANG Zhaocheng and ZHANG Wei

Center for Analysis and Prediction
State Seismological Bureau, Beijing 100036, China

Abstract

This paper provides an overview of our researches in using anomalous variation of radon content in earthquake prediction in China, including observations and studies on earthquake precursors, researches on mechanism of radon precursors, and their complexity.

In March 1966, an $M_S7.2$ strong earthquake broke out in Xingtai, North China unexpected, causing a great deal of casualty. China started earthquake prediction research ever since, so does the research in using variation of radon content to predict earthquakes. In the past 30 years, we have accumulated numerous precursory observational data, and have greatly advanced the understanding of the mechanism of radon earthquake precursors. Radon, as one of the most important items for precursory observation, has played a key role in the earthquake monitoring and prediction in China.

1. Observations on radon earthquakes precursors

1.1. Observation network

The establishment of seismo-hydrogeochemical observation network in China started from 1966 and finalized in 1990 [1, 2]. The Chinese seismo-hydrogeochemical monitoring network is divided into three parts, including essential national network, regional network and local network. They are administered at three levels by the State Seismological Bureau; Seismological Bureaus of the provinces, autonomous regions and municipalities; and Seismological Offices of the local governments; respectively. The observation stations of the network are distributed over six regions covering 29 provinces, autonomous regions and municipalities. There are totally 330 observation stations, including 68 state essential stations, 110 regional stations and 152 local stations, with 366 observational sites. The distribution of stations are shown in Table 1. Radon content observation in ground water has been carried out at all observational sites.

Table 1. Statistics of observational stations and sites in different part of China

Region	National network		Regional network		Local network		Total number	
	Stations	Sites	Stations	Sites	Stations	Sites	Stations	Sites
North China	14	17	15	21	20	20	49	58
Northwest China	16	26	11	11	55	57	82	94
Southwest China	11	14	25	25	5	5	41	44
South China	8	10	16	16	15	15	39	41
East China	12	12	33	35	19	19	64	66
Northeast China	7	11	10	10	38	42	55	63
Total	68	90	110	118	152	158	330	366

The observational stations of radon had been mainly located on the fault zones with active seismicity and in the earthquake prediction test fields. Few observational stations of radon had been set up in the areas with low seismic activities for comparison research.

1.2. Radon earthquake precursors

Since 1966, we have obtained numerous radon observational data before more than 80 earthquakes ($M_S \geq 5.0$) in the area covered by the radon observational stations in China. Among them, we have systematically studied 72 cases [1] and confirmed the existence of radon precursory anomalies before many of them, especially before five strong earthquakes of $M_S \geq 7.0$ (Table 2).

Table 2. Percentage of earthquake cases with radon precursory anomalies 1966–1990

Magnitude (M_S)	Number of earthquakes occurred in radon observation area	Earthquakes with radon precursory anomalies	
		Number	percentage (%)
≥7.0	7	5	71.4
6.0–6.9	17	8	47.1
5.0–5.9	48	16	33.3

Before some earthquakes, especially before strong earthquakes with $M_S \geq 7.0$, obvious anomalies of radon content in ground water had been observed. Fig. 1 shows the anomalous variation of radon content before and after the five strong earthquakes ($M_S \geq 7.0$), occurred from 1966 to 1976 [3]. The mid-term trend precursory anomalies appeared between half a year and more than a year before the quake. After the earthquake the radon content variation gradually returned to the normal level. For example, before the Haicheng earthquake ($M_S 7.3$) the obvious radon anomalies had been detected at the Panshan and Jinxian observational stations, which are 80 km and 200 km away from the epicenter, respectively (Fig. 1). Fig. 2 gives the curves of monthly average of the absolutely daily derivative of radon content in ground water at those stations for 10 years [4]. The radon content at both stations changed distinctly before the M_S 7.3 earthquake, and anomalies at both stations appeared simultaneously in 1974–1976, and the curves became smooth after the last strong $M_S 5.9$ aftershock (May 18, 1978). Since the aftershock, there was no anomaly appeared up to now and also no such strong seismicity like Haicheng and Tangshan earthquakes occurred in this region as well.

Before a strong earthquake, imminent abrupt anomalies can appear near and far from the future epicenter [3]. Fig. 3 shows the abrupt changes of radon content in ground water before three strong

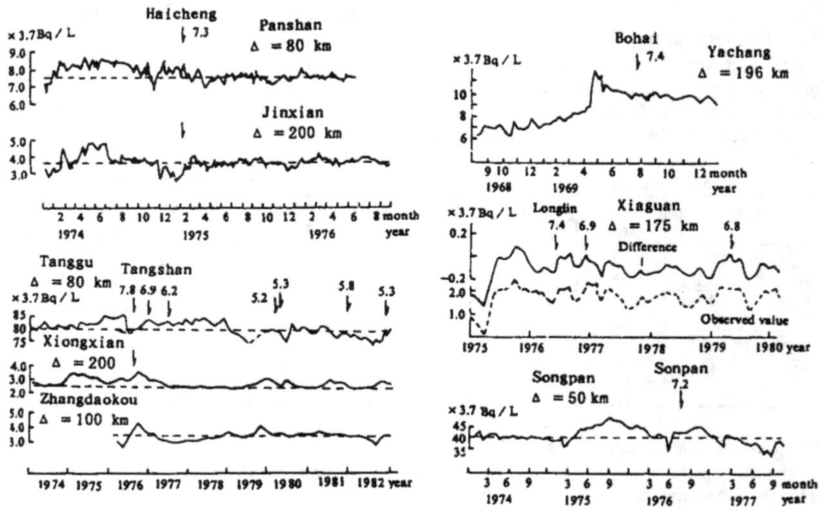

Fig. 1 Monthly variation in radon content observed before and after the five strong earthquakes

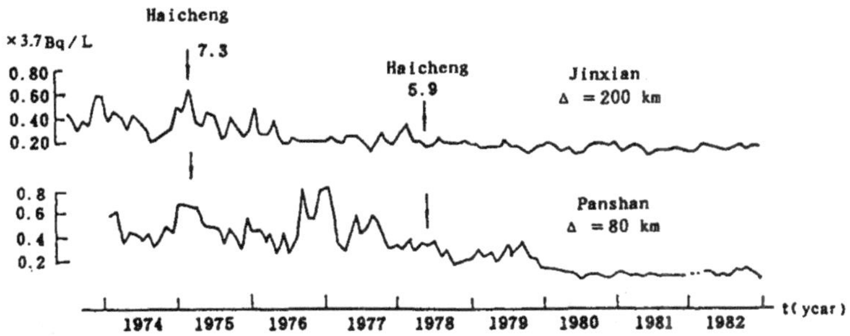

Fig. 2 Monthly average of the absolutely daily derivative of radon content in ground water at Panshan and Jinxian stations

earthquakes in North China. For example, the abrupt changes of radon content in Tanghe well appeared half an hour before Haicheng earthquake with maximum amplitude about 10 times greater than normal fluctuation. Although an abrupt change of radon content at a single site may be random, the abrupt changes at several sites increase the reliability of anomalies. Fig. 4 gives the daily frequencies of abrupt increases in radon content in the several wells of Beijing–Tianjin area during the period of half a year before the Bohai earthquake and at 50 observational sites in North China before the Tangshan earthquake. It shows that the daily frequency of abrupt changes of radon content increased distinctly 10 days before those strong earthquakes.

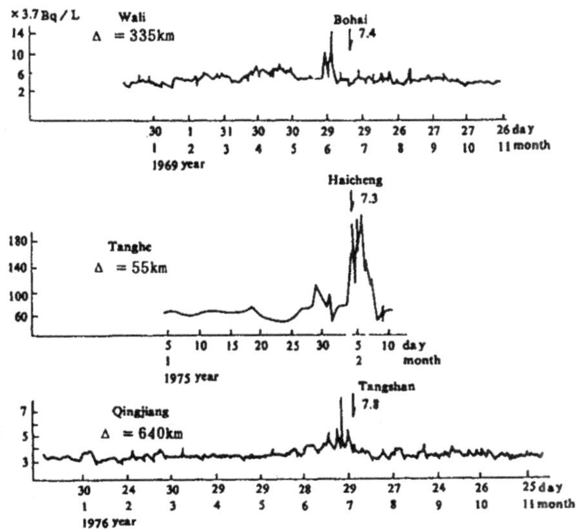

Fig. 3 Imminent abrupt changes of radon content in ground water before three strong earthquakes in North China

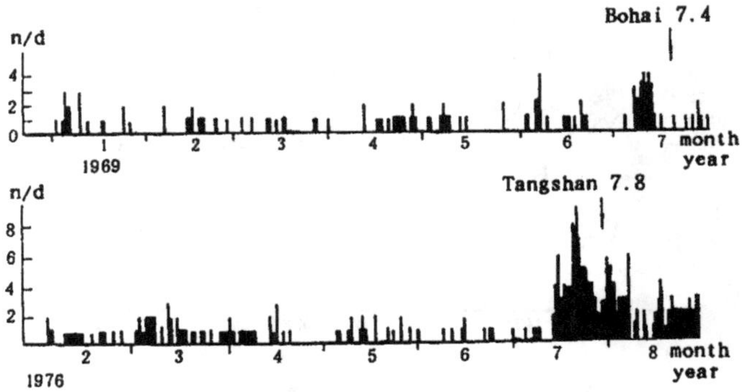

Fig. 4 The daily frequency (n/d) of abrupt radon content changes in the wells in North China before Bohai and Tangshan earthquakes

2. Researches on the mechanisms of radon precursors

For many years Chinese seismologists have carried out investigation on the mechanisms of radon precursors from many aspects [3]. Many factors can cause radon anomalous variation. In our opinion, the stress accumulation in rock stratum, the changes in stress state, and the development of deformation are possibly the main factors causing the variation in radon content before earthquakes. It can be confirmed by following:

2.1. Experimental study results

A series of experimental studies have been carried out under the laboratory conditions; such as fracturing under pressure, sonic vibration, high frequency vibration, pressure solution; and in the field conditions; such as explosions for ground loosening in industry, explosion for mineral exploitation in underground mines, nuclear explosion and large scale test of hydraulic fracturing; for simulating hydrogeochemical effect, including the anomalies of radon content, caused under instant or constant stress. Fig. 5 shows some results of those experimental studies. Abundant experimental data indicate that the stress-strain changes of rock stratum can cause anomalies of radon content.

2.2. Correlation of radon variation with seismic activity

We have studied radon anomalies, comparing with seismic activity before and after strong earthquakes in North China [3-6].

It was noted that the seismic activity of $M_S \geq 3.0$ before and after the three strong earthquakes in North China was closely corresponded to the fluctuation of the trend anomalies in radon content (Fig. 6). The monthly earthquake frequency before Bohai earthquake (July 18, 1969; $M7.4$) rose gradually with the radon content in ground water increasing, which indicates a good correlation between them. Similar pictures had also been demonstrated before the Haicheng (Feb. 4, 1975; $M7.3$) and Tangshan earthquake (July 28, 1976; $M7.8$). While the crustal stress within a large area further strengthened before these strong earthquakes [7], the earthquake frequency increased, and the radon content also increased in the same area. This indicates that the variation in radon content is closely related to the crustal stress.

We have calculated the monthly frequency of earthquakes with $M_L \geq 2.0$ (excluding the aftershocks of Bohai earthquake) from January 1973 to July 1976 in Beijing–Tianjin–Tangshan area (38°–41°N,

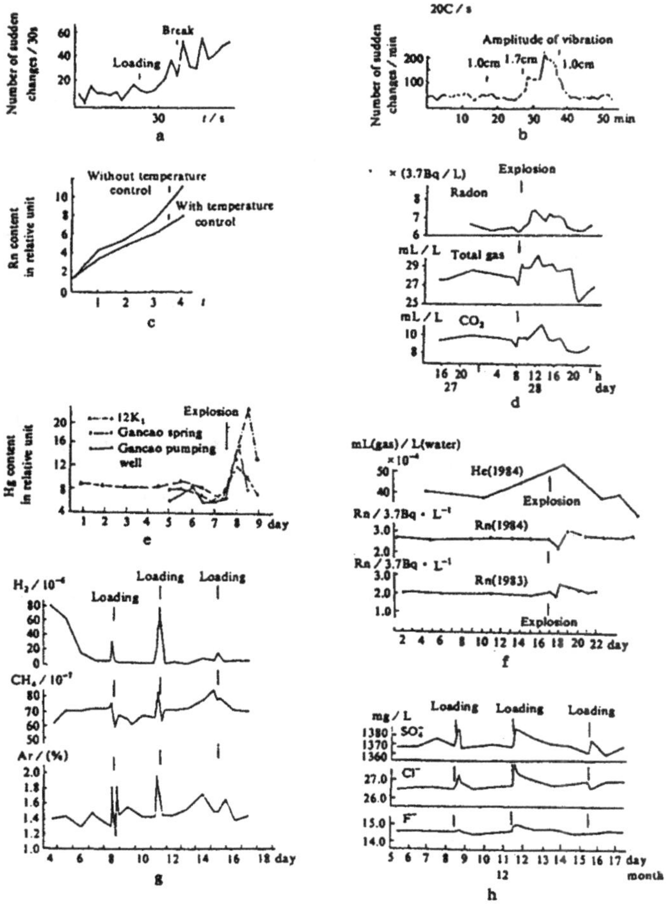

Fig. 5 Some results of investigation on the mechanism of the hydrogeochemical precursors

a. The experimental result of radon emission by fracturing under pressure for sample rock;

b. Acoustic vibration and radon emission for the granite;

c. Amount of radon emission and the number of ultrasonic vibration;

d. Variation in radon, total gas and CO_2 before and after an explosion in Anhui province;

e. Variation in Hg before and after a nuclear explosion;

f. Variation in Rn and He before and after a nuclear explosion;

g. Variation in H_2, CH_4, and Ar before and after the hydrofracturing test;

h. Variation in $SO_4^{=}$, Cl^- and F^- before and after the hydrofracturing test

and 116°–120°E). It has been noted that the monthly earthquake frequency had a certain correlation with the monthly average radon content in ground water in the Angezhuang and Tiantan wells which located near the epicenters of Tangshan $M_S7.8$ earthquake (epicentral distance of 40 km and 50 km, respectively) and its strong aftershock with magnitude of $M_S7.1$ (epicentral distance of 10–25 km). The statistics (Table 3) showed a good correlation between the monthly earthquake frequency and the

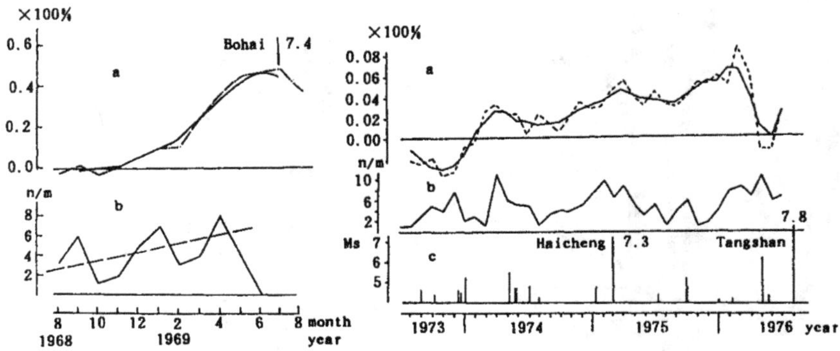

Fig. 6 The correlation between the radon content trend anomalies and the seismic activity before Bohai, Haicheng and Tangshan earthquakes

a. Compounded monthly rate of radon content from 6 sites in the north part of North China;

b. Monthly earthquake frequency (n/m) with $M_S \geq 3.0$ in North China;

c. Earthquake sequence with $M_S \geq 4.0$ in North China

Table 3. Correlation coefficient γ between monthly earthquake frequency N and monthly average radon content A

Names of wells	γ with the confidence level $\alpha = 0.05$	Correlation coefficient γ between N and A		
		between N and A in a same month	between N in the month and A one month before	between N in the month and A one month after
Angezhuang	n = 38* $\gamma_{0.05} = 0.321$	0.045	0.431	0.046
Tiantan	n = 41* $\gamma_{0.05} = 0.308$	0.207	0.337	0.056

* n represents the number of months involved in statistics.

monthly average radon content one month before (i.e. correlation coefficient γ is higher than the theoretical value with a confidence level of α=0.05). It is of precursory interest since radon anomalies appear ahead of seismic activity.

The above data show that there may be some geophysical relations between the changes of radon content and seismic activity. It has been indicated that the variation of radon content before earthquakes reflected the changes of stress-strain of earth crust.

2.3. Radon and earth solid tide

Earth solid tide is the deformation phenomenon of which we can not only calculate out in advance but also observe and record every day. If the radon change in groundwater is caused by the rock stress changes, the earth solid tide should also cause the radon change in rock-groundwater system. Therefore, whether or not the variation of radon content in rock-groundwater system has the characteristics of earth solid tide is a strong proof for the cause-effect relationship between the rock stress and radon content [8].

To demonstrate the correlation between variations of radon content and stress-strain of crust, we selected Dongsanqi well in Beijing to carry out studies on the relationships between radon and earth solid tide in 1988. While a set of Sacks-Evertson borehole strainmeter exists on the bottom of the well, we installed the automatic radon content measurement instrument of model SD-1 on the top of the well for comparative observation and research. Fig. 7a, 7b and 7c are the figures of time series spectrum of escaping radon, theoretical volumetric stress spectrum and observed volumetric stress spectrum, respectively. The spectral distributions of these three sets of data are the same. It means that radon variation in rock-groundwater system is a kind of tide change, caused by the change of sun and moon tide forces.

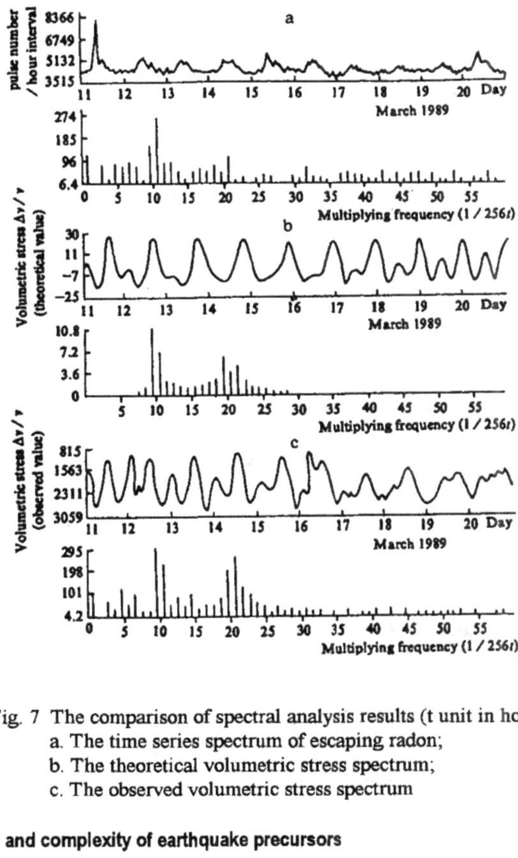

Fig. 7 The comparison of spectral analysis results (t unit in hour)
a. The time series spectrum of escaping radon;
b. The theoretical volumetric stress spectrum;
c. The observed volumetric stress spectrum

3. The general features and complexity of earthquake precursors

Zhang Zhaocheng and other 45 researchers have re-sorted the multidisciplinary data of 60 earthquake cases of $M_S \geq 5.0$ from 1966 to 1985 in Chinese mainland, with the emphasis on the precursors of these earthquakes [9-13]. 927 multidisciplinary precursory anomalies were observed before them, including 127 radon content anomalies before 36 earthquakes. Radon is one of the most important items for precursory observation. In summary, various precursors showed dynamic pictures of development with several phases in time and uneven distribution in space in a large area with relative concentration near the future epicenters. Fig. 8 shows the relation of the epicentral distance (Δ) with appearing time (T_0) of the radon anomalies. These data substantiated that the precursors do exist prior to most earthquakes although they could be quite complicated.

As mentioned earlier, we have also studied observational data of more than 80 earthquakes obtained in the area where radon observational stations exist in China. Those data also confirm the

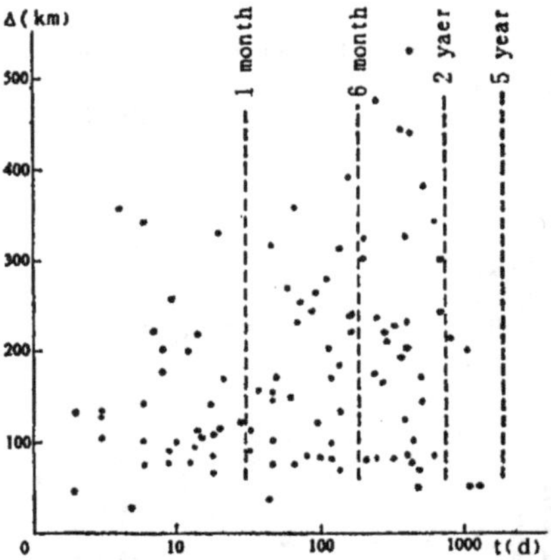

Fig. 8 Diagram of Δ–T_0 relations of radon anomalies before 36 earthquakes with $M_S \geq 5.0$ in China

above mentioned general features of anomalous variation. In the distance of 0–640 km from epicenter and for several years before strong earthquakes, the precursory anomalies of radon content appeared and relatively concentrated within 200 km near the epicenter of future earthquake (Fig. 8). The epicentral distance and duration of anomalies for earthquakes with magnitude of 5–6 (M_S) are much smaller than that for strong earthquakes. Those spatial-temporal process obviously related with earthquakes, and may be closely linked with the process of earthquake preparation.

However, the anomalous variation of radon content, we observed, belongs to earthquake precursors in a broad sense [7]. We believe there are two types of precursors. The broad scale anomalies are observed over a large area (500 km or greater) and appear to reflect a general increase tectonic activity, including seismic activity. These are useful in identifying a large area where there is an increased probability of earthquake occurrence. The small scale anomalies (200 km or less) are more closely associated with an impending earthquake. They could contain the information from a single seismic focus. At present we may detect anomalies, but it has been difficult to distinguish among them the anomalies caused by tectonic movement (they are referred to as "tectonic precursors") and the information from a concrete seismic source (called as "focal precursors"). Therefore, the relationship between precursors and concrete earthquake have a certain indefiniteness, and earthquake precursors are very complicated. One of the important scientific challenge for further study is to identify tectonic precursors from focal precursors.

4. Conclusions

Our thirty years of studying earthquake precursors have thus led us to some important conclusions:

(1) Anomalies of radon content in ground water have been observed before a lot of moderate strong earthquakes. They are related with earthquakes.

(2) At least some earthquakes, especially strong earthquakes, possess precursors, although they can be quite complicated.

(3) These precursors can be used to predict earthquakes in some cases.

(4) There has been certain progress, but there is a long way to go.

6. References

[1] Zhang Wei (chief editor), Department of Scientific Programing and Earthquake Monitoring of State Seismological Bureau of China, Hydrogeochemical Observation for Earthquake Precursors and Prediction, Seismological Press, 3–83, 1992 (in Chinese).

[2] Zhang Wei and Li Xuanhu, A survey of the hydrogeochemical observation network for earthquake prediction in China, Earthquake research in China, Edited by State Seismological Bureau, China, ALLERTON PRESS, ING./NEW YORK, Vol. 8, No. 3, 377–386, 1994.

[3] Zhang Wei, Wang Jiyi et al., Hydrogeochemical Principles and Methods in earthquake prediction, Educational Research Press, 158–185, 212–214, 252–285, 1988 (in Chinese).

[4] Zhang Wei, Xing Yuan and Xing Ruying, Variation of dispersion of radon content in the groundwater and short-term and imminent precursors of earthquakes, ACTA SEISMOLOGICA SINICA, Vol. 1, No. 3, 93–99, 1988.

[5] Zhang Wei and Lin Yiyao, Preliminary study of the application of hydrogeochemistry to earthquake prediction, Earthquake prediction, Proceedings of the international symposium on earthquake prediction, TERRAPUB, Tokyo, 539–552, 1984.

[6] Zhang Wei, Research on hydrogeochemical precursors of earthquakes, The China-Russian jointly sponsored English edition magazine "Journal of Earthquake Prediction Research", Vol. 3, No. 2, 170–182, 1994.

[7] Zhang Zhaocheng, Zheng Dalin and Wang Guixuan, Precursors to earthquakes in Chinese mainland, The China-Russian jointly sponsored English edition magazine "Journal of Earthquake Prediction Research", Vol. 3, No. 4, 528–539, 1994.

[8] Shi Yong and Zhang Wei, The correlation between radon variation and earth solid tide change in rock-groundwater system – the mechanical foundation for using radon change to predict earthquake, The China–Russian jointly sponsored English edition magazine "Journal of Earthquake Prediction Research", Vol. 4, No. 3, 423–430, 1995.

[9] Zhang Zhaocheng (chief editor), Earthquake cases in China (1966–1975), Seismological Press, Beijing (in Chinese with English synopsis and compilers' remarks), 1–222, 1988.

[10]Zhang Zhaocheng (chief editor), Earthquake cases in China (1976–1980), Seismological Press, Beijing (in Chinese with English synopsis and compilers' remarks), 1–421, 1990.

[11]Zhang Zhaocheng (chief editor), Earthquake cases in China (1981–1985), Seismological Press, Beijing (in Chinese with English synopsis and compilers' remarks), 1–294, 1990.

[12]Zhang Zhaocheng, Zheng Dalin et al., Preliminary research on the precursory data from "Earthquake Cases in China", Earthquake, No. 5, 9–24, 1990 (in Chinese).

[13]Zhang Zhaocheng, Zheng Dalin, Luo Yongsheng and Jia Qing, Studies on earthquake precursors and multidisciplinary earthquake prediction in China mainland, The China–Russian jointly sponsored English edition magazine "Journal of Earthquake Prediction Research", Vol. 1, No. 2, 191–205, 1992.

Monitoring of radon activity in the tunnel and out-flow water from mining pit of former Ogoya Copper Mine and radon anomaly related to earthquake of September 10, 1996 at Komatsu area, Japan

Kazuhisa Komura, Akira Toguchi and Seiji Yamazaki

Low Level Radioactivity Laboratory, Kanazawa University
Wake, Tatsunokuchi, Ishikawa 923-1224, JAPAN

Abstract

Radon in tunnel air has been measured in a 546 m tunnel of former Ogoya Copper Mine. Radon activity was found to be correlated inversely with the wind velocity and its variation pattern could be reproduced well by a simple cylindrical model using apparent radon exhalation rate of ~0.01 Bq m^{-1} s^{-1}. A radon monitoring project for out-flow water from a mining pit of this mine was proposed to obtain premonitory symptoms of earthquake. In addition, observation of earthquake-related radon anomaly, which began 10 days before small earthquake (M=3.2) occurred on September 10, 1996 at Komatsu area in Japan, was also presented.

1. Introduction

Radon is considered to be one of the most useful natural tracers not only from the view point of geochemistry but also of geophysics and/or environmental radioactiviy. This paper consists of three articles on radon measurements carried out at Low Level Radioactivity Laboratory (LLRL), Kanazawa University. At first presented is the measurement of radon in a tunnel air and simulation of its variation pattern by a cylindrical model. Secondly, monitoring project of radon is presented for the out-flow water from a mining pit of former Ogoya Copper Mine as one of the most promising premonitory symptoms of an earthquake. Third one presents radon anomaly observed prior to a small scale earthquake (M=3.2) occurred in September 10 of 1996 near our laboratory.

2. Measurement of radon activity in a 546 m tunnel of former Ogoya Copper Mine and simulation of its time variation using simple cylindrical model

To measure extremely low level radioactivity, it is very important to reduce background count in the measurement. It is well known that utilization of deep underground laboratory is quite effective to reduce background derived from high energy cosmic-ray components. We have been constructing an underground laboratory at 290 m points in a 546 m tunnel of former Ogoya Copper Mine closed 40 years ago after more than 300 years of operation (Fig. 1).

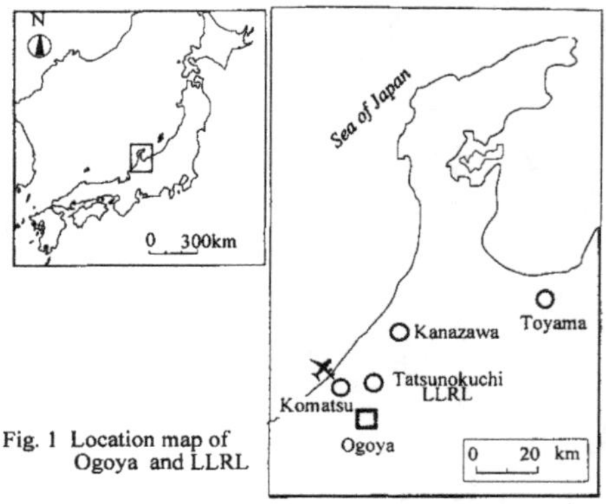

Fig. 1 Location map of Ogoya and LLRL

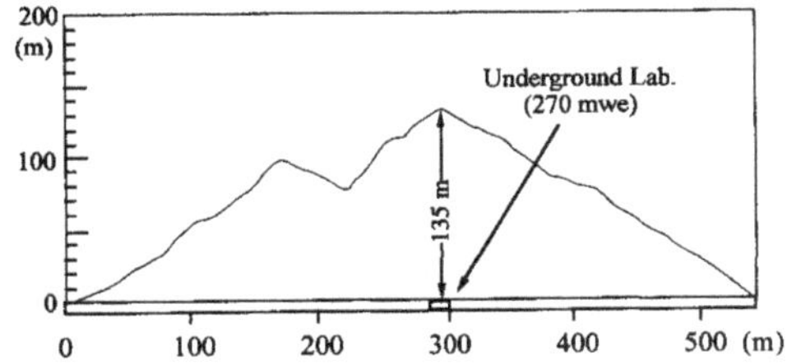

Fig. 2 Cross sectional view of Ogoya tunnel.

High energy cosmic-ray components were measured by using 3in.ϕ x 3in.NaI(Tl) detector at various points in the tunnel. Contribution of high energy cosmic-ray components (mainly muons) was found to be 1/200 of ground level at the depth of 270 mwe (meters water equivalent) point, where underground laboratory was built (Fig. 2). On the contrary, background counts due to daughter nuclides of radon (^{222}Rn) in tunnel air increased 2~5 times higher than overground measurement. Background counts due to radon daughters, however, could be successfully reduced by filling up the free space around the head of Ge detector with air-pack (or air-cup) for packing-aid and also by introducing N_2 gas from liquid nitrogen Dewar of the Ge detector[1].

In order to understand most responsible factor, which governs radon activity in the tunnel, wind velocity and air temperature has been measured simultaneously with radon. Radon activity was measured at 30 minute intervals by a portable radon monitor Pyron AB-5 with 272 ml Lucas cell near the center of the tunnel. On the other hand, wind velocity and air temperature were measured at 10 minute intervals using Canomax Anemomaster (Model 6621) placed adjacent to the radon monitor. During 2 years of monitoring, radon activity was measured to be ranging from several Bq m^{-3} to over 100 Bq m^{-3} with an average value of ~20 Bq m^{-3}. Wind velocity was zero to 8 m s^{-1} with an average of 0.4 m s^{-1} and air temperature was 8 to 14 °C with an average of 12 °C, respectively. It was found that variation of radon activity was governed by wind velocity in the tunnel. Correlation coefficient between radon activity and wind velocity was -0.8~-0.9.

To make more clear the variation pattern of radon, a simple cylindrical model developed by Miki and Ikeya[2] was applied to reproduce radon activity in the tunnel. The outline of the model is illustrated in Fig. 3. The model consists of 3 terms as given by Eq. (1), in which radon activity $C(x)$ at point x from the entrance is given as the functions of wind velocity W, exhalation rate E from the wall of the tunnel with radius of r. To correct roughness of tunnel wall, apparent exhalation rate Ea is introduced instead of E.

$$dC/dt = 2Ea/r - \lambda C - WdC/dx \qquad (1)$$

First and second terms express the exhalation of radon from the tunnel wall and radioactive decay of radon, respectively. Third one shows dilution of radon activity by air into the tunnel. By integrating the Eq. (1), radon activity $C(x)$ at a given point x is obtained by Eq.(2).

$$C(x) = 2Ea[1 - \{1 - \lambda rC(0)/2Ea\}\exp(-\lambda x/W)]/\lambda r \qquad (2)$$

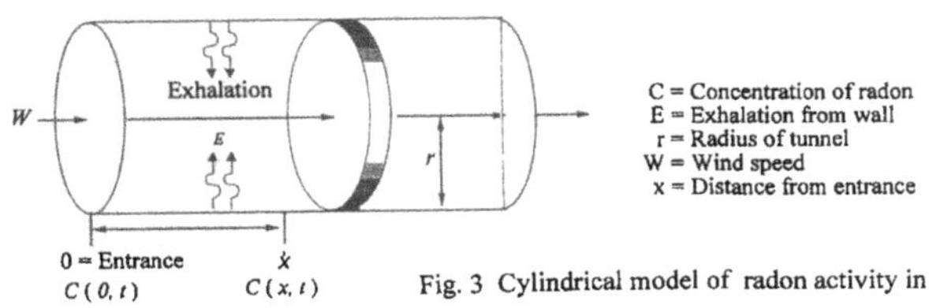

Fig. 3 Cylindrical model of radon activity in tunnel.

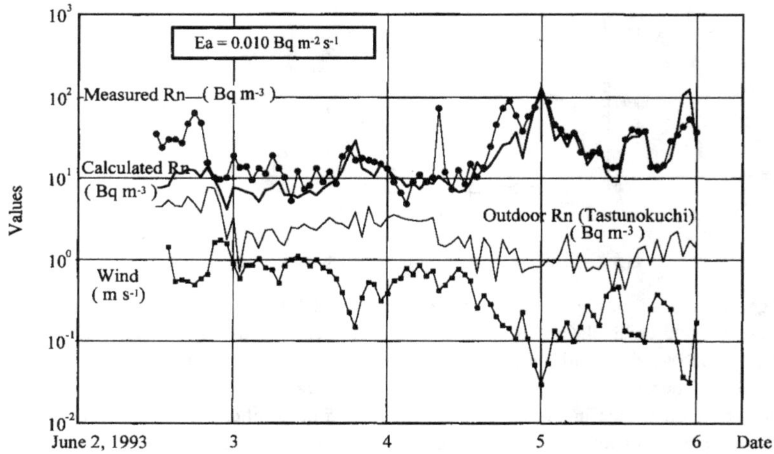

Fig. 4 Example of simulation of radon activity in Ogoya tunnel based on sylindrical model.

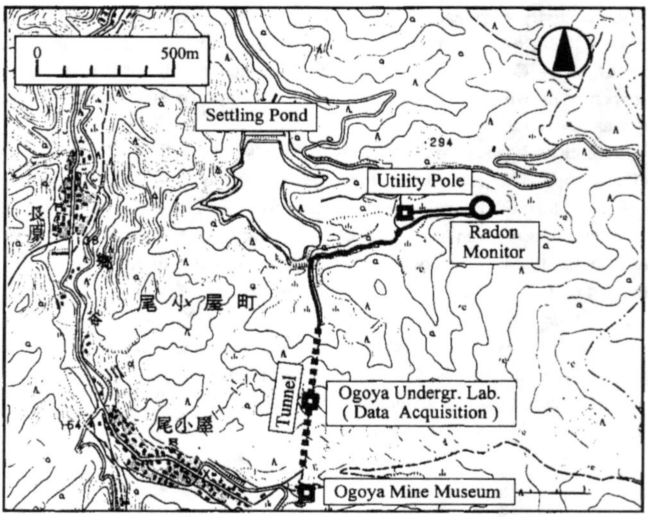

Fig. 5 Location map of Rn monitoring point at former Ogoya Copper Mine

Among the values seen in Eq.(2), $C(0)$, $C(x)$ and W are measured values. Instead of using radon activity outside of tunnel, the values measured at LLRL was adopted as $C(0)$ because the number of available radon monitor was insufficient.

Example of simulation is shown in Fig. 4, in which measured radon activity is given by closed circles and calculated radon activity is shown by bold line. As recognized from Fig. 4, variation pattern of radon activity can be well reproduced by the model, when 0.01 Bq m^{-2} s^{-1} is adopted for an apparent exhalation rate of radon. In all cases, temporal variation of radon activity could be reproduced by this model using Ea value of 0.008~0.012 Bq m^{-2} s^{-1}. According to the model, highest radon activity expectable in Ogoya tunnel is 5000 Bq m^{-3} when tunnel is completely closed for long time[3].

Radon activity in the tunnel may be arbitrarily regulated by controlling wind velocity using a large fan. If wind velocity is kept to be 1 m s^{-1}, which is equivalent to 6 h^{-1} of air ventilation, radon activity in the tunnel becomes < 10 Bq m^{-3}, which is favorable for the low level radioactivity measurement.

3. Monitoring project of radon in the out-flow water from a pit of Ogoya Copper Mine

From September of 1996, a seismometer was set at 340 m point in Ogoya tunnel by Research Center for Seismology and Volcanology, Nagoya University for the analysis of seismic wave caused by the earthquake occurred at far from Japan. Furthermore, monitoring of earth-current has been started as one of the most promising methods for the prediction of earthquakes by the joint group of the Institute of Physical and Chemical Research (RIKEN), Tokai University and Toyama University. Measurement of atmospheric electric wave has also been started near the entrance of Ogoya tunnel by the group of Nagoya Institute of Technology.

Because various kinds of geophysical observations related to earthquake are made at the same area, Ogoya is considered to be one of the best radon monitoring point to investigate premonitory symptoms of an earthquake. We decided to promote radon monitoring project at Ogoya Copper Mine.

Through the courtesy of " Hokuriku Kozan" mining company, it becomes possible to set a radon monitor at one of the pit shown in Fig. 5. The specially designed radon monitor equipped with PIN diode detector will be inserted directly in the out-flowing water from a pit. Unfortunately manufacturing of radon detector is not in time, only a schematic diagram of radon monitoring system is shown in Fig. 6. The signal from radon monitor is send through 1200 m coaxial cable to a 4k channel MCA set in underground laboratory. Integral alpha-ray count during every 30 or 60 minutes is recorded in a MCA by multi-scaler mode. The MCA is connected to a computer in Ogoya underground laboratory so that the data on the MCA can be monitored at LLRL through a modem device via telephone line. The radon monitoring will be started from April of 1998.

4. Observation of anomalous radon activity related to the earthquake occurred in September 10, 1996 at Komatsu area, Japan

Correlation between earthquake and radon anomaly has long been discussed by many authors[5-16]. In order to predict " Tokai Earthquake" considered to be occurring within a few tens of years, intensive radon monitorings have been conducted by Laboratory for Earthquake Chemistry, University of Tokyo along the Pacific coast of Honshu Island since mid 1970s[17]. In Japan, only a few radon anomaly were explained to be precursor of earthquake such as " Izu Oshima Earthquake" in January of 1978[7] and " Hanshin Earthquake" in January of 1995[15,16]. In the case of " Hanshin Earthquake" , concentration of radon in well water began to increase several months prior to the earthquake and reached its maximum of 10 times higher than normal level at one week before the earthquake[15]. Then radon activity once decreased to normal value. The earthquake occurred just after the radon activity tended to increase again.

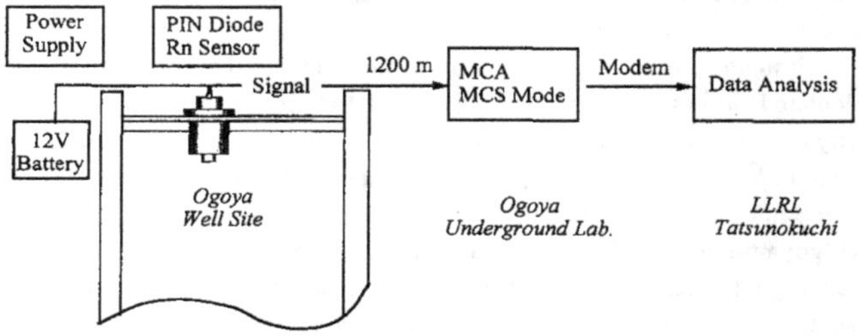

Fig. 6 Schematic diagram of radon monitoring system at Ogoya.

Increase of atmospheric radon was observed several months before Hanshin Earthquake by the group of Kobe Pharmaceutical University[16].

At our laboratoy, concentration of radon in the surface air has been measured since November of 1991 using electrostatic radon monitor developed by Iida of Nagoya University[18]. The air sucked at 1 meter above the ground was passed through a 0.45 μm Nucleopore filter to remove radon daughters and then through a desiccant box packed with P_2O_5 to remove the moisture. Alpha rays from radon daughters collected on a negatively charged collector were measured by a ZnS scintillation detector.

Seasonal variation of radon activity at Tatsunokuchi was found to be higher in summer-autumn season and lower in winter-spring season. Such a variation pattern is commonly observed at the coastal area of the Sea of Japan but opposite to the variation pattern observed in the area facing to the Pacific Ocean.

A small scale earthquake (M = 3.2) occurred at Komatsu area on 7:29 a.m. of September 10, 1996. According the data issued from Disaster Prevention Research Institute, Kyoto University, hypocenter of this earthquake was only about 9 km of distance from LLRL and 4.5 km in depth.

Figure 7 shows airborne radon activity during August to October in 1996 measured at LLRL in Tatsunokuchi. As known from this figure, in August 31 radon activity had suddenly increased upto more than 60 Bq m^{-3}, which is 3 times higher than the maximum level of normal period. Such an abnormal value had not been observed during 6 years of measurement. The high radon activity continued until September 13 and then decreased to the normal level. Peak concentration of radon is 3 to 4 times higher than the peak value of normal period, however, diurnal variation of radon activity, i.e. peaking in the morning, rapid decreasing at sun-rise, reaching to minimum after the noon and then increasing monotonically until next morning, was maintained with the large variation of radon activity. We consider that this radon anomaly is closely correlated with the earthquake though such a radon anomaly had not been observed during 6 years of observation even when earthquakes occurred near Komatsu area.

It is well known that radiation level correlates with airborne radon activity, wind velocity and precipitation. Fortunately environmental radiation level and some meteorological parameters such as wind velocity, wind direction, solar radiation and precipitation, are continuously monitored at LLRL by Ishikawa Prefectural Institute of Health Physics. All of the monitoring instruments are set on the roof of LLRL building (about 10 m above the ground). Radiation level, wind velocity and precipitation are also plotted in Fig. 7. Contrary to the expectation, no correlation was found between radiation level and airborne radon activity measured by electrostatic radon monitor. It closely correlated with the inverse of wind velocity even in the period of radon anomaly. This may be explained that layer of high radon activity was very thin and occurred quite near the building, therefore, gamma rays from radon daughters could not be detected by a NaI(Tl) detector set on the roof due to the shielding effect of 3 stories building.

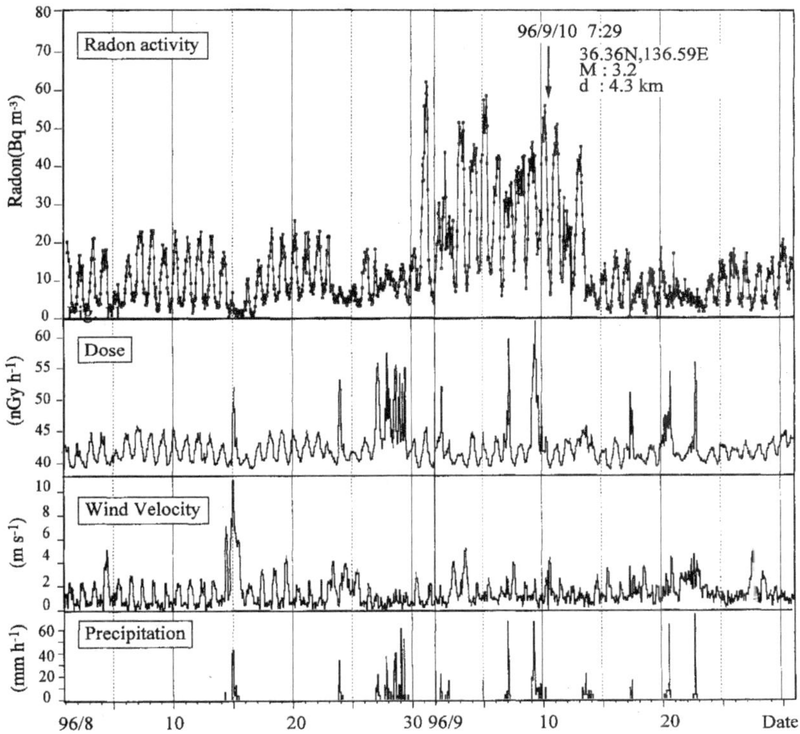

Fig. 7 Radon anomaly observed at Tatsunokuchi during Aug. 31 to Sep. 12. Time of earthquake is shown by an arrow. Radiation dose, wind velocity and precipitation are shown for comparison.

5. Conclusions

(1) Concentration of radon in 546 m tunnel air of former Ogoya Copper mine was measured to be varying from several Bq m^{-3} to over 100 Bq m^{-3}. Most relevant factor, which controlled radon activity in the tunnel, was found to be the wind velocity of air inflowing into the tunnel. The temporal variation of radon could be reproduced by a simple cylindrical model proposed by Miki and Ikeya using apparent exhalation rate of radon from tunnel wall being 0.008~0.012 Bq $m^{-2} s^{-1}$.

(2) Radon monitoring project was proposed for the water from a mining pit of Ogoya Copper Mine, where geophysical data such as earth current and seismic wave measured simultaneously with radon activity are available.

(3) Radon anomaly observed at LLRL was presented as one of premonitory symptoms of earthquake occurred on September 10, 1996.

References

[1] K. Komura, Importance of low level radioactivity measurement, Jpn. J. Med. Phys., 48 (4), 34-45 (1996).

[2] T. Miki and M. Ikeya, Atmospheric radioactivity in calcite caves at Akiyoshi Karst Plateau, J. Atom. Energy Soc. of Japan, 21 (4), 341-343 (1979) (in Japanese).

[3] A. Toguchi, Master's thesis of Kanazawa University (1992) (in Japanese).

[4] M. Okabe, Some forerunning phenomena of earthquake considered to be related with radon and exoelectron, " Environmental Radon" , Proc. '91 Radon Symposium held at Kumatori Aug. 8-10, 1991, ed. M. Shimo and T. Tsujimoto, pp. 74-85 (1992) (in Japanese).

[5] V. I. Ulmov and B. Z. Mavashev, The Tashkent earthquake of 26 April 1996, Akad Nauk Uzbek SSR FAN, 188 (1971) (in Russian).

[6] G. Igarashi and H. Wakita, Groundwater radon anomalies associated with earthquakes, Tectonophysics, 180, 237-254 (1990).

[7] H. Wakita, Y. Nakamura, K. Notsu, M. Noguchi and T. Asada, Radon anomaly : A possible precursor of the 1978 Izu-Oshima-Kinkai earthquake, Science, 207, 882-883 (1980).

[8] H. Amano, K. Kobayashi, H. Hikawa and K. Naruto, Continuous measurement of Rn in tunnel air for the prediction of earthquakes, " Environmental Radon" , Proc. '91 Radon Symposium held at Kumatori Aug. 8-10, 1991, ed. M. Shimo and T. Tsujimoto, pp. 42-52 (1992) (in Japanese).

[9] S. Tasaka and Y. Sasaki, Observation of radon family at Kamioka underground station, " Environmental Radon" , Proc. '91 Radon Symposium held at Kumatori Aug. 8-10, 1991, ed. M. Shimo and T. Tsujimoto, pp. 53-63 (1992) (in Japanese).

[10] H. Ui and I. Fujii, Radon discharge from shear zone of active faults and its relation to seismic activity. " Environmental Radon" , Proc. '91 Radon Symposium held at Kumatori Aug. 8-10, 1991, ed. M. Shimo and T. Tsujimoto, pp. 86-95 (1992) (in Japanese).

[11] H. Hikawa, K. Naruto, T. Mori, K. Kobayashi and A. Amano, The variation of radon concentrtion in the tunnel air of Matgsushiro and the relation to the nearby seismic activity, " Environmental Radon" , Proc. '91 Radon Symposium held at Kumatori Aug. 8-10, 1991, ed. M. Shimo and T. Tsujimoto, pp. 96-103 (1992) (in Japanese).

[12] C.-Y, King, W. Zhang and B.-S. King, Radon anomalies on three kinds of faults in California, Pageoph, 141 (1), 111-124 (1993).

[13] S. Danali-Cotsaki et al, ^{222}Rn in Greek spa waters : Correlation with rainfall and seismic activities, Health Phys., 64 (6), 605-612 (1993).

[14] H. S. Virk and B. Singh, Radon recording of Uttarkashi earthquake, Geophys. Res. Lett., 21 (8), 737-740 (1994).

[15] G. Igarashi, T. Saeki, N. Takahata, Y. Sano, K. Sumikawa, S. Tasaka, Y. Sasaki, and M. Takahashi, Groundwater radon anomaly before the Kobe earthquake in Japan, Science, 269, 60-61 (1995).

[16] Y. Yasuoka and M. Shinogi, Anomaly in atmospheric radon concentration : A possible precursor of the 1995 Kobe, Japanm earthquake, Health Phys., 72, 759-761 (1997).

[17] M. Noguchi and H. Wakita, A method for continuous measurement of radon in groundwater for earthquake prediction, J. Geophys. Res., 82, 1353-1357 (1977).

(18] T. Iida, An electrostatic radon monitor for continuous measurement of environmental radon. " Atmospheric radon families and environmental radioactivity" , ed. S. Okabe, Atom. Energy Soc. Japan, pp. 65-73 (1985) (in Japanese).

Variation in radon exhalation from the ground on the active fault in Kobe.

Y. Yasuoka and M. Shinogi

Kobe Pharmaceutical University
4-19-1, Motoyamakita-machi, Higashinada-ku, Kobe 658, JAPAN

ABSTRACT

Since 27 January 1997,the measurements of radon (Rn-222) exhaled from the ground have been made continuously by the use of PICO-RAD detector (Packard instrument Co.) at monitoring stations on Ashiya active fault. The fault may have been slipped by the Kobe earthquake (magnitude 7.2, 17 January 1995). The variation of relative radon exhalation on the fault was large. We guessed the large variation of relative radon exhalation on the fault was caused by not only the influence of meteorology but also the influence of other factors.

1.INTRODUCTION

The Kobe earthquake (magnitude 7.2, depth 14km) occurred beneath the sea between Kobe and Awaji Island on 17 January 1995.

We have reported already [1,2] that the remarkable increases in the atmospheric radon concentration were observed about 2 months before the Kobe earthquake (Fig.1).

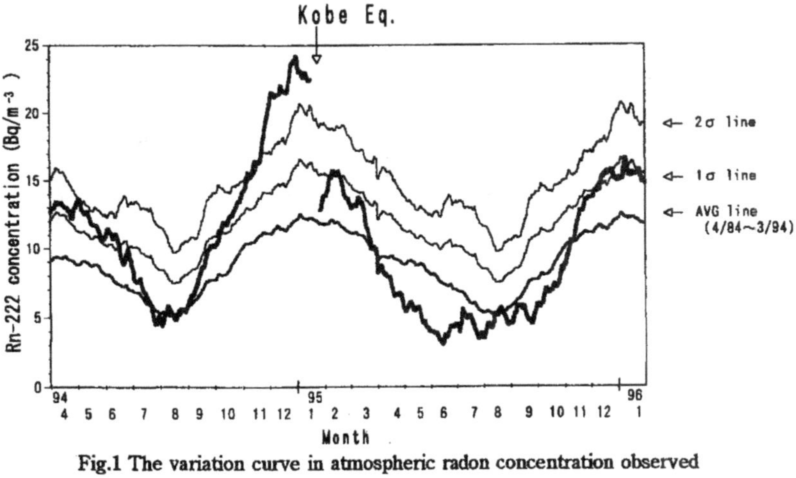

Fig.1 The variation curve in atmospheric radon concentration observed between April 1994 and January 1996 (wide solid line). The curve was obtained through the exponential smoothing method. Using the atmospheric radon concentration observed between April 1984 and March 1994,the mean radon concentration, mean $+1\sigma$ variation range and mean $+2\sigma$ variation range were created. The smoothed mean radon concentration (AVG line), mean $+1\sigma$ variation range (1σ line) and mean $+2\sigma$ variation range (2σ line) are also shown for comparison.

The monitoring station (C site) of atmospheric radon concentration is located near Ashiya active fault and about 20km away from the epicenter (Fig.2). The fault may have been slipped by the Kobe earthquake. According to Katase [3] , the increase in the atmospheric radon concentration is associated with the larger exhalation of radon from the fault.

This area is located on granite bedrock. A granite is enriched in uranium and thorium, and consequently enriched in decay products of the uranium series and thorium series.

Since 27 January 1997, the measurements of radon exhaled from the ground have been made continuously by the use of PICO–RAD detector at monitoring stations shown in Fig.2.

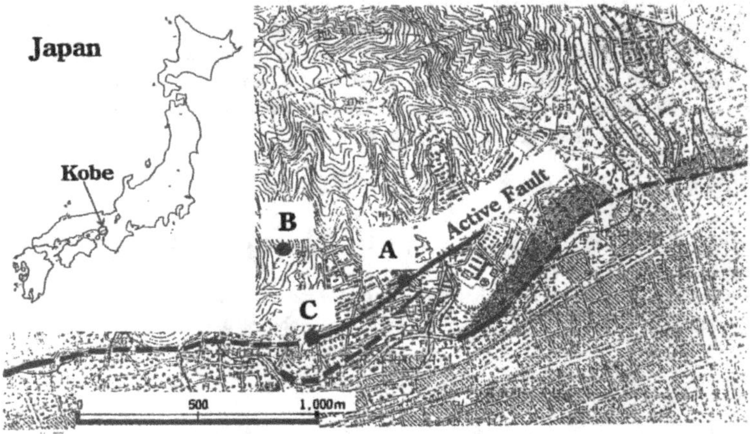

Fig.2 Location of the radon monitoring stations
A site: Monitoring station of relative radon exhalation on the active fault
B site: Monitoring station of relative radon exhalation deviated from the active fault
C site: Monitoring station of atmospheric radon concentration near the active fault

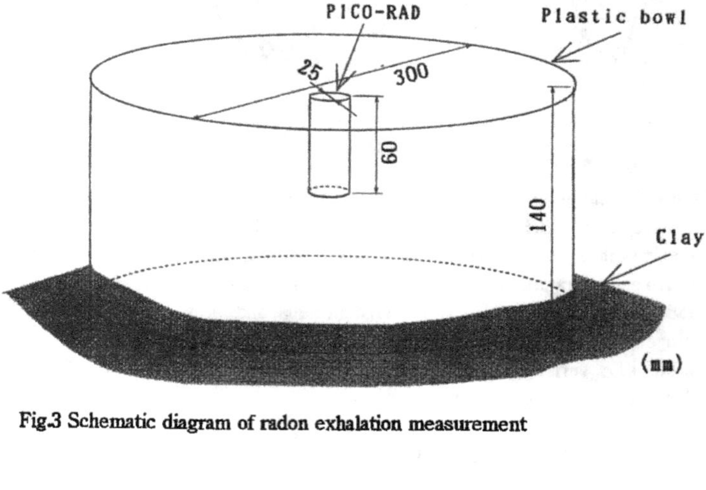

Fig.3 Schematic diagram of radon exhalation measurement

2.MATERIALS AND METHODS

PICO–RAD detector attached to a bottom of bowl (ϕ300 ×150mm) was exposed to air above the ground for 24 hours (Fig.3). The radioactivities of radon absorbed in fine activated carbon particles in PICO–RAD detector were measured with a liquid scintillation spectrometer. The measurements of radon exhaled from the ground have been made continuously by the use of PICO–RAD detector at 2 monitoring stations. One monitoring station is located on Ashiya active fault. Another monitoring station, which is control station, is located about 500 m away from the fault.

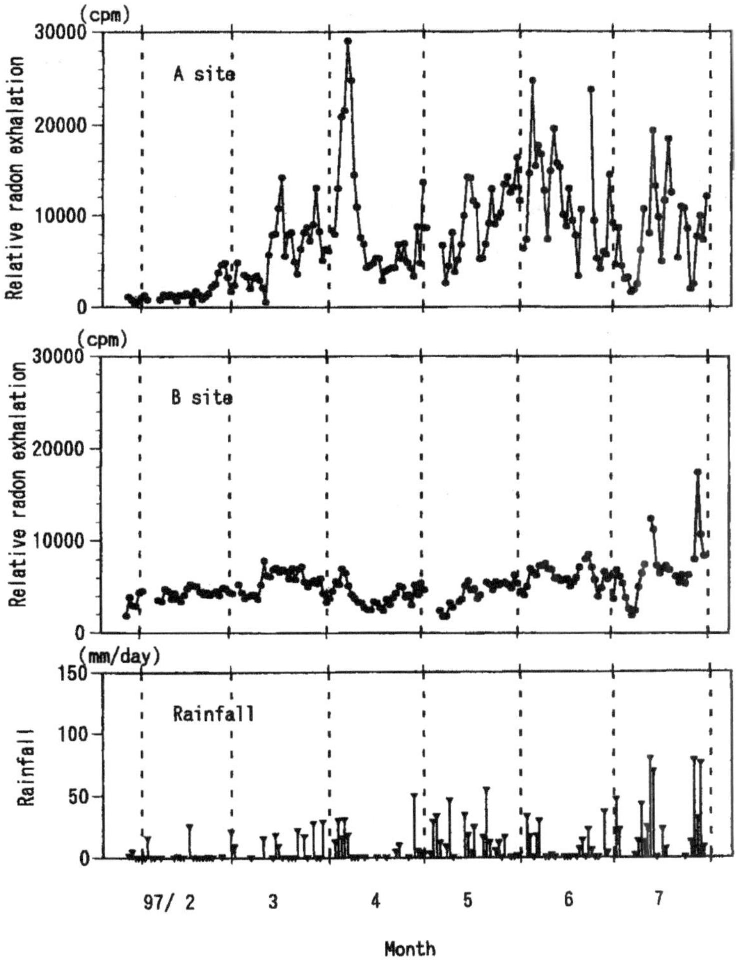

Fig.4 The variations of relative radon exhalation at A site and B site and the rainfall between 27 January 1997 and 31 July 1997.

3.RESULTS

Fig.4 shows the variations of relative radon exhalation at A site and B site and the rainfall between 27 January 1997 and 31 July 1997.

The atmospheric radon concentration shows the seasonal variation (Fig.1). The seasons which show the highest and the lowest concentration are winter and summer, respectively. While, in the variations of relative radon exhalation, up to this time, the remarkable seasonal variation is not recognized (Fig.4).

The variations of relative radon exhalation at A site were larger than them at B site. The variations of radon exhalation were related with the meteorology, especially rain and atmospheric pressure. We assumed that the increases of relative radon exhalation at B site in July were caused by the influence of rain. We guessed the large variation of relative radon exhalation at A site was caused by not only the influence of meteorology but also the influence of other factors.

4.REFERENCES

[1] Yasuoka, Y., Shinogi, M., The variation of atmospheric ^{222}Rn concentration in Kobe., Radioisotopes 43, 688–694, 1994.

[2] Yasuoka, Y., Shinogi, M., Anomaly in atmospheric radon concentration: A possible precursor of the 1995 Kobe, Japan, earthquake., Health Phys. 72, 759–761, 1997.

[3]Katase, A ., Prediction of the earthquake from the change of radon concentration in air., Mem. Coll. Univ. Towa 22,189–196,1996. (in Japanese)

Has radon emission any relation to the formation of atmospheric preseismic phenomena?

Shigeru Okabe and Tsuguo Nishikawa*

Radon Science Research Laboratory, Wakamatsudai, Sakai,Osaka
*Faculty of Engineering, Fukui University, Bunkyo, Fukui

Abstract

Mukuhira arc and earthquake cloud which are considered by some people in Japan, as the phenomena preceding earthquakes. Reliability of these phenomena are studied in relation to radon emission.

1. Introduction

The relation between the earthquake and the anomalous variation of radon concentration in the ground water or the atmosphere is rather well known. However, this relation was found usually after earthquake. Foreseeing of the earthquake with radon variation has been successful only in a few cases until now.

Mukuhira arc and earthquake cloud are known, by some people, in Japan, as the atmospheric phenomena preceding earthquake.

But the reliability of these phenomena are uncertain, their mechanism being tentatively supposed. We have tried to find the relation between these atmospheric phenomena and radon anomalous emission in order to check the reliability of these phenomena.

2. Mukuhira arc

2 - 1 Mukuhira arc

Mukuhira arc was proposed by K.Mukuhira(1930). It is a kind of halo observed a few days before earthquake in specific directions at Miyazu and Tanabe, in Japan. It was said that he had succeeded in the prediction of many famous earthquakes. However, most people have a doubt on his prediction nowadays.

Fujiwhara [1], Miyamoto [2] and Miki [3] had an interest in this problem. Efforts of Miki (1970) to record Mukuhira arc with an automatic camera for one year turned out to be a failure. Mukuhira died in 1992.

2 - 2 Earthquake cloud

Earthquake cloud is believed to appear preceding the earthquake one or two days before. The typical shape of the cloud is a long band lying over the blue sky as shown in Fig.2. It is believed that the earthquake will happen in the extended direction of the cloud or perpendicular to it. The cloud keeps its shape, staying at the same position for a few hours.

It was reported in Japan by Kagita (1966) [4] and Manabe. In China it was studied by Lu (1977) [5]. He has made earthquake prediction with the support of othe related phenomena, measuring it by laser-stress gauge and earth-electric current meter, etc., around the

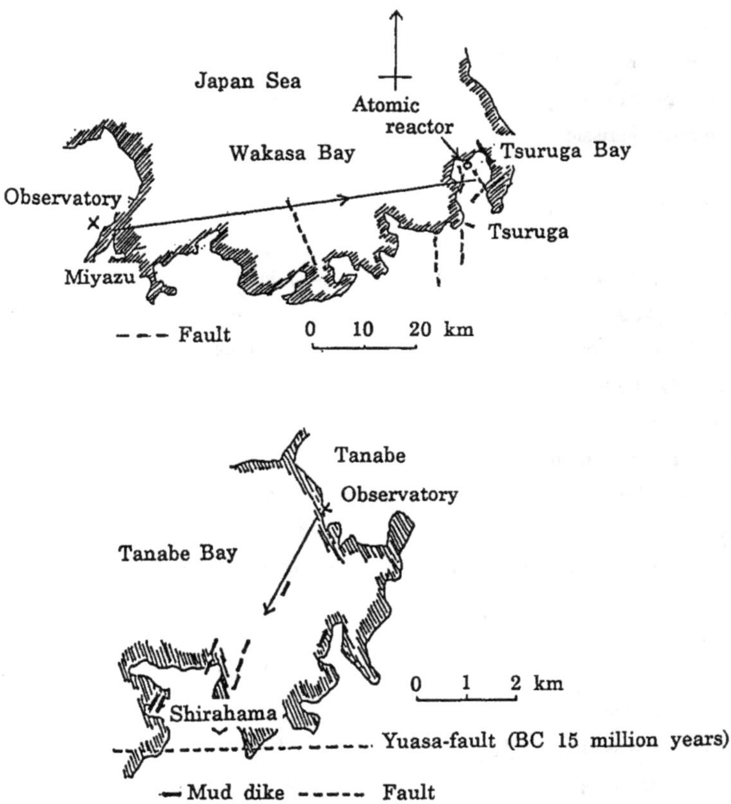

Figure 1 Map of Miyazu and Tanabe, directions, in which Mukuhira arc belived to be appeared are shown with allows. Faults and mud dikes are also shown.

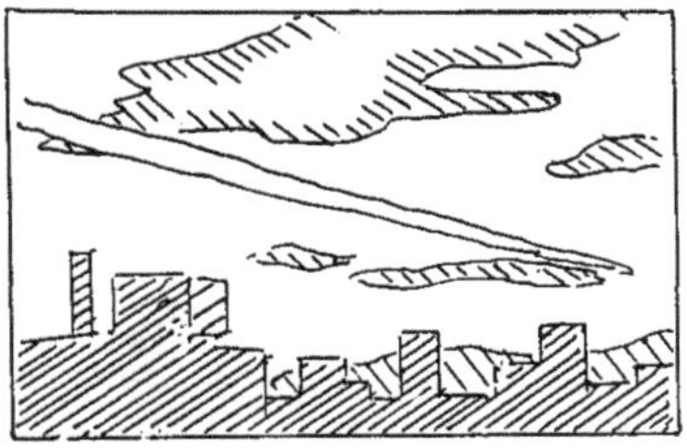

Cloud at 12, Jan., 1978
Earthquake at 14, Jan.,
300 km, M 7.0 Japan

Cloud at 24,Aug., 1979
Earthquake at 26, Aug.,
160 km, M 5.3 China

Figure 2 Sketches of some typical earthquake clouds (by photographs of Kagita,C and Lu Da-Jiong)

same place. Appearances of the earthquake clouds were the same in both cases, but assumed causes of their generation were quite different. Lu took automatically about 10^4 photographs in the morning and evening every day, for one year (1980) and got only a few photographs of possible earthquake cloud. Kagita died in 1995.

3. Investigation of the relation between radon and atmospheric phenomena preceding earthquakes

We have made a report in 1991 on the atmospheric preseismic phenomena [6] and now re-studied it with new results on the assumption that the appearance of Mukuhira arc and earthquake cloud have some relation with anomalous radon emission from a special sensitive place affected by press on earth 's crust. In such a place, many small cracks would be generated in the rock, and radon, electrons and charged particles would be emitted in ground water or porous spaces. It might be a kind of dilatancy and internal microfracto-emission.

3 - 1 Mukuhira arc

Mukuhira arc is expected to appear at a special place, in a specified direction and height.

Recently, Yamasaki et al. [7] have found a place with anomalously high radon emission in the atmosphere (10^5 times of normal case) near the expected place of Mukuhira arc at Miyazu Tsuruga peninsula line. Moreover at this point, observed variation of atmospheric radon content was large in midday and small in midnight. This tendency was completely contrary to the normal variation of the other place and it means intense blow up of radon from the ground. At Tanabe, in the specified direction that connects Tanabe and Shirahama, we have found a big old fault and increased radiation on the fault line. The fault and dike are shown in Fig.1. Furthermore, Mukuhira said that the appearance of the arc was disturbed by the experimental nuclear explosion and the operation of the atomic power plant near the expected place.

From these results, radon and/or radon doughters combined aerosol could be considered to have a relation with Mukuhira arc. However, the time correlation of each phenomena was not confirmed by observation. One of the authors, Okabe, has got three post cards from Mukuhira which predicted earthquake, but the time reliability was not enough.

3 - 2 Earthquake cloud

Kagita proposed earthquake cloud as a foreseeing phenomenon, and Lu Da-Jiong also studied it at about the same time in China. On the other hand, earthquake fog and mist were discussed by Tributsch [8] and Hoenig [9] as phenomena preceding earthquakes.

Kagita and Lu reported many successful cases of earthquake prediction by clouds. The proposed mechanism to produce earthquake cloud is a kind of electromagnetic induction, in the case of Kagita, and charged aerosol arising from the well or the fault in the case of Lu and Tributsch. Earthquake cloud formation by aerosol seems somewhat reasonable.

Recently, we have checked many colored photographs and found that some of them are lying higher than middle layer clouds, which are running in directions different to upper earthquake clouds. This means that the middle layer wind cuts the way of aerosol rising toward earthquake clouds. Many reports and photographs related on earthquake clouds of 17, Jan. 1995, Kobe earthquake (M 7.2) were also checked, but we could not found preseismic one, clearly. Estimation of the direction of earthquake generation with cloud direction is also incredible. Kagita's earthquake prediction seemed as a mixture of his intuition and the appearance of cloud. In 7, Aug. 1991, Okabe have got a telephone from Kagita which predict an earthquake but the prediction was failed.

4. Conclusion

The relation between anomalous radon emission and Mukuhira arc seems to be reasonable to some extent, because Mukuhira arc were observed right in the sky over the fault and/or anomalous radon emission places, but the existence of the arc itself is still problematic.

The existence of earthquake cloud is also doubtful though many colored photographs were shown. Does this cloud appear only before the earthquake? Most reliable preseismic phenomena related radon might be an earthquake fog which Tributsch suggested.

At the present stage, we have not been able to find any consistent mechanism leading to these preseismic atmospheric phenomena, especially with respect to correlation with time and place.

Recently (1997) Japanese earthquake prediction authority committee announced that the earthquake prediction is difficult nowadays, and it will also be impossible in the near future.

5. References

[1] Fujiwhara,S., On the so-called Mukuhira's arc as the foreshadow of an earthquake, Bull. Earthq. Res. Inst., Univ. of Tokyo,15, 706, 1937.

[2] Miyamoto,S., On Mukuhira arc, Jishin (地震). 7, 136, 1954, in Japanese.

[3] Miki,H., Mukuhira arc (1,2,3), Private document, 1971-1976, in Japanese.

[4] Kagita, C., This is the Earthquake cloud, Chunichi Shinbun, 1980, in Japanese.

[5] Lu Da-Jiong (吕大炯) ,Earthquake signs due to cloud and mist (震兆雲霞) 学林出版社 (1984), in Chinese.

[6] Okabe,S., Some forerunning phenomena of earthquake considered to be related with radon and exoelectron, Proc. '91 Radon Symp., Kumatori,. Kyoto Uni., 74. 1992, in Japanese.

[7] Yamasaki,K., et al., A mysterious spot on the out door concentration of radon and thoron, Environment International, 22,Suppl., 1,S205, 1996.

[8] Tributsch,H., Do aerosol anomalies precede earthquake?, Nature,276,606,1978.

[9] Hoenig,S.,A., Aerosol anomalies preceding earthquake, Nature, 279, 169, 1979.

Radon measurements in association with earthquakes

C. Papastefanou, M. Manolopoulou, S. Stoulos, A. Ioannidou and E. Gerasopoulos

Aristotle University of Thessaloniki, Nuclear Physics Department
Thessaloniki 54006, Greece

Abstract

A network of three radon stations has been established in the Langadas basin, North Greece. Newly made devices with plastic tubes are in operation with solid state nuclear detectors (SSNTDs) in registering α-particles from radon and radon decay products exhaled from the ground every two weeks, starting from December 1996, by using LR-115 type II non strippable Kodak films. Simultaneous measurements are made by using Lucas α-scintillation cells for instantaneous measurements of radon in soil gas before and after setting the SSNTDs at the radon stations. The new devices used have the advantage of no use of heating systems and no need of electrical power in the nearby area of the stations. Radon flux registrations ranged between 507 and 85880 tr cm^{-2} or 1.5 and 255.6 tr $cm^{-2}h^{-1}$ in the period of measurements, while radon concentrations in soil gas ranged between 528 and 35095 Bq m^{-3} at the same time.

1. Introduction

There have been many reports dealing with the measurements of radon concentration in soil gas emanating from the ground along active faults which may provide useful signals before seismic events [1-7]. Anomalous changes in subsurface radon concentrations may be expected prior to large earthquakes according to the dilatancy-diffusion model for earthquake occurrence [8]. Several investigators currently monitor radon by using a variety of techniques such as charcoal trap technique [9], nuclear track-etch technique [10] and the aerosol filtration technique [11].

Previous radon measurements in Greece in association with earthquakes were carried out at the Stivos fault in the Langadas basin, North Greece, in the period November 1982 - June 1985 [12] and at the Thessalian fault, Central Greece, in the period August 1992 - April 1994 [13]. At the Stivos fault in the Langadas basin, large earthquake events have been registered since the beginning of this century (M_L=6.5 on 5 July 1902; M_L=7.0 and 7.5 on 4 April 1904; M_L=6.5 on 8 October 1905 and M_L=7.5 on 8 November 1905 [14]).

This paper reports data obtained from field measurements over one year period, starting from December 1996, at the Langadas basin by continuous monitoring of radon emanating from the ground by using solid state nuclear track detectors (SSNTDs) as probes (integrated measurements) on one hand, and on the other hand by using Lucas α-scintillation cells for instantaneous measurements in soil gas before and after the SSNTD records in order to explore the possible relation between radon changes and earthquakes.

2. Instrumentation and Siting

The radon monitors that have been developed in our laboratory for integrated measurements were based on the registration of α-particle tracks on cellulose nitrate films. The α-particles originated mainly from the decay of radon and radon decay products and less from thoron and thoron decay products. It has been shown experimentally that a very small number of tracks of thoron α-particles (less than radon α-particles by a factor of 1000) were registered [15]. Hence, appropriate corrections have been made in our measurements. Care had been taken to prevent humidity and temperature problems on the α-particle track registration by insulation of the instrument vault which eliminates all of the diurnal variations of the above mentioned meteorological parameters. The detectors (cellulose nitrate films, LR-115 type II non strippable films) were exposed for two weeks. The retrieved films were chemically etched (NaOH 2.5 N at 60° for 2.5 h) and then the etch pits (tracks) were counted under a microscope. The radon concentration was expressed in terms of tracks per cm^2 per hour (radon flux).

Three radon monitoring stations with two devices each were installed in boreholes 1 m deep in the Langadas basin along active faults at the sites of Sholari, Stivos and Gerakarou (Fig. 1). The station at Sholari had three devices, as one of them was needed to be pretty close to a drilling for water irrigation. Each device was consisted of a plastic tube, 44 mm inner diameter, 50 mm outer diameter and 300 mm in length, and the detectors were held appropriately on the top of the tube. The sites were preselected by seismologists of the Laboratory of Geophysics, Aristotle University of Thessaloniki.

Instantaneous radon measurements were performed by Lucass α-scintillation cells (PYLON type 110 and 300 A) linked to an appropriate data acquisition system by a TEL detector. These measurements were carried out at two or three different points near the radon stations every two weeks, before and after the SSNTD records.

3. Results and Discussion

In Fig. 2, the radon flux in terms of α-particle tracks per cm^2 per hour is shown for three different radon stations for the period of measurements. Since no significant seismic events which were larger than M_L=3.0 occurred at the Langadas basin during that period, no clear radon peaks were observed in the spectra of Fig. 2. Same spectra were obtained from the data of the instantaneous measurements of radon in soil gas in terms of becqurels per m^3 as shown in Fig. 3. The data of Figs 2 and 3 show the trend of increase or decrease of radon depending on some parameters, such as soil temperature, atmospheric pressure and rain precipitation.

The data obtained from the station at Stivos fault shows significantly high radon records compared with the data obtained from the other stations (Fig. 2 and 3). It must be noted that in 20 June 1978 a ML=6.5 earthquake occurred at Stivos fault (40° 65'N, 23° 30'E, 28km long and 10km deep) and 17 years later (4 May 1995) a M_L=5.8 earthquake occurred at the same fault.

Radon peaks were observed in the previously obtained data shown in Fig. 4. Large seismic (larger than ML=3.0) events followed radon anomalies during an almost 3-year period November 1982 - March 1985 at the Stivos fault.

Radon monitoring at various sites in a seismic area (radon mapping) may be used as a tool in defining the location of faults associated with the seismic activity where large magnitude earthquakes occurred or might occur in the future.

Acknowledgements

This work was sponsored by the General Secretariat of Research and Technology of Greece and by the European Community under the contract ERBIC 15CT 960214.

4. References

[1] King C.Y., Radon emation on San Andreas Fault, Nature, 271, 516 - 519, 1978.

[2] King C. Y., Episodic radon changes in subsurface soil gas along active faults and possible relation to earthquakes, J. Geophys. Res., 85, 3065 - 3078, 1980.

[3] Birchard G. F. and Libby W. F., Soil radon concentration changes preceding and following four magnitude 4.2-4.7 earthquakes on the San Jacinto Fault in southern California, J. Geophys. Res., 85, 3100 - 3106, 1980.

[4] Shapiro M. H., Melvin J. D., Copping N. A., Tombrello T. A. and Whitcomb J. H.: In natural radiation environment III (U. S. Department of Energy, Washington D. C.) pp 137 - 153, 1980.

[5] Shapiro M. H., Melvin J. D. and Tombrello T. A., Automated radon monitoring at hard-rock site in the southern California Transverse Ranges, J. Geophys. Res., 85, 3058 - 3064, 1980.

[6] Shapiro M. H., Melvin J. D., Tombrello T. A., Mendenhall M. H., Larson P. B. and Whitcomb J. H., Relationship of the 1979 southern California radon anomaly to a possible regional strain event, J. Geophys.Res., 86, 1725 - 1730, 1981.

[7] Talwani P., Moore W. S. and Chiang J., Radon anomalies and microearthquakes at Lake Jocassee, South Carolina, J. Geophys. Res., 85, 3079 - 3088, 1980.

[8] Scholz C. H., Sykes L. R. and Aggarwal Y. P., Earthquake prediction: A physical basis, Science, 181, 803 - 810, 1973.

[9] Megumi K. and Mamuro T., A method for measuring radon and thoron exhalation from the ground, J. Geophys. Res., 77, 3052 - 3056, 1972.

[10] Fleischer R. L., Price P. B. and Walker R. M.: in Nuclear Tracks in Solids, Principles and Applications (Univ. California Press, Berkeley) 1975.

[11] Melvin J. D., Shapiro M. H. and Copping N. A.: Nucl. Instrum. Methods, 153, 239 - 251, 1978.

[12] Papastefanou C., Manolopoulou M., Savvides E. and Charalambous S.: Nucl. Geophys. 3, 49 - 56, 1989.

[13] Papastefanou C., Manolopoulou M., Stoulos S. and Ioannidou A.: Nucl. Geophys. 9, 481 - 486, 1995.

[14] Galanopoulos A.: Ann. Geol. Pays Helleniques 30, 647 - 724, 1981.

[15] Savvides E., Manolopoulou M., Papastefanou C. and Charalambous S.: Int. J. Appl. Radiat. Isot. 36, 79 - 81, 1985.

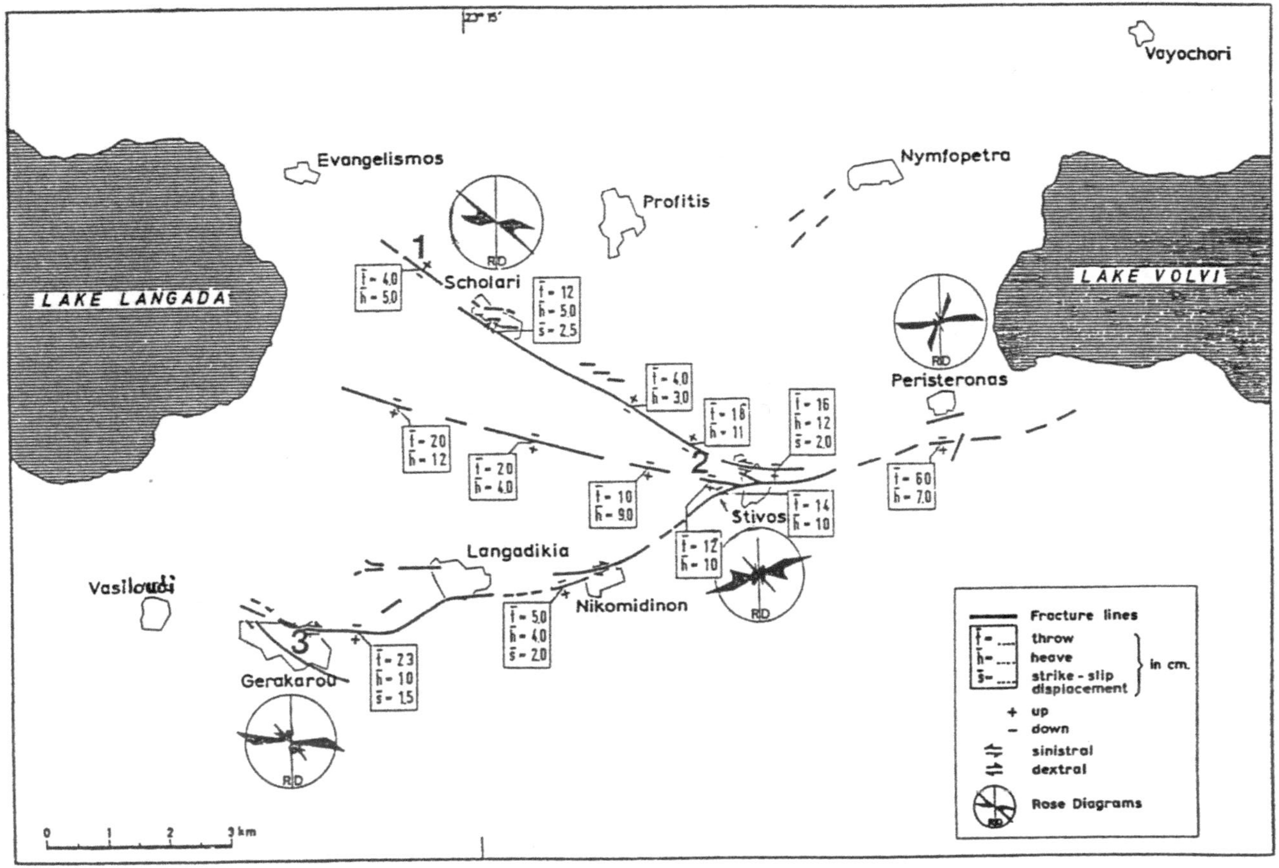

Fig. 1. Fracture lines in the epicentral area of the 20 June 1978 earthquake in the Langadas basin

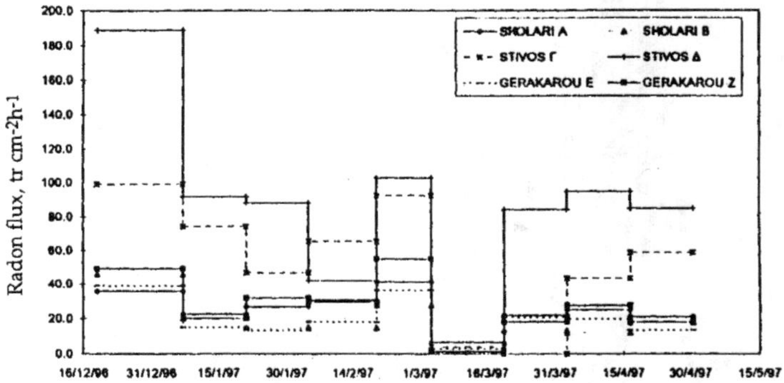

Fig. 2. Radon flux measurements in the Langadas basin

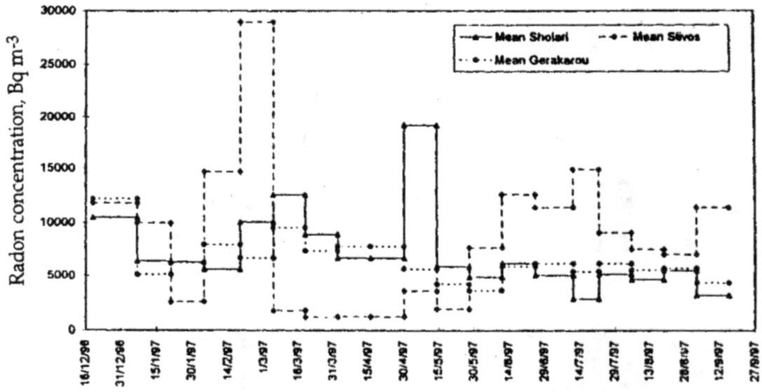

Fig. 3. Radon concentration in soil gas in the Langadas basin

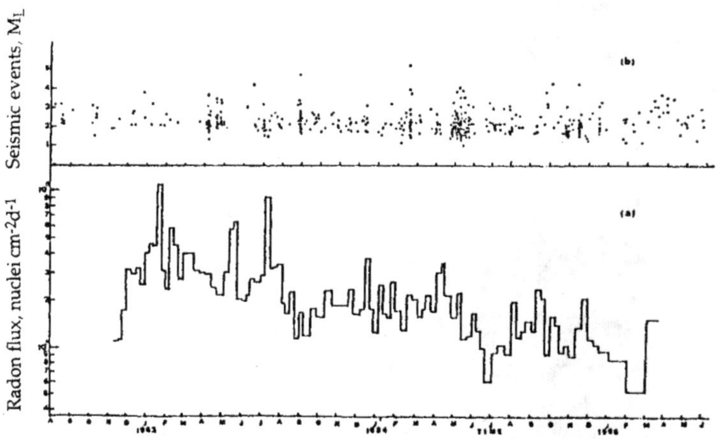

Fig. 4. Radon flux (a) and seismic events (b) at the Stivos fault for the period November 1982 - 1985

EPOS1- a multiparameter measuring system to earthquake prediction research

T. Streil 1), J. Heinicke 2), U. Koch 2),V. Oeser 1) and J. Wiegand 3)

1) SARAD GmbH, Dorfplatz 1, D-01705 Pesterwitz/Dresden, Germany
2) Saxon Academy of Sciences at Leipzig, Research Group Freiberg, Germany
3) University GH Essen, FB 9 - Geology, Essen, Germany

Abstract

The approach to earthquake prediction by geophysical, geochemical and hydrological measurements is a long and winding road. Nevertheless, the results show a progress in that field (e.g. Kobe). This progress is also a result of a new generation of measuring equipment. SARAD has developed a versatile measuring system (EPOS1) based on experiences and recent results from different research groups. It is able to record selected parameters suitable to earthquake prediction research. A micro-computer system handles data exchange, data management and control. It is connected to a modular sensor system. Sensor modules can be selected according to the actual needs at the measuring site.

1. Introduction

Research into earthquake prediction by means of seismological, geophysical and geochemical parameters has continued for more than five decades. Up to now, the use of precursors in earthquake prediction topics is to a large extent empirical, due to the many difficulties that still exist in understanding the physics of earthquakes.
A large number of investigations concerning geochemical and hydrological measurements (e.g. CO_2, Rn, He, H_2, Hg, N_2 or several ionic concentrations as well as fluctuations of the groundwater head, flow rate) and the electrical conductivity has shown possible starting points for a correlation between the tectonic stress built up in the crust and anomalous effects of the observed parameters [1-14].

On this basis, an improved measuring system EPOS1 was developed, which is able to transform the requested scientific tasks in combination with the scientific background into a reliable and applicable observation system. The EPOS1 (EPOS - Earthquake Precursor Observation System) is the innovative realization of a multiparameter station with up to 17 parameters.

2. Methods

The proposed measuring methods are the most promising parameters for the construction of a measuring system which will be chosen on the basis of a large number of international results based on this subject /1-14/ and the results of the SAW group [15,16].
The SAW group started their activities to earthquake prediction research in 1989 with first continuous measurements of the radon concentration in the spring water of the Radon-Quelle in Bad Brambach, Vogtland area - SE-Germany. Besides these data, numerous meteorological as well as hydrological parameters from different springs and wells are registered. Since the end of 1992, these parameters have been completed by the observation of additional parameters [CO2 and Rn concentration of the spring gas, water temperature, gas flow rate and electrical conductivity at the Radon- and Eisen-Quelle (springs)] by means of improved automatic measuring stations, which should be the "prototypes" of the EPOS system. For example, an interesting recorded effect is the significant correlation between CO2 and radon during some days, which is presented in Fig. 1. One important result of these long-term investigation is, that a correlation and anticorrelation between different parameters (geochemical, hydrological and meteorological) exits due to a comprehensive complex of influences, disturbances, links and side effects. Thus a multiparameter station is the basis of a reliable and statistically sufficient recording of environmental parameters, particular related to earthquake prediction research.

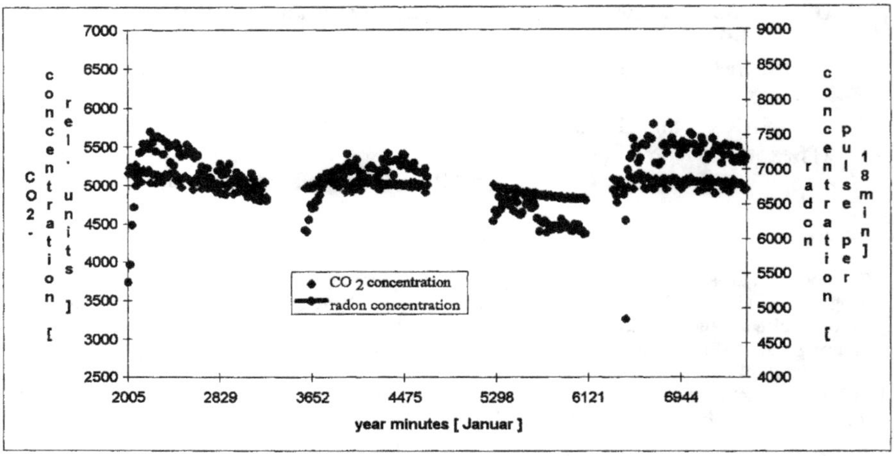

Fig. 1: Example of the correlation between CO_2 and radon in the gas phase at the Radon-Quelle, Bad Brambach

The new system EPOS1, developed by SARAD GmbH, for up to 17 registered parameters has to be installed in regions with active tectonic faults and uprising fluids. These fluids are possible transmitters for information about the deep processes due to an active stress-strain field with impacts to the open crack system. The resulting impacts onto the fluid system depend on the geological, hydrological and geochemical conditions and will be different for each location. Therefore, it is necessary for the observation of precursory phenomena to start with a comprehensive investigation of all suitable parameters at the selected sites. The numbers of sensors could be reduced after a period of basic research (background knowledge). The EPOS1 with a modular sensor system is able to fit the sensor program concerning the individual tasks. The sensors are interfaced by a fast CAN-Bus system. The sensors are grouped and between these groups a distance up to 1 km is possible to connect.
The devices work automatically for recording the data continuously and sending the results to the data storage unit. The data will be processed and interpreted by the responsible institute. The combination of geochemical, geophysical and hydrological measurements includes the following parameters:
Rn-, CO2-, N2-, CH4-concentration in the gas phase,
Eh-value, electrical conductivity, pH-value, Rn-concentration in water,
water temperature, discharge observations, water level fluctuations
the exhausting rate of the spring/well,
self-potential signals, micro-seismicity.
A further set of sensors records air temperature, barometric pressure and rel. humidity. The Fig. 2 shows a schematic overview of this system and Fig. 3 the realized system without the climatic sensor box.

Fig. 4 shows the block diagram of the electronic network which is used for the EPOS1 system. Fast radon gas transfer membranes consisting of a special plastic material (thick walled, mechanical stable, watertight) guarantee diffusion times of lower than 3 min [17]. Using this method for radon measurements in water, time resolutions of 10 min are possible. The radon is measured by a humidity corrected alpha-spectrometric monitor system using charge collection for a high efficiency [18].

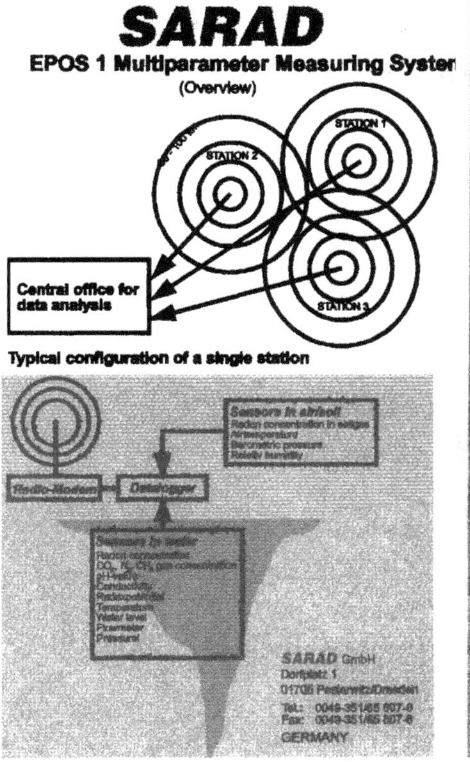

Fig. 2 Schematic view on the EPOS 1 system

Fig. 3 View of the realized system (without climatic box)

3. Field procedure

At a first stage, we want to start the investigations by selecting and organizing the measuring sites. The measuring sites will be selected according to the hydrogeological settings and the chemical and isotopic composition of the fluids. Preliminary geophysical and isotope analyses, which are reliable pathfinders at a measuring site, show where a suitable site at the seismic active area is.

The EPOS1 stations will be set-up with the selected sensor equipment for each site concerning the results of the preliminary investigations. The meteorological parameters will also be recorded simultaneously with this station at each site. The basis for the seismological data acquisition should be the existing local networks.

The second stage is characterized by the long-term field measuring period, the data analyses and the attempt to clear the possible connections between the recorded data and the local seismicity. The important task in this stage is the interpretation of the data series to improve the knowledge of anomaly generation and their possible relation to earthquake precursor phenomena. Advanced statistical methods will be applied to evaluate the observed phenomena.

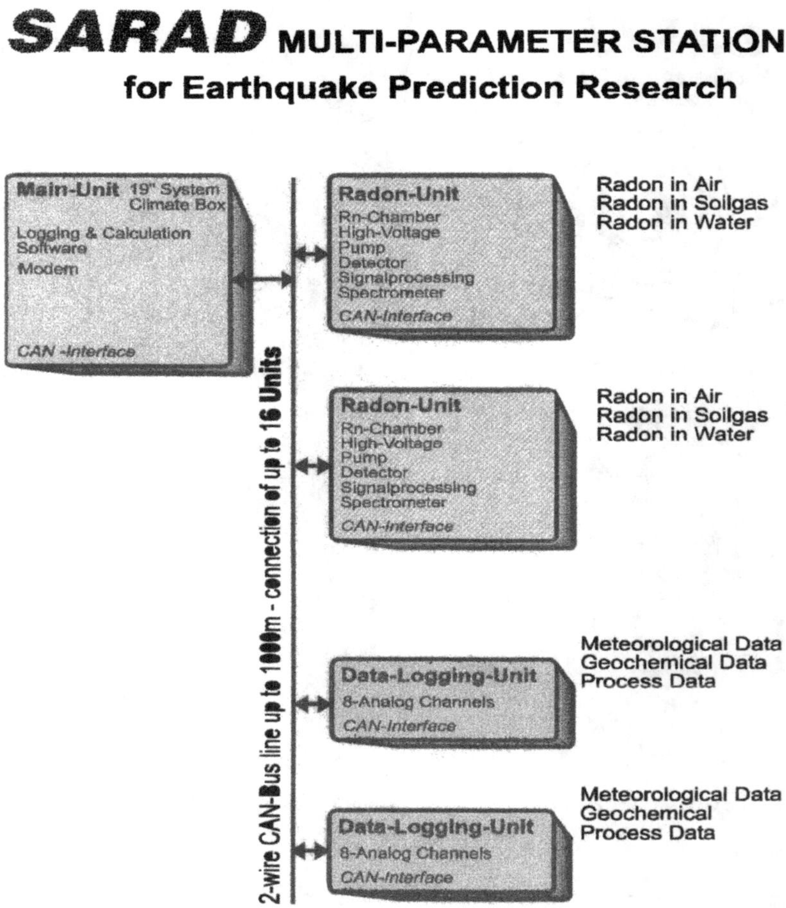

Fig. 4 Block diagram of the electronic network of the EPOS 1 system

4.References

[1] RIKITAKE, T. (ed.) (1982).
Earthquake forecasting and warning. Center of Academic Publications of Japan, Tokyo, 402 pp.

[2] WYSS, M. (1991).
Evaluation of proposed earthquake precursors. M. Wyss, ed., American Geophys. Union, Washington, 94 pp.

[3] LOMNITZ, C. (1994). Fundamentals of earthquake prediction. John Wiley & Sons, Inc., New York, 325pp.

[4] KING, C.-Y. (1990).
Gas geochemistry applied to earthquake prediction: an overview. J. Geophys. Res., 91, 12269-12281.

[5] KISSIN, I.G. and A.O.GRINEVSKY (1990).
Main features of hydrogeodynamic earthquake precursors. Tectonophysics, 178,277-286.

[6] ROELOFFS, E. (1988).
Hydrological precursors to earthquakes: a review. Pure Appl. Geophys., 126,177-209.

[7] ROELOFFS, E. (1994).
Poroelastic techniques in the study of earthquake-related hydrological phenomena. In: Advances in geophysics, Academic Press, San Diego, 37,135-195.

[8] SHI HUIXIN and C.ZUHANG (1986).
Geochemical characteristics of underground fluids in some active fault zones in China. J. Geophys. Res., 91,B12,12282-12290.

[9] TENG, T., and L.-F. SUN (1986).
Research on groundwater radon as a fluid phase precursor to earthquakes. J. Geophys. Res., 91,B12,12305-12313.

[10] THOMAS, D. (1988).
Geochemical precursors to seismic activity. Pure Appl. Geophys., 126, 241-266.

[11] WAKITA, H., Y.NAKAMURA and Y.SANO (1988).
Short-term and intermediate-term geochemical precursors. Pure Appl. Geophys. 126, 267-278.

[12] WU, K.T., M.S.YUE, H.Y.WU et al. (1978).
Foreshocks to the Haicheng Earthquake of 1975, certain characteristics of the Haicheng earthquake (M=7.3) sequence. Chinese Geophys., AGU, 1,289-308.

[13] IGARASHI, G., S. SAEKI, N.TAKAHATA et al. (1995).
Ground-water radon anomaly before the Kobe earthquake in Japan. Science, 269, 60-61.

[14] TSUNOGAI, U. and H.WAKITA (1995).
Precursory chemical changes in ground water: Kobe earthquake, Japan. Science, 269, 61-63.

[15] HEINICKE, J., U.KOCH and G.MARTINELLI (1995).
CO2 and radon measurements in the Vogtland area (Germany) - a contribution to earthquake prediction research. Geophys. Res. Lett., 22,7,771-774.

[16] KOCH, U. and J.HEINICKE (1996).
Earthquake prediction research by radiometric-geochemical methods in the Vogtland-NW Bohemian region: new results and aspects. In: Seismology in Europe - Proc. of the XXV. General Assembly of the ESC, Reykjavik, 9.-14.9. 1996,241-246.

[17] SURBECK, H. (1996) Int. Conf. Technol. Enhanced. Naturally Radioactivity (TENR) Caused by Non- Uranium Mining , Szyrk, Poland, October 16-19,

[18] STREIL, T., R.KLINKE, W. GRUNDKE, H. SURBECK and A.C. JAMES (1996), Int. Conf. Technol. Enhanced. Naturally Radioactivity (TENR) Caused by Non- Uranium Mining , Szyrk, Poland, October 141-146

Vibrations in the urban environment controlling ^{222}Rn migration in soils

Jens Wiegand

University of Essen, Department 9 - Geology, 45141 Essen (Germany)

Abstract

Comparable to investigations looking for a connection of ^{222}Rn and earthquakes, this study shows the influence of subsurface vibrations on the ^{222}Rn concentration of the soil-gas in urban environments. Generally, the ^{222}Rn concentration increases through vibrations induced by trains, street-traffic and activities at project sites. The spatial radius of the ^{222}Rn increase due to vibrations reach highest values at project sites where piled foundations or metal panels are rammed into the ground (> 60 m). Along railway tracks the radius is wider (> 30 m) than along heavy traffic roads (< 25 m). The average increase of ^{222}Rn concentrations in soil-gas due to vibrations is the highest at project sites (53 %). Along heavy traffic roads the increase of ^{222}Rn concentrations by motor vehicle traffic is higher (37 %) than that by railway traffic (11.5 %). The maximum increase of 400 % was observed in a distance of 1 m from a railway track. In the vicinity of railway tracks a difference of the vibration influence according to unconsolidated rock (11.1 %) or solid rock (11.8 %) was not noticed. Beside this vibration effect, the overall ^{222}Rn level decreases with increasing distance to the vibration source, but only at locations laying above solid rocks.

The observation of the increase of ^{222}Rn concentrations can be explained by a "pump effect": the mechanical vibration of soil and mineral particles leads to an upward motion of the whole volume of soil-gas. Therefore, ^{222}Rn is pumped out of the soil to the atmosphere and as a result the upward transport is increased.

1. Introduction

Radon is a radioactive noble-gas. There are three naturally occurring isotopes, the most stable being ^{222}Rn with a half-life of 3.8 days. It is the direct product of ^{226}Ra in the ^{238}U-decay series. Contrary to the other two radon isotopes with very short half-life's (^{219}Rn 4 seconds, ^{220}Rn 56 seconds), ^{222}Rn can move several meters within the geological environment. The two possible mechanisms of ^{222}Rn transport are diffusion and convection [1, 2]. If ^{222}Rn reaches the joint or pore space in the near surface, it can migrate from the ground into the atmosphere (^{222}Rn exhalation).

Similar to the investigations presented in this study, many investigations are conducted for trying to understand the influence of seismic waves on ^{222}Rn concentrations in soil-gas or groundwater in the sense of earthquake prediction. Already in the twenties of our century, Shiratoi [3] discussed a possible correlation of ^{222}Rn and earthquakes. Especially in Northeast Asia and North America many investigations were realised under this topic during the last 30 years [4, 5, 6, 7].

In the beginning of the nineties it was observed at a project site in Cologne (Germany) that the efficiency of soil venting systems is higher when the ground is stimulated by beating down a piled foundation (personal communication Prof. Dr. Kreuser, not published). Therefore, subsurface vibrations should have an effect on ^{222}Rn in soil-gas as well. Wiegand [8, 9] is one of the first who started to study the influence of urban vibrations on the ^{222}Rn potential. This study will describe the influence of different kinds of subsurface vibrations in the urban environment and compile the results. The investigations are conducted within the dense populated area of the Ruhr district in the western part of Germany.

2. Measuring Techniques

The ^{222}Rn concentration of the soil-gas is determined by using soil-gas samples taken from a depth of 0.8 m. The measuring points are placed usually along a profile with continuously increasing distance to the vibrations source. After a probe is rammed 0.8 m into the ground, soil-gas is sampled using a gastight syringe. The sample is transferred into an evacuated Lucas cell and after a waiting period of 3.5 hours it is measured scintillometrically [8, 9, 10]. Lucas cells with a volume of 125 ml and 600 ml are used. The reproducibility (1 σ counting error) of the ^{222}Rn measurements depend on the concentrations and the duration of the measurements. Within the range of the shown ^{222}Rn concentrations the counting error of a sample is better than 1 Bq l^{-1} (2 - 5 %). The overall statistical error is higher (approx. 11 %).

From the same depth the soil-gas was sampled (i. e. 0.8 m), soil was sampled to determine the soil moisture content gravimetrically. Using the same soil sample the uranium concentration was measured by X-ray fluorescence spectroscopy [11].

As a source for subsurface vibrations railway and motor vehicle traffic, as well as activities at project sites are used. Such vibrations produce two types of seismic waves in the ground: body waves and ground rolls. Longitudinal waves (P-wave) and shear waves (S-wave) belong to the body waves, which can go through the interior of the earth [12]. Rayleigh waves and Love waves belong to the ground rolls, which move only near the surface of the earth [13].

For the vibration measurements a block of aluminium is put halfway into the soil with care to ensure a good contact of the surrounding soil with the block. Aluminium has a similar density (2.699 g cm^{-3}) to the soil. Two oscillating receivers are placed on the block to measure the horizontal and vertical part of the vibrations [14]. One oscillating receiver measures longitudinal body P - waves and ground waves, mainly Rayleigh waves. The second receiver measures the vertical components, the body shear waves and the vertical component of Rayleigh waves.

The signals are amplified and recorded by a DAT-recorder. The analyse of the data was done in the Department of Mechanical Techniques in the University of Essen. Probable failure of the vertical oscillating receiver prevented this wave component from being used in our analysis.

3. Different vibration sources

All the investigations presented in this study are conducted within the dense populated area of the Ruhr district in the western part of Germany. The Ruhr district is build up by several large cities (Bochum, Bottrop, Dortmund, Duisburg, Essen, Gelsenkirchen, Recklinghausen) and is probably the largest urban area of Europe. Within the Ruhr district different vibrations sources are used to check the influence of subsurface vibrations on the ^{222}Rn concentrations in soil-gas. The influence is determined near heavy traffic roads, railway tracks and project sites.

3.1 Motor vehicle traffic

In the city of Bottrop the ^{222}Rn concentration of the soil-gas was measured before and during the rush-hour near to a heavily travelled road (Fig. 1) [14]. These measurements were conducted twice in a depth of 0.5 m. The ^{222}Rn concentrations shown in Fig. 1 are the average of two measurements. The soil consists of a fill (medium grained sands). The measuring point to the west is situated 25 m off the street and is located above an unconsolidated bedrock (alluvial sand), which acts as a reference site to see if there are any diurnal influences. As it can been seen, the ^{222}Rn concentrations at this point do not change. Therefore, the change of the ^{222}Rn concentrations at the three other measuring points are due to the traffic vibrations. With the beginning of the rush-hour the ^{222}Rn concentrations increase and reach their maximum during the beginning or with the peak of the rush-hour. The mean increase of the radon concentrations reach a factor of 1.37 at this site (both campaigns, all measurements(n=7) except at 25 m). Because of the

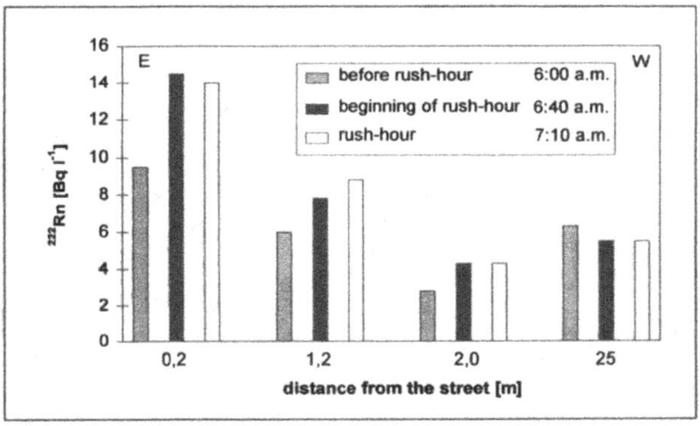

Fig. 1: ^{222}Rn concentrations in the soil-gas at the test site "Forsthaus Specht".

small amount of measurements the percental increase of the ^{222}Rn concentrations (37 %) may vary with a greater amount of measurements. Beside this, Fig. 1 shows that the radius of this influence is smaller than 25 m at this site. The general decrease of the ^{222}Rn concentrations with increasing distance from the street (0.2 m → 2.0 m) is the result of another urban effect on the ^{222}Rn potential: the sealing of soils by the impermeable asphalt. Due to the sealing ^{222}Rn is focused to the rims of the sealing were the concentrations are increased consequently [8, 9, 15]. The radius of this influence on the ^{222}Rn concentration lies within 3 m [8]. It is possible that the sealing effect intensify the vibration effect. This can be suggested from the ^{222}Rn increase just 0.2 m away from the road, which is higher than the increase at the other two measuring points.

3.2 Project sites

At a project site in the centre of Essen metal panels were rammed into the ground, which resulted in very strong vibrations. The bedrock consists of cretaceous sediments, which are overlain by a back-fill. Fig. 2 shows the result of two measuring campaigns. The first measurement was done during the metal panels were rammed into the ground, the second direct after. To see the influence of the vibrations more clearly, the ratio of both ^{222}Rn concentrations was calculated. Because of the high degree of soil sealing in such urban areas, the spacing between the measuring points is irregularly. Fig. 2 shows that 10 m off the vibration source the increase of the ^{222}Rn concentration due to the vibration reach a maximum of 80 %. Gradually, the ratio of the ^{222}Rn concentrations decrease with increasing distance to the vibration source showing the decrease of the influence of the vibrations. Nevertheless, even at a distance of 60 m the influence of the vibrations can been seen. The mean value of the ^{222}Rn increase of all four measuring points lies at 53 %.

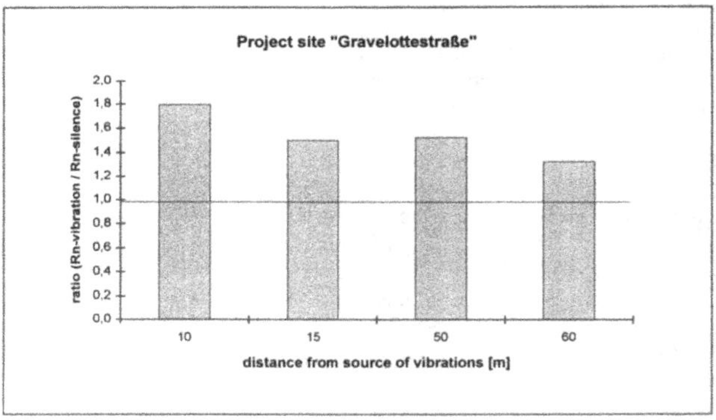

Fig. 2: Ratio of ^{222}Rn concentrations at project site „Gravelottestraße"

3.3 Train

In the cities of Essen, Bottrop and Bochum the influence of trains on the ^{222}Rn concentration in soil-gas was examined in the vicinity of five different railway tracks. At three test sites the measuring profiles are placed above solid rock, and at two sites the profiles are above unconsolidated rock. At all locations, the first few meters of the profiles consists of anthropogenic fill material. The most investigated location is named "Tossens - Büschken" (Fig. 3, Fig. 5), which lies in the East of Essen. The rock consists of Upper Carboniferous sediments, which are overlain by reassorted loess with a thickness of about 1 to 2 m. Up to a distance of 8 m the soil consists of a soil aggradation. The passing trains have an approximate speed of 50 km h^{-1}.

The test site was examined six times. Two times the ^{222}Rn concentrations of only the southernmost point was measured periodically (time interval between two measurements: 1 minute). Fig. 3 shows the increase of the ^{222}Rn concentrations each time the trains are passing, and its decrease after the passage. The increase of the ^{222}Rn concentrations due to the passage of trains reach an average of 15 %. Between the measurements of the 40. and 45. minute, two trains were passing one after the other. This results in a wider peak of the ^{222}Rn anomaly than usual. There is a similar peak around the 5. minute, which is not

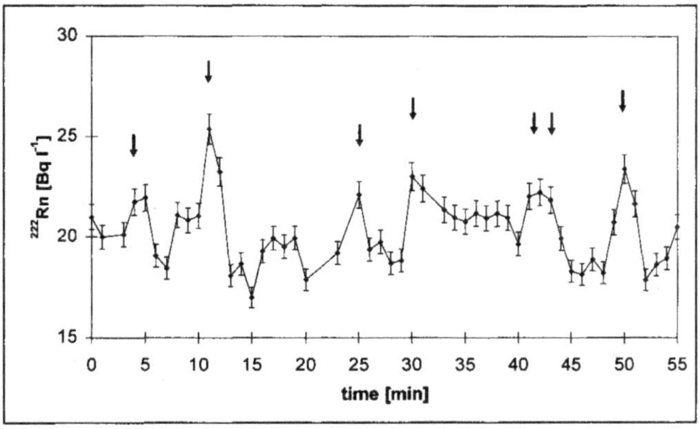

Fig. 3: ^{222}Rn concentrations and 1 σ counting errors in soil-gas (0.8 m depth) in a distance of 5 m from a railway track at the location "Tossens-Büschken". The subsoil consists of a soil aggradation. Arrows are marking the passage of trains.

explainable. Beside this, the slow decline after the "30 minute train" is not consistent with the sharp drop of the other peaks. A second measuring campaign at this location is not showing such a smooth decline.

The correlation of the measured magnitude of the vibrations with the "net effect" of the ^{222}Rn concentrations caused by the vibrations is shown in Fig. 4. The "net effect" is the difference of the ^{222}Rn concentrations between the measurement when a train is passing and the measurement immediately before. This calculation is not possible for the train at the 43. minute and therefore not shown in this diagram. Fig. 4 shows a relatively good correlation between the two parameters. Only the "30 minute train" (lower corner to the right) is not following the trend.

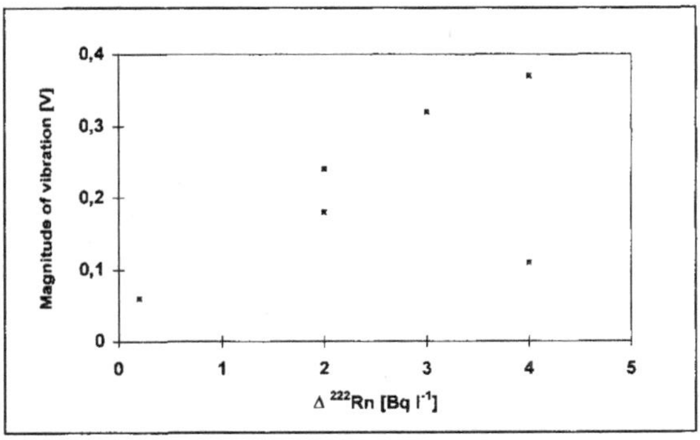

Fig. 4: Correlation between the magnitude of vibration and the difference of the ^{222}Rn concentrations when a train is passing and the measurement of pre passage immediately before.

This short-termed increase of ^{222}Rn concentrations due to the vibrations is also confirmed by the other four measuring campaigns at this site. During these campaigns the ^{222}Rn concentrations were measured along a profile consisting of six measuring points, which show an increasing distance to the railway track (Fig. 5). At each measuring point the ^{222}Rn concentration was measured three times: before, during and immediately after a train passes by. Like in Fig. 3, each time a train is passing the ^{222}Rn concentrations are rising. Beside this "short-term" influence, the radius of the influence of the vibrations could be analysed. The ^{222}Rn concentrations decrease with increasing distance from the railway track, as long as the subsoil is homogeneous. Within the soil aggradation to the south the ^{222}Rn concentrations decrease with increasing distance (only two measuring points), and also in the undisturbed soil to the north (four measuring points).

Such measuring campaigns were conducted at other test sites with a varying distance of the measuring points to the railway tracks, which lies between a minimum of 1 m and a maximum of 30 m. Just 1 m off the railway track the ^{222}Rn concentrations in soil-gas increase dramatically by about 400 % when a train is passing. Even in a distance of 30 m the short-term influence of the vibrations was observed, which shows that the radius of the influence of this kind of vibrations can exceed 30 m.

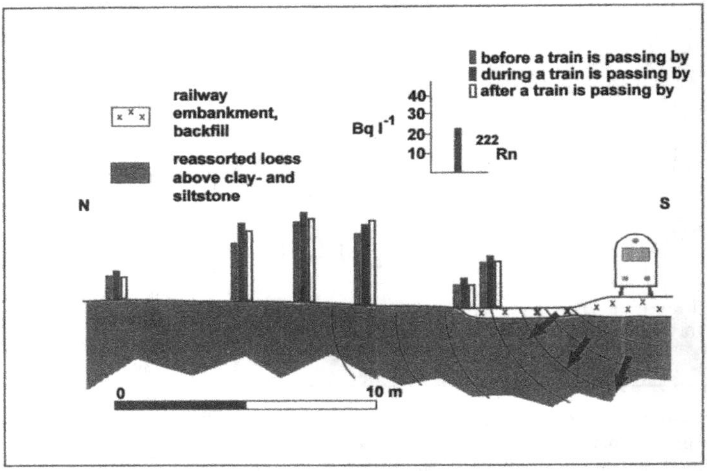

Fig. 5: Spatial and temporal distribution of ^{222}Rn concentrations in soil-gas (0.8 m depth) perpendicular to a railway track at the location "Tossens - Büschken" The measuring point of Fig. 3 is the southernmost point.

Although the "short-term effect" was observed at each location, the decrease of ^{222}Rn concentrations with increasing distance from the vibration source is not very marked, and was only found at locations, which are placed above solid rock. It is not found at locations placed above unconsolidated rock. No difference in respect to the vibrations were observed between natural soil or backfill as long as they are above the same bedrock.

Compiling the "short-term" increase of all measurements along railway tracks (n = 84) the increase of ^{222}Rn concentrations above solid rock show an average of 11.8 %, and the increase above unconsolidated rocks reach nearly the same average of 11.1 %.

At 24 measuring points soil samples were taken to verify if there is any influence of other parameters on the distribution of ^{222}Rn concentrations. Beside the uranium concentration (U_{total}), the content of soil moisture was determined. The uranium concentrations show no correlation to the ^{222}Rn concentrations [14]. The marked influence of the soil moisture content on the ^{222}Rn emanation rate is well known [2, 8, 16]. An increasing content of soil moisture results in an increase of the ^{222}Rn emanation rate, which increase the ^{222}Rn concentration of the soil-gas as well. Only at the location "Tossens - Büschken" a slight correlation between the two parameters was observed [15]. For example, the low level of the ^{222}Rn concentrations within the soil aggradation (Fig. 5) can be explained by the soil moisture content, which is 40 % lower than within the undisturbed soil to the north.

4. The model

Both observations, the "short-term" increase of the ^{222}Rn concentrations and the decrease of concentrations with increasing distances to the vibration source, can be explained by a "pump effect": Due to the subsurface vibrations mineral grains start to vibrate, whereby soil-gas is slightly blown out of the soil into the atmosphere (Fig. 6). This results in an upward movement of soil-gas. Therefore, when soil-gas is sampled during a train is passing by, soil-gas is sampled which comes from deeper layers. In the depth the measurements are done, the ^{222}Rn gradient in the soil-gas is very steep, which means that the ^{222}Rn concentrations increase pronounced with depth. Fig. 7 shows this increase of ^{222}Rn concentrations with increasing depth at the measuring point of Fig. 3. The marked differences in ^{222}Rn concentrations between Fig. 7 and Fig. 3 are due to an extreme low soil moisture content during the sampling for the analyse of the ^{222}Rn gradient of the soil-gas. We made the observation that there is a more or less good positive correlation between ^{222}Rn and soil moisture when soil and soil-gas is sampled from same sites at different times, but not from different sites at same times. As it can easily be calculated out of Fig. 7, the observed increase of the ^{222}Rn concentrations (Fig. 3) can be explained by a rise of soil-gas no more than 5 - 10 cm.

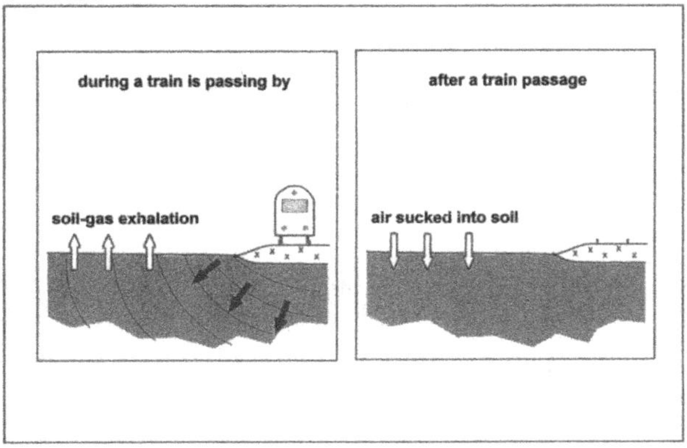

Fig. 6: The pump-effect: "breathing" of the soil due to vibrations.

It is more difficult to explain the decrease of ^{222}Rn concentrations with increasing distances from the vibration source. In response to the loss of ^{222}Rn to the atmosphere the ^{222}Rn gradient in the soil becomes steeper, and therefore, the upward directed ^{222}Rn migration is intensified. Because the loss of ^{222}Rn should decrease with increasing distance from the source of vibrations the ^{222}Rn concentrations of the soil-gas do the same. As mentioned before, the spatial decrease of the ^{222}Rn concentrations was only observed above solid rocks. Probably, the faster and greater attenuation of the vibrational energy within the unconsolidated rocks can explain this observation.

5. Conclusions

The strongest subsurface vibrations were observed at project sites where the spatial radius of the vibration influence on ^{222}Rn concentrations is > 60 m. The vibrations due to railway traffic are stronger, but last a shorter period than those of motor vehicle traffic. Therefore, the spatial radius of the influence is wider with railway traffic (> 30 m) than with motor vehicle traffic (< 25 m). Because the continuous motor vehicle traffic is stimulating the ground permanently, the increase of ^{222}Rn concentrations is higher (37 %) than by the railway traffic (11.5 %). In all three cases (project site, car and train traffic) the "pump effect" increases the upward migration due to the ^{222}Rn loss to the atmosphere and the steeper ^{222}Rn gradient in the soil.

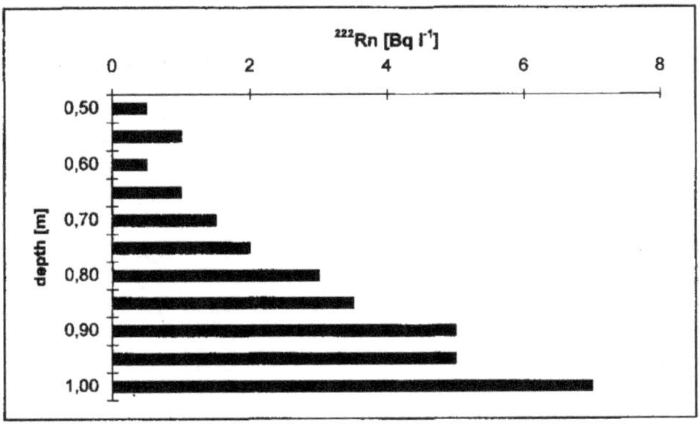

Fig. 7: ^{222}Rn concentrations in soil-gas from different sampling depths. This measuring point is the same as in Fig. 3. The values are an average of two samples taken consecutively.

The "pump effect" can influence the radiation exposure of the population in the vicinity of vibration sources. Due to the vibrations ^{222}Rn may be "pumped" into houses where it can accumulate. Because of the temporal character of project sites, such locations will be not so important in the view of radiation protection. On the other hand, the "pump effect" can play an important role along railway tracks. Because at heavy traffic roads, the increase of the ^{222}Rn concentrations persists for a longer period, an accumulation of ^{222}Rn seems even more likely. Moreover, through the intensified upward migration the subsequent delivery of radon from the subsoil is generally increased. This results in a "long-term effect", which will increase the ^{222}Rn potential in the vicinity of vibration sources.

6. Acknowledgements

I owe many thanks to Mrs. Simone Schmid, who did an excellent field work and preparation of soil samples. Prof. Dr. Volker Hans and Dipl. Ing. Heinz Wolf from the department of mechanical techniques of the University of Essen are thanked for their help with the vibration measurements and the analysis of these measurements.

7. References

[1] Tanner, A. B., Radon migration in the ground: a review. In: Adams, J. A. S.; Lowder, W. M., eds. The natural radiation environment. Chicago: University of Chicago Press; 161 – 190, 1964.

[2] Tanner, A. B., Radon migration in the ground - a supplementary review.- In: Gesell, T. F., Lowder, W. M., eds. The natural radiation environment III. Proceedings of the 3rd International Symposium on the Natural Radiation Environment. Washington, DC: U.S. Department of Energy; 5 – 56, 1980.

[3] Shiratoi, K., The variation of radon activity of hot springs. Sci. Rep. Tohoku Imp. Univ., Ser. **3, 16,** 614-621, 1927.

[4] Ulumov, V. I. and Mavashev, B. Z., A precursor of a strong tectonic earthquake. Dokl. Akad. Sci. USSR, Engl. Transl., Earth Sci. Sect. **176, 1-6**, 9-11, 1968.

[5] Fleischer, R. L. and Mogro-Campero, A., Association of subsurface radon changes in Alaska and the northeastern United States with earthquakes. Geochem. Cosmochim. Acta **49, 4,** 1061-1071, 1985.

[6] King, C. - Y., Gas geochemistry applied to earthquake prediction: an overview. J. Geophys. Res. **91, B 12**, 12269-12281, 1986.

[7] Igarashi, G. and Wakita, H., Groundwater radon anomalies associated with earthquakes. Tectonophysics **180**, 237-254, 1990.

[8] Wiegand, J., Radon in urbanen Gebieten - geogene und anthropogene Einflüsse auf das ^{222}Rn-Potential am Beispiel des Ruhrgebietes. In: Siehl, A., eds. Umweltradioaktivität; Geologie und Ökologie im Kontext. Berlin, Ernst & Sohn, 223-232, 1996.

[9] Wiegand, J. and Büchel, G., Geogene und urbane Einflüsse auf die Radon-Verfügbarkeit am Beispiel Großraum Essen.- Geowissenschaften + Umwelt, Springer Verlag, Berlin Heidelberg, 65-76, 1997.

[10] Lucas, H. F., Improved low - level alpha - scintillation counter for radon. Rev. Sci. Instrum. **28, 9,** 680–683, 1957.

[11] Schmid, S. and Tulke, S., Das Erstellen von Meßprogrammen für die Analyse von Haupt- und Spurenelementen geologischer Proben mit einem Sequenz-Röntgenspektrometer. Essen: unpub. diploma – thesis, 75 p., 1996.

[12] Kertz, W., Einführung in die Geophysik. Mannheim / Wien / Zürich: B.I. – Wissenschaftsverlag, 1969.

[13] Militzer, H. and Weber, F. Angewandte Geophysik Band 3 Seismik. Wien / New York: Springer-Verlag / Berlin: Akademie Verlag, 1987.

[14] Schmid, S., Beeinflussung des Radonpotentiales des geologischen Untergrundes durch verkehrsbedingte Vibrationen. Essen: unpub. diploma – thesis, 140 p., 1996.

[15] Tulke, S., Der Einfluß der anthropogenen Bodenversiegelung auf das Radonpotential im geologischen Untergrund. Essen: unpub. diploma - thesis; 1996.

[16] Washington, J. W. and Rose, A. W., Temporal Variability of Radon Concentration in the Interstitial Gas of Soils in Pennsylvania. Journal of Geophysical Research **97, B6,** 9145-9159; 1992.

Radon in soil gas and groundwater

Susumu Nishimura

Kyoto Institute of Natural History, Yoshidakawaramachi 14,
Sakyoku, Kyoto 606, Japan

Abstract

Disequilibrium in the ^{238}U decay series is well known. Concerning groundwater, some springs have large concentrations in *Rn*, but little in *Ra*. Often these springs are associated with granite, particularly with fracture zones. The *Ra* content of rock samples taken from the vicinity of such springs is usually revealed to be normal, but weathered rock is rather richer in *Ra* than fresh rocks of the same kind. Fault clay or fault breccia is often impregnated with *Ra*. These occurrences of ^{238}U decay series near the surface of ground are discussed.

1. Introduction

Since radioactivity was discovered, many rock types have been tested for this property. Strutt [1] was the first to determine the *Ra* content of various types of rocks, including sedimentary ones, in 1906. From his investigations, as well as those of other researchers, it was clear that *Ra* is widely disseminated, though very small amounts in all kinds of rocks. However, it is usually richer in igneous rocks than in sedimentary rocks, and also in acid rocks than in basic rocks. Because *Ra* is the parent of *Rn*, these differences in *Ra* content with rock types provide the means by which *Rn* radioactivity can be used for mapping of geologic structures. Moreover, different rates of *Rn* migration can also be used for subsurface exploration.

The use of radioactivity in prospecting dates back to the beginning of this century when von dem Borne [2] tested it in a *U* mine in 1905. In 1918, he [2] found a notable increase of radioactivity in soil samples at the intersection of buried faults in the vicinity of Blankenburg, Germany. He also found this to be the case over mineral veins. These results led him to the idea that *Rn* survey might be applicable in locating faults, mineral veins and boundaries of different rock types.

Hatuda [3] was first to use this method in *Rn* geological exploration. He [3] discussed the techniques and principles of *Rn* prospecting, described a method which he developed, and gave examples of tests and practical applications in the field. His discussion of diffusion of *Rn* and the location of hidden faults is particularly useful. Following this, his students developed several radioactive prospecting techniques which have been applied to various problems [4, 5]. γ-ray measurements using a multichannel analyzer with *NaI (Tl)* scintillation counter were carried out to reveal a coincidental increase of γ-radiation due to ^{214}Bi and ^{40}K over a fault plane [6-8].

Concerning the use of *Rn*-detection methods in earthquake prediction, it is well known that the soil gas and groundwater which fill the pores of the soil have a tolerable quantity of *Rn*. The *Rn* content shows considerable changes near the ground and water surfaces, though it gradually increases with increasing depth approaching asymptotically a finite value. Concentration depends not only on the *Ra* content and emanating power of the soil or rocks, but also on the changes in aeration caused by variable isolation and changes of atmospheric pressure and precipitation.

2. Radioactive Equilibrium and Disequilibrium

Radioactive disequilibrium in rocks, soils and groundwater is caused by the loss or addition of member nuclides in the radioactive decay series. Considering their half-lives and chemical characters, members of the ^{232}Th, ^{235}U and ^{238}U decay series were classified into a number of groups which are likely to display internal equilibrium, but which can be in disequlibrium with other groups [9]. We [10] also modified this system and proposed six groups: (i) *uranium*, (ii) ^{230}Th, (iii) ^{226}Ra, (iv) ^{222}Rn, (v) *protoactinium*, and (vi) ^{232}Th.

Table 1 The contents of radioactive elements in inland- and sea-waters, and sea-bottom clay.

Nucleide	Inland-water (g/g)	Sea-water (g/g)	Sea-bottom clay (g/g)
^{238}U	1.3×10^{-9}	3.0×10^{-9}	1×10^{-6}
^{230}Th	1.5×10^{-14}	$<3.0\times10^{-16}$	2×10^{-9}
^{226}Ra	0.7×10^{-18}	1.0×10^{-16}	4.0×10^{-12}
^{222}Rn	$\sim10^{-17}$	6.3×10^{-22}	2.5×10^{-17}
^{40}K	—	4.2×10^{-8}	—
^{232}Th	2.0×10^{-11}	$<2.0\times10^{-11}$	5×10^{-8}

Disequilibrium in the ^{238}U decay series is well known. For example, there [11, 12] were presented data on disequilibrium in the ^{238}U series in recent volcanic rocks, and discussed the origin in terms of the fractionation occurring during magmatic processes.

In the magma reservoir, ^{238}U (half-life: 4.5×10^9y) is a state of + 4 change, but in the later stage, especially pneumatolytic stage, it oxidized to +5, +6 charges. In this case, UO_2^{+2} is occured and makes $UO_2(CO_3)_2^{-2}$ and $UO_2(CO_3)_3^{-4}$, and this ion is easily dissolved to water in over pH 5. But *thorium* +4 could not change in these stage and could not moved from magma reservior. ^{234}U is more active than ^{238}U, then the activity ratio of $^{234}U/^{238}U$ is higher than 1. In the case of seawater, *thorium* observed in clay in the state of hydroxide. We can use it for dating of sea bottom clay (^{230}Th dating). ^{226}Ra (half-life 1622 y ; ionic radius 1.52 Å) is easily dissolved in water, but easily trapped in clay. In the clay, it can be observed that the activity ratio of $^{226}Ra/^{238}U$ is high, as 2-10 times (Table 1). ^{222}Rn (half-life 3.8229 day) and ^{220}Rn (half-life 51.5 sec) are in state of gas and easily moved in groundwater and soil gas.

Concerning groundwater, some springs have large concentrations of *Rn*, but little *Ra*. Often these springs are associated with granite, particularly with fracture zones and contacts. The *Ra* content of rock samples taken from the vicinity of such springs is usually revealed to be normal, but weathered rock is richer in *Ra* than fresh rocks of the same kind. Fault clay or fault breccia is often impregnated with *Ra* [13, 14] (Table 2).

Table 2 ^{226}Ra content of rocks, "Choseki-ore"* and clay

	^{226}Ra content	$(10^{-12}$g(Ra)/g)		
Sample	Fresh	weathered	Choseki-ore	Fault Clay
Daimonji granite (Kyoto)	1.42	1.54	--	--
Gyojayama granite (Kameoka)	1.46	--	--	--
Rokko granite (Kobe)	1.57	1.49	--	2.57
	mean of 8 samples (1.49-2.57)			
Rokko liparite (Kobe)	1.20	--	--	4.18
Awaji granite (Hyogo)	0.62	--	--	1.40
Awaji lamprophyre (Hyogo)	0.25	0.77	--	--
Shishitobi granite (Otsu)	2.63	--	1.79	19.0
	mean of 6 samples (1.42-3.86)		mean of 10 (0.67-3.98)	mean of 13** (2.72-68.8)

* "Choseki ore" is a part of leucocratic granite.

** Spring waters along this fault contained 0.0417×10^{-12} g(Ra)/l and had a ^{222}Rn activity of 6.62×10^{2} Bq/l (1.79×10^{-8} Ci/l) (mean value of 92 springs in Hiratsu Choseki Mine, Shiga Prefecture, Japan).

We [15] determined the *Ra* content of granitic rocks, autometamorphosed granite ('Choseki' ore) and clay collected in Hiratsu Choseki Mine, in the southern part of Shiga Prefecture, southwest Japan (Table 2). The greater part of the clays, a collective sample of fine, pale-yellow micas included in the ore, shows an even greater abundance of *Ra* than that of most of the granitic rocks. From our results, we suggested that a considerable amount of *Rn* in the water of radioactive springs has been derived from *Ra* contained in the granitic rocks and manganiferous substance, especially from *Ra* in the fine micas and clays.

The source of *Rn* in radioactive springs was also studied by many researchers. It is pointed out that the *Rn* in Masutomi hot springs, northwest Japan, is produced only about ten minutes before discharge [16]. This information was obtained from measurements of decay products of ^{222}Rn, such as ^{218}Po, ^{214}Pb and ^{214}Bi in the water. There [17, 18] are also discussion concerning the rate of diffusion of *Rn* in soil, and Katsura, K. [14] succeeded in finding out that the time of final supply of *Rn* from depth is very short, less than 30 minutes; that is, the depth of the *Rn* source can be found in terms of time by tracing with use of a whole body counter belonging to the Central Radioisotope Division of Kyoto University (Fig. 1).

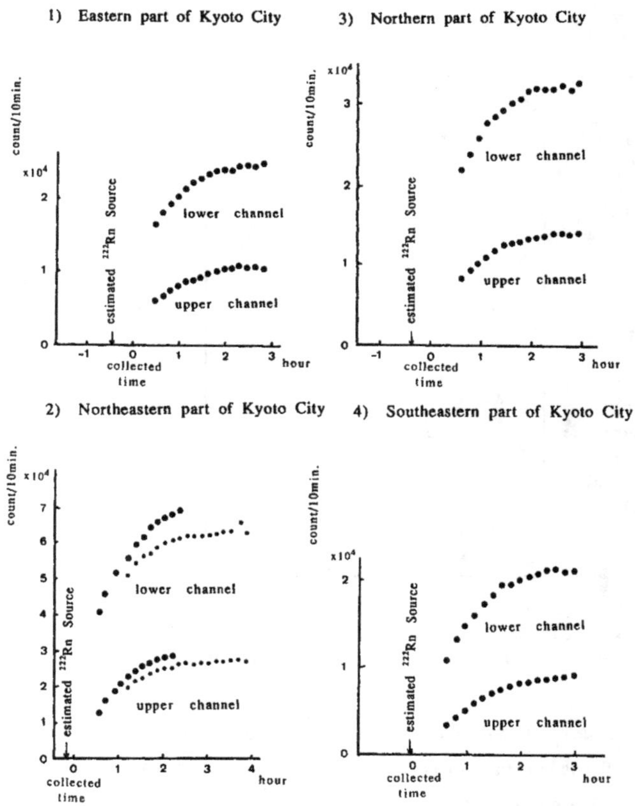

Fig. 1 Estimated ^{222}Rn source in some wells in Kyoto City (modified from Katsura, 1976).

3. Distribution of Radioactivity

Radioactive measurement is based on the fact that the *U*, *Th* and *K* contents are characteristic of rock types, and that the concentrations of their decay products are determined by the geological environment. For example, if the *Ra* content of one rock is different from that of the adjoining one, the concentration of *Rn* contained in the soil overlying the rocks differs accordingly.

From the results of measurement [9] on an ore vein of *Pb*, *Zn* and *Cu* in Honnef, those on the *Fe* ore vein [2] at Ilfield in Harz, and [20] on the same kind of ore in Japan, it is clear that there is often a relationship between radioactivity and mineralization. This is now known to be due to the fact that hydrothermal solutions contain and deposit it on the outside of ore minerals.

The distribution of radioactivity in granitic bodies and across contact zones has been studied by many researchers [21-27] . The results obtained have shown that *U* and *Th* tend to concentrate towards the periphery of intrusive masses (Fig. 2).

It is pointed out that in the vicinity of faults, there takes place an increase in the permeability of soil due to disturbances in 1908. In 1921, it [2] also stated that there will be easier circulation of solutions and gases through disturbed material so that radionuclides would be transported and accumulated at their outcrop. In this context, the experiment made [28] during drilling at Annaberg in Erz Gebirge in Germany is of great interest. He measured the *Rn* content of water and the *Ra* content of crushed rocks taken from the well bottom, and found that the *Rn* maxima always occurred at depths one or two meters shallower than the points where the *Ra* maxima appeared where the well was crossed by fissures or veins. This would imply that *Rn* is persistently ascending through rocks towards the surface of the earth; the crushed zone of a fault is providing an easier path for *Rn* ascent.

The diffusion coefficient of *Rn* in air is shown to be less than 0.07 cm^2 sec^{-1} at room temperature and have a pressure of 760 mmHg [29] . This is exceedingly low when compared with the diffusion coefficient of the other gases in air. As interdiffusion is inversely proportional to the pressure of a mixed gas, the diffusion of *Rn* is slower at a great depth than near the surface of the ground. However, the increasing temperature acts to increase diffusion. The diffusion coefficient of *Rn* in water at 18°C is 0.99-0.92 cm^2 day^{-1} and 14°C is 0.82 cm^2 day^{-1}, which are about four orders of magnitude lower than the diffusion coefficient in air.

The atomic weight of *Rn* (mono-atom gas ^{222}Rn) is greater than the molecular or atomic weight of the other gases usually found in the earth, hence the diffusion of *Rn* coexisting with other mobile gases will be so slow that much of it will be left behind. Consequently, the diffusion of *Rn* through compact rocks is practically negligible; it is only possible through cracks or fissures. The short half-life of this nuclide would seem further to emphasize this conclusion. This is one of the reasons why *Rn* concentrations occur in the neighborhood of a fault, as in its vicinity the rock is usually severely cracked and fissured.

In addition, gases and aqueous solutions can also ascend through the disturbed zone of a fault and will deposit minerals along the cracks and fissures and also in the soil overlying the fault. It is

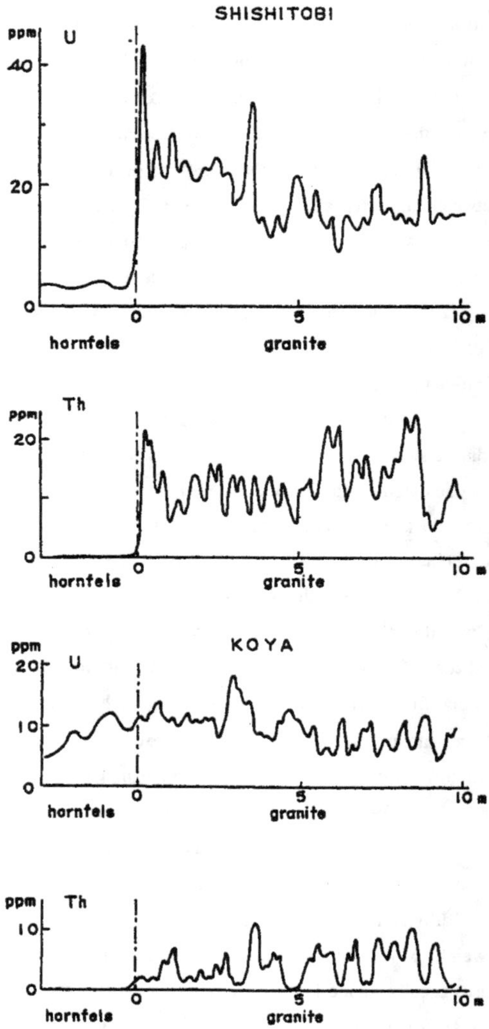

Fig. 2 The distributions of U and Th contents near the contacts of upper part and lower part of granitic intrusions.

Shishitobi : Shishitobi, southern part of Otsu City.

Koya : Koya, Watsuka-cho, Kyoto Pref.

(Nishimura, 1965)

very likely that *Ra* and *K* salts will be associated with these deposits. In consequence, the concentration of *Rn* becomes greater near faults than in other places.

A γ-survey using a multichannnel analyzer with *NaI (Tl)* scintillation and Ge semiconductor detectors is operated in similar way to the γ-scintillation counter method. After calibration of the channels with standard radioactive sources, the natural radioactivity can be identified and the γ-ray count from a specified nuclide determined.

A comparative study has been made of the different methods at Shugakuin, in the northeastern part of Kyoto City. The measurements of photo-plate, *Rn* in soil gas, Geiger-Müller counter and *in situ* ionization -chamber methods [3] were already reported, as shown in Fig. 3.

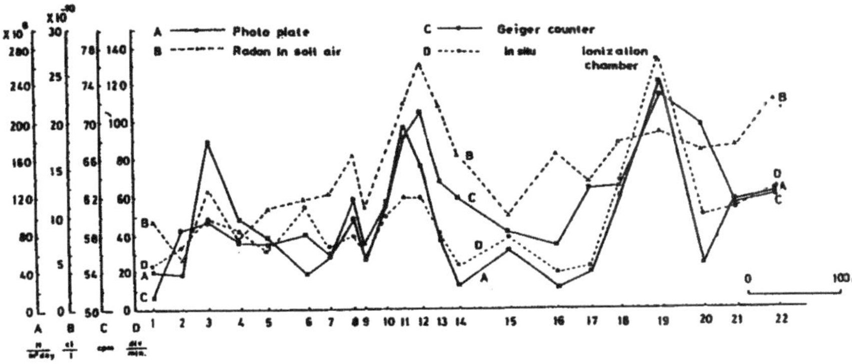

Fig. 3 Comparative study of the four different methods across a part of Hanaori active fault at Syugakuin, northeastern part of Kyoto City (modified from Hatsuda, 1954).

As shown in Fig. 3, the results obtained by the four different methods are in fair agreement. Above all, the photo-plate method is in good agreement with the *in situ* ionization-chamber method. Thus the applicability of nuclear emulsions to radioactive exploration has been proved by this experiment.

A similar comparative study has been made for *NaI (Tl)* scintillation counter and a multichannel analyzer with *NaI (Tl)* detector methods, across the Obaku active fault at Uji City, south of Kyoto City. The results [30] are shown in Fig. 4. Fault location by geological, seismic, electrical resistivity and γ-ray methods are in good agreement.

4. Conclusion

Radioactive surveys are very effective in revealing hidden geological structures, such as lithologic boundaries. Often the truth founded on *Rn* survey is all that is necessary, but where there is no preferential migration of *Rn*, another survey must be added.

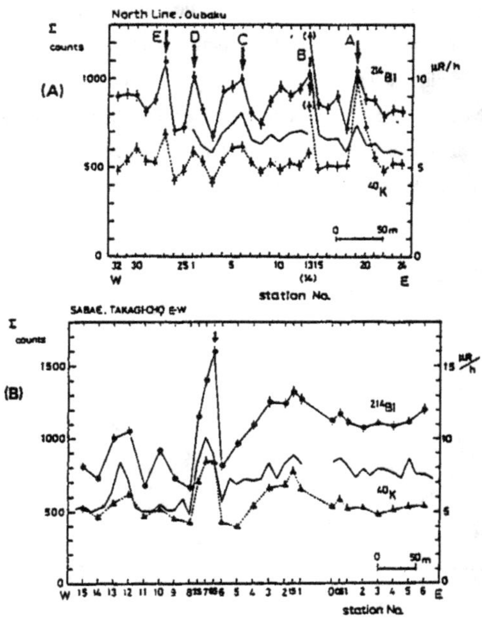

Fig. 4 Comparative study of γ-scintillation detector with multichannel spectrometer across the Obaku active fault at Uji City, south of Kyoto City (A) (modified from Katsura et al., 1986), and across active fault at Sabae, Fukui Prefecture, north of central Japan (B) (modified from Mino, 1986). The thin solid line is the data measured by γ-scintillation counter, in units of μR/h. Solid circle and triangle are the data by multichannel spectrometer, in units of counts per 5 minutes (live time). Arrows indicate fault positions, but A and B of (A) are doubtful because of very strong ^{40}K peak relative to the increase of ^{214}Bi peak.

Although it is dangerous to conclude that the possibility of successful earthquake prediction is increased from observing changes of *Rn* activity in soil gas and in groundwater, these may possibly he helpful in future earthquake prediction research. However, measurements must be corrected for changes in meteorological conditions and artificial factors.

5. References

[1] Joly, J., Radioactivity and Geology. Constable, London, 80p., 1909.

[2] Ambronn, R., Elements of Geophysics (translated by M. Cobb). MaGraw-Hill, N.Y., p. 125-128, 1928.

[3] Hatsuda, Z., Radioactive method for geological exploration. Mem. Coll. Sci., Univ. Kyoto, Ser. B, 21, No. 2, p. 231-271, 1954.

[4] Sadahiro, T. and K. Mino, The γ-ray intensity around active faults (I). Zishin (J. Seism. Soc. Japan), Ser. 2, 33, p. 51-70, 1980.

[5] Mogi, T., K. Mino and S. Nishimura, A tectonic zone in the northwestern part of Kinki district, southwest Japan (3) Yagi-Yabu, Mitoke fault zone. Zishin (J. Seism. Soc. Japan), Ser. 2, 38, p. 577-585, 1985.

[6] Katsura I., S. Nishimura, T. Sadahiro, K. Mino, T. Mogi, and J. Nishida, γ-ray survey of the Obaku Fault, Uji, Japan-Multichannel pulse-height analysis. Zishin (J. Seism. Soc. Japan). Ser. 2, 39, p. 267-275, 1986.

[7] Mino, K., A tectonic zone in the southern part of the Hokuriku district (2) Sabae fault. Zishin (J. Seism. Soc. Japan), Ser. 2, 39, p. 567-577, 1986.

[8] Katsura, I. and S. Nishimura, Discrimination of γ-ray source ^{40}K or ^{214}Bi, using γ-ray multichannel pulse-height analyzer. Zishin (J. Seism. Soc. Japan), Ser. 2, 40, p. 117-119, 1987.

[9] Rosholt, J. N.. U.N.I.C., PU.A.E., Z, Geneva, Proc., Paper OIC 183, U.N., 722, p. 230-236, 1958.

[10] Nishimura, S., Z. Hatuda and T. Asayama, A trial method of determining uranium and thorium content in rocks by radioactive measurements. Radioisotopes, 15, No. 6, p. 289-294, 1966.

[11] Somayajulu, B.L.K., M. Tatsumoto, J.N. Rosholt and R.J. Knight, Disequilibrium of the ^{238}U series in basalt. Earth Planet. Sci. Lett., 1, p. 387-391, 1966.

[12] Nishimura, S., Disequilibrium of the ^{238}U series in recent volcanic rocks. Earth Planet Sci. Lett. 8, p. 293-300, 1970.

[13] Nishimura, S., The relationships between the petrology and the radioactivity of some granitic bodies (1). Mem. Coll. Sci., Unive. Kyoto, Ser. B, 29, p.263-290, 1962.

[14] Katsura, K., Radioactive springs in the Kinki district and their radon sources (Part 2) Onsen-Kagaku (J. Balneological Soc., Japan), 22, p. 136-146, 1976.

[15] Asayama, T., S. Yagi, S. Nishimura and Y. Masuda, The radon content of the spring waters in the Hiratsu "Choseki" Mine, Shiga Prefecture, Japan. Onsen-Kagaku (J. Balneological Soc., Japan), 32, No. 2, p. 31-41, 1981.

[16] Kuroda, K., Radium isotopes in the radioactive springs. Jpn. J. Chem., 69, p. 76-77, 1948.

[17] Kikkawa, K., The sources of radon to the radioactive springs. Jpn. J. Geophys., 9, p. 105-110, 1954.

[18] Okabe, S., On some relations between the hot springs and radioactivity. Mem. Coll Sci., Univ. Kyoto, Ser. A, 28, p. 39-71, 1956.

[19] Müller, F., Radioactivitätsmessungen als geophysikalish sufschlussmethode. Z. Geophys., 3, p. 330-336, 1927.

[20] Nishimura, S., The continuous observation of radon concentration of self-spouting hot spring at Odoi, in the Izu Peninsula, Japan. Zishin (J. Seism. Soc. Japan), Ser. 2, 38, p. 587-596, 1985.

[21] Hatuda, Z. and S. Nishimura, A trial method of determining uranium and thorium contents in rocks by means of an α-ray scintillation counter and a radioscope. Mem. Coll. Sci., Univ. Kyoto, Ser. B, 31, p. 215-226, 1965.

[22] Hatuda, Z. and S. Nishimura, Variation in radioactivity across igneous contacts (The 2nd Report). Mem. Coll. Sci., Univ. Kyoto, Ser. B 25, p. 115-123, 1958.

[23] Nishimura, S., Variation in radioactivity across igneous contacts (The 3rd report). Mem. Coll. Sci., Univ. Kyoto, Ser. B, 28, p. 225-264, 1961.

[24] Nishimura, S., Variations in radioactivity and chemical elements across igneous contacts. Mem. Coll. Sci., Univ. Kyoto, Ser. B, 28, p. 265-284, 1961.

[25] Nishimura, S., Variation in the content of uranium and thorium across igneous contacts. Mem. Coll. Sci. Univ. Kyoto, Ser. B, 31, p. 227-238, 1965.

[26] Nishimura, S. and K. Katsura, Uranium, thorium and potassium content of Rokko granitic bodies. J. Japan. Assoc. Min. Petr. Econ. Geol., 62, p. 44-48, 1969.

[27] Katsura, K. and S. Nishimura, Uranium, thorium and potassium content of Kunikamiyama and Mihara granitic bodies. J. Japan. Assoc. Min. Petr. Econ. Geol., 62, p. 50-53, 1969.

[28] Aeckerlein, G., Die Erforschung des Erdinnern durch Emanationsmessung. Phys. Zeit., 38, p. 362-370, 1937.

[29] Hevesy, G. and F.A. Paneth, A manual of radioactivity. London 227p., 1938.

[30] Nishimura, S. and I. Katsura, Radon in soil gas : Applications in Exploration and Earthquake Prediction, in Geochemistry of Gaseons Elements and Compounds, ed by Durrance, E. M. et al., p. 497-533. THEOPHRASTUS Publ. S.A.Greece, 1990.

Radon measurements in soil and water and its relation with geology, Garhwal Himalaya, India

Vinay M. Choubey

Wadia Institute of Himalayan Geology,
33, General Mahadeo Singh Road, Dehra Dun, 248001, India

Abstract

Radon concentrations were measured in soil and spring water in the Garhwal Himalaya using radon emanometer. The measurements were made in the Berinag, Bhatwari and Munsiari Formations. The main rock types in the area are sheared granitic gneiss, porphyritic gneiss, chlorite schist, mica schist, mylonite, slate, phyllite, quartzite and metabasic. The radon concentrations were found to vary from 1.2 kBq/m^3 to 56.5 kBq/m^3 in soil and 0.4 Bq/l to 887 Bq/l in water. The results suggest that the radon emanation is controlled not only by uranium content of the rock and soil but also by structural zones (thrust, fault, etc.) which help in the easy migration of radon from the deeper parts of the earth crust.

1. Introduction

^{222}Rn (half life of 3.82 days) is derived from the parent radionuclide ^{226}Ra, which is a member of uranium-238 series and a long-lived alpha-emitter with a half life of 1620 years. Hence, ^{222}Rn is typically found in rocks with high uranium content. Radon generally accumulates in underground mines, caves and buildings as a result of diffusion from natural radioactive source in soil and rocks[23]. Several attempts have been made to correlate soil gas radon and groundwater radon concentration with lithology, geological structures and chemistry and mineralogy of the rocks[14, 24, 11, 10, 6, 7]. Radon being an inert and water soluble gas, its concentration in groundwater is mostly controlled by hydrodynamic factors, radium content of the reservoir rocks[2] and regional tectonics. Like many other solutes, its concentration is commonly found to be lognormally distributed in groundwater[26]. Radon concentration have been used as a diagnostic tool for predicting earthquake events and locating sub-surface fracture zones and even for estimating fracture apertures[18].

The present radon study in the Garhwal Himalaya attempts to understand the pattern of radon emanation in soil and spring water along and across the different tectonic planes and to correlate it with the structures, lithology and the associated uranium mineralization in the rocks.

2. Geology of the area

Geology of the Garhwal Himalaya have been extensively studied by several researchers [19, 1, 25, 22]. The area is characterised by polyphase deformation resulting in superimposed folding and repeated faulting and thrusting. The two main lithotectonic zones recognized in the Garhwal Himalaya are the Higher Himalaya and the Lesser Himalaya; both are further subdivided into different lithotectonic zones separated by regional scale thrust planes. Our study area falls in the Lesser Himalaya, where the lowest exposed litho-unit is the Berinag Formation (Fig. 1) which comprises the huge succession of massive and coarse grained quartzite with bands of amphibolite and meta-volcanics. This unit is overlain by Bhatwari Formation, which includes sheared mylonite, porphyritic granitic gneiss, granite gneiss, diorite, chlorite schist, phyllite and quartzite. Uranium mineralization has also been reported from this formation[17]. The rocks of the Bhatwari Formation are overlain by Munsiari Formation composed of chlorite schist, bands of mylonitized biotite rich porphyroblastic gneisses, augen gneiss with schist and sheared amphibolite dykes.

2.1 Geohydrology : Being a zone of compressive stresses[16], the consequent crustal deformation in the Himalaya has produced secondary porosity and permeability in the rocks. As a results majority of the springs are controlled by fractures and joints related to faulting and thrusting. Further, the faulting and fracturing of the rocks initiated development of various colluvial and fluvial deposits which serve as good reservoirs for the ground/subsurface water. The groundwater in the study area emanates in the form of springs and seepages. Mainly two types of springs are observed in the area, i) fault-fracture-lineament related springs (FFL) and ii) colluvial-fluvial related springs (CF).

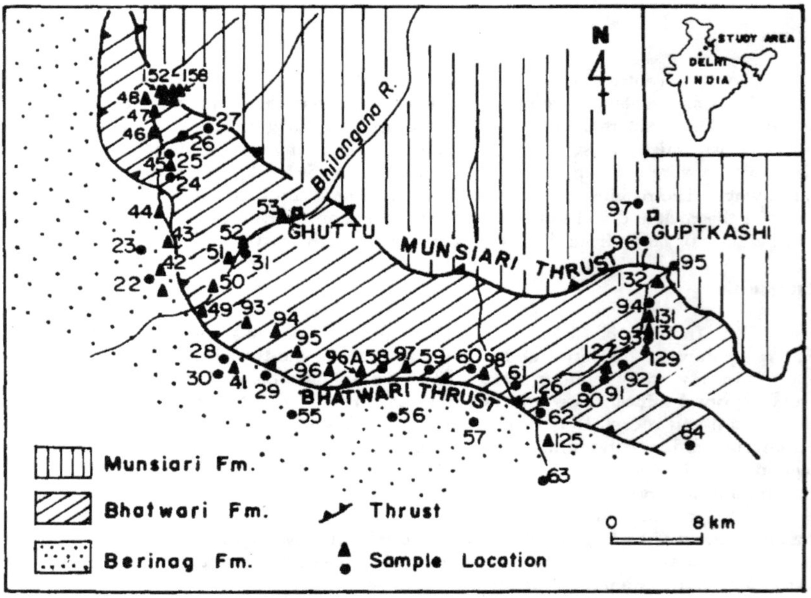

Figure 1. Geological map of a part of Garhwal Himalaya showing different lithotectonic units and the radon sample locations (solid circles indicate soil-gas samples and filled triangles spring water samples)

3. Experimental Procedure

In soil radon emanometry the auger holes, each 50 cm in depth and 6 cm in diameter, were left covered for 24 hours in order to stabilize the radon and thoron. The soil-gas probe was fixed in the auger hole and formed an air tight compartment[12]. The rubber pump, soil-gas probe and alpha detector were connected in a close circuit (Fig. 2). Air was circulated through a ZnS coated chamber for a period of 15 minutes till the radon formed a uniform mixture with the air. The alpha particles produce scintillation on the sample chamber which are sensed by photomultiplier tube and converted to the electronic pulses counted on a digital display. The detector was then isolated by clamping both the ends and observations were taken after four hours. The resulting number of the alpha counts is converted into Bq/m^3 by using the calibration factor 1 count/min. = 0.11 kBq/m^3, determined under similar conditions[21].

The radon measurements in the present study were carried out during May-June, when the weather was quite stable and the environmental changes such as temperature variation, snow fall and the moisture content were minimum.

The apparatus used for the measurement of radon in spring water is shown in Fig.3. 750 ml of water sample was taken in air tight reagent bottle of 1 lt. capacity connected in a close circuit with a ZnS coated detection chamber through a hand operated rubber pump and a glass bulb containing $CaCl_2$ to absorb the moisture. Air was then circulated in a close circuit for a period of ten minutes till the radon formed a uniform mixture with the air and the resulting alpha activity was recorded. The calibration factor 1 count/min. = 0.11 Bq/l was used to convert the recorded alpha counts in Bq/l [21].

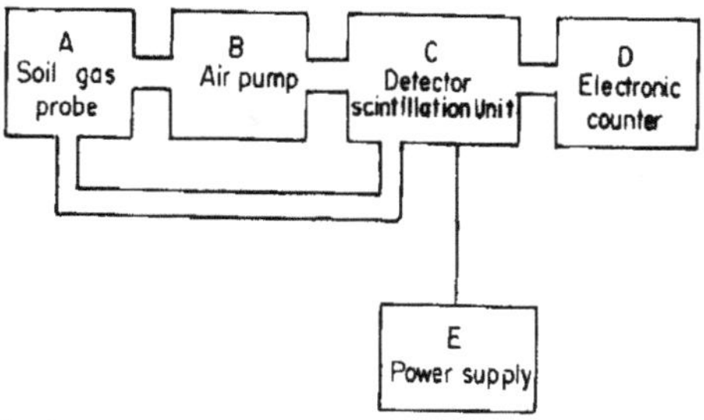

Fig. 2

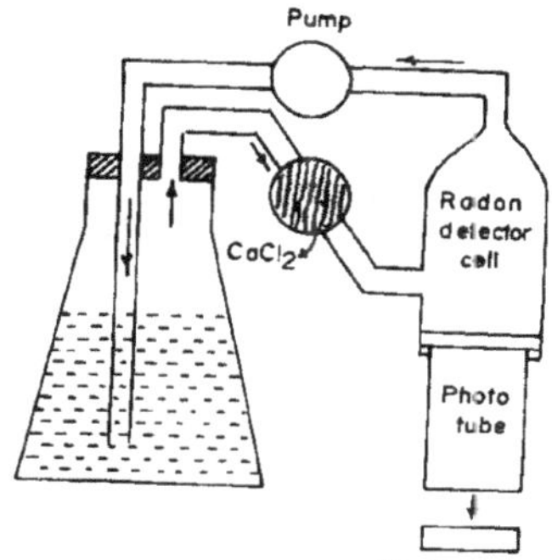

Fig. 3

Figure 2. Block diagram of close circuit technique for radon measurements in soil .
Figure 3. Apparatus for the measurement of radon in water.

4. Results and discussion

The results of the radon measurements in soil-gas and spring water are given in the Table 1 and 2, respectively and their locations are shown in Fig.1. Twenty seven samples of soil-gas and 37 samples of spring water were analysed from Berinag Formation, Bhatwari Formation and Munsiari Formation.

4.1 Radon in soil: The soil composition generally reflects the composition of the underlying rocks. The thin soil cover in the Himalaya is due to the rugged topography and fast erosion. The soil thickness in the study area vary from few cm to maximum of 2m.

The soil-gas values vary from 1.2 to 56.5 kBq/m^3 with an average of 16.7 kBq/m^3 in the Berinag Formation. The high soil-gas radon values of 56.5, 19.3, 18.1, 13.2 and 7.9 kBq/m^3 were measured close to the Bhatwari thrust. As reported in the literature the Late Archaean quartzite of Barinag Formation[4, 5] and its strike extension e.g. Manikaran quartzite from Himachal Pradesh contain uranium mineralization[9]. The presence of uranium and location of samples in the vicinity of Bhatwari thrust may be one possibility for high emanation of radon in Berinag Formation. Also, the Berinag Formation is overlain by Bhatwari Formation; the latter contains structurally controlled uranium mineralization [17]. It is possible that the uranium mineralized shear/fault extends beneath the Berinag quartzites at depth, resulting in the increased soil-gas radon emanations.

Measurements of seventeen radon soil-gas samples from Bhatwari Formation vary from 2.7 to 29.4 kBq/m^3 with an average of 12.8 kBq/m^3. Lithologically this formation comprises the sheared mylonitised granite, porphyritic granitic gneiss, chlorite schist, phyllite, etc. It has been suggested that the mylonitized shears zones with uranium mineralization can produce sufficient radon [13]. The structurally controlled uranium mineralization is reported in this formation with U_3O_8 varying from 0.047 to 0.78% in the form of pitchblende, uraninite and fluorapatite associated with the sericite-biotite granite, pelitic gneisses and sheared mylonites. The weathering of mylonites occurs along the foliation, exposing the surface of uranium even further and the foliation developed during mylonitization increases the permeability of the rocks, thereby enhancing the radon escape[13]. The higher radon values (29.4, 28, 18.1, 16.5 and 13.2 kBq/m^3) in this formation is possibly due to radioactive source and its diffusion to the adjoining area through fractures, joints, shears, thrusts and faults. Three samples of soil-gas radon were also measured from the Munsiari Formation and the values are 38.3, 15.1 and 12.4 kBq/m^3. These samples are mostly located very near to the Munsiari thrust.

Table 1. Soil radon variation in different lithotectonic units, Garhwal Himalaya

Lithotectonic units	Sample numbers	Radon concentration (kBq/m^3)
Munsiari Formation	95,96,97	38.8,12.4,15.1
---Munsiari Thrust---		
Bhatwari Formation	24,25,26,27,31 58,59,60,61,62 84,85,90,91 92,93,94	11.6,29.4,20.1,13.2,18.1 14.2,9.6,11.7,8.7, 28.0 16.5,8.2,2.7,3.9 11.5,4.0,7.5
---Bhatwari Thrust---		
Berinag Formation	22,23,28,29,30 55,56	56.5,13.2,19.3,7.9,18.1 2.1,1.2

4.2 Radon in spring water: In order to understand the geohydrological control over radon concentration in springs, water radon measurements were made in 37 springs from Berinag and Bhatwari Formation and the data are given in Table 2. In this study mostly perennial springs were selected which are normally used by villagers for drinking purpose.

In Berinag Formation the measurements were made in six springs. Four among those related to fault-fracture-lineament (FFL) show radon variation between 8.5 to 160.3 Bq/l with an average of 63.6 Bq/l, and the two colluvial-fluvial (CF) related springs show low radon concentration, i.e. 0.5 and 11.9 Bq/l. Among the 31 springs from Bhatwari Formation, 24 were related to FFL; the radon concentration in these springs vary from 4.8 to 887.3 Bq/l with an average of 238.4 Bq/l. The rest 7 springs are related to CF and show 0.4 to 135.6 Bq/l, with an average of24.3 Bq/l.

As is evident the overall high water radon concentration was found in FFL-related springs in comparison to CF-related springs. The FFL related springs are located close to the thrust and the uranium mineralisation is also reported in area along shear planes at few places. Rocks are highly fractured and jointed this possibly increases of the ratio of the rock surface area to water volume, as a result of which the emanation efficiency of radon increases[15]. Besides, the water gets enriched in radon by dissolving radon which emanates from deeper parts of the crust through deep seated thrust and faults[8].

The CF related springs were found mostly along the lower valley slopes or at the confluence of two streams in the form of terraces and associated with old landslides. The low radon emanations from the CF-related springs could possibly be related to the high water carrying capacity and high porosity and permeability of the CF deposits. The turbulent flow within such deposits is likely to cause natural de-emanation of gases which find easy pathways through high porosity of the deposits for movement into the atmosphere

The results of soil and spring water radon measurements suggest that beside, natural radioactive source, the degree of fracturing or faulting in a rocks can significantly affect the radon potential. The more extensive the discontinuities in a formation, the faster the migration of Radon[10]. However, the other factors are equally important like lithology, porosity and permeability etc.

Table 2. Spring water radon variation in different lithotectonic units, Garhwal Himalaya

Lithotectonic units	Sample numbers	Radon concentration (Bq/l)
Munsiari Formation		
---Munsiari Thrust---		
Bhatwari Formation	40,45,46,47,48	4.8,120.7,27.4,887.3,557.8
	49,50,51,52,53	87.9,167.8,125.4,34.2,135.6
	93,94,95,96,96a	1.2,1.3,0.4,10.2,174.3
	97,98,126,127,128	94.3,419.9,6.5,52.0,1.8
	129,130,131,132	110,18.5,72.8,14.5
	152,153,154,155	814.9,188.4,816.6,145.1
	156,157,158	122.1,292.8,386.9
---Bhatwari Thrust---		
Berinag Formation	41,42,43	35.9,49.9,11.9
	44,92,125	160.3,8.5,0.5

Acknowledgment

I am to thankful to the Director, Wadia Institute of Himalayan geology, Dehra Dun for providing the necessary research facilities to carry out this investigation. Critical reading of the manuscript and comments given by Drs. K.K.Sharma, M.I.Bhat and S.K.Bartarya are gratefully acknowledged.

6.References

[1] Agarwal, N.C. and Kumar, G. 1973. Geology of upper Bhagirathi and Yamuna valleys, Uttarkashi Dist. Kumaun Himalaya. Him. Geol., 3, 1-23.

[2] Ball, T.K., Cameron, D.G., Colman, T.B. and Robert, P.D. 1991. Behaviour of radon in the geological environment : a review. Q. Jour. Eng. Geol. London, 24, 169-182.

[3] Banks, D., Royset, O., Strand, T. And Skarphagen. H. 1995. Radioelement (U,Th,Rn) concentrations in Norwegian bedrock groundwaters. Environmental Geology, 25, 165-180.

[4] Bhat, M.I., Claesson, S., Dubey, A.K. and Pande, K. 1997. Sm-Nd of the Garhwal-Bhowali volcanics, western Himalayas: vestiges of the Late Archaean Rampur flood basalt province of the northern Indian craton. Precam. Res. (In press).

[5] Bhat, M.I. and Le Fort 1992. Sm-Nd age and petrogenesis of Rampur metavolcanic rocks, NW Himalaya Late Archaean relics in the Himalayan belt. Precam. Res. 56., 191-210.

[6] Choubey, V.M., Ramola, R.C. and Sharma, K.K. 1994. Soil gas and indoor radon studies in Doon Valley, India. Nucl. Geophys., 8(1), 49-54.

[7] Choubey, V.M., Sharma, K.K. and Ramola, R.C. 1997. Geology of radon occurrence around Jari in Parvati Valley, Himachal Pradesh, India. Jour. Environ. Radioactivity., 34(2), 139-147.

[8] Choubey, V.M. and Ramola, R.C. 1997. Correlation between geology and radon levels in groundwater, soil and indoor air in Bhilangana Valley, Garhwal Himalaya. Enviromental geology (in press)

[9] Das, G.R.N., Parthasarathy, T.N.and Taneja, P.C. 1972. Uranium mineralisation in the pelitic schist in the Kullu Himalaya and its probable origin. Proc. Indian Academy Sciences, 37, 267-276

[10] Ganas, M.J. Schuring, J.R. and Raghu, D. 1989. Radon concentration in dwellings. Environ. Studies., 32, 247-260.

[11] Gates, A.E. and Gundersen, L.C.S. 1989. Role of ductile shearing in the concentration of radon in the Brookneal zone, Virginia. Geology, 17, 391-394.

[12] Ghosh, P.C. and Bhalla, N.S., 1981. On the behaviour and measurement of thoron (220Rn) in soil. Ind. J. Earth Sci., 8, 1-9.

[13] Gundersen, L.C.S. and Gates, A.E. 1989. The concentration of uranium and radon in mylonite Zones from Pennsylvania, Maryland and Verginia: A function of chemical and deformational processes. EOS, 70(15), 499.

[14] Lapwood, E.R. 1948. Convection of a fluid in a porous medium. Proc. Camb. Philos. Soc., 44, 508-521.

[15] Lawrence, E., Poeter, E. And Wanty, R. 1991. Geohydrologic, geochemical and geologic controls on the occurrence of radon in ground water near conifer, colorado, U.S.A. Jour. Hydrology, 127, 367-386.

[16] Le Fort, P. 1975. Himalaya : The collided range. Present knowledge of the continental arc. Am. Jour. Sci., 275A, 1-44.

[17] Nashine, S.K., Dhanraju, R., Bhatnagar, G.R. and Narayan Das, G.R. 1982. Uranium occurrences close to Main Central Thrust around Sileth, Dhargaon, Chamyala Balganga Valley, Tehri Garhwal (U.P.). Him. Geol. 12., 305-316.

[18] Nelson, P.H., Rachiele, R. and Smith, A. 1983. Transport of radon in flowing boreholes of stripa, Sweden. Jour. Geophys. Res., 88(B3), 2395-2405.

[19] Saklani, P.S. 1972. Lithostratigraphy and structure of the area between the Bhagirathi and Bhilangana rivers, Garhwal Himalaya. Him. Geol., 2, 342-355.

[20] Sharma, Y.C., Bhatnagar, G.C. and Das, N. 1983. Shear zones mineralised with uranium: solid state nuclear track detectors techniques, in Himalayan Shears, Saklani P.S.(ed). Himalayan Books, New Delhi, 84-94.

[21] Singh, M., Singh, N.P., Singh, S and Virk, H.S. 1986. Calibration of radon detectors. Nucl. Tracks and Radiat. Meas., 12, 739-742.

[22] Rao, P.N. and Pati, U.C.1982. Geology and tectonics of Bhilangana Valley and its adjoining parts, Garhwal Himalaya with special reference to Main Central Thrust. Him. Geol., 10, 220-233.

[23] Tanner, A.B. 1980. Radon migration in the ground : A supplementary Review, in Natural Radiation Enviroment III, Symp. Houston, Tex., Thomas, F. Gesell and Wayne M. Lowder (eds.) Vol. 1, 5-56.

[24] Tanner, A.B. 1986. Geological factors that influence radon availability. in Indoor Radon, Proceedings of Air Pollution Control Association, Pittsburg, SP-54, 1-12.

[25] Valdiya, K.S. 1980. Geology of Kumaon Lesser Himalaya. WIHG publication. 1-287.

[26] Zikovsky, L. And Chah, B. 1990. The lognormal distribution of radon in groundwater. Ground Water, 28, 673-676.

Prediction of the earthquake in the western Japan from the change of radon concentrations in air

Akira Katase and Yuzuru Matsumoto*

Tohwa Institute for Science, Tohwa University, Fukuoka 815, Japan
*Kyushu Teikyo Junior College, Ohmuta 836, Japan

Abstract

The use of the increase of radon concentrations in air is discussed for the prediction of earthquake at active faults. The crust in the western Japan is pressed to the west by the moving ocean plates. The stress increases in some region and induces the slippage at an active fault in the crust. This produces the earthquake. Recently the increases of radon concentrations for some period have been observed before the earthquake at three measurement sites.

The increase of the average radon concentrations in air is induced primarily by the lager radon exhalation rate from the earth surface. The diffusion of radon atoms in the soil does not contribute for bringing more radon atoms from the underground to the earth surface to exhale from it. The ground water transports radon atoms upwards from the deep depth, where it is mixed with the pore water of high radon concentrations from rocks in the crust.

Some conditions are proposed to make the measurements of radon concentrations in air effective for the earthquake prediction.

1. Introduction

The correlations have been initially observed between the change of radon concentrations in air and the earthquake by Hatuda[1] and Okabe[2]. Afterward many reports[3,4] have been appeared in journals showing possible correlation between them. Radon concentrations in the underground water also have been seen to change in coincidence with the earthquake or prior to it[5-8]. Most of the results obtained by now showed pulse like variations of radon concentrations. Recently, however, three events [9-12] have been observed in the western Japan. Radon concentrations in air increased during a certain period accompanied with the earthquake near its end. The mechanism of pulse like changes of radon concentrations is difficult to be understood. In this report the increase of radon concentrations of some days' duration will be discussed in relation to the earthquake. The requirements on the method of measurement are proposed in order to use this increase for the prediction of the earthquake.

2. Earthquakes in the western Japan

Seismic phenomena are categorized into two kinds of the plate and the crustal fault earthquakes in the western Japan[13]. The hypocenter of the former one is located near the boundary between the Pacific Ocean Plate and the Eurasian Plate. Its depth is deeper than 50 km. The latter earthquake happens at the active faults in the crust of the main land, and its hypocenter is shallower than 20 km. This type of earthquake will be discussed.

The crust of the western Japan is constantly pressed toward the west by the Pacific Ocean Plate moving to the west about 10 cm per year and by the Philippine Sea Plate proceeding to northwest by about 4 cm per year. The upper part of the mantle is flexible, but the crust lying on the mantle is hard for deformation. The difference of the deformation between two parts induces the stress in the crust. This stress is released by fracturing in the crust producing the earthquake and faults. The crust has the history of about two million years in Japan. At an early stage after the formation of Japanese main land, new faults would be produced accompanying the earthquake. In the later phase it is more plausible that the stress in the crust is released by the slip of land at the past faults. Really periods of one to ten thousand years have been observed for many active faults to bring about the earthquake. The active faults, which are quiet for more than some hundred years without inducing the earthquake, are more probable candidates for seismic phenomena in future.

3.Experimental observations of the radon concentration increase supposed to be associated with the earthquakes

Recently three events have been observed being considered to be correlated with the earthquake. These are listed in Table 1.

Table 1. The increase of radon concentrations in air as a foreboder of the earthquake

Site of event	Period of increased radon concentrations (days)	Magnitude and date of earthquake		Depth of hypocenter (km)	Distance of epicenter (km)	Reference
Nagano	80	M 6.8	Sept.17, 1984	3.8	65	[9]
Kobe	150	M 7.2	Jan. 17, 1995	14	20	[10,11]
Kanazawa	11	M 3.2	Sept. 10, 1996	9	9	[12]

At the Nagano site, Ui et al[9] have measured radon concentrations by the cup method at six different measurement points near the active faults. Plastic detectors were replaced every one week. The detectors showed clearly the increase of radon concentrations at all points simultaneously in coincidence with the earthquake. The precursor time interval was obtained as 80 days.

Results at Kobe are more distinctive as shown in Fig. 1. The increment of radon concentrations was estimated every half month from the results obtained by Yasuoka et al [10,11]. They were monitoring the radiation level of their radioisotope laboratory continuously. The increments were obtained by subtracting the average concentrations measured far before the earthquake from those measured on dates around the earthquake. The variation of the increment with date is plotted in Fig.1 before and after the earthquake.

The increase of radon concentrations in Kanazawa has been obtained by Komura et al [12]. They were also monitoring the radiation level of their radioisotope laboratory. Their monitor showed the regular diurnal variations of radon concentrations. However, on a day, its amplitude suddenly increased and the larger swing of diurnal variations continued for about ten days. This abrupt change of amplitude can be easily recognized in their results. The earthquake happened near the end of the period of abnormal swing. After it, the monitor showed the normal concentrations.

Scholz[14,15] obtained a linear relation between the logarithmic values of precursor time interval and the magnitude of the earthquake. For three seismic phenomena given in Table 1, the expected precursor time intervals are deduced by using this relation. These values are 3.1×10^3, 5.7×10^3, and 9.1 days for Nagano, Kobe, and Kanazawa earthquakes, respectively. The value for Kanazawa fits very well to that expected from the line of Scholz. However, there are large differences in the value of the period between those for the other two phenomena and the expected ones.

4.Transportation of radon atoms in the crust

The increase of average radon concentrations in air is caused by the larger exhalation of radon atoms from the earth surface near the measurement site. More radon atoms than those on usual days are transported to the earth surface from the inside of the crust. Where do these extra radon atoms come from?

The diffusion coefficient of radon in air is estimated to be 1.2×10^{-5} m^2s^{-1} [16]. The diffusion length is about 10 m. The effective diffusion coefficient in the earth is about 10^{-6} m^2s^{-1}. This value gives the diffusion length of about 1 m in the ground. In some active faults the gouge is more porous than the soil surrounding the fault. The radon concentrations in air are measured to increase just across such a fault. The diffusion length is expected to be larger than 1 m. The length, however, can not be larger than that in air and is estimated to be less than about 10 m. Therefore, when the number of radon atoms increases at the deeper place than 10 m in the crust, this increase does not appear as the larger radon exhalation. Most of radon atoms decay to polonium atoms before to reach the earth surface. What can transport more radon atoms to the ground surface? The underground water is assumed to play the role of the possible radon carrier.

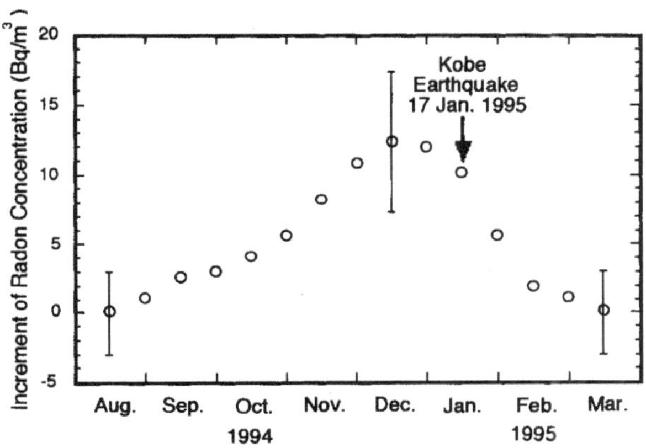

Fig.1. Variations of the increment of radon concentrations in air at Kobe Pharmaceutical University before and after the Kobe earthquake. This figure was obtained from the results of Yasuoka et al[10]. The bars of the increments in August and March are a standard error of the average concentrations of 10 Bq/m^3 measured without the effect of the earthquake. The bar in December roughly shows the fluctuation of the measured values.

In some area, there is a layer which contains the ground water. The water in this layer is supported below it by another layer of non penetrable for water. It is hard for radon atoms coming up from the lower place to penetrate this layer and enter the water. Although the behavior and the situation of water in the deep depth in the crust is not clear, a following mechanism is supposed to cause the increase of the number of radon atom concentrations in water.

The crust in the western Japan is constantly pressed westward by the ocean plates as described in section 2. The crust moves by the average rate of 2 mm per year estimated from the slippage length of active faults. The crust is not uniform. The stress can be concentrated into some region. The rocks in the crust in this region are compressed more strongly than those in other region. Rocks in the deep crust are in the environment of the high pressure of more than 10^7 Pa. Pores in the rocks are filled with water. Radon atoms are produced from ^{226}Ra atoms in rocks. Parts of radon atoms are recoiled out from the pore surface into the pore water. Its radon concentrations are larger than that of river or ordinary underground water.

When the rocks are compressed, the pore water is squeezed out the rocks or transported to the lower pressure region. There, the water is mixed with the underground water, and goes upward through fissures in rocks or more porous media produced by the past slippage of the crust. Radon atoms in the water transfer into the air at the upper level of the water near the ground surface. The rate of radon transfer depends on the behavior of the subterranean water from the deep depth to the ground surface. Whether the flow of water is fast or slow, radon exhalation will increase.

The pore fluid pressure was suggested to drop near final instability of the crustal fault due to the effect of dilatancy [17]. If the fluid pressure drops, the squeezing rate of pore water containing more radon atoms decreases and the increment of radon concentrations in air diminishes. In Fig.1 the earthquake is seen to have happened after passing the peak increment of radon concentrations, though the fluctuation of the increment is large. This phenomenon is in accord with the expected one.

5.Conditions for effective measurements for earthquake predictions

From the above discussions, necessary conditions are given for the measurement of radon concentrations to be effective for the earthquake predictions.

1. One of the measurement sites should be selected near an advisable fault. There the earthquake is expected to occur in near future from the period of past events. The period at an active fault is more than some hundreds years as explored from digging the fault. Therefore, at an active fault, where the event happened recently, its slippage will not occur within a few hundred years. The measurements near such a fault do not give an useful results.
2. Radon concentrations should be measured at other points in the region of the above site. One of them is not near the active fault. By comparing the radon concentrations with that of other points, changes of concentrations are more easily detected.
3. The gouge of the active fault should be porous. Its porosity can be tested by measuring the radon concentrations across the fault. If the measured concentrations increase at the fault, its gouge is considered to be porous. When it is not porous, radon atoms transferred upward from the depth with the water can not diffuse to the earth surface.
4. The ground water at the fault should have connections to the subterranean water deeper than1km. Radon atoms produced in rocks are transported to the fault by the ground water.
5. Radon concentrations in air have diurnal and seasonal variations. An averaging method is desirable to cancel these effect and to obtain the true change of the radon exhalation.

When the above conditions are satisfied, the measured results are expected to be used for the prediction of the earthquake. Furthermore, the results of some plural measurements in the region are useful to examine if the radon concentrations change for earthquake of far distance from the concerned fault.

6.Conclusions

The increase of radon concentrations in air is expected as a precursor phenomenon of the earthquake. In this case, the diffusion of radon atoms through the soil does not play the important role for the increase. Radon atoms in the ground water of deep depth are considered to contribute to the larger exhalation from the earth surface.

Some conditions were proposed for the measurement to be effective for the earthquake prediction. However, more experiments are required to estimate if the proposed method is useful for it.

7.References

[1] Hatuda, Z., Radon content and its change in soil air near the ground surface, Mem.Coll.Sci.Univ. Kyoto Ser. B, 20, 285-306, 1953.

[2] Okabe, S., Time variation of the atmospheric radon-content near the ground surface with relation to some geophysical phenomena, Mem. Coll. Sci. Univ. Kyoto Ser. A, 28, 99-115, 1956.

[3] King, C., Episodic radon changes in subsurface soil gas along active faults and possible relation to earthquakes, J. Geophys. Res., 85, 3065-3078, 1980.

[4] King, C., Gas geochemistry applied to earthquake prediction: an overview, J. Geophys. Res., 91, 12269-12281, 1986.

[5] Hauksson, E., Radon content of groundwater as an earthquake precursor: evaluation of worldwide data and physical basis, J. Geophys. Res., 86, 9397-9410, 1981.

[6] Hauksson, E., and J. G. Goddard, Radon earthquake precursor studies in Iceland, J. Geophys. Res., 86, 7037-7054, 1981.

[7] Teng, T., Some recent studies on groundwater radon content as an earthquake prediction, J. Geophys. Res., 85, 3089-3099, 1980.

[8] Igarashi, G.,S. Saeki, N. Takahata, K. Sumikawa, S. Tasaka, Y. Sasaki, M. Takahashi, and Y. Sano, Ground-water radon anomaly before the Kobe earthquake in Japan, Science, 269, 60-61, 1995.

[9] Ui, H., and I. Fujii, Radon discharge from shear zone of active faults and its relation to seismic activity, Environmental radon, 86-95, 1992. (in Japanese)

[10] Yasuoka, Y., and M. Shinogi, A possible precursor of Kobe earthquake detected by air monitor,

Isotope News, No 503, 74-76, 1996. (in Japanese)
[11] Yasuoka, Y., and M. Shinogi, Anomaly in atmospheric radon concentration: a possible precursor of the 1995 Kobe, Japan, earthquake, Health Physics, 72, 759-761, 1997.
[12] Komura, K., T. Wataguchi, S. Yamasaki, private communication, 1996.
[13] Katase, A., Mem. Tohwa Univ., No 22, 189-195, 1996. (in Japanese)
[14] Scholz, C.H., L.R.Sykes, and Y.P. Aggarwal, Earthquake prediction: a physical basis, Science, 181, 803-181, 1973.
[15] Scholz, C.H., The mechanics of earthquakes and faulting, Cambridge Univ. Press, 1990.
[16] Nazaroff, W.W., B.A.Mod, and R.G. Sexto: in W.W. Nazaroff and A.V. Nero, Jr. ed., Radon and its decay products in indoor air, p57, John Wiley & sons, New York, 1988.
[17] Rice, J.R., and J.W. Rudnicki, Earthquake precursory effects due to pore fluid stabilization of a weakening fault zone, J. Geophys. Res., 84, 2177-2193, 1979.

C Environmental radon and thoron

Modeling radon flux from the earth's surface

S. D. Schery and M. A. Wasiolek

Geophysical Research Center and Department of Physics
New Mexico Institute of Mining and Technology
Socorro, NM 87801 USA

Abstract

We report development of a ^{222}Rn flux density model and its use to estimate the ^{222}Rn flux density over the earth's land surface. The resulting maps are generated on a grid spacing of $1^0 \times 1^0$ using as input global data for soil radium, soil moisture, and surface temperature. While only a first approximation, the maps suggest a significant regional variation (a factor of three is not uncommon) and a significant seasonal variation (a factor of two is not uncommon) in ^{222}Rn flux density over the earth's surface. The estimated average global flux density from ice-free land is 34 ± 9 mBq m^{-2} s^{-1}.

1. Introduction

The exhalation of radon (^{222}Rn) from the earth's surface has been a subject of interest for many years [1,2,3,4,5,6]. Radon, its isotope thoron (^{220}Rn), and their decay products are a major source of atmospheric ions near the earth's surface. Particularly near the earth's surface, radon thus plays a role in the electrical properties of the atmosphere and in processes such as thunderstorm electrification and charging of aerosol particles. Due to convenient half-lives and chemical properties, radon gas and its decay products have been popular tracers for studying many atmospheric processes. Examples are tropospheric-stratospheric air exchange [5] and the residence time of aerosol particles in the troposphere [7].

During the past decade, there has been renewed interest in this subject of radon exhalation from the earth's surface for several reasons. One is the health issue that arises from recognition that radon gas and its decay products are the major source of ionizing radiation dose to the public. Although this exposure primarily occurs indoors, infiltration of radon gas from the soil has been identified as one of the main mechanisms influencing indoor radon levels in many buildings. It has become important to understand the processes controlling radon entry into buildings from the soil. It is also important to identify what locations and soil conditions are most likely to favor entry of high levels of radon gas into buildings. There is therefore interest in identifying regions of the earth's surface likely to have a strong potential for entrance of high levels of radon gas into buildings [8,9,10,11]. The scientific approach of these studies differs somewhat from the traditional study of radon flux from the earth's surface because pressure driven flow (advection), and permeability of soil, are more important for radon entry into buildings than for the diffusion-dominated flux to the outdoor atmosphere. However, in common with the situation for direct, diffusion-dominated radon flux to the atmosphere, the radium (^{226}Ra) content of the soil, and the fraction of radon gas released to the pore space (emanating fraction), remain central factors controlling the amount of radon transported.

Another reason for renewed interest in radon exhalation relates to global pollution and global climate change. Even though radon gas is not involved directly, a better understanding of its exhalation from the earth's surface has turned out to be important. For one thing, its transport involves some of the same processes controlling the soil/air exchange of important greenhouse gases such as CH_4, NO_2, and CO_2. The source of radon gas is relatively well known (the radium in soil) and radon gas does not undergo complicated chemical reactions. Study of radon gas is therefore able to provide information on the pure transport component of gaseous exchange between soil and the atmosphere.

A more immediate connection with global pollution and climate change studies involves the validation of global chemistry and transport computer models. Due to radon's well understood chemical behavior (inert gas), a convenient half-life (3.82 days), and a source term of the land's surface, atmospheric measurements of radon gas are popular for validating these models [12, and references therein]. However, as mentioned in several recent papers [13,14], development of these models has progressed to a point where improved knowledge of the radon source term (i.e., the amount and distribution of radon flux from the earth's surface) is needed for more accurate validation. It is this need that provided the catalyst for the research reported in this paper.

Traditionally, global models for radon exhalation from the earth's surface have been limited to selection of an average flux density for all land, such as 21 mBq m^{-2} s^{-1} ($\approx$1.0 atom cm^{-2} s^{-1}), with perhaps a much lower value for the oceans, such as 0.10 mBq m^{-2} s^{-1}. Sometimes, a separate flux density is assigned to tundra or polar regions. Most of these averages have been based upon very limited flux density measurements (often numbering in the hundreds or fewer [15]) or are inferred indirectly from atmospheric radon gas measurements or fallout of ^{210}Pb [4]. Such simplified radon source terms are still being used in many calculations [12].

Unfortunately, at present there appears nowhere near enough radon flux measurements distributed over time and locations to directly map radon exhalation, even using interpolation. Even within the limits of use of an average flux density for land, there remains a question of the accuracy of the chosen values. For example, although in principle a few hundred radon flux density measurements randomly distributed in time and location should be adequate to determine an average flux density, in practice most measurement sets used for averages appear to represent inadequately certain parts of the globe and certain times of the day and the year.

We report here results of an effort to improve the estimate of the distribution and amount of radon flux from the earth's surface. Our approach is first to use porous media transport theory to formulate an idealized model for radon flux density in terms of the controlling environmental variables. We then attempt to identify variables, or substitute variables, that are most important for controlling flux under typical soil conditions, and for which there exist more global data than exist for radon flux density itself. We then reformulate the model in terms of these variables, calibrate it with measurements of radon flux density, and apply it to generate a global map of radon flux density from the earth's surface. This approach follows many of the recommendations of a 1993 workshop on this subject [16].

2. An idealized model for radon flux density from soil

We summarize here an idealized, one-dimensional, steady-state model for radon flux density from semi-infinite, homogeneous soil that includes a soil temperature and moisture dependence. More complex models are possible. However, given the limited availability of appropriate global data, the present simplification seems a reasonable starting point. We base our model on the porous media theory and recommendations outlined by Holford [17].

We consider radon diffusion in a semi-infinite, homogenous soil in which the pores are partially filled with water. From conservation of mass (the continuity equation) we have

$$D_a\frac{d^2C_a}{dx^2} - \lambda(1-S+S\kappa)nC_a + \phi = 0, \quad (1)$$

where D_a is the effective radon diffusion coefficient, x is the distance from the soil's surface (positive direction downward), C_a is the radon concentration in the air of the pore space, λ is the radon decay constant, S is the fraction of pore space filled with water, κ is the Henry's Law solubility coefficient for radon gas in water, n is the total (dry) volumetric porosity of the soil, and ϕ is the production rate of radon gas into the pore space. From Fick's law

$$F = -D_a\frac{dC_a}{dx}, \quad (2)$$

where F is the flux density of radon per unit bulk area, and boundary conditions of zero radon concentration at the earth's surface and no flux density at great depths, the solution of eq. (1) substituted into eq. (2) gives

$$|F| = \frac{D_a^{1/2}\phi}{[\lambda(1-S+S\kappa)n]^{1/2}} \tag{3}$$

for the magnitude of the flux density at the soil's surface. Now

$$\phi = R\rho_b\lambda E, \tag{4}$$

where R is the radium activity concentration in soil, ρ_b is the soil bulk density, and E is the emanating fraction. Following [17], as an approximation we can take

$$D_a = D_0(T/273)^{3/2}n(1-S)^3, \tag{5}$$

to account for an increase in kinetic diffusion in pore air due to higher temperatures, or a decrease in diffusion due to pores that are partially blocked by water (soil moisture). In eq. (5), D_0 is a pore-structure-dependent constant for a reference temperature of T = 273 K. The combination of eqs. (3), (4), and (5) gives

$$|F| = \frac{R\rho_b E(\lambda D_0)^{1/2}(T/273)^{3/4}(1-S)^{3/2}}{[1-S+S\kappa]^{1/2}}. \tag{6}$$

We note that eq. (6) gives a direct dependence of flux density |F| on both the radium content of soil R and the emanating fraction E. Equation (6) also indicates that flux density should increase with decreasing water fraction S in the pore space and with increasing soil temperature T. The increase in flux density with soil temperature is due to both increased kinetic diffusion in air and less radon dissolved in the water phase since κ is a decreasing function of increasing temperature. The factor E also has a moisture dependence which will be discussed later.

The idealized nature of eq. (6) must be kept in mind in making predictions using environmental data. It assumes soil properties are the same at all depths, which is rarely true for actual soils. Up to a point, this difficulty can be corrected for by using average values for the parameters in eq. (6). Equation (6) assumes a semi-infinite soil, which in practice means a layer deep compared with the radon diffusion length, which is of the order of a meter or less for most soils. Thinner soil layers will give less flux density than predicted by eq. (6), but the reduction factor is not linear with depth. Given that many soils are sufficiently deep for the model, and global data delineating the thickness of soils may not be adequate, as a first approximation we have chosen not to include the depth correction factor. Equation (6) contains no corrections for flux density variation due to temperature-dependent sorption on the soil grains or advective flux density induced by atmospheric pressure variation. Estimates for these effects can be added. We have chosen to neglect sorption because of evidence for a minimal temperature-dependent effect in many soils due to the presence of moisture [18]. On a regional or larger basis, particularly with water tables near the surface, there is some evidence the average effect of pressure is small [6].

3. Sensitivity of flux density measurements to environmental data

Equation (6) is too idealized to be used directly with available environmental data. Specific data are not available for such variables as the emanating fraction or even the fraction of the pore space filled with water. In this section, we discuss screening of environmental data to find those most useful for predicting flux density. Our principal technique is correlation analysis between radon flux density measurements and environmental data.

We analyzed 101 measurements of radon flux density made with the fast emanometer (accumulator) system of Whittlestone and collaborators at the Australian Nuclear Science and

Technology Organisation. Data selected were mostly from mainland Australia [6], supplemented with some of the data from Tasmania and Hawaii [19,20]. Measurements were selected based on a consistent calibration and measurement methodology, and a sampling protocol designed to characterize average regional diffusive flux density. The latitude and longitude were known for all measurements. Accompanying field data were available for some variables such as soil moisture, temperature, and radium content at 10 to 20 cm depth, soil classification, surface total gamma exposure rate, vegetation density under the emanometer, and rate of change of atmospheric pressure. The values of radon flux density ranged from <1 to 128 mBq m^{-2} s^{-1}. The ranges of some accompanying field measurements for soil properties at 10 to 20 cm were as follows: radium content, 1.8 to 66 Bq kg^{-1}; temperature, -1 to 34 ^{0}C; water content as water mass divided by dry soil mass, 0.004 to 0.469.

In addition to screening variables from accompanying field measurements, we studied correlations with other sources of global environmental data. These included variables available in [21], soil diffusivity reported by [22], and soil moisture reported by [23]. Most of these global data were specified on a $1^0 \times 1^0$ grid or a $½^0 \times ½^0$ grid.

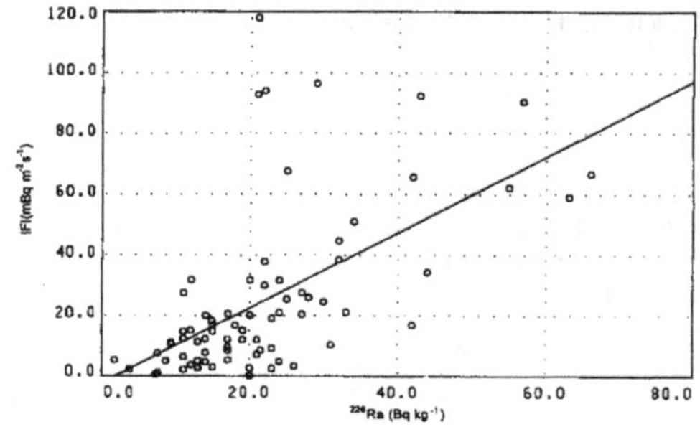

Figure 1. Measured flux density versus measured soil radium concentration.

Linear correlation analysis was run between measured radon flux densities and various environmental variables from the above sources. Variables studied included, but were not limited to, soil radium content, surface gamma radiation, monthly average air temperature, monthly average precipitation, soil type (various classification schemes), soil moisture (various classification schemes), clay content in various soil horizons, soil diffusivity, atmospheric pressure change, and vegetation indices. Correlations were typically weak. One difficulty was that we were often comparing nearly-instantaneous point measurements with regionally and/or time-averaged variables. The clearest correlation (r = 0.59) was between flux density and soil radium (Fig. 1). There was also a similar correlation (r = 0.54) between flux density and surface gamma radiation exposure rate, which is not unexpected since soil radioactivity is responsible for a significant portion of surface gamma radiation. Since surface gamma radiation measurements were available for more of the 101 flux density measurements, subsequent analysis focused on correlations with the calculated variable RNGAMMA = [flux density F divided by (gamma radiation - 30 nGy/hr)] and other variables, where 30 nGy/hr is an estimate of the residual gamma exposure rate in air when no radium is present in the underlying soil. In terms of global datasets, the remaining strongest correlations were for monthly precipitation (rain and snow) (Fig. 2, r = -0.32) and monthly air temperature (r = 0.30), which were also correlated with each other (r=-0.77). Visual inspection of accompanying plots also

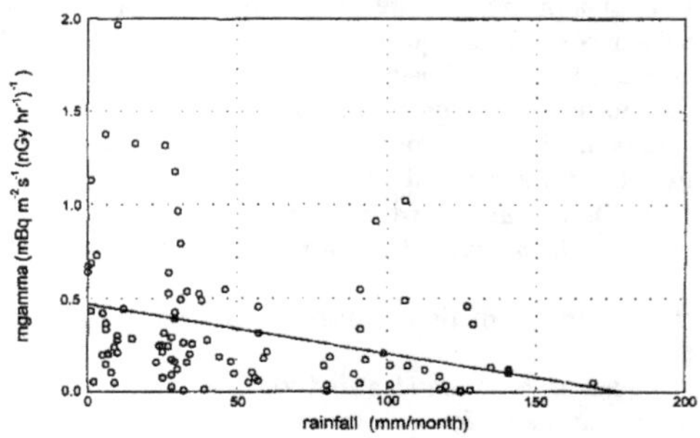

Figure 2. The variable RNGAMMA (flux density divided by net gamma radiation exposure rate) versus monthly rainfall.

indicated a possible non-linear weak correlation of stronger flux density for soils with a smaller average grain size. No, or minimal, statistically significant correlation was apparent with vegetation indices, vegetation under the emanometer, or short-term atmospheric pressure change. We note that Nielson et al. [11], in a study of flux density data for the state of Florida, identified major correlations with surface gamma radiation (positive) and soil moisture (negative), results compatible with our findings, assuming precipitation is a substitute for soil moisture.

4. A reformulated model for radon flux density

In this section we use information from sections 2 and 3 to construct a practical, reformulated model that can be used to predict global radon flux densities from land.

Equation (6) indicates that flux density should be proportional to surface soil radium content and discussion in section 3 pointed out that it gave the most important correlation. Therefore, our reformulated model should retain a proportionality to soil radium content. Because of the central importance of this factor, we include it even though global data for radium are incomplete and will have to be dealt with by estimates. A correlation was identified with average monthly air temperature (a substitute for soil temperature), and global data are available for it (Leemans and Cramer in [21]) . Therefore, we will include average monthly air temperature as an estimate for soil temperature T in eq. (6). Since the correspondence between average soil temperature and monthly air temperature is not one for one, and eq. (6) may not exactly model the full dependence of radon flux density on soil temperature, we will replace soil temperature T (in Kelvin) by $K_T T_a + 273$, where T_a is the average monthly air temperature in Celsius and K_T is a proportionality factor to be determined from the flux density measurements. A correlation was identified with monthly precipitation. We interpret this as related to the dependence of eq. (6) on soil moisture. Over the world, due to different soil characteristics and temperatures, rainfall in itself is not a reliable indicator of soil moisture. Therefore, based on the underlying science and the findings of [11], we use the global soil moisture estimates of Willmot et al. [23] for the root zone instead of monthly precipitation as an estimate for S in eq. (6), even though, for the limited number of flux density measurements mentioned in section 2, Willmott's data did not provide as strong of correlation (r=-0.23). (Part of the explanation for the weaker correlation can be traced to Willmott's use of a coarser grid and limited reference stations in Australia.) Since Willmott's soil moisture data, which are in units of millimeters of water, are not a direct measure of S, we replace S in eq. (6) by $K_S d/d_{max}$, where d is the available moisture at the given location, d_{max} is the maximum moisture possible (150 mm), and K_S is a parameter to be determined from the flux density measurements.

Equation (6) indicates a direct proportion between flux density and the emanating fraction E. Despite local studies indicating some correlation between E and soil properties such as clay content, we found no useful correlations in the limited global data we studied. However, E is known to depend on soil moisture, and following [17], we include an estimate (in a slightly more convenient functional form) for this dependence. We assume E has a non-zero value at zero moisture and approaches a higher constant value E_m at a moderately low value of S. Studies of some of the remaining factors in eq. (6), such as soil density, indicate that they do not vary as greatly from soil to soil as do radium content and soil moisture. In any case, in our initial screening, we could not identify a suitable approach to represent them in terms of available global data, so we group them as a proportionality constant reflecting average conditions. In summary, our reformulated model takes the form

$$|F| \rightarrow \frac{K_F R (E/E_m)(T/273)^{3/4}(1-S)^{3/2}}{[1-S+S\kappa]^{1/2}}, \tag{7}$$

$$\frac{E}{E_m} \rightarrow (1-K_{E1}e^{-K_{E2}S}), \tag{8}$$

where K_F is an overall normalization constant, $T \to K_T T_a+273$, $S \to K_S d/d_{max}$, and the values of constants K_F, K_T, K_S, K_{E1} and K_{E2} are determined from the flux density data.

The constants in eqs. (7) and (8) were determined from the flux density measurements described in section 3. Since more flux density and moisture data (99 values) were available for the calculated variable RNGAMMA than for radium, it was first used to determine K_{E1} and K_{E2}. The constants K_{E1} and K_{E2} were then fixed at these values, and the values of K_F, K_T, and K_S were determined by a least squares fit of eq. (7) to the flux density data using the measured data for radium (about 75 values). The temperature dependence of the solubility coefficient κ was taken as $0.1048+0.405\exp(-0.0500K_T T_a)$, which provides a good fit to the experimental radon solubility data over the range 0 to 100 ^{0}C for $K_T = 1$. If the temperature T_a was less than 0 ^{0}C, κ was fixed at its value for 0 ^{0}C (0.51). The resulting values of the constants were $K_F = 1.48$, $K_T = 3.17$, $K_S = 0.491$, $K_{E1} = 0.324$, and $K_{E2} = 20.4$ for $|F|$ in units of mBq m^{-2} s^{-1} and R in units of Bq kg^{-1}. Figure 3 shows a plot of measured flux densities against model predictions with (7). The linear correlation coefficient is $r = 0.587$. Curved lines indicate the 95% confidence range for a linear fit.

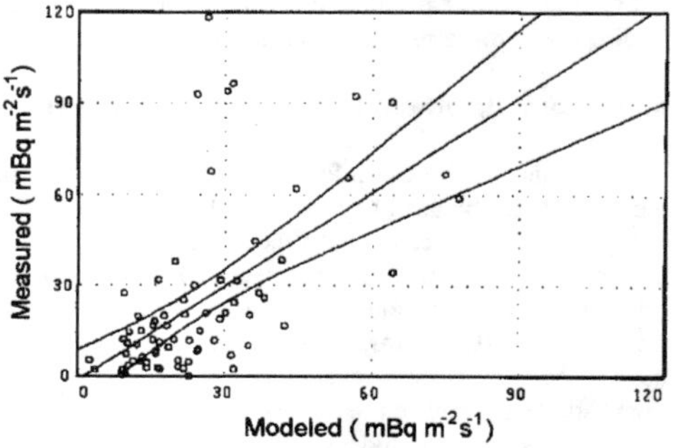

Figure 3. Measured versus modeled radon flux density.

Least squares mathematical procedures can be particularly sensitive to outlying points. As one check, we removed the three highest and lowest points from Fig. 3 (perpendicular distance from the linear regression line) and recalculated K_F. The result was a reduction in K_F by 5.6%. We concluded outlier points were not a major source of error in our procedures. The merit of the use of eq. (7) for fitting the data of section 3 is that it is based on a physical description of the process and should be useful for extrapolating flux density predictions for combinations of soil radium, temperature, and soil moisture beyond the range of the data that accompanied the calibration flux density measurements. The fact that K_T is greater than 1 and K_S is less than 1 suggests a stronger temperature dependence and weaker moisture dependence than implied by eq. (6). With these values for the constants, eq. (7) cannot then be interpreted literally in terms of its original physical basis but must be viewed more as a potentially useful mathematical parameterization. The functional form of eq. (7) puts many constraints on prediction of the flux density data, and considered as a pure mathematical exercise does not allow as much flexibility for predicting the experimental data as the number of adjustable parameters might indicate. For comparison, we fit the flux density measurements with a simpler, flexible mathematical expression involving soil radium multiplied by a second degree polynomial combination of the temperature and monthly precipitation with five adjustable parameters, which we call the quadratic model. The correlation coefficient was similar (r=0.527). However, the resulting expression gave unrealistic results at certain values of temperature and rainfall that would be encountered at other locations of the earth's surface. We concluded that eq. (7) was a superior model for predicting flux densities.

5. Global predictions of radon flux density from the reformulated model

Before making predictions with eq. (7), it was first necessary to have estimates of the radium concentration in surface soil on a global basis. The calibration of eq. (7) from the data described in section 3 is based on measurements of radium in surface soil (20 cm depth). Only in a few regions, such as China, are suitable data from direct measurements available. We therefore used other procedures to estimate surface radium for most countries. Details will be given in another paper, but we will summarize our procedures here. For some countries, aerial gamma radiation data (usually from uranium exploration surveys) were used to estimate surface radium with a calibration factor determined from surface radium measurements in the same country. For some countries, maps were available of exposure rate from gamma radiation. The deduction of surface radium from exposure rate depended on an average conversion factor from experimental measurements. This use of an average conversion factor does not allow for variation in cosmic radiation or in relative concentration of the thorium and uranium series. For some countries, data were available for uranium in stream sediments which were also converted to radium with an experimentally determined average factor. For some countries, a single number for average radium in the country was available. Where none of the above approaches were possible, we assigned soil an average radium concentration of 30 Bq kg^{-1} based on the world estimate of the NCRP [25]. Table 1 summarizes the procedures for various countries and regions. Figure 4 shows a map of the resulting radium concentrations grouped according to below average (<29 Bq kg^{-1}), average (29-30 Bq kg^{-1}), and above average (>30 Bq kg^{-1}) concentration. The world-averaged radium concentration for analyzed regions in Fig. 4 is 30 Bq kg^{-1}, which is within one percent of the NCRP value used for unanalyzed regions, although our procedures did not force this result.

Table 1. Procedures for assigning surface radium concentration

COUNTRY	PROCEDURE
United States of America (including Alaska)	aerial gamma, normalized to surface radium (Myrick et al. [24] , 48 state average of 41 Bq kg^{-1})
Canada	aerial gamma, stream sediments, country average elsewhere (24 Bq kg^{-1})
China	direct surface measurements, country average 36 Bq kg^{-1}
Western Europe	aerial gamma, exposure rate, direct surface measurements, stream sediments
India	exposure rate
Japan	exposure rate
Australia	country average (27.9 Bq kg^{-1})
rest of world	world average (30 Bq kg^{-1})

Although it is not possible to accurately illustrate on a map with the grey-scale resolution of Fig. 4, the underlying data indicate that it is not uncommon to have a radium concentration variation between regions by as much as a factor of three, such as occurs between parts of southeast and northwest China. Since in our model the radon flux density is proportional to the radium concentration, there is a comparable factor affecting radon flux density.

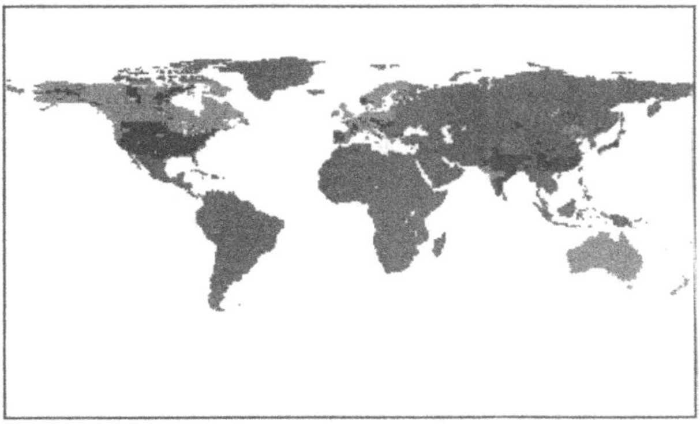

Figure 4. Surface radium concentrations assumed in flux density predictions. Light grey (below average), dark grey (average), black (above average).

Monthly-averaged estimates for the global radon flux density were then generated on a $1^0 \times 1^0$ grid using, in eq. (7), the estimated surface radium data described above, the Willmott soil moisture data, and average monthly air temperature from Leemans and Cramer in [21]. For oceans and ice-covered regions, such as Antarctica, a fixed flux density of 0.14 mBq m^{-2} s^{-1} was assigned [26]. Figure 5 shows a map of the annually-averaged radon flux density, grouped according to low, medium, and high flux densities, obtained from an average of the twelve monthly predictions. The average global flux density for ice-free land for the data used to generate Fig. 5 is 33.8 mBq m^{-2} s^{-1} (1.61 atoms cm^{-2} s^{-1}). If the surface of the earth is assumed smooth (no surface

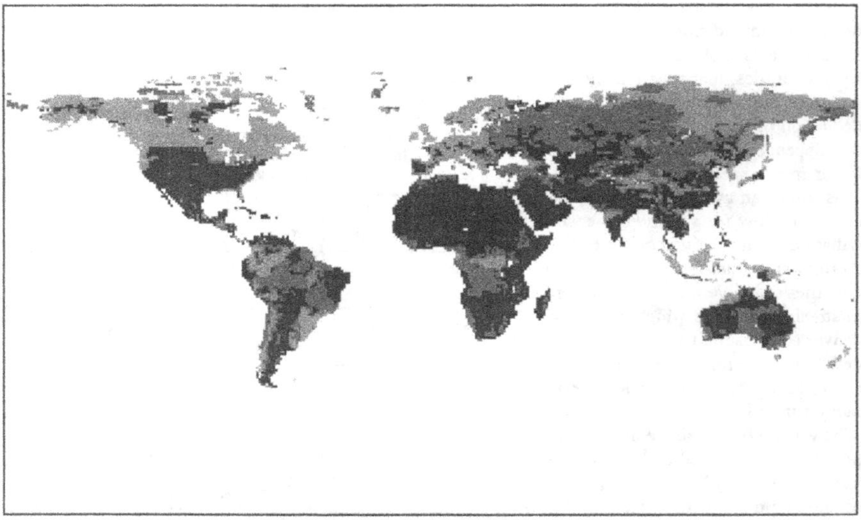

Figure 5. Annually-averaged radon flux density predicted with the reformulated model eq. (7). Data are displayed as a lower third (light grey), a middle third (darker grey), and an upper third (black).

roughness correction) integration of the radon flux density over the full globe's surface (including oceans) gives an annually-averaged radon emission rate of 25.1 kg yr^{-1}. On a monthly basis the global emission rates are surprisingly similar (within 16% or less of one another), with the highest value for July and the lowest for April. However, regionally, in places such as, for example, the central part of South America, the predicted monthly fluxes can vary more, with a seasonal variation of up to a factor of two not uncommon. Figure 6 shows low, middle, and high flux density areas for South America for the months of January and July. Since a constant surface radium concentration was assumed for this continent, patterns shown reflect variation of soil moisture and temperature only. The shift of the low flux density region from the north central in January further to the north in July corresponds to the shift of a high soil moisture region in Willmott's data. This shifting of the low flux density region appears to be consistent with observations of Pereira et al. [27] for the Amazon basin.

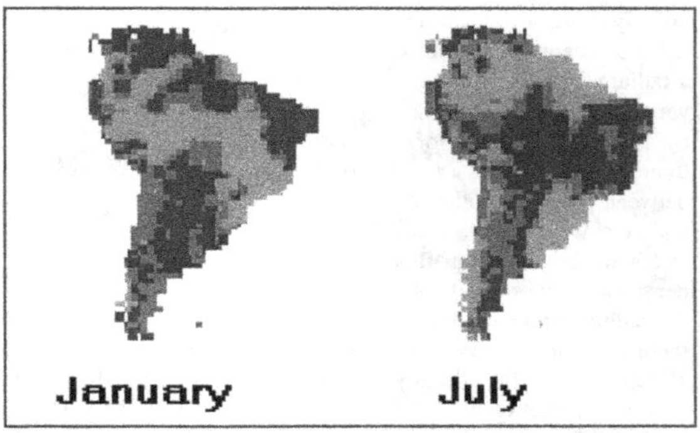

Figure 6. Radon flux density predictions for January and July for South America. The flux density levels are indicated as lowest third (light grey), middle third (dark grey), and highest third (black).

6. Discussion

Compared with previous estimates of radon flux density from the earth's surface, such as a constant value of about 21 mBq m^{-2} s^{-1}, the present modeling indicates a significant spatial, and, to lesser extent, temporal variation in flux density. Predictions for the USA and China are the most detailed due to more complete information on soil radium in these countries. Results from the model indicate that variation in soil radium is the single largest factor causing spatial variation in the radon flux density followed by temperature and soil moisture. Temperature and soil moisture change account for major seasonal regional variations, such as predicted for central South America and the southeastern United States of America. The model suggests a higher average flux density from land (of the order of 34 mBq m^{-2} s^{-1}) than generally assumed (except see [28]), although this statement has to be qualified by the error analysis discussed below. As noted elsewhere [6], some flux density measurements employing the accumulator method used for earlier estimates may have been susceptible to a back diffusion problem which could cause measurements to underestimate true flux density. Our higher result for average annual radon flux density may be more difficult to reconcile with recent analysis of ^{210}Pb fallout [13]. The model predicts only small seasonal variation in the globally-averaged monthly flux density (16% or less), apparently because increasing soil moisture or temperature in one part of the globe is compensated by decreasing soil moisture or temperature in another part.

Completely satisfactory error analysis for our model is not possible given the lack of sufficient measurements of radon flux density. However, we can make a few comments. The quadratic model mentioned in section 4, when applied to the globe, gave an average flux density 12% less than the prediction of eq. (7) for regions where the quadratic model did not require extrapolation beyond the range of the calibration data (82% of the land surface). We conclude that eq. (7) is doing a reasonably good job of interpolating the flux density data over the range of the calibration data for soil moisture and temperature; greater uncertainty exists for very low temperatures (such as Arctic regions) and very high and low soil moistures. The absolute normalization of eq. (7) is tied to the accuracy of the flux density measurements (about ±15%) and the accompanying soil radium measurements (about ±10%). Comparison of literature estimates for average global radium content of soil suggests an uncertainty of at least ±15%. Combining these uncertainties in quadrature (independent errors), including a model fitting uncertainty of ±12%, gives an overall uncertainty of about ±26% in the normalization of the present prediction of average global flux density.

There are very little independent experimental data for radon flux density in a form suitable for comparisons with predictions of the present reformulated model. Currently, only rough comparisons seem possible. The present model gives an annually averaged flux density for the USA of 44 mBq m^{-2} s^{-1}, which compares favorably with an estimated range of 40 to 48 mBq m^{-2} s^{-1} from a study of ^{210}Pb-^{226}Ra disequilibrium in soils [28]. On the other hand, Wilkening [15] gives a statewide average flux density for New Mexico, USA, of 30 mBq m^{-2} s^{-1}, which is noticeably lower than the prediction of 56 mBq m^{-2} s^{-1} from the present model. Part of this difference may be due to inadequate statistical sampling of the state, and due to Wilkening's experimental procedures underestimating flux density, but a bigger part may have to do with the accuracy of our radium concentration values for New Mexico and normalization of the model based on radium data from Australia. The present model gives an annually averaged flux density for the state of Florida, USA, of 27.6 mBq m^{-2} s^{-1}, which compares with a geometric mean of 14.8 mBq m^{-2} s^{-1} and an arithmetic mean of 50 mBq m^{-2} s^{-1} from 882 measurements reported in [11 and private communication]. The arithmetic mean value is probably the more suitable value for comparison and suggests a modest disagreement in the opposite direction from that for New Mexico: the present model predictions are lower than the measurements. We previously mentioned qualitative agreement with seasonal variation for the Amazon Basin of South America. Similarly, Dörr and Münnich [29] indicate higher radon flux densities in summer than winter for a soil of high clay content site near Heidelberg, Germany. Our model predictions for this grid sector have a ratio for summer (July) to winter (January) of 1.62.

The modeling results presented here should only be viewed as a first effort at this approach to predicting global radon flux. A number of additional steps are possible to improve predictions. The first is to obtain a larger set of flux density data covering a wider variety of conditions, such as lower temperatures, very low and high soil moistures, and different soil types. Such flux density measurements require careful, traceable calibrations. At the very least, the precise location and time of the measurements need to be recorded. Better global data for controlling variables need to be developed. Improved estimates of surface radium from both direct measurements and aerial gamma radiation data are particularly needed; there already exist much more data than we were able to properly evaluate and process in this initial effort. Improved predictions of soil moisture on a finer grid are probably becoming available and should be used. More work needs to be done to correlate soil geology with radium content and emanating radon fraction. More attention must be paid to proper statistical interpretation of data: are mean or median values being reported, do data represent point measurements or regional averages? Although there is lingering argument in the literature, we are convinced that on a global basis plant transpiration is an insignificant factor for exhalation of radon to the atmosphere [30]. However, depth of soil should be considered more carefully, and for regions with a low water table, atmospheric pressure variation might be important. For estimating integrated radon flux to the atmosphere, the roughness of the earth's surface needs consideration. The diffusion length for radon in soil is typically from a few tens of centimeters to a meter, so roughness on this scale may be important. We have made estimates with crude data for soil slope that suggest total global radon flux might be enhanced by as much as 10% due to this roughness effect.

7. Conclusions

The following summarizes the main conclusions of this paper:

○ Predictions from a model based on porous media transport theory and calibrated with experimental radon flux density data indicate a significant regional variation (a factor of three is not uncommon) and a significant seasonal variation (a factor of two is not uncommon) in radon flux density from soil over the earth's surface. The most significant factor causing regional variation in flux density is soil radium content, followed by soil temperature and moisture. Soil moisture and temperature variation are important for seasonal variation in radon flux, typically at comparable levels.

○ The model predicts an average annual global flux density from ice-free land of 34 ± 9 mBq m^{-2} s^{-1} that is higher than many previous estimates. This number is very sensitive to surface radium estimates which need to be determined more carefully.

○ More, carefully-calibrated flux density measurements over a greater variety of conditions are needed for better validation of flux density models such as the present one.

○ More extensive global data for controlling environmental variables are needed, particularly for soil radium and/or aerial gamma radiation.

8. Acknowledgments

This research was supported in part under USA Department of Energy grant DEFG03-94ER6178.

9. References

[1] Israel, H., Radioactivity of the atmosphere, Amer. Meteo. Soc., Boston, 155-161, 1951.

[2] Ikebe, Y., Variation of radon and thoron concentrations in relation to the wind speed, J. Meteor. Soc. of Japan, 48, 461-468, 1970.

[3] Wilkening, M. H., W. E. Clements, and D. Stanley, Radon-222 flux measurements in widely separated regions, in Proceedings of the Natural Radiation Environment II, edited by J. A . S. Adams et al., pp. 717-730, U. S. A. Dept. of Comm. Rep. Conf-720805, National Technical Information Service, Springfield, VA, 1972.

[4] Turekian, K. K., Y. Nozaki, and L. K. Benninger, Geochemistry of atmospheric radon and radon products, Ann. Rev. Earth. Planet. Sc., 5, 227-255, 1977.

[5] Lambert, G., G. Polian, J. Sanak, B. Ardouin, A. Buisson, A. Jegou, and J. C. Le Roulley, Cycle du radon et de ses descendants: application à l'etude des échanges troposphère-stratosphere, Ann. Géophys., 38, 497-531, 1982.

[6] Schery, S. D., S. Whittlestone, K. P. Hart, and S. E. Hill, The flux of radon and thoron from Australian soils, J. Geophys. Res., 94, 8567-8576, 1989.

[7] Moore, H. E., S. E. Poet, and E. A. Martell, ^{222}Rn, ^{210}Pb, ^{210}Bi, and ^{210}Po profiles and aerosols residence times versus altitude, J. Geophys. Res., 78, 7065-7075, 1973.
[8] Nazaroff, W. W., Radon transport from soil to air, Rev. Geophys., 30, 137-160, 1992.
[9] Gundersen, L. C. S., and R. R. Schumann, Mapping the radon potential of the United States: examples from the Appalachian, Environ. Internat., 22, Suppl. 1, S829-S837, 1996.
[10] Kemski, J., R. Klingel, and A. Siehl, Classification and mapping of radon-affected areas in Germany, Environ. Internat., 22, Suppl. 1, S789-S798, 1996.
[11] Nielson, K. K., V. C. Rogers, and R. B. Holt, Measurements and calculations of soil radon flux at 325 sites throughout Florida, Environment International, 22, Suppl. 1, S471-S476, 1996.
[12] Jacob, D. G., et al., Evaluation and intercomparison of global atmospheric transport models using ^{222}Rn and other short-lived tracers, J. Geophys. Res., 102, 5953-5970, 1997.
[13] Balkanski, Y. J., D. J. Jacob, G. M. Gardner, W. C. Graustein, and K. K. Turekian, Transport and residence times of tropospheric aerosols inferred from a global three-dimensional simulation of ^{210}Pb, J. Geophys. Res., 98, 20573 - 20586, 1993.
[14] Allen, D. J., R. B. Rood, A. M. Thompson, and R. D. Hudson, Three-dimensional radon 222 calculations using assimilated meteorological data and a convective mixing algorithm, J. Geophys. Res. 101, 6871-6881, 1996.
[15] Wilkening, M., Radon in the Environment, Elsevier, Amsterdam, Chapt. 5, 1990.
[16] Kritz, M. A., Report of the Natural Radionuclides Group in Report of the Fourth Global Radon Emissions Inventory Activity Workshop, International Global Atmospheric Chemistry Project, 1994.
[17] Holford, D. J., Rn3D: A Finite Element Code for Simulating Gas Flow and Radon Transport in Variably Saturated, Nonisothermal Porous Media: User's Manual, Version 1.0, PNL-8943/UC-402, NTIS, Springfield, VA, USA, 1994.
[18] Schery, S. D., and S. Whittlestone, Desorption of radon at the earth's surface, J. Geophys. Res., 94, 18297-18303, 1989.
[19] Whittlestone, S., W. Zahorowski, and S. D. Schery, Relationships between radon flux, radiometric contours and soil pore radon concentrations in Australia, Third International Workshop on Geological Aspects of Radon Risk Mapping, October, Prague, 1996.
[20] Whittlestone, S., S. D. Schery, and Y. Li, Thoron and radon fluxes from the Island of Hawaii, J. Geophys. Res., 101, 14787-14794, 1996.
[21] NOAA-EPA Global Ecosystems Database Project, Global Ecosystems Database Version 1.0. User's Guide, Documentation, Reprints, and Digital Data on CD-ROM. USDOC/NOAA National Geophysical Data Center, Boulder, CO, 1992.
[22] Potter, C. S., E. A. Davidson, and L. V. Verchot, Estimation of global biogeochemical controls and seasonality in soil methane consumption, Chemosphere, 32, 2219-2246, 1996.
[23] Willmott, C. J., C. M. Rowe, and Y. Mintz, Climatology of the terrestrial seasonal water cycle, J. of Climatology, 5, 589-606, 1985.
[24] Myrick, T. E., B. A. Berven, and F. F. Haywood, Determination of concentrations of selected radionuclides in surface soil in the U. S., Health Physics, 45, 631-642, 1983.
[25] National Council on Radiation Protection and Measurements Report 50, NCRP, Washington, D. C., USA, 1976.
[26] Appleby, P. G., and F. Oldfield, Application of Lead-210 to sedimentation studies, in Uranium-series Disequilibrium, eds. Ivanovich and Harmon, Clarendon Press, Oxford, 1992.
[27] Pereira, E. B., D. J. R. Nordemann, S. C. Wofsy, and S. Trumbore, Vertical radon concentration profiles over the Brazilian Amazon basin during the wet season, Eos Trans. AGU, 69, 321, 1988.
[28] Graustein, W. C., and K. K. Turekian, Radon fluxes from soils to the atmosphere measured by ^{210}Pb-^{226}Ra disequilibrium in soils, Geophys. Res. Lett., 17, 841-844, 1990.
[29] Dörr, H., and K. O. Münnich, ^{222}Rn flux and soil air concentration profiles in West-Germany. Soil ^{222}Rn as tracer for gas transport in the unsaturated soil zone, Tellus 42B, 20-28, 1990.
[30] Schery, S. D., Is transpiration an important mechanism for the release of ^{222}Rn to the atmosphere?, Health Physics, 57, 1038-1039, 1989.

Modelling of radon transport in porous media

E.R. van der Graaf, W.H. van der Spoel, R.J. de Meijer

Environmental Radioactivity Research and Consultancy Group
Kernfysisch Versneller Instituut
Zernikelaan 25, 9747 AA Groningen, the Netherlands

Abstract

This paper aims to describe the state of the art of modelling radon transport in soil on basis of multiphase radon transport equations. Emphasis is given to methods to obtain a consistent set of input parameters needed for such models. Model-measurement comparisons with the KVI radon transport facility are evaluated and suggestions are made for the direction of future research to resolve some of the pertinacious discrepancies.

1 Introduction

One of the central questions of radon research is: How to predict the indoor radon concentration of a building using details about its construction, local climate, underlying soil type etc.? In principle, given sufficient information such a prediction is possible, using models based on well-known physics. However, due to the complexity of the situation, caused by the interdependence of the processes and the multitude of input parameters needed, often all dependent on both position and time, a prediction seldomly exceeds the level of a guess. Studying subsystems with reduced complexity (such as soil and building materials) seems a logical way to proceed. Characterising entry rates of radon originating from these sources involves modelling of radon transport in porous media. It should be emphasized that not only a physical sound description of the processes involved is needed (diffusion, advection, adsorption etc.) but also consistent measurement methods for the input parameters (characterizing the porous medium).

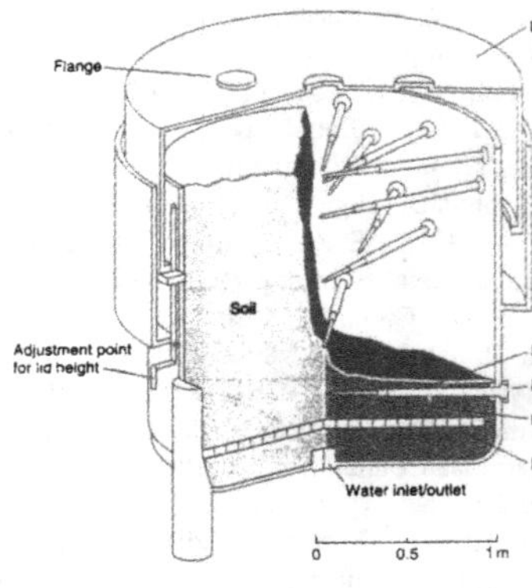

Figure 1: *KVI radon vessel.*

Over the years many studies have focussed on measuring and modelling radon entry from the soil. The geometries in which these studies were conducted can roughly be divided into three categories: 1) test houses [1]; 2) small *in situ* structures [2] and 3) laboratory facilities [3,4]. The test house and small structure studies have resulted in qualitative knowledge about correlations of indoor radon concentration with ambient parameters. However, although in some of these studies detailed modelling was undertaken, the model-measurement comparisons were often hampered by lack of knowledge of input parameters and/or uncontrollable ambient conditions.

For validation of models for radon transport in soil, laboratory facilities provide bench marks for model-measurement comparisons. Such a facility is available at the KVI and consists of a cylindrical vessel (height and diameter 2 m) filled with sand (Figure 1). At different depths probes are radially inserted for soil-gas radon concentration measurements. To study advective radon transport a circular perforated box is placed in the sand at a depth of 163 cm below the sand surface. An air flow through the sand may be induced by (de)pressurizing this box. The vessel is also equipped with a height-adjustable lid to simulate a crawl space and a water in- and outlet to set a water level. This facility is used to measure radon profiles with a systematically increase of complexity of the situation. These measurements are used to validate transport models on basis of independently measured input parameters.

In this paper first the theory of radon transport in porous media will be presented and then methods to obtain the input parameters will be discussed. Thereafter some model-measurement comparisons for the KVI facility will be evaluated with emphasis on situations were modelling fails to describe the experimental results. Finally suggestions will be made for the direction of future research to resolve some of the remaining discrepancies.

2 Theory

For a description of radon transport through a porous medium information is needed on: 1) the structure of the medium; 2) the processes causing the transport; 3) the interaction of radon with the medium and 4) radon generation.

A porous medium is a solid material traversed by usually irregular shaped and positioned pores (Figure 2). This implies that an exact description of the medium based on structural information on pore-scale is unfeasible and that structural quantities averaged over a larger volume should be defined. In most cases the pore space is filled with air and/or water, leading to a multiphase system with a mass balance equation for each phase. Parameters to quantify the structure and phase composition of a porous medium are the porosity ϵ (volume fraction of medium occupied by pores) and the fraction of water saturation m (volume fraction of pores occupied by water). In soil and building materials diffusion and advection are responsible for radon transport. The radon activity bulk fluxes ($\mathrm{Bq\,m^{-2}\,s^{-1}}$) follow from Fick's and Darcy's Laws:

$$\vec{J}_{d,a} = -(1-m)\epsilon\tau_a D_a \nabla C_a; \quad (1)$$

$$\vec{J}_{d,w} = -m\epsilon\tau_w D_w \nabla C_w; \quad (2)$$

$$\vec{J}_{a,a} = -C_a \frac{K}{\mu} \nabla P. \quad (3)$$

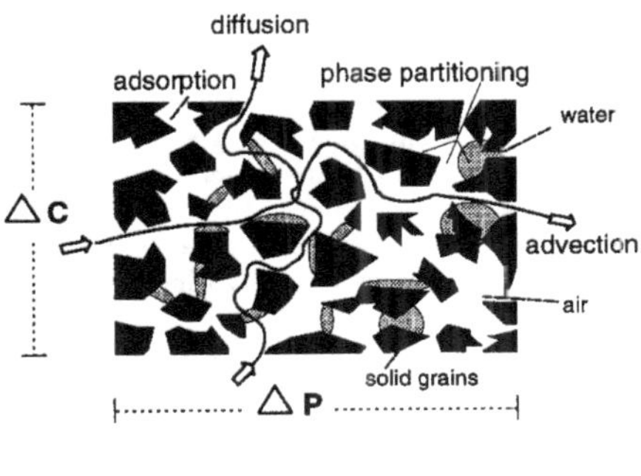

Figure 2: *Schematical presentation of porous medium.*

In these equations the subscripts d, a on the fluxes denote diffusion and advection and subscripts a and w refer to the air and water phase. The quantity τ is the tortousity, an empirical factor accounting for the elongated path of transport due to twisting and turning of the pores; D_a and D_w are diffusion coefficients ($\mathrm{m^2\,s^{-1}}$) in air and water; K (m^2) is the intrinsic permeability; μ (Pa s) is the dynamic viscosity of air, C_a and C_w ($\mathrm{Bq\,m^{-3}}$) are radon concentrations in the air and water phase and P (Pa) is the pressure disturbance field (relative to absolute pressure).

Due to adsorption and desorption of radon on solid surfaces and solution in the liquid phase radon will distribute itself between the different phases (air, water, solid surfaces). These processes are described by transfer coefficients, T_{ij} ($\mathrm{Bq\,m^{-3}\,s^{-1}}$) defined as the rate of transfer of radon activity per unit volume from phase i to phase j (denoted by the subscripts a: air; w: water or s: solid surfaces).

Inside the porous medium radon will be generated with a bulk production rate S ($\mathrm{Bq\,m^{-3}\,s^{-1}}$):

$$S = \eta\rho_b\lambda C^{Ra} \quad (4)$$

where η is the sum of fractional emanation coefficients into the air-, water-, and adsorbed phase ($\eta = \eta_a + \eta_w + \eta_s$); ρ_b is the bulk dry density of the porous medium ($\mathrm{kg\,m^{-3}}$), λ is the decay constant of radon (s^{-1}) and C^{Ra} is its radium activity per unit dry mass ($\mathrm{Bq\,kg^{-1}}$).

The mass balance equations for each phase are given by:

$$(1-m)\epsilon\frac{\partial C_a}{\partial t} = \nabla\cdot((1-m)\epsilon\tau_a D_a \nabla C_a) + \frac{K}{\mu}\nabla P\cdot\nabla C_a - (1-m)\epsilon\lambda C_a + \eta_a\rho_b\lambda C^{Ra} - \sum_{i\neq a}(T_{ai} - T_{ia}); \quad (5)$$

$$m\epsilon\frac{\partial C_w}{\partial t} = \nabla\cdot(m\epsilon\tau_w D_w\nabla C_w) - m\epsilon\lambda C_w + \eta_w\rho_b\lambda C^{Ra} - \sum_{i\neq w}(T_{wi} - T_{iw}); \quad (6)$$

$$\rho_b\frac{\partial C_s}{\partial t} = -\rho_b\lambda C_s + \eta_s\rho_b\lambda C^{Ra} - \sum_{i\neq s}(T_{si} - T_{is}). \quad (7)$$

The terms on the right hand side of equation 5 represent the rate of change due to diffusion, advection, decay, production and exchange with other phases, respectively. Similar terms occur in equations 6 and 7 except that for the water phase, advection and for the solid phase, diffusion and advection is assumed to have a negligible contribution to radon transport. In equation 7, C_s ($\mathrm{Bq\,kg^{-1}}$) is the radon concentration adsorbed per mass unit of dry material. Equations 5 to 7 are coupled through the transfer coefficients. In first order these transfer coefficients are given by,

$$T_{ij} = \alpha_{ij}\Gamma_i C_i \qquad i,j\in\{a,w,s\} \quad (8)$$

where the α_{ij}'s are exchange rates ($\mathrm{s^{-1}}$) between the phases and the combination $\Gamma_i C_i$ gives the total radon activity per unit volume in phase i (Γ_i is $(1-m)\epsilon$, $m\epsilon$ and ρ_b for i is a, w, s, respectively). Following an approach introduced by Rogers and Nielson [5] we will try to combine these equations to arrive at a single equation for the radon activity concentration in the air phase. For this we assume: 1) adsorption to wet surfaces is negligible; 2) exchange between water and air phase occurs at a time scale much shorter than time scales typical for radon transport; and 3) exchange between air and adsorbed phase is fast or adsorption to dry surfaces is negligible.

The first assumption provides most likely a good approximation for most common soil types because water has a much larger affinity to most mineral surfaces than radon.

Typical exchange times between the water-air phase are estimated to be in the order of 0.1 to 10 seconds for water layers of 10 to 100 μm [6]. Time scales for concentration changes due to diffusion and/or advection are typically much larger, of the order of hours to days, this substantiates the second assumption.

The last assumption is most uncertain. It has been established that for sand, adsorption is small [5] but for soil types like clay ad- and desorption may be important.

The consequence of the first assumption is $T_{ws} = T_{sw} = 0$; the other two assumptions give balanced transfer between the phases involved and thus $T_{as} = T_{sa}$ (=0 in case adsorption is negligible) and $T_{aw} = T_{wa}$. Using equation 8 we derive,

$$C_w = \frac{\alpha_{aw}(1-m)}{\alpha_{wa}m}C_a = LC_a; \quad \text{and} \quad C_s = \frac{\alpha_{as}(1-m)\epsilon}{\rho_b\alpha_{sa}} = k_a C_a \quad (9)$$

where L and k_a are Henry's law type coefficients, the former is often refered to as the Ostwald coefficient while the latter is the radon surface adsorption coefficient. Using equations 9 we can combine the three single phase equations to one equation for radon in the air phase;

$$\beta\frac{\partial C_a}{\partial t} = \nabla\cdot(D\nabla C_a) + \frac{K}{\mu}\nabla P\cdot\nabla C_a - \beta\lambda C_a + S \quad \text{with} \quad (10)$$

$$\beta = (1-m+mL)\epsilon + \rho_b k_a; \quad (11)$$

$$D = ((1-m)\tau_a D_a + m\tau_w D_w L))\,\epsilon; \quad (12)$$

$$S = \eta\rho_b\lambda C^{Ra}; \quad (13)$$

where β, D and S are the partition-corrected porosity [2], bulk diffusion coefficient ($\mathrm{m^2\,s^{-1}}$) and radon production rate per unit bulk volume ($\mathrm{Bq\,m^{-3}s^{-1}}$), respectively. To completely determine the problem one also needs an equation for the pressure field P. In almost all practical situations the air density can be assumed constant so that the pressure field can be calculated from Laplace's equation, viz. $\nabla^2 P = 0$.

3 Input parameters

To apply equation 10 for predicting radon transport in practical situations, values for the parameters β, D, K and S are needed. In this section we will describe methods to obtain these parameters, discuss possible misinterpretation of results and give values of these parameters for the sand in the KVI vessel.

3.1 Partition-corrected porosity

As can be seen from equations 11 to 13 all input parameters dependent on the moisture content m of the porous medium and although not explicitely shown also K is moisture dependent. In the case of soil *in situ* the moisture content will be a function of position above the water table. This information is contained in a so-called water-retention curve [7] that is measured by placing a sample of the porous medium on a box of water saturated very fine-grained sand (Figure 3: *left*). The moisture tension in the sample can be set by adjusting the height of a reservoir connected to

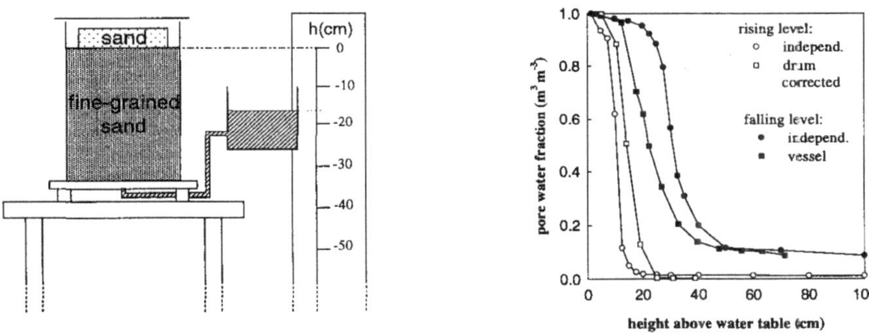

Figure 3: left *Set up for measuring water-retention curves.* right *Water-retention curves for sand in the KVI vessel; circles denote data measured with instrument in left part of the figure. Squares denote data measured by analyzing slices of drill cores taken* in situ *from the KVI vessel and of sand placed in a large diameter drum. Open en closed symbols indicate measurements with rising and falling water level, respectively.*

the box. Figure 3 (*right*) shows the water-retention curve for the sand in the KVI vessel. These curves clearly show hysteresis most likely due to irregularity of pore shapes (e.g. 'ink-bottle' shape) and different contact angles for advancing and receding water fronts [8]. From the water content in the sample at saturation the porosity of the sand was measured at 0.340 ± 0.010. Figure 3 also shows that deviations between water-retention curves measured with an independent set up and curves measured *in situ* might be large. This is most likely related to differences between packing of grains in small samples and in larger soil structures and indicates that moisture content should preferably be measured in situations close to that to be modelled.

As k_a for the sand (room dry) in the radon vessel was measured as to be only $(9.4 \pm 0.6) \cdot 10^{-7}$ $m^3\,kg^{-1}$ [9] radon adsorption can be neglected for all situations considered. This implies that for the vessel sand the partition corrected porosity is given by $\beta = (1 - m + mL)\epsilon$.

3.2 Radon diffusion coefficient

Terminology and notation of the diffusion coefficient is not consistent in the literature. This is caused by the fact that both radon concentration and flux may be expressed on basis of bulk or pore volume cq. surface area. This means that one has to be on the alert in comparing reported diffusion coefficients. Diffusion coefficients are often calculated from measurements on basis of equation 10 written in the following form,

$$\frac{\partial C_a}{\partial t} = D_e \nabla^2 C_a - \lambda C_a + S' \tag{14}$$

were $D_e = D/\beta$ and $S' = S/\beta$. In this, D_e is usually referred to as the effective [10], interstitial or pore diffusion coefficient [11]. S' is the radon production rate per partition corrected pore volume. In case of dry material we have $D_e = D/\epsilon$, and from equation 12 we can derive $D = \epsilon \tau_a D_a$. So for dry granular material on can often predict the diffusion coefficient on basis of a measured value of ϵ taking $D_a \approx 1 - 1.2 \cdot 10^{-5}\,\mathrm{m^2\,s^{-1}}$ the diffusion coefficient of radon in air and setting $\tau_a = 0.6 - 0.7$ which are typical values for dry sand (for the sand in the KVI vessel $\tau_a = 0.64$ [12] was measured by means of an electrical conductivity measurement). Situations with $m \neq 0$ are

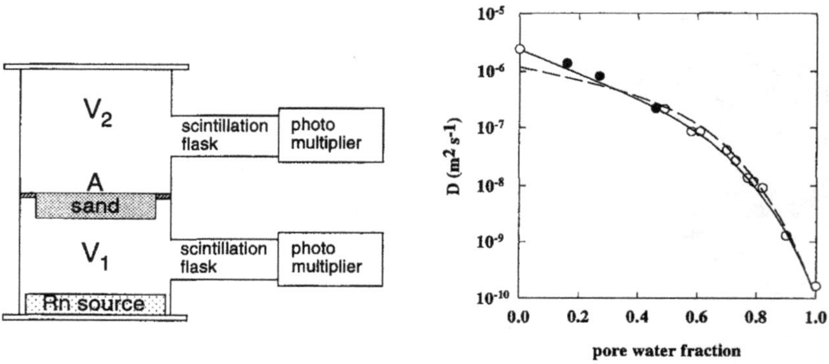

Figure 4: left *Set up for measuring diffusion coefficients.* right *Measured bulk radon diffusion coefficients; open and closed circles indicate measurements with increasing and decreasing pore-water fraction. Dashed and solid line represent correlation function from [13] and correlation function used in the KVI radon vessel experiments, respectively.*

more complex and require measurements of D as a function of m. For the sand of the KVI radon vessel D-values were measured with the sample container of the water retention set up enclosed between two compartments (Figure 4: *left*), the lower containing a radon source. Prior to a measurement upper and lower compartment are flushed and thereafter the ingrowth of the radon concentration is monitored (with scintillations flasks) in both compartments. From these curves the diffusion coefficient is derived. Figure 4 (*right*) shows the results of 14 measurements together with two correlation curves, the dashed line is a curve ($D = \epsilon\beta D_a \exp(-6m\epsilon - 6m^{14\epsilon})$) proposed by Rogers and Nielson [13] and based on more than 1000 measurements of soil samples, the solid line represents a curve ($D = (1-m)\epsilon\tau_a D_a \exp(a_1(m+m^5))$) used for modelling experiments with the KVI vessel. The free parameter $a_1 = -4.99 \pm 0.04$ results from optimizing the curve to the data.

3.3 Intrinsic permeability

Soil permeability is a parameter often routinely measured in surveys of soil radon potential [14]. *In situ* one usually measures the permeability by measuring the pressure difference between a perforated probe head inserted in the soil and ambient as a function of an induced air flow rate. For homogeneous media this will result in a linear relation from the slope of which the permeability can be calculated if the so called shape factor of the probe is known [15]. In practice it is very difficult to characterize *in situ* flow properties of a soil by permeability measurements because of the large variations within short spatial distances related to inhomogeneous moisture distribution and/or the occurance of cracks. Attempts to predict soil gas entry rates on basis of measured permeabilities and pressure differences have shown underestimation of these entry rates with a factor 2-3 [16]. The moisture dependence of the permeability (normalised with respect to the permeability of dry sand $K(0)$) of the sand in the KVI vessel is shown in Figure 5. These measurements were performed on the same sand sample as in the water retention and diffusion experiments. In these experiments the pressure difference across the sample was measured as a function of flow rate and the permeability was calculated from the slope of the linear relation. The results show an almost linear decreasing permeability to about 70% water saturation of the pore volume. This dependency has also been observed for other soil types [17]. For modelling in the KVI vessel the following correlation curve is used: $K(m) = K(0)\exp[(1.373m - 1.0057m^3)/(m-1)]$ shown by the solid line in Figure 5.

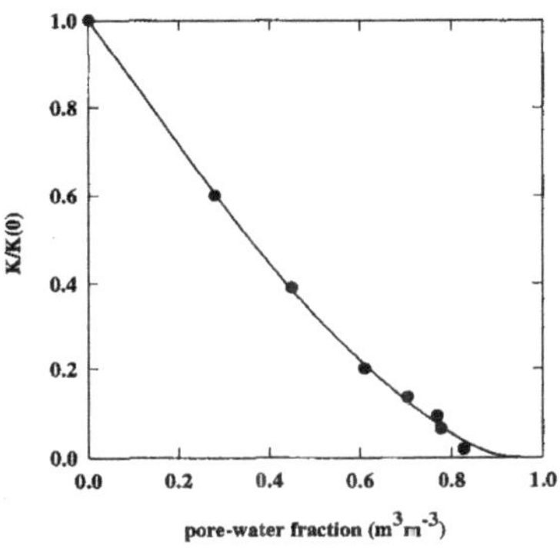

Figure 5: *Intrinsic permeability of the sand in the KVI radon vessel as a function of moisture content. Solid line represents fitted correlation function.*

3.4 Radon production

Usually radon production (emanation) is measured by using a closed can totally filled with porous material and analysing an air sample taken from the can when the produced radon activity is balanced by radon decay ('closed can method'). In this steady state situation we have $S = \beta\lambda C_a$. Thus to calculate the radon production, the partition correction porosity has to be known. Often dry samples ($m = 0$) are measured and S is calculated by setting $\beta = \epsilon$. This leads to underestimation of the production rate if adsorption is significant. Permanently flushing enclosed samples with radon free gas and collect the activity in the outgoing gas flow on an radon absorber ('flush and adsorb method')[18] will yield proper production rates.

Besides adsorption also the moisture content has to be considered in assessing radon emanation of a porous material. If a radon atom is produced inside a grain of a porous material it can escape from this grain by at least two mechanisms: 1) due to recoil the radon atom receives momentum which enables it to travel a certain distance through a material. This recoil range is about 65 μm in air and 35 nm in clay [19]; 2) radon atoms not escaping the grain by recoil may still be able to leave the material by diffusion. As this involves diffusion through a solid structure diffusion coefficients will be small and only radon atoms close to the surface will stand a chance to escape before decaying.

The following effect of moisture on the first process has been reported. Radon released from the grain by recoil rejection may not be stopped in the pore space but becomes imbedded in adjacent grains and is no longer available for transport. This effect is reduced by a water film adhering to the grains effectively stopping recoiling radon atoms in the pore water [20]. After transfer to the air phase this atom becomes available for transport. Another effect of moisture is that a sudden increase in humidity may result in a transient increase of radon emanation due to adsorption of water molecules on the grain surfaces, these molecules will compete with radon atoms for adsorption sites and thus lead to an increase in emanation [21]. Besides moisture also temperature has been reported to influence radon emanation. The amount of radon at the solid surfaces is determined by the interplay of adsorption and desorption. If the temperature of the material is increased the probability for desorption will increase and thus the total amount of radon adsorbed will decrease. A sudden change in temperature may thus result in a short extra release of radon [21].

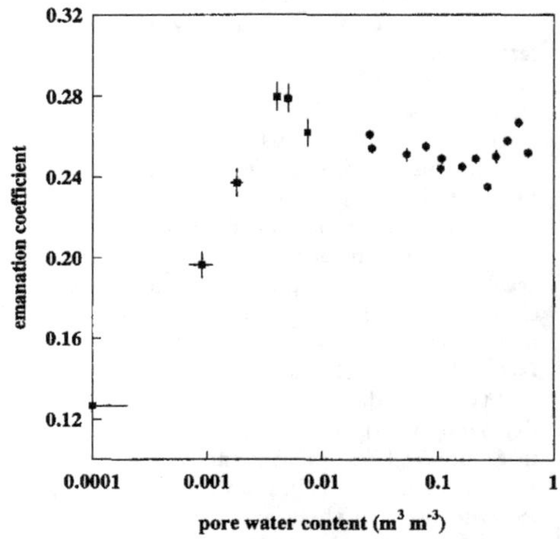

Figure 6: *Emanation coefficient of sand in KVI radon vessel as function of moisture content. Results from 'closed can method' (circles) and 'flush and adsorb method' (squares).*

All these effects stress the fact that emanation measurements are far from trivial and that for soil characterization the emanation coefficient should be measured as a function of moisture content and at ambient temperatures. Figure 6 shows the emanation coefficient (at room temperature) of the sand in the KVI vessel as a function of moisture content (η was calculated from a measurement of S and independently measured values of C^{Ra} and ρ_b). This curve shows an increase of more than a factor of two in the range $m = 0 - 0.01$, for $m \geq 0.01$ variations are less pronounced. This typical shape has also been observed by Menetrez *et al* [22] for the sandy soil in the EPA soil chamber.

4 Model-measurement comparisons

Model validation with the KVI radon vessel occurs in a framework in which starting from a well known initial situation (simple geometry, dry soil and no lid on the vessel) the complexity of the situation is increased step by step. Experiments with dry sand in the vessel on both diffusive and combined diffusive and advective transport with and without a crawl space but without a water level have shown satisfactory agreement between measurements and models (both analytical and numerical) based on equation 10 [12,23]. A typical example is shown in Figure 7 where the results of an experiment are given for a 50 cm crawl space and at different flow rates through the perforated box. For these dry sand situations, models and measurements always agreed within 10%. In these calculations values of the input parameters for $m = 0$ could be used except for the emanation coefficient. As the sand was actually only 'room-dry' ($m = 1.2 \cdot 10^{-3}$) η was deduced from Figure 6 as $\eta = 0.22$.

To proceed further with the model validation, various water levels were set in the sand. An unfortunate aspect of introducing water in the set up is that the increase of complexity is not a single step anymore because, as discussed before, several parameters are strongly dependent on the water content and hence vary simultanously. In these experiments a crawl space of 50 cm (air ventilated; $0.25\,l\,\text{min}^{-1}$) was used and water levels were set at consecutive depths of 160, 120, 80, 40 and 0 cm below the sand surface (rising water level) and thereafter the sequence was reversed (falling water level). At each level equilibrium radon concentration were measured at the probes above the water level. Figures 8 and 9 show the results for the rising and falling water level, respectively. At each water level the measured concentrations for the rising level were always lower than for the falling level probably reflecting the hysteresis in the water-retention curve. The left parts of Figures 8 and 9 also shows the (numerical) model calculations based on equation 10 that used as input the independently measured (sand box method) water retention curve to calculate m and $D(m)$ at different heights in the vessel. For both the rising and falling water level $m \geq 0.01$ at all positions above the water level in the vessel and using Figure 6 we deduced $\eta = 0.25$ independent of position. In both situations (rising and falling) model calculations substantially overestimate the experimental values. The disagreement is maximal (relative difference 80%) for the rising water level with water table at depth of 40 cm under the sand surface. This in contrast to the high water tables for the falling level were the agreement is reasonably good. Figure 6 shows that the emanation coefficient is rather sensitive to small variations in water content. As the model seems to overestimate the data it is likely that a lower emanation coefficient would give a better result. The lowest value we could reasonably choose, viz. the value for 'room-dry' sand $\eta = 0.22$, would improve the agreement considerably (right parts of Figures 8 and 9). As it was unclear if the disagreement between model and experiments stems from either insufficient knowledge on the emanation coefficient or on the water distribution it was decided to conduct an experiment with a large radon source in the crawl space (50 cm height) with the water table (falling level) at 160 cm depth. Steady-state radon concentrations were measured and corrected for the radon production in the sand. Figure 10 shows that calculations on basis of the water-retention curve measured in the vessel gives better agreement than the calculations on basis of the independently measured curve. Indicating that the former gives a better representation of the situation in the vessel. Although the *in situ* measured curves improve the agreement with the data, still deviations up to 30-40% remain. Various other attempts were made to improve the agreement, but improvement for one case caused deteriorations in other

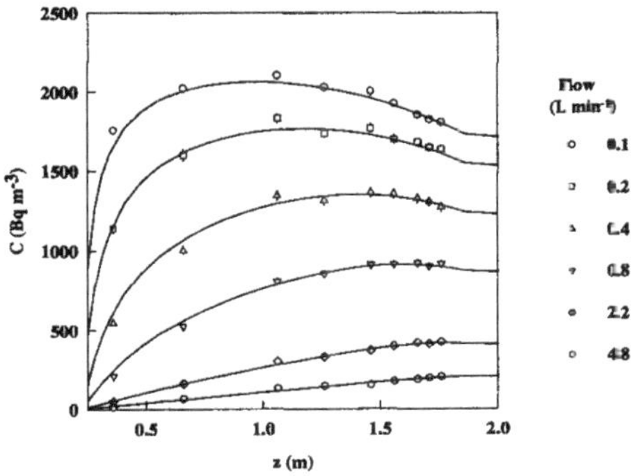

Figure 7: *Radon concentration C as a function of height z in the vessel ($z = 0$ is bottom; $z = 1.83m$ is sand surface) for different values of the air flow J for experiments with combined diffusive and advective transport with a 50 cm crawl space, air flow in the upward direction and room-dry sand. Solid lines represent numerical model calculations.*

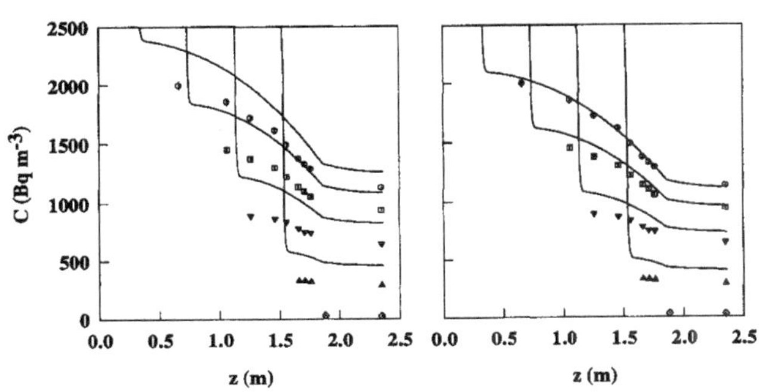

Figure 8: *Calculated radon concentration C (lines) using independently measured water retention curve (rising) and measured concentrations at different depths (markers) for a rising water level. Calculations with* left: $\eta = 0.25$; right: $\eta = 0.22$.

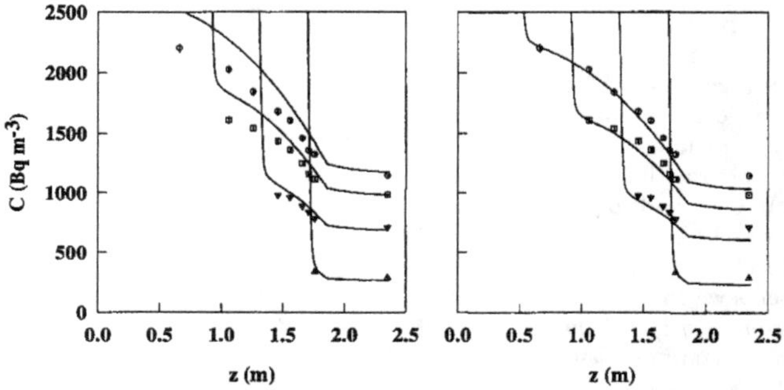

Figure 9: *Calculated radon concentration C (lines) using independently measured water retention curve (falling) and measured concentrations at different depths (markers) for a falling water level. Calculations with* left: $\eta = 0.25$; right: $\eta = 0.22$.

cases. So far no consistent set of quantities could be obtained for all experiments.

5 Conclusions and outlook

It is concluded that for situations with dry sand, model calculations and experiment are in agreement within 10%. However most experiments with a water level could not be reproduced within this limit using either a set of independently measured parameters or a set containing some *in situ* measured parameters. The most probable cause for this discrepancy is that although the sand in the laboratory facility is very homogeneous, compared to actual soils it is still to heterogeneous to obtain sufficient precise information on the input parameters from independent measurements. Another option is that some of the assumptions, made to merge the three phase equations into a single equation for the air phase, are not justified. This may for instance mean that ad- and desorption effects play a larger role than anticipated. As this study also shows that using *in situ* measurements of the parameters may improve the agreement, we propose that future research should be directed towards development and application of techniques to measure these parameters as a function of position in the soil under study. This will also have the further benefit that translation of the results to cases of real soils under actual houses will be facilitated.

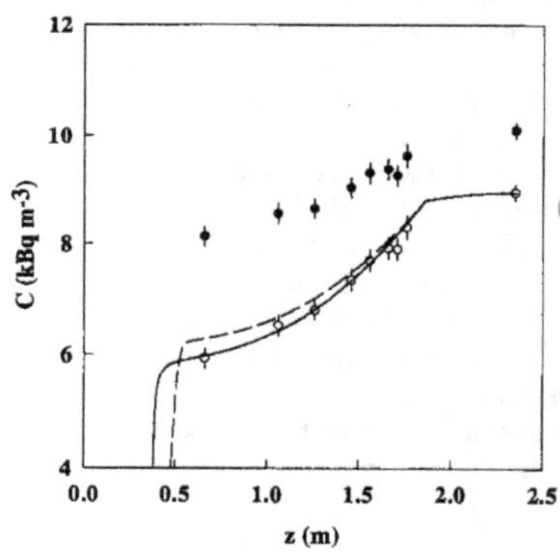

Figure 10: *Measured (filled markers) concentrations for the experiment with water table at 160 cm depth and radon source in the crawl space. Open markers are contribution of source only. Lines represent model using measured moisture content in the vessel (solid) and independently measured distribution (dashed).*

6 References

[1] Aldenkamp, F.J., Stoop, P. *Sources and transport of indoor radon: measurements and mechanisms.* PhD thesis, University of Groningen, 1994.

[2] Andersen, C.E. *Entry of soil gas and radon in houses.* PhD thesis, Risø National Laboratory, Risø-R-623(EN), 1992.

[3] Mosley, R.B., Menetrez, M.Y., Snoddy, R., Brubaker, S.A. *The influences of diffusion and advective flow on the distribution of radon activity within USEPA's soil chamber.* Environmental International, 22, Suppl. 1, 521-533, 1996.

[4] Van der Graaf, E.R., Heijs, S., de Meijer, R.J., Put, L.W., Mulder, H.F.W.M. *A facility to study transport of radon under controlled conditions.* Radiat. Prot. Dosim., 45, 223-226, 1992.

[5] Rogers, V.C., Nielson, K.K. *Multiphase radon generation and transport in porous materials.* Health Physics, 60, 807-815, 1991.

[6] Nazaroff, W.W., Moed, B.A., Sextro, R.G. *Soil as source of indoor radon - Generation, migration, and entry.* in: Radon and its decay product in indoor air. Editors Nazaroff, W W, Nero, A.V., 57-112, John Wiley & Sons, 1988.

[7] Koorevaar, P., Menelik, G., Dirksen, C. *Elements of soil physics.* in: Developments in soil science, 13, Elsevier, 1983.

[8] Gregg, S.J., Sing, K.S.W. *Adsorption, surface area and porosity.* Academic Press, London, 1967.

[9] Van der Spoel, W.H. *Radon transport in sand: A laboratory study.* PhD thesis, Eindhoven Technical University, the Netherlands, in preparation, 1997.

[10] Landman, K.A. *Diffusion of radon through cracks in a concrete slab.* Health Physics, 43, 65-71, 1982.

[11] Nielson, K.K., Rogers, V.C., Gee, G.W. *Diffusion of radon through soils: a pore distribution model.* Soil Sci. Soc. Am. J., 48, 482-487.

[12] Van der Spoel, W.H., van der Graaf, E.R., de Meijer, R.J. *Diffusive transport of radon in a homogeneous column of dry sand.* Health Physics, 72, 766-778, 1997.

[13] Rogers, V.C., Nielson, K.K. *Correlations for predicting air permeabilities and ^{222}Rn diffusion coefficients of soils.* Health Physics, 61, 225-230, 1991.

[14] Scott, A.G. *Site characterization for radon supply potential: a progress review.* Health Physics, 62, 422-428, 1992.

[15] Damkjaer, A., Korsbech, U. *A small-diameter probe for in situ measurements of gas permeability of soils.* Radiat. Prot. Dosim., 45, 85-89, 1992.

[16] Andersen, C.E., Søgaard-Hansen, Majborn, B. *Soil gas and radon entry into a simple test structure: Comparison of experimental and modelling results.* Radiat. Prot. Dosim., 56, 151-155, 1994.

[17] McCarthy, K.P., Brown, K.W. *Soil gas permeability as influenced by soil gas-filled porosity.* Soil. Sci. Soc. Am. J., 56, 997-1003, 1992.

[18] Van Dijk, W., de Jong, P. *Determining the ^{222}Rn exhalation rate of building materials using liquid scintillation counting.* Health Physics, 61, 501-509, 1991.

[19] Bossus, D.A.W. *Emanating power and specific surface area.* Rad. Prot. Dos., 7, 73-76, 1983.

[20] Morawska, L., Phillips, C.R. *Dependence of the radon emanation coefficient on radium distribution and internal structure of the material.* Geochim. Cosmochim. Acta, 57, 1783-1797, 1993.

[21] Gan, T.H., Mason, G.C., Wise, K.N., Whittlestone, S., Wyllie, H.A. *Desorption of ^{222}Rn by moisture and heat.* Health Physics, 50, 407-410, 1986.

[22] Menetrez, M.Y., Mosley, R.B., Snoddy, R., Brubaker, S.A. *Evaluation of radon emanation from soil with varying moisture content in a soil chamber.* Environmental International, 22, Suppl. 1, 447-453, 1996.

[23] Van der Spoel, W.H., van der Graaf, E.R., de Meijer, R.J. *Combined diffusive and advective transport of radon in a homogeneous column of dry sand.* Accepted for publication in Health Physics, 1997.

Airborne radon concentrations in different environments in Korea

Yoon-Shin Kim and Takao Iida1)
Institute of Environmental & Industrial Medicine
College of Medicine, Hanyang University
Seoul 133-792, Korea

1)Department of Nuclear Engineering, School of Engineering
Nagoya University, Nagoya, Japan

Abstract

Little information has been presented average radon concentrations in the living spaces in Korea to date although radon has become a greate concern in recent years to the Korean population due to association with the lung cancer risk. Since radon related health risks depends on long-term exposure, it is vital to know the relationship between these short-term measurements and long-term radon concentrations in the different living spaces. Previous studies have shown that many factors may contribute to the temporal and geographic variation in radon entry and retention in livng spaces. These factors may differ from one region of the country to another and over time. Thus, it is important that comparisons of different measurement protocols include samples drawn from a broad range of radon sources, different types of environments, and lifestyles. The present work compares the results of short-term and long-term surveys of radon measurements conducted in various areas in Korea that contains average radon-strength sources, different types of living spaces, and different seasons. Most of average airborne radon concentrations using track etch radon detectors were measured in different environments such as houses, subway stations, underground stores, indoor and outdoor office buildings. In addtion to these studies, a series of radon surveys using passive integrating radon cup monitors were undertaken in different types of dwellings in Seoul for one year since April 1996, while the same survey using radon discriminative dosimeters were conducted in major six cities in Korea. Radon concentrations in basements in the selected dwellings were higher than those levels measured in the first floor, while indoor radon concentrations were significantly higher than the corresponding outdoor levels at six survey sites. The results suggest the need to more definitely assess sources of radon concentrations as well as to provide more information about technical measurements of different radon monitors.

1. Introduction

The importance of the contribution of environmental radon to the natural radiation dose has been the subject of relatively recent studies (1-2). It is considered that the main contribution to the increase in radon levels indoors is due to the type of adjacent soil (3). Other possible sources may contribute significantly are building materials and drinking water (4). The numerous studies on measurement of indoor Rn concentrations have dealt with considerable types of indoor enviornments such as homes and public buildings where people most spent indoors (5-6). In most of these measurement, different types of sampling techniques were used with a different sampling time. Elevated levels of radon have been observed in dwellings in most developed countries, and there is concern that elevated levels of radon may contribute to an increased risk of lung cancer (7). It is estimated that the health risks are due to the inhalation of airborne high levels of radon and its progency. rather than to its direct ingestions (8-9).

In Korea, some limited studies (10-12) on the concentrations of 222Rn in homes have been carried out over short integration periods and for groups of the population chosen for convenience, rather than as representative of the whole population. In most of these studies, alpha Track Etch monitors with a sampling time of 3 months were used. However, to date little information has been published comparing nationwide screening surveys with average 222Rn concentrations in the living spaces. This paper presents a review of results of radon measurements undertaken in Korea from 1987 to 1996 and reports a preliminary results of radon measurements using passive radon dosimeters in major Korean cities.

2. Materials and Methods

The important characteristics of the measurement protocols in different surveys in Korea are summarized in Table 1.

1) Surveys of Rn levels in houses and other public buildings.

Most of measurements for airborne Rn concentration in Korea were undertaken between 1987 and 1993 in different indoor environments such as houses, public buildings, and other living spaces using alpha Track Etch detectors (13-15). The Track Etch technique has been developed and calibrated for use in large indoor surveys (16). These detectors were maily placed on the first floor and basement of each house or indoor environment for periods ranging from 3 to 4 month.

2) Surveys of Rn levels in Seoul subway stations (1989-91).

The airborne Rn levels in the Seoul metropolitan subway stations had been measured for 83 stations in four lines during summer 1989 to winter 1991 on a real-time basis by a pulsed ion chamber radon monitor (17). The monitor was intentionally placed about 0.3 m from the ground in the middle of each floor. After 5 min stabilization of the monitor, it usually took 30 min to count a radon level at a site and move to the next site. Each monitoring duration varied from 1-4 weeks. The monitor was sent to Femo-Tech Inc. every 6 month in order to be calibrated fro multipoint precision. Measurements of radon with this type of monitor were described for elsewhere (18).

3) Surveys of Rn levels in different cities (1996-97).

For a pilot radon survey in major six cities in Korea, newly developed passive electrostatic integrating Rn and Rn cup monitors were used to measure Rn concentrions. Both electrostatic Rn and Rn cup monitors have been constructed by mounting a stainless steel hemisphere with radius 50 mm on an aluminum plate bottom, that has a 16 mm inner diameter circular window with a CN film in the center. The electrostatic Rn monitors were placed inside and outside university building in six major cities since March 1996 and CN films were collected in every two months, while Rn cup monitors were placed in different environments in the Seoul area. The choice of sampling locations was often influenced by convenience considerations. Further measurements of radon with this type were described in detail elsewhere (19-20).

3. Results and Discussion

Table 1 shows the summary of the radon concentration in different sampling locations and methods. From this table, the following points may be established:

1) Radon levels in homes

Indoor radon concentrations in basements of the selected homes were several times higher than those levels measured in the first floor. Also the radon levels of the first floor in the energy efficient homes were significantly higher than the conventional homes (13). Indoor radon oncentrations of the selected homes in rural are seemed to be higher than the corresponding levels in urban areas. Approximately thirteen percent of the study homes exceeded 4 pCi/l of the radon levels of the USEPA's recommended limit. The indoor radon levels in the homes are seemed to be correlated with house location relative to geologic formation.

A pilot study on indoor and outdoor airborne radon concentration of the selected homes near nuclear power plants in two areas was undertaken to compare the radon levels (14). Average indoor radon levels in selected homes were 1.4 times higher than the outdoor radon levels. Indoor radon levels in this study seemed to be varied significantly by different types of housing, construction materials, and other chacteristics. This also suggested that indoor radon sources were not from outdoor sources, but from intrusion gas of indoor sources.

2) Radon levels in the Seoul Metropolitan subway stations

The geometric mean radon concentrations at 74 subway statioins in the Seoul Metropolitan area were 86 Bq/m3 in the upper hall and 93 Bq/m3 with a wide range of radon levels (17) which were significantly higher than the mean radon levels, 10.9 Bq/m3, of the Tokyo Metropolitan subway stations (21). Results of the surveys indicated that several stations exceeded the USEPA recommended levels of 150 Bq/m3 during the study periods. The radon levels in winter tended to be slightly higher than that in summer. It was observed that nine percent of the total stations measured in winter and eight percent in summer exceeded 300 Bq/m3. It was also observed that slight daily variation occurred in a station having elevated radon levels. During the study on daily variation, there was a real positive correlation between relative humidity and radon concentration. It was likely that possible radon sources were radon intrusion from the constuction materials and the leaking groundwater.

3) Indoor and outdoor radon levels of office buildings in major cities.

Table 3-4 shows variation of indoor and outdoor radon concentration in the university office buildings using passive eletrostatic radon monitor during February 1996-March 1997 in six major Korean cities. Table 5 gives summary statistics for the measured indoor and outdoor concentrations of radon, and indoor/outdoor ratios between summer and winter times. The mean indoor radon concentration in Taejon showed the highest levels among the cities, while outdoor radon concentrations in Kwangjoo showed the higher than any other cities. The mean indoor radon levels are higher than the corresponding levels measured in Beijing and Nagoya using the same monitor (22).

Figure 1 shows the geographic distribution of the indoor and outdoor radon levels in major cities. The pattern of this results seems to be one in which there are relatively low concentrations in the coastal region (Pusan) and more elevated values inland (Taejon), perhaps as a consequence of Rn-free breezes from the sea or of differences in geology between coastal and inland areas. The mean indoor levels in major cities were 1.5-3.6 times higher than the outdoor levels, while the indoor/outdoor ratios in winter were in general higher than the corresponding ratios in summer except for Taejon city. This means that radon levels can be affected by seasonal variation.

4) Indoor radon levels in different environments

Table 6-7 shows variation of mean radon concentrations in the basement and the first floor in different environments using passive radon cup monitor during February 1996-March 1997. Table 8 gives summary statistics for mean radon concentration in the first floor and basement, and the first floor/basement ratios between summer and winter periods. The mean radon levels in basement in each different environment were generally higher than the first floor levels except for apartment.

Previous studies undertaken in Korea focused on measurement for indoor and outdoor radon concentrations in houses, pubic buildings, underground environment, and other specific indoor environments. These has shown significant variation between several types of short-term Rn measurements and longer-term concentrations. Many factors may contribute to the temporal and geographic variation in Rn entry and retention in houses. These factors may differ from one region of the country to another and over time. Thus, it is important that comparisons of different measurement protocols include samples drawn from a broad range of Rn sources, housing types, climates, and lifestyles.

4. Conclusions

In the review of radon studies undertaken in Korea during 1987-1997, approximately threee to eleven percent of homes surveyed were found to have average radon concentrations in excess of the USEPA recommended limit. If this proportion of high-concentration homes is propoerly representative of the country as a whole, about 30,000-100,000 Korean homes would be expected to exceed the level.

Radon concentrations tend to be higher in underground environments than those of homes. Radon concentrations also tended to be enhanced in different environments that are poorly ventilated or that rely on air-conditioning for ventilation. Significant temporal variations in indoor radon concentrations in individual environments suggest that, at the present time, the only way to obtain an accurate assessment of the radon potential in a living spaces is to make a measurement that spans several seasons. Further work is planned to identify the factors that contribute to the highest Rn concentrations in particular indoor environments revelaed in the survey.

5. Acknowledgements

The authors would like to thank Dr. J.H. Lee of Choongnam National University in Taejon, Dr. S.O. Baek of Yeungnam University in Taeku, Dr. K.C. Choi of Dong-A University in Pusan, Dr. D.Y. Shin of Chosun University in Kwangjoo, Dr. B.S. Kang of Jeju National University in Jejoo for their assistance in connection with the sampling work involved in this survey. The authors are also grateful to Mr. H. Zuo of Nagoya Univesity for analytical work and Mr. J.D. Lee of Hanyang Univesity for his data collection and analysis.

6. References

[1] International Commission on Radiological Protection, Principles for limiting exposure for the public to natural sources of radiation, Oxford, Pergamon Press, ICRP Publication No. 39, 1984.

[2] United Nations Scientific Committee on the Effects of Atomic Radiation, Sources, effects and risk of ionizing radiation, New York, United Nations, 1988.

[3] Enviornmental Protection Agency, A citizen's guide to 222 Rn. What it is and what to do about it, Washington, DC. USEPA Report 520/1-86-004; 1986.

[4] Nero, A.V., Radon and its decay products in indoor air: An overview. In:Nazaroff, W.W.; Mero, A.V., eds. Radon and its decay products in indoor air, New York, NY, Wiley and Sons; 1-53, 1988.

[5] Cohen, B.L., D.R. Kulwicki., K.R. Warner., and C.L. Grassi, Radon concentrations inside public and commercial buildings in the Pittsburgh area, Health Phys., 47, 399-405, 1984.

[6] Fleischer, R., and L. Turner, Indoor radon measurements in the New York capital district, Health Phys., 46, 999-1011, 1984.

[7] Harley, N.H., and J.H. Harley, Potential lung cancer risk from indoor radon exposure. Ca-A Cancer J. Clinicians, 40, 265-275, 1990.

[8] Cross, F.T., N.H. Harley., and W. Hofman, Health effects and risk from 222Rn in drinking water. Health Phys. 48, 649-670, 1985.

[9] Crawford-Brown, D., Analysis of the health risk from inggested radon. In: Cothern, C.R., P.A. Rebers, eds. Radon, radium and uranium in drinking water. Chelsea, MI, Lewis Publishers Inc, 17-26, 1990.

[10] Kim, Y.S., Exposure to indoor pollutants in selected microenvironments, Bull. Inst. Environ. University, 1988.

[11] Kim, Y.S., Indoor radon concentrations in the Seoul area, Kor. J. Environ. Hlth. Soc., 15, 11-18, 1989.

[12] KESA, Report on "A survey of air pollution in underground spaces," Kor. Environ. Sci. Assoc., 1989.

[13] Kim, Y.S., Indoor radon levels in four different areas in Korea. Kor. J. Enviorn. Hlth.Soc., 16, 1-7, 1990.

[14] Kim, Y.S., Indoor and outdoor radon concentrations in selected homes near nuclear power plants. J. Kor. Public Hlth., 17, 60-66, 1991.

[15] Seo, J.M., Measurement of radon concentration and annual exposure rate in the underground shopping mall. Hanyang University Master Thesis. Dec. 1994.

[16] Alter, H.W., and R.L. Fleischler, Passive integrating radon monitor for environmental monitoring. Health Phys., 40, 693-702, 1981.

[17] Kim, D.S., and Y.S. Kim, Distribution of airborne radon concentrations in Seoul Metropolican subway stations. Hlth. Phys.,65,12-16, 1993.

[18] Field, R.W., B.C. Kross. Field comparison of several commercially available radon detectors. Am. J. Publ. Hlth. 80, 926-930, 1990.

[19] Guo. Q., T. Ida, and K. Okamoto, Measurement of thron concentration by passive cup method and its application to dose assessment, J. Nucl. Sci. Technol. 32, 794-803. 1995.

[20] Iida, T., Y. Yukimasa., and K.Tojo., An electrostatic radon monitor for measurement of environmental radon. Res. Lett. Atmos. Electr., 11, 55-59, 1991.

[21] Doi, M., and S. Kobayashi, Surveys of concentration of radon isotopes in indoor and outdoor air in Japan. Environ. Int. 22, 1-7, 1996.

[22] Jin, Y., Y. Ikebe., T. Iida., et al., A survey of outdoor and indoor 222Rn concentrations by passive methods in China. Hoken Butsri, 26, 341-349. (in Japanese). 1991.

Table 1. Comparison of major measurement protocols for Rn surveys undertaken in Korea.

Radon Monitor	Sampling Subject	Sampling Time	Location
Alpha Track Etch detector	House Public building	3 mo	first floor/ basement upper/lower
Pulsed ion chamber (Femto-Tech, Inc. Model R210F)	Subway station	real-time	upper/lower level
Passive electrostatic integrating Rn monitor (Aloka Co.; GS-201B)	Univ. office building	2 mo	Seoul, Pusan, Taejon, Taekoo Kwangjoo, Jeju
Passive cup monitor	House	2 mo	Indoor/outdoor

Table 2. Studies of ^{222}radon measurement results undertaken in Korea (1987–1997).

Author (year)	Sampling location	Sampling period	Sampling monitor	Sampling site(N=size)	Mean concentration
Y. S. Kim (1988)	Seoul	June–Aug 1987	Alpha Track Etch detector	Homes(N=30)	Basement : 2.49 pCi/ℓ First Floor : 0.86(living room) Approx. 5% of the homes exceeded the 4pCi/ℓ
Y. S. Kim (1989)	Seoul	Sep–Dec 1988	Alpha Track Etch detector	Homes Energy-efficient (N=20) Non-Energy-efficient(N=14) Total(N=34)	Basement / Firstfloor 3.34 / 1.96 2.84 / 1.36 2.64 / 1.71
KESA (1989)	Seoul	Jun–Dec 1988	Alpha Track Etch detector	Underground Environment Shopping center(N=3) Subway station(N=4) Underspace(N=2) Parking lot(N=2) Tunnel(N=2)	 1.5pCi/ℓ 1.6pCi/ℓ 2.1pCi/ℓ 1.3pCi/ℓ 1.6pCi/ℓ
Y. S. Kim (1990)	Seoul Songtan Dogo Kumsan	Dec1988 –April 1989	Alpha Track Etch detector	Homes Seoules(N=20) Songtan (N=19) Dogo(N=17) Kumsan(N=19) Total (N=75)	Basement (pCi/ℓ) / Firstfloor (pCi/ℓ) / Exceeding percent 3.40 / 2.72 / 5 4.23 / 2.81 / 16 4.03 / 2.79 / 18 2.84 / 2.65 / 14 3.58 / 2.74 / 13 Bedroom(N=57) : 2.86 Living room(N=75) :1.25 Kitchen(N=75) :2.43 Energy-efficient : 3.01 Non-Energy-efficient : 2.10
Y. S. Kim (1991)	Yeongkwang Uljin Total	Nov 1990 –Feb 1991	Alpha Track Etch detector	Homes near nuclear power plant (N=27) indoor outdoor Homes near nuclear power plant (N=29) indoor outdoor Total Homes(N=56) indoor outdoor Brick Concrete	Only 3% fo homes exceeded 4pCi/ℓ 2.76 1.69 1.57 1.33 2.10 1.53 2.59 1.41
D. S. Kim and Y. S. Kim (1993)	Seoul	Jul–Aug 1989 Jan 1990 Feb 1991	Pulsed ion Chamber (Femto-Tech, Inc.)	Subway Station(N=74) upper ticketing office(N=150) lower platform (N=161)	 86(Bq/m³) 96(Bq/m³) 12% of stations exceeded the 150 Bq/m³

<table>
<tr><th rowspan="2">Author (year)</th><th colspan="4">Sampling</th><th rowspan="2">Mean Concentration</th></tr>
<tr><th>location</th><th>period</th><th>monitor</th><th>site(N=size)</th></tr>
<tr><td>W. P. Hong (1989)</td><td>Taejon</td><td>Jan-Oct 1989</td><td>CR-39</td><td>Laboratory room</td><td>
<table>
<tr><td>Rn conc.</td><td>0.87(pCi/l)</td></tr>
<tr><td>Rn-Daughter conc.
Po-218
Pb-214
Bi-214</td><td>
0.53pCi/l
0.35pCi/l
0.26pCi/l</td></tr>
<tr><td>Ft</td><td>0.40</td></tr>
<tr><td>WL(×10^{-3})</td><td>3.33</td></tr>
</table>
</td></tr>
<tr><td>KAERI (1991)</td><td>Ulsan
Kwangju
Taejon
Donghae
Sokcho
Seosan Dangjin
Inchon
Nonsan
Seoul
KAERI Indoor
Chunchon
Suwon</td><td>Apr - Oct 1990</td><td>KAERI CR-39 Radon Cup</td><td>Home (N = 340)</td><td>
<table>
<tr><th rowspan="2">Area</th><th>Mean</th><th colspan="2">Median</th><th rowspan="2">Median (Bq/m^3)</th></tr>
<tr><th>(pCi/l)</th><th>(pCi/l)</th><th>(Bq/m^3)</th></tr>
<tr><td>Ulsan</td><td>1.63</td><td>1.43</td><td>52.91</td><td>21.16</td></tr>
<tr><td>Kwangju</td><td>1.14</td><td>1.04</td><td>38.50</td><td>15.40</td></tr>
<tr><td>Daejon</td><td>1.32</td><td>1.18</td><td>43.66</td><td>17.46</td></tr>
<tr><td>Donghae</td><td>1.34</td><td>1.26</td><td>46.62</td><td>18.61</td></tr>
<tr><td>Sokcho</td><td></td><td></td><td></td><td>24.00</td></tr>
<tr><td>Seosan</td><td>1.79</td><td>1.62</td><td>60.00</td><td></td></tr>
<tr><td>Dangjin</td><td>2.17</td><td>1.68</td><td>62.16</td><td>24.86</td></tr>
<tr><td>Inchon</td><td>1.13</td><td>1.07</td><td>39.60</td><td>15.84</td></tr>
<tr><td>Nonsan</td><td>2.31</td><td>1.66</td><td>61.42</td><td>24.57</td></tr>
<tr><td>Seoul</td><td>1.72</td><td>1.39</td><td>51.43</td><td>20.57</td></tr>
<tr><td>KAERI Indoor</td><td>0.98</td><td>0.914</td><td>33.82</td><td>13.53</td></tr>
<tr><td>Chunchon</td><td>1.68</td><td>1.45</td><td>53.65</td><td>21.46</td></tr>
<tr><td>Suwon</td><td>1.56</td><td>1.44</td><td>53.28</td><td>21.31</td></tr>
<tr><td>Total Avg.</td><td>1.61</td><td>1.32</td><td>48.90</td><td>19.56(~20.00)</td></tr>
</table>
</td></tr>
<tr><td>J. M. Seo (1994)</td><td>Daejon</td><td>Jan - Mar 1994</td><td>Alpha Track Etch detector</td><td>Underground store (N = 30)
Department store (N = 30)
Public buidingl (N = 30)
Daejon Police (N = 30) Station</td><td>25.2(Bq/m^3)
25.1(Bq/m^3)
25.2(Bq/m^3)
36.3(Bq/m^3)</td></tr>
<tr><td>Y. J. Bae (1994)</td><td>Seoul</td><td>Jan - Sep 1994</td><td>Charcoal canister</td><td>Underground store (N = 5)
Laboratory room (N = 5)
Office room (N = 10)</td><td>65.6(Bq/m^3)
41.6(Bq/m^3)
28.1(Bq/m^3)</td></tr>
<tr><td>Y. S. Kim (1997)</td><td>Seoul
Taegoo
Daejon
Kwangju
Pusan
Cheju</td><td>Sep 1996 - Apr 1997</td><td>Rn Cup Monitor/
Rn Electrostatic Monitor</td><td>Indoor/Outdoor office building, Home</td><td>
<table>
<tr><th colspan="3">Rn Cup Monitor(96.9-97.4)</th><th colspan="3">Rn Electrostatic Monitor(96.9-97.4)</th></tr>
<tr><th>Area</th><th>under-ground C_{Rn} (Bq/m^3)</th><th>First floor C_{Rn} (Bq/m^3)</th><th>Area</th><th>under-ground (Bq/m^3)</th><th>Firstfloor (Bq/m^3)</th></tr>
<tr><td>Hospital</td><td>30.1</td><td>24.6</td><td>Seoul</td><td>28.9</td><td>7.4</td></tr>
<tr><td>Home</td><td>15.0</td><td>59.6</td><td>Taegoo</td><td>19.3</td><td>8.1</td></tr>
<tr><td>Office building</td><td>66.9</td><td>34.8</td><td>Daejon</td><td>37.5</td><td>10.7</td></tr>
<tr><td>Apart-ment</td><td>25.2</td><td>23.9</td><td>Kwangju</td><td>23.3</td><td>10.7</td></tr>
<tr><td></td><td></td><td></td><td>Pusan</td><td>9.2</td><td>6.3</td></tr>
<tr><td>Shopping store</td><td>36.8</td><td>23.6</td><td>Cheju</td><td>26.2</td><td>7.2</td></tr>
<tr><td>Mean</td><td>34.8</td><td>33.3</td><td>Mean</td><td>24.2</td><td>8.4</td></tr>
</table>
</td></tr>
</table>

Table 3. Variation of indoor ^{222}Rn concentration($Bq \cdot m^{-3}$) using Rn electrostatic monitor in major cities (Feb 1996 - Mar 1997).

City	1996						1997
	Feb~Mar	Apr~May	Jun~Jul	Aug~Sep	Oct~Nov	Dec ~ Jan	Feb~Mar
Seoul	-	24.2	31.7	26.6	39.9	44.7	6.6
Taegu	-	15.1	-	0.8	-	30.3	32.9
Taejon	38.0	-	28.7	37.1	45.6	-	38.4
Kwangju	-	-	16.6	25.0	-	25.1	26.5
Pusan	14.9	10.6	-	-	-	9.1	8.7
Jeju	-	-	-	-	-	27.0	25.3

Table 4. Variation of outdoor ^{222}Rn concentration($Bq \cdot m^{-3}$) using Rn electrostatic monitor in major cities (Feb 1996 - Mar 1997).

City	1996						1997
	Feb~Mar	Apr~May	Jun~Jul	Aug~Sep	Oct~Nov	Dec ~ Jan	Feb~Mar
Seoul	-	6.4	7.1	10.1	10.4	10.2	0.5
Taegou	-	6.3	-	0.5	-	13.0	12.5
Taejon	9.7	-	6.2	15.2	10.0	-	11.1
Kwangju	-	-	12.2	8.2	-	10.9	11.6
Pusan	6.2	5.7	-	-	-	7.3	6.3
Jeju	-	-	-	-	-	6.7	7.6

Table 5. Mean indoor and outdoor ^{222}Rn concentrations($Bq \cdot m^{-3}$) in Korea(Feb 1996 - Mar 1997).

City	Mean ^{222}Rn concentration($Bq \cdot m^{-3}$)						Indoor/Outdoor ratio	
	Indoor			Outdoor				
	N	Mean	S.D.	N	Mean	S.D.	Summer	Winter
Seoul	6	29.0	12.3	6	7.5	3.5	3.4	4.4
Taegu	4	19.8	12.9	4	8.1	5.1	1.6	2.3
Taejon	5	37.6	5.4	5	10.4	2.9	4.6	3.5
Kwangju	4	23.3	3.9	4	10.7	1.5	1.4	2.3
Pusan	4	10.8	2.5	4	6.4	0.6	NA	1.7
Jeju	2	26.2	0.9	2	7.2	0.5	NA	3.6

N : Number of samples
S.D : Standard Deviation
N.A : Not Available

Table 6. Variation of the first floor ^{222}Rn concentration($Bq \cdot m^{-3}$) using Rn cup monitor in different environments (Feb 1996 - Apr 1997).

Location	N	1996			1997	
		Feb～Jul	Sep～Nov	Dec ～ Jan	Feb～Apr	Mean
Shopping store	8	42.0	24.1	26.4	20.2	28.2
Office building	8	23.2	38.6	37.1	29.8	32.2
Apartment	8	65.4	23.0	31.0	18.5	34.5
Hospital	8	40.6	24.3	28.5	21.1	28.6
House	8	156.5	14.9	18.0	12.6	50.5

N : Number of samples

Table 7. Variation of basement ^{222}Rn concentration ($Bq \cdot m^{-3}$)using Rn cup monitor in different environments (Feb 1996 - Apr 1997).

Location	N	1996			1997	
		Feb～Jul	Sep～Nov	Dec ～ Jan	Feb～Apr	Mean
Shopping store	8	14.0	34.7	40.5	35.4	31.6
Office building	8	12.1	73.4	72.3	55.1	53.2
Apartment	8	33.9	15.3	27.3	20.9	24.4
Hospital	8	17.3	29.4	33.4	27.3	29.4
House	8	175.4	63.2	62.7	53.0	88.6

N : Number of samples

Table 8. Mean ratios of the first floor/basement radon concentration($Bq \cdot m^{-3}$) (Feb 1996 - Apr 1997).

Location	Mean ^{222}Rn concentration($Bq \cdot m^{-3}$)						First floor/Basement ratio	
	First floor			Basement				
	N	Mean	S.D.	N	Mean	S.D.	Summer	Winter
Shopping store	8	28.2	8.3	8	31.6	10.2	0.3	1.5
Office building	8	32.2	6.2	8	53.2	24.8	0.5	1.9
Apartment	8	34.5	18.5	8	24.4	6.9	0.5	0.9
Hospital	8	28.6	7.4	8	29.4	6.5	0.4	1.2
House	8	50.5	61.2	8	88.6	50.3	1.1	3.5

N : Number of samples
S.D : Standard Deviation
N.A : Not Available

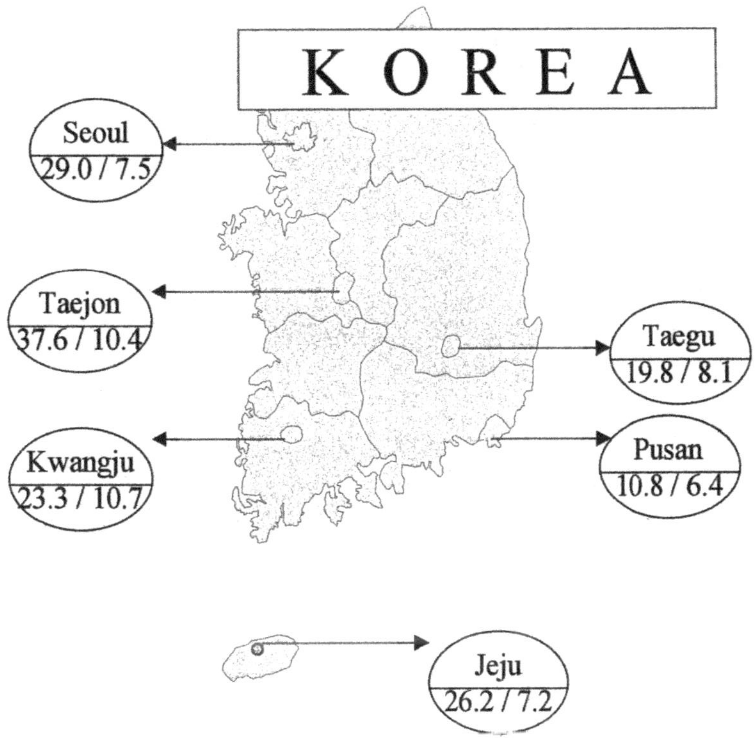

Figure 1. Mean indoor/outdoor concentration (Bq · m^{-3}) major cities

The Exhalation Rate of Radon in the Atmosphere and the Influencing Factors

H.Kojima

Dep.Liberal Arts, Faculty of Science and Technology, Science University of Tokyo, Noda Chiba, 278, JAPAN

Abstract

To clarify the mechanism of radon transport from soil to surface, the long-term and continuous measurements of the radon exhalation were performed at one site, together with radon concentrations in soil gas, soil temperatures, pressure difference, soil moisture and meteorological parameters. Based on the some preliminary results over a 8 month period, the possible influence of meteorological parameters on the exhalation and the temporal variation were discussed.

1. Introduction

The relatively extensive studies on exhalation of radon(^{222}Rn) from soil have been reported. However, as most of the studies are performed based on grab sampling methods, there are very few reports on the factors affecting exhalation and on the temporal variations. Also, there is a great confusion and much argument about the results obtained, especially regarding the influence of meteorological and soil parameters.

In the present work, the long-term and continuous measurements of the radon exhalation were performed at one site, together with radon concentrations in soil gas, soil temperatures, pressure difference between the surface and 100 cm depth, soil moisture and meteorological parameters. Based on the some preliminary results over a 8 month period, the possible influence of meteorological parameters on the exhalation and the temporal variation is discussed.

2. Experimental Methods

The experimental arrangement(Fig.1) of radon exhalation consists of a close circuit system with the following components, a) a flow-through accumulator, b) an air pump to

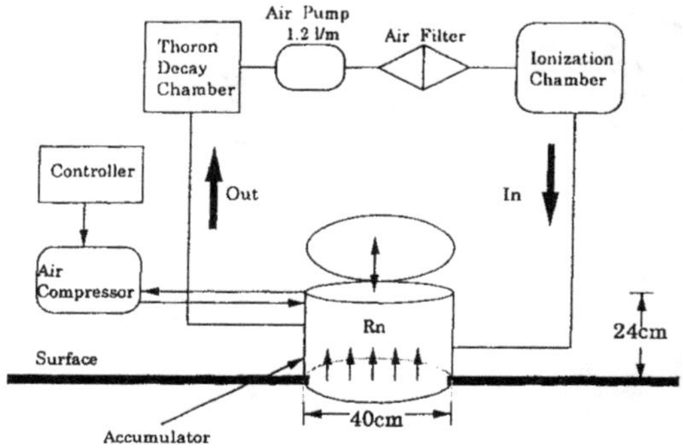

Fig.1 The experimental arrangement of radon exhalation

circulate the air through the system, c) an ionization chamber of 1.5 l for the measurements of radon concentration, d) a membrane filter, which traps radon and thoron daughters and e) thoron decay chamber.

The flow-through accumulator for unattended repetitive measurements is made of a stainless steel with diameter of 40 cm and height of 29 cm. The open of bottom 5 cm diameter of the accumulator was buried in the soil. The lid of the accumulator was automatically opened and closed by a controlled air compressor. At first, the lid of the accumulator was opened for introducing open air into the chamber and after 1 hour it was closed for accumulating the radon from the ground surface for 3 hours. The radon concentration of the sucked air from the closed accumulator was always monitored by an ionization chamber with counting intervals of 10 minutes. To eliminate the fluctuation of data, the averages of 60 readings per one datum were adopted for calculation. The radon exhalation rate was determined by the linear increase of the counting rate up to three hours; within the accumulation time, back diffusion did not lead to a deviation from the straight increase of radon concentration. The data from the automated monitor were cross checked by comparing with grab sampling data.

Measurements of the radon concentration in soil(Fig.2) were made from tubes inserted at three depths 50, 100, and 150 cm. Two vertical tubes for access were connected to a horizontal tube (8 cm in diameter and 60 cm in length), buried to the interested depth, which had many holes on the surface for collection of soil radon by diffusion. On the ground, the vertical tubes were connected to a thoron decay chamber, an air pump, an air filter and an ionization chamber of 1.5 l. In the measurement system, as the air is always circulated at the flow rate of 1.2 l/min., radon concentrations in soil air are continuously monitored.

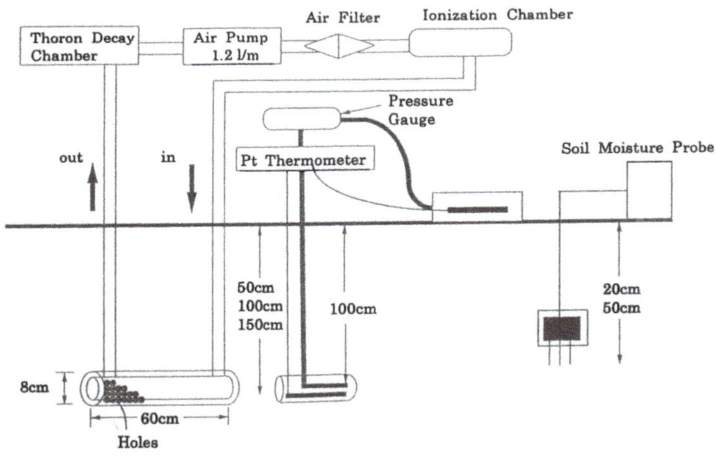

Fig.2 Experimental arrangement ofl radon and environmental factors in soil

Soil temperatures at 50,100, and 150 cm depth and soil moisture contents (volume fraction for saturation) at 20 and 50 cm depth were measured with platinum sounding probes and soil moisture probes respectively. Also, the pressure difference between the surface and 100 cm depth was monitored with a precious pressure gauge. On the ground, the meteorological data (wind direction, velocity, temperature , relative humidity, rain and radiation) are always measured.

Measurements were carried out at a campus of the Science University of Tokyo, located in the suburbs of Tokyo. The soil is a fine loam, Kanto Loam Layer. The climate has a rainy season during June to July, a humid and hot summer monsoon during July to September and a cold and windy winter monsoon during December to February. To eliminate the influence of digging the ground up, data for 6 months after the landfill were not used.

3. Results

1) Factors affecting instantaneous radon exhalation rates.

It is well-known that barometric pressure variation can significantly affect the radon exhalation (Clements and Wilkening [1], Shery and Gaeddert [5] Shery et.al [6], Duenas and Fernandez [2]). To verify this fact, temporal variations of the radon exhalation and barometric pressure, measured during the period of 10 days in Feb. 1997, are plotted in Fig. 3(a). Also, wind velocity, the pressure difference between the surface and 100 cm depth (P(0)-P(100)), the concentration ratio between radon at 50 and 100 cm depth (R(100)/R(50)) are shown in Fig.3(b) and (c). The pressure clearly changes with a period of 2 or 3 days. However, the exhalation rate does not show a change accompanied by the pressure. The same trend also has been obtained from measurements of the other series. It is seen, from the results, that there is little or no correlation between radon exhalation and pressure.

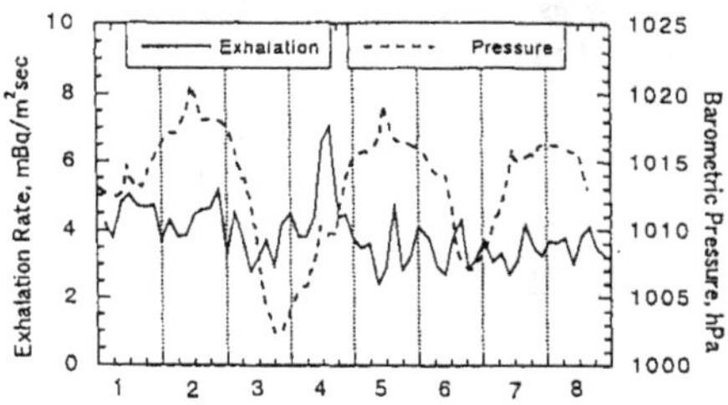

Fig.3(a)

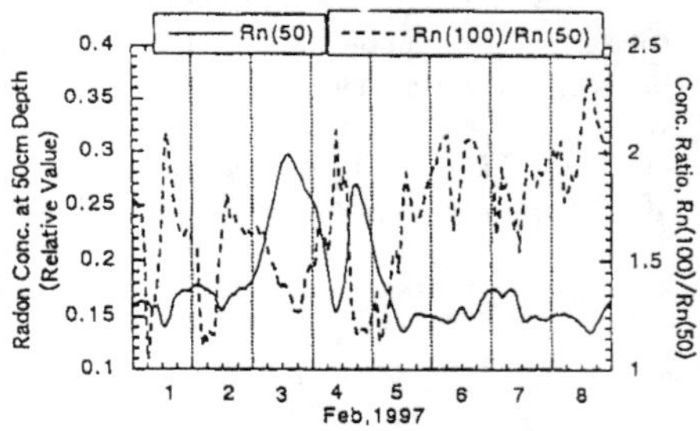

Fig.3(c)

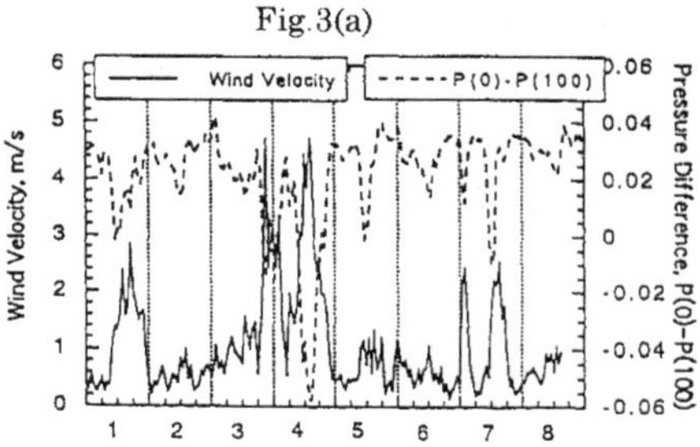

Fig.3(b)

Fig.3 Time variations of (a) radon exhalation and barometric pressure, (b) wind velocity and pressure difference and (c) radon concentration at 50 cm depth and radon concentration ratio between 50 cm and 100 cm depth.

On the other hand, Figure 3 gives a significant correlation in relation to the exhalation rate among the wind velocity, pressure difference and the radon concentration ratio. A statistically significant correlation between the exhalation and the pressure difference is shown in Fig.4. Meanwhile, the increase of wind velocity larger than about 4 m/sec on Feb 4 in Fig.3(b), corresponds to the sharp increase of the exhalation rate(Fig.3(a)). From Fig.3(b), also it is easily found that the wind velocity is negatively correlated with the pressure difference, (P(0)-P(100)). The existence of a significant pressure difference effect is not surprising as negative pressure difference will make upward flow of radon in the soil. As the wind velocity is an important factor affecting the pressure difference, the wind velocity will be closely related to the radon exhalation. The pressure difference is also negatively correlated with the radon concentration ratio (R(100)/R(50)) in soil. This shows that decreasing the pressure difference raises the concentration ratio; this will result in increasing the radon exhalation by diffusion process.

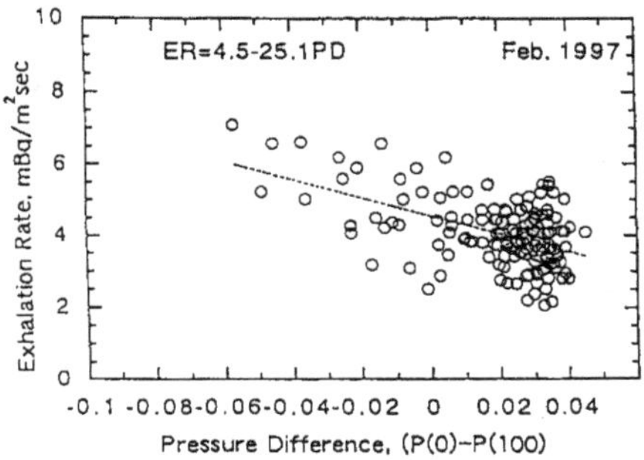

Fig.4 Relation between radon exhalation and pressure differnce.

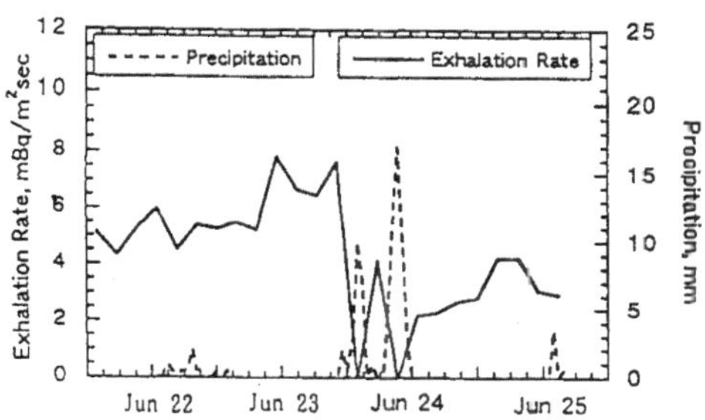

Fig.5 Radon exhalation rate and rainfall intensity over a period of 4 days.

In our experiments, the effect of barometric pressure is typically small in comparison with effects attributed to the wind velocity and the pressure difference. The result differs greatly from the other researcher's ones. In this relation, we have extensively examined further data.

Figure 5 shows the radon exhalation rate and the rainfall intensity for a period of 4 days in June of 1997. On June 24, the exhalation rate was extremely reduced by a heavy rainfall larger than 10 mm and after stopping rainfall the exhalation rapidly recovered. On the other hand, in the cases of light rainfall of the amount 1 to 3 mm on June 22 and 25, the exhalation rates were scarcely changed. The dependence of radon exhalation on the amount of rainfall, obtained here, will be consistent with the experimental results obtained by Megumi[4].

In Fig.6, the relation between the soil moisture content at 50 cm depth and the radon exhalation rate, measured during April to June, is shown. With increasing moisture content, the exhalation rate will certainly decrease. This may be due to two effects: a capping effect at ground surface during the rainfall and the reduction of air permeability in soil. From Fig.5. the exhalation seems to have a maximum at the moisture content from 0.38 to 0.385. To confirm this fact, we will need to have more data smaller than 0.38 in moisture content.

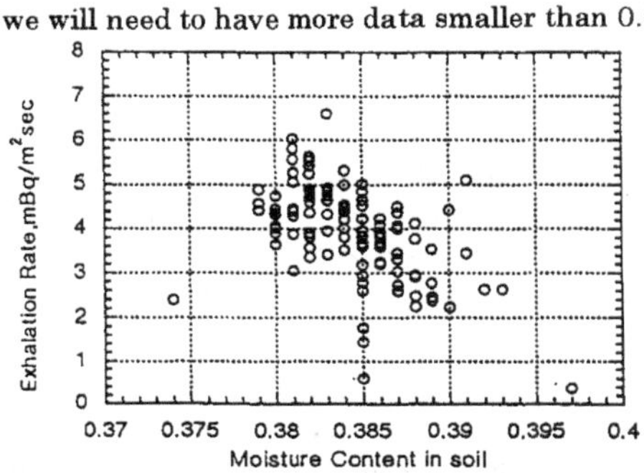

Fig.6 Relation between exhaltion rate and soil moisture at 50 cm depth

2) Temporal variations of radon exhalation

Figure 7 shows typical diurnal variations of radon exhalation, together with barometric pressure, pressure difference and concentration ratio of soil radon, measured in February. The radon exhalation shows a pattern of high level in the afternoon and low level in the early morning. But, the difference is relatively small and is of the order of about 20 %. The diurnal pattern of exhalation is inconsistent with the result conventionally predicted by the pressure variation shown in Fig.7, although there is an agreement in the hour occurring a maximum of exhalation and a minimum of barometric pressure. On the other hand, due to pressure difference effect, it is found that a maximum of exhalation in the afternoon corresponds to the diurnal drop in pressure difference, although there is a bit of delay; a reduced exhalation in early morning corresponds to the high level in the pressure difference. This results may show the existence of systematic change of upward flow between the day and the night.

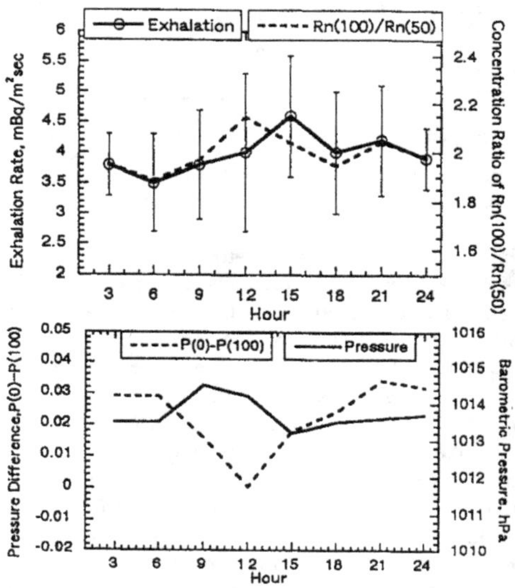

Fig.7 Typical diurnal variations of radon exhalation rate, together with barometric pressure, pressure difference and radon concentration ratio.

Several authors have proposed the effects of convection processes bearing on the radon transport in soil(e.q. Fleischer 1979). In the convection processes radon exhalation is enhanced at night when the top layers of soil are cooler than lower layers and at day it varies inversely. It will be possible to make a conclusion from our experimental results that the exhalation rates are not influenced by convective transport processes.

In Fig.8, the monthly means of radon exhalation rates during from February to September are shown together with the corresponding values for the soil moisture content of 20 and 50 cm depth and the pressure difference. The variations show a slight low level in February and September and a high level in August. From the data from June to September, it is evident that there is a good correlation between the exhalation rate and the soil moisture content in 20 cm depth. On the other hand, the pressure difference shows low level in summer and high level in early spring. From the experimental results, it can be inferred that the seasonal variation of radon exhalation is determined by a balance between the pressure difference corresponding to the force for vertical flow of soil air and the soil moisture content affecting the air permeability in soil.

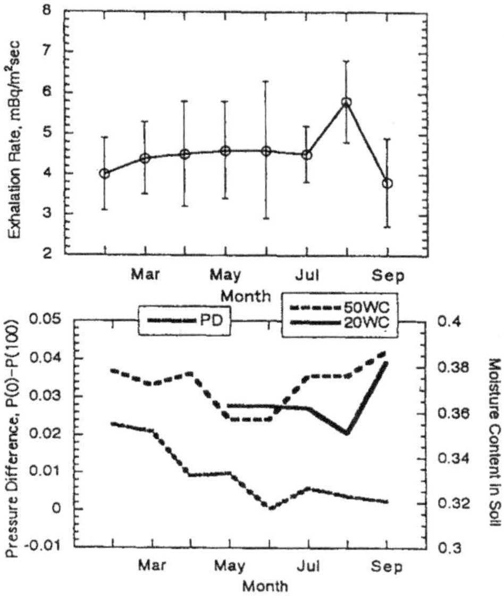

Fig.8 Monthly means of radon exhaltion rate, soil content at 20 cm and 50 cm depth.

4. Refernces

[1] Clements,W.E. and M.H.Wilkening, Atmospheric pressure effects on ^{222}Rn transport across the earth-air interface, J.Geophys.Res. **79**, 5025-5029, 1974.

[2] Duenas,C., and M.C.Fernandez, Dependence of ^{222}Rn flux on concentrations of soil gas and air gas and an analysis of the effects produced by several atmospheric variables, Ann.Geophys. **5B**, 533-540, 1987.

[3] Fleischer,R.L. and A.Mogro-Campero, A radon enhancements on the earth: evidence for intermittent up-flows?. Geophys.Res.Lett., **6**, 361-364, 1979.

[4] Megumi,K. and T.Mamuro, Radon and thoron exhalation from the ground, J.Geophys.Res.,**78**, 1804-1808, 1973.

[5] Shery,S.D. and D.H.Gaeddert, Measurements of the effect of cyclic atmospheric pressure variation on the flux of ^{222}Rn from the soil, Geophys.Res.Lett., **9**, 838-1982.

[6] Shery,S.D.,D.H.Gaeddert and M.H.Wilkening, Factors affecting exhalation of radon from a gravely sandy loam, J.Geophys.Res., **89**, 7299-7309, 1984.

Vertical Distribution of Radon Concentrations in the Atmosphere

Takao Iida[*1], Katsumi Yamada[*1], Hideyuki Miyachi[*1, *4], Yasunobu Iwasaka[*2], Katsuji Matsunaga[*2] and Takashi Ishizaka[*3]

*1 Graduate School of Engineering, Nagoya University, 464-01, Japan
*2 Solar-Terrestrial Environment Laboratory, Nagoya University
*3 Institute for Hydrospheric-Atmospheric Sciences, Nagoya University
*4 Present address: Chubu Soft Engineering Inc.

Abstract

Vertical distributions of radon (^{222}Rn) concentrations were measured up to 5 km in the atmosphere around central Japan. About 50 L air was sampled in a Tedlar® (polyvinyl fluoride, PVF) bag by using an aircraft. The ^{222}Rn concentrations in the bags were measured with electrostatic radon monitor. The detection limit is 0.10 $Bq \cdot m^{-3}$ for measuring time of 6 h. The measured concentrations were corrected by considering ^{222}Rn decay rate and the atmospheric pressure at the sampling altitude. The ^{222}Rn concentration decreases rapidly with increasing altitude. The high ^{222}Rn concentrations were observed at about 500 and 1000 m under atmospheric boundary layer over the land areas. The ^{222}Rn concentrations sampled over the sea were lower due to no ^{222}Rn exhalation from the sea surface. The vertical distribution could be explained fairly well by using the analysis of air mass trajectory and one dimensional diffusion model.

1. Introduction

Radon-222 that is natural radioactivity in the atmosphere could be used as the tracer of air pollutants originating from the continental surface. The measurements of ^{222}Rn vertical profiles clarify the behavior of not only radon but also pollutants in the upper atmosphere.

2. Sampling locations

To measure the vertical distribution of ^{222}Rn concentrations up to 5 km in the

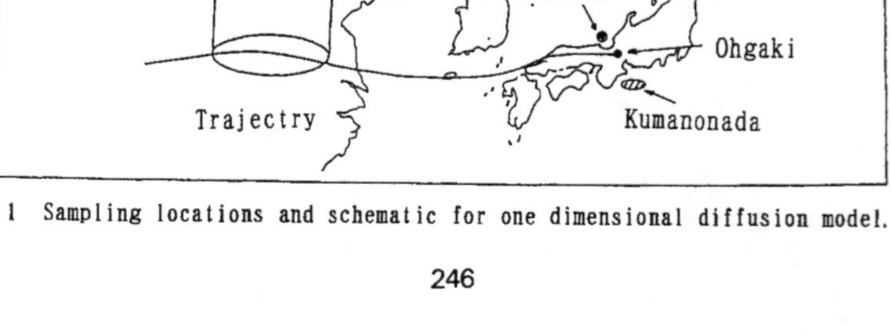

Fig. 1 Sampling locations and schematic for one dimensional diffusion model.

atmosphere around central Japan, air samples were collected by an aircraft.

Two flights were carried out over Wakasa (3 and 10 March 1994), five flights over Ohgaki (29 July 1994, 21 and 24 November 1994, 23 February 1995, and 23 April 1996) and seven flights around Kumanonada (27, 28 and 29 September 1996, 23, 24 and 27 December 1996 and 19 February 1997). The sampling locations are shown in **Fig. 1**. Two to seven air samples were collected during one flight.

3. Measurements

3.1 Electrostatic ^{222}Rn monitor

Figure 2 shows the schematic diagram of the electrostatic ^{222}Rn monitor for measuring the ^{222}Rn concentrations in grab sampling. The ^{222}Rn monitor consists of a cylindrical vessel of stainless steel which is able to be evacuated. The cylindrical vessel has a volume of about 32 L. The sampling air is passed through a membrane filter and a vapor trap to the 32-L cylindrical vessel. Then, aerosols, ^{222}Rn daughters and water vapor were removed in the air. Radon-222 decays by emission of an α-particle to ^{218}Po, and most ^{218}Po atoms are positively charged [1]. The positive ^{218}Po ions are collected electrostatically on the electrode of aluminized Mylar (0.9 mg·cm^{-2}) coated with ZnS(Ag) phosphor of 10 mg·cm^{-2}. Alpha-particles emitted from the ^{222}Rn daughters are incident on the underlying ZnS(Ag) through the Al Mylar. The scintillations due to α-particles are detected by a 52 mm photomultiplier tube optically coupled to a light pipe. The scintillation pulse which is amplified and processed is fed into a personal computer (NEC PC-9801RX) through the interface. Using the personal computer, the ^{222}Rn concentrations are calculated from α-counts at one-hour intervals on the basis of Batemann equation [2]. The obtained ^{222}Rn concentration data are stored on a floppy disk.

The detection limit depends on the background count rate and measuring time. Using the typical background count rate (2.7±1.9) counts·h^{-1} and the calibration factor (53.7±0.9) counts·h^{-1}·(Bq·m^{-3})$^{-1}$ between measured α-counts and ^{222}Rn concentration, the detection limit defined by Currie [3] is found to be 0.10 Bq·m^{-3} for measuring time of 6 h.

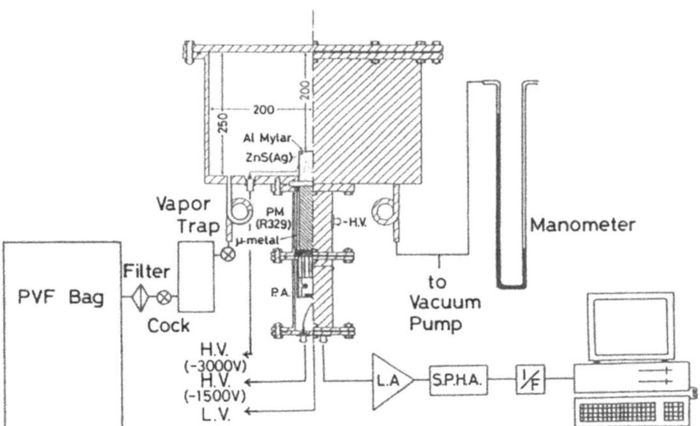

Fig. 2 Electrostatic ^{222}Rn monitor.

3.2 Measuring method

About 50 L air was sampled in a Tedlar® bag by using an aircraft. Radon-222 does not permeate through the PVF film of the Tedlar® bag. Measuring time needs more than 6 h per one air sample. Due to the preparation time and about 3 h for required to attain equilibrium between ^{222}Rn and its daughters, only two air samples could be measured a day. The half-life of ^{222}Rn is 3.82 d. Therefore, it is reasonable that six to eight air samples are collected during one flight. The measured concentrations were corrected by considering ^{222}Rn decay rate and the atmospheric pressure at the sampling altitude.

4. Results

Table 1 provides the summary of the measured ^{222}Rn concentration for the air sampled over Wakasa, Ohgaki and Kumanonada.

Figure 3 shows the vertical profiles of ^{222}Rn concentrations for the air masses sampled over Wakasa during the 4 and 10 March 1994 flights. The ^{222}Rn concentration decreases with altitude.

Table 1 Results of ^{222}Rn concentrations

Sampling time	Altitude (m)	Concentration (Bq·m^{-3})
Mar. 4, 1994	(Wakasa)	
15:15	3510	0.04±0.03
15:33	1680	3.44±0.20
15:42	305	2.55±0.09
16:27	395	2.44±0.15
16:36	1680	1.93±0.13
16:45	3050	0.61±0.07
Mar. 10, 1994	(Wakasa)	
16:30	3050	0.04±0.03
16:36	1680	1.96±0.10
16:42	305	3.20±0.17
16:48	1680	2.17±0.11
16:54	3050	0.10±0.03
Jul. 29, 1994	(Ogaki)	
11:52	2000	1.63±0.13
12:03	1000	2.35±0.10
15:17	4420	0.11±0.05
15:39	4420	0.12±0.07
15:59	457	4.16±0.17
16:12	3200	0.26±0.07
Nov. 21, 1994	(Ogaki)	
12:22	4570	0.41±0.07
12:55	4570	0.39±0.07
13:09	3180	0.62±0.10
Nov. 24, 1994	(Ogaki)	
12:00	2290	0.09±0.05
12:15	1070	2.74±0.18
12:21	457	3.51±0.21
Feb. 23, 1995	(Ogaki)	
11:25	5180	0.42±0.06
11:45	3960	0.62±0.07
12:16	2290	1.24±0.12
12:43	1370	0.77±0.11
12:57	910	4.13±0.26
13:11	457	3.54±0.22
Dec. 27, 1995	(Ogaki)	
12:45	4420	0.19±0.07
16:54	2743	0.18±0.07
Dec. 28, 1995	(Ogaki)	
09:54	5182	0.13±0.05
10:18	3658	0.43±0.09
10:40	3048	0.07±0.08
10:58	2286	−0.18±0.06
11:07	1829	−0.02±0.10
11:25	1372	0.04±0.10
11:34	914	0.20±0.14
11:51	457	0.79±0.18
Apr. 23, 1996	(Ogaki)	
08:55	5486	0.22±0.06
09:50	4267	0.65±0.09
10:43	3048	0.98±0.11
13:35	2438	1.00±0.14
14:15	1829	2.18±0.19
14:45	1219	2.41±0.25
15:17	610	3.36±0.28
Sep. 27, 1996	(Kumanonada)	
13:27	457	1.92±0.20
13:45	914	1.44±0.16
14:10	2286	0.34±0.09
15:15	914	0.92±0.16
Sep. 28, 1996	(Kumanonada)	
12:59	762	0.96±0.16
14:23	4367	0.01±0.06
14:45	1981	0.53±0.13
15:10	610	0.92±0.17
Sep. 29, 1996	(Kumanonada)	
14:40	610	0.41±0.13
15:10	2743	1.14±0.16
Dec. 23, 1996	(Kumanonada)	
11:55	5029	0.40±0.05
12:23	3658	−0.04±0.03
12:42	3658	−0.05±0.03
13:11	1067	0.09±0.06
Dec. 24, 1996	(Kumanonada)	
14:08	3658	0.21±0.06
14:23	2743	1.81±0.15
14:36	1829	0.63±0.11
14:49	914	0.55±0.12
Dec. 27, 1996	(Kumanonada)	
13:50	2896	0.38±0.06
14:09	1981	1.41±0.13
14:38	1981	1.24±0.12
15:04	610	1.79±0.17
Deb. 19, 1997	(Kumanonada)	
10:40	500	1.31±0.12
10:47	1000	1.65±0.17
10:53	1500	1.80±0.15
10:59	2000	0.89±0.12
11:07	3000	0.08±0.05
11:15	4000	0.15±0.04
11:25	5000	0.04±0.03

The results sampled over Ohgaki are shown in **Fig. 4**. The high concentrations of ^{222}Rn were observed at about 500 and 1000 m, since the air mixing is relatively high under atmospheric boundary layer. The ^{222}Rn concentration sampled on 23 February 1995 fluctuated with altitude from 0 m to 3000m. It is considered that the air mass of high ^{222}Rn concentration have been brought with ascending current. Japan is covered with maritime air mass in summer. However, the ^{222}Rn profile on 29 July 1994 was similar to those in other seasons.

The vertical profiles of ^{222}Rn concentrations around Kumanonada are shown in

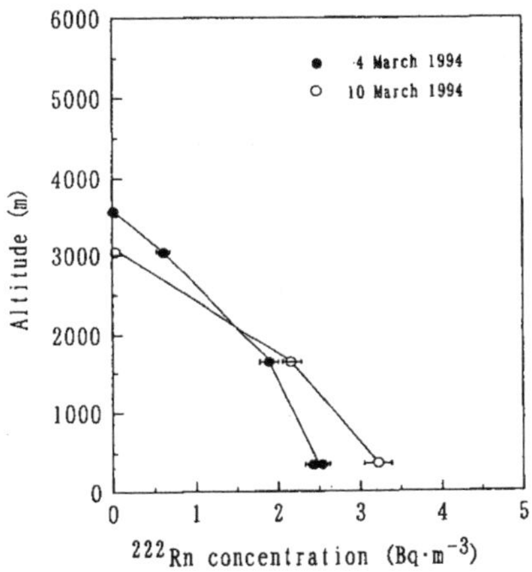

Fig. 3 Vertical profiles of ^{222}Rn concentrations over Wakasa.

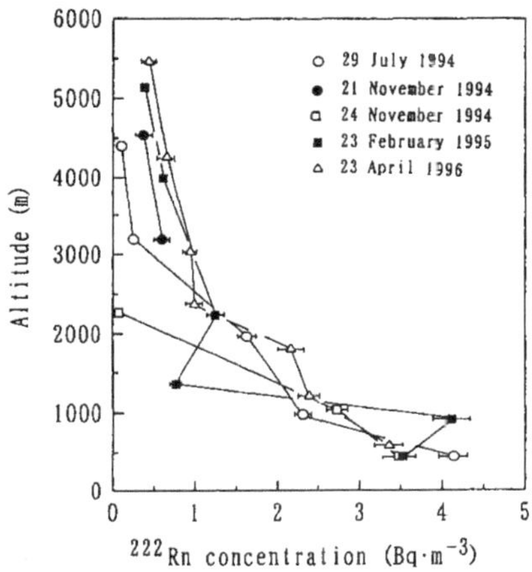

Fig. 4 Vertical profiles of ^{222}Rn concentrations over Ohgaki.

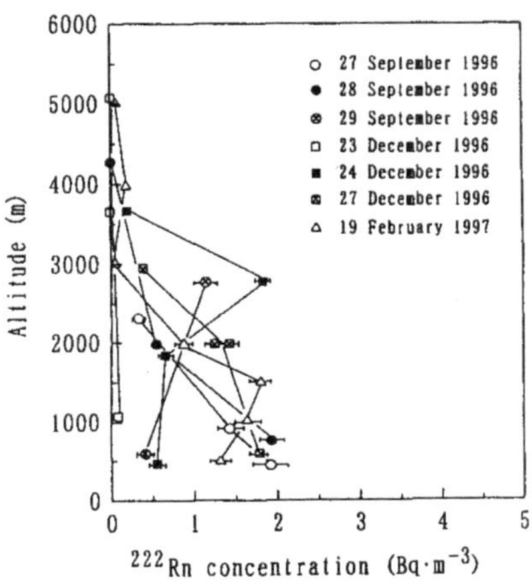

Fig. 5 Vertical profiles of ^{222}Rn concentrations around Kumanonada.

Fig. 5. The ^{222}Rn concentrations under atmospheric boundary layer are lower than those observed over Ohgaki. The results on 24 December 1996 show very high ^{222}Rn concentration at the altitude of about 3000m. The vertical profile observed on 19 February 1997 is also shown in **Fig. 5**. The ^{222}Rn concentration was almost 0 $Bq \cdot m^{-3}$ over 3000 m. The low concentration at lower altitude is due to no ^{222}Rn exhalation from the sea surface.

5. Discussions

5.1 Analysis using air mass trajectory

The trajectories presented in **Fig. 6** were constructed backwards in one week from the positions where the ^{222}Rn concentrations were measured on the 23 February 1995 flight. The trajectories were constructed at the isobaric surface of 850hPa, 700hPa, and 500hPa which correspond to the altitude of about 1450m, 3000m, and 5600m in the standard atmosphere, respectively. The trajectories originated in the continent. This is the reason why the ^{222}Rn concentrations are high over atmospheric boundary layer.

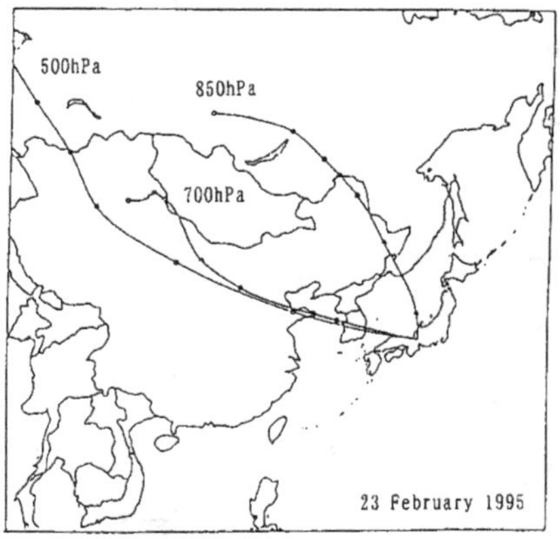

Fig. 6 Air mass trajectories constructed for the 23 February 1995 flight.

5.2 Trajectory plus one dimensional diffusion model

The concept of trajectory plus one dimensional model is shown in **Fig. 1**. It has been supposed that the vertical column of air moves along the trajectory at the isobaric surface of 850hPa. In the column, ^{222}Rn emits only from the surface on the land areas, and diffuses to the upper atmosphere. We used the vertical diffusion profiles to classify the unstable condition for day-time and stable condition for night-time, referring to the vertical profiles of Jacobi and Andre [4]. The calculated vertical distribution of ^{222}Rn is shown in **Fig. 7**. The calculated profile agrees fairly well with observed profile. However, close agreement was not obtained at the upper atmosphere. The reason is considered to be as follows; (1) the reliability of the trajectory, (2) the assumption of the uniform ^{222}Rn emission rate on the land area, (3) the assumption of two vertical diffusion profiles a day, (4) the boundary condition of the calculation, and (5) the limit of one dimensional diffusion model. Moreover, this model could not simulate the rapid ascend of ^{222}Rn to the upper atmosphere around clouds.

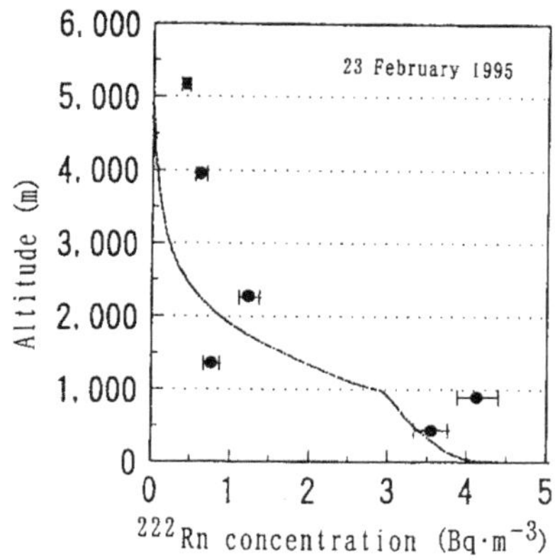

Fig. 7 Comparison of measured and calculated ^{222}Rn concentrations.

6. Conclusion

The vertical distributions of ^{222}Rn concentrations up to 5 km in the atmosphere were observed around central Japan by using an aircraft. The ^{222}Rn concentration decreases rapidly with increasing altitude over atmospheric boundary layer. The vertical profile could be evaluated fairly well by using the analysis of air mass trajectory and one dimensional diffusion model. The subjects for a future study are the evaluation of the ^{222}Rn emission rate on the continent and the assumption of correct vertical profiles for diffusion coefficient.

The authors developed the numerical model of three dimensional atmospheric transport [5]. The ^{222}Rn concentration levels and the diurnal variations simulated at Nagoya agree well with the observed concentrations. We will apply the numerical model to explain the vertical profile of ^{222}Rn concentrations.

7. References

[1] Dua, S. K., P. Kotrappa, and P. C. Gupta, Influence of relative humidity on the charged fraction of decay products of radon and thoron, Health Phys., **45**, 152-157, 1983.

[2] Iida, T., An electrostatic radon monitor for the continuous measurement of environmental radon, in Atmospheric Radon Progeny and environmental Radioactivity, edited by S. Okabe, Atomic Energy Society of Japan, Tokyo, pp. 65-73, 1985 (in Japanese).

[3] Currie, L. A., Limits for qualitative detection and quantitative determination, Anal. Chem., **40**, 586-593, 1968.

[4] Jacobi, W., and Andre, K., The vertical distribution of Radon 222, Radon 220, and their decay products in the atmosphere, J. Geophys. Res., **68**, 3799-3814, 1963.

[5] Sakashita, T., T. Murakami, T. Iida, et al., The numerical model on the three dimensional atmosphere transport with application to the transport of ^{222}Rn, J. Atmos. Electr., **14**, 57-62, 1994.

The Vertical Profile of ^{222}Rn Concentration in the lower Atmospheric Boundary Layer at Shimane Peninsula

Katsuhiro Yoshioka

Public Health and Environmental Science Institute of Shimane Prefecture

582-1, Nishihamasada, Matsue, 690-01 Japan

Abstract

To estimate local meteorological condition by studying the behavior of atmospheric ^{222}Rn, we measure the time variation of the vertical profile of ^{222}Rn concentration in the lower atmospheric boundary layer. The measurements of ^{222}Rn vertical profile were continuously obtained simultaneously at the four locations, where two mountaintops have different elevation (h=150,520m) and our labolatory (h=30m) is sited at Shimane peninsula and Oki island lies in Japan Sea.

A diurnal change is clearly shown in ^{222}Rn profiles at 30m and 150m high due to the temperature inversion layer in the surface layer through the nighttime. But this change is rarely shown in ^{222}Rn profile at 520m high. It is considered that the top of surface inversion layer could not reach on the mountaintop of 520m high. And then, these ^{222}Rn profiles are clearly different in the long term variation. It is definitely considered that the atmospheric stability depends on the altitude in the atmospheric boundary layer.

^{222}Rn concentration in Oki is lower value on the whole and it rarely shows a diurnal change and its magnitude is small and, Oki is a seagirt island. Consequently, we would regard ^{222}Rn concentration in Oki as the background level in this area.

1. Introduction

In Japan of varied and complicated topography, to estimate the vertical and/or horizontal distributions of meteorological elements, it is necesary to prepare several high observation towers and many meteorological instruments and to observe at many locations through the long period. But such observation is very difficult.

The atmospheric ^{222}Rn concentration would be easily measured everywhere using a small instrument. We would observe a vertical distribution of ^{222}Rn to study the turbulent transfer at some mountaintops with different altitude in place of high observation tower in Shimane peninsula with varied topography.

2. Measurement method

2.1 Electrostatic radon monitor

In the natural environment, the range of atmospheric ^{222}Rn concentration is between 0.1 and 100 Bq/m^3, so that the available method of measurement is limited in time series observation. To meet the above range, a ^{222}Rn monitor is required to have the following characteristics : 1) a capability to measure continuously variable and lower level with high accuracy and sensitivity : 2) a stable efficiency over a long period : 3) an easy maintenance.

We would adopt the electrostatic continuous radon monitor which has been developed by T. Iida et. al. [1]. The schema of this radon monitor is shown in Fig. 1 and four monitors are made.

2.2 Comparison with four electrostatic radon monitors

Four radon monitors are perfectly adjusted and are given each calibration factor by the sensitivity test using ^{222}Rn gas. The results of comparison with four radon monitors are shown in Fig. 2 (a) and (b). Only small quantity of difference can be seen among between measurements of four monitors.

To compare directly with each measurement, measurements of ERM-1, ERM-2 and ERM-3 have been normalized to ERM-0 using the compensation factors of which have been calculated by the least square method using the above result.

2.3 Measurement locations and elevations

Four measurement locations are shown in Fig.3. Misakayama and Fukada are each mountaintop of 520m and 150m high and Hamasada is on our laboratory roof of 30m high. These locations are within roughly 10km area of the eastern Shimane peninsula. Oki lies 75km away from Shimane peninsula in the Japan Sea and this elevation is 90m high.

3. Results and Discussion

The atmospheric ^{222}Rn concentrations have been continuously observed at the above locations since December 1996.

3.1 Profiles of atmospheric ^{222}Rn concentration

Monthly means of ^{222}Rn concentration at four locations are shown in Fig.4. ^{222}Rn concentration is affected by the meteorological condition of each location due to height or topography, so that each profile difinitely differ from others.

On the whole, the mean ^{222}Rn concentration at Misakayama (h=520m) is approximately –45% value of

Hamasada (h=30m) and −42% value of Fukada (h=150m). Fukada is approximately −6% value of Hamasada.

It is well known that the surface inversion layer is grown by the radiation cooling through the nighttime. It is assumed that Hamasada and Fukada are lower than the top of temperature inversion layer, but Misakayama is higher than the top of this layer.

The profile of Fukada is extremely changeful. It is considered that this height (h=150m) is nearly a border line between the surface boundary layer and the lower Ekman layer. In February and March 1997, ^{222}Rn concentration of Fukada is lower than this of Hamasada. It is considered that the top of temperature inversion layer has not frequently reached the height of Fukada in this period.

In May, June and July 1997, ^{222}Rn concentration of Misakayama is very lower level but these of Hamasada and Fukada are very higher level. These values of Misakayama are between −70% and −86% values of Hamasada and/or Fukada. It is considered that the stable stratification has been developed below the height of Misakayama through this period.

^{222}Rn concentration in Oki(h=90m) is nearly equal to Misakayama except May, June and July 1997. ^{222}Rn concentration in Oki is considered a lower level, this is because the exhalation of ^{222}Rn from Oki island is a little quantity due to small islands and that from Japan Sea is quite negligible. ^{222}Rn from Eurasian Continent and/or Japanese Islands are sufficiently mixed and diffused through the way of transportation.

Consequently, ^{222}Rn concentration in Oki would be the background level in this area.

3.2 Diurnal change and Daily variation

In calm days, ^{222}Rn concentration shows a regular change due to the rotation of the earth in such Fig.5 (a) and (b). The time lag is seen among the three diurnal changes due to the different elevation within 10km area. In addition, the changes of Misakayama and Oki are barely shown and these quantities are small.

The daily variation of ^{222}Rn concentration is shown in Fig.6 between (a) and (b) and these variations are irregular and the quantities are large in Hamasada and Fukada and small in Misakayama and Oki. And these quantities are small in summer and large in winter and spring. Due to the above variations, it is considered that the meteorological condition is unstable when the atmospheric boundary layer is low and/or in winter and spring.

Acknowledgement

The author wishes to express heartly thank to Dr. T. Iida of Nagoya univ. for his help in calibration.

4. Reference

[1] T. Iida et. al., Continuous measurements of outdoor radon concentrations at various locations in east Asia, Environ. Int., 22, s139 -- s147, 1996

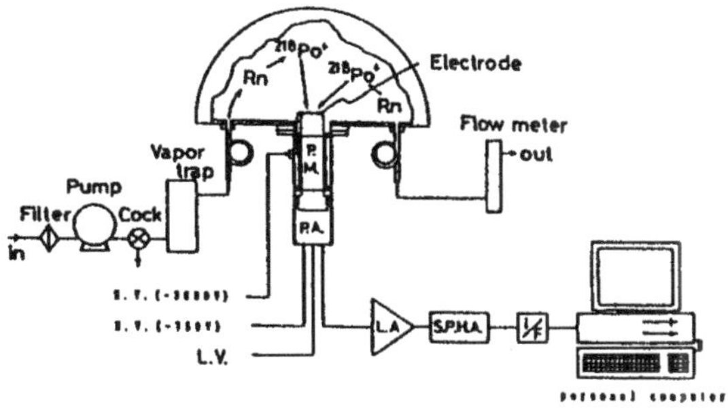

Fig. 1. Schema of electrostatic radon monitor

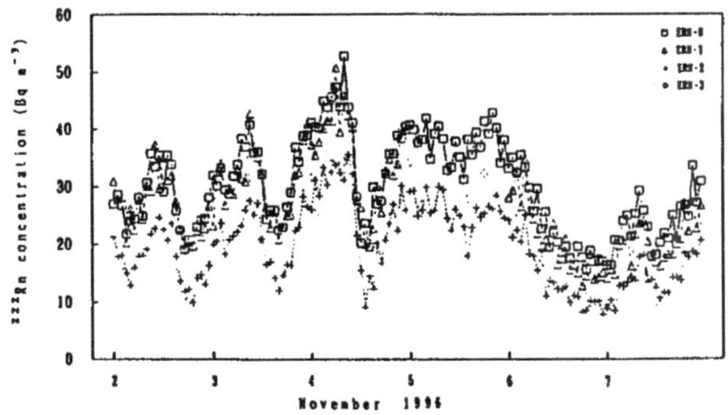

Fig. 2 (a). Comparison of ^{222}Rn concentrations with ERM-0, ERM-1, ERM-2 and ERM-3

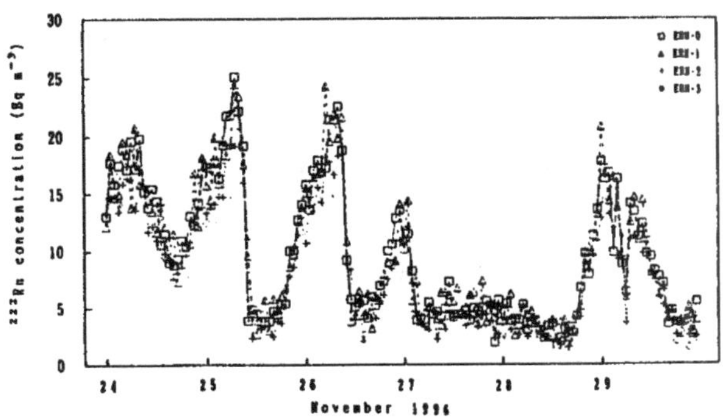

FIg. 2 (b). Comparison of ^{222}Rn concentrations with ERM-0, ERM-1, ERM-2 and ERM-3

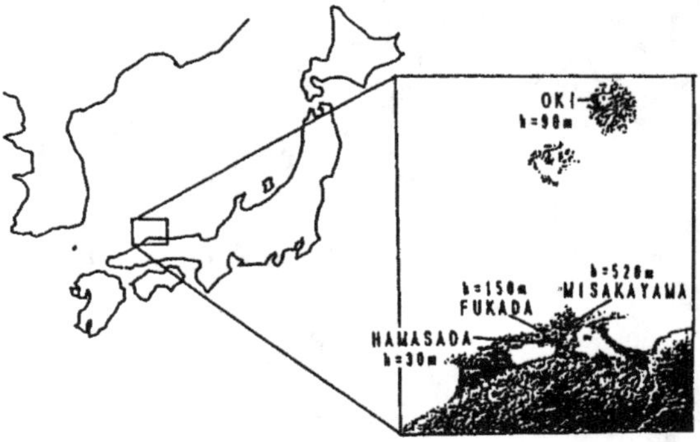

Fig. 3. Map of measurement locations

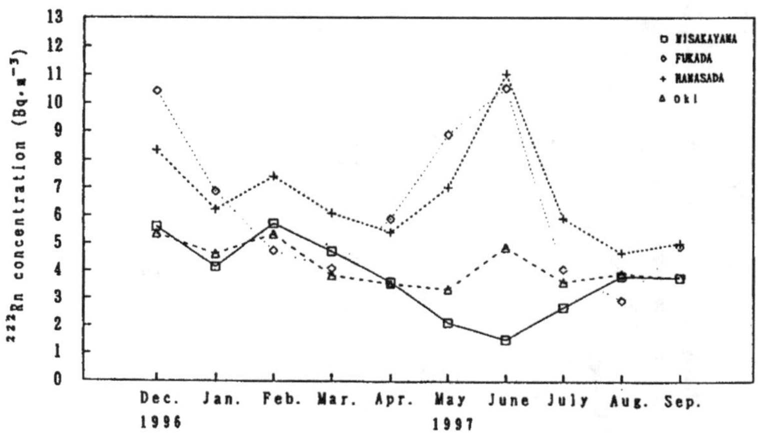

Fig. 4. Monthly means of ^{222}Rn concentration

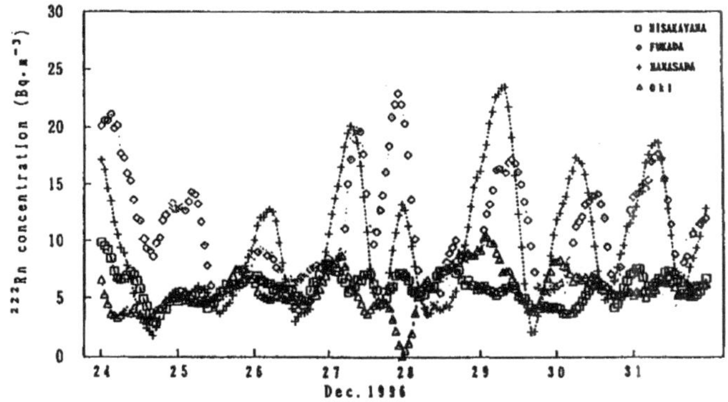

Fig. 5 (a). Diurnal changes of ^{222}Rn concentration

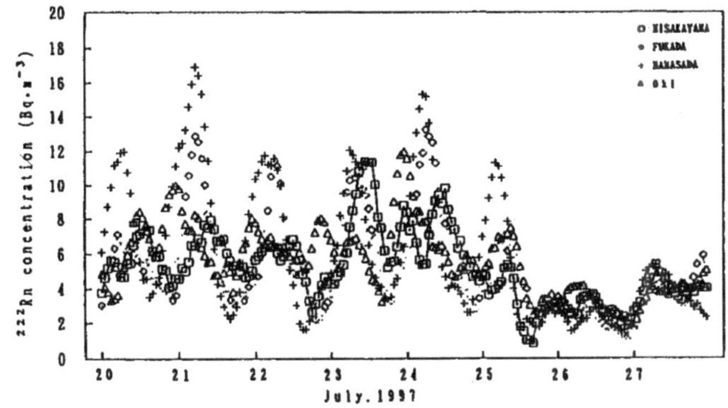

Fig. 5 (b). Diurnal changes of ^{222}Rn concentration

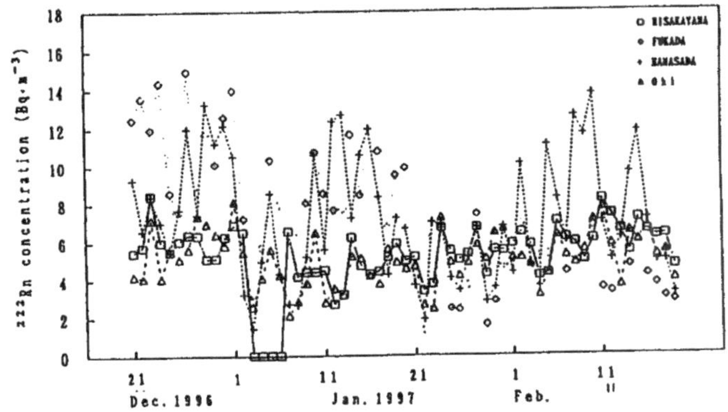

Fig. 6 (a). Daily variations of ^{222}Rn concentration

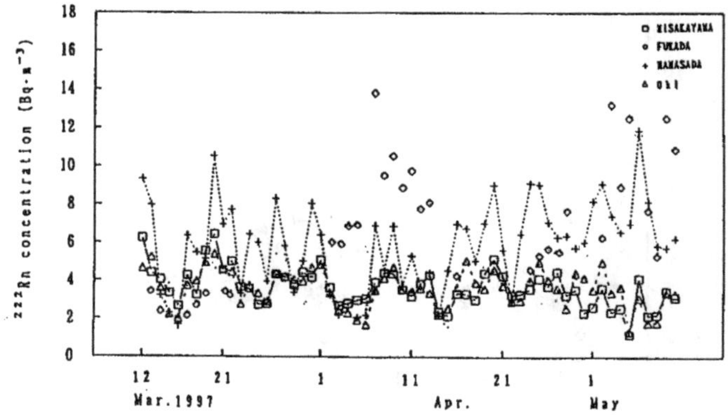

Fig. 6 (b). Daily variations of ^{222}Rn concentration

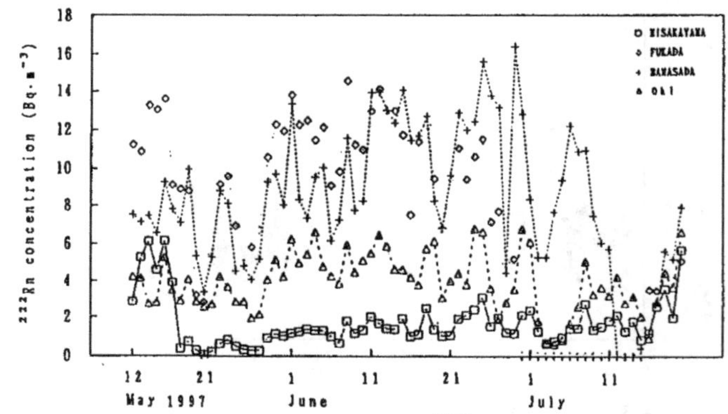

Fig. 6 (c). Daily variations of ^{222}Rn concentration

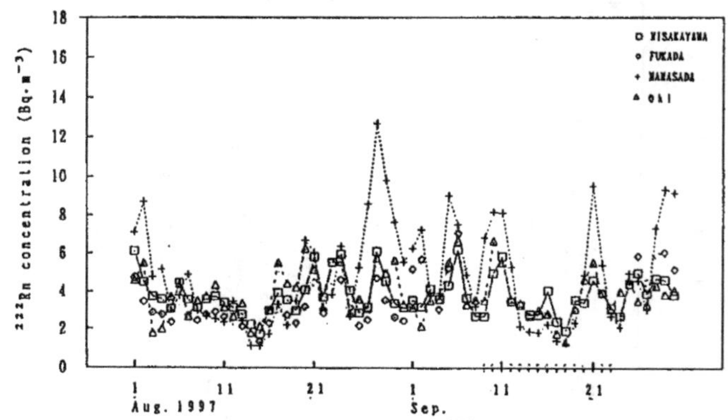

Fig. 6 (d). Daily variations of ^{222}Rn concentration

Estimates of seasonal variations of ^{222}Rn from different origins by using the correlation between ^{222}Rn and ^{212}Pb concentrations in air

Takatoshi Hattori and Takeshi Ichiji

Central Research Institute of Electric Power Industry
2-11-1, Iwatokita, Komae-shi, Tokyo 201, Japan

Abstract

Continuous measurements of ^{222}Rn ($T_{1/2}$ = 3.82 d) and ^{212}Pb ($T_{1/2}$ = 10.6 h) concentrations in outdoor air were carried out in Tokyo from December 1993 to November 1996. Positive correlations were found in the relation between ^{222}Rn and ^{212}Pb concentrations. Using the intercept values obtained from the regression line, a new method is proposed to separate the seasonal variation of monthly mean ^{222}Rn concentration by its origin. The estimated concentration of ^{222}Rn originating from distant regions in 1994 agreed with the calculations by the long-range transport model, which verified the method reliability. The present method would be useful as a means to study the cause of the seasonal variations of ^{222}Rn concentration.

1. Introduction

Radon-222 with a half-life of 3.8 days is available as a tracer for continental air masses. This is because ^{222}Rn is an inert gas unaffected by chemical interactions in transit and is released from land surface at about one hundred times the rate of ocean release. Its radioisotope, ^{220}Rn, has almost the same properties, but the half-life of 55 s is too short to use it as a tracer. On the other hand, ^{212}Pb, the decay product of ^{220}Rn, exists in air attached to aerosol particles with a half-life of 10.6 h. There have been many studies which utilized ^{212}Pb as a tracer to estimate the influence from local land [1-5].

In the present paper, the seasonal variation of monthly mean ^{222}Rn concentration is separated by the origin by using the intercept values obtained from the regression line of the relation between ^{222}Rn and ^{212}Pb concentrations. The reliability of this method is verified through comparison of the estimated concentration of ^{222}Rn originating from distant regions and the calculated one by the long-range transport model.

2. Experimental Methods

Measurements of ^{222}Rn and ^{212}Pb concentrations in air were carried out at Komae city, a suburb of Tokyo, during the period from December 1993 to November 1996. Outdoor air sampling point was at a height of about 1.5 m above the ground.

Hourly measurements of ^{222}Rn were made with an electrostatic monitor based on the method reported by Iida et al. [6]. Sampling air was pulled into the 15 l container of the monitor through the membrane filter at a flow rate of 0.5 $l.min^{-1}$. Radon-222 in the sampling air decays to ^{218}Po by emission of an alpha particle. The positive ^{218}Po ions are collected electrostatically on the electrode of aluminized Mylar coated with ZnS(Ag). The scintillations due to alpha particles emitted from the ^{222}Rn progeny are detected by a photomultiplier tube. The scintillation pulse is amplified and fed into a personal computer.

The sampling air is dehumidified to below 1 % relative humidity before the filter as the collection efficiency depends on the humidity in the container. The mean concentration of ^{222}Rn for each hour are calculated by alpha counts at one-hour intervals compensated for counts accumulated during past 5 h. Sensitivity of the monitor is about 1 $Bq.m^{-3}$ which is the concentration level when the relative statistical error is 50 %. The flow rate of the sampling air is low enough to remove the effect of ^{220}Rn, but the ^{222}Rn concentration in the container follows the outside one within a delay of 30 min if the outdoor concentration changes remarkably. In the case of decrease of flow rate or increase of humidity, the monitor stops measurements and indicates an alarm message.

The concentration of ^{212}Pb in air was measured with continuous monitor [7] and recorded every four hours. The monitor is able to measure the concentrations of ^{218}Po, ^{214}Pb and ^{214}Bi, and ^{212}Pb assuming the equilibrium with ^{212}Bi. Radon progeny are collected on a roll type membrane filter for 10 min every four hours. Ten seconds after sampling, two energy spectra for alpha particles are observed for 10 min and 38 min. This combination of sampling time and counting times are fixed so that ^{218}Po can be estimated most accurately. The sensitivity of the monitor for ^{212}Pb is about 0.013 $Bq.m^{-3}$ which is the concentration level when the relative statistical error for counts is 50 %. The mean concentration of ^{212}Pb for 10 min every four hours are calculated from alpha counts in the region of interest which is for an alpha particle of 8.8 MeV except in the region lower than 7.7 MeV in the second energy spectrum. The alpha counts are corrected by the sampling flow volume and by the collection and counting efficiencies. In the case of decrease of flow rate or increase of suction pressure near the filter, the monitor stops the measurement automatically.

The simultaneous measurements of ^{222}Rn and ^{212}Pb concentrations were completed in about 86 % of all measuring periods without any operational troubles such as suspension of electric power supply and automatic measurements halts.

3. Experimental Results

The geometric mean values of ^{222}Rn and ^{212}Pb concentrations for each month are shown in Fig. 1. The error bar shows the scattering of the geometric standard deviation. The concentration of ^{212}Pb was quite low about 1 % of ^{222}Rn. It can be seen that the ^{222}Rn concentrations were high in winter and low in summer. The same tendency was observed in ^{212}Pb concentration.

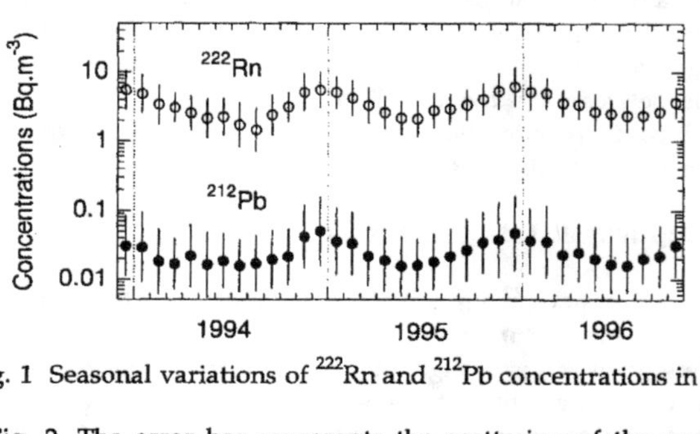

Fig. 1 Seasonal variations of ^{222}Rn and ^{212}Pb concentrations in air.

The diurnal variations of ^{222}Rn and ^{212}Pb concentrations obtained from all the data in 1994 are given in Fig. 2. The error bar represents the scattering of the geometric standard deviation. Both the concentrations tended to be high from night to morning with a minimum level occurring around 16 h. The reason why such tendencies were observed is that ^{222}Rn and ^{212}Pb originating from the near measuring site are accumulated in the lower atmosphere in the nighttime and diffused into the upper air during the daytime. This indicates that ^{222}Rn and ^{212}Pb originating from near site would show the similar behavior.

Figure 3 shows examples of the relation between ^{222}Rn and ^{212}Pb concentrations observed in January and August in 1994. A value of about 0.8 in correlation coefficient was found in the relations. The reason for the good correlation would be that the behavior of these nuclides was similar to each other as was suggested from the diurnal variations.

Comparing the two correlations carefully, it can be seen that the intercept value obtained from the regression line in January is higher than that of August. In the present study, we defined that the intercept value consists of the components due to ^{222}Rn originating from distant regions where the half-life of ^{212}Pb is too short to be transported to the measuring site. Whether the definition is adequate or not will be discussed in the following section. Using above the method, the concentration levels of ^{222}Rn originating from distant regions were evaluated for the 36 monthly correlations in the period from December 1993 to November 1996. The results are shown in Fig. 4. In the figure, the concentration level of ^{222}Rn that originated near measuring site, which were obtained by subtracting those from distant regions from the monthly arithmetic mean concentrations, is also given. It can be recognized that the seasonal variation of ^{222}Rn concentration, which were high in winter and low in summer, was influenced by both the components. It can be also found that the variation of ^{222}Rn from near the site was clearer and the concentration level, except for in winter, was lower than those from distant regions.

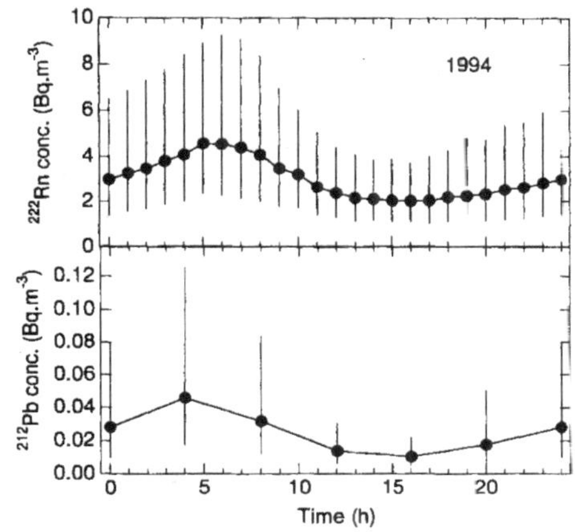

Fig. 2 Diurnal variations of ^{222}Rn and ^{212}Pb concentrations in air.

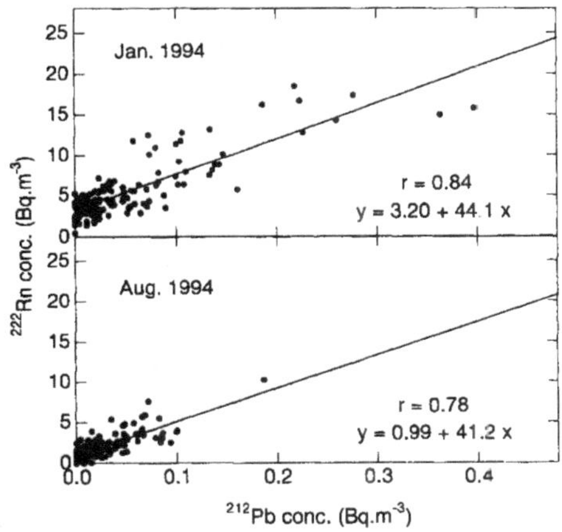

Fig. 3 Relations between ^{222}Rn and ^{212}Pb concentrations in air

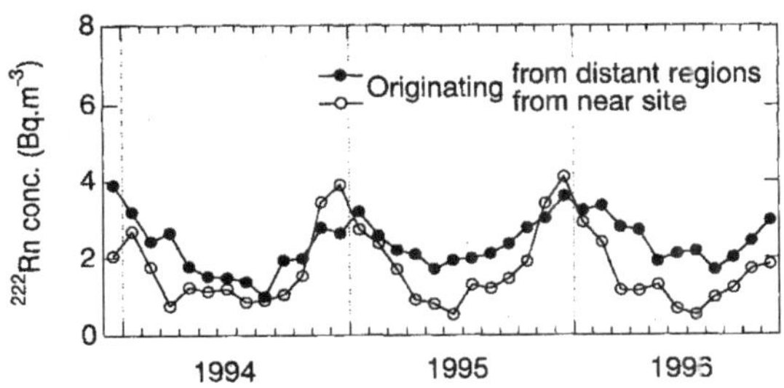

Fig. 4 Separated components of ^{222}Rn concentration in air.

4. Discussion

In the above section, it was assumed that the intercept value consists of the components of

^{222}Rn that originated from distant regions where the half-life of ^{212}Pb is too short to be transported to the measuring site. In order to know the accuracy of the present method, the following two effects on the correlations between ^{222}Rn and ^{212}Pb should be clarified; 1) washout of ^{212}Pb due to precipitation and 2) variation of the production ratios of ^{222}Rn and ^{212}Pb which can reach to the site, namely the seasonal or locally variations of the percentage of the exhalation rate of ^{222}Rn to that of ^{220}Rn which is parent nuclide of ^{212}Pb. Rainout of ^{212}Pb due to precipitation was not discussed here because it can be considered that scavenging in the clouds at an altitude of several hundred or thousand meters would scarcely affect the ^{212}Pb concentration in air at a height of 1.5 m above the ground.

The influence of "washout" on aerosol particles due to rain was evaluated by Slinn et al. [8] along with estimates of the effect of Brownian motions, inertial contributions and thermophoresis. It can be recognized from their results that the washout coefficient was at a minimum in the range between 0.05 and 0.5 μm in particle radius and its value was at most $3x10^{-6}$ - $3x10^{-8}$ s^{-1}. On the other hand, the radius of aerosol particles which were attached by ^{212}Pb was about 0.2 μm [9]. From the above reasons, it can be concluded that there was little effect of washout on the relation between ^{222}Rn and ^{212}Pb.

Radon-220 is released from underground as well does ^{222}Rn. The depth of soil from where the exhalation occurs is less than a few centimeters due to its short half-life, 55 s [10]. Therefore, the variation of some parameters such as soil temperature, soil moisture and atmospheric pressure might affect the exhalation of ^{220}Rn more than that of ^{222}Rn. The exhalation rates of ^{222}Rn and ^{220}Rn at one measuring site in Osaka were observed by Megumi et al [11]. They reported that no appreciable seasonal difference was found in the exhalation rate of ^{222}Rn, the exhalation of ^{220}Rn depended strongly on the moisture of a thin surface soil layer and ^{220}Rn exhalation rate in summer was about twice its value in winter. This conclusion implies that the experimental correlations between ^{222}Rn and ^{212}Pb concentrations in air might have been influenced by the seasonal variations of the percentage of the exhalation rate of ^{222}Rn to that of ^{220}Rn. On the other hand, Schery et al. [12] measured the exhalation of ^{222}Rn and ^{220}Rn at a variety of locations throughout Australia in winter and concluded that the exhalation rate of ^{222}Rn and ^{220}Rn were strongly correlated. It might be recognized from these two reports that whether the percentage of the exhalation rate of ^{222}Rn to that of ^{220}Rn changes or not is dependent on the location, but there was no reports about the large-scale survey of the exhalations of ^{222}Rn and ^{220}Rn around Tokyo. Therefore it is difficult to estimate the effect of the above 2) in this paper. Further study on the exhalation is necessary to estimate the accuracy of the present method.

The trajectory analysis of the air mass has been used as a means to explain the variation of ^{222}Rn originating from distant regions by many authors [2, 13-17]. In this paper, in order to verify the reliability of the present method, the estimated concentrations of ^{222}Rn originating from distant regions were compared with those calculated by the long-range transport model with the back trajectories.

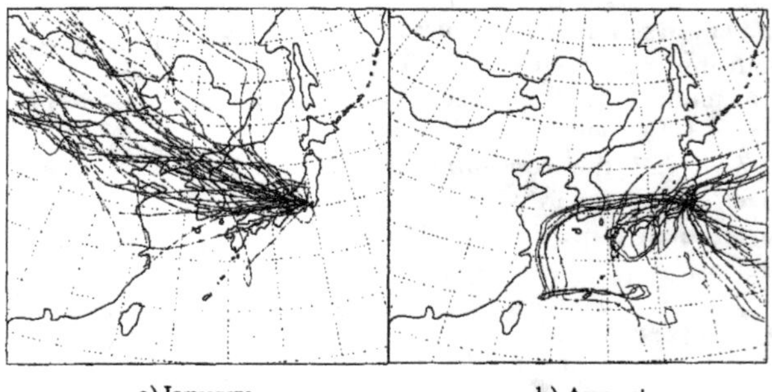

a) January b) August

Fig. 5 Examples of back trajectories ending at Komae in 1994.

The back trajectories for air masses in East Asia were analyzed for 365 days in 1994. It was assumed that ^{222}Rn was transported by the wind in the upper air at a pressure of 850 hPa. The wind data were observed at 20 sites in Japan and 34 sites in the foreign countries and were obtained from Japanese aerological data and weather charts. The observation time was 0 h and 12 h in Greenwich Mean Time. The trajectories ending at Komae city were chased for 10 days at one-hour intervals. In the case that the trajectory ran out of the region, east-west 4160 km and north-south 4400 km, the calculation was stopped. The interpolations of the time of the wind data were linear. The interpolations of the locations of the wind data were carried out with the objective analysis weighting by an inverse square of the distance. Examples of the trajectories for one month are shown in Fig. 5. The arrow represents the trajectory of air masses for every six hours. It can be seen that almost all the trajectories in January and August were from the Continent of Asia and the Pacific Ocean, respectively.

It was assumed that ^{222}Rn was supplied from ground surface at a constant exhalation rate of 15 mBq.m^{-2}.s^{-1} [18] and never from the sea. Horizontal diffusion of ^{222}Rn was not considered. The vertical distribution of ^{222}Rn was considered to be uniform in the mixing layer at a height of 1500 m without mixing into the upper layer. In order to calculate the components due to ^{222}Rn originating from distant regions where the half-life of ^{212}Pb is too short to be transported to the measuring site, the equivalent concentration of ^{222}Rn from distant regions were evaluated by subtracting the ^{222}Rn concentration calculated on the assumption that its half-life was 10.6 h.

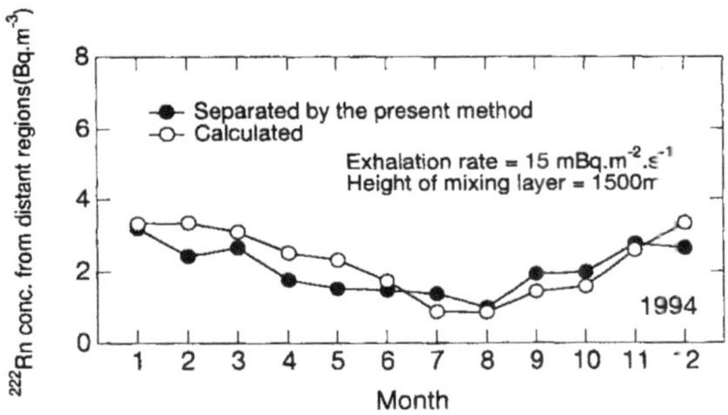

Fig. 6 Comparison of ^{222}Rn concentrations estimated by the present method and calculated by the transport model.

The monthly mean values of the calculated concentrations of ^{222}Rn are given in Fig. 6. In the figure, the concentration of ^{222}Rn from distant regions estimated by our experimental method are also plotted. The estimated concentrations agreed well with the calculation results by long-range transport model. This implies that the assumption that the intercept value consists of the components due to ^{222}Rn originating from distant regions would be reliable. The application of the long-range transport model might make it possible to estimate the contribution of ^{222}Rn originating from distant regions in the various districts in Japan.

5. Conclusions

A new method was proposed to separate the seasonal variation of monthly mean ^{222}Rn concentration by the origin using the intercept value obtained from the regression line of the correlation between ^{222}Rn and ^{212}Pb concentrations in air. The estimated concentration of ^{222}Rn originating from distant regions in 1994 agreed with the calculations by the long-range transport model, which verified the method reliability. The present method would be useful as a means to study the cause of the seasonal variations of ^{222}Rn concentration. Further study is necessary to estimate the accuracy of the present method.

6. References

[1] Lopez, A., Guedalia, D., Servant, J. and Fontan, J., Advantage of the use of radioactive tracers ^{222}Rn and ^{212}Pb for the study of aitken nuclei within the lower troposphere, J. Geophys. Res., 79, 1243-1252, 1974.

[2] Polian, G., Lambert, G., Ardouin, B. and Jegou, A., Long-range transport of continental radon in subantarctic and antarctic areas, Tellus, 38B, 178-189, 1986.

[3] Duenas, C., Fernández, M. C. and Senciales, M., Usefulness of Rn, decay products of Rn, and ThB to study diffusion in the lower atmosphere, Atmos. Environ., 24A, 5, 1255-1261, 1990.

[4] Whittlestone, S., Schery, S. D. and Li, Y., Pb-212 as a tracer for local influence on air samples at Mauna Loa Observatory, Hawaii, J. Geophys. Res., 101, D9, 14777-14785, 1996.

[5] Iida, T., Miyachi, H., Yamada, K., Kurihara, O., Ikebe, Y., Tanaka, H. and Nagao, I., Continuous measurements of radon concentrations and radon and thoron daughters concentrations at Ogasawara Hahajima Island and analyses of the data., 3a-II-4, The 34th annual meeting on radioisotopes in the physical sciences and industries, 1997 (in Japanese).

[6] Iida, T., Ikebe, Y. and Tojo, K., An electrostatic radon monitor for measurements of environmental radon., Res. Lett. Atmos. Electr., 11, 55-59, 1991.

[7] Hattori, T. and Ishida, K., A continuous monitor for radon progeny and its unattached fraction., Radiat. Prot. Dosim., 55, 113-122, 1994.

[8] Slinn, W. G. and Hales, J. M., A reevaluation of the role of thermophoresis as a mechanism of In- and Below-Cloud Scavenging, J. Atmos. Sci., 28, 1465-1471, 1971.

[9] Becker, K. H., Reineking, A., Scheibel, H. G. and Porstendörfer, J., Radiat. Prot. Dosim., 7, 147-150, 1984.

[10] Guedalia, D., Laurent, J. L., Fontan, J., Blanc, D. and Druilhet, A., A study of radon-220 emanation from soils, J. Geophys. Res., 75, 357-369, 1970.

[11] Megumi, K. and Mamuro, T., Radon and thoron exhalation from the ground., J. Geophys. Res., 78, 1804-1808, 1973.

[12] Schery, S., Whittlestone, S., Hart, K. P. and Hill, S. E., The flux of radon and thoron from Australian soils., J. Geophys. Res., 94, D6, 8567-8576, 1989.

[13] Wilkniss, P. E., Larson, R. E., Bressan, D. J. and Steranka, J., Atmospheric radon and continental dust near the antarctic and their correlation with air mass trajectories computed from nimbus 5 satellite photographs, J. Appl. Meteorol., 13, 512-515, 1974.

[14] Ikebe, Y., Kojima, S. and Shimo, M., On the origin and transport of Rn-222 in the atmosphere., Res. Lett. Atmos. Electr., 3, 51-54, 1983.

[15] Downey, A., Jasper, J. D., Gras, J. L. and Whittlestone, S., Lower tropospheric transport over the southern ocean., J. Atmos. Chem., 11, 43-68, 1990.

[16] Whittlestone, S., Robinson, E. and Ryan, S., Radon at the Mauna Loa observatory: transport from distant continents., Atmos. Environ., 26A, 251-260, 1992.

[17] Miyake, H., Sasaki, J., Iwao, T., Oda, K., Yamauchi, T. and Inoue, T., Nucl. Instr. and Meth., B75, 282-286, 1993.

[18] Wilkening, M. H., Clements, W.E., and Stanly, D., Radon-222 flux measurements in widely separated regions, 717-730, CONF-72085, 1975.

Radon Daughters In Precipitation

Tsuguo NISHIKAWA[1], Yoichi TAMAGAWA[1] and Shigeru OKABE[2]

1. Faculty of Engineering, Fukui University,
Bunkyo 3-9-1, Fukui 910, Japan
2. Radon Science Research Laboratory, Wakamatsudai 3-10-10,
Sakai, Osaka 590-01, Japan

Abstract

The continuous measurement of the radon daughter concentration in precipitation have been carried out repeatedly for a long time. The concentration varied largely in short time interval when the precipitation rate is low. The quantities of radon daughters brought to the ground surface by the precipitation in each 15 minutes show almost a linear correlation with the amounts of precipitation. Sometimes, the slope of this linear correlation changes suddenly before and after the light precipitation or the temporary stop of the precipitation even in one precipitation period. This phenomenon is confirmed to be related to the passage of the front or the air mass boundary over the observation point.

1. Introduction

The precipitation brings the radon daughters from the upper air to the ground surface by its scavenging effect. The radon daughters brought to the ground surface increase the environmental gamma-radiation near the ground surface. The increase of the total gamma-rays and the gamma-radiation dose on rainy and/or snowy period is several to several dozen percent of its average in the case of no precipitation [1]. Furthermore, the primary gamma-rays emitted from the radon daughters increases to several-fold value of its normal level [2]. Such large variation of the environmental gamma-radiation is a troublesome disturbance for its monitoring carried out in and around the nuclear facilities. Therefore, it is important for a reliable monitoring of the environmental gamma-radiation to investigate the variation of the radon daughter concentration in precipitation and its relation with the meteorological conditions.

The variation of the radon daughter concentration in precipitation is considered to be influenced by the characters of the upper air masses such as a continental or maritime one and the efficiency of the scavenging of precipitation to the radon daughters in the atmosphere. The scavenging effect of precipitation was investigated using the rainout-washout model [3] and the four-phase model [4]. The studies using the measured data of the radon daughter concentration in precipitation were carried out by several researchers [5-8]. However, there were scarcely any systematic studies of the time variation of the concentration for a long time period.

The present authors developed the automatic measuring instrument for the radon daughter concentration in precipitation [1] and have been measuring the concentration continuously for a long time period. This paper shows the results of this measurement and the discussion on the obtained data and the relation of them to the meteorological condition as the front.

2.Experiment

The continuous measurements of the radon daughter concentration in precipitation are carried out using the automatic measuring instrument developed by the authors [1]. Figure 1 shows the schematic diagram of this instrument. It is equipped with the well-type NaI(Tl) scintillator as its detector so as to measure the gamma-radioactivity of precipitation in a liquid state without any additional treatments such as condensation and extraction. The amount of precipitation is measured simultaneously using the rainfall intensity recorder. This instrument can detect a low concentration of about 0.5 Bq/ml in the case of light rain of about 0.5 mm/hr within a short time interval of 15 min, and has a sufficient sensitivity to the radon daughters in precipitation in practical conditions.

The continuous measurements of the radon daughter concentration in precipitation are carried out at the campus of Fukui University located in the coastal region of the Sea of Japan in the Central Japan. The continuous measurements of the radon daughter concentration in precipitation have been carried out repeatedly under the various meteorological conditions.

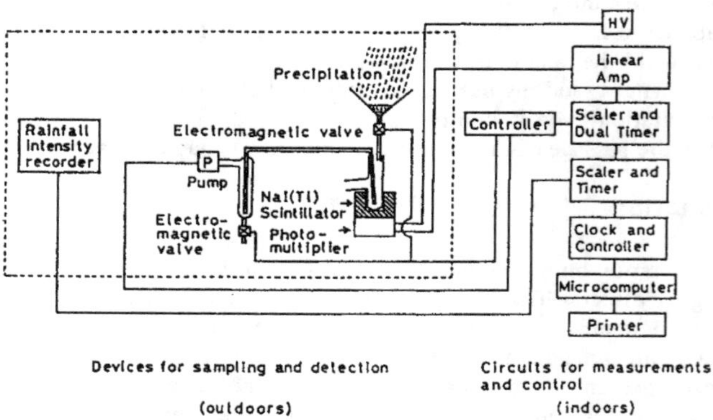

Figure 1 Schematic diagram of the automatic measuring instrument for the radon daughter concentration in precipitation.

3.Results and discussion

Figure 2 shows an example of the time variation of the amount of precipitation, the radon daughter concentration in precipitation and the dose rate of the environmental gamma-radiation. The plotted data show the values of every quarter of an hour. The amount of precipitation and the radon daughter concentration in precipitation show large variations in a short time interval. As the result of these time variations, the dose rate of the environmental gamma-radiation shows the complicated time variation.

Figure 3 shows the relation of the radon daughter concentration in precipitation to the amount of precipitation using the data shown in Fig. 2. The concentrations are distributed in a wide range when the amount of precipitations are little. In the case of the high precipitation rate, the concentrations tend to show almost constant value.

Figure 4 shows the relation of the quantities of radon daughters brought to the ground surface by the precipitation, that are the products of the radon daughter concentration in precipitation and the amount of precipitation, to the amounts of precipitation using the data shown in Fig. 2. The former quantities have almost a linear correlation with the latter quantities. The slope of the line representing this linear correlation sometimes changes before and after the time when the amount of precipitation becomes little or the precipitation stops temporarily even in one precipitation period as shown in this figure. The tenfold value of the slope of this linear correlation is almost the average concentration of the radon daughters in precipitation. Therefore, the change of the slope is considered to be caused by the phenomenon that the the air mass boundary passes over the observation point and the character of the air mass bringing the precipitation changes.

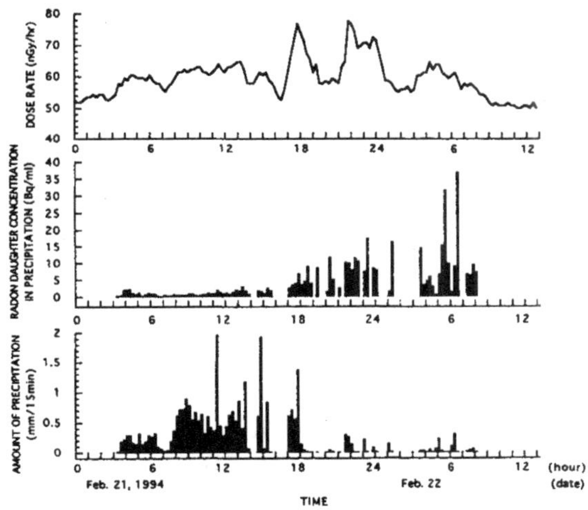

Figure 2 Example of the time variation of the radon daughter concentration in precipitation and the amount of precipitation. (Measured on February 21 - 22, 1994 at Fukui.)

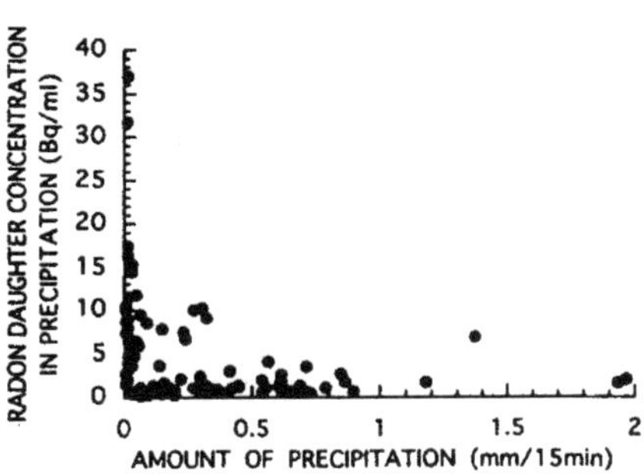

Figure 3 Relation between the radon daughter concentration in precipitation and the amount of precipitation shown in Fig. 2.

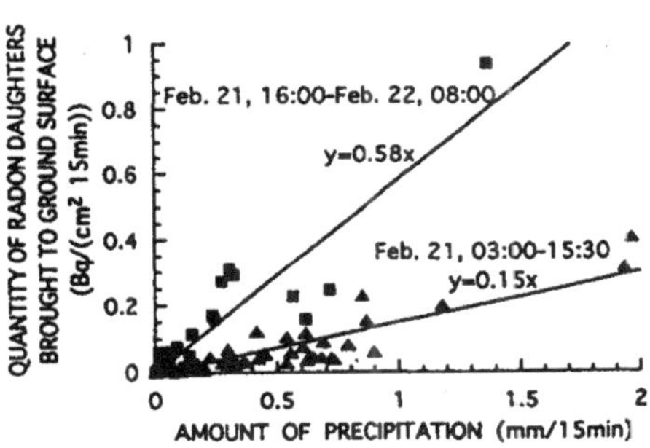

Figure 4 Relation between the quantities of radon daughters brought and the ground surface by the precipitation to the amounts of precipitation shown in Fig. 2.

Figures 5 show an example that the change of the slope of this linear correlation clearly corresponds to the passing of the front. Figure 5a shows the other correlation as shown in Fig.4 and Fig. 5b shows the weather maps of that time. From the weather maps, it is shown that the cold front passes over the observation point at about midday of November 20, 1992. The slope of the line becomes steep suddenly and this change shows the sudden increase of the average concentration of the radon daughters in precipitation. At this observation point, before the passage of the cold front, the precipitation is brought by maritime air mass which has low concentration of the atmospheric radon daughters. The other hand, after that time, the continental air mass of which concentration is high spreads over the observation point and brings the precipitation. Therefore, the change of the slope agrees with the passage of the cold front.

Figure 6 shows an example that the changes of the slope of the line are shown three times even in one precipitation period and the passage of the front is not recognized in the weather map. From this fact, the changes of it is a useful indicator of the air mass boundaries even in the case that the clear front is not recognized from the weather map.

(5a)

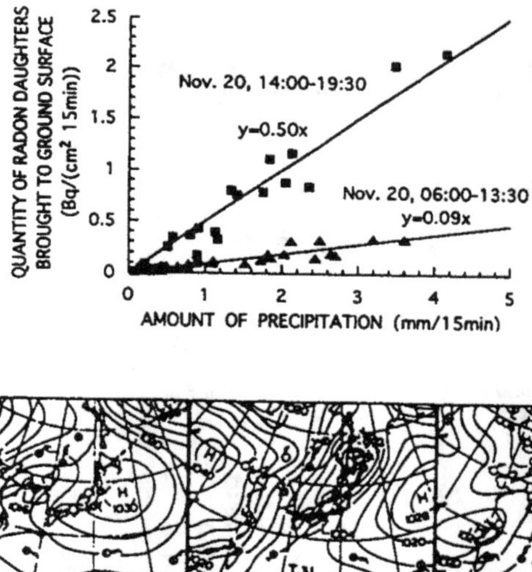

(5b)

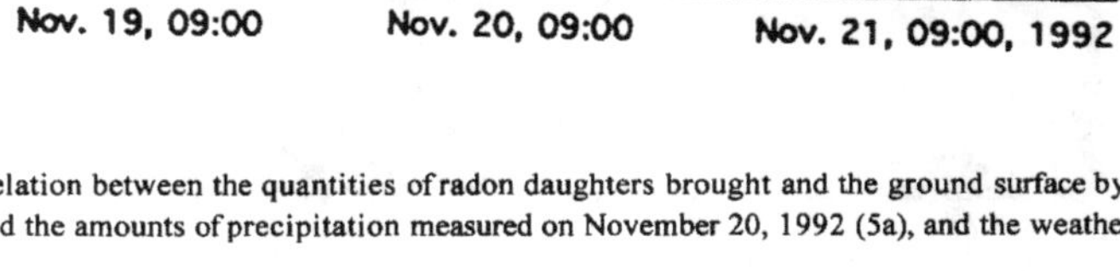

Figure 5 Relation between the quantities of radon daughters brought and the ground surface by precipitation and the amounts of precipitation measured on November 20, 1992 (5a), and the weather maps (5b).

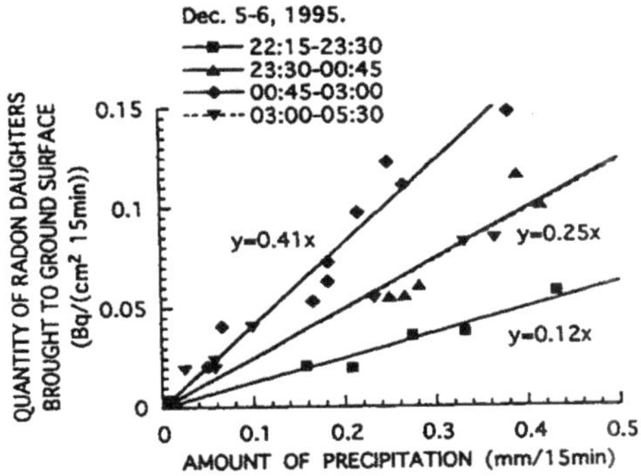

Figure 6 Relation between the quantities of radon daughters brought to the ground surface by precipitation and the amounts of precipitation measured on December 5 - 6, 1995.

4.Conclusion

The radon daughter concentration of the precipitation have been measured continuously for a long time under the various meteorological conditions. When the precipitation rate is low, the concentration varied largely in a short time interval. The concentration has the tendency to show nearly constant value when the precipitation rate is high. The linear correlation is shown between the amounts of precipitation and the quantities of radon daughters brought to the ground surface by the precipitation using the data obtained in each 15 minutes. Sometimes, the slope of this linear correlation changes suddenly before and after the time when the precipitation rate becomes low or the precipitation stops temporarily even in the one precipitation period. This phenomenon is mostly considered to correspond to the passage of the front over the observation point that is shown in the weather map. In the other cases, although the front is not shown clearly in the weather map, this phenomenon is considered to be caused by the passage of different air masses. It is suggested that the change of the slope of the line is a clear indicator of the passage of the air mass boundary.

5.References

[1] Nishikawa, T., Aoki, M. and Okabe S., Automatic measuring instrument for radon daughters concentration of precipitation, J. Nucl. Sci Technol., **23**, 1001-1007, 1986.
[2] Katase, A., Narahara, Y., Ishihara, Y., Tanaka, K. and Matsuyama H., Variation of intensity of environmental gamma-rays measured with Ge(Li) spectrometer, J. Nucl. Sci. Technol., **19**, 918-927, 1982.
[3] Takeuchi, N. and Katase, A., Rainout-washout model for variation of environmental gamma-ray intensity by precipitation, J. Nucl. Sci. Technol., **19**, 393-409, 1982.
[4] Ikebe, Y., Saijo, K., Iida, T., Nagai, H. and Takeda, T., A model for precipitation scavenging of short-lived Rn-222 daughters, Res. Lett. Atmos. Electr., **5**, 17-21, 1985.
[5] Damon, P. E. and Kuroda, P. K., On the natural radioactivity of rainfall, Trans. Am. Geophys. Union, **35**, 208-216, 1954.

[6] Bhandari, N. and Rama, Study of atmospheric washout processes by means of radon decay products, J. Geophys. Res., **68**, 3823-3826.
[7] Ikebe, Y., Aigyo, K. and Nakashima, Y., Measurements of radioactivity of rain water (in Japanese), Proc. Soc. Atmos. Electr. Japan, No. **20**, 33-38, 1978.
[8] Takeyasu, M., Yamasaki, K., Tsujimoto, T., Ogawa, Y. and Urabe, I., Radon-222 progeny in precipitation, Proc. Asia Cong. Radiat. Prot., 618-621, 1993.

Time series analysis of atmospheric ^{222}Rn concentrations and meteorological factors in JAPAN

YAMADA Katsumi[1], IIDA Takao[1], IKEBE Yukimasa[1], MIYACHI Hideyuki[1], NAGAO Ippei[2] and KOMURA Kazuhisa[3]

[1]Dept. of Nuclear Eng., Graduate School of Eng., Nagoya University
Furo-cho, Chikusa-ku, Nagoya, 464-8603 Japan

[2]The Institute for Hydrospheric Sciences, Nagoya University
Furo-cho, Chikusa-ku, Nagoya, 464-8601 Japan

[3]Low Level Radioactive Laboratory, Kanazawa University
Tatsunokuchi-machi, 923-1224 Japan

Abstract

The aim of this study is to examine a phenomenon of material transports in the atmosphere. To consider the meteorological factors contributing to the variation of atmospheric ^{222}Rn concentration, time series analysis (1) which contained the trend model, the power spectrum and the correlation function has been applied to the relationship of ^{222}Rn concentrations observed at Kanazawa, Nagoya and Ogasawara-Hahajima with meteorological factors. The analyzing period was from December 8 to 17, 1995. During this period two low pressures passed above Japan, and we observed then rapid increases of ^{222}Rn concentrations at Ogasawara-Hahajima. We confirmed from the time series analysis at Nagoya that the ^{222}Rn concentrations correlated with atmospheric pressure, temperature and wind speed. In addition, relative humidity shows a similar cyclic variation. In addition to this analysis, we examined the isentropic backward trajectory of air mass at Ogasawara-Hahajima. It gives a good account of the relation between the ^{222}Rn concentrations and transport direction of air mass. The trajectory analysis enables us to discuss the long distance transport of ^{222}Rn in relation to snow depth on the continent.

1. Introduction

Radon-222 is a noble gas, and has comparatively long half life of about 3.82 day. Radon-222 has been observed as a useful natural radioactive tracer. One of the significance for ^{222}Rn observation is to study the transport of materials in the atmosphere. A great deal of effort has been made on atmospheric ^{222}Rn and its progeny transports. By numerical analysis, we can be fairly certain that the ^{222}Rn concentrations consists of remote source and local components. It has been explained that the variation of remote source component results from the movement of air mass, and the variation of local component results from local factors, for example, the variations of temperature, wind speed, and the growth of mixing layer or the decrease of the exhalation rate due to rain and snow. Meteorological factors are necessary for analyzing the variation of ^{222}Rn concentrations in detail. However, it had been difficult to obtain the meteorological factors, then unless we measure them simultaneously. Since 1995 we could easily obtain Japanese meteorological factors, provided by Japan Meteorological Agency (JMA) as a CD-ROM. Further recently we can obtain the world meteorological factors via internet. Up to now, only a scatter plot has been made in analyzing observed ^{222}Rn data to estimate correlation coefficients of ^{222}Rn concentrations and some meteorological factors. Sakashita (2) had argued the effect of variation by time series analysis. In this paper we introduce a better method to examine the relationship of cause and effect for the variation of ^{222}Rn concentrations. The purpose is to consider more the meteorological factors contributing to variation of atmospheric ^{222}Rn concentrations using useful numerical statistic method, time series analysis, and many meteorological data.

2. Methods

We continually measured atmospheric ^{222}Rn concentrations every one hour at a height of 1m above the ground with ERM (Electrostatic Radon Monitor) (3) at three stations, Kanazawa, Nagoya and Ogasawara-Hahajima, from January 1995 to October 1996. Useful meteorological factors are surface observation data, aerological data and ASM (Asia Spectral Model) provided by Japan Meteorological Agency, and surface observation data, especially snow depth, provide by the National Climate Data Center (NCDC, USA). At Nagoya there were vertical profiles of temperature and wind provided by Aichi Prefecture.

The observed ^{222}Rn concentrations contained inevitable fluctuation. Furthermore they also reflect error of preceeding collecting terms. For the reason stated above, it was difficult to estimate the correct error of the concentrations. To estimate the error, first, we calculated the trend of time series data using Kalman filtering and smoothing, from which we estimated σ using the maximum likelihood method. Secondly, we examined power spectrum from the trend. The spectrum represents a magnitude of various periodical components included in the variation. Thirdly, we calculated the correlation functions, an application of scatter plot, between ^{222}Rn concentrations and surface meteorological factors.

In order to draw the isentropic backward trajectory, first we decided the potential temperature at the height of 850hPa in a pressure altitude from aerological data, then we calculated grid point of wind along the isentropic surface and traced the route of plume by Pickering's (4) method.

3. Results and discussions

Figure 1 shows the trends of observed ^{222}Rn concentrations at the 3 stations. Most of the observed hourly data are within $\pm 3\sigma$ of the trend component. Diurnal variation of ^{222}Rn concentrations, which was high level at night and low at daytime, had been often observed at Nagoya and Kanazawa. It is clear that the variation depends on the diurnal variation of solar radiation. However, no diurnal variation of ^{222}Rn concentration had been observed at Ogasawara-Hahajima, because of a maritime climate. Rapid increases had been observed many times at Ogasawara-Hahajima when a low pressure with cold front had passed. We can say that the ^{222}Rn at Ogasawara-Hahajima is mainly composed of remote source component, which traveled with migratory atmospheric pressure.

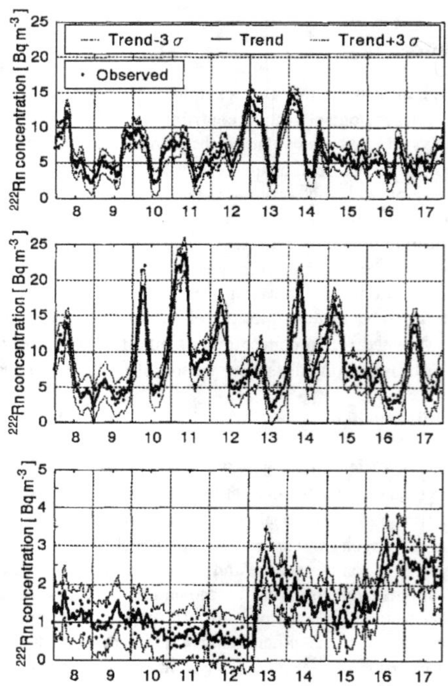

Fig. 1. The trends of ^{222}Rn concentrations at three stations on December 8-17, 1995. Above: Kanazawa, middle: Nagoya, bottom: Ogasawara-Hahajima

Figure 2 shows the power spectra and autocorrelation functions at the 3 stations. The highest peak of spectrum at Kanazawa was 24.0h cycle, that at Nagoya was 22.9h. These peaks account for the diurnal cycle. The peak of 96.0h at Nagoya may account for a variation of 4-day cycle. The rest peaks were also higher harmonics of 24.0h and 22.9h. Figure 3 shows the correlation functions at Nagoya between ^{222}Rn concentrations and meteorological factors (atmospheric pressure, solar radiation, relative humidity, wind speed, sheer of potential temperature and sheer of wind speed). There is a good evidence to suggest the reason for 4-day cycle in the spectrum of ^{222}Rn concentrations at Nagoya. Autocorrelation function of ^{222}Rn concentrations was similar to each autocorrelation function of relative humidity, wind speed and sheer of potential temperature. It is clear that a diurnal variation of each factor results from solar radiation and a variation of 3 or 4 days cycle results from atmospheric pressure. Here, we notice that atmospheric pressure affects these four factors. We can say that 4-day cycle indicated in the spectrum results from atmospheric pressure. Next, cross-correlation functions of ^{222}Rn concentrations with relative humidity, sheer of potential temperature and sheer of wind speed were similar to auto-correlation function of ^{222}Rn concentrations. It is well-known that the sheer of potential temperature and wind speed considerably correlate with ^{222}Rn concentrations, and in fact they are parameters of vertical diffusion coefficient in the dispersion model Level 2 of Mellor and Yamada (5). Relative humidity is the ratio of the amount of vapor in a unit volume to the density of saturation vapor which depends on a temperature. The amount of vapor mainly varies with a movement of air mass. We suggest that variations of ^{222}Rn concentrations and relative humidity may be controlled by a common condition. It is clear from cross-correlation functions that wind speed is in inverse correlation with ^{222}Rn concentrations. We notice that ^{222}Rn concentrations were delayed a few hours, compared with these correlated parameters. It is said that ^{222}Rn is also delayed to variation of methane (6). However, we have not cleared the reason for the delay.

When low pressure with a cold front went through Ogasawara-Hahajima in the midnight on December 13, the direction of air mass changed and ^{222}Rn concentrations increased. Figure 4 shows the results of trajectory analysis at Ogasawara-Hahajima and distribution of snow depth in east Asia. The air mass on December 11 and 12 originated from Siberia, although on December 13 and 14 originated from China. It is expected that the exhalation rate was low in Siberia because snow existed on a surface. Namely, it is suggested the air mass of low concentration due to existence of snow traveled to Ogasawara-Hahajima before the cold front went through.

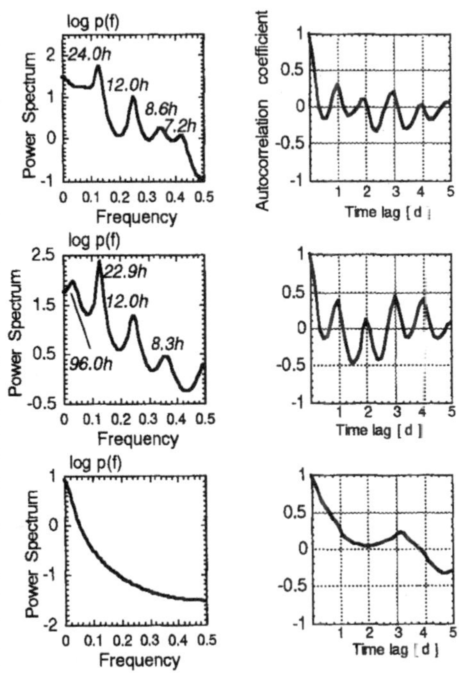

Fig. 2. Power spectra and autocorrelation functions
Above; Kanazawa, middle; Nagoya,
bottom; Ogasawara-Hahajima

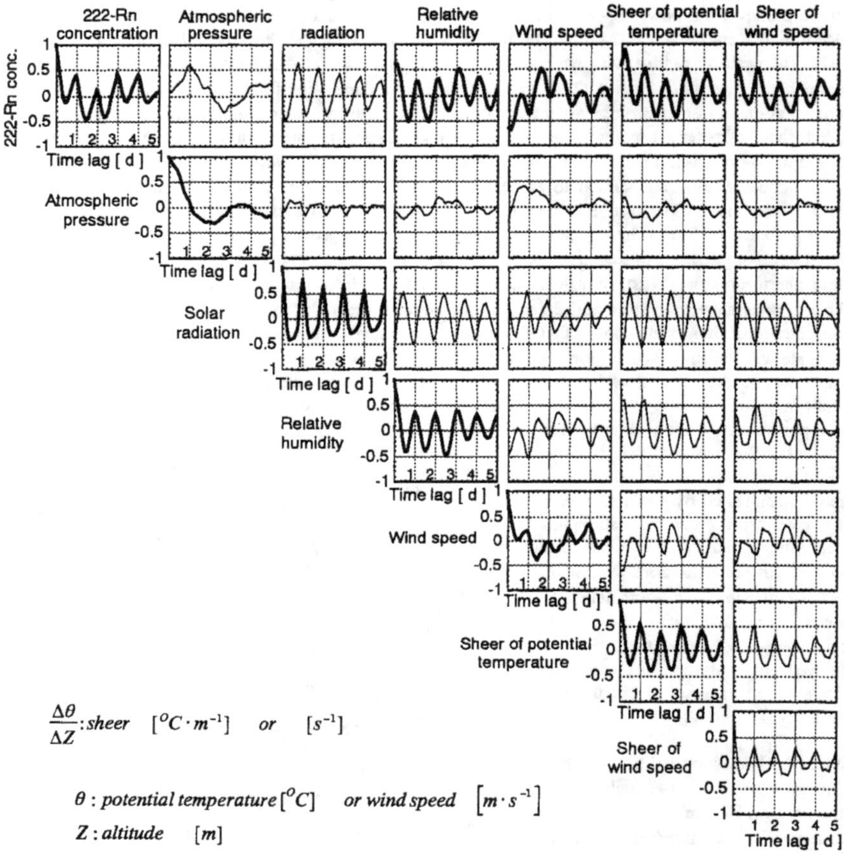

Fig.3 Cross-correlation functions between ^{222}Rn concentration and meteorological factors at Nagoya

4. Conclusions

We continually measured atmospheric ^{222}Rn concentrations at the 3 stations, Kanazawa, Nagoya and Ogasawara-Hahajima, from January 1995 to October 1996. We could examine the relationship of the ^{222}Rn concentrations with a lot of meteorological factors in detail by using time series analysis. Especially, it was reconfirmed that the atmospheric pressure, solar radiation and wind speed have correlations with ^{222}Rn concentrations. Furthermore it was suggested that ^{222}Rn concentrations and relative humidity may be dominated by a common meteorological condition. The correlation functions provided considerably good numerical hints. It was also suggested by using trajectory that remote source component correlated with snow depth on the continent.

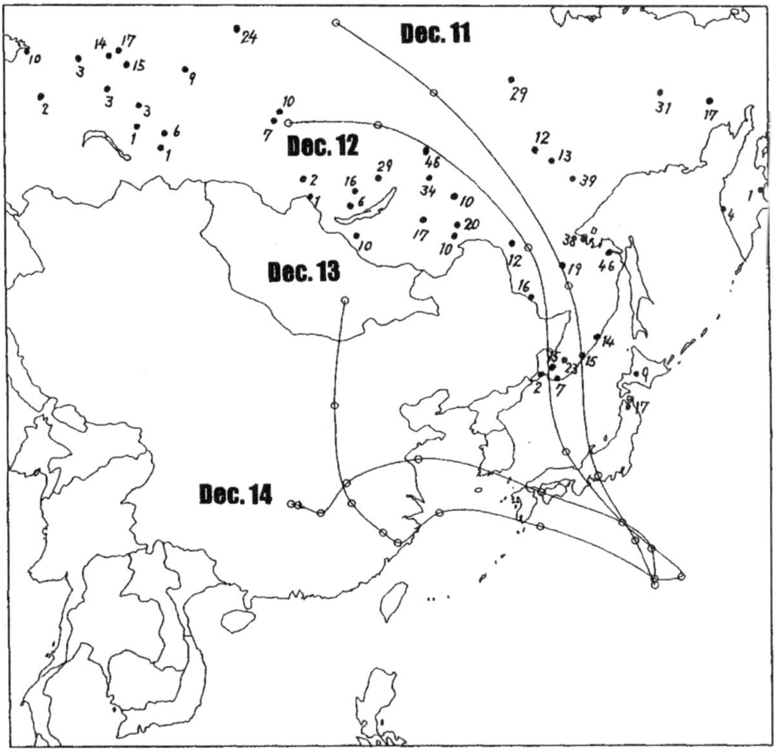

snow depth [cm]

Fig. 4. The trajectory of plume to Ogasawara-Hahajima and the distribution of snow depth on the beginning of December 1995

5. Acknowledgment

I am indebted to a number of my colleagues at Nagoya University, especially to Dr. Sakashita , who gave me valuable advices for computer programming.

6. References

[1] Kitagawa, G.: "Programming of Time Series Analysis", Iwanami Press, 245 (1993), [in Japanese].
[2] Sakashita, T., et al.: J. of Nucl. Sci. and technol., **34**, 63 (1997).
[3] Iida, T., et al.: Res. Lett. Atmos. Electr., **11**, 55 (1991).
[4] Pickering, A. E., et al.: Mon. Wea. Rev., **122**, 864 (1994).
[5] Mellor, G., Yamada, T.: J. Atmos. Sci., **31**, 1791 (1974).
[6] Mase, Y.,: Nagoya Univ. M. Paper, (1996), [in Japanese].

A subnationwide survey of outdoor and indoor ^{222}Rn concentrations in China by passive method

Y. Jin
Fujian Institute of Radiation Health Protection,
282 Guangrong Road, Fuzhou 350001, China

T. Iida and Y. Ikebe
Graduate School of Engineering, Nagoya University,
Furo-cho, Chikusa-ku, Nagoya 464-01, Japan

Z. Wang
Laboratory of Industrial Hygiene, Ministry of Public Health,
2 Xinkang Street, Deshengmenwei, Beijing 100088, China

S. Abe
National Institute of Radiological Sciences, 4-9-1,
Anazawa, Inage-ku, Chiba 263, Japan

Abstract

Survey of outdoor and indoor ^{222}Rn concentrations in China were made with 2-month integrating times from Nov. 1988 to Jan. 1990, and from Feb. 1991 to Mar. 1993. In the first survey, the annual mean outdoor ^{222}Rn concentrations at Beijing, Xi'an, Nanjing, Shanghai, Fuzhou, and Guiyang were 8.0, 11.2, 6.4, 4.8, 9.5, and 12.2 $Bq\cdot m^{-3}$, respectively. The results of the second survey at Changchun, Huhehaote, Shanghai, Wuhan, Fuzhou and Guiyang were 10.1, 8.6, 4.3, 14.6, 10.2, and 11.8 $Bq\cdot m^{-3}$, respectively. The annual mean indoor ^{222}Rn concentration was 21.8 $Bq\cdot m^{-3}$ in typical buildings constructed with concrete and brick. The highest indoor ^{222}Rn concentration of 38.1 $Bq\cdot m^{-3}$ was observed at Guiyang, and the lowest of 11.3 $Bq\cdot m^{-3}$ was at Huhehaote. The annual effective dose equivalent resulted from outdoor and indoor ^{222}Rn and its daughter products was estimated to about 0.67 $mSv\cdot a^{-1}$ in China.

1. Introduction

Many types of passive ^{222}Rn monitors have been used for the survey of time-averaged indoor ^{222}Rn concentrations[1,2,8]. To evaluate lung dose due to the inhalation of ^{222}Rn and its daughters precisely, it is also necessary to measure the mean concentrations in various outdoor environments. However, measurements of regional distribution of outdoor ^{222}Rn concentration have seldom been carried out due to lack of a passive ^{222}Rn monitor with sufficiently high sensitivity[6]. Information about the regional distribution of ^{222}Rn concentration and its time variation is also important in studying ^{222}Rn in relation to ^{226}Ra concentration in soil[10], ^{222}Rn exhalation, geographical features, and its behavior in the atmosphere.

Iida et al. studied and developed a new type of passive electrostatic integrating ^{222}Rn monitor (EIRM) that is capable of measuring outdoor ^{222}Rn concentration accurately[4]. It had been successfully used to measure the regional distribution of ^{222}Rn concentration around

Nagoya in Japan[9]. Cooperated with Japan, a survey of outdoor and indoor ^{222}Rn concentration in China were carried out by using the EIRMs, and the results are described in this paper.

2. Instruments and Methods

2.1 Electrostatic integrating ^{222}Rn monitor: The principle and structure of the monitor (EIRM) had been reported by Iida et al.[4]. It has a volume of about 2.26 L, and the measured exchange rate was 0.67 h^{-1}. Since the electrostatic collection of $^{218}Po^{+}$ atoms depends on the humidity in air, the inside of the EIRM was dehumidified with a P_2O_5 drying agent which is powerful and does not absorb ^{222}Rn gas. According to Currie's definition[3], the detection limit is found to be 0.4 Bq•m^{-3} for an exposure time of two months. Therefore, the outdoor ^{222}Rn level can be measured with high accuracy.

2.2 Monitoring duration and monitor arrangement: Surveys of outdoor and indoor ^{222}Rn concentrations in China were carried out about 2-month integrating times from Nov. 1988 to Jan. 1990, and from Feb. 1991 to Mar. 1993. The 9 central cities in six different regions of China: Changchun, Huhehaote, Beijing, Xi'an, Nanjing, Shanghai, Wuhan, Fuzhou, and Guiyang were selected as survey locations. Gaoxiong in Taiwan and Nagoya in Japan were selected as control cities. The location are shown in Fig.1. The first survey involved 102 measuring sites located in Beijing, Xi'an, Nanjing, Shanghai, Fuzhou, and Guiyang; the second included 81 sites in Changchun, Hugehaote, Shanghai, Wuhan, Fuzhou, and Guiyang. In principle, the surveys were carried out by setting the monitors in indoor and outdoor simultaneously at each site. The EIRMs set in outdoor ranged from 1 to 1.5 m high above the ground. The indoor monitors were set in apartments and offices constructed with concrete and brick, which are typical buildings in China.

Typical apartments in China have 1 to 6 stories. To evaluate variations in outdoor ^{222}Rn concentrations with height in the living environment, monitors were also set at the balconies of 1 to 6 stories in Xi' an and Fuzhou.

3. Results and Discussion

3.1 Distribution of outdoor ^{222}Rn concentrations in China: The average outdoor ^{222}Rn concentrations at 9 cities of mainland China from Nov. 1988 to Nov. 1989 and from Apr. 1991 to Apr. 1992 are shown in Tables 1 and 2, and Fig.1. As represented in Tables 1 and 2, Shanghai, Fuzhou, and Guiyang participated in both surveys. Except for the indoor ^{222}Rn concentrations at Guiyang, the results of the first and second surveys agreed fairly well for both outdoor and indoor ^{222}Rn concentrations at the 3 cities. The difference of indoor ^{222}Rn concentrations at Guiyang may be explained as that the second survey was carried out in only one building.

The annual mean outdoor ^{222}Rn concentration of the 9 cities was 9.3 Bq•m^{-3}. This is almost the same level as the average of 10 Bq•m^{-3} in continental areas indicated in the UNSCEAR report[7]. Among the 9 cities, the highest and lowest annual mean ^{222}Rn concentration were 14.6 Bq•m^{-3} (8.4～23.9Bq•m^{-3}) at Wuhan and 4.8 Bq•m^{-3}(3.6～6.3Bq•m^{-3}) at Shanghai, respectively. The ^{222}Rn concentrations in China were higher in southern parts than those of northern parts, and they were lower in seashore areas than that of inland areas.

The mean outdoor ^{222}Rn concentrations for the 2-month intervals at each city are shown in Fig.2. The pattern of seasonal variations at the 9 cities had a summer minimum and a winter

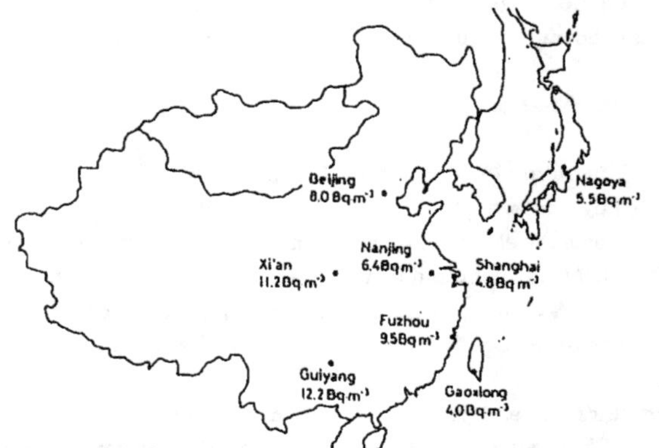

Fig. 1a. Average outdoor ^{222}Rn concentrations at 6 cities in China from Nov.1988 to Nov.1989

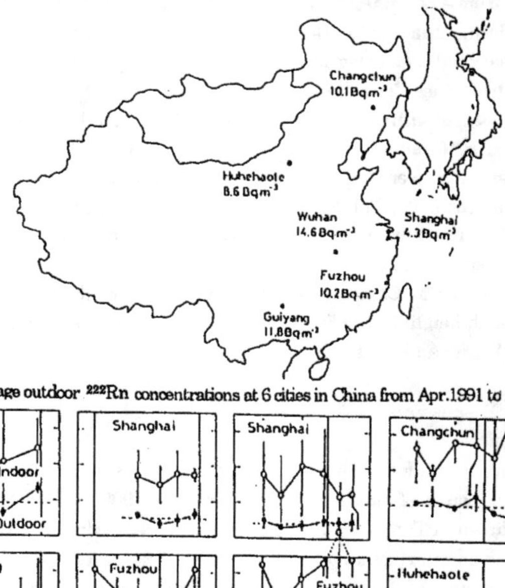

Fig. 1b. Average outdoor ^{222}Rn concentrations at 6 cities in China from Apr.1991 to Apr.1992

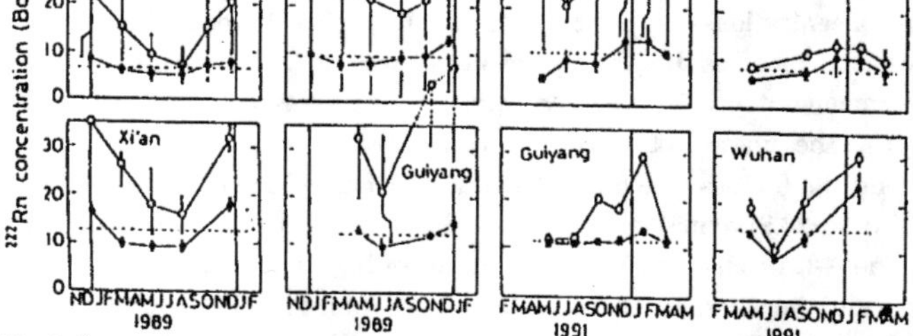

Fig. 2. Seasonal variation of outdoor and indoor ^{222}Rn concentrations at 9 cities in China
The error bars indicate the standard deviations of each datum

maximum, and this was similar to that of the control cities and most areas in the world. Table 3 represents the ratios of the maximum to the minimum. They ranged from 1.5 to 2.2 with an average of 1.8 at eight cities except for Wuhan.

Fig.3 shows the relationship between outdoor ^{222}Rn concentrations and ^{226}Ra concentrations in soil[10]. A good correlation coefficient of 0.88 was obtained for the data at Nanjing, Shanghai, Fuzhou, Guiyang, Nagoya and Gaoxiong. Changchun, Huhehaote, Beijing, Xi'an and Wuhan were excluded from the calculation of the correlation coefficient. Because it is suggested that the high outdoor ^{222}Rn levels at these places are due to the high exhalation rate of ^{222}Rn from dry soil of low ^{226}Ra concentrations. In addition, some of the high ^{222}Rn concentrations may be attributed to the high atmospheric stability and frequent occurrence of inversions in the inland.

The low ^{222}Rn levels at Nanjing and Shanghai correspond to the low exhalation rate of ^{222}Rn from soil due to the low ^{226}Ra concentrations in soil and high soil moisture. The high ^{222}Rn levels at Fuzhou and Guiyang result mainly from the high ^{226}Ra concentrations in soil. A clear seasonal variation at Wuhan situated in an inland area may be caused by high atmospheric stability and low soil moisture in winter.

3.2 Comparison between the two sides of Taiwan Straits: As shown in Table 3, the ratios of maximum outdoor ^{222}Rn concentrations in winter to the minimum in summer ranged from 1.5 to 2.2, with an average of 1.8, at the eight cities except Wuhan in the mainland. However, the ratio of Gaoxiong in Taiwan was up to 3.5. This result suggests that ^{222}Rn in airflow over Taiwan was largely affected by the air mass over the mainland. The ^{226}Ra concentration of 40 Bq•kg^{-1} in the soil of Gaoxiong is similar to the average of 38 Bq•kg^{-1} at the 9 cities in mainland China. So it can be considered that about 50% (the excess ratio 1.7) of the outdoor ^{222}Rn concentration at Gaoxiong was contributed by the airflow of the mainland in winter. To estimate by the same way, about 35% of the outdoor ^{222}Rn concentration at Nagoya in Japan was derived from China in winter. The various degrees of influence between Gaoxiong and Nagoya are due to the different movement distances of airflow.

3.3 Height distribution: The annual mean outdoor ^{222}Rn concentrations measured at the balconies of 1 to 6 stories in Xi'an and Fuzhou are shown in Fig.4. There was no clear variation from the height of 1 m to 30 m in Xi'an. While in Fuzhou, the outdoor ^{222}Rn concentration was slightly higher at 1 m.

3.4 Indoor ^{222}Rn concentrations: Annual means and seasonal variations of indoor ^{222}Rn concentrations are shown in Tables 1 and 2, and Fig.2. The highest of 38.1 Bq•m^{-3} was observed at Guiyang in the first survey, and the lowest was 11.3 Bq•m^{-3} at Huhehaote. The seasonal pattern of the indoor ^{222}Rn concentrations was the same as that of the outdoors. As shown at the right column of Tables 1 and 2, the ratios of indoor to outdoor ^{222}Rn concentrations in China ranged from 1.3 to 3.4, and they were the same as those of Japan[5].

3.5 Estimation of annual effective dose equivalent: It is necessary to consider many factors for estimating lung dose from inhalation of ^{222}Rn decay products. According to the estimating model recommended in the UNSCEAR report[7], the annual effective dose equivalents were calculated using the annual mean outdoor and indoor ^{222}Rn concentrations. The results are represented in Table 4. The annual mean effective dose equivalent was 0.67 mSv•a^{-1}, ranging from 0.40 mSv•a^{-1} at Huhehaote to 1.11 mSv•a^{-1} at Guiyang.

As described above, the annual mean outdoor ^{222}Rn concentrations in China were the same as the average in continental areas of the world, while the effective dose equivalents were lower than that of the world, as shown in Table 4. This may be mainly due to the high ventilation rate

Table 1. Average outdoor and indoor ^{222}Rn concentrations in China (Nov.1988 – Nov.1989)

Location	Outdoor: No. of monitors	Outdoor: Mean	Outdoor: S.D.	Indoor: No. of monitors	Indoor: Mean	Indoor: S.D.	Indoor/Outdoor
	Average ^{222}Rn concentration (Bq m^{-3})						
Beijing	10	8.0	2.5	11	22.6	3.4	2.8
Xi'an	14	11.2	3.0	9	23.9	7.7	2.1
Nanjing	10	6.4	1.3	10	13.7	5.3	2.1
Shanghai	5	4.8	1.6	5	16.6	4.8	3.4
Fuzhou	10	9.5	1.9	10	24.5	5.2	2.6
Guiyang	4	12.2	2.0	4	38.1	13.9	3.1
Mean		8.7	2.8		23.2	7.7	2.7
Gaoxiong[a)]	1	4.0		1	15.5	9.1	3.9
Nagoya[b)]	6	5.5		6	11.3	2.3	2.1

S.D. - standard deviation
a) - Feb. 1988 to Feb. 1989.
b) - Dec. 1985 to Dec. 1986.

Table 3. The ratios of maximum outdoor ^{222}Rn concentrations in winter to the minimum in summer during the 2-month intervals

Location	Ratio	Location	Ratio
Changchun	1.7	Wuhan	2.8
Beijing	2.2	Fuzhou	1.7
Huhehaote	1.7	Guiyang	1.5
Xi'an	1.9		
Nanjing	1.6	Gaoxiong	3.5
Shanghai	1.9	Nagoya	2.6

Table 2. Average outdoor and indoor ^{222}Rn concentrations in China (Apr.1991 – Apr.1992)

Location	Outdoor: No. of monitors	Outdoor: Mean	Outdoor: S.D.	Indoor: No. of monitors	Indoor: Mean	Indoor: S.D.	Indoor/Outdoor
	Average ^{222}Rn concentration (Bq m^{-3})						
Changchun	7	10.1	2.0	4	26.4	4.9	2.6
Huhehaote	4	8.6	1.9	4	11.3	2.0	1.3
Shanghai	9	4.3	0.8	7	15.2	3.2	3.5
Wuhan	11	14.6	5.9	15	19.8	4.9	1.4
Fuzhou	9	10.2	2.8	9	31.8	6.5	3.1
Guiyang	1	11.8	1.1	1	17.3	6.3	1.5
Mean		9.9	3.4		20.3	6.9	2.1

S.D. - standard deviation

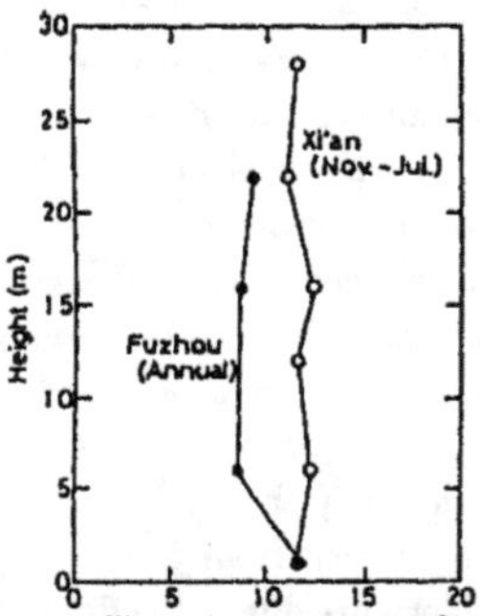

Fig. 4. Height distribution of ^{222}Rn concentration in outdoor air

Table 4. Calculated annual effective dose equivalent

Location	H_{eff} (mSv a^{-1}): Outdoor	Indoor	Total
Beijing[a)]	0.10	0.57	0.67
Xi'an[a)]	0.14	0.60	0.74
Nanjing[a)]	0.08	0.34	0.42
Shanghai[a)]	0.06	0.42	0.49
Fuzhou[a)]	0.12	0.62	0.74
Guiyang[a)]	0.15	0.96	1.11
Changchun[b)]	0.13	0.67	0.80
Huhehaote[b)]	0.11	0.29	0.40
Shanghai[b)]	0.05	0.38	0.43
Wuhan[b)]	0.18	0.50	0.68
Fuzhou[b)]	0.13	0.80	0.93
Guiyang[b)]	0.15	0.44	0.59
Mean	0.12	0.55	0.67
Gaoxiong	0.05	0.39	0.44
UNSCEAR (1993)	0.13	1.05	1.18

a) Annual means for Nov. 1988 to Nov. 1989.
b) Annual means for Apr. 1991 to Apr. 1992.

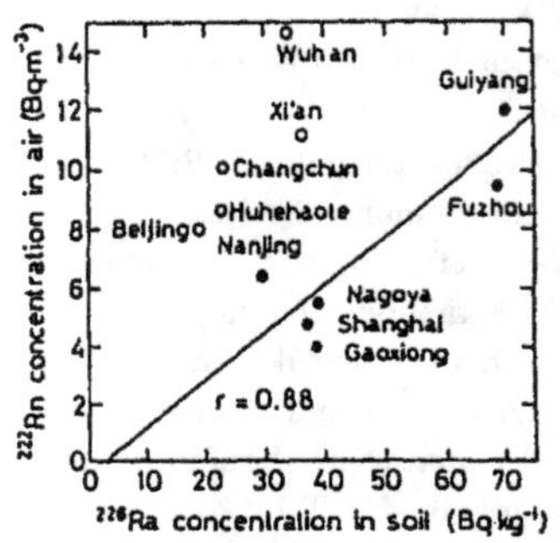

Fig. 3. Relationship between outdoor ^{222}Rn concentrations and ^{226}Ra concentrations in soil. Changchun, Huhehaote, Beijing, Xi'an and Wuhan were excluded in the calculation.

that results in low indoor ^{222}Rn concentration in Chinese buildings, which causes the indoor annual mean effective dose equivalent of 0.55 mSv•a^{-1}, far less than the mean dose of 1.05 mSv•a^{-1} of the world[7].

4. Conclusions

Surveys of outdoor and indoor ^{222}Rn concentrations at 9 cities in China were carried out by China and Japan from Nov. 1988 to Jan. 1990 and Feb. 1991 to Mar. 1993. The annual mean outdoor and indoor ^{222}Rn concentrations in China were 9.3 Bq•m^{-3} and 21.8 Bq•m^{-3}, respectively. The ratios of the indoors to outdoors ranged from 1.3 to 3.4. The annual effective dose equivalent due to the outdoor and indoor ^{222}Rn was 0.67mSv•a^{-1}(0.40~1.11 mSv•a^{-1}).

The variation of ^{222}Rn concentrations could be analyzed by the factors of geological conditions, geographical features, and meteorological conditions. To quantitatively analyze the ^{222}Rn concentrations in more detail, the study of ^{222}Rn exhalation rate from soil and continuous monitoring of ^{222}Rn concentrations should be performed.

Acknowledgment—The authors would like to express their appreciation to Ms. H. Chen, Jilin Sanitary and Epidemic Prevention Station, and Ms. L. Wu, Inner Mongolia Radiation Hygiene & Protection Institute, and Prof. Q. Zeng, Wuhan Radiation Protection Institute, and Prof. K. Du, Shanghai Medical University, and Mr. S. Li, Guizhou Radiation Protection Institute, for their kind help in carrying out this survey.

5. References

[1] Alter, H.W.; Fleischer, R.L. Passive integrating radon monitor for environmental monitoring. Health Physics. 40: 693-702; 1981.

[2] Bartlett, D.W.; Gilvin, P.J.; Dixon, D.W.; Solanki, H.L.; Miles, J.C.H. The performance of the NRPB radon personal dosimeter. Radiat. Prot. Dosim. 17: 139-142; 1986.

[3] Currie, L.A. Limits for qualitative detection and quantitative determination. Anal. Chem. 40: 586-593; 1968.

[4] Iida, T.; Ikebe, Y.; Hattori, T. et al. An electrostatic integrating ^{222}Rn monitor with cellulose nitrate film for environmental monitoring. Health Phys. 54: 139-148; 1988.

[5] Iida, T.; Ikebe, Y.; Yamanishi, H.; Abe, S.; Jin, Y. A survey of outdoor and indoor ^{222}Rn concentrations by passive method in Japan and China. Proc. int. conf. on Radiat.Eff. & Prot., Mito, Japan; 1992. Tokyo: The Japan Atomic Research Institute;1992: 364-369.

[6] Put, L.W.; de Meijer, R.J. Variation of tim-averaged outdoor radon concentrations over a distance of about 50 kilometers. Radiat. Prot. Dosim. 24: 97-99; 1988.

[7] UNSCEAR(United Nations Scientific Committee on the Effects of Atomic Radiation). Sources and effects of ionizing radiation. New York: UN; 1993: 33-89.

[8] Urban, M.; Piesch, E. Low level environmental radon dosimetry with a passive track detector device. Radiat. Prot. Dosim. 1: 97-109; 1981.

[9] Yamanishi, H.; Iida, T.; Ikebe, Y.; Abe, S.; Hata, T. Measurements of regional distribution of ^{222}Rn concentration. J. Nucl. Sci. Technol. 28: 331-338; 1991.

[10] Zhang, S.; Pan, J.; Li, R.; Xu, C.; Zhu, C.; Wang, X.; Lou, R. Levels and distributions of radionuclides in soil in China. Chn. J. Radiat. Med. Prot. 8: 1-15; 1988.

Environmental effects of radon and its progeny from uranium waste rock piles: part 1 - measurements by passive and continuous monitors

Yuu ISHIMORI, Kimio ITO, Sada-aki FURUTA

Ningyo Toge Works, Power Reactor and Nuclear Fuel Development Corporation
Kamisaibara, Okayama, 708-06 Japan

Abstract

The radon concentration in atmosphere on and around the uranium waste rock pile sites has been measured by integrating passive monitors since 1989. In fiscal 1996, except for the Katamo-shimo 1, the average concentration of radon on the sites, around the sites and in control areas, ranged from 11 to 194 Bq/m^3(average: $45Bq/m^3$), from 8 to $75Bq/m^3$ (average: $26Bq/m^3$), and from 9 to $77Bq/m^3$ (average: $30Bq/m^3$), respectively. Additionally, the typical daily and seasonal variations of radon concentration, radon progeny concentration and radon exhalation rate are observed with continuous or automatic monitors. According to the measurement results by passive monitors and continuous monitors, the environmental effects of radon and its progeny from the waste rock pile sites are estimated small in residential regions around the sites.

1. Introduction

Ningyo Toge Works of Power Reactor and Nuclear Fuel Development Corporation (PNC) carried out research and development projects related to the uranium exploration from 1967 to 1988. "The waste rock pile sites" which pile up the rocks left by the exploration activities have been retained following the Japanese Mine Safety Law. These waste rock piles are located both in Okayama prefecture and in Tottori prefecture (Fig.1). The measurements of the radon concentration on and around the sites were started in 1989, when the Japanese Mine Safety Law was revised according to ICRP Publication 26.

In order to evaluate the behavior and the environmental effects of the radon and its progeny from the waste rock piles, two way approaches take place: environmental monitoring and dispersion model use.

The limitations against radon are determined with equilibrium equivalent radon concentration (EECRn) in current Japanese laws, including the Japanese Mine Safety Law. The Law requires lower exposure dose than 1mSv/year out of the site, exceeding natural background level. 1mSv/year is equivalent to 9 Bq/m^3 of EECRn. Although it is necessary to obtain the long-term average concentration of radon progeny, for example over a month, most of the commercial radon progeny concentration monitors are expensive and require the AC power source. Therefore, it is necessary to determine the typical equilibrium factor (F) by grab sampling methods on and around sites in order to estimate the radon progeny concentration from the radon concentration obtained by integrating passive monitors. Furthermore, the measurements of radon concentration, radon progeny concentration and radon exhalation rate have been carried out by continuous or automatic monitors for several years, in order to complement grab sampling data.

Radon and its progeny exist naturally, and the daily and seasonal variations of their concentration are in a large scope. Thus, it is necessary to estimate the environmental effects of the radon and its progeny which are distinguished from natural component. The effects have been evaluated by dispersion model with the data of radon exhalation rate. Additionally, the dispersion experiments using the SF_6 tracer gas were carried out three times in order to examine the validity of the model.

In this study, the environmental effects of radon and its progeny from uranium waste rock piles are illustrated in three parts: Part 1, the measurement results obtained by integrating passive monitors and continuous monitors; Part 2, the measurement results obtained by grab sampling methods and the summary of the results by environmental monitoring; Part 3, the evaluation by dispersion model use and the tracer gas experiments.

2.Method

2-1 Integrating passive radon monitor

The radon concentration on and around waste rock pile sites has been measured by electrostatic integrating passive monitors, which was developed by Iida et al.[1] and is produced by Aloka Co. (Japan). The monitor is used in the naturally ventilated case, the height of which is about from 1 to 1.5 meters. The data are mainly obtained quarterly. The calibration of the monitors is carried out every year with the radon test chamber developed by PNC[2]. The test chamber has the control system of temperature, humidity and air pressure. The volume of the chamber is about 7.5m^3. The radon concentration in the chamber is determined by ionization chamber.

2-2 Continuous monitors for radon and its progeny concentration

The radon concentration and its progeny concentration are measured by PMT-TEL produced by the PYLON Co.(Canada) and by the continuous radon progeny monitor developed by PNC[3][4][5], respectively. The radon progeny monitor has a long roll of membrane filter for collecting radon progeny and also has a vacuumed cell for alpha spectrometry by SSD. The measurements are carried out on the boundary of Katamo-1,2 waste rock pile site(R-5) and in the Japanese pear orchard in the vicinity of the site. Measuring data is obtained every hour.

2-3 Automatic radon exhalation rate monitor

The radon exhalation rate is measured by the automatic monitor developed by PNC[6]. The monitor has a scintillation cell for detecting radon and also has a "Hat" -a chamber for accumulating radon released from ground surface. The Hat is only closed while accumulating radon; this means that the change of the meteorological conditions, an important parameter for radon exhalation, directly affects the ground surface. Thus, the daily and seasonal variation ranges of radon exhalation rate are able to be observed. Measuring data is obtained every six hours.

3.Results and Discussion

3-1 Radon concentration by integrating passive monitors

The radon concentration obtained by passive monitors in fiscal 1996 is shown in Fig.2, Fig.3 and Table 1. The location of measuring points in the Katamo area is also shown in Fig.4. Except for the Katamo-shimo 1, the annual average concentration of radon on the sites, around the sites and in control areas ranged from 11 to 194 Bq/m^3(average: 45Bq/m^3), from 8 to 75Bq/m^3 (average: 26Bq/m^3) and from 9 to 77Bq/m^3 (average: 30Bq/m^3), respectively. Although the relatively high radon concentration is observed on the sites, the levels of radon concentration both around the sites and in control areas are the same. Thus, the effects of radon from the sites should be estimated small in residential regions around the sites compared with natural background level. Fig.3 shows that the radon concentration in the Japanese pear orchard, around R-18 or R-19, is almost at natural background level in the district. Thus, the effect of radon from the Katamo-1,2 site should be also estimated quite small in the residential region. Table 1 shows the seasonal variations of radon concentration in residential regions. The radon concentration during summer is higher than that during winter. It is the typical variation in the Sea of Japan side[7].

The annual average concentration of radon at the Katamo-shimo 1 in front of a closed mouth of level is 366Bq/m^3. The experimental evaluation concerning the Katamo-shimo 1 is examined in part 2.

3-2 Radon and its progeny concentration by continuous monitor

The radon concentration and its progeny concentration obtained by continuous monitors are shown in Table 2 and Fig.5. The seasonal variation, such as relatively high concentration is observed during summer, agrees with the results by using integrating monitors.

According to the measurement results by integration monitors, the radon and its progeny concentration in the Japanese pear orchard should be regarded as the background level. Thus, on the boundary of the Katamo-1,2 site (R-5), the radon and its progeny concentration from the site probably is equivalent to the difference between the concentration at R-5 and that in the orchard.

From 1993 to 1996 fiscal year, the annual average concentration of radon and its progeny on the boundary of Katamo-1,2 site ranged from 52 to 88 Bq/m^3 and from 7.0 to 9.1 Bq/m^3, respectively. And those of radon and its progeny in the orchard ranged from 17 to 23 Bq/m^3 and from 4.4 to 5.0 Bq/m^3, respectively. The results meant that the annual average concentration of radon progeny on the boundary of the Katamo-1,2 site was probably lower than 9Bq/m^3 except background level even if the concentration in the orchard is not regarded as the background level.

The daily and seasonal variation of radon progeny concentration at R-5 is shown in Fig.6. The relatively high radon progeny concentration is observed in summer morning. The phenomena can be explained by the influence of wind shift rather than that of radiative cooling.

3-3 Radon exhalation rate by automatic monitor

The example result of the radon exhalation rate monitoring is shown in Fig.7. It was obtained at the PNC Ningyo Toge Works site in 1994. The soil of the measurement point is the decomposed granite soil. Radon exhalation rate lowers rapidly at the beginning of rain. And soil recovers its radon exhalation rate gradually after the end of rain. The recovery of exhalation rate requires about 3 days at the site. Therefore, the monthly average of radon exhalation rate depends on the number of rainy days, though the maximum of the radon exhalation rate is almost fixed.

The estimation by dispersion model will be relatively excessive, because it is calculated from the exhalation rate obtained by grab sampling method on the days without raining.

4. Conclusion

The radon concentration in atmosphere on and around the uranium waste rock pile sites has been measured. Although the relatively high radon concentration is observed on the waste rock pile sites, the environmental effects of radon and its progeny from the sites are estimated small in residential regions around the sites. The estimation is examined further in part 2 and part 3.

5. Reference

[1] Iida, T.; Ikebe, Y.; Hattori, T.; et al. An Electrostatic Integrating ^{222}Rn Monitor with Cellulose Nitrate Film for Environmental Monitoring. Health Physics, 54: 139-148; 1988.

[2] Ito, K.; Furuta, S.; Matsuo, M. Establishment of Radon Test Chamber. Proceedings of the 27th Annual Meeting of the Japan Health Physics Society (in Japanese), Akita Japan, June 11-12; 1992: 4.

[3] Ito, K.; Ishimori, Y.; Furuta, S.; Kashimura, O. Development of Radon Progeny Continuous Monitor. Proceedings of the 29th Annual Meeting of the Japan Health Physics Society (in Japanese), Tsuruga Japan, May 19-20; 1994: 10.

[4] Furuta, S.; Ito, K.; Ishimori, Y. Development of Continuous Monitor for Radon Progeny . Atmospheric Radon Families and Environmental Radioactivity III. Tokyo: Japan Atomic Energy Society; 1995: 22-24.

[5] Furuta, S.; Ito, K.; Ishimori, Y. Development of Continuous Monitor for Radon Progeny . PNC Technical Review, 93: 62-67; 1995.

[6] Ishimori, Y.; Ito, K.; Furuta, S. Automatic Radon Exhalation Rate Monitor. Proceedings of the 31st Annual Meeting of the Japan Health Physics Society (in Japanese), Sapporo Japan, May 31-June 1; 1996: 47.

[7] Furuta, S.; Ito, K.; Matsuo, M. Study on Environmental Radon Measurements. PNC Technical Review, 81: 122-125; 1992.

Table 1 Radon concentration obtained by integrating passive monitors in residential regions in fiscal 1996.

LOCATION	Rn Concentration Bq/m^3 1st Period	2nd Period	3rd Period	4th Period	Annual Average
KAN'NOKURA	11	15	15	13	13
ASABATAKE	11	15	12	8	11
KATAMO	12	17	15	8	13
KAWAKAMI	9	10	9	8	9
SUGEGAWARA	53	83	53	45	58
YENDANI	16	21	18	14	17
HIROSE	22	28	27	22	25
Okayama Univ. Branch Hospital	12	22	19	15	17
PNC Kurayoshi Dormitory	11	12	13	10	12
ANAGAMO	13	20	14	13	15
NAKATSUGO-1	14	21	24	19	20
NAKATSUGO-2	22	38	26	19	26
NAKATSUGO-3	11	11	11	12	11
OMBARA	10	12	10	-	11
NINGYO	14	9	7	-	10

"-" means that no data was obtained because of much snow.

Table 2 Summary of the radon and its progeny concentration obtained by continuous monitors on the boundary of the Katamo-1,2 waste rock pile site (R-5) and in the Japanese pear orchard in the vicinity of the site: "N" means the number of data.

LOCATION	Fiscal Year	Rn Bq/m3 Mediar	Ave.	Std	N	EECRrBq/m3 Mediar	Ave.	Std	N	F
R-5	1993	26.0	54.3	76.0	8219	4.0	9.1	17.3	5998	0.168
	1994	33.0	51.9	55.3	7669	4.2	7.0	8.2	7826	0.134
	1995	36.3	70.9	83.2	7682	4.6	8.3	9.9	8342	0.117
	1996	55.8	88.4	90.6	8089	4.4	7.9	10.1	8573	0.089
	WHOLE DATA	36.7	66.5	79.0	31659	4.3	8.0	11.4	30739	0.120
Pear Orchard	1993	13.4	17.3	15.9	8733	3.4	4.4	3.9	7499	0.253
	1994	12.8	19.5	21.0	6674	3.7	4.9	4.0	6464	0.253
	1995	18.0	23.1	19.4	6929	3.8	4.9	4.0	7111	0.212
	1996	17.4	21.7	16.7	7990	3.8	5.0	4.3	7487	0.232
	WHOLE DATA	15.5	20.3	18.3	30326	3.7	4.8	4.1	28561	0.237
Difference	1993		37.0				4.8			
	1994		32.5				2.0			
	1995		47.8				3.4			
	1996		66.7				2.9			
	WHOLE DATA		46.2				3.2			

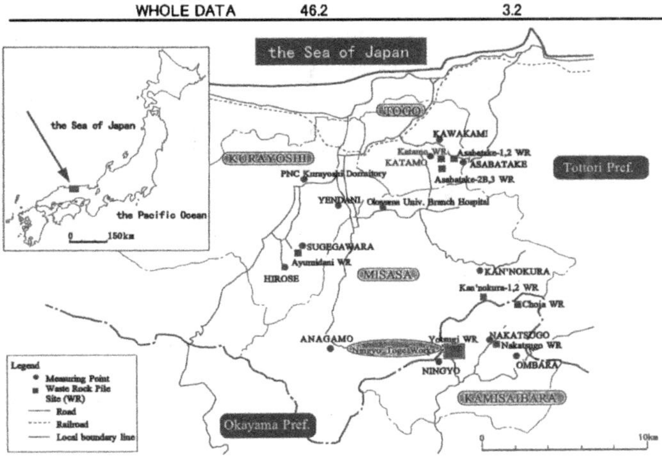

Fig.1 Location of the measuring points on and around the waste rock pile sites.

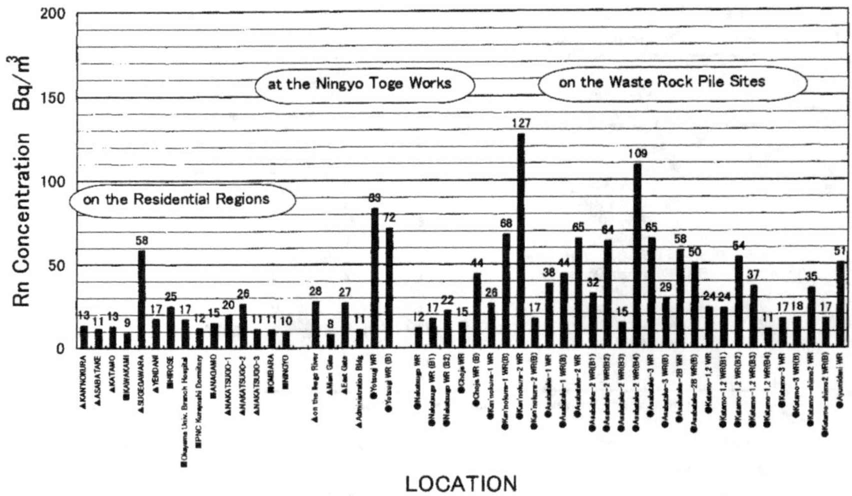

Fig.2 Radon concentration on and around the waste rock pile sites in fiscal 1996: "WR" and "(B)" means the measuring point on the waste rock pile site and on the site boundary, respectively.

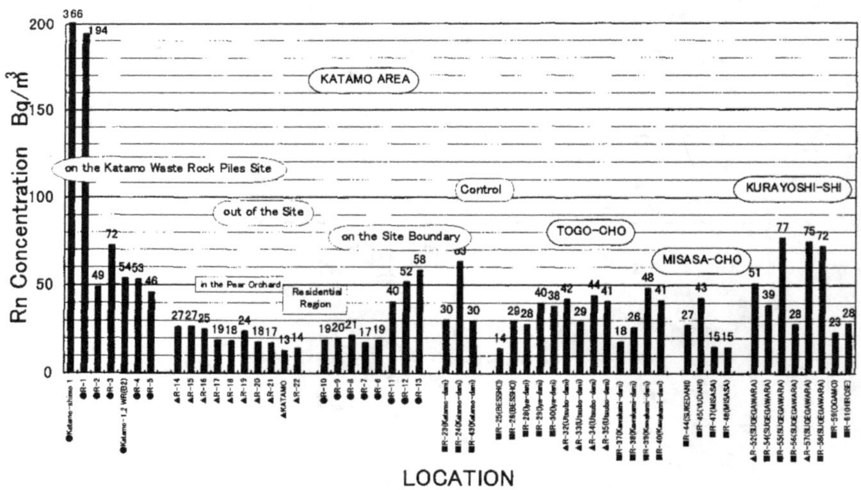

Fig.3 Radon concentration on and around the Katamo waste rock pile site in fiscal 1996.

Note: In Fig.2 and Fig.3, ●, ▲ and ■ means the measuring point on the waste rock pile site, around the site and in the control area, respectively.

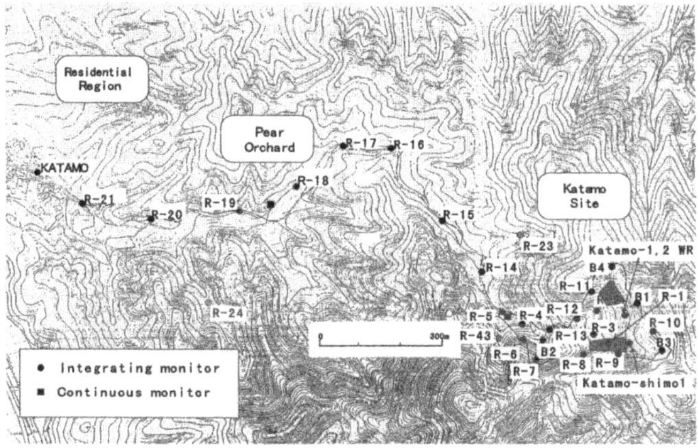

Fig.4 Measuring points in the Katamo area.

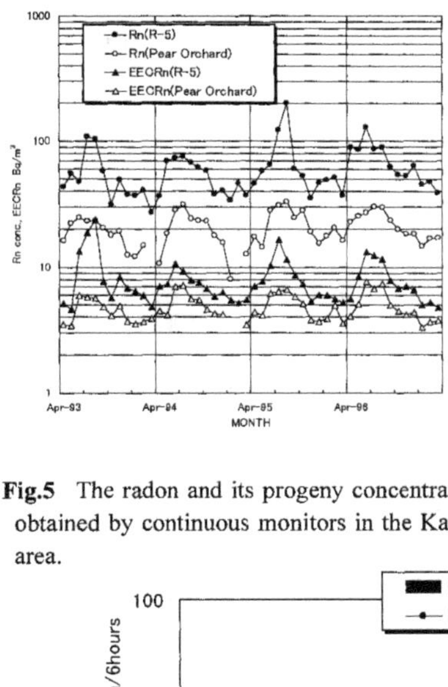

Fig.5 The radon and its progeny concentration obtained by continuous monitors in the Katamo area.

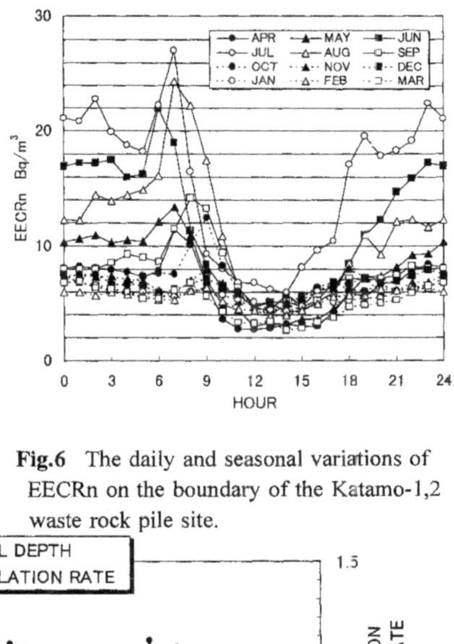

Fig.6 The daily and seasonal variations of EECRn on the boundary of the Katamo-1,2 waste rock pile site.

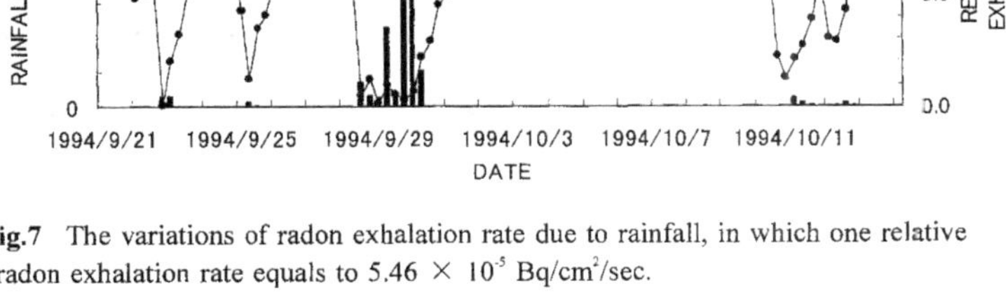

Fig.7 The variations of radon exhalation rate due to rainfall, in which one relative radon exhalation rate equals to 5.46 × 10^{-5} Bq/cm²/sec.

Environmental effects of radon and its progeny from uranium waste rock piles: part 2 - measurements by grab sampling methods

Kimio ITO, Yuu ISHIMORI, Sada-aki FURUTA

Ningyo Toge Works, Power Reactor and Nuclear Fuel Development Corporation
Kamisaibara, Okayama, 708-06 Japan

Abstract

In the uranium mine waste rock pile sites and in the nearby control area, the radon concentration, equilibrium equivalent radon concentration, and radon exhalation rate have been measured by a grab sampling method. The measurement results of radon concentration and radon exhalation rate showed a large fluctuation by the measurement sites and the radon concentration and radon exhalation rate were almost around 100 Bq/m^3 and 10～100×10^{-6}Bq/cm^2/sec respectively. While, in the control area, they were around 10 Bq/m^3, 1～3×10^{-6}Bq/cm^2/sec, respectively. The equilibrium equivalent radon concentration was mostly 1～10 Bq/m^3, and there was not a large difference between the waste rock pile sites and the control area.

The estimated equilibrium equivalent radon concentration was discussed with the equilibrium factors which were obtained from these measurements and the radon concentration obtained by passive monitors (see part 1), and it was suggested that the contribution of a natural background was important to the equilibrium equivalent radon concentration.

1. Introduction

The radon concentration, equilibrium equivalent radon concentration in atmosphere and radon exhalation rate from the ground surface are measured both in the waste rock pile sites and in the control area by the grab sampling method.

In the present act of Japan that includes the Japanese Mine Safety Law, the regulation for radon has been defined by the equilibrium equivalent radon concentration that is actually radon progeny concentration. As reported in part 1, PNC has carried out the integral measurement of the radon concentration in the waste rock pile sites and in the nearby control area for several years. Therefore, the measurement by the grab sampling method is very important in order to obtain the equilibrium factor for estimating the equilibrium equivalent radon concentration from the integrated radon concentration, and from the results obtained until now. The equilibrium factor in the waste rock pile sites have been proven to be considerably lower than the value reported by other researchers.

Furthermore, the radon exhalation rate is also important as a factor of release rate in the simulation of the radon dispersion from the waste rock pile sites, which is shown in part 3.

2. Method

2-1 Radon concentration

The measurement of the radon concentration in the grab sampling method is based on the measurement by a gas charging type ionization chamber. The volume of the ionization chamber is 1500cm^3 and a radon concentration of about 5Bq/m^3 is measurable by the electric charge storage method for ionization current measurement. The conversion factor from the ionization current is obtained from the comparison between the ionization current and the radon concentration in which the radon gas enclosed from the radium solution, which the radium concentration is already known, and it is also confirmed that the data obtained by our method agreed well with the results of other research institutes within error, through various inter-comparison measurement that were carried out in domestic and overseas facilities[1].

In actual measurement, at first, air is sampled in a plastic bag for 1 hour*1, and then the average radon concentration for 1 hour is obtained by measuring the air in the bag, because the equilibrium equivalent radon concentration is an average concentration for the 1 hour.

2-2 Equilibrium equivalent radon concentration

The equilibrium equivalent radon concentration is measured by a portable measuring instrument (WLM-200plus made by TracerLab. Germany). This measuring instrument is an equipment which makes outputs of the measurement result by continuously sampling air on a fixed filter with flow rate of about 100 liter/hour. A silicon semiconductor detector is used for the detection of each alpha particle from Po-218 and Po-214. The equilibrium equivalent radon concentration is calculated from the result of measured Po-218 and Po-214, and 20 ratios of each activity recorded in the CPU of the instrument.

The calibration of the instrument is not carried out at present, because the standard source of radon progeny is not established. It has been confirmed that the data obtained by our instrument agreed well with the results of other organizations within error in various inter-comparison measurement carried out in the domestic and overseas facilities[1].

2-3 Radon exhalation rate

The radon exhalation rate from the ground surface is obtained by putting a radon storage container with the volume of about 16 liter(40cmϕ,13cmH) on the ground surface. The radon exhalation rate is calculated from the radon which is accumulated for 30 minutes in the container. For the measurement of the radon concentration, the scintillation cell (LUCAS CELL 300 made by Pylon Co. Canada) is used. The calibration of the scintillation cell is made by a comparison with the gas charging type ionization chamber.

2-4 The measurement point

The measurement points are located both in the waste rock pile sites and in the control area around Ningyo Toge (see part 1, Fig.1).

3. Result

3-1 Radon concentration

The measurement results are shown in Fig.1 and Table 1. As a whole, the radon concentration is ranged over 2 or 3 orders, and it tends to be a little higher in the waste rock pile sites. Especially, the radon concentration at Katamo-shimo1 is much higher than the other sites. This point is located in front of an old mining tunnel which is temporarily closed with sandbags. It is considered that the radon will escape from the tunnel. At Kan'nokura-2, the radon concentration seems to be high, because thick weeds may suppress the diffusion of the radon around the measurement point. The radon concentration in other waste rock pile sites are around 100 Bq/m^3 or less.

In other measurement points except for the waste rock pile sites, the radon concentrations are 10Bq/m^3 or less though high radon concentration was observed at Sugegawara in Kurayoshi City. At Sugegawara, the natural radioactivity is a little higher from the view of geology, and the diffusion of atmosphere is suppressed in the district along a narrow rill. Therefore, higher radon concentration is observed in this area.

3-2 Equilibrium equivalent radon concentration

The measurement of the equilibrium equivalent radon concentration results are shown in Fig.2 and Table 1. The results are ranged from 1 to 10 Bq/m^3 both in the waste rock pile sites except for Katamo-shimo 1 and in the control area. The variation of total results is very small compared with the radon concentration. And, the difference between the results in the waste rock pile sites and in the control area is also small. On the basis of these measurement results, it is suggested that the contribution of the natural background is important to the equilibrium equivalent radon concentration on the waste rock pile sites.

---

*1 This procedure was begun in fiscal 1992, and the data obtained before 1992 were not used in this report.

Although the data over 300Bq/m^3 which is the weekly mean concentration required to set up a radiation controlled area was once obtained at Katamo-shimo 1 after the tunnel was closed temporarily, the average concentration of 77.4Bq/m^3 is lower than 300Bq/m^3.

3-3 Radon exhalation rate

The measurement results are shown in Fig.3 and Table 2. Fig.3 does not show the range of the results for 1 measuring point, but it shows the range of all the results for many points in a waste rock pile site. The range of the fluctuation reaches 2 or 3 order. The results in the waste rock pile sites are approximately 10 $\sim$100$\times$10^{-6}Bq/cm^2/sec. The range of the results in control area is within an order in spite of the widely distributing control area in Togo town, Misasa town and Kurayoshi city, and that the range is in the order of 1$\times$10^{-6}Bq/cm^2/sec.

4. Discussion

4-1 Radon concentration and equilibrium equivalent radon concentration

The equilibrium factor for each grab sampling measurement was obtained from the results of the radon concentration and equilibrium equivalent radon concentration. The result is shown in Fig.4 and Table 1. According to UNSCEAR report[2], 0.8 of the averaged equilibrium factor in normal environment is estimated. However, lower equilibrium factor than 0.5 is often observed in a certain environment[3,4,5].

The equilibrium factors obtained from our measurements are generally low and they are about 0.4 even in the control area and ranged from 0.001 to 0.4 in the waste rock pile sites. The difference of the equilibrium factor originates from the difference of the radon concentration at the measurement points, because there was not much difference of the equilibrium equivalent radon concentration by the measurement points except for Katamo-shimo1, as it was described in the preceding chapter. Therefore, the radon concentration is relatively high at a place where the equilibrium factor is small.

These measuring points are roughly discriminated into two groups. In the first group with the equilibrium factor over 0.1, the radon concentration is relatively low and obtained mainly on the boundary of the waste rock pile sites or the control area. The equilibrium factor is approximately over 0.1 also in Nakatsugo-WR and in Choja-WR where a remediation for the waste rock pile site has already finished. In the second group with the equilibrium factors under 0.1, the contribution of "fresh radon" which is released from the waste rock pile is increased and it is considered that the radon progeny sufficiently has not grown yet.

As to the residence time of the radon, it needs about 30 seconds or about 5 minutes to reach to 0.01 or 0.1 of the equilibrium factor respectively, if it was assumed that the radon would originate only from the waste rock pile. The waste rock pile site is not a closed space although it is covered with small trees. The radon can not remain 5 minutes on the waste rock pile. Therefore, it is suggested that the contribution of natural background for the radon progeny is important even in the waste rock pile sites if the equilibrium factor is over 0.1. However, the contribution of the radon from the waste rock pile seems to increase at such situation as the equilibrium factor becomes under 0.01 although the waste rock pile site is a limitted area.

At the point of R-5 in Katamo-1,2 WR, the radon and radon progeny concentration are measured continuously, as it was introduced in part 1. The equilibrium factor obtained from these results was 0.12, which agreed well with 0.15 obtained by the grab sampling measurement. According to these results, it is considered that the equilibrium factor obtained in the grab sampling measurement is regarded as the first approximation of the equilibrium factor at the measuring point.

4-2 Estimated equilibrium equivalent radon concentration

The equilibrium equivalent radon concentration was estimated both from the radon concentration obtained by passive monitors in fiscal 1996 which is described in part 1 and from the equilibrium factor obtained by the grab sampling measurement.

The estimated results are shown in Table 1. The equilibrium equivalent radon concentrations obtained by the grab sampling method and the concentration estimated from the radon concentration obtained by passive monitors agreed comparatively well in the waste rock pile sites except for Katamo-shimo1, and they agreed within factor two in control area except for Sugegawara. At Katamo-shimo1 and Sugegawara, they are greatly different. At Katamo-shimo1, the estimated value is lower than the average value, which is lower than 300Bq/m^3, by the grab sampling measurement. Consequently, for a long term average the estimated value seems to be appropriate.

All of the estimated value except for the results of Katamo-shimo1 and Sugegawara are 10Bq/m^3 or less and no clear difference between the waste rock pile sites and the control area is observed. Therefore, the contribution of natural background seems to be important as described in 3-2.

4-3 Radon concentration and radon exhalation rate

It is reported that the radon in atmosphere consists of a component from the ground in the vicinity and component carried from a far field by wind[6], and it seems to be mainly attributed to the first component. The relationship between radon exhalation rate and radon concentration was derived by assuming vicinity data can represent the measured radon exhalation rate. The correlation of the radon concentration and the radon exhalation rate is shown in Fig.5. The averaged value at each measuring site was used for this figure. The data, which is shown by a circle spot on the upper position, is for Katamo-shimo 1 and it is suggested that there is another release source of radon as described in 3-1.

From this figure, the correlation coefficient between both data except for Katamo-shimo1 is about 0.8, which indicates that there is clearly a positive correlation and that the contribution of the vicinity component is qualitatively important. The parameters on the diffusion situation of atmosphere are complicated with the radon concentration and radon exhalation rate, and a more detailed examination seems to be necessary.

5. Conclusion

In the uranium mine waste rock pile sites and in the nearby control area, the radon concentration, equilibrium equivalent radon concentration, and radon exhalation rate have been measured by the grab sampling method. The measurement results of the radon concentration, equilibrium equivalent radon concentration and radon exhalation rate in the waste rock pile sites except for Katamo-shimo1 are around 100Bq/m^3, 1～10 Bq/m^3 and 10～100×10^{-6}Bq/cm^2/sec respectively. The results in the control area except for Sugegawara are 10Bq/m^3, 1～10 Bq/m^3 and 1～3×10^{-6}Bq/cm^2/sec respectively.

The point of Katamo-shimo 1 in the Katamo1,2 waste rock pile site is a very special place where the concentrations of the radon and equilibrium equivalent radon are high and about 4800, 77 Bq/m^3 respectively. However, the latter value is much lower than 300Bq/m^3.

The estimated equilibrium equivalent radon concentration was discussed and the results show similar conclusions between that obtained from the measurement results by the grab sampling method and that from passive monitors.

The positive correlation between the radon concentration and the radon exhalation rate was observed and it is estimated that the vicinity component of radon is the main contribution to radon in air.

6. Reference

[1]M. Shimo, K. Fujitaka, S. Furuta et al, Radon intercomparison at EML, U.S.DOE, Japan Health Physics Vol.32 pp.285-294(1997)

[2]UNSCEAR. 1988 report, Sources, effects and risks of ionizing radiation (1988)

[3]Neil Whyte, Jacques Dubeau, David M. Smith, Instrumentation for automated field measurement of radon progeny, Int. conf. on radiation safety in uranium mining (1992)

[4]J.Bigu, Radon, thoron, and their progeny in working, living, and natural environments, Int. conf. on radiation safety in uranium mining (1992)

[5]Keizo Yamazaki, Jun Saegusa, Tadashi Tsujimoto, et al, A mysterious spot on the outdoor concentration of radon isotopes

[6]Y. Ikebe, H. Yamanishi, K. Tojo, T. Iida, Relationship between radon exhalation from ground and

atmospheric radon concentration, J. Atomic Energy Soc. Jpn. vol.35, pp.735-738 (1993)

Table 1 The summary of the results of radon and equilibrium equivalent radon concentration

Site	Radon [Bq/m^3]				EECRn [Bq/m^3]				F-value [-]				Estimated
	Ave.	Max.	Min.	N	Ave.	Max.	Min.	N	Ave.	Max.	Min.	N	EECRn
Asabatake-1WR	119.4	754	15	13	4.3	10.4	1.3	16	0.095	0.29	0.006	12	3.6
Asabatake-1WR(B)	96.2	184	18	10	3.7	11.9	0.9	10	0.051	0.11	0.008	10	2.2
Asabatake-2WR	85.0	330	8	37	8.1	12.6	3.9	45	0.085	0.65	0.012	37	5.5
Asabatake-2WR(B)	77.4	371	7	34	3.7	22.9	0.5	44	0.087	0.39	0.013	34	9.5
Asabatake-2BWR	64.8	155	13	8	3.6	10.4	1.0	9	0.093	0.23	0.013	8	5.4
Asabatake-3WR	146.4	280	68	9	3.3	8.0	0.7	12	0.032	0.088	0.016	9	2.1
Katamo-1,2WR(s1)	4786.7	66300	14	48	77.4	339	0.6	49	0.067	0.29	0.001	47	24.5
Katamo-1,2WR	20.0	57	6	24	3.6	7.9	0.7	25	0.27	0.65	0.037	24	6.4
Katamo-1,2WR(R-5)	28.7	130	6	33	3.2	10.8	0.9	37	0.15	0.63	0.032	32	6.8
Katamo-1,2WR(B4)	21.3	56	9	8	2.9	6.5	1.1	9	0.17	0.33	0.039	8	1.9
Katamo-S2WR	54.7	176	9	7	3.7	9.7	0.9	8	0.085	0.17	0.038	7	3.0
Katamo-3WR	37.3	67	14	6	2.0	2.9	1.3	7	0.071	0.14	0.021	6	1.2
Ayumidani-WR	22.9	48	6	11	3.4	5.9	2.0	12	0.18	0.35	0.071	11	9.1
Kan'nokura-1WR	44.3	107	6	9	2.2	4.6	0.6	9	0.092	0.35	0.022	9	2.4
Kan'nokura-1WR(B)	83.1	135	37	9	3.6	5.5	0.8	9	0.043	0.073	0.022	9	2.9
Kan'nokura-2WR	437.7	1210	15	9	7.7	26.1	0.9	9	0.047	0.25	0.008	9	6.0
Kan'nokura-2WR(B)	59.2	297	6	6	3.5	12.0	1.1	8	0.22	0.37	0.040	6	3.8
Nakatsugo-WR	11.7	21	5	10	2.4	5.8	0.2	32	0.29	0.67	0.049	14	4.1
Nakatsugo-WR(B)	10.8	20	5	9	2.9	5.4	1.5	12	0.35	0.78	0.15	10	5.0
Choja-WR	23.0	91	5	10	1.8	3.4	0.6	15	0.19	0.49	0.012	10	2.8
Choja-WR(B)	41.7	127	5	6	2.0	3.1	0.6	8	0.18	0.62	0.018	6	7.9
Sugegawara	14.7	42	5	37	3.7	12.0	0.4	47	0.27	0.67	0.031	37	15.8
Controls(others)	22.9	191	5	26	2.8	8.4	0.2	60	0.36	0.70	0.031	28	5.0

Note1: "EECRn" means equilibrium equivalent radon concentration

Note2: "N" means the number of data.

Note3: "Estimated EECRn" were obtained from the integrated radon concentration in fiscal 1996.

Table 2 The summary of the results of radon exhalation rate

[$\times 10^{-6}$ Bq/m^2/sec]

Site	Ave.	Max.	Min.	N
Asabatake-1WR	109.3	1540	1	117
Asabatake-2WR	105.0	2600	2	335
Asabatake-2BWR	18.3	80	1	20
Asabatake-3WR	143	543	8	23
Katamo-1,2WR(s1)	170.5	1190	0.4	67
Katamo-1WR	70.3	449	2	191
Katamo-2WR	146.7	808	6	44
Katamo-1,2WR(B1)	3.5	9	0.4	18
Katamo-S2WR	22.1	66	0.4	13
Katamo-3WR	17.8	78	2	9
Ayumidani-WR	14.9	120	0.4	109
Kan'nokura-1WR	73.5	755	1	44
Kan'nokura-2WR	74.3	406	1	35
Nakatsugo-WR	20.2	182	0.4	166
Choja-WR	60.4	495	1	70
Sugegawara	4.0	66	0.4	52
Control(others)	1.8	7	0.4	54

Note: "N" means the number of data.

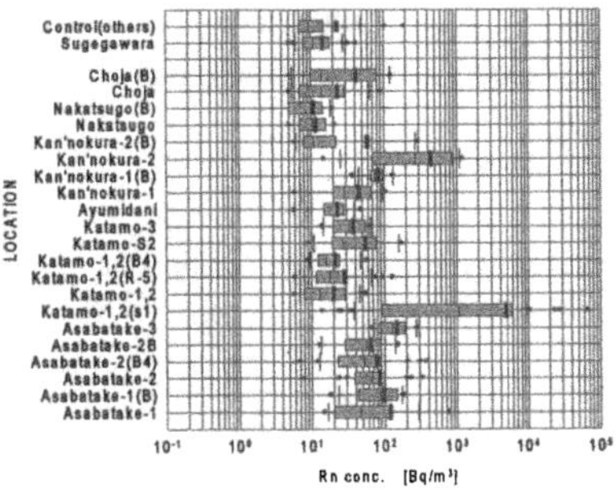

Fig.1 the variation of radon concentration

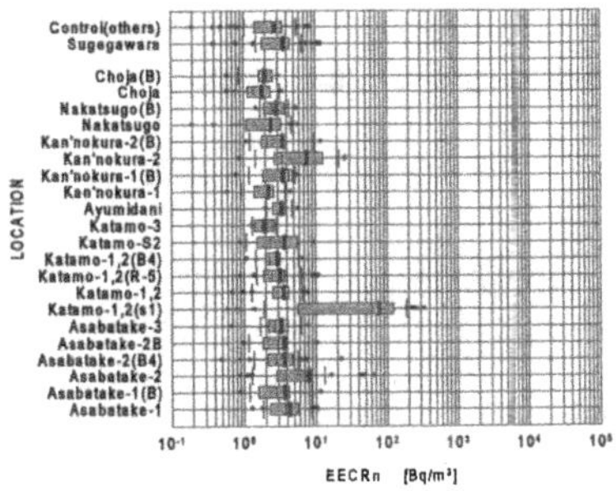

Fig.2 the variation of equilibrium equivalent radon concentration

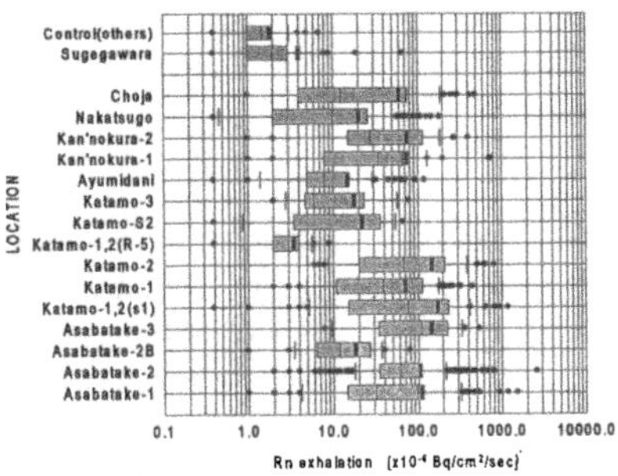

Fig.3 the variation of radon exhalation rate

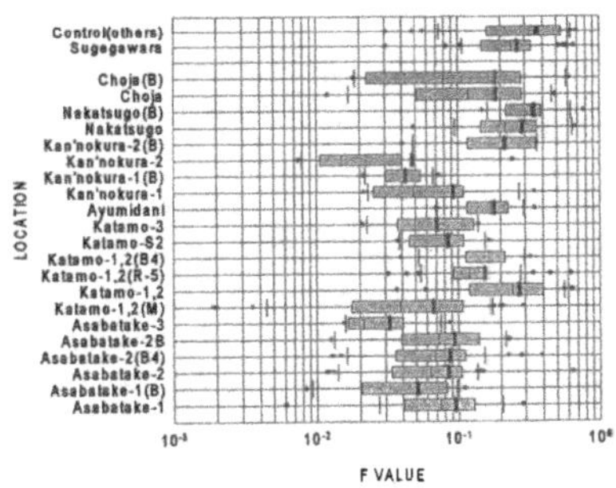

Fig.4 the variation of equilibrium factor

Note for Fig.1,2,3,4: a figure of box means the median and the range from 25th to 75th percentile; a line with error bar, from 10th to 90th percentile; thick line, the arithmetic mean and black dots, other data

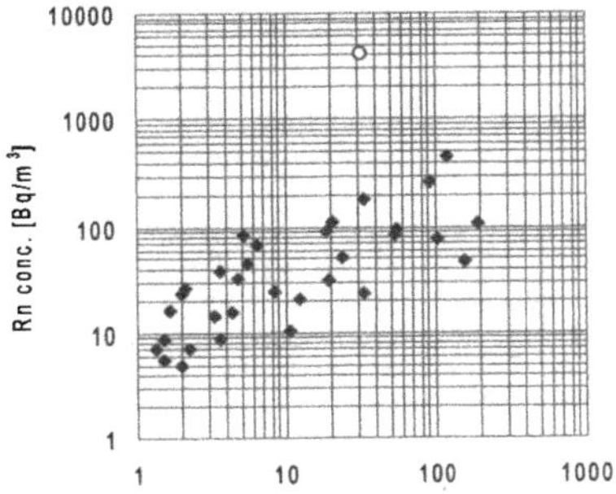

Fig.5 the correlation between radon concentration and radon exhalation rate

Note: the spot circle (○) means the data for Katamo-shimo 1, and others (◆) for each measuring site

Environmental effects of radon and its progeny from uranium waste rock piles: part 3-estimation by dispersion models and experiments

Sada-aki FURUTA, Kimio ITO, Yuu ISHIMORI

Ningyo Toge Works, Power Reactor and Nuclear Fuel Development Corporation
Kamisaibara, Okayama, 708-06

Abstract

For the purpose of evaluating the effects of radon and its progeny on the environment from the waste rock piles around Ningyo Toge Works of Power Reactor and Nuclear Fuel Development Corp. (PNC), the radon emanated from the surface of the waste rock piles were measured, and radon concentration, equilibrium equivalent radon concentration, and effective dose out of the sites were estimated by the dispersion calculation. The tracer gas dispersion experiments at the typical waste rock site were carried out, and the validity of the plume model calculation was examined. Consequently, it was confirmed that the effective doses caused by the waste rock piles are less than 1mSv/year, the public effective dose limit, out of the site boundary.

1. Introduction

The uranium exploration in Japan was started in 1956 and succeeded to PNC in 1967. Although the uranium exploration was closed by 1987, the waste rock piles are remained in the mining areas around Ningyo Toge. Therefore, the measurements of radon in the environment and the evaluation of the exposure have been necessary by the Japanese regulation since 1989 according to the ICRP Publication26. However, as the radon exists naturally and the concentration changes greatly in different atmospheric conditions, it is very difficult to judge whether the effect from the mining activity or from natural component by measurement. Therefore, for the purpose of evaluating the environmental effect from the waste rock piles, the radon emanated from the surface of the piles were measured, and radon concentration, equilibrium equivalent radon concentration, and effective dose out of the site were obtained by the dispersion calculation. The effects were also evaluated by the tracer gas dispersion experiments at Katamo area, and the validity of the plume model calculation was examined.

2. The condition of the calculation

In this study the evaluation was carried out with representative emanation data which are average measured emanation data at each waste rock pile. When there are multiple measured emanation data at the same point, the data were averaged at the each point, and further averaged at the each pile. The radon emanation from the mouth at Katamo-shimo 1 which has been temporarily closed by the sandbags was also measured, and it was added to the evaluation. It was assumed that only the radon itself emanated from the ground surface of the waste rock piles, and the progeny nuclides were calculated according to the Betheman equations. The deposition of the progeny nuclides on the ground surface or the obstacles during the dispersion was disregarded in order to evaluate conservatively. In Japan the equilibrium equivalent radon concentration is regulated, but the exposure by the radon gas itself is not regulated. However, in this study, as the equilibrium factor was small and the exposure by radon gas can not be disregarded either, the evaluation was carried out including the effective dose by radon gas.

3. Dispersion model

Though the one-dimensional dispersion model from soil to atmosphere by W.Jacobi et al. is well known[1], its application is difficult in the finite source like the waste rock piles. The model in which the basic dispersion formulation assumed the Gaussian distribution is well used in report[2] to evaluate radon effect from the uranium mill tailing piles. In this evaluation, we assumed that the emission height and the calculation points are ground surface for using the plume model, since the plume model with the Gaussian distribution is widely adopted by the many guidelines or manuals[3] as a simple model. The land form of the waste rock piles was assumed to be a plane rectangle, which has the same size as the waste rock pile area. Since the lower limit of the usable wind velocity of the plume model is empirically 1m/s, in the manual as a reference, the wind velocity of 1m/s was also adopted in this calculation. As for the wind direction, it was conservatively assumed that the wind blows only in one direction throughout a year, which means that the wind direction distribution was disregarded . Therefore, the directions of the calculation points were along the short and long axes of the rectangle, and the shortest distance from the edge of the waste rock pile to the site boundary was used in the calculation. The model is shown in figure 1. Since the purpose of the calculation is to obtain the annual mean, as for the atmospheric stability, neutral D was adopted because the frequency of D accounts for more than a half of the annual total and more unstable category than D holds 80% according to the weather observation.

Because the source of radon is a plane, it can be approximated as a cluster of point sources. The area of the waste rock pile was divided into 1m × 1m meshes for the calculation, and the numerical integration was carried out by assuming that the radon gas was emitted from the each center.

4. Calculation result and discussion.

The test calculations were carried out with the plume model (atmospheric stability=D,wind velocity=1m/s) to evaluate how the modeled shape of the piles affects the calculation results. A 3600m^2 pile area with unit emission(1Bq/m^2/s) was assumed. The calculated effective doses for various combinations of short and long sides of the same rectangle area are shown in figure 2. The modeled of the pile is given by "A(width) × B(length)" in this figure, and the wind blows along the B side. On this figure, it was confirmed that longer B pile brings larger dose, and that the difference of the shapes can be disregarded from more than a few hundreds meters to about ten kilometers on down wind side. It is also indicated that the effective doses decreases monotonously in every case. Therefore, we suggest that the assumed wind direction along the longer side of the rectangle and the shortest calculation point are appropriate for the conservative evaluation.

For each site, the results of radon concentration, equilibrium equivalent radon concentration, and effective dose calculated by the plume model (atmospheric stability=D, wind velocity=1.0m/s) are shown in table 1. In this table, the results from the multiple waste rock piles within the same site are simply summed, and it was confirmed that the effective doses do not exceed 1mSv/year at each site boundary. In addition, the similar results were obtained even at a wind velocity of 0.5m/s, assuming that the dispersion is weak at night and dawn, using not the shortest distance but realistic one from the wind direction estimated by the measured radon concentrations described in Part 1.

The calculation results were compared with the radon concentration and equilibrium equivalent radon concentration reported in Part 1 and Part 2. As for the radon concentration, calculated value and measured value agree well, considering that the measurement points are near the boundary within the site, the radon concentration of the background is about 20Bq/m^3, and the ununiformity of emanation brings about localized radon concentration. However, as for the equilibrium equivalent radon concentration between calculated and measured value, poor agreement is observed with one order difference, because the measured values are not determined by long term measurements but by grab samplings. The calculated results seems to be too small to discriminate from the background level in the normal environment reported in other reports [4] [5] [6].

5. The dispersion experiment

The dispersion experiments were conducted three times by using SF_6 tracer gas in 1992 and 1994 in order to obtain the radon influence from the waste rock piles to the residential area at Katamo as a demonstration field shown in figure 3,. And, the radon transport speed was also estimated by the puff emission in 1994. The waste rock pile site is located in mountain about 1.5km away from the residential area along the small valley surrounded by forest, and the area a half way from the residential area to the site is utilized as pear orchards. In case of such land form, it is known that the mountain breeze (nocturnal drainage flow) arises at night. In order to know how much of radon is flowing to the residential area, the tracer gas was emitted from the pile while the mountain breeze was observed at night, and the air samples from the site to the residential area were collected by automatic samplers and analyzed by gas chromatography. In addition, the radon concentrations in the collected samples were also measured by ionization chambers, and the effect from the waste rock piles was examined with the radon emanation.

6. Experiment result and discussion

It is observed that the result of the dispersion experiments by the continuous emission has a good agreement with the estimated concentration on the down wind axis by the plume model (atmospheric stability=D, wind velocity=0.5,1m/s), as shown in figure 4. Although several experimental concentrations which exceeded the model results were observed, those were only temporal phenomena which occur when the mountain breeze was broken up in the morning. And the transport speed of tracer gas toward residential area was estimated to be about 0.5m/s by the puff emission experiment. It can be said that the dispersion experiments were carried out conservatively considering that the mountain breeze is not always observed even at night. Therefore, the weather conditions of the model, such as atmospheric stability D, 1m/s wind velocity, and one wind direction, seem to be valid for the purpose of annual evaluation. And, it seems to be also adequate to assume atmospheric stability D and 0.5m/s wind velocity from transport speed by the puff emission experiment as the calm condition at night. On the basis of three time experiments, the maximum effect of radon concentration on the site boundary was estimated to be about 20 Bq/m^3 from the radon samples collected out of the site at dawn as shown in figure 5, assuming that the behavior of the radon is the same as the tracer gas.

7. Conclusion

By the dispersion calculation of the plume model, it was confirmed that the effect of radon out of the site would not exceed 1mSv/year, and that the dispersion model explained well by the dispersion experiment in Katamo area. In addition, the effects from the piles seems to be insignificant considering the natural background radon.

From Part 1 to Part 3, the environmental effects of radon and its progeny from the waste rock piles have been investigated. Although the relatively high concentrations are observed within the site, the results out of the site indicate that the effects are small compared with the natural background level, and the effective doses are less than 1 mSv/year based on both measurement and calculation out of the site boundary.

8. References

[1] W.Jacobi and K.Andre, The Vertical Distribution of Radon222, Radon220 and Their Decay

Products in the Atmosphere, Journal of Geophysical Research, Vol.68, No13, (1963)

[2] W.E.Clements, S.Barr, and M.L.Marple, Uranium Mill Tailings Piles as Sources of Atmospheric Radon-222, Natural Radiation Environment Ⅲ, p1559-1583 (1980)

[3] Control manual of the total amount of NOx, the Environment Agency (Rev.2 1988) (in Japanese)

[4] T.Yamasaki, T.Iida, M.Shimo, Y.Ikebe, Continuous Measurement of Outdoor Radon and Its Progeny Concentrations, Hokenbutsuri, Vol.30, No.2 pp149-154 (1995)

[5] H.Hoetzl and R.Winkler, Long · term variation of outdoor radon equilibrium equivalent concentration, Radiat Environ Biophys, Vol.33, No.4 pp381-392 (1994)

[6] G.Keller, K.H.Folkerts, and H.Muth, Activity concentrations of ^{222}Rn, ^{220}Rn, and their decay products in German dwellings, dose calculations and estimate of risk, Radiat Environ Biophys, Vol.20, No.4 pp263 · 274 (1982)

Table 1. The results of radon concentration, equilibrium equivalent radon concentration and effective dose calculated by the plume model (atmospheric stability =D, wind velocity =1.0m/s)

waste rock	conditions of the estimation				results of the estimation			
	size of waste rock area	waste rock area	down wind distance	radon emanation	radon concentration	equilibrium equivalent radon concentration	effective dose	total of the site
Nakatsugo	120 m × 100 m	12,000 m^2	2 m	0.22 $Bq/m^2/s$	12.7 Bq/m^3	0.157 Bq/m^3	0.033 mSv/y	0.033 mSv/y
Choja	12 m × 26 m	312 m^2	14 m	0.78 $Bq/m^2/s$	11.4 Bq/m^3	0.111 Bq/m^3	0.026 mSv/y	0.026 mSv/y
Kan'nokura-1	39 m × 130 m	5,070 m^2	2 m	0.47 $Bq/m^2/s$	27.7 Bq/m^3	0.363 Bq/m^3	0.075 mSv/y	0.075 mSv/y
Kan'nokura-2	72 m × 89 m	6,408 m^2	2 m	1.10 $Bq/m^2/s$	57.9 Bq/m^3	0.567 Bq/m^3	0.134 mSv/y	0.134 mSv/y
Asabatake-1-1	33 m × 43 m	1,419 m^2	2 m	0.59 $Bq/m^2/s$	24.3 Bq/m^3	0.137 Bq/m^3	0.044 mSv/y	0.086 mSv/y
Asabatake-1B	27 m × 34 m	918 m^2	4 m	0.76 $Bq/m^2/s$	21.9 Bq/m^3	0.137 Bq/m^3	0.042 mSv/y	
Asabatake-2	25 m × 80 m	2,000 m^2	5 m	0.77 $Bq/m^2/s$	30.0 Bq/m^3	0.352 Bq/m^3	0.076 mSv/y	0.076 mSv/y
Asabatake-2B	16 m × 25 m	400 m^2	4 m	0.30 $Bq/m^2/s$	7.5 Bq/m^3	0.039 Bq/m^3	0.013 mSv/y	0.013 mSv/y
Asabatake-3	32 m × 50 m	1,600 m^2	4 m	0.50 $Bq/m^2/s$	17.1 Bq/m^3	0.138 Bq/m^3	0.036 mSv/y	0.036 mSv/y
Katamo-1	22 m × 91 m	2,002 m^2	5 m	0.71 $Bq/m^2/s$	28.9 Bq/m^3	0.366 Bq/m^3	0.076 mSv/y	0.294 mSv/y
Katamo-2	29 m × 59 m	1,711 m^2	2 m	0.71 $Bq/m^2/s$	32.7 Bq/m^3	0.233 Bq/m^3	0.065 mSv/y	
Ore storage area	16 m × 24 m	384 m^2	32 m	0.04 $Bq/m^2/s$	0.3 Bq/m^3	0.006 Bq/m^3	0.001 mSv/y	
Storage-area	11 m × 26 m	286 m^2	45 m	0.017 $Bq/m^2/s$	0.1 Bq/m^2	0.002 Bq/m^3	0.000 mSv/y	
Shimo-1 mouth	point	-	130 m	824 Bq/s	59.9 Bq/m^3	0.704 Bq/m^3	0.151 mSv/y	
Katamo-3	21 m × 24 m	504 m^2	4 m	0.21 $Bq/m^2/s$	5.1 Bq/m^3	0.026 Bq/m^3	0.009 mSv/y	0.009 mSv/y
Katamo-shimo2	20 m × 65 m	1,300 m^2	4 m	0.32 $Bq/m^2/s$	12.2 Bq/m^3	0.116 Bq/m^3	0.028 mSv/y	0.028 mSv/y

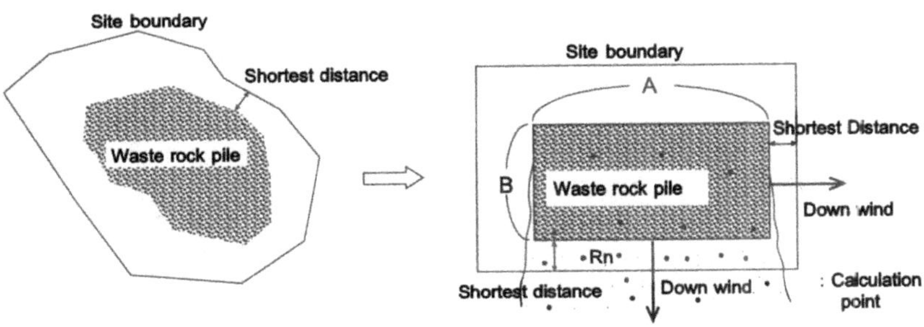

Real land form of the site Model of the site

Fig.1. Waste rock pile site modeled for calculation

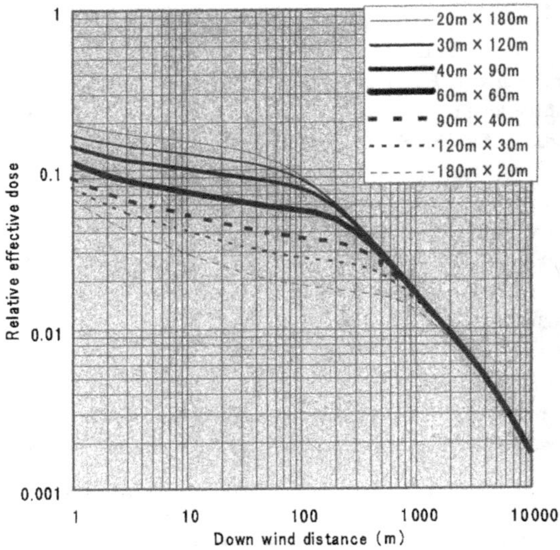

Fig.2. The effective doses calculated for various parameter sets of the same area rectangle

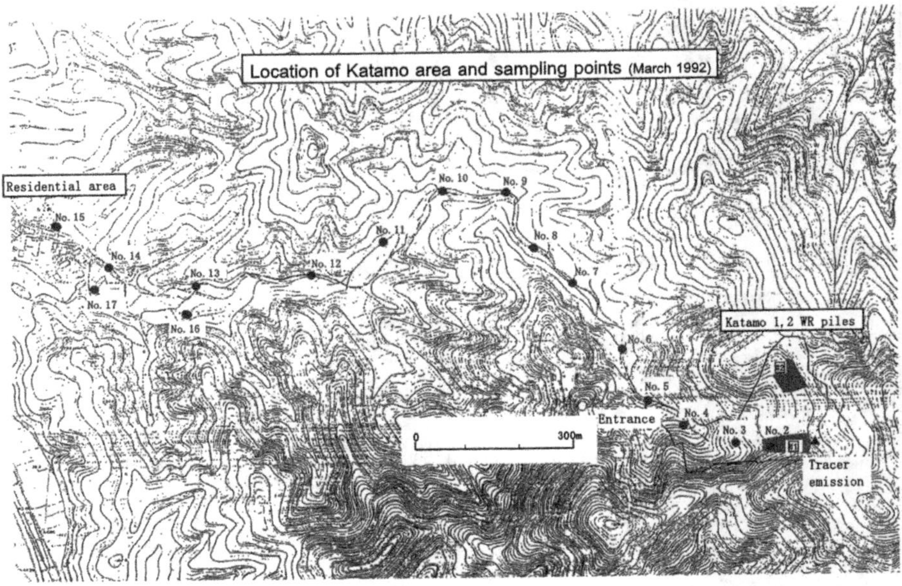

Fig.3. Location of Katamo-1,2 waste rock piles, residential area, and tracer gas sampling points

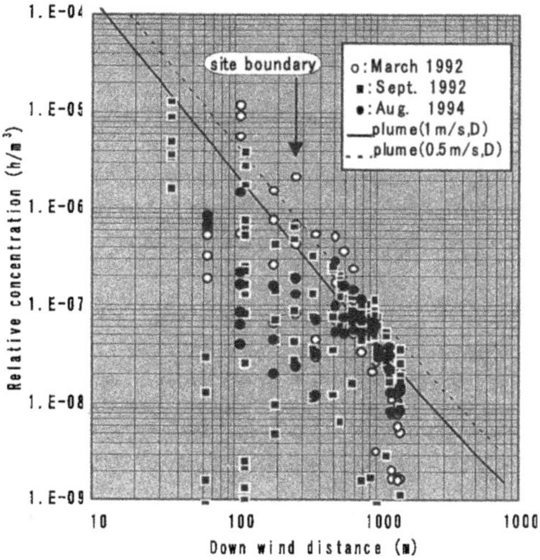

Fig.4. The results of dispersion experiments

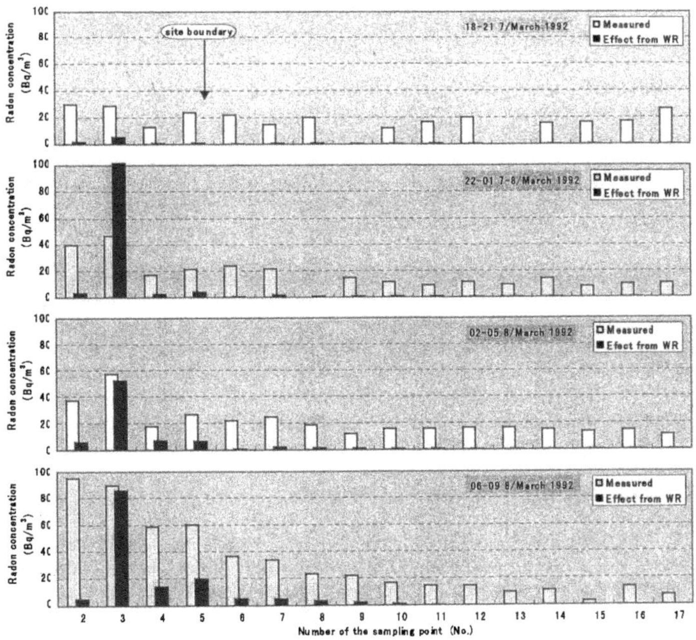

Fig.5. Measured radon concentration and estimated effects from waste rock piles by the dispersion experiments

Radon measurements on uranium mining and mill tailing in Kalan West Kalimantan Indonesia

Mudiar Masdja and Sutomo Budihardjo

Nuclear Minerals Development Center
National Atomic Energy Agency
Jakarta 12060, Indonesia

Abstract

Radon measurements on uranium mining and mill tailing in Kalan , West Kalimantan, Indonesia have been performed using SC-6 and 58-1 models scintillation cells, SAC-R5 counter and Ludlum 1000 scaler. The purpose of this report is to present the mode and technique of radon measurements on uranium mining and mill tailing in Kalan , West Kalimantan . Radon measurements (inhalation) in an atmosphere tunnel is conducted by instantaneous mode, it shown in the working area was 0.16 ± 0.02 WL. Radon emanation from drilling hole surface Long Year - 65 is carried out by semi-integrating mode, it was between 0.1 to 122.6 Bq/m^2-sec. Radon realized (exhalation) of 40 cm tailing has been determined by semi-integrating mode, it was 32.8 x 10^{-2} Bq/cm^2 - second. The soil single layer cover 50 cm, it was realized radon with 33.9 attenuation factor. The compacting soil, 48 gr./cm^2 and 15 cm, it was released radon with 113.4 attenuation factor. Radon of uranium mining and mill tailing in Kalan , West Kalimantan were achieved by the method selected for measurement of the actual concentration of radon gas, they were the instantaneous and semi-integrating modes, and it depends upon measurement facilities, the scope of measurement, existing equipment and supporting financial.

1. Introduction

Radon is an inert, odorless and colorless gas. Under natural condition in such low concentration that there is no conventional chemical analytical techniques with sufficient sensitivity to measure it. Radon must be determined using technique that depend on the radioactive properties of radon and its decay products. The radioactive decay sequence from ^{222}Rn to stable ^{206}Pb shows, ^{218}Po and ^{214}Po are α emitter, while ^{214}Pb and ^{214}Bi are β and γ emitters [1].

The radon measurements, especially in the atmosphere of uranium mining and mill tailing are most important. Uranium ore milling is the process extracts uranium from ore, here internal contamination is dominantly than radon measurement. In operational uranium milling, the large quantities of tailing produces and they contain ^{226}Ra and its daughter product ^{222}Rn. For environment protection, the measurement of radon exhalation from tailing mill must be conducted [2]. There is thus a need to ensure that health risk of the worker and environment from radon of uranium mining and mill tailing are reduced to acceptable level. Most of instruments detecting α particles and γ rays have been developed [3].

2. Radon measurements and methods

Two of the three basic mode for radon collection and measurement [3] have been applied in radon measurement of uranium mining and mill tailing in Kalan, West Kalimantan, and they are instantaneous and semi-integrating (continuous real time). The fully integrating (time averaging) has not been developed yet. The instantaneous mode has been conducted for radon inhalation determining in the atmospheric tunnel, and the semi-integrating mode has used for radon emanation from the drilling hole LY-65 surface and radon exhalation from mill tailing. The instantaneous mode for radon concentration is determined of radon in the order of ten minutes. In this short time result of the radon concentration in the tunnel has been determined.

By the semi-integrating mode, the radon are allowed to reach a steady state equilibrium, where the rate of radon gas production is equal to the rate of decay, this occurs within a time frame of the order of a few hours. The cross section of the Remaja tunnel and the location of the drilling hole LY-65 is shown in Fig. 1.
The measurement technique by scintillation cell has been selected for radon inhalation, emanation and exhalation determinations, it is constructed of plastic, coated with ZnS powder as a scintillation , material and in varies model and volume. The equipments of measuring of the radon inhalation, emanation and exhalation [4] are shown in Fig. 2.

Radon inhalation is the radon in the atmosphere that is inhaled intake in man. The radon inhalation is calculated by radon concentration equation :

$$C = \frac{Cpm - Bg}{60 \times V} \times e \quad (1)$$

where, C = radon concentration (Bq/l)
Cpm = count rate per minute
Bg = background per minute
V = scintillation cell volume (liter)
e = calibration value of scintillation cell (Bq/Cpm)

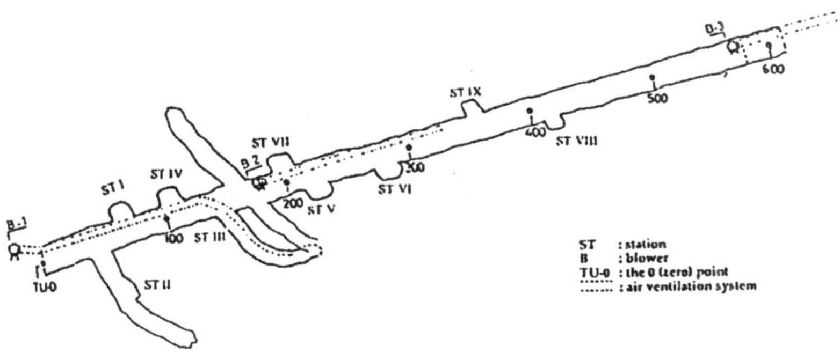

Fig. 1 The cross section of the Remaja tunnel, Kalan, West Kalimantan.
(ST : the location of drilling bors LY-65 holes)

The methods of measuring of radon inhalation, the scintillation cell SC-6 model, 1.4 liter volume is vacuumed and put in the Remaja tunnel. The valve of scintillation cell is opened, radon gas will get into, for 5 minutes and then valve is close tightly. Radon inhalation is counted within 20 minutes outside of the tunnel and counted by SAC-R5 counter and measured by Ludlum 1000 scaler.

The radon emanation and exhalation are calculated by radon flux equation :

$$E = \frac{\left[\left(Cpm \times Fk\right) - Bg\right] e.h}{V.t} \quad (2)$$

where, E = emanation and exhalation rate (Bq/m^2.sec.)
Fk = correction of ^{222}Rn activity for decay during counting
h = height of container (m)
t = time of radon gas collecting (second)
V = scintillation cell volume (m^3)

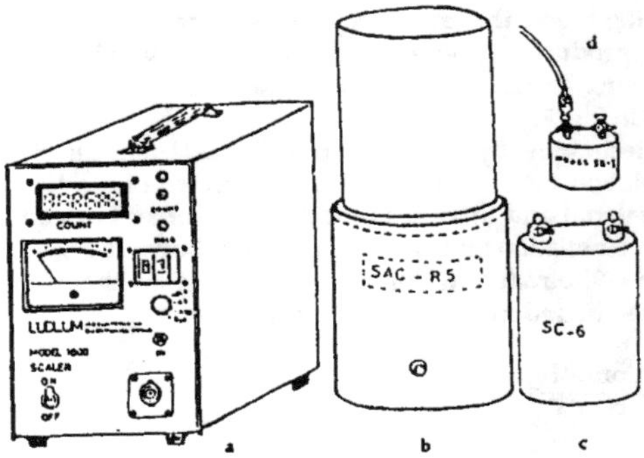

Fig. 2. The equipments of measurement of radon inhalation, emanation and exhalation. (a : Ludlum 1000 scaler; b : SAC-R5 counter; c : Scintillation cell SC-6 model and d : the vacummed scintillation tube)

The technique of measuring of the radon emanation is shown in Fig. 3. The radon emanates from radioactive solid/ore that may or not may retained within the pore space of the solid phase. Radon gas from the drilling LY-65 surface at the Remaja tunnel is collected for 60 minutes by using the PVC pipe, diameter 3.81 cm and 125 cm in length. Scintillation cell SC-6 model, 1.4 liter volume is vacuumed and connected to the PVC pipe [5]. The other scintillation cell valve is opened slowly and the pressure in the cell is equal to air pressure, for while the radon gas will enter into, and then the valve is closed tightly. The scintillation cell is held within 3.5 hours and counted by SAC-R5 counter and measured by Ludlum 1000 scaler.

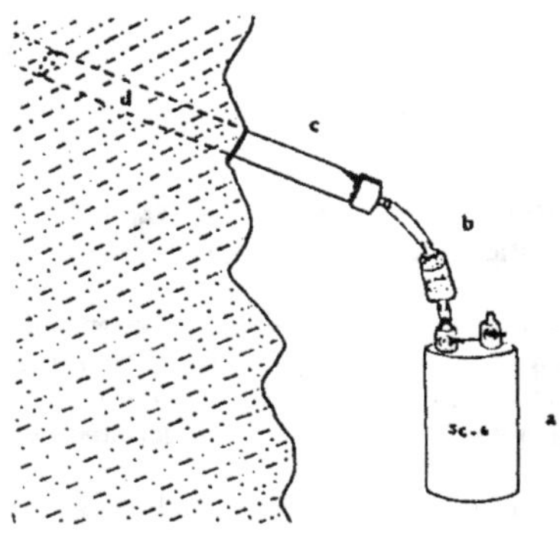

Fig. 3. The technique of measuring of the radon emanation from the drilling hole LY-65 (a : scintillation cell SC-6 model; b : PVC pipe; c : filter and d : the drilling hole)

Radon exhalation is radon releases from bulk solid phase into the atmosphere. The measurement of radon exhalation rate of tailing is characterized by the radon flux rate. Radon exhalation rate from 40 cm uranium mill tailing was collected for 30 minutes by container [6]. The scintillation cell 58-1 model, 0.5 liter volume was vacuumed and connected by valve to container, like figure is shown in Fig. 4. Radon gas was moved from container to scintillation cell by opening slowly the other valve. The scintillation cell was kept within 3.5 hours and counted by SAC-R5 counter and measured by Ludlum 1000 scaler.

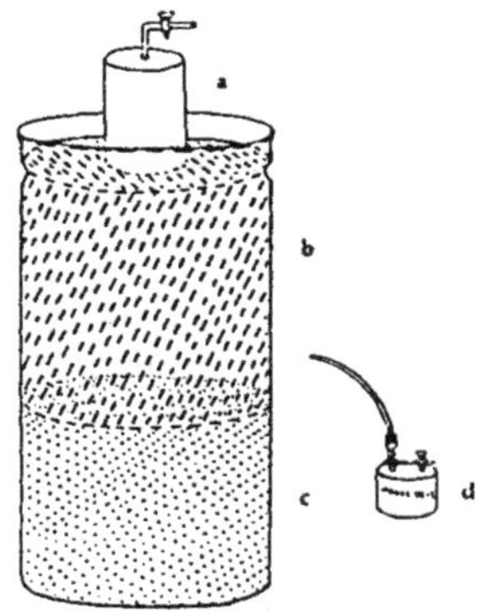

Fig.4. The measurements of radon exhalation for 40 cm uranium mill tailing. (a : concave cover; b : solid or other material cover layer; c: uranium mill tailing layer and d : the vacuumed scintillation tube).

3. Result and discussion

Radon inhalation measurement in the Remaja tunnel was conducted by instantaneous mode, radon was counted within 20 minutes. Correction factor of ^{222}Rn activity for decay during counting is assumed 1. The radon concentration and working level in the Remaja tunnel must be determined within a few minutes, consequently the instantaneous mode has been selected. If the radon concentration more than 1.1 Bq/l or 0.3 WL, the ventilation must be operated for 30 minutes. The workers are allowed inter the Remaja tunnel when the working condition has been adjusted [2]. The radon concentration or working level in the Remaja tunnel was between 0.008 ± 0.002 and 0.16 ± 0.02 WL or the maximum radon exposure was 27.2 ± 3.4 WLM.

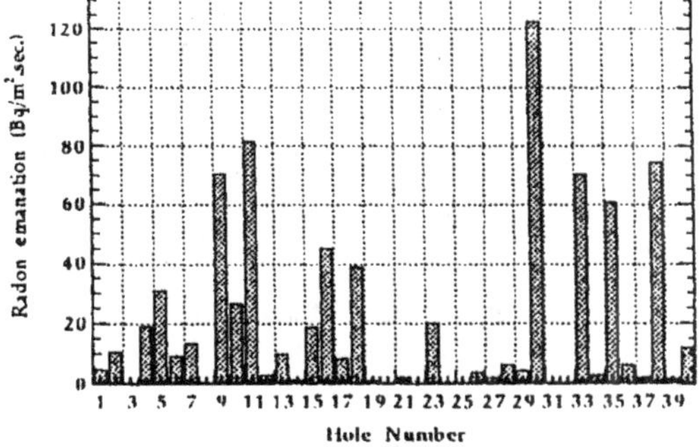

Fig. 5. The radon emanation from the 40 drilling holes LY-65 in the Remaja tunnel.

The radon emanation from drilling hole LY-65 surface has been measured by semi-integrating mode. By this mode the reducing the statistical counting accuracy of the sample will be eliminated. In the general case, suitable ages for counting are 3 - 18 hours for the first counting [3], it will give the correction of ^{222}Rn activity for decay during counting equal to 1. In the Remaja tunnel, the radon emanation of 40 drilling holes LY-65 surface have been determined and the emanation rate varied between 0.10 to 122.6 Bq/m^2.sec. These values are 1500 times greater than the radon emanation rate from the earth surface, that is 0.033 to 0.078 Bq/m^2.sec.[7]. The result of radon emanation measurement from the drilling holes LY-65 surface is shown in Fig. 5, that some of the drilling holes surface must be shut off by cement.

The radon exhalation measurement from 40 cm tailing layer shown 32.9 ± 0.37 Bq/m^2.sec., this value 3.29 times greater than radon exhalation measurement in the Andujar, Spain and Western area in USA. The radon exhalation rate in the some site were reported between 0.03 to 10.2 Bq/m^2.sec.[2]. The radon exhalation rate will be 0.97 ± 0.10 Bq/m^2.sec., if the tailing layer was covered by 50 cm soil covered layer, it means the covered layer gave 33.9 attenuation factor. The attenuation factor of the 15 cm compact soil covered layer on the radon exhalation measurement of tailing shew 113.4, so the attenuation factor of the 15 cm compact soil layer is greater than 50 cm soil covered layer. This concept will be adopted in reclamation to improve the appearance after close out, more than the 50 cm soil covered layer could be applied for vegetation and the more than 15 cm compact soil layer for the road.

The ionization chamber can be used for radon measurement, its detection limit as low as 4 Bq/m^3. Activated charcoal detector are used extensively for determination of in door radon concentrations with a sensitivity of about 1 Bq/m^3 and an error of 50 %. Nuclear track detector is utilized for long term radon concentration such as several month to one year. The solid state surface barrier detector is applied for measurement of radon exhalation rate from tailing, high cost and the care and the maintenance requirements and impractical relatively. Thermoluminescent dosimeter are used for environmental radon monitoring with lower limits of detection of the order of 1 Bq/m^3 of ^{222}Rn for one week sampling. The electric is a device which acts like a capacitor, is used to determine the radon concentration during the period of exposure [2]. Among those measurement techniques, the scintillation cell with varied mode and volume has been selected for measuring the radon of uranium mine and mill tailing in Kalan , West Kalimantan since 1990. The scintillation cell techniques has been selected because of base on the technical report series no. 333 [2], it is relative simplicity, raggedness, versatility, sensivity, reliability and low costs.

4. Conclusions

The result of radon inhalation, emanation and exhalation measurements in Kalan, West Kalimantan shows that :

1. Radon measurements of uranium mining and mill tailing in Kalan , West Kalimantan have been conducted by scintillation cell mode either inhalation radon concentration in the tunnel, radon emanation rate from the drilling hole LY-65 surface and radon exhalation rate from mill tailing.
2. The instantaneous, mode of measurement, has been adopted for inhalation radon determination while the semi-integrating for radon emanation and exhalation rate. The mode for radon collection and measurement selection are based on convenience and time available for sample collection, existing equipment, cost, sensitivity and the supporting facilities and financial base.
3. Working condition and acceptable value for radon inhalation in the Remaja tunnel are gained by operating the ventilation and the radon emanation rate from the drilling hole LY-65 surface is reduced by cementation of the hole surface. The reducing of the exhalation radon rate from mill tailing is performed by soil covered layer.

The fully integrating time has not been applied yet. In the near future this mode will be developed for radon environment monitoring in Kalan , West Kalimantan as secondary measurement.

5. Acknowledgment

We would like to acknowledge to Dr. Ir. Karyono, HS, Head of Nuclear Minerals Development Centre, National Atomic Energy Agency for the permition and support. Further aknowledgement go to all of the KKL (Work Safety and Environment Division) staffs, especially for Mr. Ngatino, who help us to collect data and reconstruct some of the pictures.

6. References

[1]. EPA, Radon Reference Manual, United States Environmental Protection Agency, EPA 520/1 87-20, 1987.

[2]. IAEA, Measurement and Calculation of Radon Releases from Uranium Mill Tailing, Technical Report Series No. 333, 1992.

[3]. IAEA, Current Practice for the Management and Confinement of Uranium Mill Tailings, Technical Report Series No. 235, 1992.

[4]. Eberline Instrument Cooperation, Technical Manual for Scintillation Alpha Counter Model SAC-R5, Scintillation Cells Model, CS-3, SG-5, SC-6, 1982.

[5]. Santosadjaja, B. et al, Measurement of the Emanation Rate of Radon Gas from the Drilling Hole Long Year-65 Surface, PPBGN/PUS/LAP/II/1991 (Internal Report), 1991, and its references, in Indonesian

[6]. Heryati, T, Tailing Covering Layer of Uranium Ore Processing, PPBGN/PUS/LAP/II/1991 (Internal Report), 1991, and its references, in Indonesian.

[7]. Greenberg, A.E., Apha et al, Standard Method for the Examination of Water and Waste Water, American Public Health Association, American Water Works Association, Water Environment Federation, 18th Edition, 2 - 20, 1992.

[8]. National Atomic Energy Agency-BATAN, Safety in the Mining and Milling of Radioactive Ores, Director General Decree No. 574/DJ/XII/1993, 1993.

[9]. National Atomic Energy Agency-BATAN, Technical Addendum of Safety in the Mining and Milling of Radioactive Ores, Director General Decree No. 610/DJ/XII/1993, 1993.

Is radon a potential risk in geothermal fields?

M. Balcázar, L. Tavera and A. López

Instituto Nacional de Investigaciones Nucleares
Apartado Postal 18-1027, México D. F. 11801 México

1. Abstract

Some environmental groups in Mexico have shown concern, by the fact that natural radon emanation in geothermal fields could be considered as a potential risk for human health.

As part of the studies recommended by the Ministry of Energy and to provide technical answers to the request of environmental groups, a study of the radon emanation and its impact to human health was carried out at the Primavera Mexican geothermal energy field under prospecting.

It was found that radon concentration decreases with temperature of geothermal fluids and increases with the enthalpy of the well. In any case, using the appropriate corrections, radon values were similar to those of ground water. Radon values in the atmosphere of the field show no potential risk for human health.

1. Introduction

The geothermal installed power capacity in Mexico is nearly 800 MW[1], the program of the Mexican Ministry of Energy states to gradually increase the use of the other energy sources, including geothermal; it also recommends to perform studies on the emissions and the necessary associated equipment to take care of the environment.

The geothermal energy field La Primavera under prospecting is situated at the west coast of México in Jalisco State on the so called Neovolcanic Belt.

Radon (Rn-222) is contained in the steam and discharge into the atmosphere in concentrations of 3,700-78,000 Bq/kWh[2]. At Larderello geothermal field (Italy) the radon concentration in air at ground level is 5.5 Bq/m^3 and at the Geysers (California) ranges[3] from traces to 6 Bq/m^3. By comparison the average levels of radon in air elsewhere are around 3 Bq/m^3. There is little evidence that radon concentration is raised above the background level by geothermal emissions. However, some environmental groups in Mexico demanded technical answers to the Federal Electricity Board, in charge of prospecting geothermal field in Mexico, relating to what is the potential risk of radon from the geothermal field to human health.

2. Methodology

Water and vapour samples from two prospecting wells were collected for radon assessment in a scintillation detector. Additionally, several plastic nuclear track detectors were placed in the field 70 cm deep below the earth's surface to evaluate radon emanation from the soil and at 20, 50 and 100 cm high over the surface to determine the radon dispersion in air.

Two different conditions at the wells were found: PR-9 well, Figure 1, was on a maintenance program, which required to have partially open valves to reduce its working pressure. Besides, there was a failure in a valve producing a continuos fluid relief. On the contrary well PR-11, Figure 2, was totally closed with a small discharge from a partially open valve and connected to a vapour relief tank to low down the noise.

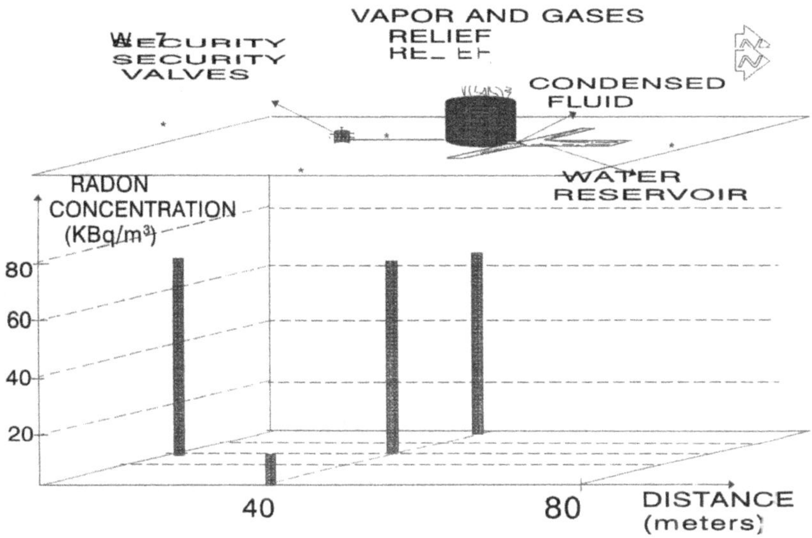

Figure 1.- PR9 well, on maintenance program, required to have partially open valves; values of radon concentrations from the ground, are represented by the four bars.

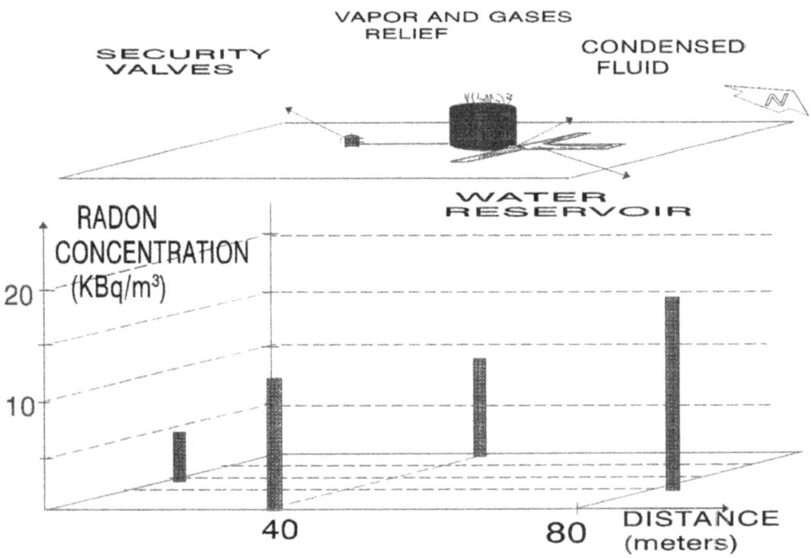

Figure 2.- PR11 well, totally closed, with small discharge from a partially open valve and connected to a vapour relief tank to low down the noise. This well presents higher radon concentration values than PR9.

Liquid samples.

Measurements of radon concentration in liquid samples, require a reproducible sampling procedure. Collection and representative of the samples are the main source of variation in the results. Considering this, a careful sampling method was performed; the fluid was passed through a narrow elliptical-shape tube as indicated in Figure 3, the tube was inside a 200 l tank filled with cool water. The condensed water from the cooling system was contained in a 1l cup.

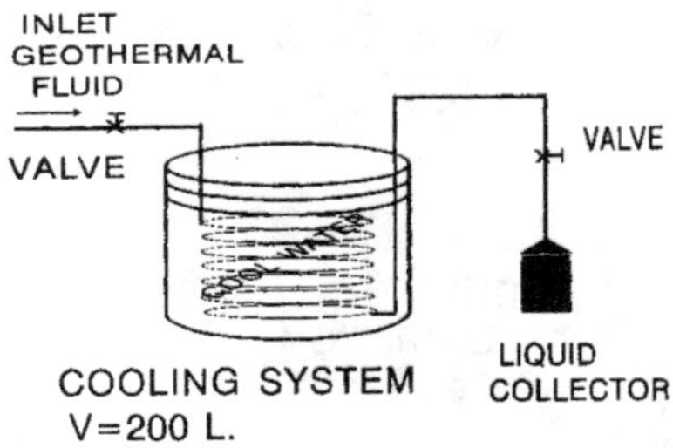

Figure 3 Sampling method to determine radon in geothermal fluids.

The samples collected were identified as follows:

WELL	CODE SAMPLE	VOLUME (l)
PR9	PR9-I	1
PR9	PR9-II	1
PR11*	PR11-I	0.5

* For well PR11, two more samples were taken; one after the relief tank and one more in the reservoir, see Figure 2.

Environmental radon and radon gas exhalation.

In order to evaluate radon concentrations in air and soil, two devices were employed. Both of them were a cellulose nitrate plastic track detector LR115 type for α detection.

For the environmental radon concentration measurements a cup with two detectors, one inside at the bottom base and the other one outside were used; the cup is sealed by a thin plastic which permits radon to diffuse into the cup[4]. Three cups were fixed on a long vertical holder, at 20, 50 and 100 cm from the ground, in order to monitor the radon dispersion in the air, as it is shown in Figure 4.

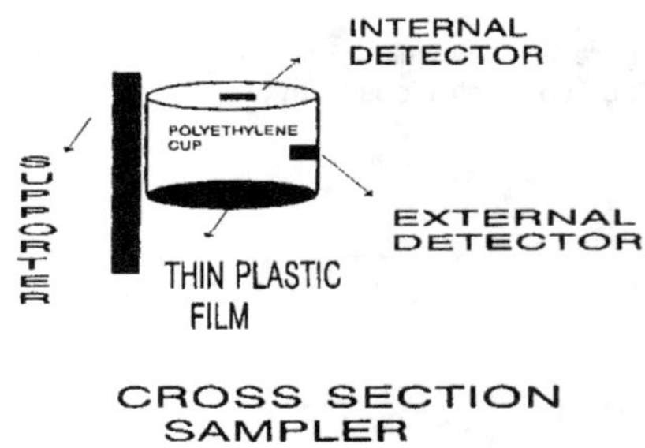

Figure 4 Radon determination in air. The sampler consists of three polyethylene cups placed at 20, 50 and 100 cm on vertical supporter

For radon gas exhalation a monitor device displayed in Figure 5, was buried in the ground. The monitor consists of a PVC tube one meter long which allocates inside a second PVC 30 cm long with a LR115 detector at its upper part, in this way the undesirable radon 220 decays before reaching the detector. Five points were monitored with both kind of devices: one point was at the centre of the well and other four points were chosen 50 m from the well towards each of the cardinals points.

After 20 days of exposure, detectors were picked up for their evaluation. A known chemical solution was used (10% NaOH at 56 °C, during 90 min) to etch the tracks; α-track counting was performed using a spark counter device and radon concentration was calculated with an appropriate calibration factor[5].

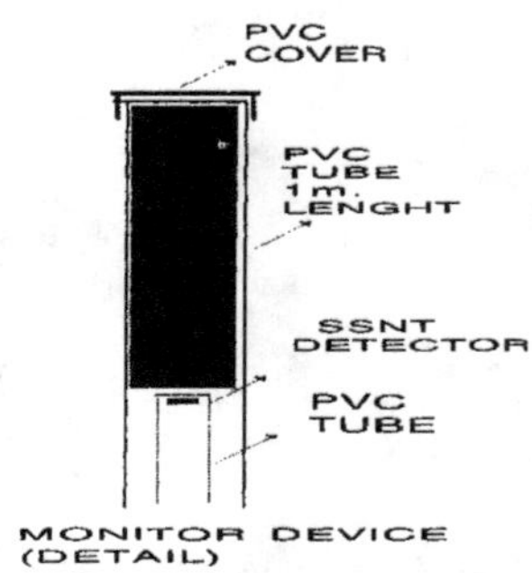

Figure 5 Device for radon evaluation in soil.

3. Results and Discussion

Radon concentrations for liquid samples of both wells are tabulated in table I. In Table II, a mean radon concentration as a function of height, for all the monitored points are presented. The values obtained for radon gas exhalation are presented in Table III.

Radon in fluids. Radon specific activities obtained in fluids, have relative low values compared to that reported for hot water wells[5], and for acquifer water and drinking water. Degasification of springs provoke a decrease in radon concentration within the fluids.

Underground radon. The results obtained in this work, are consistent with values reported in other experiences in geothermal field[1], where high radon concentrations are commonly found, due to active faults and to internal high temperatures.

Table I.- Radon specific activity in liquid samples.

WELL -SAMPLES	SPECIFIC ACTIVITY (Bq/m^3) ± 30%
PR9-I	3,357
PR9-II	3,500
PR11-I	940
PR11-II	3,357
PR11-III	3,285

Table II.- Environmental radon

HEIGHT (cm)	MEAN CONCENTRATION (Bq/m^3) ± 50%
20	80
50	80
100	38*

* At the centre of the well the mean concentration is about 128 Bq/m^3.± 40%

Environmental radon. Radon concentrations in air, obtained in this work are around environmental values reported in other part of the world. Radon levels are under the values recommended by the Regulatory Agencies as investigation levels in houses.

4. Conclusions

Geothermal fluids is one of the natural contributors to the atmospheric radon, among the oceans, natural gas, volcanic gases, caves, mines, and coal combustion; the main contributor to atmospheric radon concentration is the terrestrial surface.

Radon concentrations measurements obtained in this work present:

a) levels in fluids are in-between of those reported for geothermal fields.

b) The values obtained for environmental radon are under the investigation leve s or levels recommended by Regulatory Agency (IAEA)[6].

Table III.- Underground radon emanation around the wells.

WELL POSITION	CONCENTRATION (Bq/m^3) ± 20%
PR9 NORTH	20,480
PR9 SOUTH	3,540
PR9 EAST	9,484
PR9 WEST	8,793
PR9 CENTRE	-
PR11 NORTH	-
PR11 SOUTH	80,000
PR11 EAST	10,516
PR11 WEST	73,472
PR11 CENTRE	80,000

5. References

[1] Balcázar, M., Radon and geothermal energy production, Radon Measurements by Etched Track Detectors, World Scientific, 345-362, 1997.

[2] Layton, D. W., Anspavgh, L. R., and O´Banion, K. D., Health and environmental effects document on geothermal energy, 1981.

[3] Barbier, E.,Nature and Technology of Geothermal Energy: A Review Renewable & Sustainable Energy Reviews, 1(1997) 1-69.

[4] Balcázar, M., Chávez, A., Tavera, L., and Camacho, M. E., Anomalous radon in a research reactor, Nucl. Tracks Radiat. Meas. (in press).

[5] Balcázar, M., Santoyo, E., and González, D., Radon measurements in heat-producing geothermal wells, Nucl. Tracks Radiat. Meas. part D V19(1991) 283-287.

[6] Basic safety standards for protection against ionising radiation and for the safety of radiation sources, Safety Series No. 115, IAEA, 1996.

Miners annual exposure and the outlet air

Dariusz Kluszczyński, Tomisław Domański

Institute of Occupational Medicine, ul. Św. Teresy 8, 90-950 Łódź, Poland

Abstract

The air in all underground mines is contaminated by radioactive gas radon. The presence of it is the potential danger for the health of workers. The measure of this danger is the annual exposure of miners. It may be evaluated using several techniques, but all of them base on measurements carried out in the work-places. The measurements lead in this way, form the problems in choosing the optimal measurements' strategy. It is known that mines are large and dynamic (in the physical sense) objects. The air contamination change both in time and places as well and the changes are in the order of magnitude frequently. Authors were searching for a new method which can average the air contamination's level and which is simple in use. The proposed method bases on measurements carried out in outlet air stream in mines. These streams are sum of many local streams, so they carry information on average radiation in the air on work-places. The main idea of the method is to correlate parameters of the contamination's distribution measured in the outlet air streams with the appropriate parameters of contamination's distribution measured in the work-places. It was proved that the geometric mean obtained from the measurements in the outlet air can be correlated with the geometric mean and the geometric standard deviation of the measurements' distribution obtained in the work-places. The determination coefficient is 0.87 and 0.59 respectively. The idea of the method may be useful in all situations, where the air contamination frequently vary and is no possibility to take continuous measurements in all work-places.

1. Introduction

The number of underground miners in Poland amounts to 300,000. They work under conditions of varying exposure to technologically-enhanced natural radiation[1]. According to the legally-binding regulations [2] such working conditions make the employer responsible for assessing miners' radiation exposure and maintaining a relevant monitoring system. Previous reports underline that owing to their characteristics, radiation measurements in the mine are difficult to carry out and supervise. Thus there is an urgent need for developing such a measuring strategy that would make it possible to estimate miners' exposure and to verify the estimates. The strategy would also have to be adjusted to the actual conditions of a given mine and to the capacities of the measuring unit. The strategy in question could be based on environmental measurements performed at specific points of the mine ventilation network rather than at particular workposts. The locations for the measurements could be all the available sites in the major outlet air streams of the mine. The air flow at these points is the total of all the air flows within a specific ventilation area. Consequently, the mean radon daughters concentration there can be expressed as a function of the mean radon daughters concentration at the workposts. In the diagram of the ventilation system (Fig. 1) the main outlet air streams have been marked with P. For technical reasons it may sometimes be impossible to make the measurement in the every ventilation channel. The air flow in such channel may amount to 300-500 m^3/s. The air from one ventilation area may also be let out through a number of channels. In this case the measurements should be carried out at the sites (points P_1) where the channels with the contaminated air meet the ventilation channel. Fig 1. displays also the exploration regions (R) where increased radon emanation (R_1, R_2, R_n) can be expected as well as the fresh air area (I) and contaminated air stream area (O).

2. Methodology

In order to determine the procedure for a group risk assessment of miners radiation exposure two simultaneous measuring strategies, the study and the reference one, were applied. The reference strategy consisted in providing a representative group of miners with passive dosimeters. The strategy under study was the one referring to measurements carried out in the main outlet air streams. In both cases OC-1 passive dosimeters with Kodak LR-115 track-etched detectors were used [3,4,5,6,7,8,9]. A single measurement lasted one month. The measurements made it possible to calculate the mean monthly concentration of the radiation potential energy of radon daughters.

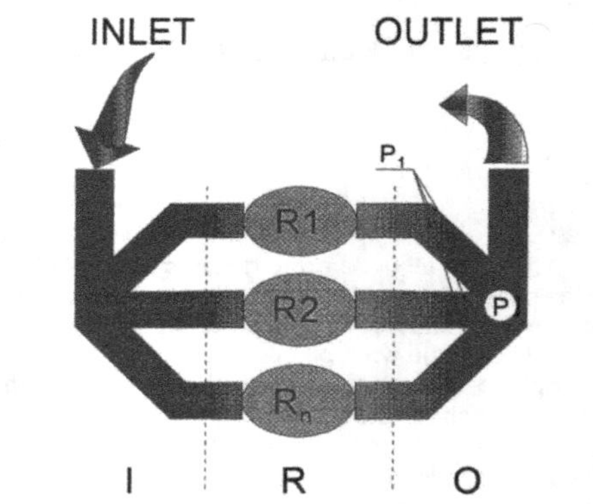

Figure 1 The simplification of the ventilation network (see text)

The work on the present project was undertaken among others with a view to developing a measuring strategy that would be applicable in any mine in Poland regardless of its type. Accordingly, seven mines of different type were selected for the study. These were 2 chemical raw materials mines, 4 metal ore mines and 1 coal mine. The mines varied also with respect to the mileage of headings, number of staff and potential radiation hazard. Totally, 7744 measurements of the mean monthly radon daughters concentration have been carried out. In the "Piast" coalmine, employing about 7000 underground miners, 408 measurements altogether were performed at the workposts and in the outlet air stream. Owing to technical obstacles the measurements could not be limited only to the main outlet air streams. Consequently, each measuring cycle required 7 measurements of the local outlet air streams which were necessary to cover the area with the contaminated air transported through ventilation network from the mine to the surface. As for the metal ores the studies were carried out in two copper ore mines ("Rudna" and "Lubin") and in two lead and zinc ore mines ("Pomorzany", "Trzebionka"). The mines employ approx. 12,000 underground miners altogether. From three to six measurements were carried out during a single measuring cycle. Only in the "Trzebionka" mine the number of measurements in the outlet air streams corresponded to the number of outlet shafts. In the others the number of performed environmental measurements was higher than that of the outlet shafts. For the chemical raw materials the studies were performed in the "Boguszów" and "Wiry" mines. Both the mines are rather small with respect to the heading mileage and number of workers (100 and 20 for "Boguszów" and "Wiry", respectively). Although each of the mines had just one outlet shaft we could not, for technical reasons, set an only one measuring point in the main outlet air stream.

3. Data analysis

Our study was to investigate the parameters and correlations between the results of the measurements carried out in the outlet air streams and those referring to the worksite measurements. The searching for the above was based on an assumption that the distribution of results of measurements in the outlet air streams may yield one parameter which will help determine at least two parameters of the distribution of radon daughters concentration at the workposts. Such an assumption is well-grounded as:

1) the distribution of radon daughters concentrations at workposts is a log-normal one,
2) to determine a log-normal distribution two parameters are necessary,
3) there is a relationship between the mean concentration at the workposts and result dispersion for these workposts.

The density function for the log-normal distribution can be expressed using the following formula:

$$f(y) = \begin{cases} \dfrac{1}{y\sigma\sqrt{2\pi}} e^{-\frac{1}{2}\left(\frac{\ln y - \mu}{\sigma}\right)^2} & y>0 \\ 0 & y \le 0 \end{cases} \tag{1}$$

The function depends on two parameters "μ" and "σ" defined as:

$$\mu = \frac{\sum_{i=1}^{n} y_i}{n} \qquad \sigma = \sqrt{\frac{\sum_{i=1}^{n} (y_i - \mu)^2}{n-1}} \tag{2}$$

The measurements revealed that these values are related. The correlation coefficient was found to be 0.88. In the log-normal distribution the mean arithmetic value ($\bar{y}$) and the standard deviation (SD_y) can be described using the following formulas:

$$\begin{aligned} \bar{y} &= e^{\mu + \frac{\sigma^2}{2}} \\ SD_y &= e^{2\mu + \sigma^2}\left(e^{\sigma^2} - 1\right) \end{aligned} \tag{3}$$

Thus the correlation between the arithmetic mean for the distribution of workpost concentration and the standard deviation indicates the existence of the correlation between μ and σ parameters of the density function. This finding makes us assume that only one of these parameters is an independent variable which is correlated with the parameter of the distribution of measurement results for the outlet air stream. Considering the simplicity of application two parameters of the result distribution were investigated: the arithmetic mean (AM) and the median (Me). The arithmetic mean can be calculated from the following formula:

$$AM = \frac{\sum_{i=1}^{n} C_i}{n} \tag{4}$$

where C_i is the I-th result of the concentration measurements in the outlet air stream, and n stands for the number of measurements performed. In the log-normal distribution the median value can be expressed as:

$$Me = e^{\mu} \tag{5}$$

where

$$\mu = \frac{\sum_{i=1}^{n} \ln C_i}{n} \tag{6}$$

All the remaining notations followed those from the formula (4). Considering (5) and (6) we have obtained:

$$Me = \left(\prod_{i=1}^{n} C_i\right)^{\frac{1}{n}} \tag{7}$$

The application of (7) has some limitations when the measurements reveal the zero value, e.g. when the meter's sensitivity is not high enough. In this case, for formula (7) to be applicable the zero value should be excluded or replaced with an arbitrary value, or eventually, the third parameter can be introduced into the density function. In this study the median was determined according to its definition. In order to determine the median the results should be ordered in a nondecreasing series and the middle result should be chosen.

In order to select proper parameter of the result distribution of measurements in the outlet air streams, a number of correlations were investigated. The determination coefficient was the main criterion for selecting the parameter under study The results of the analyses have been presented in Table I. As evidenced by the data from the Table the equations with arithmetic mean reveal weaker

Table I The correlation between the distribution parameters

parameter		function shape	determination coefficient R
outlet air (x)	workposts (y)		
arithmetic mean	arithmetic mean	$y = a x^b$	0.58
arithmetic mean	median	$y = a \ln(x) + b$	0.88
median	arithmetic mean	$y = a \ln(x) + b$	0.71
median	median	$y = a \ln(x) + b$	0.89

correlations (in term of the determination coefficient) then those with the median values. In the asymmetric distribution the high values of the measurement results affect the arithmetic mean to a higher extend than do the median. As the result the primary parameter in the measuring strategy we have developed was the median value from the result distribution of measurements in the outlet air streams (Me_s). A correlation was then examined between Me_s and the median determined for the result distribution for the workpost measurements (Me_h). An additional parameter that can be calculated from the measurement results is the geometric standard deviation (GSD). All the correlations discussed above have been presented in the diagram in Figure 5. The diagram also displays the 95% confidence limit for the regression curve and the 95% confidence limits for the median and GSD, respectively. The correlations between Me_s,Me_h and GSD variables can be expressed using the following functions:

$$Me_h = 0.94 \cdot \ln(Me_s) + 0.94 \tag{8}$$

$$GSD = 0.86 \cdot (Me_s)^{-0.44} + 2.23 \tag{9}$$

where Me_s, Me_h and GSD have been expressed as [μJm^{-3}].
The correlation coefficient for both (8) and (9) was found to be 0.89 and 0.59, respectively. An interesting characteristic of the latter correlation is that an increase in radon daughters concentration in the outlet air streams leads to a decreasing dispersion of results of the measurements performed at the workposts, down to the asymptotic value of 2.23 (formula (9)).

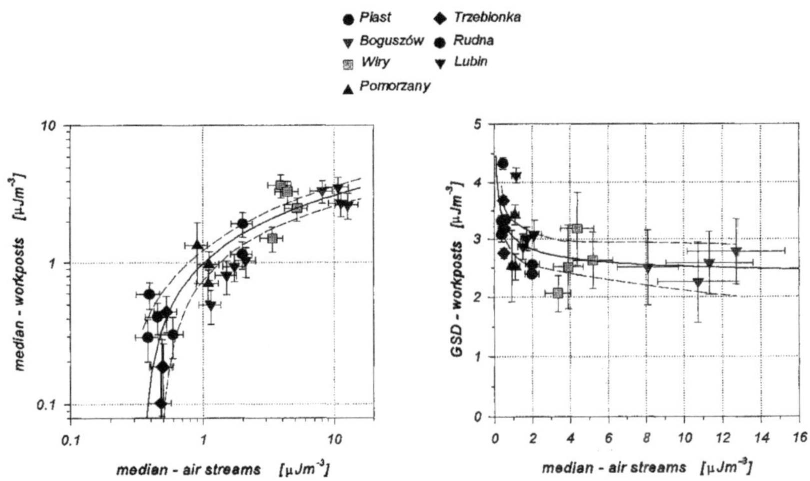

Figure 2 The correlation between the results of measurements in the outlet air streams and the workposts carried out from 1991 to 1994.

Figure 3 presents the correlation between the arithmetic mean values obtained from the workpost measurements and those carried out in the outlet air streams, calculated using formulas defined in log-normal distribution. The correlation coefficient R=0.91 indicates a high consistence of the empirical data with the model we have designed.

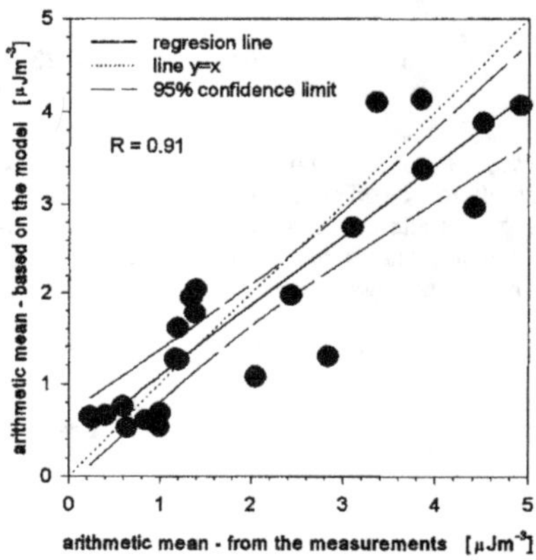

Figure 3 The correlation between the mean potential energy concentration of radon daughters determined from the workpost measurements and the respective concentration calculated from the outlet air stream measurements

4. Conclusions

The strategy developed in this study allows group risk assessment of miners exposed to radon daughters. The strategy employs measurements of radon daughters concentration in the main outlet air streams. The assessment is based on two calculation procedures:

1) determining the distribution of radon daughters concentration at the workposts on the grounds of the results of measurements carried out in the outlet air streams
2) converting the distribution of radon daughters concentration to miners' annual exposure. The procedure has been reported in Kluszczyński [10].

The use of this strategy limits the number of the necessary measurements. In the case of passive dosimetry the number of the measurements can be reduced by approximately ten times.

5. References

[1] Radiation protection of workers in mines. ICRP 47, 1985

[2] Polish atomic law. 1987 (in Polish)

[3] Kodak cellulose nitrate materials for radiation recording applications. Eastman Kodak Co., 1986.

[4] Chruścielewski W., Orzechowski W., Domański T., Świątnicki G.: Measurement of exposure to radon and its progeny using Kodak LR-115 Type II foil; II. Calibration of the detector. Proc Radon Spec Meeting, NEA, 3-8.03.80, Roma, Italy, RD Press Univ Utah, 30-38, 1982.

[5] Chruścielewski W., Domański T.: System of personal dosimetry introduced in Polish nonuranium mines. Proc Internat Conf Rad Hazard Mining, US Dept Labor, 4-9.10.81, Golden, USA, 589-596, 1982.

[6] Domański T., Chruścielewski W., Hofman M.: Monitoring of exposure to radon decay products in the air of mines using 'passive' track detectors. Health Phys, 40, 211-217, 1981.

[7] Żórawski A., Hawryński M., Kluszczyński D.: The computer system of automatical microscope analysis of miners' individual dosimeters. Rad Prot Practice, Sydney, 10-17.04.88, vol I, 396-397, 1988.

[8] Żórawski A., Czerski B.: Method of calibration and analysis of track detectors. Rad Prot Practice, 10-17.04.88, Sydney, 398-399, 1988.

[9] Domański T., Chruścielewski W., Kluszczyński D., Olszewski J.: Polish radon daughters dosimeters in international intercomparison. Nukleonika, 38(4), 19-26, 1993.

[10] Kluszczyński D.: Transformation of field monitoring results into miners annual exposures to radon progeny. Pol J Occup Med Envir Health, 5(1), 77-86, 1992.

Differential radon measurements for outburst prediction in coal mines

Wlodek Zahorowski and Stewart Whittlestone

Australian Nuclear Science and Technology Organisation (ANSTO)
PMB1 Menai, NSW, 2234, Australia

Abstract

In some areas, coal mines are subject to occasional outbursts consisting mainly of carbon dioxide or methane and coal dust. These vary widely in severity, but can cause loss of life. There are several techniques which may permit prediction of the outbursts. These include monitoring acoustic signals or continuous measurement of methane, carbon dioxide or radon concentrations. While it is not expected that any one will be adequate on its own, each technique is being studied so that a combination of all can be evaluated. This paper presents an experiment in which the concentration of radon in underground air is measured with high precision upon entry to and leaving a workface. This was considered the most practicable way to detect any radon flux anomalies preceding an outburst in a working mine. There are many practical difficulties in mounting an experiment in an active mine because of the dusty, wet conditions and because of the need for rigorous safety requirements in an environment which may contain an explosive mixture of air and methane. Details of the development of suitable equipment are given, and an analysis made of the first two months of operation.

1. Introduction

Radon-222 has been used as a naturally occurring tracer of some processes in geological formations which are accompanied by a change in the permeability and/or the effective grain size. The best known applications are in earthquake research and hydrology [1,2].

Relatively little has been published on radon monitoring as a method of prediction of strata events such as instantaneous outbursts or rock burst in underground mining operations. Studies conducted by McDonald [3] and Lebecka et al. [4] have demonstrated that before rock bursts and instantaneous outbursts of coal and gas in mines, radon concentration anomalies do occur and can be utilised as a hazard indicator. In the case of instantaneous outbursts the radon method was claimed to be more accurate than the prediction methods currently used, including those based on gas pressure and desorption intensity [4]. These studies, although successful in demonstrating the feasibility of using radon as a prediction tool in mining, have had some basic limitations. First, the techniques employed were labour-intensive because they required specially drilled boreholes to collect radon or prone to adverse effects from fluctuating ventilation and other environmental parameters. Second, the monitoring was based on week long, manually collected samples.

The method presented in this paper addresses the limitations of the previous studies. It allows for radon changes in ventilation air in large areas of the mine (as in [3]) by monitoring radon concentrations changes simultaneously in intake and return air of a current workface face. The resulting signal difference is free from major environmental fluctuations. Moreover, the method incorporates an automated, continuous monitoring system capable of measuring relevant changes of radon flux from the face with a time resolution of less than 1 hour.

2. Methodology

The applied method (Fig.1) relies on monitoring the change of radon flux from a current workface. This is done by measuring concentration in ventilation air which passes along a roadway towards the face (intake air) and then back to the main headings (return air). Two radon detectors are positioned at the low-hazard site of the stopping in a cross-cut, typically a few hundred meters away from the face. The first detector samples the intake air (with the radon concentration C_1) and the second, through a sampling header mounted in the stopping samples return air radon concentration $C_2 = C_1 + C$. Here, C is the radon concentration change caused by the contact of the ventilation air with radon exhaling surfaces which occurs during the passage of air between the two sampling points. By subtracting C_1 from C_2, the detection system directly monitors the change in radon flux which occurs in the monitored area and thus any change due to an impending outburst.

A major question in the planning phase of the project was to decide what detector sensitivity was required. Since no preliminary measurements of radon concentration in underground air had been made it was assumed that the average concentration will be slightly above the environmental levels, up to about 20 Bq m^{-3}. In recent years ANSTO has developed and deployed very high sensitivity radon detectors with the lower detection limit down to few tens of mBq m^{-3} [5]. However, the higher the desired sensitivity the bigger the size of the detector's delay chamber (Section 3) and hence a compromise had to be made between a practicable instrument and the best possible sensitivity.

An estimate of the required sensitivity was made as follows. If J is the radon exhalation rate (flux) from the surface of a roadway, f is the circumference of the roadway's cross-section and q the ventilation air flow rate, then the increase of radon concentration in the air passing through a section of the roadway of length x is equal to $C(x)= J\ f\ x\ /\ u$. The formula does not take into account a small radon concentration decrease caused by radioactive decay. Assuming that the change in radon flux will be tenfold (from an average of 15 mBq $m^{-2}s^{-1}$) and limited to an area very closed to the workface (x=15m and f=15m) the resulting radon concentration increase gives the value of about 1 Bq m^{-3}. Transport conditions are known to cause elevated radon exhalation rates. For example, the authors have measured a radon exhalation rate 100 times higher than normal coming from a dragline spoil at an open cut mine. The elevated exhalation rate was limited to an area with a radius of several metres only and was due to a local change in radon-to-surface transport conditions.

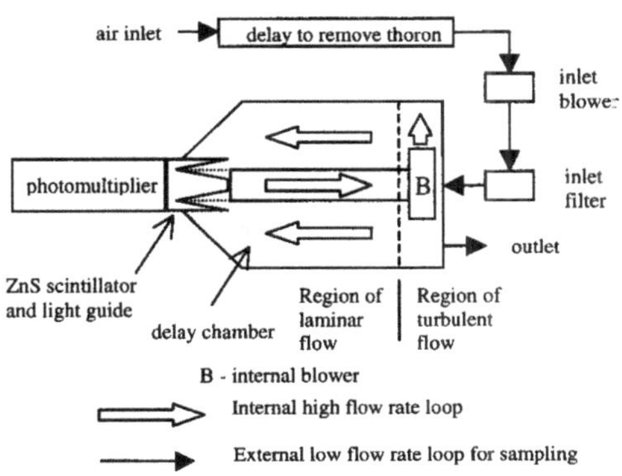

Fig. 1. Schematic diagram of dual flow loop two filter radon detector

Based on the above estimate a detector with the sensitivity of about 1000 counts h^{-1} per Bq m^{-3} (with a corresponding volume of about 750L) was chosen. Consequently, the pair of radon detectors constructed for the project could resolve radon concentration changes larger than 0.5 Bq m^{-3} in an ambient level of 10 Bq m^{-3}. This figure is based on a one hour counting period. The resolution reflects the ability of the system to detect the changes derived from a signal which was obtained by subtracting two relatively large concentrations recorded by the individual detectors.

3. Experiment

The radon detectors used in the experiment are examples of the so called two filter detector where the analysed air stream is directed through a set of two filters separated by a delay chamber. The first filter removes all radon and thoron decay products ("daughters"), then some new daughters are produced in the chamber. Subsequently, some of the daughters are collected on the second filter. The number of the collected daughters is proportional to the radon and thoron concentrations. The detector type used in this project, called dual loop two filter radon detector, departs from the above principle by using two air flows and a wire screen as the second filter [5]. The inlet loop takes ambient air at a low flow rate through a small delay volume to remove thoron and into the delay chamber (Fig. 1). The flow rate is set to obtain an air exchange time of less than 30 minutes, so that the overall time resolution of the detector is 45 minutes (defined as a time to reach 50% of maximum count rate after step increase in radon concentration at the inlet). The internal loop uses a high flow rate to maximise the collection of the unattached daughters on the wire screen. To avoid plateout and loss by decay of Po-218 the internal loop should circulate the air in less than 1 minute. The use of wire screen ensures that the internal flow impedance is low but requires counting alpha particles from both faces of the screen. This is achieved using a detector-screen assembly (Fig. 1) where the screen is inserted into a specially shaped light guide which has the surfaces facing the screen lined with a ZnS scintillator foil. The light guide is attached to a photomultiplier tube (PM). A diffuser separates the region of turbulent flow from the region of almost laminar flow. A detailed discussion of the dual lop two filter detector can be found elsewhere [5].

The detectors constructed for the underground measurements had to meet some stringent safety requirements. Construction materials were restricted to stainless steel, brass, fire resistant antistatic (fras) tubing and gaskets and polycarbonate for the light guide. The return air detector was fitted with a methane detector to shut down the unit whenever methane concentration in the sampled air exceeded a preset value.

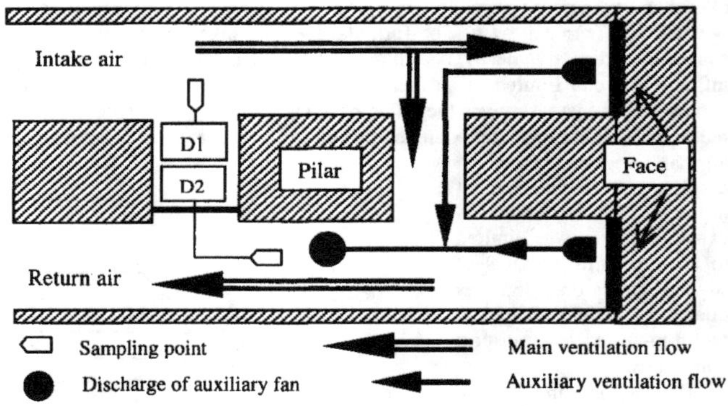

Fig. 2. Two radon detectors are positioned at the low-hazard site of the stopping in a cross-cut, typically a few hundred metres away from the face. The detectors sample the intake and return air. A sampling header mounted in the stopping reaches a point about 20 m down flow from the discharge point of an auxiliary fan.

The detectors' delay chamber volume was 730L, flow rates were set to 40L/min and 800 L/min for the low and high flow rate loops and the thoron delay volume was 200L.

Calibration of the detectors was performed by injecting radon from a dry radon source (model RN-25 by Pylon Electronics Corp. Canada) into the inlet flow. The nominal radon output was used in the calculations. The resulting sensitivity was about 1000 counts h^{-1} per Bq m^{-3}. No attempt was made to evaluate the instrumental background. Similar detectors, with a slightly lower sensitivity, at about 700 counts h^{-1} per Bq m^{-3}, had their instrumental background estimated to be equal or less than 40 mBq m^{-3} [6].

Typically, detectors stayed in a given cross-cut for about 10 days, before being moved to a next position in order to follow the advancing workface. During the experiment, the detectors were put in five consecutive cross-cuts. The distance between adjacent cross-cuts was 150 metres. Further in the paper, the detection positions are referred to by number from 1 to 5, with position 5 being the latest. The mine's ventilation system maintained an air flow at 30 $m^3\ s^{-1}$ in the roadways to and from the two workfaces (Fig. 2). There was also an auxiliary fan drawing air from each of the two mined faces at about 20 $m^3\ s^{-1}$. A discharge of the auxiliary fan was located in the return air roadway (Fig. 2). Intake air was sampled close to the low-hazard end of the current cross-cut and return air was sampled at a point 20 metres down flow from the discharge. The relative sampling configuration was the same for each of the five positions of the detectors.

The two detectors were linked in a master-slave configuration. Data, including detector counts, detector flow indicators, temperature and PM tube biases were sampled every 15 minutes and stored in a logger. The logger was connected to a computer located in an above ground control room using a standard serial port and a data line supported by a pair of line drivers. The computer uploaded the data at the same frequency as they were recorded by the logger to prevent data loss in case of power cut in the investigated area. The other major function of the computer was to query the status of the logger to resume the monitoring as soon as the power was restored to the area.

4. Preliminary results and discussion

The investigated workings were known to have experienced no outbursts in the past. None occurred during the monitoring period, either. Nevertheless, the acquired data offer some first estimates of the performance of the instrumentation, the employed methodology and the magnitude of signals involved.
The trial period lasted 53 days. The data recovery was equal to 40%, 35%, 26%, 31% and 20% in detection position 1 to 5, respectively. In absolute terms, these numbers corresponded to 111, 108, 68, 49 and 55 hours

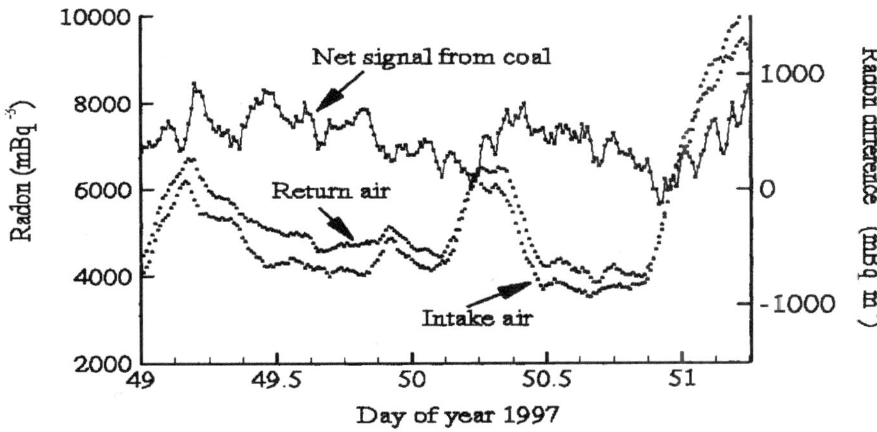

Fig. 3. An example of radon concentrations measured in intake and return air. The signal from the coal face is the difference between the two.

of monitoring. Such a low data recovery was mainly due to frequent breaks in the mining activity in the investigated area. Most of the recorded data are grouped in several 2-3 day long periods of mostly uninterrupted measurements. There are also few shorter uninterrupted periods, each representing no less than 1 shift (8 hours).

A sample of a 2 day long period of 15 minutes data smoothed by the 4-point moving average is shown in Fig. 3. The net radon signal is determined with the specified precision and time resolution. The sensitivity is considered adequate even for the gateroad development where the workface area is relatively small (about 15 m^2). The detectors' dynamic range is adequate to accommodate diurnal changes in radon concentration, evident in the intake and return air signals, much bigger than the net signal from the coal face. The instrumental performance has validated the choice of construction materials required for underground environment. Fig.4 shows minima, means and maxima of the net signal distributions in the five detection positions.

It is perhaps necessary to point to some independent evidence that the recorded net signal corresponds to a real effect. For instance, one can expect that the radon intake concentration should exhibit some dependence on the detection position. Such a dependence can be demonstrated by analysing the minima of the radon intake concentrations corresponding to each of the five detection positions. Figure 4 shows these minima. It is clear that each time the intake air sampling point was moved by another 150 metres down the intake air roadway there was a shift in the minimum of the intake air radon distribution. The rising minimum value can be attributed to an extra radon flux from an additional 150 metres long section of the roadway. The increase of the minimum, resulted from the four consecutive relocations of the detectors, is equal to 660, 670, 1260, 880 mBq m^{-3}. The average value of the increase corresponds to a radon flux equal to about 10 mBq m^{-2} s^{-1} which is probably a good estimate of the true value.

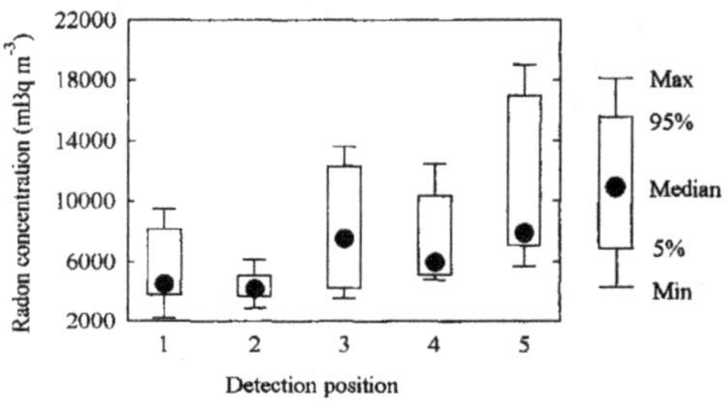

Fig. 4. Distribution of radon concentration in intake air in function of detection position.

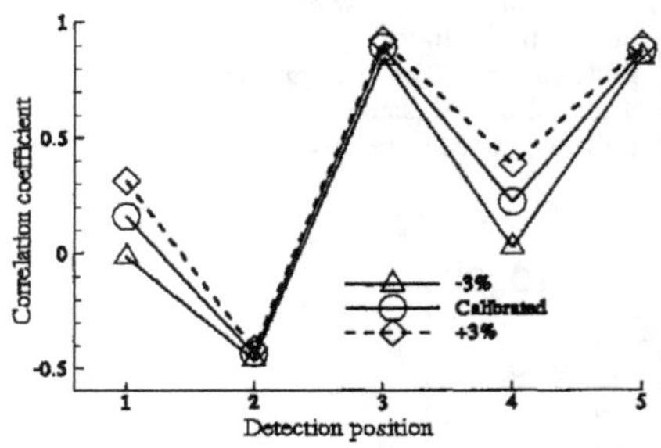

Fig. 5. Correlation between the intake air radon concentration and the net signal.

One can also expect that the net radon concentration should be largely independent of the intake air radon concentration. This is illustrated in Fig. 3 where the net signal is relatively independent of a significant increase in the intake air radon concentration. A detailed examination of the data set returns a more complex answer to this proposition. Fig. 5 shows correlation coefficients between the intake air radon concentration and the net radon signal for all detection positions (points connected by a solid line). In some positions (3 and 5), there is a relatively strong correlation, in others (1, 2, and 4) there is no correlation at all. Qualitatively, the above result is not sensitive to changes in the relative calibration of the two detectors. This is illustrated by the dotted and dashed curves in Fig. 5 which indicate correlation coefficients corresponding to the relative calibration varied by ±3% from the one found experimentally. Large variation of the correlation coefficient is attributed to changes in mine ventilation rate which has not been recorded.

The data enabled the authors to estimate the approximate value of an increase in the net radon concentration during the workface advancement. The resulting figure, expressed as an increase per unit length of the path between the intake and return air sampling points, vary from about 4 to 20 mBq m^{-3} per metre. This is a preliminary figure, since only few brief periods were available where concentrations were recorded during the face advancement. The lower figure corresponds to a radon flux of about 10 mBq m^{-2} s^{-1}, consistently with the flux estimate based on the shifting minimum of the distributions. The higher figure suggests that the upper estimate for a contribution from the workface is about 20 mBq m^{-3} per metre of the face. This results from a radon flux of about 50 mBq m^{-2} s^{-1}.

No correlation between the measured radon concentration and surface pressure has been found.

5. Conclusions

A new method for prediction of instantaneous outburst in underground coal mines has been described. It is characterised by a non-disruptive, continuous and fully automated analysis of intake and return air radon concentrations in the monitored workings with time resolution of less than one hour. As such it is most suitable for continuous mining operation. The method does not require the detectors being positioned in the high risk area close to a work face. The instrumentation is easily redeployable and with minimal maintenance requirements.

The results of a preliminary underground monitoring have validated the authors' assumptions made in the planning phase of the project concerning expected absolute and incremental radon levels in underground air and hence the choice of the size of the radon detector. The sensitivity proved to be adequate even for the gateroad development where the workface area is relatively small. The detectors' dynamic range fully accommodates environmental changes in radon concentration and the instrumental performance has validated the choice of construction materials required for underground environment.

The preliminary results seem to be consistent. They lead to some plausible estimates of radon fluxes from roadway walls. Based on these estimates, an approximate value of the radon flux from the workface was found to be equal to 50 mBq m^{-2} s^{-1}.

It is clear that the method can be applied to prediction of rock failure leading to methane emission. Such an application shows a good measure of an early success bearing in mind the frequency of methane emissions in longwall mining and much bigger face area than in gateroad developments which will lead to correspondingly higher net radon signals.

It is also clear that much longer monitoring periods than the present one will be required to confirm the reported observations and, above all, to record changes in the net radon concentration attributable to outbursts.

Acknowledgments. The authors wish to thank Mr. R. Phillips and Mr. G. Relf of Bellambie Collieries Pty. Ltd. for their support during the underground experiment. Mr. B. Stenhouse of ANSTO built the detectors with the assistance of Mr. M. Hyde. The project was partly financed by ACARP under Grant C5035.

6. References

[1] Isotopic and geochemical precursors of earthquakes and volcanic eruptions, Proceedings of an Advisory Group Meeting held in Vienna, 9-12 September 1991, IAEA-TECDOC-726, 1993.

[2] Selected Problems in Radon Measurements for Earthquake Prediction, Proceedings of the 2nd Workshop on Radon Monitoring in Radioprotection, Environment and/or Earth Science, Trieste (Italy), 25 Nov.-6 Dec., 1991, World Scientific 1993, Singapore, Furlan, G., and L. Tommasino (Eds).

[3] McDonald, A. J., Radon gas emissions and seismicity in deep-level gold mines. Proceedings of the 1st International Congress on Rockbursts and Seismicity in Mines, Johannesburg, 1992, SAIMM, Johannesburg, 1994, Gay, N. C., and E. H. Wainwright (Eds)

[4] Lebecka, J. M., Wysocka M., and S. Chalupnik, Radon monitoring for the prediction of mine outbursts. Proceedings of the 2nd Workshop on Radon Monitoring in Radioprotection, Environment and/or Earth Science, Trieste (Italy), 25 Nov.-6 Dec., 1991, World Scientific 1993, Singapore, Furlan, G., and L. Tommasino (Eds).

[5] Whittlestone, S., W. Zahorowski, and P. Wasiolek, High sensitivity two filter radon/thoron detectors with a wire or nylon screen as the second filter. ANSTO E- report, ANSTO /E718, December 1994.

[6] Whittlestone, S., and W. Zahorowski, Baseline radon detectors for ship-board use: Development and deployment in ACE-1, to be published in Journal of Geophysical Research, ACE-1 Special Issue, 1998.

ROLE OF ENVIRONMENTAL VARIABLES ON RADON CONCENTRATION IN SOIL

H. CLIMENT, M. BAKALOWICZ, M. MONNIN

Géofluides, Bassins, Eau. UMR 5569 CNRS- UM II, cc 57
2, Place E. Bataillon, F-34095 Montpellier Cedex 5, France

ABSTRACT

In the frame of an European project, radon concentrations in soil and measurements of environmental variables such as the nature of the soil or climatic variables were monitored. The data have been analysed by time-series analysis methods, i.e. Correlation and Spectrum Analysis, to point out relations between radon concentrations and some environmental variables. This approach is a compromise between direct observation and modelling. The observation of the rough time series is unable to point out the relation between radon concentrations and an environmental variable because of the overlapping of the influences of several variables, and the time delay induced by the medium. The Cross Spectrum function between the time series of radon and of an environmental variable describes the nature of the relation and gives the response time in the case of a cause to effect relation. It requires the only hypothesis that the environmental variable is the input function and radon concentration the output function. This analysis is an important preliminary study for modelling. By that way the importance of soil nature has been pointed out. The internal variables of the medium (permeability, porosity) appear to restrain the influence of the environmental variables such as humidity, temperature or atmospheric pressure.

KEYWORDS

Radon, Transport, Environmental Variables, Correlation and Spectrum Analysis, Soil.

INTRODUCTION

In the frame of the European project "Criteria for indoor radon concentration-an experimental study considering especially the Leipzig-Halle brown coal area" (Jonsson G. et al., 1995; Baixeras et al., 1996) the monitoring of soil radon concentrations has been carried out during 15 months on a karstic site in southern France. The variation of atmospheric pressure, temperature, humidity, wind speed and water table level have been measured simultaneously. These data have been analysed by Correlation and Spectrum Analysis (Mangin, 1984) in the aim to find a relation between radon transport in the soil and the influence of internal and/or environmental parameters (Climent, 1996). The Cross Spectrum function between radon concentration time series and an environmental variable evidences the type of relation and gives the response time in the case of an input-output relation. Moreover, the cross correlation function is the unit impulse response in the case of a random input function.

EXPERIMENTAL FIELD AND MONITORING

The data have been collected in a karstic area in southern France. Radon concentration, atmospheric pressure, temperature, humidity, wind speed and water table level were monitored during 8 months every one hour. The radon concentrations were measured in two different soils S1 and S2 distant of 40 m. S1 is more clayey than S2 which is a well drained out soil. Radon concentrations were measured with Clipperton II electronical radon probes (Morin, 1992, Jonsson et al., 1996). Figure 1 shows the field equipment with radon electronical probes. The climatic parameters were monitored with a weather station type Weather Monitor II (Davis Inc.) and the water table level with an IRIS probe (BRGM, Durand, 1992).

INTERPRETING METHOD

The time series are analysed on the basis of Fourier Transform. The method is commonly used in hydrology to determine unit hydrograph which is the unit response of hydrological systems (Mangin, 1994). In the first step (simple analysis), the Fourier Transform gives the spectrum (in frequency) which characterises radon and environmental variables time-series. The time-series are compared by the autocorrelation function (in time). In the second step (cross analysis), the following is assumed : the environmental variable is considered as the input function, or the cause, while radon is considered as the output function or the effect. The cross correlogramme evidences how environmental variables operate on radon. Figure 2 describes the different type for cross correlogrammes. Figures 2.a and 2.b show a cause to effect relation (where t is the response time), the input determine the output and if the input is a random function the cross correlogramme is the transfer function. On figure 2.c the two functions have a common cause and the cross correlogramme is symmetrical. On figure 2.d the two functions are independent variables.

RESULTS INTERPRETATION

The aim of this study is to point out the importance of the soil nature on the influence of external parameters. Figures 3 to 7 show the cross correlation function between radon time series as the output and respectively, as input, temperature, atmospheric pressure, wind speed, humidity and water table level at sites S1 and S2. The two sites are 40 m distant. Site S1 is located close to an open barn which more or less protects it from the wind. Therefore, S1 and S2 are subjected to the same environmental conditions, except for the wind.

Description of the results

At site S1, radon variations are related to temperature, while radon variations at site S2 are directly led by temperature variations with a complex response of 42 and 60 days (figure 3).

At site S1, radon and pressure variations are independent. At site S2, radon variations are correlated to pressure variations with a complex response time of 7,5 and 58 days (figure 4).

At site S1, radon variations are led by wind speed variations with a response time without delay. At site S2 radon and wind speed variations are independent variables (figure 5).

At site S1, radon variations are not related to humidity variations even though at site S2 radon variations are directly led by humidity variations with a complex response time of 4 and 120 days (figure 6).

At site S1, radon variations are related to the water table level variations while there is no relation between radon variations at site S2 and those of the water table level (figure 7).

Results interpretation

External temperature variations produce a variation of the temperature gradient between the top and the bottom of the external PVC tube (figure 1). The gradient variation generates a gas flow in the tube which causes a pumping effect on soil gases when the soil is not compact enough. This is the case for the soil at site S2 where radon variations are led by temperature variations. At site S1, the soil is more clayey and more compact than at site S2. Therefore the pumping effect does not appear and radon and temperature variations are not directly correlated but are the effect of a common cause which is not yet identified.

In the same way, atmospheric pressure variations create a pumping effect in the case of a non compact soil. When pressure decreases, soil gases are going upward and radon concentrations increase. This effect takes place at site S2 where radon and pressure variations are inversely correlated with a short time effect of 7,5 days. At site S1, the soil is more compact and the pumping effect is not appreciable. Therefore radon and pressure variations at site S1 are independent.

The pumping effect generated by temperature or pressure variations is shown by the cross correlogrammes at site S2 where soil is less compact. Radon variations are led by temperature and pressure variations at site S2.

The wind is sweeping the upper layers of the soil and depletes it of radon. The time for recharging the depleted zone in radon depends on the compacting rate of the soil. If the soil is not compacted, the recharge time is immediate, radon and wind speed variations are independent as shown on figure 3 by the cross correlogramme. If the soil is compacted, an increase of wind speed corresponds to a decrease of radon concentrations. Wind speed and radon variations are inversely correlated at site S1.

When soil and air humidity increase because of the rain, the emanation factor and thus radon concentrations increase. It affects a permeable soil such as at site S2 . The complex response is made up by a short term effect (4 days) due to the increase of the emanation factor and by a longer term effect (120 days) due to rainfall and infiltration. At site S1, the soil is more impermeable, the increase of air humidity as well as rainfall have a very weak influence on soil humidity. Radon variations at both sites and rainfall are independent because it rained very little during the experiments (January to August 1995).

The variations of the water table level produce two effects : a degassing effect and a piston effect. In both cases, the increase of the water table level corresponds to an increase of radon concentrations. At site S1 these effects are noticeable; however, radon and water table variations are not correlated but have a common cause. The piston and/or the degassing effect are dramatically reduced when the soil is not compacted. In this case the volume of the soil involved is more important, the gases are easily homogenised and radon and water table level variations are independent (site S2).

Comparison to literature

Authors as Rudakov (85) find an inverse correlation between radon concentrations and temperature. Others like Stranden et al. (84) point out a correlation. Several explanations have been advanced : the difference between soil and air temperature (Klusman & Jaacks, 87), the dilatation of the soil when air temperature increase should increase radon concentrations. In any case, it seemed difficult to define rules governing radon migration versus temperature.

The results published about the link between pressure and radon concentrations are also contradictory. Clements & Wilkening (74) found a correlation while Ball et al. (83) point out both a positive and a negative correlation between radon and pressure variations. It seemed that other environmental variable influences prevent the observation of the role of the only pressure.

Kraner et al. (64) point out a radon dilution in the upper part of the soil swept by the wind. Singh et al. (88) found also an inverse correlation between radon concentrations and wind speed. But the influences of temperature, pressure or rainfall were more important and could hide the wind speed effect.

Fleischer (87) showed the correlation between radon concentrations and humidity variations. Kraner et al. (64) took into account soil characteristics such as porosity and/or permeability.

The variety of the results and their contradictions seem to be due to the overlapping of the influences of several environmental variables at different time scales. The intrinsic characteristics of the medium are rarely taken into account as well as the response time of radon concentrations to the influence of the variation of an environmental variable.

Correlation and Spectrum analysis allows to determine the type of relation between radon concentration and one environmental variable. It only assumes that radon time series variations is the output function and the environmental variable the input function. In the case of a cause to effect relation, it gives the response time. This study evidences the fact that the nature of the soil restrains the effect of the external variables on radon concentration variations. The response can be complex and delayed. This means that intrinsic parameters of the medium are playing a leading role in radon transport from the soil to the atmosphere.

CONCLUSIONS

The observation of the raw data gives some indications about the influence of a set of climatic variables on radon variations but no information is given about the response time. Modelling, on the other hand, requires many assumptions, generally very difficult to verify. Correlation and Spectrum Analysis are an in between way to analyse the data. The only assumption is to consider an environmental variable time series as the input function and a radon concentrations time series as the output function of the system. We used Correlation and Spectrum Analysis to study radon concentration variations in two different soils S1 and S2 submitted to the same climatic conditions except wind speed. The next table gives the results of Cross Spectra at sites S1 and S2 with radon variations as output and as input respectively : temperature, pressure, humidity, wind speed and water table level (t is the time delay).

	S1	S2	Figure
Temperature	common cause	correlation (t=42 and 60 days)	3
Humidity	independent variables	correlation (t=4 and 20 days)	4
Pressure	independent variables	inverse correlation (t=7,5 and 58 days)	5
Wind Speed	inverse correlation (t≈0)	independent variables	6
Water Table	common cause	independent variables	7

Results of the Cross Spectra

The two sites present physical characteristics of their soil which are very different (S1 : clayey and compact; S2 : coarse and well drained out). Therefore, their different behaviour with respect to gas flow, as shown by radon variations, are assumed to be the consequence of their physical characteristics, e.g. gas permeability controlled by compactness and water content. The Correlation and Spectrum Analysis is an objective method for describing the existence, the type and the rate of relations between radon and environmental variables variations. This statistical analysis method is an efficient method for preparing to modelling.

REFERENCES

C. Baixeras, G.U. Bacmeister, H. Climent, W. Enge, K. Freyer, G. Jönsson, R. Ghose, M.M. Monnin, L. Font, R. Devantier & G. Sciocchetti *"Report on the first phase activity of an EU-project concerning coordinated radon measurements in five European countries."* Accepted for publication in **Environmental International (1996)**.

T. K. Ball, R. A. Nicholson & D. Peachey (1983). "Effects of meteorological variables on certain soil gases used to detect buried ore deposits." Trans. Instit. of Mining and Metallurgy **92**(b): 183-190.

W. E. Clements & M. H. Wilkening (1974). "Atmospheric pressure effects on 222Rn transport across the Earth-air interface." J. Geophys. Res. **79**(33): 5025-5029.

H. Climent, 1996, *"Radon transport at the soil/air interface : study of the influence of external parameters"*, **Thesis**, speciality Physics and Chemistry of the Earth, Montpellier II University.

V. Durand, (1992). *"Structure d'un massif karstique. Relations entre déformations et facteurs hydro-météorologiques"*. **Thesis**, option Hydrogéologie, Université de Montpellier II.

R. L. Fleischer, (1987). "Moisture and ^{222}Rn emanation." Health Phys. **52**(6): 797-799.

G. Jönsson, D. Albarracin, G.U. Bacmeister, H. Climent, R. Devantier, W. Enge, K. Freyer, L. Font, R. Ghose, M.M. Monnin, G. Sciocchetti, J.-L. Seidel & H.C. Treutler *"Comparison of radon measurements done by Solid State Nuclear Tracks Detectors and electronic devices in the frame of an European project"*. Accepted for publication in **Radiation Measurements** (18th ICNTS, EGYPT 1-5 Sept. 1996).

G. Jönsson, D. Albarracin, G.U. Bacmeister, C. Baixeras, H. Climent, G. Cotellessa, R. Devantier, W. Enge, K. Freyer, L. Font, R. Ghose, M.M. Monnin, G. Sciocchetti, J.-L. Seidel & H.C. Treutler *"Continuous measurements of soil radon under regular field conditions"*. Accepted for publication in **Environmental International**.

R. W. Klusman & J. A. Jaacks (1987). "Environmental influence upon mercury, radon and helium concentrations in soil gases at a site near Denver, Colorado." J. Geochem. Explor. **27**: 259-280.

H. W. Kraner, G. L. Schroeder & R. D. Evans (1964). "Measurements of the effects of atmospheric variables on radon-222 flux and soil gas concentrations." Nat. Radiat. Environ. **I**: 191-215.

A.Mangin, (1984). "Pour une meilleure connaissance des systèmes hydrologiques à partir des analyses corrélatoire et spectrale." J. Hydrol. **67**: 25-43.

J. P. Morin, (1992). *"Le radon traceur du transport des fluides dans le sol : simulation, modélisation et applications géophysiques."* **Thesis**, spécialité Physique Nucléaire, Université Blaise Pascal, Clermont Ferrand.

V. P. Rudakov (1985). "Nature of the seasonal variations in subsoil radon." Geokhimiya **7**: 133-135.

M., R. Singh, C. Ramola, N. P. Singh, S. Singh & H. S. Virk (1988). "The influence of meteorological parameters on soil gas radon." J. Assoc. Expl. Geophys. **9**: 85-90.

E. A. Stranden, K. Kolstad & B. Lind (1984). "The influence of moisture and temperature on radon exhalaison." Radiat. Protect. Dosim. **12**(1): 33-38.

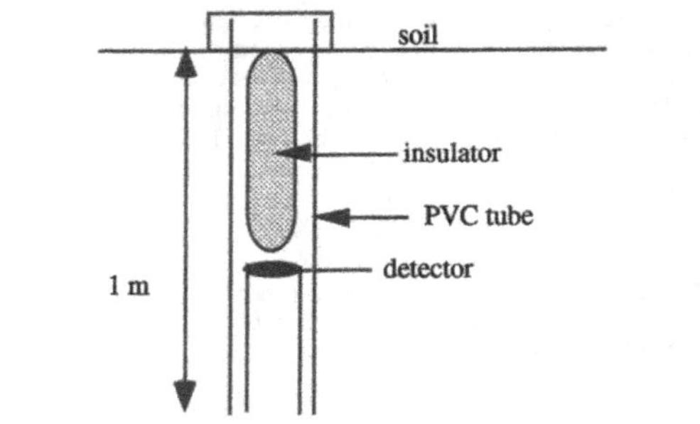

Figure 1 : Field equipping for monitoring radon in the soil

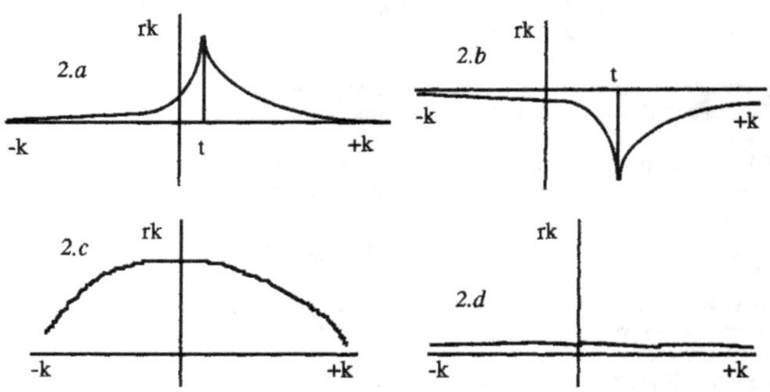

Figure 2 : Different Types of Cross Correlogramme function

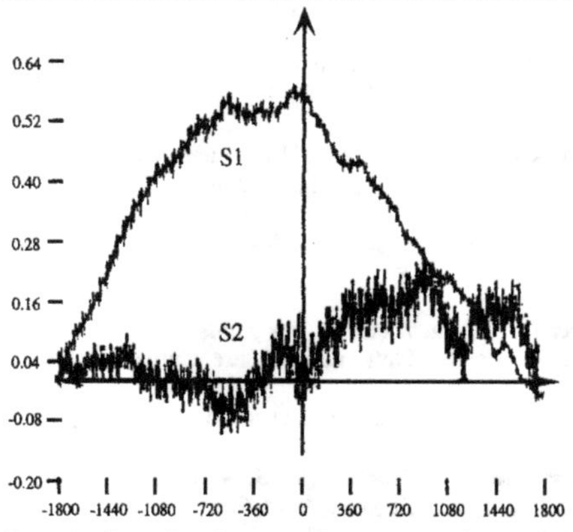

Figure 3 : Cross Correlogramme function for Radon (output) and temperature (input) time series

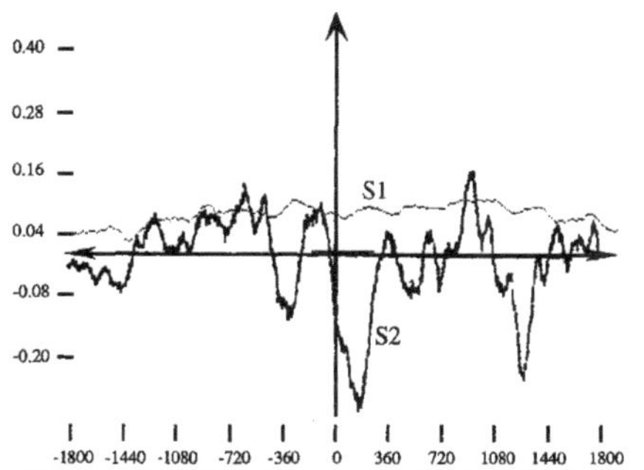

Figure 4 : Cross Correlogramme function for Radon (output) and Pressure (input) time series

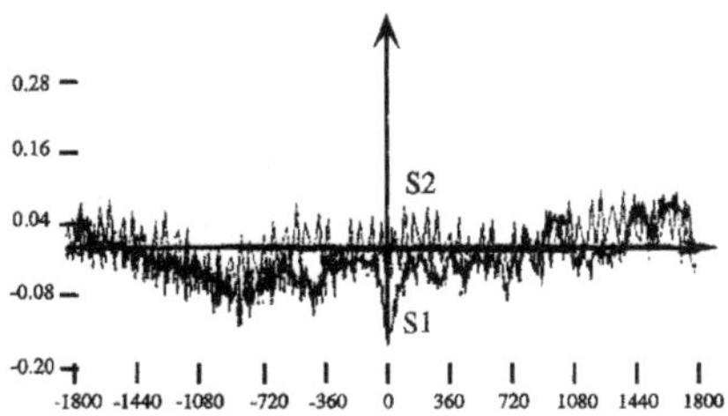

Figure 5 : Cross Correlogramme function for Radon (output) and Wind Speed (input) time series

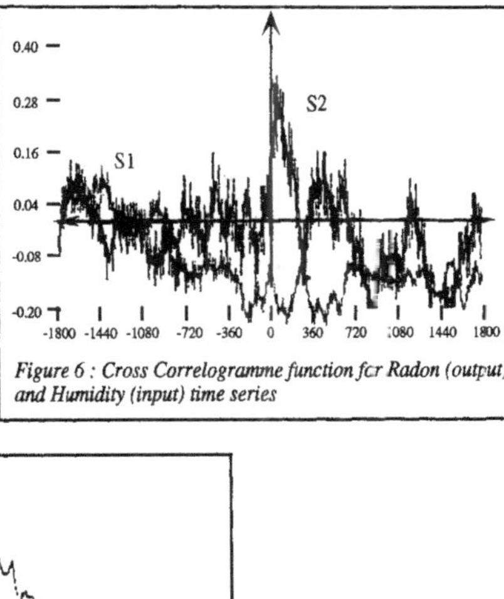

Figure 6 : Cross Correlogramme function for Radon (output) and Humidity (input) time series

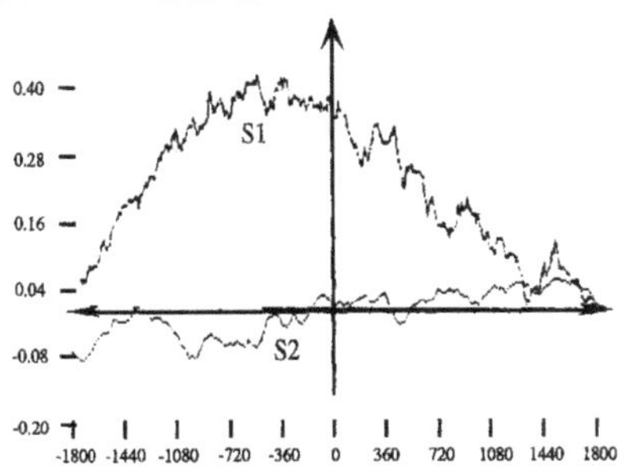

Figure 7 : Cross Correlogramme function for Radon (output) and Water Table level (input) time series

Radon movement simulation in overburden by the "Scattered Packet Method"

H. Marah, , A. Sabir*, L. Hlou , M. Tayebi

Laboratoire de Physique Nucléaire Appliquée
Faculté des Sciences - Kénitra
Ibn Tofail University , Kenitra (Morocco)

Abstract

The analysis of Radon (^{222}Rn) movement in overburden needs the resolution of the General Equation of Transport in porous medium, involving diffusion and convection. Generaly this equation was derived and solved analytically. The "Scattered Packed Method" is a recent mathematical method of resolution, initially developed for the electrons movements in the semiconductors studies. In this paper, we have adapted this method to simulate radon emanation in porous medium. The keys parameters are the radon concentration at the source, the diffusion coefficient, and the geometry.

To show the efficiency of this method, several cases of increasing complexity are considered. This model allows to follow the migration, in the time and space, of radon produced as a function of the characteristics of the studied site.

Forty soil radon measurements were taken from a North Moroccan fault. Forward modeling of the radon anomalies produces satisfactory fits of the observed data and allows the overburden thickness determination.

1. Introduction

Radon is a natural radioactive gas from the uranium family which is the main component of the annual radiation exposure of the population. Part of the radon formed in the earth's crust can migrate to the atmosphere through cracks and fissures by transport mechanisms. ^{222}Rn anomalies can occur if radon is transported from the shear zone througth the cover by both diffusive and convective processes. Different parameters as the location of radon sources, the porosity, the humidity in the soil can influence this migration and therefore the radon emanation in the atmosphere.An understanding of radon movement in overburden is acquired by establishing and solving the diffusion equation of radon. Analytical solutions of the diffusion equation such as the Monte Carlo method and the Direct Method were derived [1-2-3].In order to combine the advantages of the above two methods, a new technique, called the "Scattered Packet Method" (SPM) have been developed by J.P. Nougier and al. [4-5-6]. In the following, we develop an adaptation of the SPM to simultate radon emanation in porous medium, taking diffusion and convection mechanism into account. The method is applied to the two-dimensional fault model and used to determine fault zone parameters from a radon survey across a fault zone in Moroccan area.

2. Model calculations

General Transport Equation

The gas molecules migrate from regions of high concentration to regions of low concentration. Diffusion in an isotropic medium is based on Fick's law, which states that the amount of substance diffusing through a unit area is proportional to the concentration gradient measured normal to the section. Fick's law [7] may be written as:

$$J = -D_e \nabla N \qquad (1)$$

where J is the radon flux across the surface, De is the effective diffusion coefficient of ^{222}Rn in the porous medium; N is the concentration of the diffusing radon per unit volume; and ∇ is the gradient operator.

The advective transport of fluid througth a porous medium is described by the Darcy's law [8], which may be written as:

$$V = -\frac{K}{\mu}\nabla P \qquad (2)$$

where V is the superficial velocity vector (Darcy's velocity); K is the intrinsic permeability of the soil; m is the dynamic viscosity of the fluid and ∇P is the gradient of the dynamic pressure.

We may writte a general balance equation that accounts for the diffusion and advection effects as :

$$D_e\nabla^2 N - V.\nabla N - \lambda N + Q = \frac{\partial N}{\partial t} \qquad (3)$$

where λ is the decay constant of ^{222}Rn ($2.07 \times 10^{-6}\ s^{-1}$); and Q is the possible contribution of the soil overburden in ^{222}Rn radioactivity [9]. This equation describes the steady-state condition of 222Rn transport in a three-dimensional porous and isotropic medium, where there exists a mass flow of speed V cm/s. It supposes that permeability, porosity and diffusivity are constant in space and over time.

"Scattered Packet Method" of resolution

In order to simulate the transport of radon in a porous medium the domain of propagation is discretised in oblong mesches of Dx and Dy sides (Fig. 1). Each mesh M contains Nij atoms (carriers) and is repered in the plan (O,x,y) by its coordinates i and j The ensemble of radon atoms contained in one cell at time t are defined as the "packet" Pij. If a and b are the boundaries of the domain , the number of cells along Ox and Oy directions is simply $Nx=a/\Delta x$ and $Ny=b/\Delta y$.

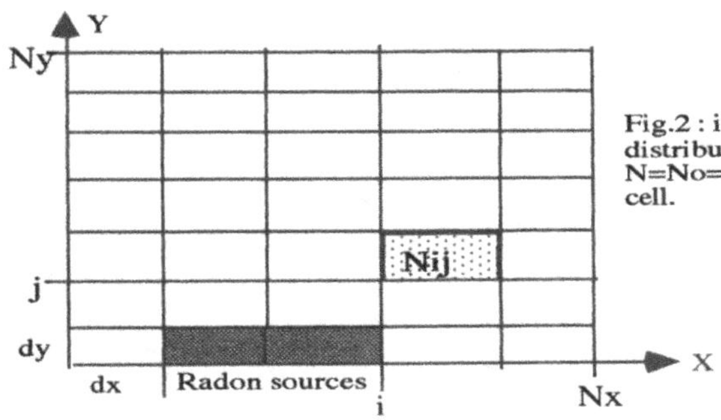

Fig.2 : initial distribution of radon. N=No=Cte in each cell.

It is assumed that under the effect of diffusion and advection the packet of carriers initially in the state M undergoes a free flight during Dt. At the end of the flight the packet is distributed in all the meshes of the domain.The probability for a carrier initialy in mesh M at time t to be in an other mesh M' at time $(t+\Delta t)$ is computed numerically. By repeating the procedure a time-dependent evolution matrix B allowing to compute $Nij(t+\Delta t)$ in each cell if Nij(t) is known, is evaluated.

$$[Nij(t+\Delta t)]=[B][Nij(t)]$$

The diffusive process can be simulated by a free flight with suitably chosen speed Vd. To determine Vd we consider a dissipative flow for which we have the two following expressions of the flux density of radon activity :

$$J = -D_e\nabla N$$

and

$$J = N.Vd$$

By identification we obtain;

$$Vd = -\left(\frac{D_e}{N}\frac{\partial N}{\partial x}i + \frac{D_e}{N}\frac{\partial N}{\partial y}j + \frac{D_e}{N}\frac{\partial N}{\partial z}k\right)$$

Under the conjugated diffusive and convective processes a packet make a free-flight of duration Δt, at $Vg=Vc+Vd$ speed. At the end of the free-flight the packet overlappe a certain number of cells (maximun 4) (Fig.2).

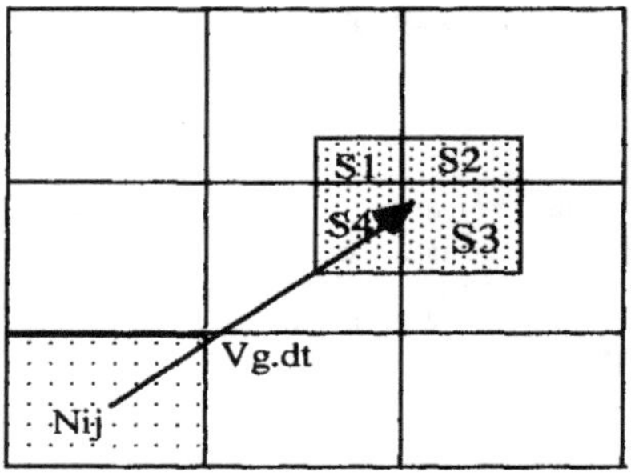

Fig. 2: The portions S1 , S2 ,S3 and S4 for initial packet Nij will be affected to the cells where they are

The content of each overlapped cell is incremented by a number of atoms, proportional to his covering with the incident packet. Once the packets spreading ends we reduce the Nij number of atoms in each mesh by a ($\lambda.N_{ij}.\Delta t$) quantity, taking radioactive decay of atoms into account. All the process is repeated by time-step of Δt, until to reach the stationary distribution function. The flow chart in Fig.3. represents the different steps of the program which is used to calculate the radon distribution function in the overburden.

To determine the distribution of ^{222}Rn in overburden the model is used under the following hypotheses:

- considering the fault , the extent is much larger than its width and the tree-dimentional problem reduces to a two-dimentional one
- the source of ^{222}Rn, of width W and concentration No, lies at the overburden-rock interface (y=0) .

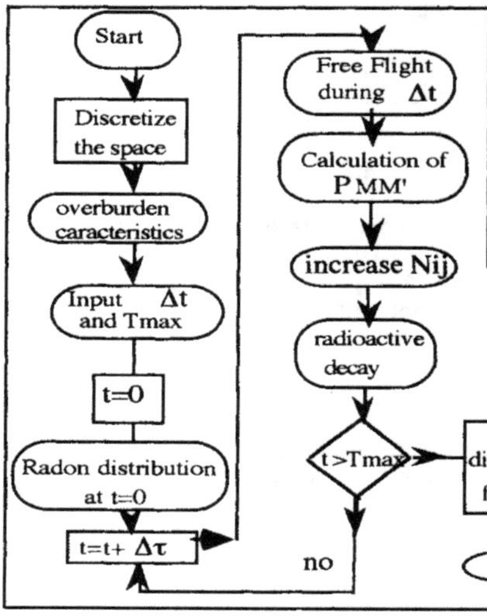

Fig. 3 : Flow chart of the program used to calcultate the stationary radon distribution function. Nij is the number of atoms in a cell at t=0; $P_{MM'}$ is the transition probability from the cell number M to the cell M', during time interval dt

3. Results

Simple case

The nonradioactive overburden (Q=0) is isotropic and homogeneous, and has a constant thickness b and a diffusion coefficient De . In order to simplify the calculation, initial concentration of radon in the cells of overburden is considered void.

Figure 4 shows a comparison between the steady state distribution function given by the SPM and the analytical methods (AM) [3], at different depths, when both mecanisms of diffusion and advection are taken into account (full curves). As we can see on the results, a full agreement is carried out between this two methods in this simple case. The emanation coefficient N/No of radon is maximal above of the source and on all its width. The contribution of the convection effect (dashed line) increases the emanation coefficient , and therefore the possibility of detecting the source from the overburden surface.

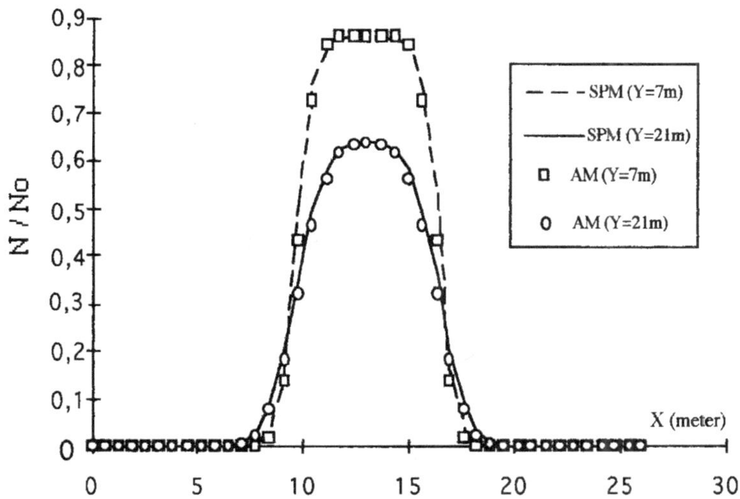

Fig. 4 : Comparison between the radon distribution function given by the SPM and the analytical methods (AM), at different depths (Y= 7 m and 14 m) , when both mecanisms of diffusion and advection are taken into account .

Complex case

The situation of a radon source covered by an homogeneous overburden is very simplistic. The reality is often more complex : several sources with different caracterisctics can exist under an inhomogeneous overburden. The schematized situation in Fig. 5 constitute an approach to this complexity. We considere three sources with the same initial concentration Nc , covered by different overburdens (different thickness and different diffusion coefficients). For example this could represent the case of a fault whose edges constitute preferential ways for the radon emanation as a result of the grinding that they have undergone.

In such situations classical methods of solution of transport equations meets with great difficulties in calculating radon distribution function. The result presented in Fig. 6. shows the suppleness and efficiency of "Scattered Packet Method".

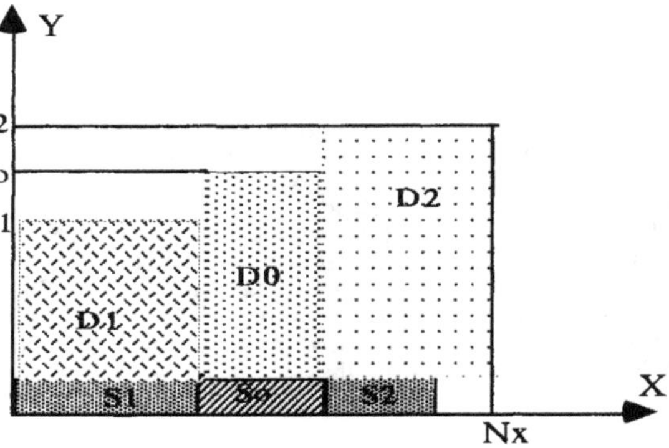

Fig.5 : Three sources S_1, So, S_2 covered by different overburdens. D_1, Do, D_2 are the diffusion coefficients of radon in the medium; b_1, bo, b_2 are the respective ticknesses; N_1, No and N_2 are the initial concentrations of the sources

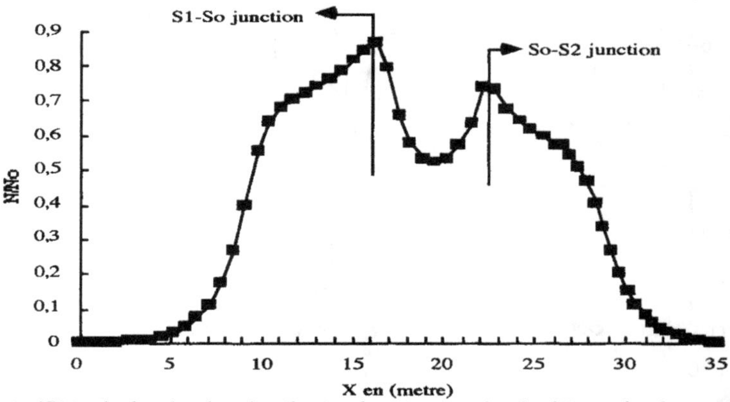

Fig. 6 : SPM calculated radon distribution function at a depth of 70 cm for three simultaneous sources covered by three kinds of overburden soils caracterized by the diffusion coefficients $D_1=5.10^{-6}$, Do= 3.10^{-6} and $D_2=10^{-5}$ m^2s^{-1} .Used parameters are No=N_1=N_2 (same initial concentrations), b_1 = 1.6 m ; bo= 2 m ; b_2=2.8 m; W_1=Wo=W_2= 6,5 m.

Fitting of experimental data

Rommani is a region approximately situated at 80 km in the South-Est of Rabat. This region presents several faults system which the most important is the fault of Oued-El-Malha (OEM). A total of nearly 40 radon measurements were taken at Rommani field area distributed into two traverse lines [10]. At sampling site it was attempted to space the individual sample points approximately 6 meters apart . Far from the fault zone the distance between stations was chosen as 20 m and the values considered as field area background level. The films were buried at a depth of 60 cm and kept in the ground for five weeks. The interpretation of the ^{222}Rn anomalies was based on the SPM developed earlier. Modelling was performed using the case of three sources covered by the same overburden. The initial concentration No of the central source is greater than N_1 and N_2. The diffusion coefficient is chosen as De= 3.10^{-6} m^2 . Figure 7 shows that there is a good fits between the observations and the model. Modeling of the selected traverse line shows that the widths of the sources at the soils-rock interface are Wo =16 m; W_1= 35 m; W_2 = 25 m. The tickness of overburden that gives the better agreement is b= 2,5 m

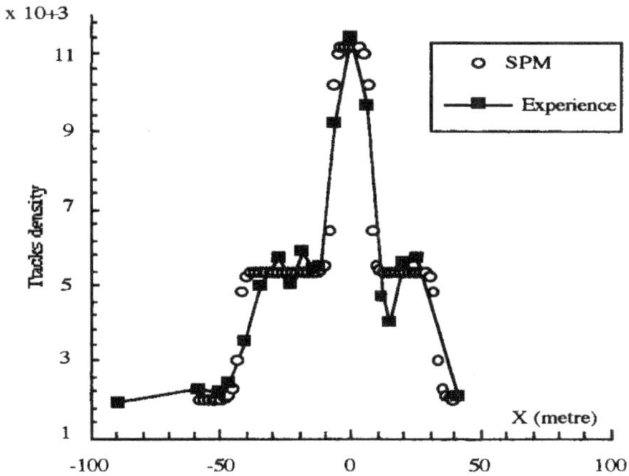

Fig. 7 : Comparison of Observed and Scattered Packet model curves for Rommani Fault.

4.Conclusion

In the radon emanation studies the SPM seems to be a suppleness and efficiency alternative method. For the simple cases a full agreement is carried out between the steady state distribution function given by the SPM and the analytical methods (AM), when both mecanisms of diffusion and advection are taken into account . For complicated ones SPM is more practicable, with a high accuracy calculation of radon distribution functions and reasonable CPU time requested. The analysis is applied to the two-dimensional fault model. Forward modeling of the radon anomalies produces satisfactory fits of the observed data and allows the widths of the sources and overburden thickness determination.

5.References

[1] Soonawala, N.M. 1976. PhD Thesis, McGill University, Montreal, 187 pp

[2] Tanner, A.B. 1964 , Radon migration in the ground, a review. In : J.A.S Adam and W.M. Lowder (Editors), Univ. of Chicago Press, Chicago, 1069 pp.

[3] Adboh A. and M. Pilkington, Geoexploration, 25 (1989) 341-354

[4] Nougier J.P., Hlou L., Houlet, P. , Vaissiere J.C. , Varani L. , Proceeding of the third International Workshop on computational Electronics, Portland, Oregon, May 18-20, 1994.

[5] Hlou L., Thèse de Doctorat es Sciences, University of Montpellier II (1993)

[6] Vaissiere J.C., Nougier J.P., Hlou L., Achachi A., 12th International conference on noise in physical Systems and l/f fluctuations, St Louis, Missouri (USA), August 16-20 (1993)

[7] Shroeder G.L., Kraner H.W. and Evans R.D., 1965, J. Geophys. Res.,70:471-474

[8] Nazaroff W., Reviews of Geophsics, 30,2/May 1992 p.137-160

[9] Soonawala N.M. and W.M. Telford, Geophysics, vol.45, n°8 (1980) p1297-1315

[10] Marah H., Thèse de Doctorat 3ème cycle, (1996) Université Ibn Tofail, Kénitra, Morocco.

D Indor radon and thoron

Radon in the United Kingdom

Gerry Kendall

National Radiological Protection Board
Chilton, Didcot, Oxon OX11 ORQ, UK

Abstract

A survey of radon in UK houses showed that indoor concentrations are distributed log-normally with a mean of about 20 Bq m^{-3}. Nevertheless concentrations vary greatly from one part of the country to another. An extensive mapping exercise has identified Radon Affected Areas and targetted measurements have been made in over 350,000 homes at greatest risk. Ten percent of these homes exceeded the UK Action Level of 200 Bq m^{-3}. The aim of UK radon policy is that such homes should be identified and remedied.

1. Introduction

The earth's crust contains small amounts of the primordial radionuclides ^{238}U and ^{232}Th. These decay through a chain of radioactive nuclides to stable isotopes of lead. Most of the decay products are isotopes of solid elements but two are gases: ^{222}Rn from the decay of uranium and ^{220}Rn from thorium. The name "radon" is sometimes used generically for all isotopes of the gas but is sometimes reserved for ^{222}Rn, ^{220}Rn being called thoron. This paper will generally refer to both isotopes together.

Although radon is ultimately derived from uranium its immediate precursor is radium. In some cases the latter may be concentrated (or diluted) by geological or geochemical processes. If radon escapes from the matrix in which it was formed it can make its way to the gas in soil beneath houses. Houses are normally at slightly lower pressure than the soil gas and the latter, including any radon, is gently sucked in (Figures 1, 2)

Uranium in ground

↓ geological process

Radium in ground

↓ faulting of rock
permeability
movement of gas
nature of overburden

Radon in soil gas

↓ connection of building to ground
underpressure in building
ventilation of building

Radon in buildings

Processes leading to radon in houses

Figure 1

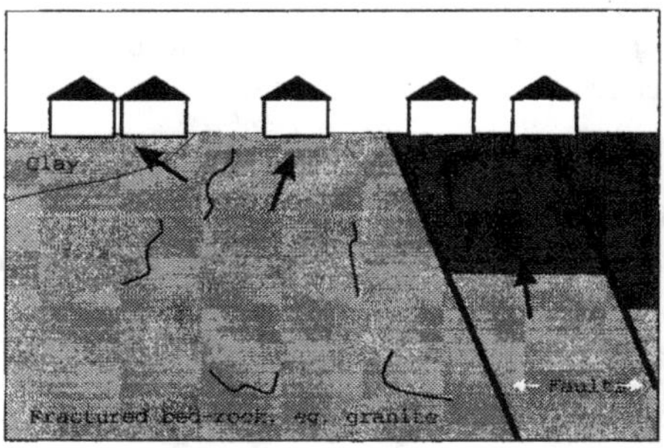

Geological factors and radon in houses

Figure 2

Other sources of radon in houses can include building materials and radon in the outside air but these are not usually important in the UK[1].

Radon in dwellings has been recognised as an important component of public exposure. Table 1, based on the 1993 UNSCEAR report[2], gives data for a number of countries around the world.

Table 1. Mean radon levels in dwellings (world-wide)

	Mean, Bq m^{-3}
Australia	12
Canada	34
China 7 provinces	24
Sechuan	19
Shenzhen	16
Egypt	9
India	60
Japan	29
United States of America	42
World average	40

2. Radon levels in UK dwellings

In the early 1980s NRPB carried out a survey to determine radon levels in UK buildings[3]. It was found that the mean indoor concentration of ^{222}Rn was about 20 Bq m^{-3} and that concentrations were distributed log normally (particularly if allowance is made for the background radon concentration in outside air, 4 Bq m^{-3})[4]. This is illustrated in Figure 3.

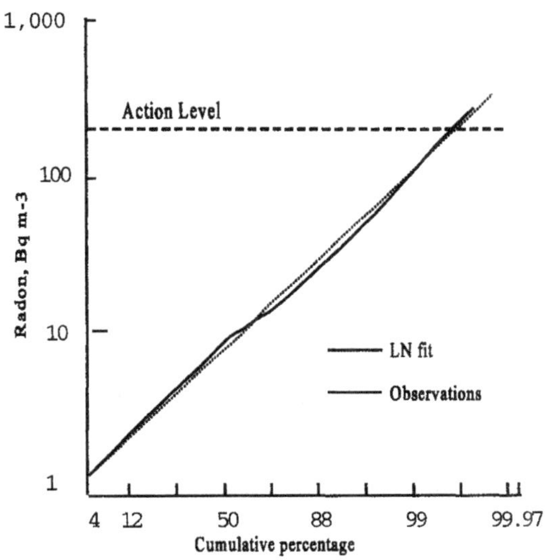

Log probability plot of UK domestic radon concentrations (corrected for background in outside air)
Figure 3.

A UK survey[5] of exposures to ^{220}Rn suggested that doses from this nuclide would exceed those from radon in only about 2% of cases and that the mean dose from ^{220}Rn is about an order of magnitude smaller than that from ^{222}Rn.

It has been estimated[6] that domestic radon exposures are responsible for about 2,000 of the 40,000 lung cancer deaths in Britain each year. Smoking is, of course, by far the dominant cause of lung cancer, but estimates suggest that radon is more important than passive smoking, thought to be responsible for a few hundred deaths each year[7].

3. Application in the UK of ICRP advice on controlling domestic radon

ICRP, in Publication 65[8], has given advice on controlling domestic exposures to radon. The approach is essentially pragmatic and it is accepted that practical constraints may determine national radon policy rather than the balancing of costs and detriment which would be the normal approach in other areas of radiation protection. Put bluntly, ICRP advise that national authorities should not choose a radon policy which cannot be implemented.

ICRP advise that two tools may be useful in defining a radon policy:

a) A radon Action Level

i.e. a mean domestic radon concentration above which remedial action should be taken; such action may be mandatory or advisory. One way of selecting an Action Level is to set it at ten times the national average radon concentration.

b) Radon-Prone Areas

High radon houses can be found anywhere in a country and ICRP advise that there is merit in focussing attention on radon-prone areas where concentrations of radon in buildings are likely to be higher than in the country as a whole. One criterion for a radon-prone area would be that 1% of houses exceed the Action Level but other criteria may also be chosen.

It is important to recognise that ICRP are not prescriptive on the actions to be taken within radon-prone areas. There is no implication that all houses must be measured or that all new buildings must be constructed with anti-radon measures. Such responses may be sensible in any particular set of national circumstances but they do not follow automatically.

The UK has implemented the ICRP advice in a very direct way[1,9]. The UK Action Level for radon in homes is set at 200 Bq m^{-3}, ten times the national average of 20 Bq m^{-3}. Radon-prone areas (termed "Radon Affected Areas") are those where more than 1% of houses exceed the Action Level.

4. The UK domestic radon programme

Programmes to assess and control the hazard from domestic exposures to radon are in place in all four parts of the United Kingdom. Common themes are mapping the geographical variation of radon concentrations and then targeting measurements on those houses most likely to be at risk. The methods used for mapping are described below.

Measurements of radon levels in dwellings are made by postal surveys using small and inexpensive track etch dosemeters[10]. These consist of a small piece of CR-39 plastic contained in a virtually sealed plastic holder which nevertheless allows radon to diffuse in and out. Two such detectors are sent to each householder taking part in the survey. One is put in the living area and one in an occupied bedroom. After three months the detectors are returned to the laboratory for assessment. This involves increasing the size of the α particle tracks in the CR-39 by etching with sodium hydroxide solution followed by counting the number of tracks with a computer controlled microscope. The number of tracks is proportional to the radon exposure. The measured radon concentrations are corrected to an annual value by taking into account the season of the year when the measurement was made. The householder is informed of the result and, if appropriate, advice on remedial measures is offered.

To date, NRPB has undertaken measurements in about 350,000 houses, 10% of which have exceeded the Action Level. Details are in Table 2.

Table 2. Statistics of the UK domestic radon programme

Country	Number of radon measurements	Number above the Action Level
England	332,000	34,000
Northern Ireland	12,000	400
Scotland	6,000	200
Wales	4,000	200
Total UK (rounded)	350,000	35,000

5. Radon Remedial Measures

If a house is found to have radon concentrations above the Action Level then the householder is advised to undertake remedial work to reduce it. The most effective method is to construct a radon sump ie, a system in which a small cavity is excavated below a ground-supported concrete floor, or below a concrete oversite under a suspected timber floor. Sumps may also be created be excavating outside an external wall and putting a pipe through the foundation wall to connect to the underfloor space. A pipe extends from the sump to a point away from the building or internally through the roof, and is terminated by a suitable fan. The exhaust from the fan must be well away from windows and doors. The fan creates a reduced pressure in the sump and draws soil gas with radon into the sump and discharges it to the atmosphere. Sumps can be installed by the householder, but a considerable degree of expertise is required to make them effective and unobtrusive.

Increasing the natural ventilation of the house, for example by trickle vents on windows, will help dilute the radon but is unlikely to reduce levels by a large factor. Positive ventilation, in which filtered loft-space air is blown into the occupied spaces by a small fan, is only marginally more effective. Natural or mechanical ventilation of the underfloor void usually give reductions similar to those from increased ventilation of the house itself.

An instinctive reaction to the knowledge that radon is entering the house through cracks and gaps in the floor is to try to seal them. This may also be attractively cheap if the householder is prepared to do the work, although it is often expensive if contractors are employed. However, a very high proportion of the openings must be sealed for this measure to be successful: if only half the openings are blocked, the radon will come in almost twice as fast through those that remain. An alternative to sealing cracks in the floor is to lay a radon-proof membrane across it. This is often ineffective, however, and can lead to serious problems with rot in timber floors if ventilation is inadequate. The method is no longer recommended.

Most houses with high radon levels can be remedied quite easily although there have been some instances where the action taken was ineffective. The procedure most likely to be successful and to show the largest reduction in radon level is the sump. Where sumps have failed the reasons have generally been that the fan was of an inappropriate type or the exhaust was discharged in the loft space or too close to windows. The efficiency of various remedial measures is compared in Table 3.

Table 3. Effectiveness of different approaches to reducing radon levels in homes[11]

Method	No of houses	Geometric mean of reduction factor	Typical cost (£)
Sump	429	9.5	700†
Positive ventilation	94	2.7	500
Additional natural ventilation of the house	115	1.9	300
Mechanical ventilation of the underfloor void	63	2.8	300
Natural ventilation of the underfloor void	171	1.9	300
Sealing	71	1.7	1000*

† £300 if undertaken by the householder
* £ 50 if undertaken by the householder

6. Making and using radon maps

Maps showing the way that radon concentrations vary across the country are useful for a number of reasons:

to understand the variation in radiation exposure
to guide measurement programmes and the remediation of existing houses
to identify areas where anti-radon measures are needed in new buildings
for certain epidemiological purposes

Most radon maps of the UK have been based on 5 or 10 km grid squares of the UK national Ordnance Survey. There are about 10,000 such 5 km grid squares. To make deductions about radon concentrations at the 1% level in a square might be thought to require about 100 measurements with a requirement of a million measurements to map the whole country. Fortunately a much more economical technique was developed[12].

Investigations showed that radon concentrations within any particular grid square could usually be taken to be log-normal. This means that only two parameters are needed fully to characterise the distribution: the geometric mean (GM) and the geometric standard deviation (GSD). The GM can be estimated if a few measurements are available for the square in question. However, it can be shown that, unless estimates are based on many measurements (say >20), they are improved by smoothing estimates of GM with those in adjacent squares. More measurements are necessary to determine the GSD than the GM and where there are insufficient measurements in a square for its direct determination, a standard GSD for the region in question can be assumed. Both the smoothing of GMs and the assumption of default GSDs improve the accuracy of mapping when limited data are available but there is inevitably some loss of precision and detail.

A detailed radon map of the whole of England has been produced in this way (Figure 4).

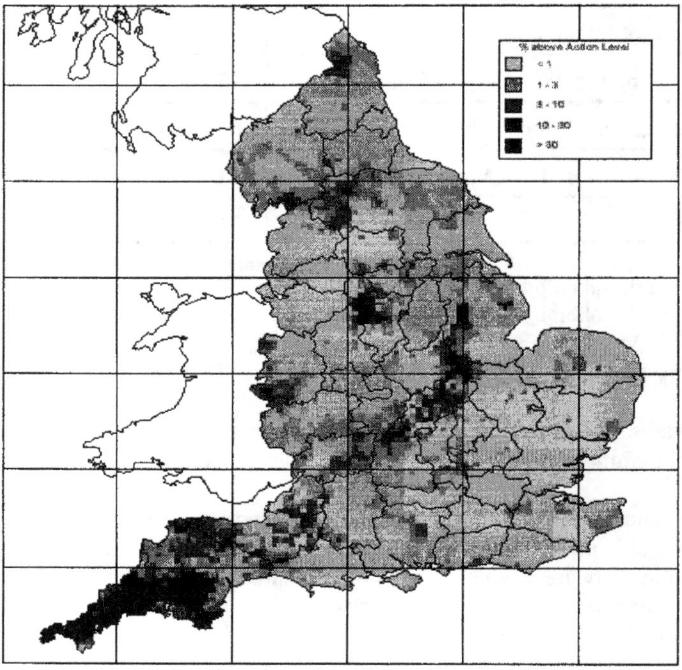

Variation in proportion of houses exceeding the radon Action Level in England

Figure 4.

This map is the basis upon which radon Affected Areas in England have been defined[9] and it is not expected that these definitions will be refined in the short or medium terms. However, there is still a demand for more detailed and precise radon maps. These can, of course, be provided using the techniques described above if the measurement density is increased sufficiently. However, there are likely to be practical constraints on the number of measurements which can be made. The most promising way of improving radon maps based on a given number of radon in house measurements is to include geological information.

As outlined in Figure 1, geological processes broadly determine the concentration of radon in soil gas. The way in which radon in soil gas enters room air depends on details of the way the house is constructed and used. This topic goes beyond the subject matter of this paper but it should be noted that two adjacent and superficially similar houses can have quite different indoor radon concentrations. Although levels of radon in soil gas are governed by geological processes these are extremely complex and abinitio modelling often has limited success[13]. A pragmatic blend of house measurement data and geological knowledge is the most promising way ahead. Simply grouping house measurement results by geological unit rather than grid square has been shown to improve predictions[14].

7. Possible future developments

The UK radon programme has been very successful in identifying areas, particularly of England and Northern Ireland, where high radon levels are likely and in finding these houses. Mapping exercises in Scotland and in Wales will shortly give similar results for these countries. Attention is being directed towards methods for encouraging householders who have high radon levels to undertake remedial measures.

8. Acknowledgements

The development of the UK radon programme has involved many organisations and individuals. A key role has been played by the Department of the Environment, the Scottish and Welsh Offices and the Department of the Environment for Northern Ireland. Practical questions about radon control frequently go to local councils, and we are especially grateful for the support of members of the Chartered Institute of Environmental Health. Colleagues at the Building Research Establishment have been responsible for developing and testing many of the radon remedial and preventative measures. We also acknowledge the many colleagues at NRPB who have contributed to the radon programme, in particular Mr M C O'Riordan, Mr D W Dixon, Dr B M R Green and Mr J C Miles.

9. References

[1] G M Kendall, J C H Miles, K D Cliff, et al. Exposure to radon in UK dwellings. NRPB-R272 (1994) London, HMSO.

[2] United Nations Scientific Committee on the Effects of Atomic Radiation. Sources and effects of atomic radiation. 1993 Report to the General Assembly. United Nations, New York, (1993).

[3] A D Wrixon, B M R Green, P R Lomas, et al. Natural radiation exposure in UK dwellings. NRPB-R190 (1988) London, HMSO.

[4] J A Gunby, S C Darby, J C H Miles, et al. Factors affecting indoor radon concentrations in the UK. Health Physics, 64, 2–12 (1993).

[5] K D Cliff, B M R Green, A Mawle, J C H Miles. Thoron daughter concentrations in UK homes. Radiat. Prot. Dosim., 45, 361–366 (1992).

[6] J Metters. Setting radon in context. Radiat. Prot. Dosim., 42, 159–164 (1992).

[7] M R Law, A K Hackshaw. Environmental tobacco smoke. Br. Med. Bulletin, 52, 22–34 (1996).

[8] International Commission on Radiological Protection. Protection against radon-222 at home and at work. ICRP Publication 65, Ann. ICRP 23 (2) (1993).

[9] National Radiological Protection Board. Radon affected areas: England. Documents of the NRPB, 7, 3–9 (1996).

[10] G D Hardcastle, C B Howarth, S P Naismith, et al. NRPB etched-track detectors for area monitoring of radon. NRPB-R283 (1996) London, HMSO.

[11] S P Naismith, J C H Miles, C R Scivyer. The Influence of house characteristics on the effectiveness of radon remedial measures. (In preparation).

[12] J C H Miles. Mapping radon prone areas by log-normal modelling of house radon data. Health Physics to be published.

[13] J Hulka, I Fojtikova, Z Borecky, et al. Indoor radon risk mapping in the Czech Republic. Proceedings of the European Conference on Protection against Radon at Home and at Work. Prague, 2–6 June 1997.

[14] J C H Miles and T K Ball. Mapping radon-prone areas using house radon data and geological boundaries. Environment International, 22, Suppl. 1, 5779–5782 (1997).

Indoor Radon Levels and Measurements in China

WANG Hengde

China Institute for Radiation Protection
P. O. Box 120, Taiyuan, Shanxi 030006, China

Abstract

Mankind has been exposed to indoor radon since the ape-men moved from trees to caves and had their indoor environment. Following the harmful health effects of radon among underground miners became known, indoor radon attracts more and more attention. This paper describes the importance of indoor radon study, radon levels in different kinds of Chinese houses, a few radon measuring methods used in China that may not be common in other countries, and finally discussion about some issues on quality assurance of radon measurement.

1. Importance of indoor radon study

As well known, exposure to radon and decay products is of most important contributors to radiation doses to members of the public, especially exposure to indoor radon and decay products. UNSCEAR[1] reported that the exposure to the public from radon and decay products accounts for 43.3% of the exposures from all sources averaged over the world, 44.75% in China[2] and 10.5% in Japan (as medical exposure takes much bigger percentage in Japan) [3], as shown in Tables 1 and 2. Therefore, it is of great significance to study sources and effects of and countermeasures against indoor radon and decay products.

Table 1 Contributions of various sources to human radiation exposures

Source	Percentage, %		
	World[1]	China[2]	Japan[3]
Natural sources except Rn and progeny	36.7 (1.1)*	48.65 (1.0)*	28.4 (1.08)*
Radon and progeny	43.3 (1.3)*	44.75 (1.1)*	10.5 (0.4)*
Medical exposure	19.7(0.59)*	4.21(0.114)**	59.2 (2.25)*
Nuclear power generation	0.002 (0.007)*		
Occupational exposures	0.0007(0.002)*		
Others (including severe accidents and nuclear tests)	0.3		1.08 (0.07)*

Note: * The number in the brackets is corresponding individual annual effective dose in mSv.

** Dose from radiotherapy in China is not included as the data is not available at the moment.

Besides, as indoor levels of radon and decay products are generally higher, maybe considerably high, up to several hundreds $Bg \cdot m^{-3}$ or more in some houses, which provide a good site to conduct epidemiological study for long-term health effects of radiation. This site could be better than underground mining workplaces as the residents spend much longer time in the houses than miners do in the workplaces, and the residents live in a relatively even radon level in a house unlike the miners work in a uneven radon level in the workplace, therefore, the dose to the residents may be determined more accurately. To conduct epidemiological study in a house with elevated radon level more efficiently, it may be necessary to reconstruct the dose to the residents in early years if it was not properly measured. C. Samuelsson and L. Johansson[4], S. Oberstedt and H. Vanmarcke[5] reported methods of retrospective radon monitoring using long-lived radon decay products as long-term radon exposure indicator in traps of readily available items in residential houses.

Table 2 Population doses in China

Source	Annual collective dose	
	Collective dose, man·Sv	Percentage, %
Natural exposure	2.55×10^6 (2.298*)	93.4
Cosmic ray	0.317*	
Territorial γ ray	0.54*	
Rn and progeny	1.101*	44.75
K-40	0.17*	
Other radionuclides	0.17*	
Occupational exposure	3.4×10^4	1.25
Nuclear industry	7×10^1	
Nuclear & radiation techniques application	2.2×10^2	
Underground coal mining	2.6×10^4	
Underground nonferrous metals mining	8.4×10^3	
Sea fishing	-6.6×10^2	
Public exposure	3×10^4	1.16
Nuclear industry	2.31×10^1	
Air travel	3.6×10^1	
Isotope applications (I-131 and I-125)	4	
Coal-fired power plants	2.5×10^3	
Coal slag application	3.5×10^3	
Other coal burning	3×10^4	
Transportation	-5.15×10^2	
Concrete buildings	-7.8×10^3	
Drinking water	-1.36×10^2	
Medical exposure	1.15×10^5	4.21
X rays	9.16×10^4	
Nuclear medicine	2.32×10^4	
Total	2.73×10^6	

Note: * These numbers are individual annual effective doses in mSv.

Furthermore, radon is being with us everywhere and in all of our time, so people do not dread it though it is a sort of substance emitting radiation. Studying on radon, especially on higher levels of radon in houses, and comparing the risk from indoor radon and decay products with that from nuclear power production and other kinds of power production, and other human activities should promote better understanding of radiation, nuclear science and technology, nuclear power and other beneficial human activities involving radiation.

2. Radon levels in houses of China

China is big in size. People living in different areas may live in different kinds of residency such as brick houses, multistory apartment buildings and semi-underground cave dwellings, where radon concentration may be different. Different contents of radionuclides in soil and building materials may also cause different radon level. In addition, some underground air-raid shelters, that were constructed during cold-war years, have been reused as shops or hotels. These underground buildings usually have higher radon concentration depending on ventilation.

Well-planed national investigation of radon and decay products has not conducted so far in China. However, some investigations have been accomplished.

Pan[2] summarized the results of indoor radon investigations that were conducted in some 20 provinces of China, as shown in Table 3, and stated that the population-weighted indoor radon-222 concentration and equilibrium equivalent radon concentration of indoor radon decay products was 23.5 and 11.5 Bq·m^{-3}, respectively, and the annual effective dose caused by radon-222 and progeny was 0.916 mSv, of which 0.786 mSv was due to indoor exposure and 0.130 mSv outdoor exposure, averaged over the 20 provinces in which 73% of Chinese population lived.

Table 3 Concentrations of radon-222 and decay products in houses

Location	Rn concentration, Bq·m^{-3} Rooms measured	Mean	Equilibrium factor	EEC, Bq·m^{-3} Rooms measured	Mean
Hubei	380	22.5	0.53	1920	10.2
Guangdong	220	19.0	0.54	220	10.3
Hebei				44	8.0
Liaoning	397	30	0.49		14.5
Beijing	537	30.3			
Anhui	856	16.2			
Henan	350	21.5			
Shandong	46	18.7	0.47	46	8.83
Hunan	78	42.8	0.51	78	21.8
Fujian	338	51.8	0.49	338	24.4
Shanxi	204	27.5		105	9.7
Jiangsu	486	17	0.42	491	8.0
Tibet	252	9.6	0.57	140	5.8
Sha'anxi	837	43.3	0.6	849	25.4
Heilongjiang	413	20.8	0.44	40	9.2
Shanghai	148	11.8	0.5		5.9
Jiling	51	8.7	0.67		5.3
Zhejiang	209	17.2	0.44	351	7.52
Inner Mog.	277	31.5	0.31	45	11.3
Leshan,Sichuan	629	10.5	0.46		4.83

In some places, some buildings are built up with coal-slag bricks, that contain more naturally occurring radionuclides, so that the radon levels in these buildings are generally higher. Table 4 gives the results of some surveys in this kind of building in Jiangxi and Hunan provinces.

Table 4 The concentrations of Rn and progeny in some coal-slag brick houses in Jiangxi and Hunan provinces

Location		Rooms measured	Rn concentration $Bg \cdot m^{-3}$	α potential energy conc. $x10^{-8}$ $J \cdot m^{-3}$	Ref.
Jiangxi	A	47	254	67.2	[6]
	B	42	300	53.9	
	C	11	169	20.0	
Hunan	A	14	295	95.7	[7]
	B	4	222	62.2	
	C	7	116	40.2	
	D	2	177	64.5	
	E	5	71.1	20.7	

In some hilly areas, some people, mainly farmers, live in cave dwellings with higher radon concentration. A survey[8] made in the east of Gansu province showed that the radon concentration in cave dwellings in Pingliang and Qingyang areas of Gansu were ranging from 74 to 267 Bq.m-3, as shown in Table 5.

Table 5 Rn-222 concentration in cave dwellings in $Bq \cdot m^{-3}$

Cave dwelling type	Pingliang	Qingyang
A	87	74
B	165	162
C	123	231
D	130	267

Table 6 Rn-222 concentration in underground constructions in $Bq \cdot m^{-3}$

Location	Usage	Rooms measured	Rn concentration	Ref.
Beijing	Subways	28	17.4	[9]
	Hotel	83	45.9	
	Basement	33	46.6	
	Recreation	26	49.6	
Kunming	Closed	12	644.5	[7]
Wuhan	Closed	2	50.0	
Guiyang	Closed	2	23.1	
Shenyang	Closed	112	293.0	

Radon concentration in underground constructions varies considerably one from another, depending mainly upon ventilation. Surveys have been made in closed underground air-raid shelters in Kunming, Wuhan, Guiyang and Shenyang cities and in underground constructions of different kinds in Beijing city. The results are summarized in Table 6.

3. Radon monitoring

Here are briefly described several monitoring methods that are used in China besides the common methods using ionization chamber, scintillation chamber, charcoal canister, CR-39 and TL.

Double filter method uses a cylinder with two filters at its two ends, respectively. The inlet filter removes radon decay products in air to be measured and outlet filter collects radon decay products that are formed by radon during passing through the cylinder. Then the radon decay products collected on the outlet filter are measured and radon concentration in the measured air is derived. To obtain higher sensitivity, the cylinder cavity is usually big, therefore it is not quite convenient to be used in a smaller room.

Balloon method uses a rubber balloon of usually 1 m^3, of which an opening is covered with two filters. The first filter removes radon decay products in inlet air when the air is forced with a pump into the balloon. Following stay in the balloon for certain period of time, the air is discharged through the filters. The second filter collects radon decay products that are formed in the balloon during waiting time. Then the radon decay products collected on the second filter are measured and radon concentration in the measured air is derived.

Electret is an isolator with electrical charges maintained stably on it. Electret method uses an electret in the ion collection chamber to provide an electrostatic field. The measured air with radon diffuses into the chamber through a filter by which radon decay products are removed. The air is ionized in the chamber by α particles produced from decays of radon and its decay products that are formed inside the chamber. Then the ions are driven by the electrostatic field and collected by the electret. The ion collection reduces the surface electrical potential of the electret. The potential reduction is a measure of radon concentration in the measured air. Therefore, the radon concentration can be derived.

4. On quality assurance of Rn measurement

Reliability of measurement of Rn and its decay products is affected by many factors such as the energy of reference source used for calibration, calibration geometry and environmental conditions. To obtain better quality of measurements, quality control measures must be taken.

Any instrument needs to be properly calibrated and it is important to maintain the same conditions in calibration and in measurement. Special attention should be paid to the following issues.

A plate reference source is usually used to determine the detection efficiency of a detector to be calibrated to α radioactivity on a filter. It is essentially important for α

measurement that the reference source should have the same α energy as the α radioactivity to be measured and the geometrical condition of calibration should be the same as that of measurement. In the case of measuring radon and its decay products, Am-241, Pu-239 and U-238 references are frequently used, however, the α energy of those references are significantly smaller than the α energy of RaA(Po-218) and RaC'(Po-214) which is 6.00 and 7.69 Mev, respectively. This may cause systematic deviation. It is desirable to use RaA+RaC' source of known activity as a reference source that can be prepared with standard radon source.

Moisture can affect the efficiency of detectors of radon and its progeny in several ways. Moisture may become condensation nuclei which radon decay products may attach to and affect deposition of radon decay products on detector surface and walls, and moisture may affect normal functioning of filter if humidity is high.

Atmospheric pressure may cause change in detection efficiency of radon monitors of ionization chamber type and scintillation chamber type as difference in pressure may cause difference in air absorption of α energy. Temperature change may cause change in the quantity of absorbed radon by activated charcoal, therefore affect the detection efficiency of activated charcoal type detector.

In addition, the following general measures might be helpful.

1. All monitoring activities must comply with national regulations and technical codes related.
2. Monitoring personnel should be appropriately trained.
3. Take part in inter-laboratory intercomparisons of radiation monitoring, and make necessary modifications according to the experience gained in the intercomparisons.
4. Reference sources should be traceable to the national standards.
5. All equipment should be calibrated regularly.
6. Sampling and measurements should be made according to pre-determined procedures.
7. All monitoring data should be compiled and kept.
8. Monitoring data should be processed according to pre-determined procedures, and the results should be checked and reviewed.

References

[1] UNSCEAR, UNSCEAR 1993 Report to the General Assembly: Sources and Effects of Ionizing Radiation, 1993.

[2] Pan Ziqiang, Preliminary Assessment of Population Dose in China (in Chinese), Radiation Protection, Vol. 17, No. 2, 1997.

[3] Japan Nuclear Safety Research Association, Radiation in the Living Environment, 1992.

[4] C. Samuelsson and L. Johasson, Long-Lived Radon Decay Products as a Long-Term Radon Exposure Indicator, Radiation Protection Dosimetry, Vol. 45, Nos. 1-4, 1991.

[5] S. Oberstedt and H. Vanmarcke, Volume Traps: A New Restrospective Radon Monitor, Health Physics, Vol. 70, No. 2, 1996.

[6] Qin Shizhong et. al., Hygienic Assessment of A Building Material Made of Stone-Like Coal, Radiation Protection, Vol. 5, No. 6, 1985.

[7] Shi Jinhua, Survey of Concentrations of Radon and (Potential Energy of Rn Daughter Products in Air in Some Regions of China (1983-1990) (in Chinese), Radiation Protection, Vol. 12, No. 2, 1992.

[8] Shang Bing et. al., Survey of Radon Concentrations in Cave Dwellings in the East of Gansu Province (in Chinese), Radiation Protection, Vol. 15, No. 6, 1995.

[9] Lin Lianqing et. al., Radon Concentration in Beijing Underground Buildings and Assessment of the Internal Dose Equivalent (in Chinese), Radiation Protection, Vol. 7, No. 2, 1987.

Assessment of radiation dose from exposure to radon and thoron progeny

Stephen B. Solomon

Australian Radiation Laboratory
Lower Plenty Road
Yallambie Melbourne
Victoria Australia 3085

Abstract

The assessment of exposure to radon and thoron is usually derived from measurements of the airborne concentrations of the radioactive gases or their decay products. Monitoring programmes and surveys can provide detailed information about temporal and spatial variations in these airborne concentrations. The International Commission on Radiological Protection (ICRP) recommends that the calculation of the inhalation dose for an individual exposed to radon progeny should be based on a single conversion factor derived from epidemiological studies of uranium mine exposure. This approach does not take into account differences in aerosol characteristics, particularly the fraction and diameter of any ultrafine (unattached) mode. The Australian Radiation Laboratory (ARL) has developed a number of systems for assessing the aerosol parameters required by the dosimetric models of radon progeny inhalation. These systems include a practical dichotomous-sampling field device and computational algorithm which has a measurement characteristic as a function of particle size that closely simulates the dependence of effective dose on aerosol size. This device can be used as a radon or thoron progeny *Effective Dosimeter*. This paper describes the application of the ARL systems to the assessment of radon progeny radiation dose in a range of Australian exposure situations and examines the applicability of the ICRP *conversion convention*. The quality of radon and radon progeny measurements can be optimised by proper calibration methods, supported by intercomparison exercises such as those organised within the International Radon Metrology Programme (IRMP). There is a need for a similar programme to intercompare radon progeny dose assessment methods.

1. Introduction

Radon gas, in both the isotopic forms ^{222}Rn (*radon*) and ^{220}Rn (*thoron*), is all pervading. The measurement of airborne radon concentration provides a practical means for assessing the potential risk from exposure to radon. There are now a large number of research activities identifying areas with elevated levels of ^{222}Rn. National surveys of radon levels in homes, buildings, caves, spas, mines and other workplaces have been carried out in many countries [1]. The International Commission on Radiological Protection (ICRP), in ICRP65, has recommended action levels for limiting exposure to environmental radon in homes and in the workplace [2]. Many countries, including Australia, have incorporated these recommendations into their national radiations standards [3, 4]. For Australia, the recommended action levels are;

- for homes 200 Bq m^{-3}
- for workplaces 1000 Bq m^{-3}

These levels form the basis for implementing remedial action, in the case of homes, or the imposition of radiation protection measures in the case of workplace exposure.

The health risk associated with radon arises from the inhalation of the short-lived decay products, i.e. *radon progeny*. The radiation dose delivered to the respiratory system, and the resultant health detriment, is a complex function of the radon progeny aerosol characteristics and the physiological parameters of the exposed individual. The magnitude of this risk has been quantified in a number of epidemiological studies, most dealing with the increased rate of lung cancer amongst uranium mine workers [5]. A limited number of studies have been made linking radon exposure in homes to increased lung cancer rates [6]. ICRP65 recommends the use of the risk factors determined in uranium mining epidemiological studies as the preferred method converting radon progeny exposure to effective dose. The ICRP, while recognizing that differences in aerosol and physiological characteristics could modify the dose conversion factor (DCF), considers the epidemiological approach to be simpler and more direct than the alternate dosimetric modelling approach.

This paper summarises a number of research programmes undertaken by the Australian Radiation Laboratory characterising radon progeny aerosol in order to assess the use of the ICRP65 *conversion convention* in a range of Australian environments.

2. Radiation Dose from Radon Progeny

The radiation dose from exposure to ^{222}Rn or ^{220}Rn progeny can be derived from a measure of the potential alpha energy concentration (PAEC), combined with a dose conversion factor (DCF or dose per unit intake) which reflects the radiation dose from the deposition of the inhaled progeny in the respiratory tract. For ^{222}Rn the approach recommended by the ICRP is to use the radon progeny *conversion convention*. The use of a single value for the DCF assumes:

- for occupational exposure, that the radon progeny particle size distribution, in particular the fraction and particle size of the ultrafine mode, is not too different from the uranium mine size distributions,

- for environmental exposure, that the differences in aerosol conditions are offset by lower breathing rates for members of the public, particularly children.

The *conversion convention* is not applicable to ^{220}Rn progeny.

A second approach to the estimation of radiation dose uses the results of the dosimetric models for inhalation of airborne radioactivity. Such models incorporate data on respiratory tract deposition, including data on deposition of ultrafine particles in nasal passages [7,8]. These models are widely used for the assessment of dose for most inhaled radionuclides. The models show that the dose per unit intake of radon progeny is dependant upon the site of deposition within the respiratory tract, which in turn is strongly dependent on the particle size of the airborne radon progeny, particularly in the ultrafine region below 10 nm in diameter, (the so-called *unattached fraction*). A weighted dose conversion factor for a particular exposure location can be derived from measurements of radon progeny size distribution, combined with the particle-size dependent dose conversion factors, calculated using one of these dosimetric models.

A range of techniques for the measurement of size distributions of radon progeny associated with both accumulation mode and ultrafine particles have been developed. These have been extensively reviewed elsewhere [9]. Many of these methods rely on the fractionation of the sub-micron radioactive particles by diffusion processes, using sets of wire screens, or by inertial and impaction processes, using cascade impactors. The latter systems are capable of resolving an additional coarse mode at particle sizes exceeding 1 μm). These systems have been used to measure radon progeny size distributions in homes, workplaces and uranium mines [10-15]. A simpler technique involving the measurement of *unattached fraction* has been widely used [16]. This method relies on the separation of the ultrafine and accumulation modes (*attached* particles in size range 50 - 500nm) using a diffusion sampler, usually a single wire screen. These samplers poorly differentiate the two modes and provide little information on the particle size of either mode.

The Australian Radiation Laboratory has developed a number of systems for characterising radon and thoron progeny aerosol. These include:

- Two-stage wire screen sampler for *unattached fraction*
- Multi-stage wire screen parallel diffusion battery for activity size distributions
- Serial graded screen array for activity size distribution of the ultrafine mode
- Two-stage *effective dosimeter* for direct measurement of radon progeny dose

The primary purpose of these systems is to measure radon progeny parameters for a range of sampling locations and exposure groups, with the aim of estimating the associated radon progeny inhalation dose using the results of dosimetric modelling, in this case the ICRP Human Respiratory Tract Dosimetric Model [7].

2.1 The ICRP Human Respiratory Tract Model

Most of the ARL studies to date have concentrated on radon progeny, i.e. the short-lived decay products of ^{222}Rn [17 - 20]. Two of these radon progeny, ^{218}Po and ^{214}Po, are alpha-particle emitters and airborne concentrations of progeny are usually specified in terms of potential alpha energy concentration (PAEC), usually measured in terms of J.m^{-3}. The historical unit for PAEC is the Working Level (WL), with 1 WL = 20.8 μJ m^{-3}. The risk arising from the inhalation of radon progeny is related to the effective dose, a measure of the equivalent whole body radiation dose derived from a weighted sum of the radiation dose to particular organs. The dose conversion factor (DCF) relating the effective dose to the radon progeny PAEC is a complex function of the physical properties of the airborne progeny and the physiological parameters associated with exposure by inhalation. The extensive dosimetric model recommended by the ICRP in Publication 66 [7] forms the basis for the derivation of radon progeny doses for this study.

Figure 1 shows the typical response functions for the dose conversion factor for an adult male with breathing rates of 1.2 m^3.h^{-1} and 0.78 m^3.h^{-1}, for occupational and environmental exposure, respectively. These curves were derived using the computer program RADEP, which implements the ICRP66 Human Respiratory Tract Model [21]. Assumptions were made for the representative activity concentration ratios for ^{218}Po : ^{214}Pb : ^{214}Bi of 0.8 : 0.02 : 0.0 and 0.8 : 0.4 : 0.2, for particle sizes < 20 nm and ≥ 20 nm, respectively. The resultant doses are relatively insensitive to the exact choice of values for the equilibrium ratios [21]. The ICRP66 model-derived values have been adjusted by a factor of 0.3 to provide consistency with the epidemiologically-derived risk estimates in ICRP65 [21]. The results from the RADEP program in Figure 1 show that the ICRP66 model produces values for the dose per unit exposure for ultrafine (*unattached*) radon progeny with an activity median thermodynamic diameter (AMTD) ~1-3 nm, that are up to 20 times higher than for *attached* radon progeny, with an AMTD in the range 100 to 300 nm.

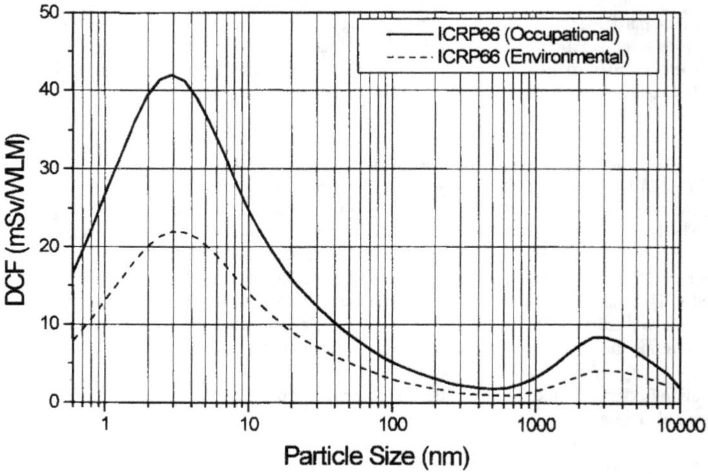

Figure 1: Effective dose per unit intake of 222Rn progeny as a function of particle size, as determined from the computer code RADEP. The RADEP-derived values have been adjusted by a factor of 0.3 to provide consistency with the epidemiologically-derived risk estimates.

2.2 Unattached fraction measurement

Originally, the *unattached fraction* was described in terms of the fraction of free molecular radon progeny atoms or ions, possibly associated with other molecules such as water. Usage of the term has expanded to cover *unattached fractions* for each daughter (f_a, f_b, f_c) and for the total PAEC (f_p). Diffusion techniques using tubes, parallel plates or wire screens are the most convenient and most common measurement type. Analysis of these measurements assumed a single constant diffusion coefficient for the unattached ^{218}Po, usually 0.054 $cm^2.s^{-1}$, based on the original estimates of Chamberlain and Dyson [22]. With the development of a wire screen penetration theory by Cheng and Yeh [23,24] and the use of a semi-empirical diffusion coefficient equation in the molecular cluster size range [25], collection efficiency curves can be estimated over both the ultrafine and the accumulation mode size ranges.

The left hand side of Figure 2 shows a typical sample configuration used by ARL for the measurement of unattached fractions. This system uses a single 4 cm diameter, 105 mesh wire screen operated at a flow rate of 1.5 lpm. The efficiency derived using the fan filtration theory of Cheng and Yeh [23]for the collection of radon progeny onto the backup filter, is plotted on the right hand side of Figure 2 This system has a d_{p50} (the particle size corresponding to the 50% collection efficiency point) of 4 nm. The backup filter collects >99% of attached radon progeny, assuming the particle size is greater than 100 nm, and <2% of unattached progeny has activity less than 1 nm. Without corrections for the presence of attached progeny on the backup filter, the single screen method will over estimate the unattached fraction for low values of *unattached fraction*. The reference filter is assumed to collect both attached and unattached progeny with 100% efficiency.

2.3 Measurement of ultrafine mode

The concept of unattached radon progeny constituting a clearly defined mode with a single particle size has been shown to be an over-simplification. Present evidence would suggest that the ultrafine mode comprises one or more peaks in the size range 0.5 to 10 nm, with the exact position and behavior of the mode dependent upon the gaseous medium. As shown in Figure 1, the bronchial dose conversion factors derived from the dosimetric models are sensitive to the diameter of the ultrafine mode. In the past few years, serial graded screen arrays (GSA) have been developed as extensions of single wire-screen techniques, to provide particle size information for the ultrafine mode [26 - 29]. The configuration of the ARL GSA is shown in the left hand portion of Figure 3. The radon progeny are sampled through a series of wire screens, and either diffuse onto the screens or are collected on the backup filter. The collection efficiency for each stage of the ARL GSA is shown on the right hand side of Figure 3. These curves were derived using the Cheng-Yeh equations [23], correcting the activity incident on each screen for the portion collected on preceding screens. The deposited radon progeny activities on the screens and the back up filter are usually determined from multiple alpha particle counts, obtained either by gross alpha counting or alpha spectroscopic methods.

2.4 Measurement of activity size distribution

The attached radon progeny are usually associated with the accumulation mode aerosol with activity median thermodynamic diameters (AMTD) of between 50 to 500 nm. Diffusion batteries utilising tubes, parallel plates, porous carbon or wire screens have all been used to measure radon daughter activity size distributions in this size range [30, 31]. Wire screen diffusion batteries have been in use since the early 1970's and have proven to be the most convenient of the available methods. For a particular set of wire screens and sampling configurations, the Cheng-Yeh equations can be applied to derive efficiency as a function of particle size for collections of particles on a filter located behind each stage of the diffusion battery. Figure 4 shows the sampling configuration and the collection efficiency curves for the ARL five-stage wire screen parallel diffusion battery with 50% efficiency cut-points of 4 nm, 16 nm, 65 nm and 250 nm, using different numbers and diameters of 105 mesh screens and a sampling rate of 1.9 lpm. For this battery, the decrease in collection efficiency, due to increasing contributions from the impaction terms, does not occur until particle sizes > 1200 nm (1.2 μm).

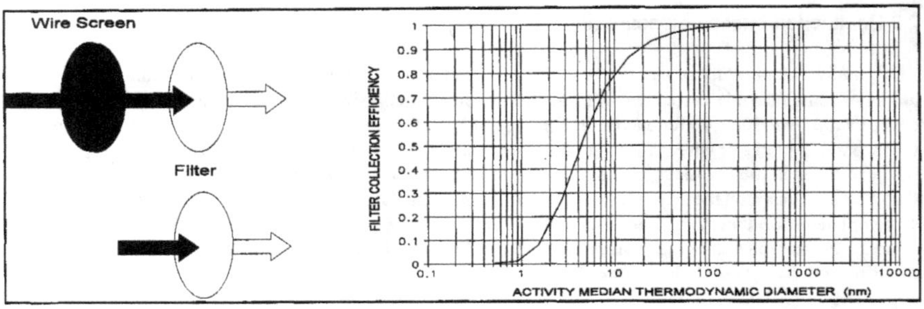

Figure 2. Sampling configuration and collection efficiency for ARL single wire screen measurement of unattached fraction.

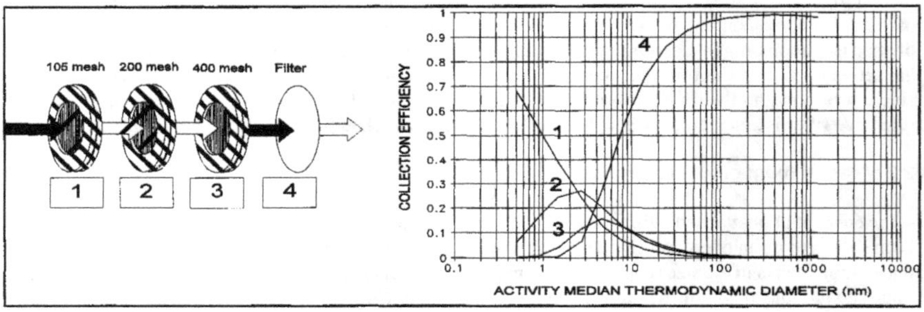

Figure 3. Sampling configuration and collection efficiency for ARL serial graded screen array for the measurement of particle size distributions in the 0.5 - 20 nm size range.

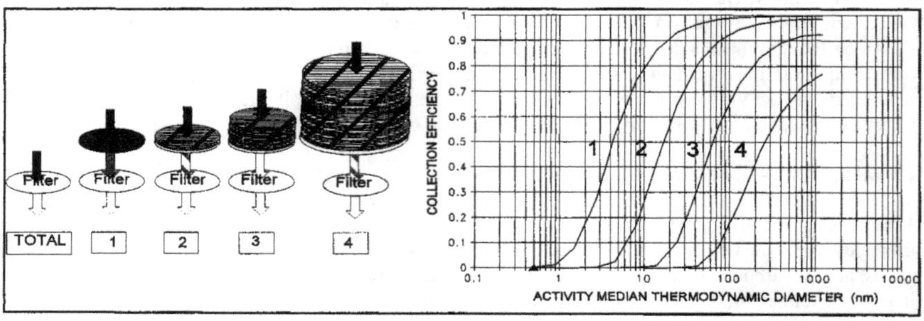

Figure 4. Sampling configuration and collection efficiency for ARL parallel multi-stage wire screen diffusion battery measurement of particle size distributions in the 0.5 - 500 nm size range.

2.5 Effective Dosimeter

The functional form of the DCF versus particle size response shown in Figure 1, is strongly affected by two sets of deposition processes. With size decreasing below 3 nm there is a dose reduction due to deposition losses in the nose (a pre-separation stage). Above 3 nm the dose decreases due to reduced deposition in the respiratory tract with increasing particle size (collection stage). This particle size response can be simulated with a sampler comprising two wire-screen collectors. A sampling system providing a close match to the DCF response function can be achieved using a wire screen pre-separator simulating the nasal deposition process and a wire screen collector matching the respiratory tract collection. The particle size response for a two screen system using a 100 mesh, 0.8 cm diameter pre-separator and a 400 mesh, 4.0 cm diameter collector at a sampling rate of $1.7 \times 10^{-5}\ m^3.s^{-1}$ (1.0 lpm) is shown by the solid curve in Figure 5. A two screen sampler with a collection efficiency, $\epsilon_{HE}(d_p)$ matched to the DCF response function $DCF(d_p)$, provides a measure of the radon progeny DCF value, irrespective of the exact form of the radon progeny particle size distribution.

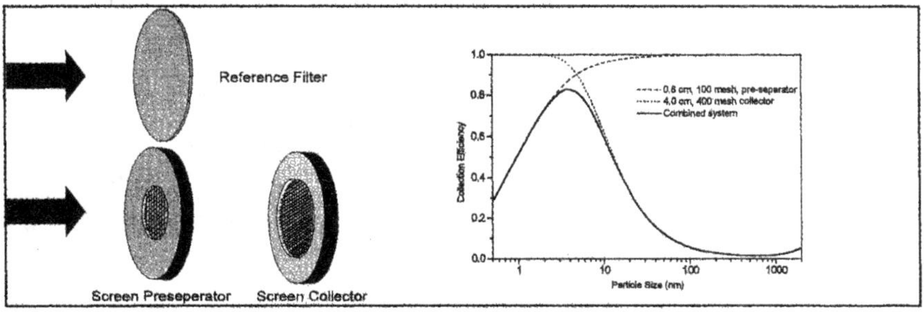

Figure 5. Sampling configuration and calculated collection efficiency of ARL *Effective Dosimeter* for a sampling rate $1.7 \times 10^{-5}\ m^3\ s^{-1}$ (1.0 lpm), a 0.8 cm diameter, 100 mesh pre-separator and a 4.0 cm diameter, 400 mesh screen collector.

Table 1 summarises five parameter sets for the ARL *Effective Dosimeter* [32], for sampling rates between $2.5 \times 10^{-6}\ m^3.\ s^{-1}$ (0.15 lpm) and $6.7 \times 10^{-5}\ m^3\ s^{-1}$ (4.0 lpm). These parameters are not unique and were chosen to provide a good qualitative match between $\epsilon_{HE}(d_p)$ and $DCF(d_p)$, the dose conversion factor for a particle with diameter d_p over the particle size range 0.6 nm to 1000 nm. For all five sampling systems a single linear scaling function can be used to derive an estimated dose conversion factor $DCF_\epsilon(d_p)$;

$$DCF_\epsilon(d_p) = DCF_{min} * (1.0 + 35.0 * \epsilon_{HE}(d_p)) \tag{1}$$

where DCF_{min} is the value of the dose conversion factor at the minimum in the particle size response function (~ 500 nm). For the RADEP implementation of the ICRP66 respiratory tract model, normalised to epidemiological data,

$DCF_{min.} = 285$ mSv/ (J h m^{-3}) (1.0 mSv/WLM) for environmental exposure (env), and
$DCF_{min} = 513$ mSv/ (J h m^{-3}) (1.8 mSv/WLM) for occupational exposure (occ).

Table 1. Selected screen parameters (pre-separator diameter d_s, collector diameter d_c) for *Effective Dosimeter* operated at sampling rates between $2.5 \times 10^{-6}\ m^3\ s^{-1}$ (0.15 lpm) and $6.7 \times 10^{-5}\ m^3\ s^{-1}$ (4 lpm).

	Inlet	Pre-separator		Collector	
Sampling Rate ($m^3\ s^{-1}$) (lpm)	Length (cm)	Mesh Number	d_s (cm)	Mesh Number	d_c (cm)
2.5×10^{-6} (0.15)	0.50	100	0.30	400	1.50
8.3×10^{-6} (0.50)	3.00	100	0.50	400	3.00
1.7×10^{-5} (1.00)	3.00	100	0.80	400	4.00
3.3×10^{-5} (2.00)	3.00	100	1.20	635	4.50
6.7×10^{-5} (4.00)	3.00	100	1.80	635	6.50

As a first approximation, the DCF is proportional to the collected fraction. For monodispersed radon progeny the DCF errors are estimated to be up to 35%. For polydispersed radon progeny distributions in the range 1 nm to 1000 nm, the DCF error was ±10% and ±20%, for occupational and environmental exposure, respectively. This error range can be reduced to ± 10% between 0.6 nm and 700 nm by using a polynomial fitting function [32]. The response to complex particle size distributions has been investigated using a data set of 70 experimental measurements taken as representative of environmental and occupational exposure. For environmental exposure, the derived mean DCF error and the root mean square (RMS) error were +3.6% and ± 4.8% for polynomial fitting functions, respectively. The corresponding values for occupational exposure were 0.5% and ± 3.1%. The simulation results show that the *Effective Dosimeter* is relatively insensitive to variations in sampling rate and in the screen parameters [32].

3. ARL Field Measurements of Radon Progeny Aerosol

In the period between 1989 and 1995, a programme of radon progeny aerosol measurements was carried out in a number of homes, a building, caves and an operational underground uranium mine [17 -20]. The sizing measurements used grab sampling with a combined 4 stage parallel diffusion battery and 4 stage serial graded screen array (PDB/GSA) and continuous sampling with a 5 stage parallel diffusion battery (CDB). The details are summarised in Table 2.The sampling locations for the under ground uranium mine were considered as two operational areas. Work areas with vehicles and equipment using diesel engines were designated as *diesel* areas, while areas with no diesel vehicles or equipment were termed *non-diesel*.

Table 2. Summary of ARL sizing measurements sampling locations during period 1990-1993.

Sampling Location	No. of Sites	No. of Samples	Period	Equipment used
House	6	75	1990-1993	PDB/GSA and CDB
Building	1	44	Nov 1993	PDB/GSA and CDB
Cave	1	2	Oct 1990	PDB/GSA
Uranium mine (*diesel*)	6	21	July 1992	PDB/GSA
Uranium mine (*non-diesel*)	1	4	July 1992	PDB/GSA

The activity size distributions obtained from these samples were analysed to determine:

- mean size-weighted dose conversion factors for occupational and environmental exposure
- average geometric mean of accumulation (attached) mode radon progeny
- average geometric standard deviation of accumulation mode
- estimated values of *unattached fraction* (f_p)for samplers with d_{p50} values of 4 nm and 15 nm.
- estimated values of the collection efficiency for the *Effective Dosimeter* (f_{HE})
- dose conversion factors for occupational (occ) and environmental (env) exposure

The analysis was performed for each of the 5 exposure types listed. The measurement results, averaged over each sampling location are summarised in Table 3.

Table 3. Average dose conversion factors (DCF), accumulation mode geometric means (GM) and geometric standard deviations (GSD), unattached fractions (f_p) and *Effective Dosimeter* efficiency from ARL measurements. Highlighted DCF values indicate values appropriate to exposure group.

Sampling Location	DCF(env) (mSv/WLM)	DCF(occ) (mSv/WLM)	GM (nm)	GSD	f_p(4nm) (%)	f_p(15nm) (%)	f_{HE} (%)
House	**5.0**	9.4	140	2.6	21.1	28.1	11.0
Building	**5.4**	**10.1**	115	2.7	18.1	26.7	11.8
Cave	**4.2**	**7.9**	170	2.5	14.2	19.6	8.6
U-mine *(diesel)*	3.2	**5.6**	270	2.6	4.0	10.7	5.5
U-Mine *(non-diesel)*	4.0	**7.1**	120	2.8	6.2	14.7	7.4

4. ARL Field measurements in show caves.

In March 1994 a 12-month study of ^{222}Rn levels in Australian show caves commenced. Radon monitors were placed in representative sampling sites in 52 cave systems in 24 locations around Australia. The measurements were restricted to established show caves and a total of 286 sites were monitored. A pair of passive, integrating radon monitors utilising CR-39 detectors were used to measure 3-monthly, seasonal, and annual radon levels at each site. The exposure period for the monitors was May 1994 to June 1995. The values for the radon concentration at each monitoring site ranged from ambient (< 20 Bq.m^{-3}), up to over 9000 Bq.m^{-3}. There was marked seasonal variability at most measurement sites; the highest value being measured during summer. Of the 67 cave tours in the study, 14 tours had time-weighted yearly average radon levels in excess of 1000 Bq m^{-3} [33]. The yearly radiation dose (annual effective dose) was calculated for 116 tour guides using the ICRP *conversion convention.* The estimated yearly radiation dose for 82 of these guides was less than 1 mSv, between 1 and 5 mSv for 30 guides, and between 5 and 10 mSv for the remaining 4 guides. The highest estimated radiation dose was 9 mSv per year.

In a follow-up study, three of the show caves, Jenolan, Buchan and Nelson, have each been monitored over a one week period using a 6 stage wire screen diffusion battery, one or more *Effective Dosimters* operated as a continuous alpha counting sampler, and a Genitron Alphagard radon monitor. A total of 19 sites were monitored to determine average values of radon, the radon progeny potential alpha energy concentration (PAEC), the equilibrium ratio, the *Effective Dosimeter* efficiency, the radon progeny DCF for occupational exposure, the effective dose rate for inhalation, derived from the product of the DCF and the effective dose rate normalised per unit radon. The measured values are summarised in Table 4.

Table 4. Summary of ARL measurements in caves during period 1996-1997.

Location	[Rn] (Bqm^{-3})	PAEC (mWL)	F	%f_{HE}	DCF(occ) (mSv/WLM)	Dose rate (μSv/h)	(μSv/h)/kBqm-3
Jenolan Caves							
Baal		190		18.2	**13.9**	15.7	
Katie's Bower		980		4.9	**5.4**	30.6	
Cathedral	1275	170	0.51	6.6	**6.6**	6.4	5.2
Queen's Canopy	3570	280	0.37	15.4	**12.1**	19.9	6.4
Nellie's Grotto	4800	460	0.37	10.0	**8.9**	23.9	5.1
Sink hole	650	130	0.78	4.9	**5.3**	4.4	6.7
Buchan Caves							
Kings Chamber		265		10.4	**9.1**	13.6	
Entrance	2340	230	0.38	8.1	**7.6**	10.3	4.5
Eastern Chamber	2400	240	0.38	9.9	**8.8**	12.6	5.3
Ivory Palace	2290	160	0.26	11.7	**8.0**	7.4	(3.2)
Amber Bank	1850	160	0.32	8.7	**6.1**	5.5	(3.0)
Princess Royal		340		6.4	**6.5**	12.8	
Skeleton	1770	210	0.45	7.6	**7.3**	9.1	5.2
Reid	1920	230	0.44	7.9	**7.5**	10.1	5.3
Font of the Gods	1970	180	0.34	11.4	**9.8**	10.4	5.3
Octopus	1290	120	0.34	8.3	**7.7**	5.3	4.1
Princess Margaret Rose Cave							
Entrance	490	61	0.51	8.1	**7.6**	2.7	6.8
Centre Section	530	66	0.47	6.4	**6.5**	2.5	4.7
End Section	510	85	0.50	6.3	**6.4**	3.2	5.0

DISCUSSION

The combined ARL results constitute a limited data set that may not be representative of Australian or international conditions. However, the data in Table 3 provides some indication of those exposure situations in which the use of the ICRP conversion convention may be inappropriate. For exposure to members of the public in Australian homes, buildings and caves, the ARL results in Table 3 are consistent with a DCF value of 5 mSv/WLM . For these locations, the higher values for *unattached fraction* are offset by lower breathing rate and the range of age groups. For occupational exposure in a uranium mine with diesel-powered equipment, the ARL dosimetric model-derived DCF value is within 10% of the *conversion convention.* Since the *conversion convention* was derived for uranium mine conditions, this provides confirmation of the normalisation value of 0.3 applied to the RADEP values in this paper.

The initial ARL data on occupational exposure for "clean air" situations in Table 3 shows that the dosimetric modelling approach leads to higher values for the DCF, up to a factor of two greater than the conversion convention values. These results are supported by the more detailed studies in Australian caves, summarised in Table 4. Since the radon levels in these caves exceed the intervention level, the tour guides would need to be considered as occupationally exposed workers, and the DCF for occupational exposure should be applied. For the three caves studied, the dosimetric-modelling leads to average DCF(occ) values in the range 6.8 - 8.7 mSv/WLM. If the dosimetric modelling results are correct then the use of the *conversion convention* for these exposure situations would underestimate the radon progeny effective dose by up to a factor of 1.7.

At the time when the ICRP 65 recommended the use of the *conversion convention* there was little data published on occupational exposures in "clean air" environments. There is now a developing body of research on aerosol conditions for these exposure situations and a number of groups are developing the use of dosimetric modelling to estimate DCF values from their data. Comparing the results of these studies is complicated by the diversity in the measurement methods and differences in the dosimetric modelling. Radon and radon progeny intercomparisons provide methods of ensuring uniformity of data for measurements of airborne concentrations. The International Radon Metrology Programme (IRMP) of the International Atomic Energy Agency (IAEA), has organised such intercomparisons both regionally and globally.

In order to ensure the consistency of pooled data, there is a need to extend these intercomparison programmes to cover methods for assessing radon progeny effective dose. The IRMP has organised a number of radon progeny aerosol intercomparisons with a limited number of participants [34 - 37], but there is presently no agreed approach to the estimation of radiation dose. The ARL *Effective Dosimeter* may provide one such method for deriving consistent radon progeny doses.

ACKNOWLEDGEMENTS

The assistance of ARL Staff, particularly members of the Health Physics Group, in the development of the measurement systems and the collection and analysis of field data. The input of Dr A.C. James and Dr A. Birchall in providing the RADEP results is acknowledged.

REFERENCES

[1] United Nations Scientific Committee on the Effects of Atomic Radiation. Sources, Effects and the Risks of Ionising Radiation, 1988 report to the General Assembly, with Annexes. New York: United Nations (1988)

[2] International Commission on Radiological Protection (ICRP) *Protection against radon-222 at home and at work.* Oxford:Pergamon Press; ICRP Publication 65, Ann. ICRP 23(2) (1993).

[3] National Health and Medical Research Council (NHMRC) and National Occupational Health and Safety Commission (NOHCS) 1995. Recommendations for Limiting Exposure to Ionising Radiation. (Guidance note [NOHSC:3022 (1995)]), Radiation Health Series; **39**. (Canberra: Australian Government Publishing Service)

[4] National Occupational Health and Safety Commission (NOHCS), 1995. National Standard for Limiting Occupational Exposure to Ionising Radiation. [NOHSC:1013 (1995)] (Canberra: Australian Government

Publishing Service)

[5] Samet, J.M. 1989. *Radon and Lung Cancer*. J. Natl.Cancer Inst. **81**:745-757

[6] Pershagen, G., Akerblom, G, Axelson,O, Clavensjo,B., Damber, L., Desai, G., Enflo, A., Lagarde, F., Mellander, H., Svartengren, M. and Swedjemark, G.A. Resedential radon Exposure and Lung Cancer in Sweden. New England Journal of Medicine. 330: 159- 164 (1994)

[7] International Commission on Radiological Protection (ICRP) *Human Respiratory Tract Model for Radiological Protection. A Report of Committee 2 of the ICRP*. Oxford:Pergamon Press; ICRP Publication 66, Ann. ICRP 24(1/4) (1994).

[8] NCRP. Deposition, *Retention and Dosimetry of Inhaled Radionuclides*. Bathesda, NCRP **125** (1997)

[9] National Research Council (NRC). 1991. *Comparative dosimetry of Radon in Homes and Mines.* Panel on the Dosimetric Assumptions Affecting the Application of Radon risk Estimates. Washington D.C. National Academy Press.

[10] Tu, K. and Knutson, E.O.. *Indoor radon progeny particle size measurements made with two different methods*. Radiat. Prot. Dosim. **24**:251-255 (1988).

[11] Strong, J.C. *The Size of Attached and Unattached Radon Daughters in Room Air*. J. Aerosol Sci. **19**:1327-1330 (1988).

[12] George, A.C., Hinchcliffe, L. and Sladowski, R.. *Size distributions of radon daughter particles in uranium mine atmospheres.* Am. Ind. Hyg. assoc. J. **36**:484-490 (1975).

[13] George , A.C. and Breslin, A.J.. *The distribution of ambient radon and radon daughters in residential buildings in New Jersy-New York area.* National Radiation Environment III, Vol. 2. CONF-780422. Oak Ridge, Tenn.: Technical Information centre, U.S. Department of Energy (1980).

[14] Tu, K., Knutson, E.O. and George, A.C. *Indoor radon progeny aerosol size measurements in urban, suburban and rural regions*. Aerosol Sci. and Technol. **15**:170-178 (1991).

[15] Reineking, A. and Porstendorfer, J.. *High-volume screen diffusion batteries and α-spectroscopy for measurement of the radon daughter activity size distributions in the environment. J.* Aerosol Sci. **17**:873-879 (1986).

[16] Hopke, P.K. 1990. *A critical review of measurements of the "unattached" fraction of radon decay products*. Radon Technical Report Series. DOE/ER-0451P. Washington D.C.: U.S. Department of Energy, Office of Health and Environmental Research

[17] Solomon, S.B. and Ren, T.. *Characterisation of Indoor Airborne Radioactivity*. Radiation Protect. Dosim. **45**: 323-327 (1992).

[18] Solomon, S.B. and Wilks, M. *Characterisation of Airborne Radioactivity in an Australian Dwelling*. Radiation Protect. Dosim. **56**: 109-11 (1994).

[19] Solomon, S.B., O'Brien, R.S., Wilks, M. and James, A.C. *Application of the ICRP's New Respiratory Tract Model to and Underground Uranium Mine*. Radiation Protect. Dosim. **53**: 119-125 (1994).

[20] Solomon, S.B. and Wang, Z. *Risk assessment for radon in an air-conditioned workplace*. In: Akber, A.A. and Harris, F. Radon and radon progeny measurements in Australia. Canberra: Supervising Scientist for Alligator Rivers Region. 83-90 (1994).

[21] Birchall, A. and James, A.C. *Uncertainty Analysis of the Effective Dose per Unit Exposure from Radon*

Progeny and Implications for ICRP Risk-Weighting Factors. Radiat. Protect. Dosim. **60**(4):321-326 (1995).

[22] Chamberlain, A.C. and Dyson, E.D. The dose dose to the trachea and bronchi from the decay products of radon thoron. Brit. J. Radiol. 29:317-325 (1956).

[23] Cheng, Y.S. and Yeh, H.C. *Theory of screen type diffusion battery.* J. Aerosol Sci. **11**:313-319.(1980).

[24] Cheng, Y.S. Keating,J.A. and Yeh, H.C. *Theory and calibration of screen type diffusion battery.* J. Aerosol Sci. **11**:549-556, (1980).

[25] Ramamurthi, M. and Hopke, P.K. *On improving the validity of wire screen "unattached" fraction Rn daughter measurements.* Health Phys. **56**:189-194, (1989).

[26] Holub, R.F. and Knutson, E.O.. *Measuring polonium-218 diffusion-coefficient spectra using multiple wire screens* In: Radon and its Decay Products: Occurrence, Properties, and Health Effects, P.K.Hopke, ed., Symposium Series 331, Washington D.C., American Chemical Society (1987).

[27] Holub, R.F., Knutson, E.O. and Solomon S.B. *Tests of graded wire screen techniques for measuring the amount and size distribution of unattached radon progeny.* Rad. Prot. Dosim. **24**:340-356, (1988).

[28] Ramamurthi, M. and Hopke, P.K. *Simulation studies of reconstruction algorithms for the determination of optimum operating parameters and resolution of graded screen array systems (non-conventional diffusion batteries).* Aerosol Sci. Technol. **12**:700-710, (1990).

[29] Solomon, S.B. and Ren, T. *Counting efficiencies for alpha particles emitted from wire screens.* Aerosol Sci. and Technol. **17**:69-83, (1992).

[30] Sinclair, D., George, A.C. and Knutson, E.O. *Application of diffusion batteries to measurement of submicron radioactive aerosols.* pp103-114 in Airborne Radioactivity. La Grange Park, Ill.:American Nuclear Society (1977).

[31] Portsendorfer, J. *Radon: Measurements Related to Dose.* Environment International **22**, Suppl.1 :PPS563-S583, (1996).

[32] Solomon, S.B. *A Radon Progeny Sampler for the Determination of Effective Dose.* Radiat. Protect. Dosim. **72**: pp71-42 (1997).

[33] Solomon, S.B., Langroo, R., Peggie, J.R., Lyons, R.G. and James, J.M. *Occupational exposure to radon in Australian tourist caves: an Australia-wide study of radon levels.* Australian Radiation Laboratory Technical Report ARL/TR119, (1996).

[34] Knutson, E.O., Tu, K.W., Solomon, S.B. and Strong, J. *Intercomparison of Three Diffusion Batteries for the Measurement of 222Rn Decay Product Particle Size Distribution.* Radiat. Protect. Dosim. **24**:261-264 (1988)

[35] Hopke, P.K., Strydom, R., Ramamurthi, M., Knutson, E.O., Tu, K.W., Scofield, P., Holub, R.F., Cheng, Y.S., Su, Y.F., Winklmayr,, W, Strong, J.C. and Solomon, S.B. *The Measurement of Activity Weighted Size Distributions of 222Rn Progeny: Methods and Laboratory Intercomparison Studies.* Health Phys. **63**:560-570 (1992)

[36] Reineking, A., Knutson, E.O., George, A.C., Solomon, S.B., Kesten, J. Butterweck, G, Portsendorfer, J. *Size Distribution of Unattached and Aerosol-Attached Short-lived 222Rn Decay Products: Some Results of Intercomparison Measurements.* Radiat. Protect. Dosim. **56**:113-118 (1994).

[37] Tu, K.W. *International Intercalibration and Intercomparison Measurements of Radon Progeny Particle Size Distribution.* Environmental Measurements Laboratory, EML-589, July 1997, New York.

Study of indoor radon levels in Garhwal Himalayas using solid state nuclear track detector

R. C. Ramola

Department of Physics, H.N.B.Garhwal University Campus
Tehri Garhwal - 249 001, India

Abstract

The measurements of radon in indoor and outdoor environment by using LR-115 type II, plastic track detector in bare mode have been carried out in four districts (Tehri, Pauri, Utterkashi and Dehradun) of Garhwal Himalaya. One year data has been obtained from each house which is distributed in four quarters viz. Winter, Summer, Rainy and Autumn. About 350 houses have been surveyed and the radon levels were found to vary from 25 to 395 Bq/m^3. The annual average radon values were found to vary from 59 Bq/m^3 to 303 Bq/m^3. The indoor radon concentration was found to vary with seasonal variation, building material and mode of the construction of house. The winter/summer ratios for different type of houses vary substantially among different places. The radon concentration was found higher in mud houses due to the typical houses construction and building material used. More that 50% values of radon concentrations in mud houses are recorded above the ICRP recommended values and there is a specific need to measure the effect to the people living there. The results show that the building materials, mode of construction of house and the seasonal variations should be taken in account to calculate annual effective dose equivalent.

1. Introduction

A major contribution to the natural radiation dose rate arises from inhalation of short-lived radioactive decay products of ^{222}Rn. The effect of much lower values usually found in homes and other building was unknown, but a linear extrapolation the results given by Sevc et al.[1] to the levels found in homes suggests that thousands of people per year could be affected in U.S. by increase in radon level that are caused by protecting structures against heat loss and public concern over radon in homes is increasing. Since then a number of measurements of radon were carried out all over the world[2-10]. Elevated levels of radon have been observed in houses in a number of countries including the United States, Sweden and the United Kingdom[11]. These elevated levels can exceed the international control guidelines and may contribute to an increased risk of lung cancer. They are found to be associated with a number of factors including soil porosity, uranium content of the soil, building material, mode of construction, ventilation and meteorological variables. It is therefore desirable and possibly necessary to measure average concentration of radon over a period of time at different places.

This paper presents some results of radon survey carried out in indoor atmosphere of Garhwal Himalaya. These measurements were made during the period from November 1993 to October 1996. Most of the houses in the area are built with soil and local stone. The ground floor and walls are coated with a thin layer of mud and ceiling is made of wood and slates. The effect of different parameters such as building materials, ventilation, seasonal variations etc. has also been observed in order to locate the actual radon rich areas.

2. Calibration of Detector

The experiment on the calibration of LR-115 type II, plastic track detector was performed at Environmental Assessment Division, Bhabha Atomic Research Centre, Bombay. The detector films (2.5 cm X 2.5 cm) were exposed in bare mode in a calibration chamber of known radon concentration. The condition was kept similar to that in the field measurements. In all 12 samples were exposed and an average value was obtained as 3.12×10^{-2} tracks/cm^{2}d = 1 Bq/m^3.

3. Method

The measurements of radon in indoor atmosphere have been carried out by using LR-115 Type II, plastic track detector. The detector film attached on a glass slide is suspended inside the room at a height of about two meters from the ground floor. After an exposure time of three months, films are removed and etched in 2.5 N NaOH solution for two hours at 60 °C in a constant temperature bath. The films are then scanned under an optical microscope for track density measurements. The recorded track density was converted in Bq/m^3 by using the calibration factor as discussed above.

4. Results And Discussion

The results of recorded radon concentration in the houses of Tehri Garhwal are given in Table. 1. In general, the indoor radon concentrations are found quite high under normal ventilation condition. The average indoor radon concentration is found to vary from 33 Bq/m^3 to 199 Bq/m^3. The results reveal that in most of the areas the radon concentration is high in winter while it is low in summer. However, the average values show mere 25% variation. This low difference may be due to cold climate of the area throughout the year and houses are not highly ventilated during summer. The major source of radon inside the house is underlying soil and building materials with possible variation due to poor ventilation.

Table 1 Radon concentration and equivalent dose in the dwellings of Garhwal Himalayas

Place	Average Radon Concentration (Bq/m^3)	EECRN (Bq/m^3)	Dose (mSv/y)
Padiyargaon	199	90	7.9
KotiColony	122	55	4.8
Tehri	110	50	4.3
New Tehri	119	54	4.7
Nail	119	54	4.7
Chamma	109	49	4.3
Dikholgaon	76	34	3.0
Thanegaon	64	29	2.5
Sabligaon	72	32	2.8
Khanda Srikot	178	80	7.0
Srinagar	75	34	2.9
Chauras	69	31	2.7
Kotdwar	33	15	1.3
Malideval	133	60	5.2
Serain	82	37	3.2
Rajgaon	118	53	4.7
Uppu	48	22	1.9
Dang	54	24	2.1

The high values of radon in the area are mainly due to the typical construction of the houses. Most of the houses in the villages (particularly old houses) are mud houses, built with soil and local stone and having poor ventilation conditions. People in this area use wood fuel inside the house and the resulting smoke may increase the circulation of radon daughters. Other major causes may be the continuous land slides, several hot water spring and many small rivers (local GADS) in the area. During rainy season, the increased flow of rivers carry the soil from mountains which is deposited at river bank at different places on its way. It is also observed that indoor radon concentration in Tehri town and surrounding places is relatively high than that in other places. This may be due to the heavy construction work in the area. Due to processes of underground digging and rock blasting, the resulting dust goes up with the wind and deposited on the window pans and doors and thus increase the indoor radon concentration.

The radon levels were measured in different seasons of the year in the houses of Garhwal Himalaya, Northern India. The periods of measurements are Winter (January-March), Summer (April-June), Rainy (July-September) and Autumn (October-December).

Based on the available data the annual average indoor radon concentration in India is calculated as 33.2 Bq/m^3 [12] against the global average of 40 Bq/m^3 [13]. Most of the indoor radon data in India is scattered and a limited work has been done particularly on the effect of seasonal variations to indoor radon levels. It is reported that there is no significant variation of indoor radon levels for different seasons in a tropical country like India. However, the situation is different for hilly regions in India as these areas have cold climate for most of the time in a year. Our study area falls in the same regions. It is therefore necessary to measure the radon for different seasons. Unusual weather condition, air condition in the dwelling and so on are to be taken in account for the extrapolation to an annual average value of the measured radon concentration. The behaviour of indoor radon in different type of houses for different seasons is discussed as follows:

Cemented Houses

In this category, two different type of houses were selected for radon study i.e residential houses and work places:

Residential Houses

The results reveal that 32% houses in the winter, 39% houses in the autumn and 29% houses both in the summer and rainy season have radon levels in the range of 101 Bq/m^3 to 150 Bq/m^3. On the other hand the corresponding number of houses are reverse for the radon levels in the range of 51 Bq/m^3 to 100 Bq/m^3. 57% houses in the summer, 43% houses in the rainy season, 29% houses in the winter and 36% houses in the autumn have the radon levels in this range. The number of houses having radon values exceeding 200 Bq/m^3 are relatively less i.e. 11% and 3.5% in the winter and the summer, respectively and 7% in both the autumn and the rainy season. Overall, the indoor radon levels are found higher during the winter and the autumn than those in the summer and the rainy season.

Work Places

The results from cemented houses as working places (office, school, etc.) show that 34% of houses in the winter, 27% houses in the autumn, 21% houses in the summer and 14% in the rainy season have recorded the radon levels in the range of 151 Bq/m3 to 200 Bq/m^3. On the other hand 38% houses in the summer have recorded the radon levels in the range of 51 Bq/m^3 to 100 Bq/m^3.

The radon levels were found higher in the winter for both categories of the houses. However, the overall average radon levels were observed higher in work places than those in residential houses. Since the work places remain closed during week ends, holidays and vacations, the radon is accumulated inside the room due to poor ventilation condition. As such higher radon levels were observed in the houses of this category.

Mud Houses

In this category of houses again two different types of houses are selected for radon study i.e residential houses and work places:

Residential Houses

From the data obtained for this category of houses, it was observed that 23% houses in the winter, 30% houses in autumn, 10% houses in the summer and 16% houses in the rainy season have radon levels in the range of 151 Bq/m^3 to 200 Bq/m^3. On the other hand the frequency of the houses increases for the range of 51 Bq/m^3 to 100 Bq/m^3, which is maximum (40%) for the summer and minimum (26%) for both the winter and the rainy season. Average value is again observed higher in the winter. However, the frequency of the houses having radon levels higher than 200 Bq/m^3 is increased. It is 30% and 20% for the winter and the autumn, respectively and 16% for both summer and rainy season.

Work Places

From the results of radon measurements for different seasons in the houses used as work places was observed that 56% houses in the winter have the radon levels in the range of 151 Bq/m^3 to 200 Bq/m^3. While the number of houses having radon levels in this range are 11%, 27% and 22% in the summer, the autumn and the rainy season, respectively. For most of the seasons, the radon levels are recorded higher in the range of 51 Bq/m^3 to 100 Bq/m^3. The frequency of houses for this range are 45% in the autumn and 44% in both the summer and the rainy season. However, only one house has recorded the radon level below 100 Bq/m^3 in the winter. The average annual radon levels were recorded higher than 200 Bq/m^3 in two houses.

5. Analysis of the Seasonal Variation

The comparisons of observed indoor radon values in different seasons are shown in Tables 2 and 3. It was observed that the average indoor radon concentration in the winter is higher than that in any other season both in cemented and mud houses. The winter/summer, winter/rainy and winter/autumn ratios were found to vary from 0.51 to 2.13 (average 0.63 to 1.64), 0.58 to 2.26 (average 0.59 to 1.56) and 0.56 to 2.29 (0.64 to 1.88), respectively, in cemented houses. In mud houses, the winter/summer, winter/rainy and winter/autumn ratios were found to vary from 0.59 to 3.47 (average 1.02 to 2.22), 0.50 to 4.24 (average 0.88 to 1.74) and 0.41 to 5.07 (0.91 to 2.22), respectively. The overall average value of winter/summer ratio was found to be maximum in both cemented and mud houses. It was found 1.16 and 1.42 in cemented and mud houses, respectively. As such the average radon level was recorded maximum in the winter and minimum in the summer. However, in few selected houses the winter/summer ratio was found less than one i.e high radon value was recorded in the summer than in the winter. This abnormal behaviour was mainly in the schools which remained closed for a long time during the summer. A few residential houses have also recorded high radon levels in summer which perhaps may be due to the same reason as the occupants of the houses may have shifted temporarily to other places for a short duration.

6. Conclusions

On the basis of the data recorded so far, it is concluded that indoor radon concentration Garhwal Himalaya is strongly dependent on the building materials and type of construction of the house. Since most of the houses in the villages of this area are constructed with local stone and soil (mud houses), the high radon values recorded in mud houses of the region have significant importance for further study. This data provides a complete distribution pattern of environmental radon in Garhwal Himalayas and resulting annual exposure dose can be calculated for the people living there.

Table 2 Seasonal variation of indoor radon concentration in cemented houses

Place	Number of measurements	Ratio of indoor radon concentrations		
		W/S	W/R	W/A
Tehri	7	1.40	1.44	1.05
New Tehri	8	1.16	1.15	1.17
Padiyargaon	1	0.63	0.59	0.64
Nail	2	1.25	1.23	1.15
Chamba	3	1.64	1.91	1.65
Dikholgaon	1	1.22	1.09	0.86
Thanegaon	1	1.10	0.89	1.02
Serain	2	1.00	1.00	0.78
Uppu	3	1.00	1.21	0.85
Chham	3	1.22	0.96	1.10
Koti Colony	2	1.49	1.11	0.87
Chauras	2	1.12	0.80	1.18
Nageshwarsaur	4	1.00	1.17	0.94
Pratapnagar	2	1.32	0.92	0.96
Nagni	2	1.37	1.51	1.27
Narendranagar	2	0.76	0.53	----
Lambgaon	1	1.34	1.11	1.88
Ghansali	4	1.08	1.56	1.27
Chamiyala	3	1.09	0.98	0.93
Ghonti	2	1.27	0.89	1.57
Average		1.16	1.05	1.11

W-Winter, S-Summer, R-Rainy, A-Autumn

Table 3 Seasonal variation of indoor radon concentration in mud houses

Place	Number of measurements	Ratio of indoor radon concentrations		
		W/S	W/R	W/A
Tehri	3	2.22	1.30	1.30
Nail	6	1.02	1.23	0.99
Manjur	3	1.09	0.98	0.92
Dikholgaon	2	1.23	1.14	1.03
Thanegain	1	1.48	1.09	1.07
Sabligaon	2	1.81	1.00	----
Malideval	3	1.80	1.36	1.23
Rajgaon	1	1.26	1.01	1.05
Uppu	2	1.12	1.09	0.91
Dang	2	1.45	0.88	1.47
Changora	4	1.46	1.47	1.40
Pratapnagar	2	1.27	1.57	1.34
Ghonti	1	1.20	1.74	2.22
Average		1.42	1.22	1.25

W-Winter, S-Summer, R-Rainy, A-Autumn

Acknowledgements - Authors are thankful to BRNS, DAE, Govt of India for providing financial assistance under research scheme No. 27/1/96-R&D-II.

7. References

[1] Sevc J., Kunz E. and Placek V. (1976) Health Phys. 30, 433.
[2] Abu Jarad F. and Fremlin J.H. (1982) Health Phys. 42, 82.
[3] Keller G. and Folkerts H.H. (1984) Health Phys. 47, 385.
[4] Nazaroff W.W. and Doyle S.M. (1985) Health Phys. 48, 265.
[5] Ramachandran T.V., Lalit B.Y. and Mishra U.C. (1986) Nucl. Tracks Radiat. Meas. 11, 245.
[6] Ramola R.C., Singh M., Singh S. and Virk H.S. (1987) Ind. J. Pure Appl. Phys. 25, 127.
[7] Ramola R.C., Singh M., Singh S. and Virk H.S. (1992) J Environ. Radioactivity 15, 95.
[8] Ramola R.C., Rawat R.B.S. and Kandari M.S. (1995) Nucl. Geophys. 9 (1995) 383-386.
[9] Segovia N. and Cejudo J. (1984) Nucl. Tracks Radiat. Meas. 8, 407.
[10] Wolfs F., Hofstede H., De Meijer R.J. and Put L.W. (1984) Health Phys. 47, 271.
[11] UNSCEAR (1988) Effects and Risk of Ionizing Radiation. New York, United Nations.
[12] Ramachandran T.V. (1996) Proc. 10th National Symp. on SSNTDs, Kurukshetra, pp. I1-I12.
[13] UNSCEAR (1993) United Nations Scientific Committee on the Effects of Atomic Radiation,New York

The Measurement of Thoron Concentrations in Australia using the Japanese Passive R-T Dosimeter

L F Toussaint[1], S Tokonami[2], M Doi[2], S B Solomon[3] and J R Peggie[3]

[1]Radiation Health Section,
Health Department of Western Australia.

[2]National Institute of Radiological Sciences,
Inage, Japan

[3]Australian Radiation Laboratory,
Melbourne, Australia.

Abstract

Previous studies have shown higher gamma radiation levels and higher soil thorium concentrations in the Darling Scarp area to the east of the city of Perth in Western Australia. It was considered desirable to measure the concentrations of the ^{220}Rn (thoron) in houses built near such areas and compare these measurements with similar measurements at other sites within Australia. A monitor developed for surveying ^{220}Rn and ^{222}Rn levels in Japan was considered most suitable for this purpose. The monitor used for the survey was a passive radon-thoron discriminative dosimeter, similar in type to that used to measure thoron concentrations in traditional Japanese houses. Houses for this study were selected from areas in the Darling Scarp. To provide data on what might be considered typical thoron concentrations associated with other soil types within Australia, a small number of houses in areas on the coastal plain of Western Australia and from the coastal areas of Victoria were also monitored. The mean thoron concentration in homes on the Darling Scarp of 14.0 Bq.m^{-3} was higher than that in the coastal plain in Western Australia (3.0 Bq.m^{-3}) but comparable to the measurements in Victoria (11.3 Bq.m^{-3}). For the radon measurements in homes, the mean Scarp concentration was typically 49.2 Bq.m^{-3} while the WA coastal plain and in Victoria mean concentrations were 11.6 Bq.m^{-3} and 15.4 Bq.m^{-3} respectively. For outdoors in the Darling Scarp, the thoron mean was 76.3 Bq.m^{-3} while the radon level was 18.2 Bq.m^{-3}. These measurements would suggest that the source of the thoron in homes on the Scarp is from the underlying soil rather than from the building materials.

1. Introduction

The city of Perth is located on the west coast of Australia. The Darling Scarp area of Western Australia (WA) lies some 30 km to the east of the city of Perth. Whereas the coastal plain consists primarily of limestone and sand-plain, the Darling Scarp area is composed of other types of soils, including clays and laterite (particularly pisolitic laterite in some areas). These soils have a higher thorium content than normal soils. The soils in the coastal area near Perth, Western Australia have radionuclide levels within typical global ranges of 10 - 50 Bq.kg^{-1} for both uranium and thorium [1]. The thorium content in the Darling Scarp is much higher than the world mean and indeed falls into the upper range quoted by UNSCEAR for soils[1]. The equivalent ranges for uranium and thorium activity concentration, measured in gravel soils in the Darling Scarp are 20 - 110 Bq.kg^{-1} and 40 - 500 Bq.kg^{-1}, respectively [2]. The question arises as to whether this elevated thorium content is reflected in increased radiation levels within dwellings in the area.

Studies carried out by Yeates and King [3] showed gamma radiation levels in the Darling Scarp which were higher than for the WA plain. Gamma levels up to 0.4 μGy.h^{-1} (and even slightly higher in some small localised areas) were measured. A later study by Toussaint [4] showed that the thorium in the soil was the main contributor to the higher gamma levels. Radon (^{222}Rn) measurements have shown that the concentrations in homes situated on the Darling Scarp are typically in the range 25 - 40 Bq.m^{-3} [5]. These values are about three times higher than for homes located on the coastal plain and other locations in Western Australia where the median indoor radon concentration was determined to be 11 Bq.m^{-3} [6,7].

There have been no corresponding surveys of thoron (^{220}Rn) concentrations in Australian dwellings. Given that higher levels of thorium are present in Darling Scarp soils, it was considered desirable to measure ^{220}Rn concentrations in air in, and outside, houses in the Darling Scarp area. This study describes the results of a small-scale survey of ^{220}Rn concentrations in a number of homes in the Darling Scarp area and compares these values with ^{220}Rn levels measured elsewhere in Australia. This study was a collaboration between the Radiation Health Section of the Health Department of WA, the Australian Radiation Laboratory and the National Institute of Radiological Sciences, Japan.

2. Methodology

Passive ^{220}Rn Monitor

While there have been several passive monitors available for measuring ^{222}Rn concentrations, it is only relatively recently that a reliable passive monitor has become available to measure concentrations of ^{220}Rn [8,9,10]. The passive monitor has the advantage that it can be kept in place for long periods of time and hence provide a more accurate indication of the long-term thoron concentration when compared with simple grab sampling.

The passive radon-thoron discriminative dosimeter used in this study was provided by the National Institute of Radiological Sciences, and is similar in type to the monitor used to measure Japanese homes. Two pieces of polycarbonate disk (Iupilon[1]), 300 μm in thickness, 50 mm in diameter are installed as solid state nuclear track detectors (SSNTD) in the two hemispherical diffusion chambers of diameter 110 mm and 70 mm respectively. The detector housing is made of anti-static polycarbonate (Carbon Black FD-9054F[2], which is a compound of Ketjenblack[3]), because of its low electrical resistance (which is less than 5.3 x 10^3 Ω cm) and negligible radioactivity. Radon and thoron enter the 110 mm-hemispheric chamber through glass fiber filter set on the fan-shaped air inlet of size 36.2 cm^2, then they diffuse into the 70 mm hemispherical chamber through a pinhole air inlet of size 1 mm in diameter.

After the exposure, the polycarbonate disk in the 110 mm hemispherical chamber was subjected to preliminary chemical etching (CE) for 3 h, which was performed at 30°C with a blend of 8.0 N KOH solution (80% volume) and C_2H_5OH (20% volume). It was then subjected to electrochemical etching (ECE) under an alternating current of 800 V (2kVrms), 2kHz. In this condition, the alpha-tracks produced by thoron and its progeny are registered more effectively than those by radon and its progeny.

---

1 Iupilon (type FE-2000 for printing) is a registered trade mark of the polycarbonate produced by Mitsubishi Gas Chemical Company, Inc., Mitsubishi Bld. 2-5-2, Marunouchi, Chiyoda-ku, Tokyo, 100 JAPAN, and 17th floor, 520 Madison Avenue, New York, NY10022, U.S.A

2 Carbon black polycarbonate plastic (type FD-9054F) is a product of Lion Cooperation, 1-2-22, Yokoami, Sumida-ku, Tokyo 130, JAPAN

3 Ketjenblack is a registered trade mark of the carbon compound produced by Ketjen Black International Company, The Netherlands.

In contrast, the polycarbonate disk in the 70 mm-hemispheric chamber is subjected to CE for 0.5 h followed by ECE for 3 h. In this condition, the registered etch pits originate predominantly from radon and its progeny.

The lower detection limits are 3.4 $Bq.m^{-3}$ for radon and 5.3 $Bq.m^{-3}$ for thoron respectively after 3 months exposure. More detailed technical information on the dosimeter is given in the papers by Doi et al. [8,9,10].

Monitoring Sites

For this study, a small number of houses from areas on the Darling Scarp were selected. To provide data on what might be considered typical thoron concentrations for other soil types within Australia, a small number of house were selected in areas on the coastal plain of Western Australia and from the coastal areas of Victoria. The monitors were placed in each of the homes for a period of 3 months, after which there were removed and returned to Japan for assessment.

Darling Scarp

Homes in the Darling Scarp area were selected on the basis that previous measurements of gamma radiation levels and soil thorium content had shown higher values than similar measurements for areas on the WA coastal plain. In the Darling Scarp area, 5 homes were selected. Four of these were from the Roleystone area (Rol.1 , Rol. 2, Rol.3 and Rol. 4) and one was from the Lesmurdie area (Les. 1). In general 3 monitors were used for each home that was tested. Two of these monitors were placed in an area representing the normal inside area used by the occupants. The third monitor was placed in an outside area, such as a garden shed or similar, where thoron was possibly emanating directly from the soil.

Perth Coastal Area and South West

To provide a comparison with the measurements in homes with high thorium soil content, two homes considered as typical of the Perth Coastal Plain were selected from areas which had not previously been classified as excessively high either in gamma radiation levels or radon levels. One home was in Nedlands (Ned.1) the other in Shelley (Shel.1). One other location on the south coastal plain north of Bunbury in Western Australia was also monitored (Bun.1). These homes were built over the sandy soils which had relatively low concentrations of thorium.

Melbourne Suburban Area

A total of 22 monitors were distributed to 11 staff members at the Australian Radiation Laboratory, for placement in the bedroom and living room in their homes. The staff were chosen because the location of their homes provided a relatively wide distribution of sites around suburban Melbourne. The exposures were carried out for a 3-month between December 1995 and March 1996, during the Australian summer.

3. Results

3.1a Coastal Areas - Western Australia

The coastal area (inside houses) was monitored commenced on 31 May 1995 for a period of approximately 3 months. The results are given in table 3.1a

Table 3.1a Radon and thoron levels in dwellings on the coastal plain.

Location	Days	Rn(Bq.m^{-3})	Tn(Bq.m^{-3})	Comments
Ned.1	114	27.9	0	Living Room
Shel.1	101	9.7	4.9	Living Room
Shel.1	101	7.5	6.3	Family
Shel.1	101	8.5	5.2	Bed Room
Bun.1	101	4.6	0.8	South Coastal
Mean		*11.6*	*3.0*	

3.1b Coastal Areas - Melbourne, Victoria

The area was monitored during December 1995 through to March 1996. The results are given in table 3.1b

Table 3.1b Radon and thoron levels in dwellings in suburban Melbourne.

		Bedroom		Living Room	
Location	Days	Rn	Tn	Rn	Tn
Viewbank	107	20.7	0	12.9	11.5
Eltham	109	25.5	6.8	38.7	7.6
W Heidelberg	112	6.8	5.6	12.1	0
Beaumaris	121	3.8	1.6	3.2	1.4
Kensington	109	7.0	3.0	2.7	6.3
St Andrews	99	53.1	43.8	42.1	39.1
Windsor	109	5.4	6.5	5.8	14.3
Camberwell	135	22.5	4.6	18.4	30.8
Ferntree Gully	135	5.8	0	2.5	0
Kinglake West	122			11.3	8.9
Northcote	106	10.0	7.5	12.3	43.4
Mean:		*16.1*	*7.8*	*14.7*	*14.5*

3.2a Darling Scarp Area (inside houses)

The Darling Scarp area (inside houses) was monitored from 31 May 1995 for a period of approximately 3 months. The results are given in table 3.2a

Table 3.2a Radon and thoron levels in dwellings in the Darling Scarp.

Location	Days	Rn (Bq.m^{-3})	Tn (Bq.m^{-3})	Comments
Rol.1	114	9.1	9.3	Downstairs
Rol.1	114	10.6	0.9	Upstairs
Rol.2	114	21.4	1.2	Living room
Rol.3	112	43.1	12.4	Living room
Rol.3	112	38.2	19.2	Bed room
Rol.4	127	19.8	8.5	Bed room
Rol.4	127	29.8	11.0	Living room
Les.1	112	125.2	31.8	Living room
Les.1	112	145.9	31.8	Bed room
Mean:		*49.23*	*14.0*	

3.2b Darling Scarp Area (outside houses)

The Darling Scarp area (outside houses) was monitored from 31 May 1995 for a period of approximately 3 months. The results are given in table 3.2b

Table 3.2b Radon and thoron levels around dwellings in the Darling Scarp.

Location	Days	Rn (Bq.m^{-3})	Tn (Bq.m^{-3})	Comments
Rol.1	114	10.4	60.6	Outside
Rol.2	114	8.5	17.0	Outside
Rol.2	114	21.8	5.4	Outside
Rol.3	112	49.8	351.5	Near rock
Rol.4	127	11.6	9.5	Outside shed
Les.1	112	6.9	13.5	Front porch
Mean:		*18.2*	*76.3*	

4. Discussion

This current study was an preliminary thoron study with a relatively small sample size. The monitoring was effective in showing the large variations in thoron concentrations, often even within the same house. These variations are primarily due to the relatively short half-life of the thoron and the monitor to thoron source distances involved. While there appears to be a difference between Darling Scarp mean thoron concentrations and the limited data for the WA Coastal Plain, there is no such difference between the Darling Scarp data and the Victorian data. The mean Scarp concentration was 14.0 Bq.m^{-3}, while the Melbourne coastal mean was 14.7 Bq.m^{-3}. Areas such as St Andrews, Northcote and Camberwell in fact have higher, or comparable thoron concentrations to the Darling Scarp. These concentrations were in the range 30-44 Bq.m^{-3}. St Andrews also had higher radon concentrations (up to 53 Bq.m^{-3}). These variations are possible due to differences in house construction rather than an indication of increase thoron production.

For the radon concentrations, the Scarp area was much higher than the coastal areas in both Western Australia and Victoria. The mean Scarp concentration was 49.2 Bq.m^{-3} while the WA coastal and Victorian coastal areas were 11.6 Bq.m^{-3} and 15.4 Bq.m^{-3} respectively. The monitoring that was done outside homes in the Scarp had higher concentrations of thoron (and lower concentrations of radon) than for inside houses in the Scarp. The mean outside thoron level was 76.3 Bq.m^{-3} while the radon mean was 18.2 Bq.m^{-3}. This data suggests that the source of the thoron and radon is primarily from the soil over which the houses are built rather than from the brick building material. Although the mean thoron concentration appears to be relatively high, it is biased somewhat by monitor Rol 3. Also, as this monitoring was generally done within 1 metre of the ground, it does not represent the normal breathing zone for adults.

As thoron concentration reduces with height above emanating soil and as out-door occupancy (20%) is usually lower than indoor occupancy (80%), the relatively higher out-door concentrations may not contribute significantly to overall adult dose. It may be desirable however, in a future study, to monitor situations where occupants could spend considerable time close (ie less than 0.5 m) to a soil from where thoron emanation is known to be high[11]. This may apply to such situations as:

- children's play areas where a child's breathing zone is closer to soil.

- caravan and camping areas where the breathing zones may be close to the soil and there isn't the normal protection of a solid building to isolate the thoron from the breathing zone.

5. Acknowledgments

The authors wish to express their gratitude to Dr Kobayashi (Director NIRS, Japan), Prof. Dr Ikebe and Dr Iida (Nagoya University) and Prof. Dr Kurosawa (Waseda University) for their guidance on the analytical details of this project.

6. References

[1] UNSCEAR (1993), Sources, Effects and Risks of Ionising Radiation. United Nations Scientific Committee on the Effects of Atomic Radiation 1993. Report to the General Assembly, with annexes (New York, NY. United Nations).

[2] Alach Z J, Broun B M, Breheny R S and Toussaint L F (1996). Radionuclide Concentrations in the Darling Scarp of Western Australia. Radiation Protection in Australia, Vol. 14, No. 2 (April),pp. 35-38. Australian Radiation Protection Society.

[3] Yeates D B and King B E (1973). Estimation of the Gamma-ray natural background radiation dose to an urban population in Western Australia. Health Physics, Pergamon Press 1973. Vol. 25, (Oct), pp. 373-379.

[4] Toussaint L F (1985). Background Radiation in Western Australia. Radiation Protection in Australia, Vol. 3, No. 4, pp. 151-155. Australian Radiation Protection Society.

[5] Toussaint L F (1994). Radon Concentrations in Western Australian Homes. In: Proc. Symposium on Radon and radon progeny measurements in Australia, Canberra, February 1994, pp. 91-100.

[6] Solomon S B, Peggie J R, Wise K N and Paix D, (1988). Radon Levels in Australian Homes. Proceedings of the Seventh IRPA International Congress, Sydney, 1988. Pergamon Press, 1988, pp. 215-218.

[7] Langroo M K, Wise K N, Duggleby J C and Kotler L H (1991). A Nationwide Survey of ^{222}Rn and gamma Radiation Levels in Australian Homes. Health Physics, Vol. 61, No. 6 (December), pp. 753-761, 1991.

[8] Doi M, Fujimoto K and Kobayashi S (1991). Etch pit formation model during chemical and electrochemical etching in polycarbonate foil. Radiation Protection Dosimetry, 37: 5-12;

[9] Doi M, Kobayashi S and Fujimoto K (1992). A passive measurement technique for characterisation of high-risk houses in Japan due to enhanced levels of indoor radon and thoron concentrations. Radiation Protection Dosimetry Vol. 45 No. 1/4 pp. 425-430. Nuclear Technology Publishing.

[10] Doi M and Kobayashi S. (1994). The passive radon-thoron discriminative dosimeter for practical use. Hoken-Butsuri, 29: 155-166;

[11] Schery S D, Whittlestone S and Hart K P, (1989). The Flux of Radon and Thoron from Australian Soils. Journal of Geophysical Research. Vol. 94 No. D6, pp 8567-8576, June 20, 1989.

Influence of ^{220}Rn on ^{222}Rn measurement in Chinese cave dwellings

Bing Shang, Zhouyuan Wang

Laboratory of Industrial Hygiene, Ministry of Health, 2 Xinkang Street, Deshengmenwai, Beijing 100088, China

Takao Iida, Yukimasa Ikebe, Katsumi Yamada

Department of Nuclear Engineering, School of Engineering, Nagoya University, Furo-cho, Chikusa-ku, Nagoya 464-01, Japan.

Abstract

Alpha track detectors (ATDs) are widely used for measurement of indoor ^{222}Rn concentration. However, some results showed there were quite differences of ^{222}Rn concentration in some Chinese cave dwellings by using different kinds of ATDs. The abnormal results had caused a special attention. Further experiments found some filter membranes used with some detectors had very large air exchange rates, that means ^{220}Rn gas can also get into diffusion chamber through filter with ^{222}Rn gas together. In this way, ^{222}Rn concentration would be overestimated for the disturbance of ^{220}Rn. Since the ^{220}Rn concentrations are very high (152 ~ 296Bq m^{-3}) in some cave dwellings north west in China , its contribution should not be neglected.

1. Introduction

Alpha track detectors (ATDs) are widely used for measuring indoor ^{222}Rn concentration. One of the advantages of ATDs is that it can integrate ^{222}Rn concentration in environment. In addition, since it is inexpensive, small, easy to handle and mail, so it is suitable for large-scale investigations.

^{220}Rn, the another isotope of radon, also exists in the human environment. But it is often ignored because of its shorter half-life ($T_{1/2}$ = 55.6s). Recently, high ^{220}Rn concentrations had been measured in some traditional Japanese houses [1,2]. Similar problem appeared in Chinese cave dwellings [3]. The influence of ^{220}Rn to ATDs has being given attention. This paper reports a preliminary result.

2. Experiment

2.1 Alpha track detectors

A pilot study of lung cancer and residential radon exposure was conducted in northwest China in 1993 [3]. In this study we used 300 Landauer (LD) ATDs to measure indoor ^{222}Rn concentrations. Meanwhile 100 ATDs from Laboratory of Industrial Hygiene (LIH) were also placed with LD side by side in some rooms. In order to know the distribution of indoor ^{222}Rn concentration , the monitors were placed at three different positions in each room (front, middle and back of the room). Heights of monitors placed were 1.8 ~ 2.0 m from floor. After 6 months, LD monitors were returned to US for evaluation, and the films in LIH monitors were etched in 6.25 mol NaOH solution for 5 h at 80 ℃, then alpha tracks were counted by using optical microscope.

The data of ^{222}Rn concentrations obtained by using LD and LIH monitors are not agree in some rooms. Table 1 lists the ^{222}Rn concentrations (C_{Rn}) measured by using the two types of ATDs in three cave dwellings. We found LD monitors which were placed near walls gave obviously higher values than those farther from wall. The data measured with LIH monitors at different positions were almost the same.

Table 1. C_{Rn} measured by using LD and LIH monitors at different positions (Bq m^{-3})

No. Cave	Type. ATDs	Distance from Wall (cm)		
		5-10(Front*)	50-70 (Middle)	5- 10(Back*)
1	LD	611, 440	104, 89	1195, 784
	LIH	76, 77	66, 74	82, 140
2	LD	495, 592	263, 159	1069, 703
	LIH		117**	
3	LD	100, 115	70, 67	107, 37
	LIH		79**	

*: near walls. **: only one monitor placed.

2.2 LD and LIH detector

LD monitor provided by Track-etch, Tech-Ops-Landuaer, Glenwood, IL.USA. It has a simple structure, and a small volume (22ml). Its material is electric conductive plastic. There are nine 0.6cm air exchange pores on the cover (see Fig 1).

LIH monitor uses KfK type diffusion cup that was designed by Karlsruhe Nuclear Research Center, Germany [4]. There is a filter-halting between filter membrane and cover, which can stop dust and rain water while measuring. Some improvements had been made at LIH. CR-39 was used instead of Makrofol, which has a wider energy response on alpha particles than other SSNTDs. We select polyethylene membrane in LIH monitor that presents a permeability relatively low to thoron and high to radon.

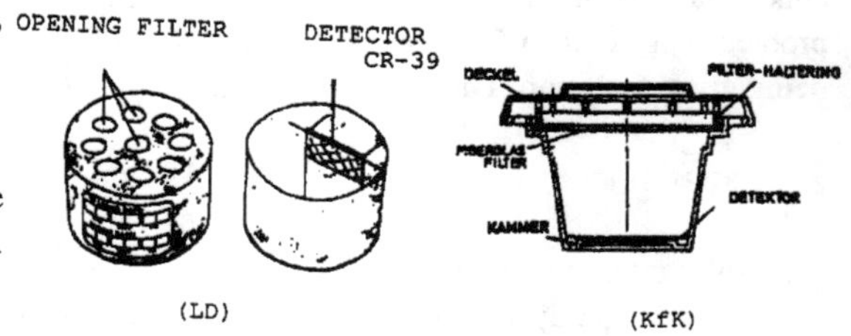

Fig.1. LD and KfK detectors

2.3 Permeability and air exchange rate equation

The exchange rate of ^{222}Rn or ^{220}Rn gas can be represented by the equation [5]:

$$r = K A / \delta V_2 \tag{1}$$

Where r is air exchange rate in h^{-1}; K is permeability of ^{222}Rn or ^{220}Rn in the filter in $cm^2 h^{-1}$, permeability equals diffusion coefficient (D_f) for a purely diffusive barrier; A is area of the filter in cm^2; δ is the thickness of the filter in cm; V_2 is the interior volume of the monitor in cm^3.

2.4 The measurement of air exchange rate (r)

The air exchange rate of ATD was measured by ionization chamber [6]. First, the monitors were remodeled into ionization chamber. ^{222}Rn gas was introduced into the ionization chamber. After 3.5 h when ^{222}Rn and its short-lived decay products reached radioactive equilibrium, starting ionization current (I_0) was observed. Then to pull off the rubber cork of the ionization chamber, ^{222}Rn gas was diffused naturally through the filter for certain time (t). Plugged the rubber cock again, ionization current (I) was measured after 3.5 h. The relationship between r and I_0, I as follows:

$$I = I_0 e^{-(\lambda + r)t} \tag{2}$$

3. Results and Discussions

3.1 D_f/δ of some filter membranes

The choice of filter membrane for ATD is very important. A suitable filter membrane should : (a) stop ^{222}Rn and ^{220}Rn decay products; (b) discriminate ^{222}Rn from ^{220}Rn. It is necessary to know the diffusion coefficient of the membrane to be used.

Table 2. D_f/δ of some membranes

ATDs	Membranes	Manufacturer	Thickness (mm)	D_f/δ ($cm\ h^{-1}$)
LD	Paper Filter (PF)	Radtrak Landauer, Inc. USA	0.157 ± 0.007	174
KfK	Glass Fiber (GF)	Glasfaser Rundfilter Germany	0.223 ± 0.012	202
LIH	Polyethylene (PE)	2Plastic factory Beijing, China	0.024 ± 0.001	0.90
Rn-Tn*	Cellulose Nitrate (CN)	Toyo Roshi co., LTD. Japan	0.136 ± 0.002	172

*: Rn-Tn monitor is developed in Nagoya University, Japan.

The measured results of some filter membranes used by ATDs are listed in Table 2. In order to avoid measuring uncertainty in thickness and uniformity of the membranes, permeability D_f/δ value was adopted to instead of D_f in common use. From Table 2 we can see the D_f/δ of PF used in LD is quite large (174 cm h^{-1}). The D_f/δ of GF and CN are also very large. The result of CN is close to PF. PE membrane has a lower diffusion ability, D_f / δ only 0.90 cm h^{-1}.

3.2 Estimation of ^{222}Rn or ^{220}Rn concentration ratio (R) inside to outside the monitors

The radon gas was diffused naturally through the filter into the monitor. If exposure time is long enough ($t >> \lambda^{-1}$), the ratio of ^{222}Rn or ^{220}Rn concentration inside to outside of the monitor can be expressed by following expression [5]:

$$R = \frac{C_{in}}{C_{out}} = \frac{r}{\lambda + r} \tag{3}$$

Where R is the ratio of ^{222}Rn or ^{220}Rn concentration inside to outside of the monitor. C_{in} and C_{out} are the concentration of ^{222}Rn or ^{222}Rn inside and outside the monitor in Bq m^{-3}. r is air exchange rate in h^{-1}; λ is the decay rate of ^{222}Rn or ^{220}Rn in h^{-1}.

Table 3 lists the results of r and R for each ATD. Since PF membrane used in LD has high permeability, the value of r calculated by D_f / δ from Eq (2) is 20h^{-1}. The r measured using ionization chamber is 24.2h^{-1}. The r for KfK is 51 h^{-1}, which consistent with 30 ~ 41 h^{-1} reported [7].

Table 3 . r and R values of some ATDs

Monitor	Filter Membrane	D_f / δ (cm h^{-1})	r (h^{-1})	R	
				^{222}Rn	^{220}Rn
LD	PF	174	24.2	0.9996	0.3497
KfK	GF	202	51	0.9999	0.5313
LIH	PE	0.90	0.23	0.9682	0.0051
Rn-Tn*	CN	209	0.157	0.9540	0.0035

*: From reference [4] .

Ward suggested that R (^{222}Rn) > 0.90, R (^{220}Rn) < 0.01 would be sufficient to prevent diffusion of ^{220}Rn and permit the rapid establishment of equilibrium between ^{222}Rn outside and inside the monitor For ATD [8]. R (^{220}Rn) of LD is 0.3497 (Table 3) which is much large than 0.01 acceptable range. Therefore, Alpha tracks recorded of this kind of monitor come from both ^{222}Rn and ^{220}Rn.

Table 3 also shows that the R of LIH monitor was less than 0.01 for ^{220}Rn as using lower permeability PE membrane. It is close to that of Rn-Tn monitor. LIH monitor can effectively prevent the interference from ^{220}Rn.

3.3 Field experiment

3.3.1 The ^{222} Rn and ^{220} Rn concentrations in cave dwellings

Generally, cave dwelling can be classified into four types by their position relative to ground and structure stile[3]. Aboveground cave dwelling (AGCD) is one type of the four types. The difference of AGCD with others is the former built with thick soil brick. The others built by tunneling into the ground or the side of a hill. The number of AGCD is on the increase recent years. The soil brick, main building material of AGCD, is made up of local soil by pressing in which ^{232}Th content is about 50 Bq kg^{-1}.

Rn-Tn ATDs were placed in six AGCDs (including three houses in pilot study). Because the limitation of number of Rn-Tn monitors, only one pair of Rn-Tn monitors placed in each house with a distance from wall of 20 cm. Exposing time was from October to December, 1996. Table 4 is preliminary results of ^{222}Rn and ^{220}Rn concentrations in the cave dwellings. The average of ^{222}Rn concentration in ADCD is 83.2 Bq m^{-3}, it shows no difference with the result (81 Bq m^{-3}) in past study [9]. However, ^{220}Rn concentrations are very high in these houses, with average of 197 Bq m^{-3}. For this reason, ^{220}Rn concentrations and behavior in cave dwellings is going to be a question of further research.

Table.4. The ^{222}Rn and ^{220}Rn concentrations in caves (Bq m^{-3})

Cave	1	2	3	4	5	6	Average
AGCD (^{222}Rn)	62	74	77	81	75	109	83.2
(^{220}Rn)	156	166	200	152	213	296	197

3.3.2 Comparison of the results by using different ATDs

LD, LIH and Rn-Tn monitors were placed in same positions (front, middle and back) in cave dwellings as pilot study. High values of ^{222}Rn from LD monitors have been found near the walls (see Fig 2). The cave dwellings made up of bare soil walls. Even decorated the house, it was only covered with some paper or coat with thin mud on the surface of the wall. As a result ^{220}Rn is easily released from soil wall and gathered near around wall. Since LD monitor has a quite large air exchange rate, ^{220}Rn gas can also enter into the diffusion chamber through the filter with ^{222}Rn gas together. This is the reason why to get high ^{222}Rn concentration values near wall with LD monitors. The results from Rn-Tn and LIH monitors have no obvious deferent at different positions.

3.3.3 Relation between ^{220}Rn and ^{222}Rn concentrations and distance from wall

Six pairs of Rn-Tn monitors were set at distance of 10,20,40,60,80,100 cm from wall. Relationship between ^{220}Rn and ^{222}Rn concentrations and distance from wall are shown in Fig 3. There is a quite high ^{220}Rn concentration near wall and it decreases obviously with the distance from wall for its short half-life. On the contrary, ^{222}Rn concentration is no big difference with distance.

The monitors were usually hanged on the wall with a nail or placed on the cabinet. The distance was about 5 ~ 20 cm from the wall. If air exchange rate of ATD is too large, ^{222}Rn concentration would be overestimated from the disturbance of ^{220}Rn. On other hand, the disturbance from ^{220}Rn on ^{222}Rn measurement using ATDs may be solved by placing the monitors with distances being away from the wall (more than 60cm to walls).

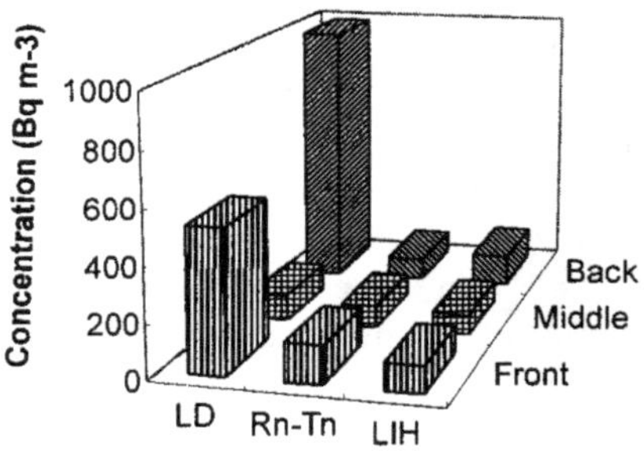

Fig.2. The comparison of the results measured by using different ATDs

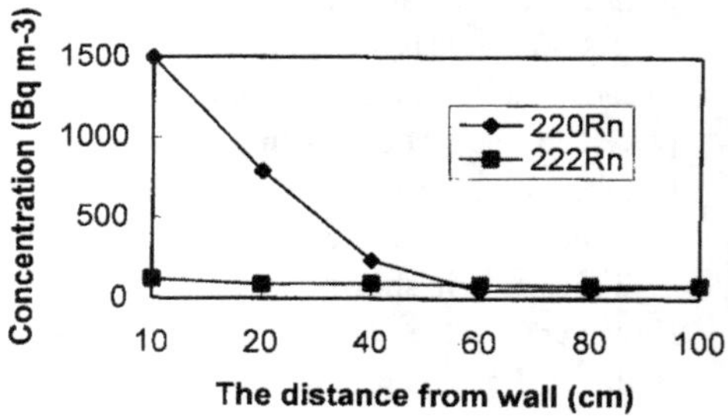

Fig.3. The relationship between ^{220}Rn and ^{222}Rn concentrations with distance from wall

4. References

[1] Iida,T., Sone,K.,and Ikebe, Y. Some Problems on the Measurement of Radon Concentrations by cup Method. Proceedings of the 25th Annual Meeting of Japan Health Physic Society (in Japanese), Tsukuba Japan, May ., 63, 24-25, 1990.

[2] Doi,M., Fujimoto, K, and Kobayashi, S, Spatial Distribution of Thoron and Radon Concentrations in the Ambient Air of a Traditional Japanese Wooden House. Health Phys., 66,43-48, 1994.

[3] Wang, Z.Y.; Lubin, J. H.; Wang, L.D, et al. Radon Measurements in Underground Dwellings from two Prefectures in China. Health Physics., 70, 192-197, 1996.

[4] Urban, M., and Piesch, E. Low Level Environmental Radon Dosimetry with a Passive Track Etch Detector Device. Radiation Protection Dosimetry, 1, 97-109, 1981.

[5] Guo, QiuJu., Iida, T, and Okamoto, K. Measurement of Thoron Concentration by Passive Cup Method and its Application to Dose Assessment. Journal of Nuclear Science and Technology. 32.694-796. 1995.

[6] Iida, T., Ikebe,Y, and Hattori, T., Hattori.T; et al. An Electrostatic Integrating ^{222}Rn monitor with Cellulose Nitrate Film for Environmental Monitoring. Health Phys, 54, 139-147,1988.

[7] Doi, M., and Kobayashi, S.; Characterization of Japanese Wooden House with Enhanced Radon and Thoron Concentrations. Health Phys., 66, 274. 1994.

[8] Ward,W.J., Fleischer,R.L and Mogro.C.A. Barrier Technique Separate Measurement of Radon Isotopes. Rev. Sci.Instrum., . 48,1440-1442, 1977.

[9] Shang,B., Wang, Z.Y,and Gao, P.Y, et al. Survey of Radon Concentration in Cave Dwellings in the East of Gansu Province. Radiation Protection (in Chinese), 15, 461-463, 1995.

Correlation between nationwide indoor radon and terrestrial gamma survey data in Japan

Kenzo Fujimoto

National Institute of Radiological Sciences
9-1, Anagawa-4, Inage-ku, Chiba, 263 Japan

Abstract

Using the results of nationwide indoor radon and external exposure surveys, a correlation between the indoor radon concentration and dose rate in air from terrestrial gamma radiation is examined although the surveys were not conducted at the same time nor the same location. The radon concentration shows a log normal-like distribution, whereas the dose rate in air shows a normal-like distribution. In the log-linear scatterplot for each pair of the indoor radon concentration and gamma dose rate in air in each municipality reveals a clear relationship than that in the linear-linear scatterplot. The regression line indicates that radon concentration increases with gamma dose rate in air. The indoor radon concentrations below the lower 95% confidence limit could be used for the estimation of outdoor radon concentration and found the average outdoor radon concentration to be around 5 Bq m^{-3} in Japan. Based on the finding in the present study a certain criterion level of air dose rate could be established for an effective survey to find out houses which might be required a remedial action.

1. Introduction

A nationwide indoor radon survey was performed at more than 7,000 houses by passive radon detectors in Japan and an annual average radon concentrations at 5,717 houses [1] were obtained as shown in Fig. 1. The concentration varies from 2 Bq m^{-3} to 313 Bq m^{-3}. In addition, our group conducted a nationwide external exposure survey at 1,115 sites throughout Japan [2]. The average dose rate in air in each municipality are shown in Fig. 2. Minimum and maximum of the average exposure rates were 34 nGy h^{-1} (3.9μR h^{-1}) and 148 nGy h^{-1} (16.9 μR h^{-1}).

Although the gamma dose rate in air arises from not only uranium series, which is a precursor of radon, but also thorium series and potassium in the soil, it is believed that the indoor radon concentration has a coarse correlation with the dose rate in air from terrestrial gamma radiation at the area. The other factor which must be considered is that the gamma dose rate is governed by the radioactivity distributed in soil at the area of around a few hundred meters with a few tens of cm depth from the ground surface, whereas the outdoor radon concentration around a few tens of kilometers with a few meters depth. In addition, variability of the indoor radon concentration is expected owing to the type of structure, infiltration rate of houses, wall materials, way of living, the air permeability of soil etc. In spite of these disturbances the correlation between the indoor radon concentration and dose rate in air from terrestrial gamma radiation is studied to find out useful relationships between the two.

2. Method

The nationwide radon survey covered 1,390 municipalities throughout Japan and provides annual average indoor radon concentrations. The external exposure survey was performed with an ionization chamber as well as NaI(Tl) surveymeter. The measurement values by NaI surveymeter were converted to those measured by the ionization chamber using the conversion factor obtained empirically from simultaneous measurements by the ionization chamber and surveymeter. Thus, all results presented in the paper contain cosmic ray components. For the analysis of the correlation between the radon concentration and gamma dose rate in air in the present study the cosmic ray component was subtracted from the original data taking into account the cosmic ray contribution in the results [3]. The derived values represent site averaged absorb dose rate in air from terrestrial gamma radiation at 1,115 sites in 768 municipalities.

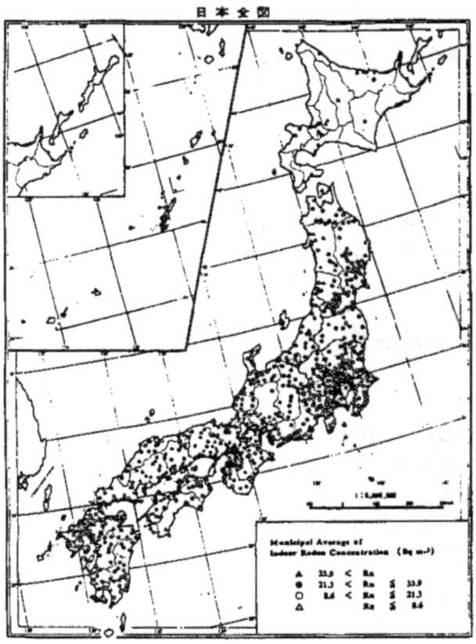

Fig. 1. Municipal average of indoor radon concentration

Fig. 2. Municipal average of dose rate in air due to terrestrial gamma radiation

Although the measurements of indoor radon concentration and gamma dose rate in air were not performed at the same location nor at the same time, two results could be compared since these data were obtained carefully as representative values at each house or each site. Large portion of outdoor radon concentrations are believed to be influenced by the radon emanated from the soil of around a few tens kilometers [4]. Thus the municipality average of gamma dose rate in air, if more than one data in a municipality are available, might be a better indicator of the radon contribution from the soil to the indoor radon in a house located at the same municipality than the individual site average of gamma dose rate in air. Therefore, the correlation was examined between the municipality average of gamma dose rate in air and radon concentration in each house at the same municipality. Both measurements were found to have been performed at 537 municipalities. The measurement results of indoor radon concentrations were obtained in 4,242 houses at these 537 municipalities.

3. Results and discussion

Fig. 3 shows the scatterplot of indoor radon concentrations in 4,242 houses at the 537 municipalities where the gamma dose rate in air was obtained by the nationwide external exposure survey. Least square regression is applied to the plot and the result of the straight-line fit is superimposed in Fig. 3. The correlation coefficient is 0.2407. A relationship with the shape of a right-angled triangle is found between the indoor radon concentration and gamma dose rate in air. The indoor radon concentration varies from the outdoor concentration to the highest concentration that is achievable at the gamma dose rate in air depending on the type of structure, infiltration rate of house, way of living and so on. When the gamma dose rate in air is low, the indoor radon concentration cannot reach to a high level. The local maximum as well as regression line of the radon concentration indicate a tendency to increase with the gamma dose rate in air.

Indoor radon concentration distribution has often been reported to have a log normal distribution. Log normal cumulative frequency distribution of indoor radon concentration is shown in Fig. 4. It tells that our result has a log normal-like distribution although the log normality test is rejected (p value is less than 0.001 by Kolmogorov-Smirnov test). On the other hand, as shown in Fig. 5 the gamma dose rate in air indicates a normal-like distribution although normality test, Shapiro-Wilk W test shows that the distribution is not accurately normal (p is 0.0332). For the analysis of correlation of these two values with different distributions, log conversion is applied to the radon concentration to make both distributions close to normal distribution.

The relationship between the two in log-linear scale graph is shown in Fig. 6 which illustrates a better correlation than that in Fig. 3. The correlation coefficient of these two variables is 0.3118, which is larger than that in linear-linear scale scatterplot mentioned above. In addition, the lowest radon concentration also demonstrates a clear increase with the gamma dose rate in air whereas it had not been clearly found in the linear-linear scale graph of Fig. 3. A regression line shown in the middle and upper and lower 95% confidence limits for individual predicted value are superimposed in Fig. 6 with the ellipsoids, which are

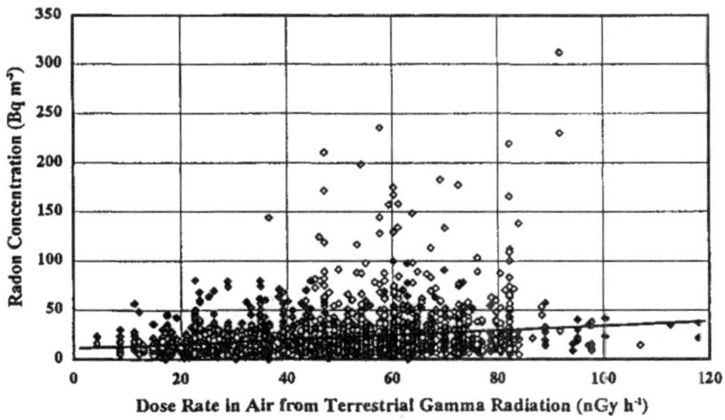

Fig. 3. Scatterplot for each pair of indoor radon concentration and gamma dose rate in air

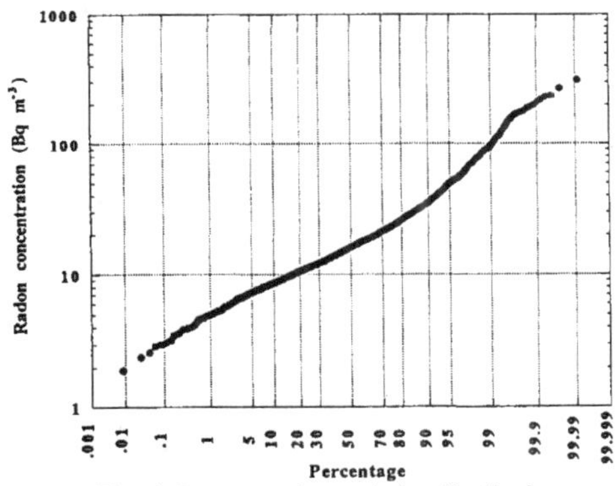

Fig. 4. Log-normal cumulative distribution of indoor radon concentrations

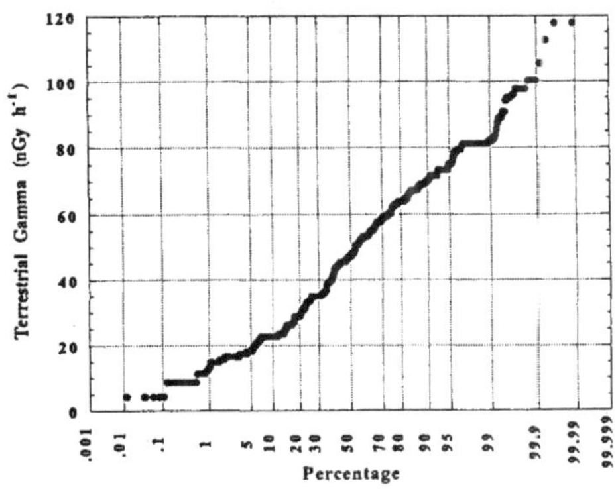

Fig. 5. Normal cumulative distribution of gamma dose rates in air

both density contours and confidence curves for 90 and 95%. Scatterplots below the lower 95% confidence limit strongly reflect the outdoor radon concentration and could be used for the estimation of outdoor radon concentration. This estimation of outdoor radon concentration is very useful since there is no nationwide outdoor radon survey data available at present in Japan. Number of houses is 44 whose radon concentrations are below 95% confidence limit. Their radon concentrations range from 1.9 to 9.1 Bq m^{-3} and the average is 4.7 Bq m^{-3}, which are all in the range of normal outdoor radon concentrations found in Japan [5,6]. When scatterplots below the lower 90% confidence limit are selected, the number of plots is 132, the range of the concentration is from 1.9 to 9.1 Bq m^{-3} and the average is 5.8 Bq m^{-3}. Based on these calculation outdoor radon concentration is estimated to be around 5 Bq m^{-3} in Japan.

The regression line shown in the middle is expressed as $Y_{Bq\ m^{-3}} = 10.3\ EXP\ (0.0101X_{nGy\ h^{-1}})$, where $Y_{Bq\ m^{-3}}$ is the expected indoor radon concentration at the municipality where the gamma dose rate in air is $X_{nGy\ h^{-1}}$. Regression lines in the two extreme group beyond the upper and lower 95% confidence limits were obtained to be $Y_{Bq\ m^{-3}} = 30.4\ EXP\ (0.0101X_{nGy\ h^{-1}})$, $Y_{Bq\ m^{-3}} = 3.46\ EXP\ (0.0101X_{nGy\ h^{-1}})$, respectively. These regression lines tell that diversity of correlation between radon concentration and dose rate in air increases with dose rate in air even in the log-linear scale. Table 1 indicates the predicted radon concentration from dose rate in air due to terrestrial gamma radiation using the regression lines shown in Fig. 6. The positive coefficient of the equation mentioned above confirms the positive correlation between the indoor radon concentration and gamma dose rate in air

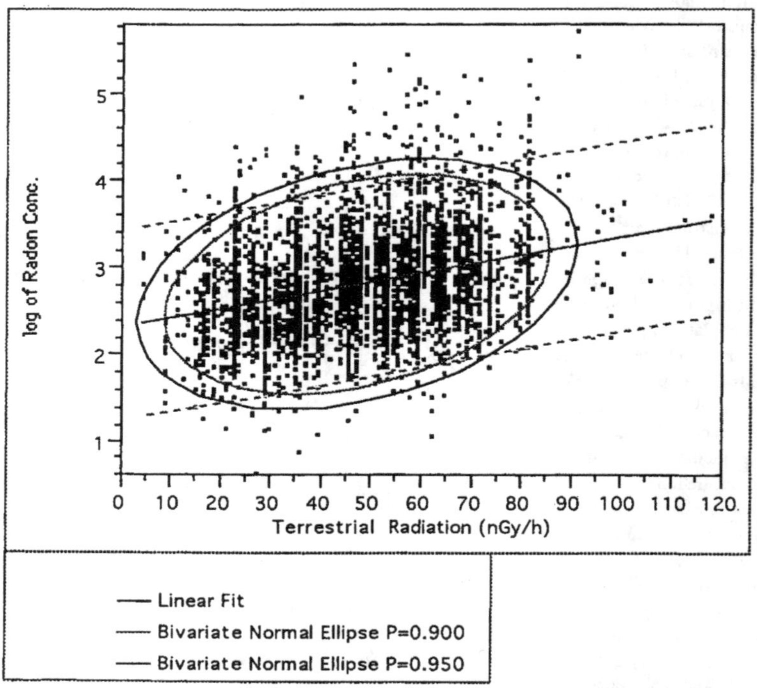

Fig. 6. Scatterplot in log-linear scale for each pair of indoor radon concentration and gamma dose rate in air

Table 1. Predicted Radon Concentration from Dose Rate in Air due to Terrestrial Gamma Radiation

Terrestrial Radiation (nGy h^{-1})	0	20	40	60	80	100	120
Upper 95% confidence limit	30	37	46	56	68	84	102
Radon Concetration (Bq m^{-3})	10	13	15	19	23	28	35
Lower 95% confidence limit	3	4	5	6	8	10	12

although the line has a slight concave curve in the linear scale. It is satisfactory to consider that the correlation between the indoor radon concentration and gamma dose rate in air is the result of geological property in the area since the gamma dose contribution due to radon and radon progeny in air is negligible judging from model calculations presented in the papers [7,8].

Judging from the relationship between the indoor radon concentration and gamma dose rate in air, the houses built on an area where the gamma dose rate in air is less than a certain level might not have a high radon concentration, even if they have an air tight structure and less frequent opening and shutting of windows. On the contrary high radon houses are expected in an area where they have more than a certain level of gamma dose rate in air. Thus, a survey to find out high radon concentration houses, which might require a remedial action, could be focussed on the area where high gamma dose rates in air are observed.

The degree of dependency of indoor radon concentration on the air dose might vary according to nations or races as well as climates. However, the dependency might be clearly found in other countries as far as relatively large number of measurements are used for the analysis.

4. Conclusion

The correlation between indoor radon and dose rate in air was studied. A good correlation was found between the indoor radon concentration and municipality average of gamma dose rate in air, although the nationwide indoor radon and external exposure surveys were not conducted at the same time nor the same location. The indoor radon concentration shows a log normal-like distribution, whereas the dose rate in air from terrestrial gamma radiation has normal-like distribution. Therefore, log-linear scatterplot demonstrates a clearer relationship than linear-linear scatterplot. The indoor radon concentrations at the area with a certain gamma dose rate in air, vary from the lowest concentration, that is close to the outdoor radon concentration, to the highest, which is possible at the gamma dose rate in air, depending on the structure, way of living etc. The average, maximum and minimum as well as regression line of radon concentration were found to increase with the gamma dose rate in air. As a whole the indoor radon concentration are governed by the gamma dose rate in air at the area. However, the indoor radon concentrations below the lower 95% confidence limit strongly reflect the outdoor radon concentration and might be close to outdoor radon concentration. This estimation of outdoor radon concentration is useful since there is no nationwide outdoor radon survey data available at present in Japan. Its range and average are 1.9 to 9.1 Bq m^{-3} and 4.7 Bq m^{-3}, respectively. Based on this assumption, outdoor radon concentration is estimated to be around 5 Bq m^{-3} in Japan.

Based on the finding in the present study a certain criterion level of air dose rate could be established for an effective survey to find out houses which might be required remedial action.

The degree of dependency of the indoor radon concentration on the air dose might vary according to nations or races as well as climates, however, the dependency might be found as clearly recognized in Japan.

The structure type is another factor which modifies the indoor radon concentration. The effect of these factors on the relationship will be analyzed in the future study.

5. References

[1] Fujimoto, K.; Kobayashi, S.; Uchiyama, M.; Doi, M.; Nakamura, Y. Nationwide indoor radon survey in Japan. Hoken Butsuri 32: 41-51: 1997.

[2] Abe,S.; Fujitaka, K.; Abe, M.; Fujimoto, K. Extensive field survey of natural radiation in Japan. J. Nucl. Sci. Technol. 18: 21-45; 1981.

[3] Fujimoto, K. Significance of measurement of natural radiation in the fields of applied nuclear energy and radiation. Thesis 1987 (in Japanese).

[4] Sakashita, T.; Murakami, T.; Iida, T.; Ikebe, Y.; Chino, M.; Suzuki, K. Simulation of Diurnal Variation of Atmospheric ^{222}Rn concentrations with three-dimensional atmospheric dispersion model. Hoken Butsuri. 31: 161-168; 1996.

[5] Yamanishi, H.; Iida, T.; Ikebe, Y.; Abe, S.; Hata, T. Measurements of regional distribution of radon-222 concentration. J. Nucl. Sci. Technol. 28: 331-338; 1991.

[6] Hattori, T.; Ichiji, T.; Ishida, K. Equilibrium factor and unattached fraction of radon progeny in outdoor air. Radioisotopes. 44: 710-714; 1995.

[7] Fujimoto, K. External gamma exposure to radon progeny in indoor air. J. Nucl. Sci. Technol. 22: 1001-1010; 1985.

[8] Miller, K. M.; George, A. C. External gamma-ray dose rates from radon-222 progeny indoors. Health Phys. 54: 203-206; 1988.

On the long term indoor radon concentration measurements the basement of Japanese hospitals

Mika Shiotsu[1], Koichi Ikeda[1], Susumu Yosizawa[2], Takao Iida[3]

[1] The Institute of Public Health, Tokyo, 108 Japan
[2] Sience University of Tokyo , Tokyo, 162 Japan
[3] Nagoya University , Nagoya, 464-01 Japan

ABSTRACT

Long term indoor radon gas measurements using a passive monitor were conducted in the basement rooms of five hospitals, which are built in the Tokyo metropolitan area. The ventilation rate averaged over all hospital was 12.3 air change per hour (ACH). The highest radon concentration was 56.3Bq/m^3. The value averaged concentration over all hospital and all season was 20 Bq/m^3. But the average of the two hospitals were twice as high as those of three others. Since the total ventilation rate integrated over a day is not different much,the reason of the concentration difference is considered to be the difference in radon emission rate from concrete wall and soil of the each hospital. Though we inquired workers about "Environmental complaints", it do not show the direct relation between radon and environmental complaints.

1.INTRODUCTION

The hospital basements are not usually residential space, but sometimes used for radiology, examination, laundering and kitchen so forth. Although many people are working, there is usually no window, therefore mechanical ventilation is only way to ventilate the space. Moreover, as the space was surrounded by soil and thick concrete walls, extremely high indoor radon concentration is expected.
Long term indoor radon gas measurements using a passive cup monitor were now being conducted in the basement rooms of five hospitals, which are built in the Tokyo metropolitan area. Ventilation rate are also measured during the monitoring pried. We also made questionnaires concerning complaints about environmental conditions in the work place to the workers in the terms of radon health risk.

2.METHODS

Indoor radon concentration measurements in 5 hospital basements in the Tokyo metropolitan area was conducted from 1995 summer to 1996 spring. Measurements of 35 pointswere performed by passive type monitor in each hospitals. Background measurement was also conducted in the each 5 hospitals. . Table 1 shows outline hospital serveyed. The duration of measurements is 3 month as one season. Figure 1 shows floor plan of the second basement of hospital D and measuring points.

Radon gas measuring instruments

Passive ^{222}Rn cup monitor method with cellulose nitrate (CN) film (Kodak Pathe, LR115 type 2) was used for measurements. Sample air enters into the monitor through a membrane filter and alpha rays emitted from radon in the monitor are detected by the CN film. We decided to put two monitors for each measuring point so that we can take average of two monitors as sensitivity of the monitors are not high enough.

Ventilation rate measurement

We measured the ventilation rates in almost all the rooms. Ventilation rates were determined by the tracer gas decay method. We used SF_6 as a tracer which was detected by photo acoustic spectroscopy method.

Calibration

Calibration of passive monitors was conducted in August 1995 and July 1996 at Dept. of Nuclear Engneering, School of Engineering, Nagoya University in Japan.

Calibration of the monitor were performed by setting fifteen or sixteen cups monitors into a 247.8L stainless steel chamber which was connected with 1.5L cylindrical ionization chamber containing high concentration ^{222}Rn gas. Relationship between integral radon concentration and the number of etch pits is shown in Figure 2. These passive monitors were also calibrated at the intercomparison held in December 1996 in Critical Assembly in Kyoto University in Osaka.

Questionnaire

We were told from some workers who worked in these hospitals that they were having anxiety about of indoor air quality, for example a feeling of stagnant air, lack of ventilation, and so on. Therefore we inquired workers about these conducting questionnaire about "Environmental complaints". The questionnaire was consisted of three items, affirmation, complaints about work place environmental and satisfaction with them.

3.RESULT AND DISCUSSION

Measurement results of radon concentration obtained from summer of 1995 to spring of 1996 in Tokyo are shown in Figure 3.

The highest radon concentration, 56.3Bq/m^3, which is almost one third of the EPA action level, 150 Bq/m^3, was observed in the summer season (June to September) in hospital A. The lowest radon concentration was, on the other hand, almost equal to lower detection limit of the monitor.

The averaged radon concentration over all hospital and all seasons was 20 Bq/m^3. The average at hospital A and B were 35.6 and 26.6Bq/m^3 respectively, which were twice as high as at the another three.

Figure 3 does not show seasonal change about concentration of radon gas, although it is well known fact that radon concentration is changing depending on the seasons and atmospheric pressure.

Ventilation

The result of ventilation measurements were also shown in Figure 4.

The ventilation rate averaged over all hospital was 12.3 ACH. The rates are ranging from 3.6 to 20.6 ACH. Although we could measure ventilation rates at some rooms the average in this measurement, we had measured them almost all room in the previous measurements[1], the average of which was 8.9 ACH, and was a little bit lower than present measurement. The ventilation systems were always running during the ventilation rate measurement,

From both results, these rooms of the hospitals were kept high ventilation rate so far as the ventilation systems are on.

However, there is anxiety that ventilation rates during the ventilation systems off was not high. In fact, the measurement during the systems off should considerably low ventilation rate, i.e., about 1.0 ACH, although the number of measurements is limited.

Figure 3 does not show seasonal change in the concentration of radon gas.

It is a well known fact that radon concentration is changing depending on the seasons and atmospheric pressure.

The reason for the concentration difference between hospital A and B, and other three is considered to be the difference in emission rate from concrete wall and soil of the each hospital. Since the total ventilation rate integrated over a day is not different much.

Questionnaire result

Tble 2 shouws 36 persons answered.

About 36 percent of the subjects answered that they are satisfied with nember of staffs and the medical facilities with which they are working. The items pointed out by the subject as "satisfactory environmental factor". were a little different from items we were expected. Some 39 percent of the subjects were complaining of the shortage of room spaces and width of the corridor in each hospital. We think that these items are deeply related to building planning. We also did not expected them as environmental factor. The workers of 36 percent complained that the work place has no window, since they could not see outside and could not introduce outdoor air. Another 31 percent did not complain of the factors related to HAVC systems, such as "Bad smell", "thermal sensation" and so forth.

But these results of questionnaire do not show the direct relation with radon concentration and health.

4.CONCLUSION

1. The highest radon concentration was, 56.3Bq/m^3, which is almost one third of the EPA action level, 150 Bq/m^3.

2. The average concentration at hospital A and B were 35.6 and 26.6Bq/m^3 ,respectively, which were twice as high asthose of three others.

3. The ventilation rate averaged over all hospitals was 12.3 ACH.

4. The result of questionnaire does not show the direct relation with health or radon concentration.

5.ACKNOWLEDGEMENT

We thank to the technical staffs of 5 hospitals (we wish we could name them personally and the name of the hospitals.) And at last we would like to thank Mr.Arakawa of Taisei oncyo co.Ltd,.

6.REFERENCE

[1]Mika Shiotsu, Koichi Ikeda, Kazukiyo Kumagai, et al., Lomg Term Radon Gas Measurement in Japanese Hospitals Using Passive Monitors, *Proceedings of INDOOR AIR '96* Nagoya(Japan), Vol.1, 111-116, 1996

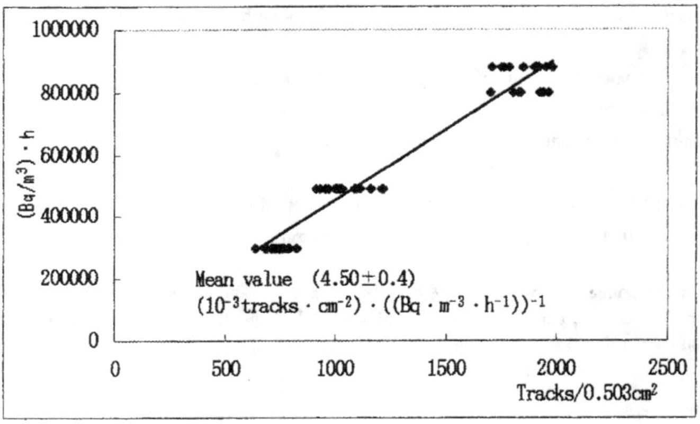

Figure 2. Calibration Curve

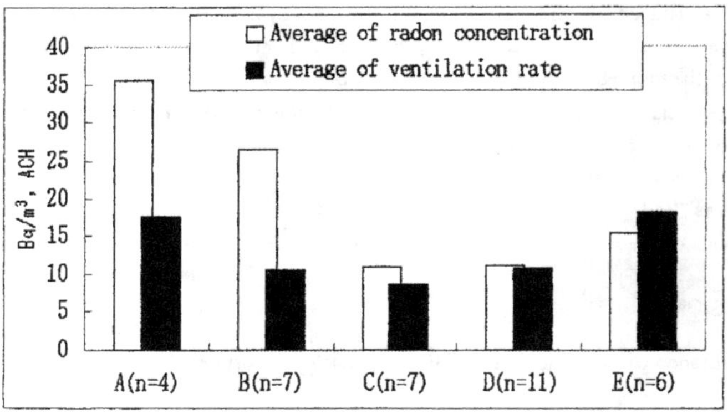

Figure 4 . Average Radon Gas Concentration and Ventilation rate

Table 2. Queationnaire Result

		Nunber of subject							
		Satisfied with				Complainng of			
Hospitals	Answer of Workers	Planninig	HAVC systems	Human among reratioship	others	Planning	HAVC systems	Human among reratioship	others
A	4	1			3	3(1)	1		
B	12	4		3	5	11(3)	2	2	1
C	9	1		2	3	6(4)	5	1	4
D	10	2			2	7(5)	3	2	
E	1	0							
Total	36	8		5	13	27(13)	11	5	5
Percentage	100	22		14	36	75(36)	31	14	14

( ):complain about no window

Table 1. Outline of the hospital surveyed

Hospital	Floor	Points
A	the first basement	5
B	the first basement	8
C	the first basement	8
D	the second basement	12
E	the fourth basement	7
	total	40

Besides these points in each hospital there is one other point for background measurement

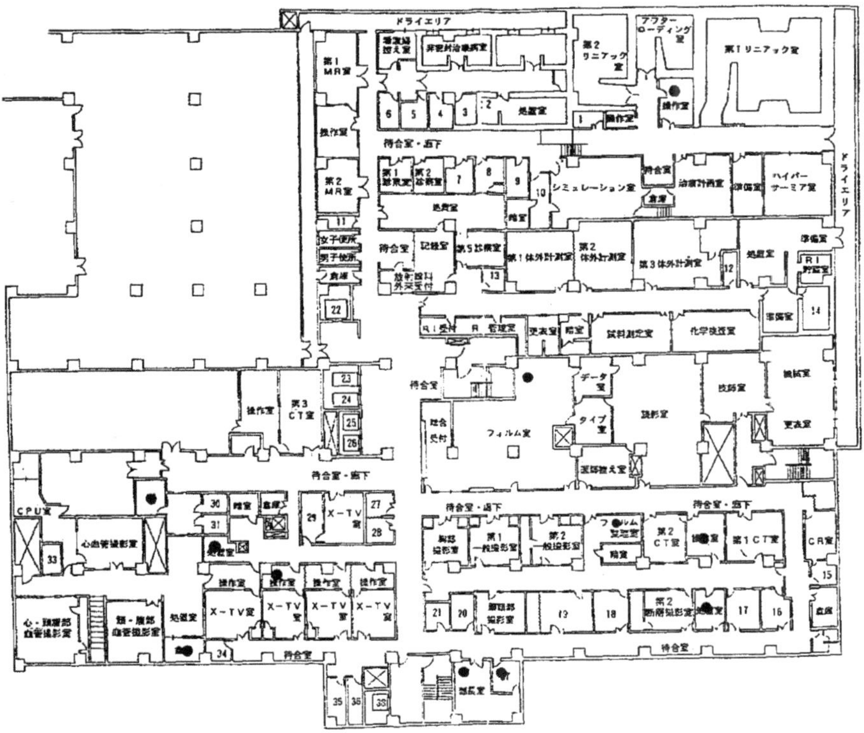

Figure 1. Floor plan of the second basement of Hospital D (● : measuring points)

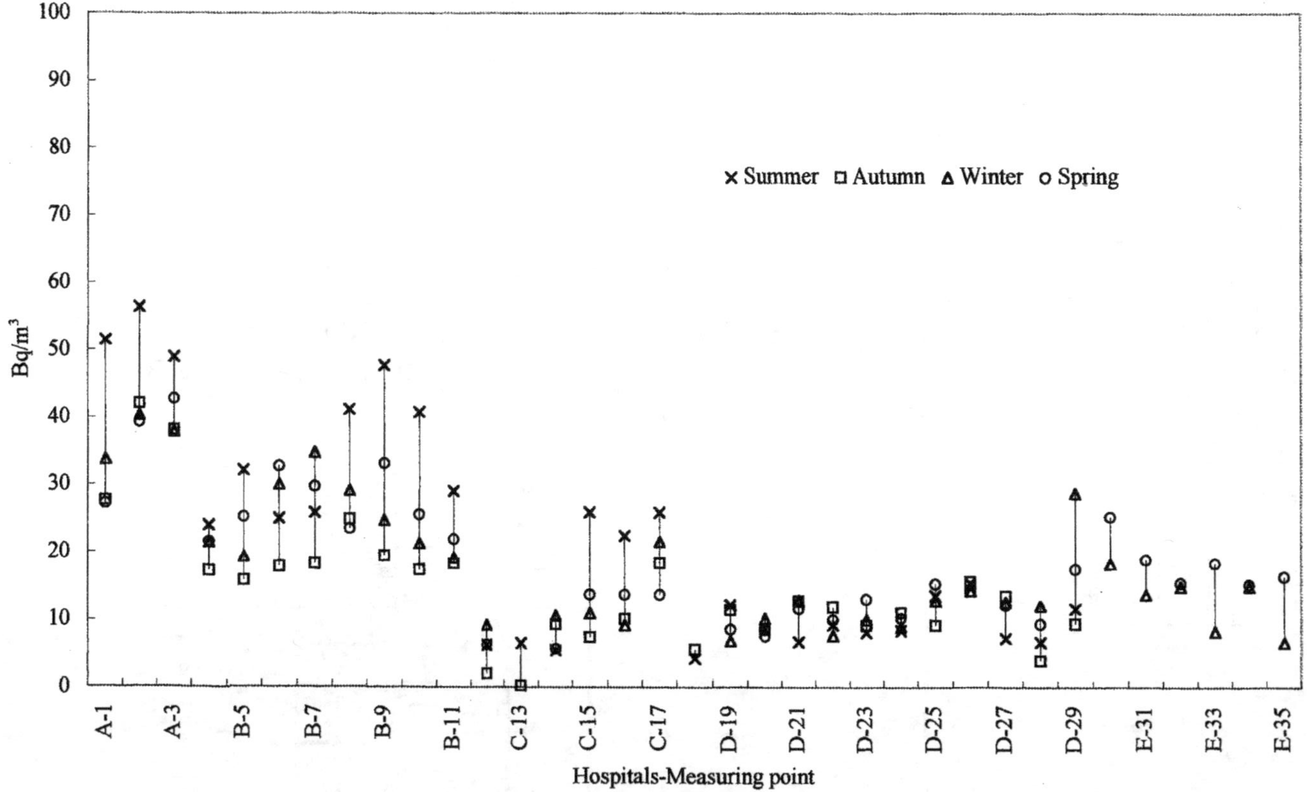

Figure 3. Seasonal Change in Radon Concentration

Radon concentrations in wooden houses

Hiromu Watanabe and Takashi Okuyama.

School of Agricultural Sciences, Nagoya University
Nagoya, 464-8601 Japan

Abstract

Japanese houses which had been adapted to the hot and humid of summer, are different from western houses regarding air tightness. Especially in the last two decades, many Japanese people live in houses having high air tightness, built with inorganic materials. Excessive usage of aluminum framed sliding-doors increases the air tightness inside wooden houses, allowing the radon emanates into the air tightened rooms from the soil under the crawl-space as well as from a kind of gypsum used for heat insulation.

In order to determine the potentially harmful environmental conditions, the diffusion behavior of radon in Japanese wooden houses was assessed and the properties of wood and wood-based materials as radon sealants were investigated.

The annual mean concentrations of radon in the crawl-space were 34 Bq m^{-3}. The concentrations in the first and the second floors were lower than that in the crawl-space. This fact shows that the main source of radon in the wooden houses is the soil under the crawl-space.

In any board thickness of Spruce, the reduction rates in the tangential direction was 90 % or more. It was made clear that wood has good properties as a radon sealant.

1. Introduction

Almost every ordinary house in Japan had wood framed windows and doors, which loose structure allowed air induction easily. In the last two decades, aluminum framed windows and doors have commonly been used for houses. Furthermore, the aluminum framed window has been used to allow a better insulation when air conditioning equipment is used. Thus, houses have become well tightened from the air. Therefore, the influence of the indoor radon emanations has turned into a serious problem.

Japanese houses, in general, have the crawl-space under the first floor with a ventilation opening to prevent the damp. Especially the dry and rainy season are clearly differentiated. Accordingly, the style of Japanese houses differ much from that of the European and American ones.

In this study, the radon emanations investigated were those coming out of the building materials used and from the soil under the crawl-space. Indoor radon concentrations were measured at various locations in Japan. The ability of wood as a radon sealant was also investigated.

2. Outline of the survey

The measurement sites were selected according to the climatological condition and the type of construction (wooden or reinforced concrete house). The cities selected were Sapporo, Morioka, Funabashi, Kawasaki, Fukuroi, Nagoya, Matsuyama and Kagoshima (Figure 1). In Hokkaido, where Sapporo is located, the effect of the rainy season is not as marked as in the rest of Japan. The elevation of a conventional Japanese house is displayed in Figure 2. It has a crawl-space under the first floor with a ventilation opening to prevent the damp. The radon concentrations were measured in the crawl-space, in the first floor and in the second floor. Furthermore, radon emanations out of the building materials used in Japanese houses were investigated, and also the ability of wood (Spruce) as a radon sealant.

3. Methods

3.1 Indoor Radon Concentrations

The cup method was employed for measuring the indoor radon concentrations. Cellulose nitrate film (CN film, Kodak LR115 type2) was used to detected alpha-particles. The cup monitor was developed by the department of nuclear engineering in Nagoya University, Japan [1]. The monitors were sent to measurement sites and the CN films were exchanged every two months. The films were etched with 2.5 N NaOH solution at 60℃. The numbers of etched tracks were counted automatically with an image analyzer (KONTRON ZEISS, IBAS Ⅱ) and a microscope. For calibration the number of tracks were counted previously using a film with a known number of tracks. Radon concentrations were computed from the number of tracks, using Equation (1).

$$
\begin{aligned}
C &= \frac{1}{T}\int_0^T C(t)dt \\
&= \frac{1}{TAC_F}\left(N - N_B\right) \qquad (1)
\end{aligned}
$$

where C is the radon concentration in Bq m^{-3}, T is exposure time in hours, C_F is the calibration factor in tracks cm^{-2} (Bq m^{-3} h)$^{-1}$, N is the number of tracks in the area of A(cm^2), and N_B is the number of background tracks in the area of A.

3.2 Radon emanations out of building materials and the soil

The building materials and the soil under the crawl space were crushed into particles (about 2 × 2 × 0.5 cm). The sample was set in a desiccator with the cup monitor to measure radon concentrations. The weight of each sample was 0.15 kg. The volume of the desiccator was 1.4×10^{-3} m^3. The dried sample was embedded in the bottom of the desiccator. The desiccator was left in a chamber at constant temperature of 24 ℃ for 60 days. The radon emanation out of the samples was computed by Equation (3) obtained from Equation (2).

$$
\begin{aligned}
C &= \frac{1}{T}\int_0^T \frac{EM}{V}\left(1 - e^{-\lambda t}\right)dt \\
&= \frac{EM}{V}\left(1 - \frac{1}{\lambda T} + \frac{1}{\lambda T}e^{-\lambda T}\right) \qquad (2)
\end{aligned}
$$

$$
E \approx \frac{\lambda CTV}{M(\lambda T - 1)} \qquad (3)
$$

where E is the radon emanation in Bq kg^{-1}, M is the weight of the soil sample (0.15 kg), V is the volume of the desiccator (1.4×10^{-3} m^3), and λ is the decay constant of radon (7.55×10^{-3} h^{-1}).

3.3 Reduction rate of radon using wood as a radon sealant

Figure 3 shows the schematic diagram of an apparatus for scintillation method. PYLON model RNC was used as the radon source. Radon concentrations were obtained by the radon monitor (PYLON, AB-5). The concentration on the side of the radon source was detected by cell PYLON 300A, and on the other side by PYLON PMT-EL. On each side the air was circulated by the built in pump onto AB-5. The specimen used was Spruce. Each specimen was cut as a circular plate whose diameter was 10 cm. The side of each specimen was coated with epoxy resin. The radon reduction rates due to wood were obtained from the following equation.

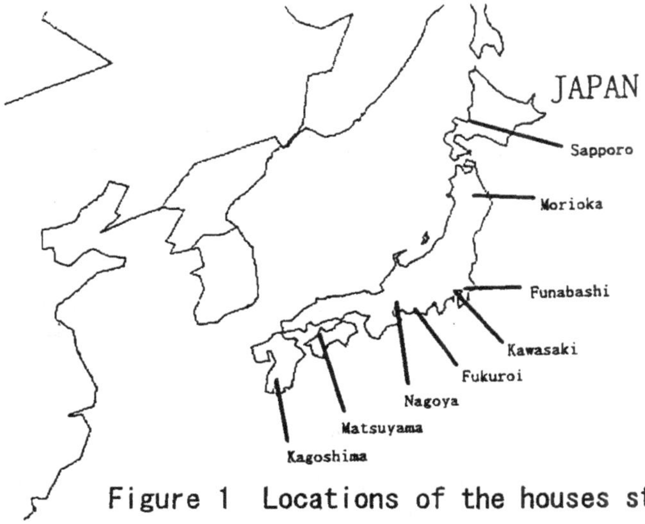

Figure 1 Locations of the houses studied.

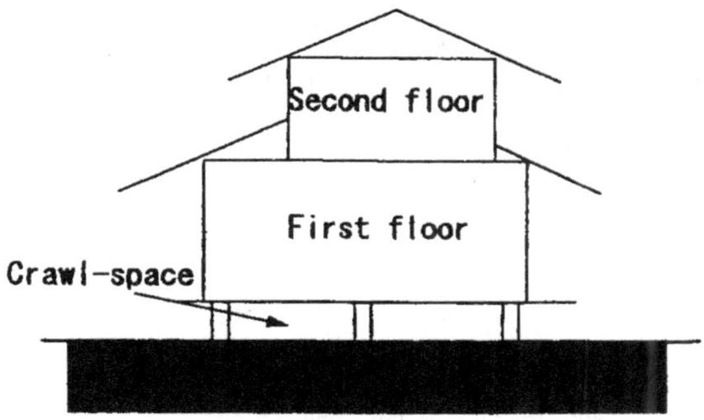

Figure 2 Outline of a typical Japanese house.

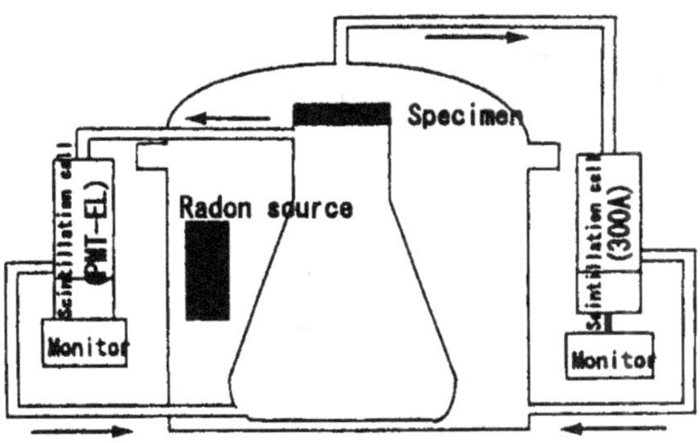

Figure 3 Schematic diagram of the testing apparatus.

$$R = \left(1 - \frac{\int n_{in} dt}{\int n_{out} dt}\right) \times 100 \qquad (4)$$

where R is the radon reduction rate of wood in %, n_{in} is the radon concentration in the desiccator on the opposite side of the radon source in Bq m^{-3}, n_{out} is the radon concentration in the desiccator on the side of the radon source in Bq m^{-3}.

4. Results and discussion

4.1 Radon emanations out of building materials and soil

Radon emanations out of building materials like cement and gypsum used in Japan, are shown in Table 1. Radon emanations out of building materials ranged from 0.1 to 29.8 Bq kg^{-1}. The radon emanates out of ply-wood, which is a wood-based material, was very little.

4.2 Indoor radon concentration

Table 2 shows the radon emanations out of the soil under the crawl-space. The emanations were over 1.0 Bq kg^{-1}. These emanations show that the main source of indoor radon is the soil under the crawl-space, especially in wooden houses. Indoor radon concentrations are displayed in Table 3. The higher radon concentration was found in the crawl-space rather than in the first and the second floors. Geometric mean (G.M.) of the radon concentrations in the crawl-space, in the first floor and in the second floor were 34.1, 17.8 and 18.4 Bq m^{-3}, respectively. These results show that radon emanates out of the soil under crawl-space and diffuses to upper floors.

The seasonal variations of indoor radon concentrations are shown in Table 4. The concentration in the crawl-space is relatively high in summer and low in winter. The concentrations in the first floor and in the second floor are higher in winter than in summer. In summer, the windows tend to be opened. So in the habitable rooms, the radon concentrations are relatively lower in summer.

4.3 Radon concentration in the crawl-space

The radon concentration in the house is shown in Figure 4. The sampled house was located in Sapporo city (43° 03'N, 141° 20'E), Hokkaido, in the northern part of the Japan. The ventilation opening had a cover, which was closed in winter because of the harsh weather. The radon concentration in the crawl-space is clearly higher in winter than in the other seasons. In the habitable rooms, which are in the first and second floor, the radon concentrations were also higher in winter than in the other seasons.

4.4 Reduction rate of radon using wood as a radon sealant

The reduction rates of the radon are displayed in Figure 5. The reduction rate of radon increased with an increment in the thickness of wood. In any thickness, the reduction rates were 90% or more. These rates show that the wood can reduce the penetration of the radon.

5. Conclusions

The radon emanations out of building materials were ranging from 0.1 to 29.8 Bq kg^{-1}. The radon emanation from ply-wood which is wood-based material, was little.

Indoor radon concentrations were higher in the crawl-space than in the first and second floor. Geometric mean (G.M.) of radon concentrations in the crawl-space, in the first floor and in the second floor were 34.1, 17.8 and 18.4 Bq m^{-3}, respectively.

The reduction rate of radon increased with an increment in the thickness of wood. In any

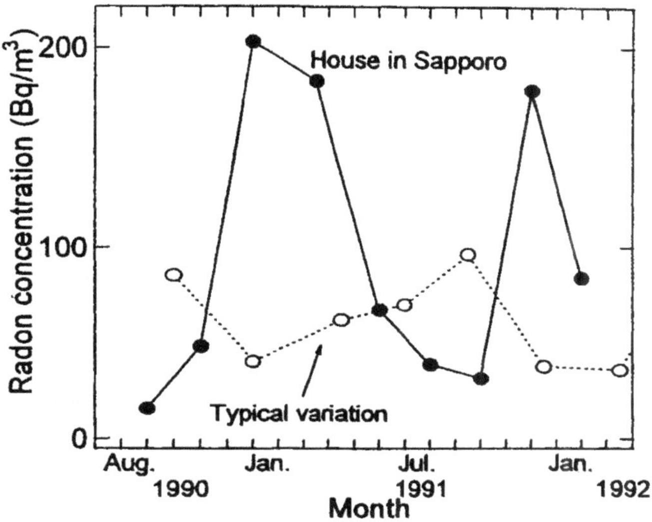

Figure 4 Seasonal variation of radon concentration in the crawl-space.

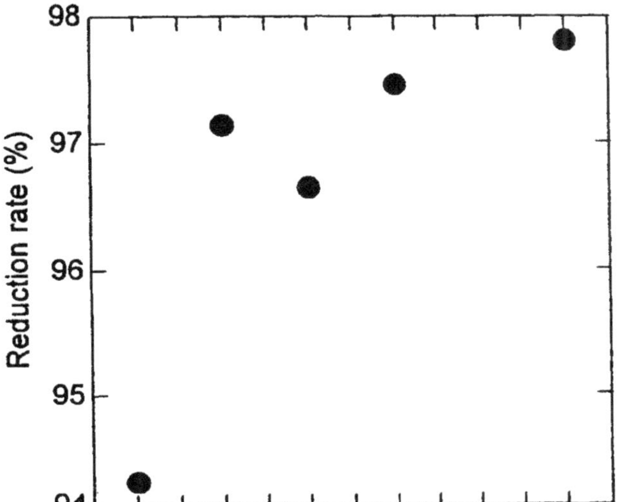

Figure 5 Reduction rate of radon using wood.

thickness, the reduction rates were 90% or more. These rates show that the wood has the function to reduce the penetration of the radon.

Acknowledgments- The authors would thank to Miss Kiyoko Sawada and Miss Yuko Tomita for their assistance with experiments. The authors would also thank to following persons for their cooperation and advice: Drs. Takanori Arima, Shinsuke Fujita, Kazuo Hayashi, Takao Iida, Koyo Kitta, Jun Ohtani, Takeshi Okano, Osamu Sawabe, Shigehiko Suzuki.

6. References

[1] Iida, T., Q. Guo and Y. Ikebe, Some problems on the measurement of ^{222}Rn concentrations by passive cup method, Health Phys., 69, 508-511, 1995.

Table 1 Radon emanations out of building materials.

Samples		Radon emanation (Bq/kg)
Inorganic building materials		
	Asbestos cement sheets	11.3 – 13.0
	Pulp cement boards	7.2 – 7.3
	Portland fly-ash cement boards	5.9 – 7.0
	Gypsum boards	6.0 – 29.8
	Cemented chip boards	3.4 – 3.6
	Autoclaved asbestos-cement calcium silicate booards	2.3 – 2.6
	Concrete blocks	1.0 – 1.2
	Light-weight concrete blocks	1.5 – 2.7
	Mud wall	3.8 – 10.5
Wood-based building material		
	Plywood	< 0.1

Table 2 Radon emanations out of soil.

Locations	Radon emanation (Bq/kg)
Sapporo 1	4.0
Sapporo 2	1.7
Sapporo 3	7.1
Morioka	6.4
Funabashi	5.3
Fukuroi 1	4.1
Fukuroi 2	4.7
Matsuyama	5.8
Kagoshima 1	1.3
Kagoshima 2	2.1
Kagoshima 3	4.6

Table 3 Radon concentrations in wooden houses.

	Radon concentration (Bq/m^3)		
	minimum	mean	maximum
Crawl-space	4	34	232
First floor	4	18	37
Second floor	3	18	49

Table 4 Radon concentrations in wooden houses.

Season	Radon concentration (Bq/m^3)								
	Crawl-space			First floor			Second floor		
	min.	mean	max.	min.	mean	max.	min.	mean	max.
Spring	16	42	167	7	20	37	4	22	49
Summer	28	76	232	10	19	29	12	20	29
Autumn	4	32	90	4	18	35	3	17	30
Winter	8	27	70	5	15	33	6	18	41

Radon and thoron daughter concentrations in Romanian houses

Constantin Milu, Raluca Gheorghe

Institute of Public Health
Str. Dr.Leonte No 1 - 3, 76256 Bucharest 35, Romania

Abstract

The indoor and outdoor radon and thoron daughter concentrations were determined in 119 Romanian dwellings, by air sampling on a membrane filter, followed by repeated gross - alpha countings. Several constructive and physical parameters have been recorded.

Knowing the radon and thoron daughters concentrations, the Potential Alpha Energy Concentration (PAEC), the Equilibrium Equivalent Concentration (EEC), the equilibrium factor (F) and the effective dose for adults (ED) have been calculated.

EEC (Rn) experimental values for indoor exposure ranged from 3 to 130 Bq/ m^3, with a mean value of 22 Bq . m^{-3}. The main observed influencing parameters were: ventilation rate, type of building material and indoor relative humidity. It was also pointed out that the contribution of the thoron daughters to the total ED is about 20 %. The results of the concentrations within an experimental house with a part of walls from phosphogypsum are presented.

1. Introduction

Man spents at least 60 % of his life indoor and this fact imposes a careful investigation of his habitual environment. Indoor exposure rises up to 50 % of the total radiation exposure of the public to natural sources [1]. The main contribution is done by the inhalation of radon and thoron and their progeny which are naturally - occuring radioactive gases that are found in rocks, soils and building materials.

It is already well recognized that the radon family increases the risk of lung cancer, and is believed to be the second most important cause of the disease after smoking.

It is also well known that the exposure to radon, thoron and their progeny comes mainly from the inhalation of the decay products of radon and thoron, which deposit inhomogeneously within the human respiratory tract and irradiate the bronchial epithelium.

With the aim to evaluate the indoor exposure of the population in Romania houses, several measurements of the radon and thoron daughters were performed.

2. Materials and method

For determination of Radon and Thoron daughter concentrations, a short air sampling (10 minutes) was used, followed immediately by 60 gross - alpha repeated countings (1 minute each) of the collecting filter and, after a delay of minimum 5 hours, by a single measurement of 50 minutes.

Airsols were collected on a membrane filter type Sympor (35 mm diameter, 0.85 micrometer porosity) at a flow rate of 20 to 30 liter/ minute.

The indoor and outdoor determinations were performed during the morning period; for indoor, after minimum 3 hours of closing doors and windows of the house.

The experimental data have been processed by applying Sima and Sonoc method [2], in order to determine each of the components: ^{218}Po (RaA), ^{214}Pb (RaB), ^{214}Bi (RaC) and ^{212}Pb (ThB).

The radioactive equilibrium between ^{214}Bi and ^{214}Po, as well as between all the Thoron daughters has been assumed. Using the determined Radon and Thoron daugters concentrations, the Potential Alpha Energy Concentration (PAEC), the Equilibrium Equivalent Concentration (EEC) and the equilibrium factor (F) have been calculated, according to the relations [1]:

PAEC (Rn) = (0.105 C (RaA) + 0.505 C (RaB) + 0.380 C (RaC)/ 3.7 (mWL)
PAEC (Tn) = 3.64 C (ThB) (mWL)
EEC (Rn) = PAEC (Rn) x 3.7 Bq. m^{-3}
EEC (Tn) = C (ThB) Bq. m^{-3}
F = EEC (Rn) / C (Rn)

^{222}Rn concentration has been approximated to be equal with ^{218}Po concentration plus 10 % assumption of the unattached fraction of ^{218}Po [1].

During the measurements of the Radon and Thoron concentrations several constructive and physical parameters have been recorded: construction tupe, constructive solution design, building materials used. heating systems inside the house, season, indoor and outdoor temperature, pressure and relative humidity, direction and speed of the wind.

The determinations were performed in 119 Romanian dwellings.

3. Results

The measured indoor radon and thoron daughter concentrations are prescuted in Table 1, expressed in median, geometric mean and standard deviation of the individual PAEC and resp. EEC values.

Table 1. Indoor Radon and Thoron short-lived daughter concentrations

	Experimental value range	Median	Geometric mean	Standard deviation
PAEC(Rn) mWL	0.8 - 34.4	5.85	4.63	1.26
EEC(Rn) Bq.m^{-3}	3.0 - 130	21.60	17.20	4.66
PAEC(Tn) mWL	0.5 - 16	5.57	5.00	1.11
EEC(Tn) Bq.m^{-3}	0.013 - 4.4	1.54	1.38	0.31

After mathematic analyses of the results, a lognormal distribution of PAEC (Rn) values was obtained, in a good agreement with data from literature [1].

The mentioned determinations have been equally distributed among seasons, so that the mean values could be considered as representative for the whole annual exposure.

The EEC (Rn) ranged from 3.0 to 130 Bq. m^{-3}.

From the presented concentration values, the individual Committed Effective Dose Equivalents (CEDE) were calculated, and are presented in Table 2.A difference between rural and urban population was found and contribution of the thoron daughters is pointed out.

Table 2. Individual Committed Effective Dose Equivalent (Heff) for adults caused by inhalation (mSv/year)

Area	Ocup. factor	EEC ind		EEC out		Heff	Heff	Heff	Heff %	
		Rn	Tn	Rn	Tn	Rn	Tn	total	Rn	Tn
Rural	0.6	29	1.17	3	0.23	1.117	0.28	1.405	84	16
Urban	0.8	17.2	1.38	3	0.23	0.914	0.307	1.221	75	25

The total PAEC values for different areas from Romania are given in Table 3, determined by several radiation hygiene laboratories in 258 houses.

Table 3. Radon and Thoron short-lived daughter concentrations (mWL) in areas from Romania (Range of experimental values and Median)

Area	Number of houses	Indoor (mWL)		Outdoor (mWL)	
		range	mean	range	mean
Arges	14	1.4 - 17.0	6.7	0.4 - 3.6	1.7
Valcea	7	2.6 - 11.2	5.8	0.3 - 1.0	0.6
Bacau	18	2.6 - 21.0	8.9	0.1 - 4.7	1.8
Brasov	15	0.2 - 48.3	4.0	0.1 - 1.5	0.5
Cluj	17	0.6 - 52.0	5.6	0.3 - 2.3	0.7
Constanta	7	1.0 - 4.1	2.3	0.4 - 1.1	0.6
Caras-Severin	20	0.5 - 9.2	1.4	0.1 - 0.9	0.5
Dolj	5	2.2 - 14.7	7.6	0.9 - 2.6	1.5
Galati	10	2.0 - 9.7	4.3	0.4 - 3.7	1.5
Braila	10	1.5 - 13.6	7.9	0.4 - 3.7	1.9

Vrancea	10	0.8 - 16.8	4.1	0.4 - 3.7	1.0
Iasi	17	0.1 - 20.2	4.5	0.1 - 1.0	0.2
Maramures	22	0.3 - 14.4	4.4	0.2 - 1.5	0.7
Prahova	5	1.2 - 9.3	2.3	0.6 - 2.1	1.3
Buzau	5	0.8 - 3.2	9.0	0.1 - 3.7	1.9
Suceava	19	1.0 - 50.2	6.5	0.3 - 2.2	0.9
Mures	5	4.2 - 9.4	6.7	0.8 - 3.7	2.1
Bucuresti	52	1.0 - 44.4	7.4	0.1 - 6.9	1.6

As the building material used was considered an important influence factor, several determinations of the natural radioactive content of different building materials used in Romania were performed, using a Ge(Li) detector with multichannel analyzer. The results obtained can be found in Table 4.(A).

Table 4. ^{226}Ra, ^{232}Th and ^{40}K concentrations (Bq/kg) in building materials

Building material	^{226}Ra			^{232}Th			^{40}K		
	min.	max.	av.	min.	max.	av.	min.	max.	av.
A) Commonly used in Romania									
red brick	17	100	36	19	53	32	319	833	493
cement	9	66	33	4	97	18	4	504	152
gravel, sand	5	9	6	4	12	8	27	611	428
lime	4	41	17	4	19	8	4	114	58
clay	17	31	25	35	67	49	114	704	402
granite stone	98	137	115	66	301	181	1300	1700	1460
natural gypsum	4	43	18	4	27	10	4	277	103
B) Industrial solid wastes									
phosphogypsum	237	970	715	11	42	23	59	167	67
ash, slag	62	218	155	74	126	89	277	945	569
sterile mass (uranium mines)	512	6800	4030	24	158	94	319	1370	780

4. Determinations within an experimental house

Since 1980, the potential use as building material of some industrial solid wastes was signalized, mainly phosphogypsum, a by-product from phosphate fertilizer industry, and slag and ash from coal-fired power plants. " Mountains " were and still are available of such solid wastes

The first our step was to check their radioactive content. As can be seen from Table 4 the mean value obtained for ^{226}Ra for samples of phosphogypsum was 40 times the content of natural gypsum.

In order to see which is the influence of this ^{226}Ra content to indoor radon level a real (experimental) detached house (with basement, ground floor, garret) was built near Bucharest. The main external walls were built using concrete and aerated concrete bricks and some internal wall made (later) from phosphogypsum.

Two sets of measurements were performed, without and with internal walls of phosphogypsum. The main results are presented in table 5. It can be pointed out an increase four times of the EEC (Rn) by use of phosphogypsum.

Table 5. EEC(Rn) in Bq.m^{-3} for an experimental house built <u>without</u> 'A' and <u>with</u> some internal walls of phosphogypsum 'B'

Building solution	Experimental values [range]	Median	Geometric mean	Standard deviation	Ratio of medians (B/A)
A	[4.7 - 32.6]	11.7	10.1	2.4	4.3
B	[7.4 - 145.4]	50.2	41.8	5.7	

This final result determined the Ministry of Health to introduce restrictions in the use of phosphogypsum as building material for social dwellings.

5. Conclusions

The Radon and Thoron daughter concentrations in Romanian houses are comparable with those observed in other European countries and considered as " normal " [1, 3, 4].

The contribution of thoron daughters to the effective dose from inhalation of radon and thoron daughters is not neglijable, its value up to max. 20 % was noticed.

The use as building material of some industrial solid wastes (phosphogypsum, ash) proved to be an important source, which increase the radon concentration in homes. As a consequence, the Romanian Ministry of Health use now the ^{226}Ra content (max. 100 Bq/ Kg) as an acceptance test of any new building material.

Within a national scientific meeting organized in 1997 in Bucharest, several unusual high level of radon were reported [5], particularly in detached houses, with basement of stone and slag and without cellar. In some of such homes the average concentration of radon was four to five times the simple action level (200 Bq . m^{-3}) recommended by World Health Organization in 1988 [6].

These recent results impose the need of a mass survey in Romania, using a simple method for direct detection of radon gas concentrations throughout the whole conntry.

6. References

[1] United Nations Scientific Committee on the Effects of Atomic Radiation, Sources and Effects of Ionizing Radiation, New York, UNSCEAR Report, 1993

[2] Sima O., Sonoc S., Dovlete C., Critical analysis of the least square method applied to the determination of the environmental radon and thoron daughters, St. Cercet. Fizica, Bucharest, 38, 464 - 495, 1986

[3] National Radiological Protection Board, Board Statement on Radon in Homes, Documents of the NRPB, Vol.1, No.1, 1990

[4] Radon Policy: Theory and Practice, Annales de l'Association Belge de Radioprotection, Vol.21, No.1, 1996

[5] Institute of Public Health, Bucharest, Summary of Papers, XXXII - nd International (70th Anniversary) Conference, 186 - 189, 8 - 9 May 1997

[6] World Health Organization, Air quality guidelines for Europe, No.23, Copenhagen, WHO, 1988

Radioactivity of some specific underground workplaces in Hungary

Pavel Szerbin

"Frédéric Joliot-Curie" National Research Institute for Radiobiology and Radiohygiene, POB 101, 1775 Budapest, Hungary

Abstract

In this study, radon and radon daughter concentrations were measured in specific underground workplaces: in vine cellars, showcaves and caves for speleotherapeutic treatment of patients. These workplaces are considered to be affected by high radon concentrations, the workers in some cases can be exposed to radiation from radon and short-lived daughters at levels, comparable to those of other underground workplaces such as uranium mines. The aim of the study was to reveal the variation of the radon and radon progeny concentration, and to estimate the dose to the workers and visitors of these workplaces from the radon daughters.

Seasonal variations of radon concentrations have been found in most of the caves investigated. The magnitude of the variation depends on the degree of isolation of the cave cavity from the outdoor atmosphere. The radon concentrations of the vine cellars were much lower than that of the caves. Because of the produced carbon dioxide gas in the process of sugar fermentation even the very old cellars are constructed well ventilated, consequently radon can not be accumulated in this underground cavities. From the measured radon and progeny concentrations cumulative potential alpha energy exposure, bronchial dose, and effective dose values were calculated for the staff, patients and visitors. The effective dose values for the cave personnel do not exceed the 10 mSv year^{-1} upper bound of the action level recommended for workplaces and even much lower for workers of cellars. The doses for visitors and patients were one or two magnitudes lower than that of the workers.

1. Introduction

Hungary is famous of the old traditional vine-growing areas. Vineyards are represented mostly by two different types: the first one is where vine is stored and treated in cellars of small farms and the second one is the huge labiriths of vine producing enterprises and distilleries. These specific underground workplaces are considered to be affected by high radon concentrations. Besides, Hungary reach of natural caves particularly in the vicinity of Budapest and in the karstic region in the North-East of Hungary. Some of these caves are used for treating patients suffering from respiratory diseaes. The atmosphere of these caves is usually stable; the temperature varies between 10-14 °C, the relative humidity is about 100 %. The vine cellars have less stable atmosphere because of the better ventilation. The source of the radon in the caves and cellars is the ^{226}Ra belonging to the ^{238}U decay series found in all rocks and soils. The radium activity in the host minerals usually varies widely, between 1-4500 Bq kg^{-1} [1]. After alpha recoil, radon atoms that enter capillaries migrate in cracks and fissures and appear in underground cavities. The soil gas radon concentration due to the continuous radium decay and radon diffusion is very high, with measured values up to thousands of Bq m^{-3} [2]. The radon concentration in soils and underground cavities depends on different factors: the radium concentration of the minerals and soils, the emanation rate of the radium containing media, its porosity and permeability, water content and, the meteorological conditions [3, 4, 5, 6, 7, 8].

2. Materials and methods

Sampling sites were chosen where the cellar workers activity is concentrated, and along the route of the guided tourist visits, and at the locations where speleotherapeutic treatments occur. Air was sampled every three hours directly using Lucas cells of 1000 cm^3 volume. Radon measurements were performed by NP-420P type (MÉV, Mecsek Mining Company) radon meter. The alpha activity was counted after three hours to permit radioactive equilibrium between radon and the daughters to be established. The Lucas cells were calibrated by a standard ^{226}Ra solution and by a Pylon (Pylon Electronic Development Inc.) RNC standard. The lower limit of detection (LLD) of the equipment was 4-7 mBq l^{-1}, depending on the counting interval and the background of the scintillation cell.

Continuous radon and radon daughter measurements were carried out by Pylon AB-5 type monitors (Pylon Electronic Development Inc.), equipped with Continuous Passive Radon Detector (CPRD) and a working level detector, model number AEP-47. The LLD for the CPRD was 11 Bq m^{-3}, for the AEP-47 it was 0.7 Bq m^{-3} equilibrium equivalent concentration (EEC).

The equipment is calibrated on a monthly basis, using Pylon 3150A, RNC, and RN-190-47 calibration standards. The original certificated calibration of the Pylon instrumentation was verified by continuous filter sampling of radon progeny in air and by continuous sampling of radon gas in air and simultaneous intercomparison with working standard instruments of reference values.

Summer mean concentrations in cellars and yearly mean radon concentrations in caves were calculated along the paths of the tourist visits and at the sites of speleotherapeutic treatment of the patients. The time spent in the underground cavity by the worker or by a visitor was assessed on the basis of personal communication with the staff and the management. From the mean radon concentrations and the known occupancy in workplaces the potential alpha energy exposure, bronchial dose, and effective dose values were calculated for the staff, patients and visitors.

3. Results and discussion

Radon and radon progeny concentrations were measured at 12 sampling sites in the cellars of Hungarovin Ltd., and at 12 sampling sites in the cellars of Promontorvin Ltd., both located in the south of Budapest. Measurements were done only in summer period, no seasonal variation data available yet. Results of continuous radon and radon progeny measurements of lowest and highest values in the Hungarovin cellars for sampling sites No.3 and No.10 are given in Figure 1. and Figure 2. respectively.

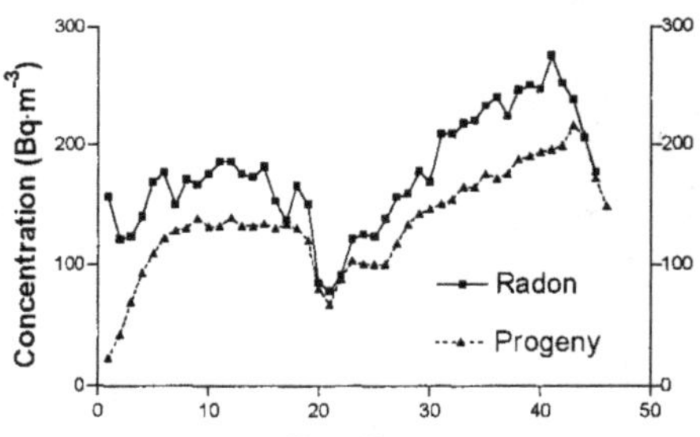

Figure 1. Radon and radon progeny concentration at sampling point No.3 in Hungarovin cellars.

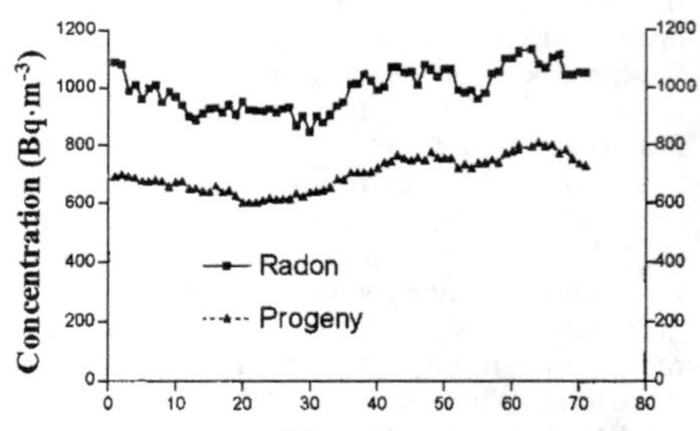

Figure 2. Radon and radon progeny concentration at sampling point No.10 in Hungarovin cellars.

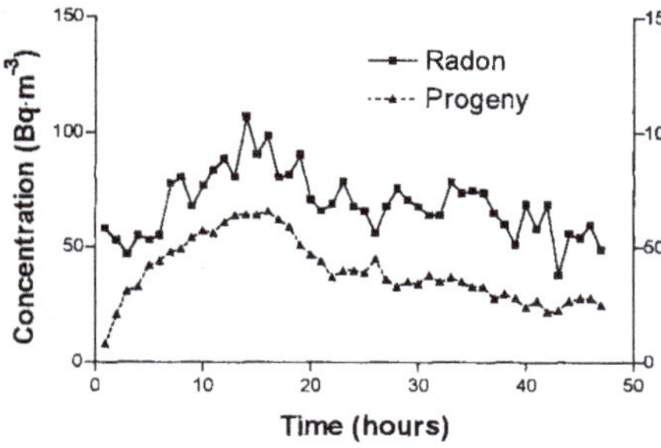

Figure 3. Radon and radon progeny concentration at sampling point No.9 in Promontorvin cellars.

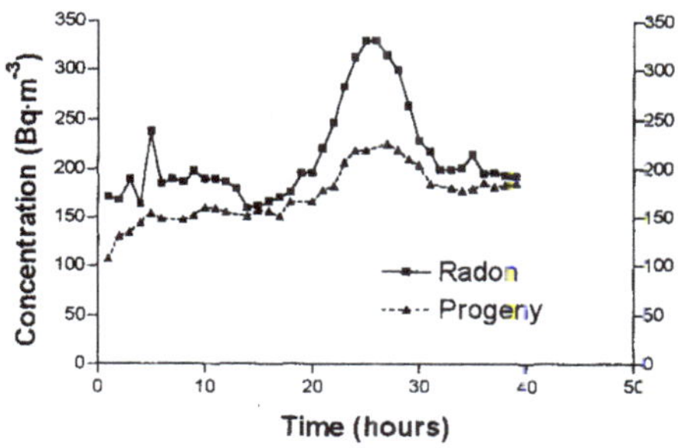

Figure 4. Radon and radon progeny concentration at sampling point No.7 in Promontorvin cellars.

The lowest and highest radon and radon progeny concentrations measured in Promontorvin cellars are given in Figure 3. and Figure 4. As it can be seen from the pictures, airborne radioactivity in the cellars is much changeable and dependent on the effect of outdoor climatic conditions. The sampling point No.10 in the Hungarovin cellars was the deepest among the sites investigated. Radon levels in Hungarovin cellars were much higher than that of the Promontorvin cellars in spite of that both located in the same region, constructed by the same technology, surrounded by the same limestone and the distance between them is not more than approximately 500 metres.

Results of radon measurements in caves are shown in Figures 5-8. Long-term measurements (Figures 5-7) revealed seasonal variation of radon concentration, which was more spectacular in shallow caves (Pál-völgyi cave and Tapolca cave) and less expressed in cave with well isolated cavity (Szemlő-hegyi cave). It is due to microclimatic reasons in the cave and its environment, and the phenomena has been described earlier [9, 10, 11]. It points to the necessity to take into account these changes in dose calculations and to use long term integrating measuring techniques. Effect of outdoor weather conditions was proved also by short-term measurements in a shallow cave. When according the weather recordings wind speed rised slowly due to a warm front, radon concentration in the cave was affected slowly and moderately. Sharp and instantenous drop of radon level was observed when cold front passed the area with strong northern wind (Figure 8).

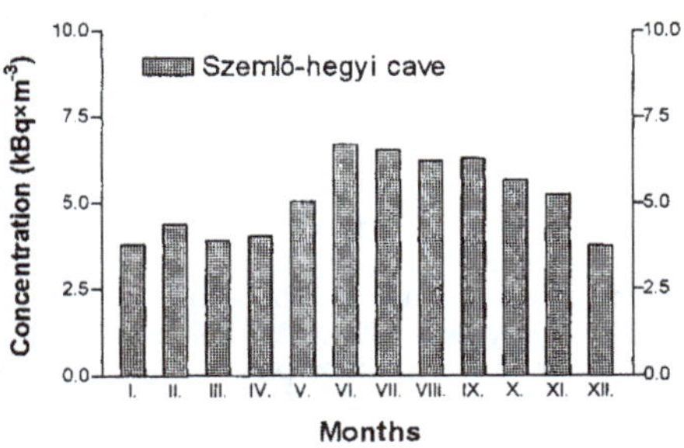

Figure 5. Seasonal variation of radon concentration in Szemlő-hegyi cave.

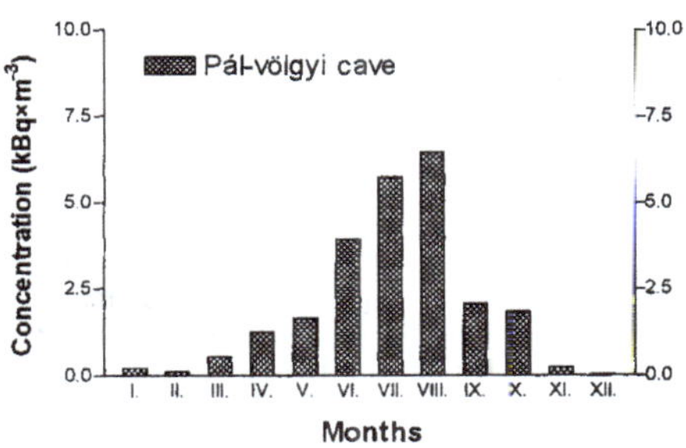

Figure 6. Seasonal variation of radon concentration in Pál-völgyi cave.

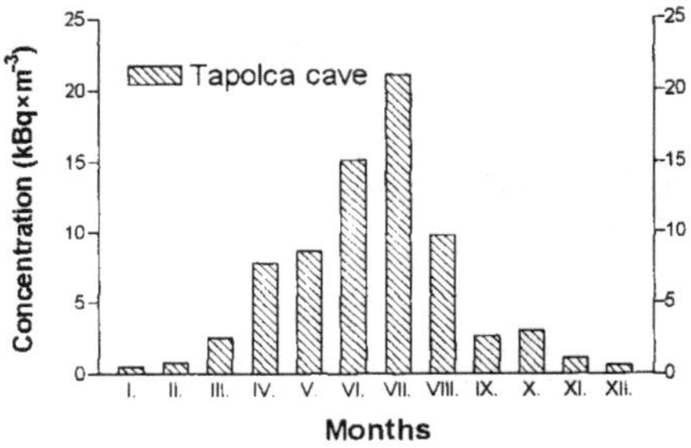

Figure 7. Seasonal variation of radon concentration in Tapolca cave.

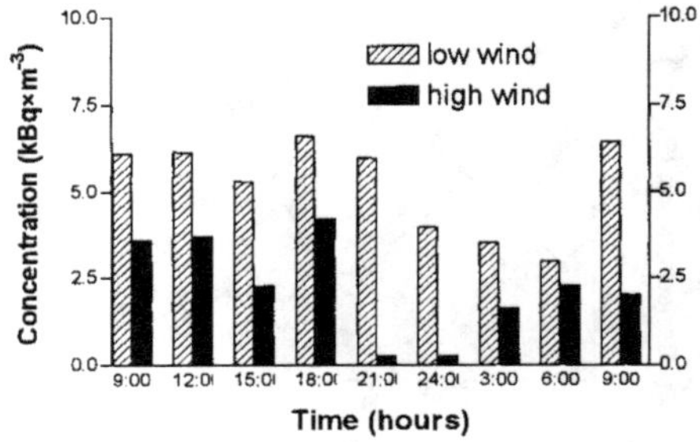

Figure 8. Effect of different climatic conditions on radon concentration in a shallow cave.

Results of simultaneous radon and radon progeny measurements are shown in Figures 1-2. It can be seen that radon and progeny concentrations more stable in caves with well isolated inner cavity. Calculated equlibrium factor for the caves was about 0.5 and this value was adopted for dose calculations. In the cellars the equlibrium factor changed more widely, the values were between 0.35 and 0.92. The F values were calculated for each sampling site separately and the corresponding values were used for dose calculations.

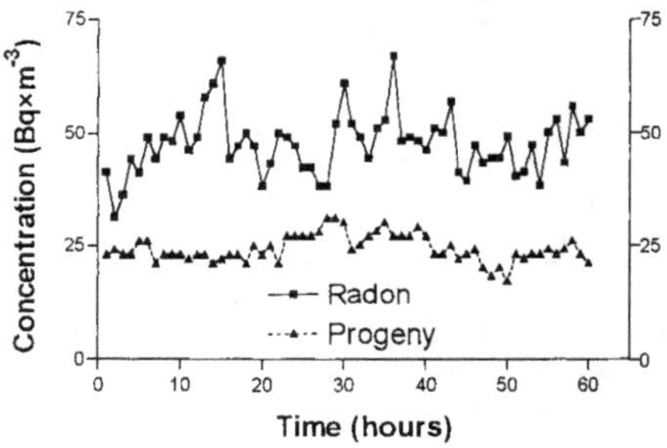

Figure 9. Radon and radon progeny concentration in a shallow cave.

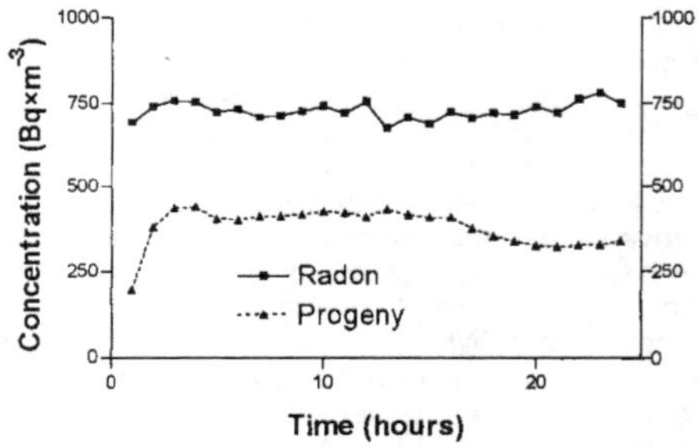

Figure 10. Radon and radon progeny concentration in a medium deep well isolated cave.

For caves the yearly mean values, for cellars the summer measurement results of radon concentrations were used for potential alpha exposure, bronchial dose and effective dose calculations, the results are shown in Table 1. Doses were calculated using conventions, published by ICRP publications [12, 13]. The occupancy in the cellars was assumed to be about 1560 hours year^{-1}, the occupancy in caves was given as 250 hours year^{-1} for tourist guides, and the speleotherapy staff. An average 60 hours year^{-1} exposure for speleotherapy patients was used. It can be concluded that doses for the workers never exceeded the 20 mSv year^{-1} dose limit, recommended by ICRP [14], and lower than the upper bound of the 3-10 mSv year^{-1} dose limit recommended by ICRP [13].. Doses to the cave visitors assuming 1 hour per year exposure were very low and varied in the range of 7 to 211 μSv year^{-1}.

Table 1. Radon concentration, the calculated potential alpha energy exposure, bronchial dose, and effective dose values for cellar, cave staff and patients of speleotherapy.

Sampling area	Person	Mean radon concentration (kBq m^{-3} ± S.D.)	Equilibrium factor (F)	Potential alpha energy exposure (mJ h m^{-3})	Bronchial dose (mSv y^{-1})[a]	Effective dose (mSv y^{-1})[b]
Hungarovin lowest	worker	0.18 ± 0.03	0.8	0.11	2 - 5	0.2
Hungarovin highest	worker	0.99 ± 0.12	0.7	0.52	9 - 25	0.7
Promontorvin lowest	worker	0.07 ± 0.02	0.6	0.03	1 - 2	0.04
Promontorvin highest	worker	0.22 ± 0.04	0.8	0.13	2 - 5	0.2
Szemlő Hill cave	worker	5.15 ± 1.13	0.5	3.68	67 - 178	5.2
	patient			0.89	16 - 43	1.3
Pál Valley cave	worker	2.01 ± 2.22	0.5	1.45	26 - 69	2.1
	patient			0.35	6 - 17	0.5
Hospital cave	worker	6.10 ± 6.60	0.5	4.39	79 - 210	6.2
	patient			1.06	19 - 50	1.5

a - calculated by ICRP Publication 32 [12]
b - calculated by ICRP Publication 65 [13]

4. Conclusions

- Diurnal and seasonal changes were observed in cellars and caves. These changes are more expressed in shallow underground cavities with good air connection between the outdoor and indoor atmosphere. Short-term atmospheric meteorological changes affect the radon levels in the cellars and caves.
- Radon progeny concentrations change in parallel with radon values. Equlibrium factors in the cellars and caves vary widely.
- The calculated effective dose values from radon and radon daughter radiation for workers of investigated cellars and caves do not exceed the dose limit and action level, recommended by ICRP [13, 14].
- Further studies need to be done to establish seasonal variation of radon concentration in cellars and to reveal possible differences between small farm cellar and enterprise cellar conditions.

Acknowledgements - Funding for this study was approved by the Polish Hungarian Aid for Reconstruction and Education (PHARE) under project 134/3 Cave Climatology and Cave Therapy, and supported by National Environmental Health Program (NEKAP) of the Ministry of Welfare, Hungary.

5. References

[1] United Nations Scientific Committee on the Effects of Atomic Radiation (UNSCEAR). Sources and Effects of ionizing Radiation. New York: United Nations, 1993.

[2] Wilkening, M.H; Radon in the Environment. Elsevier: Amsterdam-Oxford-New York-Tokyo. 1990.

[3] Megumi, K; Mamuro, T; A method for measuring radon and thoron exhalation from the ground. J. Geophys. Res. **77,** 3052-3056, 1972.

[4] Miki, T; Ikeza, M; Accumulation of atmospheric radon in calcite caves. Health Phys. **39,** 351-354, 1980.

[5] Somogyi, Gy; Lénart, L; Hakl, J; Németh, Gy. Seasonal change of radon concentration in the caves of Bükk-mountain. In: Proceedings of the 2nd national conference of karst and cave research activity of education institutes, 65-71, 1987. (in Hungarian)

[6] Davis, R.M; Watson, J.E. Influence of ^{226}Ra concentration in surrounding rock on ^{222}Rn concentration on ground water. Health Phys. **58,** 369-371, 1990.

[7] Hunyadi, I; Hakl, J; Vásárhelyi, A. Radon monitoring experiences and important facts that may affect dose estimation. In: Radiation protection in neighbouring countries in central europe, proceedings of the Austrian-Italian-Hungarian radiation protection IRPA symposium. Oberburgl, Austria. **3,** 414-417, 1993.

[8] Hakl, I.; Csige, I.; Hunyadi, I.; Várhegyi, A.; Géczy, G.; Radon transport in fractured porous media - experimental study in caves. Environ. Int.; **22,** Suppl. 1, 433-437, 1996.

[9] Wilkening, M.H; Watkins D.E. Air exchange and ^{222}Rn concentrations in the Carlsbad caverns. Health Phys. **31,** 139-145, 1976.

[10] Eheman, C; Carson, B; Rifenburg, J; Hofman, D. Occupational exposure to radon daughters in Mammoth Cave National Park. Health Phys. **60,** 831-835, 1991.

[11]Lively, R.S; Krafthefer, B.C. ^{222}Rn variations in Mystery cave, Minnesota. Health Phys. **68,** 590-594, 1995.

[12] International Commission on Radiological Protection. Limits for Inhalation of Radon Daughters by Workers. Oxford, Pergamon Press; ICRP Publication 32, 1981.

[13] International Commission on Radiological Protection. Protection Against Radon-222 at Home and at Work. Oxford: Pergamon Press; ICRP Publication 65, 1993.

[14] International Commission on Radiological Protection. 1990 Recommendations of the International Commission on Radiological Protection. Oxford: Pergamon Press; ICRP Publication 60, 1991.

The influence of the working conditions on the equilibrium factor F and the unattached fraction fp

T. Streil[1], *A. Reichelt*[2] *and A. Reineking*[3]

[1]SARAD GmbH, Dorfplatz 1, D-01705 Pesterwitz/Dresden
[2]TÜV Süddeutschland, Westendstr. 199, D-80686 München
[3]Georg-August Universität Isotopenlab., Burckhardtweg 2, D-37077 Göttingen

Abstract

We report on the influence of the working conditions on the dose estimation. Especially the equilibrium factor and the unattached fraction are influenced by the working conditions.
For instance in a cabinet-maker's shop the radon concentration is strongly influenced by the ventilation system. The factor F is influenced by dust producing work processes.

For a better knowledge of the radon dosimetry it is also necessary to measure continuously and separately the unattached fraction of the radon progeny. We present first results obtained with a SARAD EQF 3020 monitor which contains three alpha detector microsystems measuring radon in the air, attached radon daughters and unattached radon daughters. The three components are measured quasicontinuousely with a temporal resolution of 2 hours. The measuring range is 1 to 10 MBq/m^3 at a low detection limit of 0.1 Bq/m^3 equivalent radon concentration. The system was tested in buildings, caves, mines, waterworks and other places. By using silicon microsystems with alpha detector, ADC, memory and logics - which are integrated on few chips - the costs for the production of equipment are far lower than if produced by means of assembled individual components.

1. Introduction

The influence of the working conditions on the dose estimation is very important. Especially the equilibrium factor and the unattached fraction are influenced by the working conditions.
For instance in a cabinet-maker's shop the radon concentration is considerably influenced by the ventilation system. The factor F is influenced by dust producing work processes.
For a better knowledge of the radon dosimetry it is also necessary to measure continuously and separately the unattached fraction of the radon progeny. The dose factors for unattached radon daughters are considerably higher than for radon daughters attached to aerosols. Therefore dose calculations and assessments of remedial actions based only on radon gas or attached radon daughter concentrations may be misleading. Reducing e.g. the aerosol concentration by aerosol filters, the attached fraction is clearly reduced but the unattached fraction increases, resulting in a dose increase instead of a dose reduction.

One of the reasons why one rarely measures all three components (radon, attached fraction and unattached fraction) is that there are no cheap portable instruments available which can measure them all. A monitor built out of alpha spectrometric silicon microsystems [1]. The first step to use microsystems for the radiation detection was the application of special designed d-RAM cells [2]. These systems were only applicable for counting and not for spectrometry. The small size and the low power consumption of silicon microsystems also make them ideal for portable instruments. We present first data obtained only with this type of monitor containing three detectors and measuring all three components with a temporal resolution of 2 hours.

2. Methods

In a previous paper we have described a radon and radon progeny monitor allowing the simultaneous measurement of radon and radon progeny with two silicon microsystem detectors [3]. In this system the air passes an aerosol filter registered by the first detector and then flows through the radon measuring chamber. We now have added a metal screen/detector assembly that can be switched into the air flow in front of the aerosol filter to catch and measure the unattached radon daughters. A microprocessor controls the movement of the two radon daughter detectors and the screen. The cut-off factor of the screen at an air flow of 2.5 l/min and the screen geometry is 5 nm. We are thus able to measure radon progeny and attached progeny separately.

3. Results and discussion

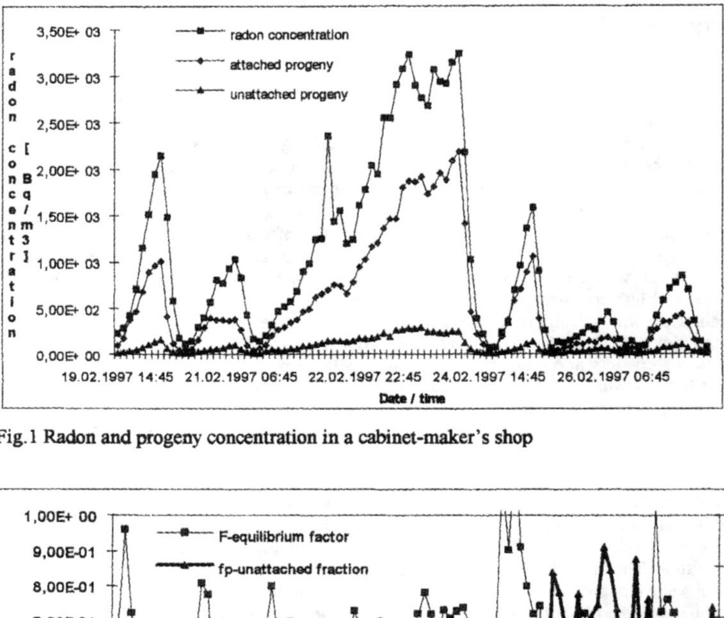

Fig.1 Radon and progeny concentration in a cabinet-maker's shop

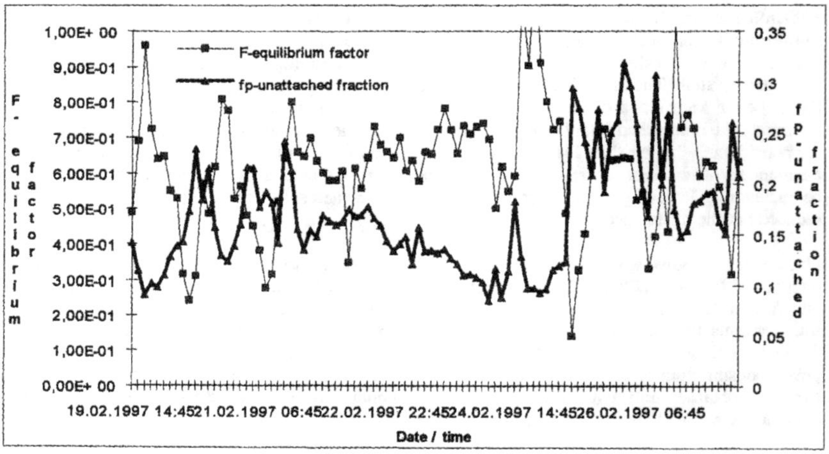

Fig.2 Equilibrium factor F and fp in a cabinet-maker's shop

Fig. 1 shows a long time measurement of the radon concentration, the attached and the unattached progeny concentration in a cabinet-maker's shop. The daily variations can be seen clearly . In the morning the ventilation system is switched on and in the evening the ventilation system is switched off. At a high ventilation rate the unattached fraction is increased but in the evening - due to the small

ventilation rate and the dust producing production processes - the attached fraction is increased. In the dusty production rooms this fact is excellently shown
(especially during the weekend) in Fig. 1. On Monday, early in the morning, the ventilation reduces rapidly the aerosol concentration, and the factor fp increases (Fig. 2). When the production process starts, the dust increases the aerosol concentration as well as the factor F.

Fig. 3 shows a measurement carried out in Dresden (Germay) concerning the equilibrium
factor F and fp of the unattached fraction at a work place in a school cellar next to the coal heating system.

The big variation of the fp factor is caused by the change of the ventilation rate and the aerosol concentration in this cellar.The radon and radon progeny concentration is shown in Fig.4. The factor fp ranging up to 0.11 shows that for the dose calculation the unattached fraction is not negligible. A long-term study in this place shows that the average value of the unattached progeny concentration is in the order of 200 Bq/m³ and the attached progeny concentration is of approximately 2,000 Bq/m³. Using the James/Birchel lung model (recommendation of ICRP 66) for the dose calculation, the exposed person (the stoker) accumulates during 2,000 working hours approx. 28mSv/a from the unattached progeny and 32 mSv/a from the attached progeny. In this way we can conclude that the unattached fraction has almost the same dose contribution like the attached fraction at this work place.

Using this new measuring technique it is possible to get more exact information of the real dose at work places or other interest places.

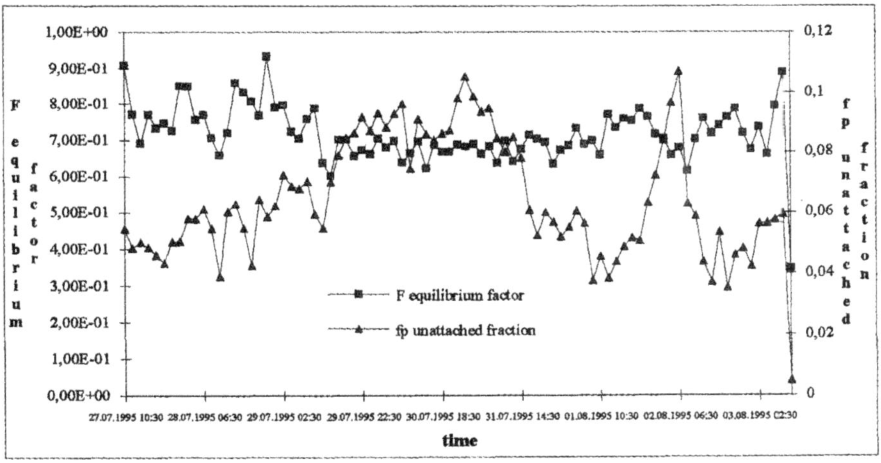

Fig.3 Measurement of the equilibriumfactor F and fp in a school cellar in Dresden

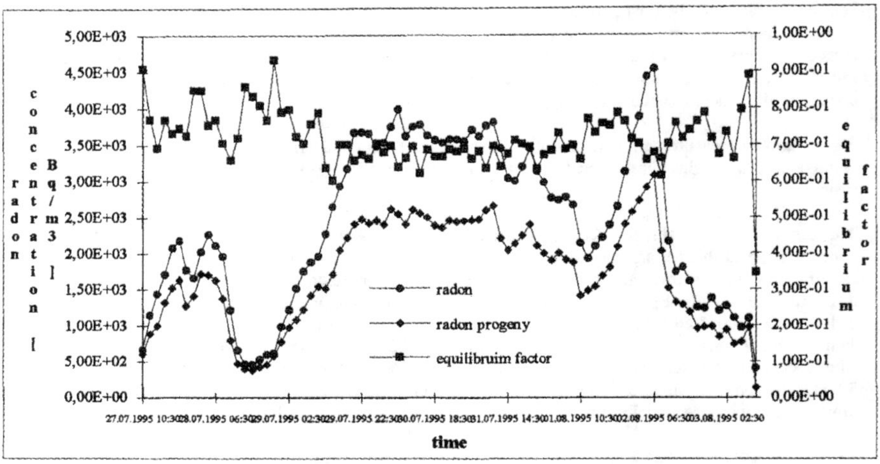

Fig.4 Measurement of radon and progeny concentration in a school cellar in Dresden

Another work place is shown in Fig.5 and 6. In collaboration with Ivan Kobal and Janja Vaupotic from the "Jozef Stefan" Institute in Lublijana (Slovenia), we measured the radon, attached and unattached progeny concentration in the Postoina show cave in several places. Especially in the place "partizanski rov" we measured a very low factor F in the order of 0.1 (Fig.6). This low factor F is caused by the extremely clean and aerosol-free air and the natural ventilation in this cave. In this case the main contribution for the dose of the tour guides resulted from the unattached progeny with approx. 30 mSv/a during 2,000 working hours.

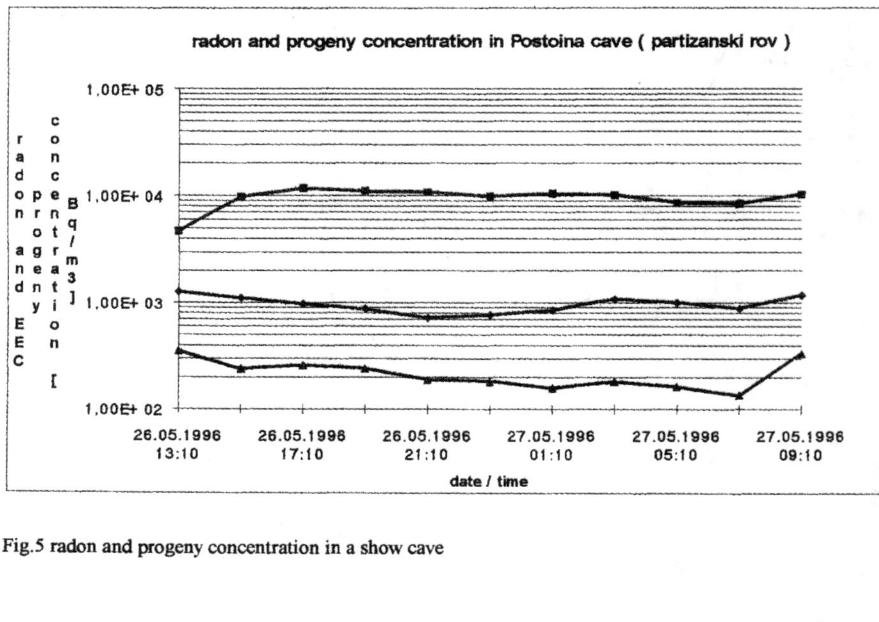

Fig.5 radon and progeny concentration in a show cave

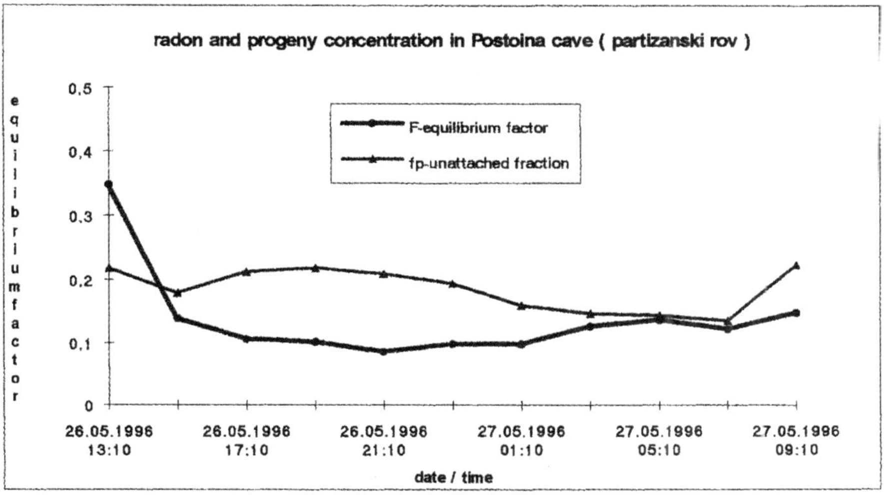

Fig.6 Equilibrium factor F and fp in a show cave

The contribution of the attached fraction is of approximately 15 mSv/a. In this cave the usual working hours for the guides are 10 hours per day and in this way the above-mentioned dose estimation reflects the almost real conditions.

This first study shows that at work places the working conditions have a strong influence on the real dose of the worker due to the radon progeny . We found places with very low factors F of approximately 0.1 and dusty places with high factors F of approximately 0.8. The unattached fraction fp varies considerably from 0.02 in waterworks [6] up to 0.3 in ventilated rooms.

4. References

[1] Streil, T., R. Klinke, A. Erlebach, P. Hübler, W. Kluge, H. Kück and G. Zimmer, Sensors and Actuators A, 41-42 (1994) 85-87

[2] Streil, T., A. Erlebach, P. Hübler, A. Scade and W. Kluge, Proc. 2nd Workshop on Radon Monitoring in Radioprotection, Environmental and/or Earth Sciences, Trieste Nov./Dec. 1991 554-559

[3] Streil, T., R. Klinke and G. Just, Indoor Air An Integrated Approach, 1995 Elsevier Science Ltd., 135-138

[4] Streil, T , R. Klinke, W. Birkholz and G. Just, Radiation Measurements, Vol. 25 Nos 1-4 (1995) 621-622

[5] Streil, T. R. Klinke, W. Grundke, H. Surbeck and A.C. James, Proceding International conference " Technologically enhanced natural radiation caused by no-uranium mining" 16.-19. October 1996 Szyrk, Poland , P.141

[6] Reichelt, A and K.H. Lehman " Antropogene Stoffe und Produkte mit natürlichen Radionukliden" Teil IIa Oktober 1994, Publikation Bayerisches Staatsministerium für Landesentwicklung und Umweltfragen

Continuous radon gas measurement along with intermittent radon- and thoron-progeny measurement for one year

Tsuneo Kobayashi and Yuji Takaku

Department of Physics, Fukushima Medical College
1 Hikariga-oka, Fukushima City 960-12, Japan

Abstract

Continuous indoor radon gas measurement along with intermittent outdoor and indoor radon- and thoron-progeny measurement in the afternoon were made for one year around the authors' office that is located in the northeastern part of Japan. Indoor radon-progeny concentrations were smaller than outdoor ones, and quite small equilibrium factors (F-value, mean value was 0.2) were observed. A definite seasonal variation was observed for continuous radon data and intermittent radon-progeny data (higher concentrations in autumn), while no significant seasonal variation was observed for indoor thoron-progeny. Statistical and time series analyses indicated several interesting results such as several-day periods of variation that was similar to wind velocity, 20- to 30-day periodicity that concerns the rotation of the sun, etc.

1 Introduction

Investigations have been made on radon (^{222}Rn) gas, radon (^{222}Rn) progeny and thoron (^{220}Rn) progeny in air. Clarifying the behavior of these radioisotopes is an issue of wide importance to radiation protection, predicting earthquakes, etc. The aim of the present research was to survey the level and behavior of these radioisotopes at the authors' college located in the northeastern part of Japan (37°45′N,140°28′E, 67.4 m above sea level). Measurements were made from April 1996 to April 1997.

2 Materials and methods

Indoor radon gas was measured with a continuous electrostatically collecting scintillation cell monitor in active mode (Pylon AB-5 PMT-TEL, cell volume 18.5 L). Air was sampled (0.5 L/min) through glass-fiber filter and desiccating agent. Data were acquired every one hour; to exchange the desiccating agent, one value was missed about once a week. Lower detection limit for radon concentration was 0.93 Bq/m^3.

A filtration method with alpha spectroscopy was employed for the measurement of progeny [1]. Pore size of membrane filters used was 0.8 μm. After 10 minutes sampling (20 L/min) of the progeny in air, the activity on the filter was detected for 50 minutes for radon progeny and additional 60 minutes for thoron progeny. By using an evacuated chamber, ^{212}Po peak (8.78 MeV) was clearly separated.

The alpha spectrum for progeny measurements was divided into three regions: ^{218}Po region (4.6 to 6.2 MeV), ^{214}Po region (6.2 to 7.9 MeV), and ^{212}Po region (7.9 to 9.3 MeV). Interference

Table 1: Basic statistical quantities for ^{222}Rn, ^{222}Rn progeny, ^{220}Rn progeny, and F-value (equilibrium factor). c_p denotes potential alpha energy concentration.

Variable	Mean	Std. Dev.	Minimum	Maximum	#
Indoor ^{222}Rn (1 day) [Bq/m^3]	7.79	1.93	3.91	13.75	326
Indoor ^{222}Rn (13:00) [Bq/m^3]	4.66	1.47	1.76	10.88	384
IndoorRaA [Bq/m^3]	2.90	1.81	0	9.75	201
IndoorRaB [Bq/m^3]	.80	.71	0	3.33	201
IndoorRaC [Bq/m^3]	.71	.57	0	3.31	201
Indoor $c_p(^{222}$Rn) [nJ/m^3]	5.54	2.56	1.59	19.03	201
Indoor $c_p(^{220}$Rn) [nJ/m^3]	1.90	1.47	0.00	8.08	201
Indoor F-value	0.22	0.07	0.07	0.42	200
Outdoor RaA [Bq/m^3]	3.44	2.14	0	11.21	198
Outdoor RaB [Bq/m^3]	2.03	1.29	0	7.85	198
Outdoor RaC [Bq/m^3]	2.12	1.30	0	6.54	198
Outdoor $c_p(^{222}$Rn) [nJ/m^3]	12.45	5.52	2.49	33.87	198
Outdoor $c_p(^{220}$Rn) [nJ/m^3]	1.60	1.71	0.00	10.77	197

among these three regions were negligible, especially because of few counts of ^{212}Po region. Potential alpha energy concentration (PAEC, c_p) of radon progeny, PAEC(Rn), was calculated from ^{218}Po, ^{214}Pb and ^{214}Bi concentrations. Thoron progeny concentration, PAEC(Tn), was estimated from the counts of ^{212}Po region. Lower detection limits were 0.4 nJ/m^3 for ^{222}Rn and 2 nJ/m^3 for ^{220}Rn progeny, respectively.

Progeny measurements were made for indoor (from 12:55 to 13:05) and outdoor (from 14:55 to 15:05) air every day except holidays; measurements at the identical times are essentially indispensable for time series analysis. Outdoor air was sampled on the fourth floor of an emergency staircase. Indoor air was sampled around the authors' office, which is on the fourth floor of a five-storied building. The college building was built in 1988. For indoor air of 13:00, F-value (equilibrium factor) was calculated.

3 Results and discussion

3.1 Basic statistics and seasonal variation

Figure 1 shows the temporal variation of daily averaged radon, radon at 13:00, outdoor and indoor radon/thoron progeny, and indoor F-value at 13:00. Table 1 shows basic statistics of various data. Indoor radon progeny was lower than outdoor progeny, while indoor thoron progeny was slightly higher than outdoor progeny.

To see the seasonal variation, one-way layout analysis was made for four groups of data, i.e., spring (March to May) summer (June to August) autumn (September to November) winter (December to February). Indoor radon, indoor radon progeny, outdoor radon and thoron progeny indicated significantly higher level in autumn, while indoor thoron progeny and F-value indicated no seasonal variation.

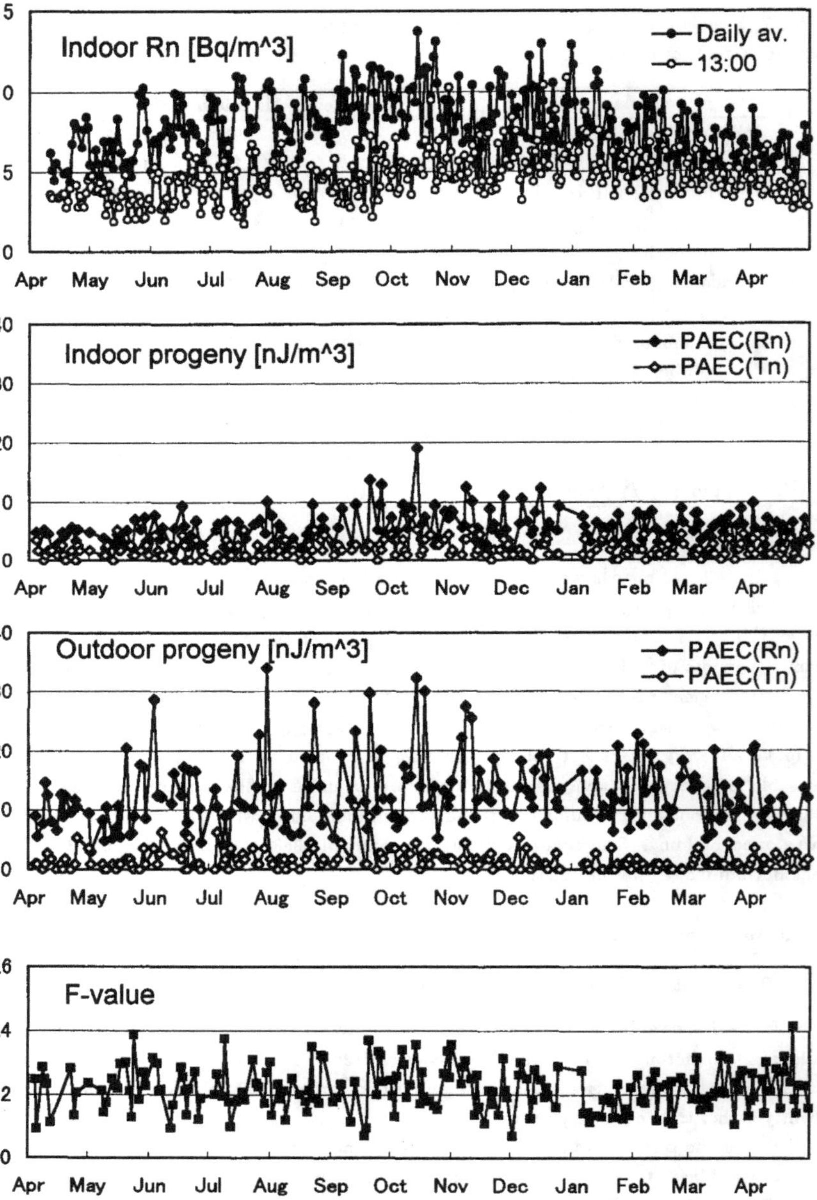

Fig. 1: Time series of radon, indoor radon- and thoron-progeny, outdoor radon- and thoron-progeny, and F-value (equilibrium factor).

3.2 Statistical distribution

From histogram of radon and progeny data, shape of the statistical distribution was examined. Higher level radon around dawn revealed log-normal distribution, while lower level radon in the afternoon revealed normal distribution. Radon progeny reveled log-normal distribution, while thoron progeny revealed Poisson-like distribution. F-value revealed normal distribution.

3.3 Correlation analysis

Correlation analysis was made for data shown in Table 1. Highly correlated combinations with significant level of 1% are enumerated below (+ and - denote positive and negative correlation, respectively). Indoor radon at 13:00 vs. atmospheric pressure (+), and wind speed (-). Indoor PAEC(Rn) vs. outdoor humidity (+). Indoor PAEC(Tn) vs. wind velocity (-). Outdoor PAEC(Rn) vs. outdoor humidity (+). Outdoor PAEC(Tn) vs. outdoor temperature (+), indoor temperature (+), outdoor humidity (+), indoor humidity (+), and wind speed (-). Outdoor thoron progeny was sensitive to meteorological data. All indoor radon data (each one hour data and daily averaged data) were highly correlated with indoor and outdoor radon- and thoron-progeny data.

Correlation between indoor radon and progeny data with outdoor data may be attributed to the ventilation system that introduces outer air from 7:30 to 23:30.

3.4 Time series (spectral) analysis

Time series analysis was made for various variables [1]. Power spectra were estimated with autocorrelation function and autoregressive model [2] that is equivalent to maximum entropy method [3]. Figures 2 and 3 show the results, where the periods are also written for main seven peaks.

Indoor radon, radon progeny, and thoron progeny had similar peaks in their power spectra. Besides remarkable one-year period, three to five day peaks were often seen. Wind speed and atmospheric pressure also indicated several-day periods, which are probably attributed to the passage of air masses. Twenty-day to thirty-day peak may be attributed to meteorological phenomena corresponding to the rotation period of the sun.

Indoor F-value indicated no periodicity and temperature indicated no significant periodicity except overwhelming one-year period.

4 References

[1] Kobayashi, T. and Takaku, Y., Intermittent measurements of ^{222}Rn and ^{220}Rn progeny in air for four years, RADIOISOTOPES, **46**, 603-614, 1997.

[2] Akaike, H., Fitting autoregressive models for prediction, Annals of Institute Statistical Mathematics, **21**, 243-247, 1969.

[3] Burg, J. P., Maximum entropy spectral analysis, paper presented at the 37th Annual International Meeting, Soc. of Explor. Geophys., Oklahoma City, Okla., Oct. 31; 1967.

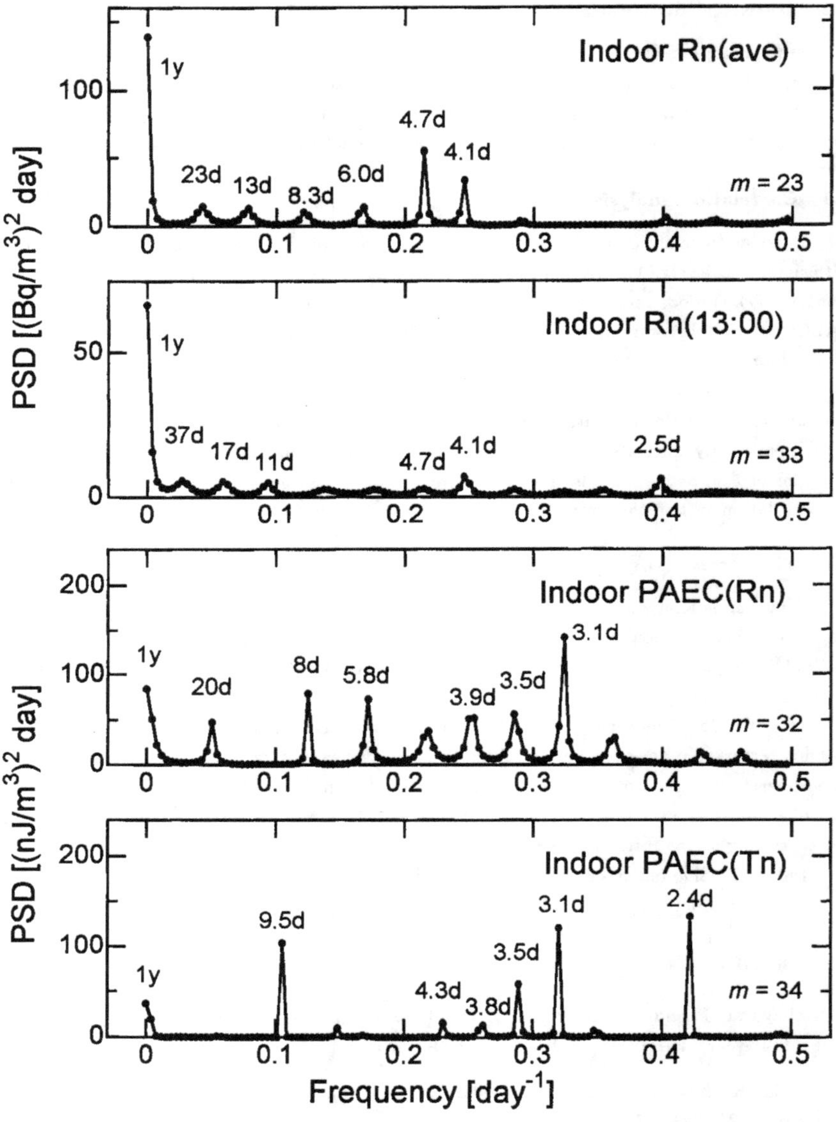

Fig. 2: Power spectra for radon concentration, radon progeny, and thoron progeny. PSD means power spectral density. Rn(ave) means the daily average radon concentration and Rn(13:00) means radon concenctration obtained from integration of 12:30 to 13:30. PAEC means potential alpha energy concentration. *m* is the degree of autoregressive model.

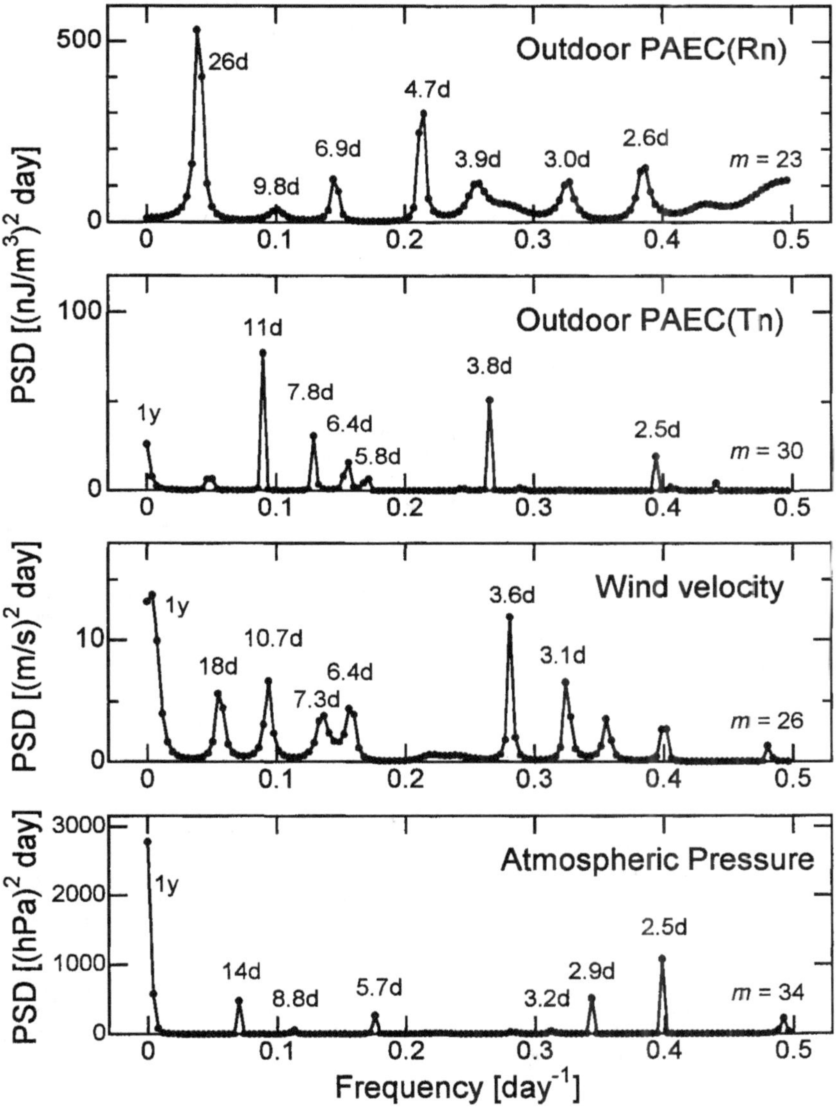

Fig. 3: Power spectra for outdoor radon progeny, thoron progeny, wind velocity and atmospheric pressure. PSD means power spectral density. PAEC means potential alpha energy concentration. *m* is the order of autoregressive model.

Comparison between indoor and outdoor measurements of Rn-220 decay product concentrations and meteorological parameters

M. Trautmannsheimer, K. Bunzl, J. Peter [†]

Institut für Stahlenschutz, GSF-Forschungszentrum für Umwelt und Gesundheit, Postfach 1129, D-85758 Neuherberg, Germany

[†] Institut für Strahlenhygiene, Bundesamt für Strahlenschutz, Postfach 1108, D-85758 Neuherberg, Germany

Abstract

Data from continuous measurements of the potential alpha energy of thoron progeny (Rn-220) in several dwellings within the same village, located in the south of Germany, show that there is a typical seasonal dependency of the Rn-220 concentration.

Additional measurements of Rn-220 outdoor concentrations in the vicinity of these houses allows testing for a correlation between indoor and outdoor concentrations. Decay product concentrations of Rn-222 at all locations are also evaluated for the comparison of exposure by thoron and radon progenies.

For collecting these data a computer controlled filter device with a silicon α-detector and a flow meter controlling the air flux was used. The peak areas of Po-214, Po-218 and Po-212 can be calculated by special procedures from the α-spectra to get the equilibrium equivalent concentration for thoron and radon decay products.

A meteorological station at a distance of not more than 500 m of each outdoor and indoor measuring location was installed to collect data of temperature, air pressure, humidity, wind velocity and wind direction. The comparison of these parameters with the thoron and radon concentrations shows the influence of the weather conditions on the radon and thoron exposure in the dwellings and outdoors.

1. Introduction

For many years various projects had been performed by the GSF to measure radon decay product concentrations at several villages indoors and outdoors [1],[2]. The main purpose of these projects was to get a detailed overview about the different situations at specific locations concerning the exposition of the population by radon and its progenies.

The last study at a small village about 50 km east of Munich, which was performed during the years 1995 to 1997, included continuous radon decay product measurements at two different locations outdoors and in several houses. A simple weather station was also installed to register temperature, huminity, air pressure, wind velocity and wind direction. We obtained a very good data set of a time range of more than one year.

Because our radon decay product monitors do full alpha spectroscopy we can also evaluate the thoron progeny concentrations from the original data using the Po-212 peak from the spectra. This allows a detailed comparison of radon and thoron decay products and an investigation of the influence of the weather conditions.

To find out whether there are specific local contributions to the thoron concentration and whether these concentrations are also typical for other parts of Germany we used data from the German meteorological service (Deutscher Wetterdienst, DWD) and from another decay product station 10 km north of Munich.

2. Experimental arrangement

For the evaluation of the radon decay product concentration a computer controlled instrument with a filter and an α-spectrometer was used. The instrument consists of a sampler for the continuous measurement of radon and thoron decay products and a central recording unit. 80 to 100 liters of air per hour are sucked through a polytetrafluoroethylene filter of 1.2 μm pore size. The alpha radiation of the decay products collected on the filter surface is recorded by a passivated implanted planar silicon detector (CAM-PIPS, Canberra Semiconductor, Olen, Belgium) with a diameter of 20 mm. The counting efficiency is in the order of 17%. The pulses of the detector are counted in 256 energy channels by a multichannel analyser (Tracerlab Instruments, Frechen, Germany). Simultaneously the air flow rate is recorded. The system can separate the peaks of Po-218 (6.0 MeV), Po-214 (7.7 MeV) and Po-212 (8.8 MeV) and determine the equilibrium equivalent concentrations every two hours [3],[4]. The pump operates 90 minutes and then it is switched off for half an hour. The data inclusive the spectra were stored on a floppy disk which has to be changed every two weeks.

Special characteristics of the devices are their small size and their silent operation.

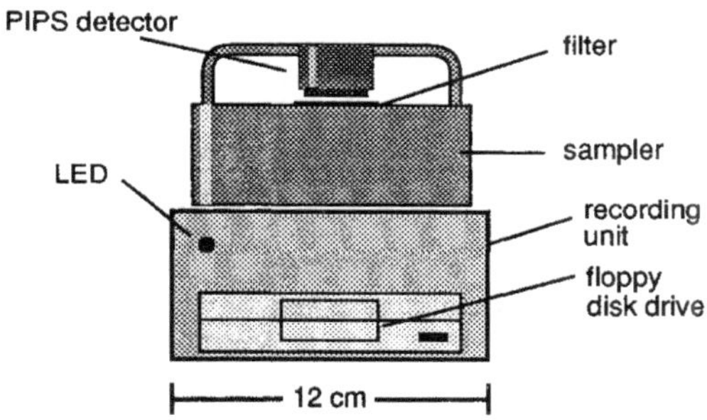

Fig. 1 The instrument developed at the GSF can measure the concentrations of radon decay products by alpha spectrometry in dwellings continuously over longer time periods.

3. Measurement locations

In South Germany about 50 km east of Munich two instruments were placed outdoors and five were used for indoor measurements in inhabited dwellings. A station to collect weather parameters like air temperature, air pressure, huminity, wind velocity and direction was installed in the vicinity of one outdoor station. All instruments were located in the same village. Data were collected from the end of 1995 till the spring of 1997, containing daily averages of the radon and thoron decay product concentrations (EEC Rn-222 and EEC Rn-220).

An additional instrument is operating outdoors at the GSF 10 km north of the center of Munich. For comparison, daily averages of Pb-212 concentrations at 35 locations throughout Germany collected by the German meteorological service (Deutscher Wetterdienst, DWD) in the framework of the IMIS System were used.

4. Results

The similarity of time dependencies measured at the GSF and a station of the DWD located about 5 km away (Fig. 4) shows that the evaluation methods at the GSF (EEC Rn-220 by α-spectrometry) and at the DWD (γ-spectrometry of Pb-212) yield the same results.

Fig. 2 shows the seasonal dependency of the EEC (Rn-220) and the EEC (Rn-222) outdoors in the year 1996. The EEC (Rn-220) and the EEC (Rn-222) measured 1.5 m above ground show pronounced seasonal variation with annual averages of 0.04 Bq/m^3 and 6 Bq/m^3 respectively. The maximum of the EEC (Rn-220) is reached in summer whereas the EEC (Rn-222) has its highest values in winter. In the case of Rn-222 this is the typical seasonal behavior also found elsewhere [5]. At the second instrument, located at a distance of about 500 m, also outdoors, the same concentrations as at the first station were measured.

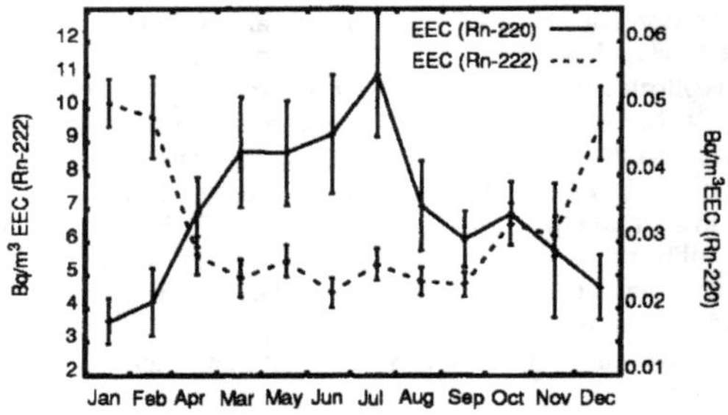

Fig. 2 Seasonal variation of the EEC Rn-220 (right axis) and the EEC Rn-222 (left axis) at outdoor location No. 2. Plotted values are monthly averages of daily measurements. With the EEC the standard deviation of the mean are given as error bars.

The instrument at the GSF and several stations of the DWD in south Germany also show their maximum thoron values in summer but with different average concentrations.

The results of the indoor measurements are shown in Fig. 3. The radon decay product concentrations show the well known seasonal variation with a typical maximum during the winter months. The thoron decay product concentrations have the same time dependency (e.g.[6]) and a good correlation with radon decay product concentrations [7]. In the dwelling presented here, the annual average of the EEC (Rn-220) is 1.2 Bq/m^3 compared to an EEC (Rn-222) of 241 Bq/m^3. The thoron decay product concentrations in these dwellings are in agreement to values measured in dwellings in east Bavaria with the same method [8].

Concentrations in the other 4 dwellings show the same seasonal pattern with annual averages from 24 to 208 Bq/m^3 EEC (Rn-222) and 0.5 to 1.6 Bq/m^3 EEC (Rn-220).

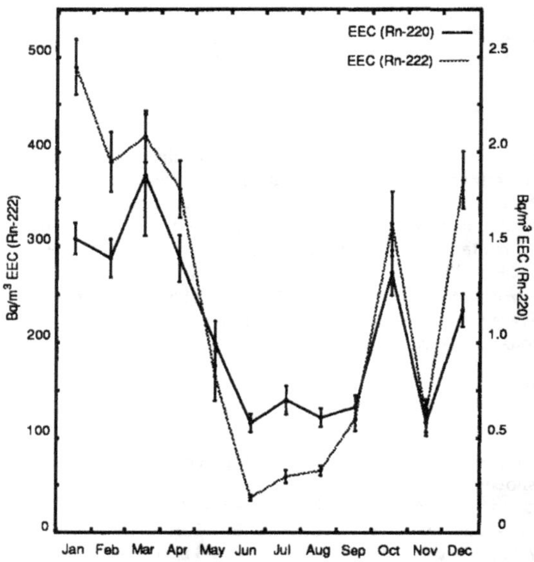

Fig. 3 Seasonal variation of the EEC Rn-220 (right axis) and the EEC Rn-222 (left axis) at one indoor location (dwelling No. 3). Plotted values are monthly averages of daily measurements. With the EEC the standard deviations of the mean are given as error bars

5. Discussion and Conclusions

In dwellings the concentration of thoron decay products follows the concentration of radon progeny. This is due to the change in ventilation in the houses between winter and summer which influences both concentrations in the same way.

Outdoors the reduction of concentrations by exchange of air between layers near the ground and higher layers affects both radon and thoron decay products in the same manner [9]. The similarity of time dependency of thoron decay product concentrations measured at the GSF and at the DWD in Munich and Regensburg (Fig. 4) indicates that local parameters influence concentrations not as strong as the general weather conditions.The difference in seasonal dependency must be caused by meteorological conditions affecting the exhalation rate from the soil. This can be explained by the fact that the half life of thoron gas is very short compared to the diffusion time of both radon and thoron gases through wet or frozen soil [10]. As a consequence thoron decay product concentrations are low during periods of wet or frozen ground.

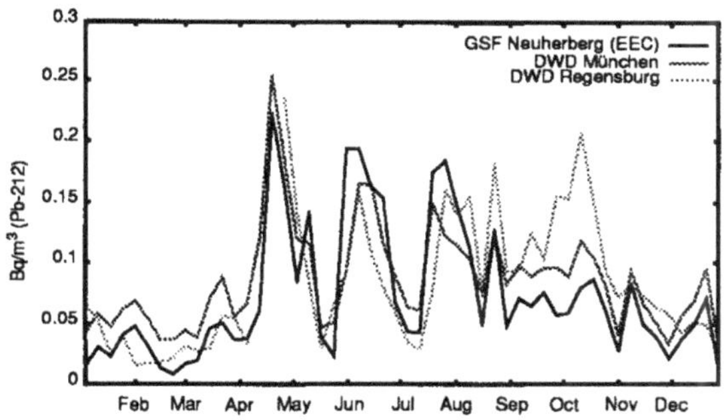

Fig. 4 Weekly averages of the concentration of Pb-212 at various locations in South Germany

6. References

[1] Trautmannsheimer M., Bunzl K.,
Untersuchungen zur radiologischen Bewertung von Schlackeablagerungen,
Bericht des Bayerischen Staatsministerium für Landesentwicklung und Umweltfragen
vom 30.4.97

[2] Trautmannsheimer M.,
Untersuchungen für eine radiologische Bewertung bestehender Bergbauhalden in Bayern,
Zwischenbericht des Bayerischen Staatsministerium für Landesentwicklung und Umweltfragen
vom 31.7.96

[3] Peter J. E.,
Analysis of radon and thoron daughter concentrations in air by continuous alpha spectroscopy.
Radiat. Prot. Dosim. 56:267-270; 1994.

[4] Peter, J. E.,
Erarbeitung von Bewertungskriterien für Messungen des Radons und seiner Zerfallsprodukte in
Häusern. GSF-Bericht 35/93. ISSN 0721-1694; 1993.

[5] Hötzl H., Winkler R.,
Long-term variation of outdoor radon equilibrium equivalent concentration,
Radiat. Environ. Biophys. 33:381-392;1994

[6] Keller G., Folkerts K.H.,
Radon-222 concentrations and decay product equilibrium in dwellings ant in the open air
Health Physics 47:385-398; 1984

[7] Schery S.D.,
Measurements of airborne ^{212}Pb and ^{220}Rn at varied indoor locations within the United States,
Health Physics 49:1061-1067; 1995

[8] Peter, J. E.,
Measurements of Rn-220 decay product concentrations in German dwellings,
IRPA9, 1996 International Congress on Radiation Protection, Vienna April 14-19, 1996

[9] Martz D.E., Falco R.J., Langner G.H.jr.
Time-averaged exposures to ^{220}Rn and ^{222}Rn progeny in Colorado homes,
Health physics, 58:705-713; 1990

[10] Singh M., Ramola R.C., Singh S., Virk H.S.,
Influence of moisture content on radon diffusion in soil,
Nuclear geophysics 4:479-482; 1990

Effect of humidity on radon exhalation rate from concrete

Hirokuni Yamanishi[1], Haruo Obayashi[1], Naruhito Tsuji[2] and Hisao Nakayoshi[2]

[1] National Institute for Fusion Science
Toki 509-52, Japan
[2] Japan Airconditioning Service Co. LTD.
Nagoya 468, Japan

Abstract

The objective of the present study is evaluation of seasonal humidity effect on radon exhalation rate from concrete. Three concrete pieces have been placed in three different fixed humidity circumstances for about a year. The three fixed humidities are selected 3, 10, 25 g m^{-3} in absolute humidity, those correspond to dry condition as control, winter and summer, respectively. Radon exhalation rate from each concrete piece is measured every one month during humidity exposure. Under the lower humidity, radon exhalation rate from concrete is small. On the contrary, radon exhalation rate is large in the higher humidity circumstance. This trend is consistent with the seasonal variation of indoor air radon concentration in low air-exchange-rate room.

1. Introduction

The experimental hall for nuclear fusion experiments is required to keep airtight in order not to leak airborne radioactivities which are induced by neutrons [1]. It is necessary to monitor air exchange rate and air quality in that room. Radon concentration in a room is mainly governed by air exchange and radon exhalation from walls. We are planning to use this relationship for monitoring air exchange rate. So, we have investigated measuring method of air exchange rate using radon (^{222}Rn) as a tracer gas, because the concentration of radon will be high in a room with low air exchange rate. Under the circumstances, it is necessary to know well about variation pattern of radon exhalation rate from concrete wall and its factors. In case of seasonal variation, radon concentration level is high in summer and fall, and low in winter and spring, when we measured continuously in the room with low air exchange rate, which is located in the building of Nagoya site, NIFS [2]. It is supposed that radon exhalation rate from wall varies with seasonal atmospheric humidity.

On the other hand, there are only a few investigations which are studied about variation of exhalation in long period exposure to controlled humidity environment [3.4], especially lack of data about high humidity environment.

The objective of the present study is evaluation of seasonal humidity effect on radon exhalation rate from concrete.

2. Experiment

Figure 1 shows the system of controlled humidity environment. Three sealed boxes (30 cm × 30 cm × 30 cm) are placed in an incubator where temperature is controlled at 30 ℃. In each box, concrete piece whose size is 15 cm diameter and 5 cm thickness has been set. They have been exposed in three different fixed humidity circumstances for about one and half years. The three fixed humidity are selected at 3, 10 and 25 g m^{-3} in absolute humidity. Those humidity conditions correspond to dry as control, winter and summer, respectively. The humidity in each sealed box is controlled by flow rate of wet air supply system and dry one. Its wet air is made by bubbling water, and dry air is made by through desiccant.

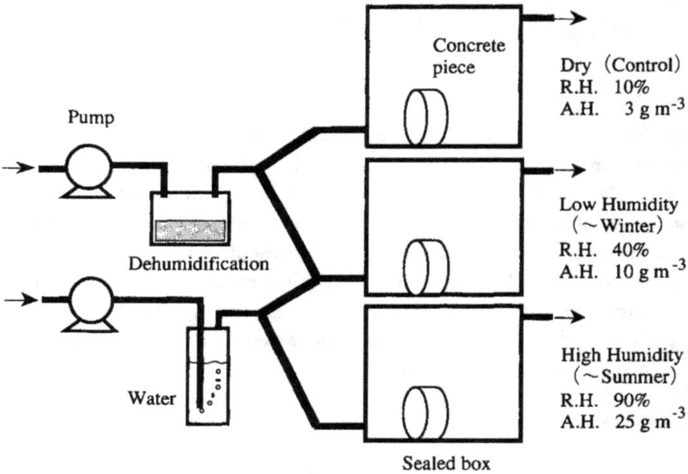

Figure 1 System of controlled humidity environment.

Before humidity exposure, first, three concrete pieces were dehydrated by heater at 105 ℃, because of dry weight measurement and initialized water content. Secondly, the sides of concrete pieces were covered with plastic film, and gaps between concrete and films were filled up with epoxy resin adhesive, because of limitation of the area for humidity permeation and radon exhalation.

Water content and radon exhalation rate from each concrete piece were measured at intervals of about 30 days. Water content of concrete piece is evaluated by the following equation.

$$W = (w_m - w_d) / w_d \qquad (1)$$

where W is water content(%), w_m is measured weight(g) and w_d is dry weight(g). Radon exhalation rate from concrete piece is measured by the following process. First, a concrete piece is confined to the vessel, which has 6.3 liter volume, 20 cm diameter and 20 cm height. Second, air in the vessel is substituted with radon free gas. Then, it is kept intact for 8 hours in order to accumulate exhalated radon. After accumulation, radon gas in the vessel is sampled to 1.5 liter ionization chamber and measured.

3. Results

Three concrete pieces are used in the present experiment, which are named for No.2, No.3 and No.4. No.2 sample has been exposed to dry circumstances. No.3 and No.4 have been set low and high humidity, respectively. After 500 days exposure, No.4 sample was moved to low humidity place.

Figure 2 shows the results. Measured water content (W) and radon exhalation rate (E) are plotted with elapsed time. W and E increase with elapsed time. In the high humidity, they increase up to 2.4 % and 0.43 mBq m^{-2} s^{-1}, respectively. In the low humidity environment, they increase up to 0.88 % and 0.13 mBq m^{-2} s^{-1}, and are saturated at early time. It takes 80 days for low humidity, 180 days for high humidity to saturate these value. After saturation of W, E follows the variation of water content. W and E of No.4 sample are decreased to 1.65 % and 0.25 mBq m^{-2} s^{-1} after placed low humidity.

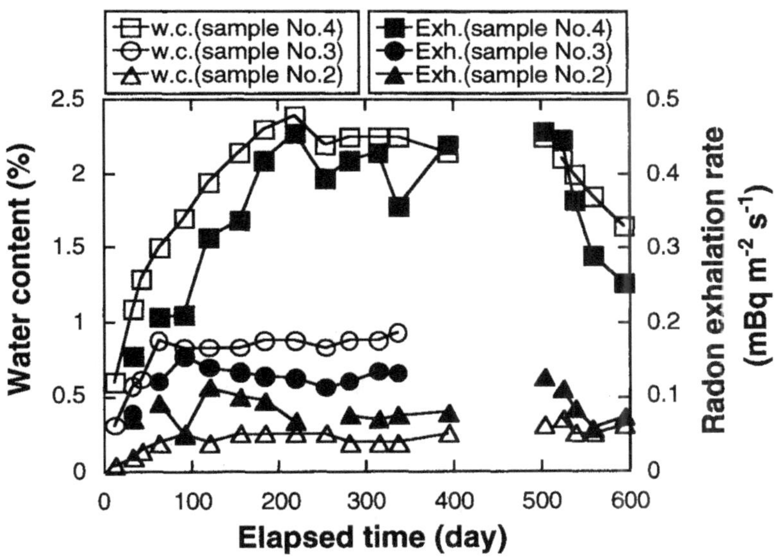

Figure 2 Time variation of water content and Rn exhalation rate.

Figure 3 shows the correlation between W and E. Radon exhalation rate increases almost linear, under 2.4% water content. The trend is similar to the result of Stranden et al [5]. The following equation is a result to fit these measured value with polynomial expression.

$$E = 0.096 - 0.051\,W + 0.085\,W^2 \qquad (2)$$

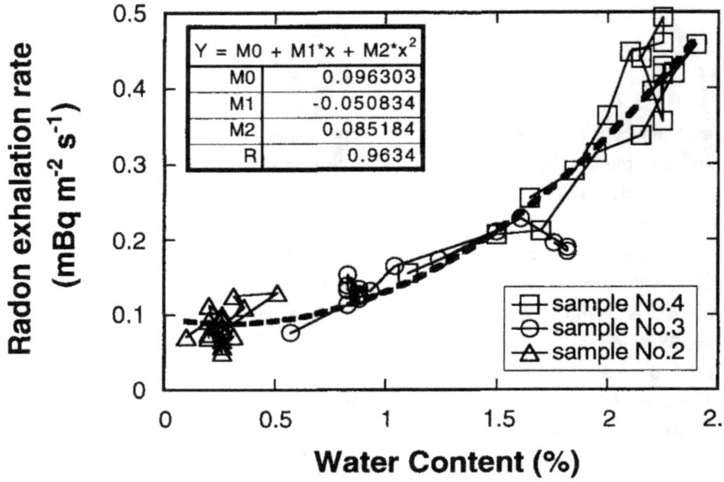

Figure 3 Relationship between water content and Rn exhalation rate.

4. Discussions

4-1 Comparison of measured value

The wall thickness of common building is about 20 cm. Since the thickness of concrete piece in this investigation is 5 cm, it is necessary to revise obtained E in order to compare the data from other studies. We applied the equation in the UNSCEAR 1993 report [6] for evaluating E. Some coefficients are quoted from UNSCEAR 1993 report. The revised values range from 0.3 to 1.5 mBq $m^{-2} s^{-1}$. In the UNSCEAR 1993 report, the average of E of building materials is 1.5 mBq $m^{-2} s^{-1}$. The value which we directly measured on wall varied from 1.4 to 2.5 mBq $m^{-2} s^{-1}$ [7]. The obtained data in the present study is smaller than others. We should obtain the own data, which are porosity and diffusion coefficient etc., in order to evaluate more realistic E.

4-2 Humidity effect to exhalation rate

The measured data which are shown in Fig. 2 were fitted with a function along exposure time as follows.

$$f(t) = (C - C^*) \exp(-a t) + C^* \qquad (3)$$

where a is an increase or decrease coefficient, C* is saturation level in the present exposed environment, C is saturation level before the present exposure. In the case of high humidity exposure, increase coefficient is estimated as 0.017 and 0.011 for water content and exhalation rate, and decrease coefficient is estimated as 0.009 and 0.010 for water content and exhalation rate. In the case of low humidity exposure, increase coefficient is estimated as 0.033 and 0.036 for water content and exhalation rate. The coefficient for E is almost the same as that for W, except on the exposure from dry to high humidity.

We try to simulate seasonal variation of E from absolute humidity in a room. First, W is derived from absolute humidity in a room by using the relationship between absolute humidity of humidity exposure system and saturation value of W. There is time delay of changing W on effect from humidity. Time delayed W is estimated by using equation (3), where C is the present W, C* is the derived W from absolute humidity of next day. In this simulation, absolute humidity which are measured in the room with low air exchange rate are used. Finally, E is estimated from time delayed W using the relationship shown in Fig. 3 or equation (2). Figure 4 shows the result. The result indicates that if air exchange rate of a room is constant, indoor radon concentration varies 2 or 3 times between seasons. Acutually, in this room, observed radon concentration varies two times under low air exchange rate. Further investigations are required to predict the variation of E.

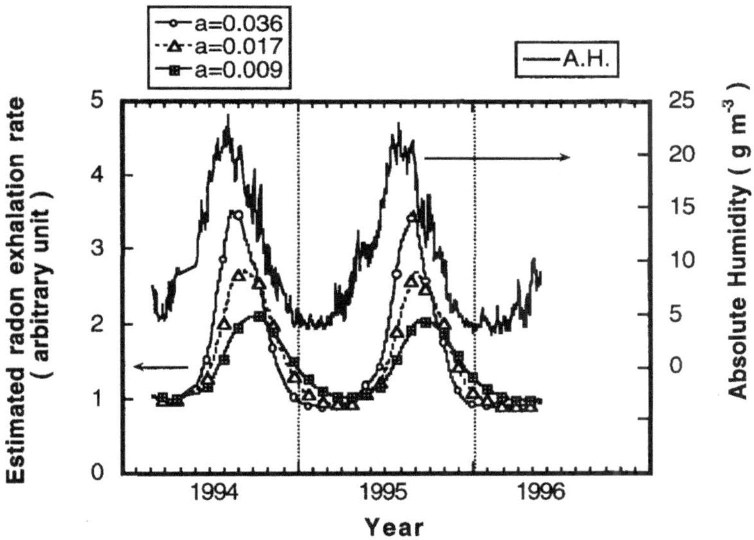

Figure 4 Simulation of the seasonal variation of Rn exhalation rate.

5.Coclusions

Three concrete pieces have been continuously exposed in each three different humidity environments for one and half years. Water content and radon exhalation rate increase with elapsed time. In the high humidity, they increase up to 2.4 % and 0.43 mBq m^{-2} s^{-1}, respectively. After saturation of water content, radon exhalation rate follows the variation of water content. The increase or decrease speed of exhalation rate is almost the same as that of water content, except on the exposure from dry to high humidity.

Radon exhalation rate is increased almost with linearity, before 2.4% water content. Under the lower humidity, radon exhalation rate from concrete is small. On the contrary, that is large in the higher humidity circumstance. This trend is consistent with the seasonal variation of indoor air radon concentration in low air exchange rate room.

Acknowledgment ; The authors acknowledge the financial support of the Japanese Ministry of Education, Science and Culture, Grant-in-Aid for Encouragement of Young Scientists program.

6. Refernces

[1] Handa, H., et al., Radiation shielding analysis of a large helical device, Fusion Eng. and Design, 28, 515-524, 1995.

[2] Yamanishi, H., et al., Variation of radon and radon progeny in a room with low ventilation rate, Proc. 6th Intern. Congress, Indoor Air '93, Vol.4, p.549-554, 1993.

[3] Dijk, W.V., Jong, P.D., Determining the ^{222}Rn exhalation rate of building materials using liquid scintillation counting, Health Phys., 61, 501-509, 1991.

[4] Roelofs, L.M.M., et al., The effect of aging, humidity, and fly-ash additive on the radon exhalation from concrete, Health Phys., 67, 266-271, 1994.

[5] Stranden, E., et al., The influence of moisture and temperature on radon exhalation, Radiat. Prot. Dosim., 7, 55-58, 1984.

[6] Report of UN Sci. Comm. on the Effects of Atomic Radiation, "Ionizing Radiation ; Sources and Biological Effects", (1993), UN, New York.

[7] Yamanishi, H., et al., Measurement of air exchange rate and radon exhalation rate from building wall, Proc. 7th Intern. Congress, Indoor Air '96, Vol.1, p.135-140, 1996.

Indoor radon levels in the environs of Guru Nanak Dev University campus

B.S. Bajwa and H. S. Virk

Deptt. of Physics, Guru Nanak Dev University
Amritsar- 143005, India.

Abstract

This paper presents the results of radon measurements carried out in and around Guru Nanak Dev University (GNDU) campus using SSNTD technique. The method used involved exposing small stripes of LR-115 Type II films in the bare mode. The estimated levels of radon and its progeny vary from 38.8 to 174.6 Bq/ m^3 and from 5.2 to 23.6 mWL respectively. For sake of comparison indoor radon measurements were also carried out using Alpha Guard and it has been observed that integrated measurements using plastic detectors gives higher values of radon concentration in the dwellngs.

1. Introduction

The inhalation of radon and its short lived daughters in dwellings is one of the most important radiation risks to human beings [1], because it constitutes more than 50 % of the total radiation dose. In India, the available data indicate that, it accounts for 52% of the total dose [2]. The exposure to high levels of Rn^{222} and its progeny in mines leads to lung cancer [3 - 4]. It is widely accepted that the primary effect is not due to Rn^{222} as such, since inhaled Rn^{222} is mostly expelled, but from the decay of alpha active daughters of radon, viz. Po^{218}, Pb^{214}, Bi^{214} and Po^{214}. These are attached to aerosols which can be lodged in the lungs when inhaled.

Indoor radon concentrations levels have been measured in various countries since seventies and confirmed the relationship between the exposure of inhabitants to high radon and an increase in lung cancer risk [5-6]. As Radon is a noble gas, it diffuses through the porous building materials and also through the sub-soil under the building and is subsequently mixed with the air inside the room. Subsoil emanation is mainly responsible for the fact that the basement of a building contains a higher radon concentration than the rooms on other floors [7-8]. Poor ventilation enhances the build-up of radon concentration in the rooms.

Passive time integrated technique using the solid-state nuclear track detector (SSNTD) is being used for radon concentration measurements inside the dwellings ,as it is the most reliable technique for integrated and long-term measurement of indoor radon [9 - 11]. As a part of the ongoing national project of mapping of the country for indoor radon, measurements have been carried out in and around GNDU campus, and the preliminary results obtained have been briefly presented in this paper.

2. Experiment

In the present work, SSNTD technique was used for the radon measurements [9 - 11] . LR- 115 (Type-II) plastic detectors, of 2x2 cm^2 in size, were fixed on glass slides which were then fixed at the same height on the walls inside the houses of the university campus. After an exposure of one month, the samples were removed and etched in 2.5 N NaOH at 60^0 c for duration of 2 hr. These detectors were then scanned under an optical microscope at a magnification of 600X for measurement of track density. From the observed track densities, actual radon gas levels have been estimated using the conversion factors generated earlier in this laboratory [12] .

3. Conclusions

Results of the radon levels measured in the different type of houses and even in various rooms under different ventilation conditions are given in Table - 1. Mean radon levels in dwellings of university campus varies from 63.1 Bq/m^3 in the drawing rooms to 116.7 Bq/m^3 in the stores . The observed mean radon concentration in the drawing rooms has been found to be much less than as compared to that in the stores. The main cause of this difference is due to a consequence of the low air exchange rate in the poorly ventilated rooms. The high ventilation rates reduce the concentration towards the ambient level, this being possible because the emanation does not supply radon fast enough to replace the amount removed [13] . So in normal dwellings, the indoor radon content are high in partially or poorly ventilated rooms. Similar results have been observed from the radon content estimations carried out in the village of Jodhpur (Table- 2), Distt. Sangrur, Punjab, where the stores are showing quite high radon concentrations as compared to the other rooms.

As is well known that kitchen in general, should have higher radon concentrations than in the stores due to the usage of LPG and other kitchen appliances [14], but in our observations, the measured radon levels in the stores were more or less equal to that in the kitchens . This may be due to the prevailing higher ventilation conditions in the kitchens as compared to that in stores .

Radon measurements have also been carried out using Alpha-Guard in the A-type houses and at some other places in the university campus (Table - 3) . It is obvious from the Table-3 that the average values of radon concentration determined by using Alpha-Guard for 24 hrs. are much below the values determined through the plastic detectors over a period of one month, which accounts for diurnal and hourly variations .

International and national recommendations for the radon limitation vary. US federal authorities recommended 150 Bq / m^3 for existing homes [15]. ICRP recommendations are 400 Bq / m^3 for existing homes and 200 Bq / m^3 for future homes [16-17] . In UK , a value of 200 Bq / m^3 has been recommended for existing homes [18]. The present preliminary survey on radon levels in the dwellings of university campus indicates that the radon levels are well within the permissible level recommended by all the above agencies. However , more data on radon levels in different seasons of the year are required to draw other definite conclusions on the indoor radon levels.

4. References

[I] UNSCEAR , United Nations Scientific committee on the effects of Atomic Radiation (1988) , Sources , Effects and Risks of Ionising radiation, Report to the General Assembly, United Nations, New York.

[2] Narayanan , K. K., Krishnan , D., and M.C. Subba Ram , Population exposure from natural and man made sources of ionising radiation , Indian Society for Radiation Physics, Kalpakkam Chapter (1991).

[3] Wagoner, J.K., Archer, V.E., Lundin, F.E., Holaday, D.A., and J.W. Lloyed , Radiation as the cause of lung cancer among uranium miners, New Eng. J. Med. , 273 , 181-185, 1965.

[4] Sevc., J. , Kunz , E. , and V. Placek , Lung cancer in uranium mines and long-term exposure to radon daughter products , Health Phy., 30, 433-437, 1976.

[5] UNSCEAR (1982) Ionising Radiation : Sources and Biological Effects , United Nations.

[6] Health Risk of Radon and other Internally Deposited Alpha-emitters, BEIR VI (1988), Committee on the Biological Aspects of Ionising Radiation , National Research Council , National Academic Press, Washington, DC .

[7] Abu-Jarad , F. , and J.H. Fremlin , The activity of radon daughters in high rise buildings and the influence of soil emanation , Environ. Int. , 8 , 37-43, 1982.

[8] Fleischer, R.L. , Mogro -Campero, A. , and L.G. Turner , Radon levels in homes in the north eastern United States energy efficient homes, Report No. 80 , CRD 88 (Schenectady, N.Y., General Electric), 1980.

[9] Frank , A.L. , and E.V. Benton , Radon dosimetry using plastic nuclear track detectors, Nucl. Track Detect. , 1 , 149-179 , 1977.

[10]Abu -Jarad , F. , and J.H. Fremlin, A working level monitor for radon measurements inside houses, Radiat. Protect. Dosimet. , 1 , 221-226 , 1981.

[11]Alter, H.W. , and R.L. Fleischer , Passive integrating radon monitor for environmental monitoring , Health Phys. , 40 , 693-702 , 1981.

[12]Singh, M. , Singh, N. P. , Singh ,S. , and H.S. Virk , Calibration of radon detectors, Nucl. Tracks Radiat. Measur. , 12 , 739 - 742 , 1986.

[13]Harley, J.H. , Environmental radon , In Proc. Symp. on Noble Gas , ed. R. E. Staneley and A. A. Moshissi , Environmental Research Centre , Las Vagas , Nevada , 109 - 114 , 1973 .

[14]UNSCEAR , United Nations Scientific committee on the effects of Atomic Radiation (1993), Sources , Effects and Risks of Ionising radiation , Report to the General Assembly, United Nations , New York.

[15] US EPA (United States Environmental Protection Agency) , A citizen's guide to radon , Washington DC, US Environmental Protection Agency , 1986.

[16] ICRP Publication -39, Lung cancer risk from indoor exposure to radon daughters , Oxford , Pergamon Press , 1984.

[17] ICRP Publication -65, Protection against radon at home and at work , Annals of the ICRP, 23 (2), 1 - 48 , 1993.

[18] NRPB (National Radiological Protection Board) , Documents of the NRPB 1, 1, 1990.

Table 1 : Radon levels in the houses of GND University campus

Location	No. of Samples	Radon Levels					
		Max	Min (Bq/m^3)	Mean	Max	Min (mWL)	Mean
Stores	9	174.6	83.1	116.7	23.6	11.2	15.8
Kitchen	8	116.4	77.6	95.6	15.7	10.5	12.9
Drawing room	9	99.7	38.8	63.1	13.5	5.2	8.5

Table 2 : Radon levels in village Jodhpur, Distt. Sangrur, Punjab

House No.	Sample No.	Location	Radon Levels	
			(Bq/m^3)	mWL
1	JD 6	Bed room	233.9	31.6
	JD 7	Bath room	216.9	29.2
	JD 15	Shed	299.2	40.4
2	JD 12	Bed room	121.6	16.4
	JD 9	Drawing room	194.3	26.3
	JD 11	Store	337.7	45.6
	JD13	Kitchen	216.1	29.2
	JD8	Front room	237.9	32.2
	JD10	Veranda	227.5	30.8
	JD14	Ist floor	127.8	17.3
3	JD18	Bed room	121.6	15.4
	JD16	Drawing room	133.0	18.0
	JD17	Store	341.9	46.3

Table 3. Indoor radon measurement using Alpha Guard

Location	Duration (hours)	Radon conc.(Bq/m^3)
Punjab		
Research Lab.	12	35±5
A-4 Store	12	36±4
A-4 Bed Room	02	40±4
A-5 Store	04	36±4
A-5 Drawing Room	05	33±3
Boys Hostel	12	29±3
Faculty House	12	27±2
Physics Library	07	27±2
Anandpur Sahib	12	32±5
Himachal Pradesh		
Pakhloh	12	150±15

^{222}Rn Concentration in some characteristic dwellings in Cluj, Romania

Ildiko Mocsy[1]*, Ilona Hunyadi[2], Krisztina Simon[3], Corina Fulea[1], Attila Vasarhelyi[2]

1* Medical Centre for Health Services and Management Cluj-Napoca, Str.Pasteur 6, 3400 Cluj-Napoca, Romania

2 Institute of Nuclear Research of the Hungarian Academy of Sciences, Debrecen, Hungary

3 Sanitary Policy and Preventive Medicine, Cluj-Napoca, Romania

Abstract

We studied ^{222}Rn concentration in indoor air in typical types of dwellings in Cluj (3500.000 inhabitants): four and ten floor block-of-flats and detached-houses. These dwellings are classified by the building materials, by type of heating, by their high from the soil and by the dimension of the rooms. ^{222}Rn concentration was determined in all the rooms under normal living conditions in the period of 1994-1996. The radon activity was measured by scintillation cells and ionisation chambers. Daily variations of ^{222}Rn concentrations due to the ventilation and season conditions were eliminated through long term measurements using Solid State Nuclear Track Detectors (CR-39).

The mean values of ^{222}Rn from the air in the detached houses were between 94-540 Bq.m^{-3} and in the blocks-of-flats 49-168 Bq.m^{-3} . The maximum values were found in detached houses in rooms having no cellars 1850 Bq.m^{-3} . From the analysis of the results follows that the most important contribution to the accumulation of ^{222}Rn in the indoor spaces is related to the type of foundation of the houses.

1. Introduction

The main source of exposure to natural radiation comes from the inhalation of the ^{222}Rn and its short-lived daughters. There is a currently growing interest in the effect of exposure to ^{222}Rn , because it became recognised as an important cause of lung cancer [1, 2].

In the 80's, this noble gas was discovered to be accumulated mostly in the indoor air of the buildings. In the industrialised countries the people today spend a great deal of their time (~80 %) within buildings (homes, work-places, public buildings), thus the greater fraction of the internal radiation exposure depending on the indoor ^{222}Rn and its daughter product concentrations. Therefore, in the last years, determining ^{222}Rn activity concentration in the indoor air became more and more important, and the presence of this radionuclide was considered to be a real problem of public health.

Sources of indoor radon are: the soil beneath the building, building materials, tap water and natural gas. The accumulation of these radionuclides in dwellings depends on some factors such as: the geochemical composition and geophysical properties of the soil where the location of the construction is, the nature and type of the building foundation, the construction techniques, the level above the ground, the ^{226}Ra content of the building materials, the dimension of rooms, the rate of ventilation, the meteorological and season parameters, the ^{222}Rn and ^{226}Ra content of the tap water, the type of heating used and living and working conditions [3, 4].

The purpose of this paper is the determination of the ^{222}Rn and its daughters' concentrations in different types of dwellings in Cluj, big city in Romania.

2. Experiment

Two types of dwellings were study in the urban settings of Cluj. The detached houses (built between 1920-1940) and four and ten floor block-of-flats (built in 1960-85) were situated in an area with "normal" radiation background. These buildings are characteristic for this city. Cluj has 350000 inhabitants and according to the statistical data from 1990 the proportion of the population living in block flats is 54 %.

The walls of the detached houses are made red brick while stone , slag and in several cases of concrete were used for the basement. There are cellars under some rooms and others no cellar of the houses taken under study. The heating of the rooms is provided by burning natural methane gas in terra -cotta stoves. Natural gas is used for cooking. The rooms were the almost same volume 5m x 5m x 3.5m. The walls of the studied block-of flats buildings are made either of red brick or prefabricated concrete elements. The foundations are made of concrete. There are cellars or garage under buildings. All the flats have central heating, and natural methane gas is used for cooking. The living rooms volume are 4m x 4m x 2.5m.

The source of the tap water was the same for both types of dwellings.

The determinations of the ^{222}Rn content were performed under normal living conditions, in 1994-1996 by measuring the alpha activity of the indoor air with scintillation cell and ionization chamber. The air samples were collected in the middle of the premises in detached houses and in the bedrooms at each floor level, at 0.5 m from the floor. In order to assess and compare the values of the indoor ^{222}Rn concentrations, measurements were performed in the outside as well, at 0.5 m from the ground, during the same period of time. The mean values of ^{222}Rn activity concentration obtained by averaging the results from scintillation cell and ionisation chamber measurements, were compared to long-term measurements performed in rooms with Solid State Nuclear Track Detectors (SSNTDs). The Solid State Nuclear Track Detectors (CR.-39) were offered, and evaluated by the Institute of Nuclear Research of Hungarian Academy of Sciences, Debrecen. The exposure time of the SSNTDs was 4 x 90 days.

We made study of ventilation and season influence in radon accumulation in rooms. As well, we study the contribution of the tap water and methane gas at the indoor ^{222}Rn concentration.

3. Results and discussion

The annual mean values of ^{222}Rn activity concentration inside and outside the detached houses are shown in Table 1.

Table 1. The annual mean values of the in detached houses

HOUSE	BASEMENT	^{222}Rn ACTIVITY CONCENTRATION* (Bq.m^{-3})
room - under it is cellar	concrete, stone, slag	142 ±112
- no cellar	concrete	245 ± 154
	stone, slag	540 ± 285
kitchen		89 ± 43
bathroom		94 ± 35
outdoor		29 ± 12

* Note: represent average resulted of the 140 - 400 measurements

In the case of the detached houses the in rooms without cellar is twice or thrice as high as in rooms with cellar beneath. The building materials ^{222}Rn concentration being identical for the buildings, the high differences between ^{222}Rn activity concentrations could be explained by the dissimilarities in their respective basements. The ^{222}Rn can easily spread and reach the interior of the rooms in the case of houses without cellar due to the high porosity of stone and slag which the basement s made of. The lower values of the ^{222}Rn concentration in rooms with cellar beneath could be explained by the fact that underground radon is temporarily accumulated in the cellar, and then easily evacuated through the windows and doors of the cellar into the open air. The indoor ^{222}Rn activity
concentration in room under it is concrete foundation being 3 - 4 times lower than those measured in the rooms with stone and slag basements. It can be explained by the presence of a concrete foundation that plays the role of a screen for the ^{222}Rn coming out from the soil. The ^{222}Rn activity concentration was 5 - 18 times higher indoor than outdoors.

The smaller mean values of the ^{222}Rn activity concentration in kitchens, could be explained by the fact that kitchens - according to their function in the household activities - are more frequently ventilated.
The smaller value of ^{222}Rn activity concentration in the bathroom, is explained by a permanently opened ventilation aperture through the chimney. Measurements in the air of the bathroom after running the shower for 20 min. proved an increase in the ^{222}Rn concentration of 35 ± 19 Bq.m^{-3}.

From the measurements performed in the initially ventilated rooms, during standard periods of 24 hours, it can be noticed that during the first 3 hours after closing the windows and doors, the speed of accumulation of the radon is 50 Bq m^{-3} h^{-1} in rooms under it is cellar and 90 Bq m^{-3} h^{-1} in room without cellar, followed by a slower accumulation, of 22 and, respectively, 28 Bq m^{-3} h^{-1} determined during a 9 hour period. Sudden decreases of ^{222}Rn activity concentration after the opening of the windows have been observed in all sampling points (Figure 1).

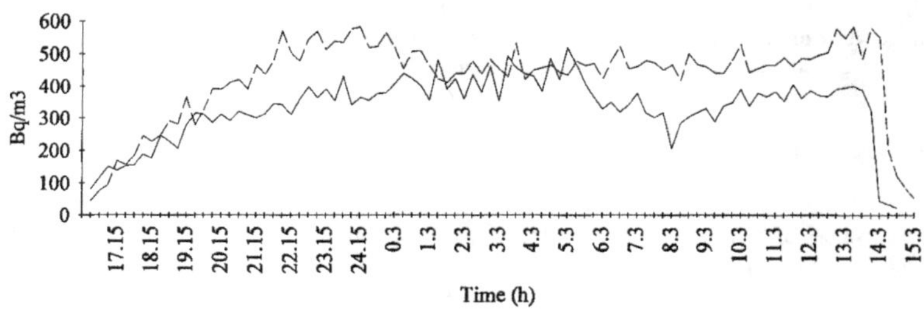

Figure 1. Accumulation of ^{222}Rn in living rooms

Table 2. The seasonal variation of the ^{222}Rn activity concentration (Bq.m^{-3}) in living rooms

HOUSES	SPRING	SUMMER	AUTUMN	WINTER
with cellar				
I	121	76	172	183
III	57	62	71	67
without cellar				
I	115	86	238	254
II	261	188	323	447
III	363	119	434	728
IV	121	75	246	248

The seasonal variation of the ^{222}Rn activity concentration in living rooms of the different detached houses are shown in Table 2. The determinations were made with SSTD. The ^{222}Rn activity concentrations in winter period are in almost all cases 2.5-3 times higher than in summer. The natural ventilation in this period is very low, therefore the accumulation of ^{222}Rn is higher than in summer when the windows are frequently open.

We selected three rooms with high ^{222}Rn level for establish the ^{222}Rn sources in these buildings. The results of the measurements performed between 3 September 1996 and 3 January 1997. The SSTD were placed in three rooms at different places and highs. The results show in table 3. Beneath the rooms

Table 3. Variation of the ^{222}Rn activity concentration with high

ROOM	NUMBER OF PLACE	HIGH ABOVE THE FLOOR (m)	^{222}Rn ACTIVITY CONCENTRATION ($Bq.m^{-3}$)
A	1	0	547
	2	0	474
	3	3	612
	4	1.5	579
	5	3	645
B	1	0	725
	2	0	1376
	3	0	1294
	4	3	696
LOFT above the A and B rooms		1	115
C	1	0	1586
	2	0	1366
	3	0	1304
	4	1.5	1339
	5	3	1297
LOFT above the C room		0.5	38

A, B and C, the high porosity pavement made of stone and slag allows this noble gas to easily spread and reach the interior of the rooms. The highest values were obtained at the 0 m distance the floor, near the windows. The outwalls in these house are cladding. The radon level decreased with hight in all cases. So, the main sources of radon are the building material of the foundation. The contribution of the natural gas to the ^{222}Rn activity concentration should be about 40 Bq/m^3 [5].

The annual mean values of ^{222}Rn activity concentration inside and outside the block-of -flats are shown in Table 4. The average of all indoor ^{222}Rn activity concentration measurements in block rooms was nearly 3 times higher than the open air. The substructure of blocks is made of concrete, and there are cellars under the ground level. The maximum values obtained in rooms at the ground level might be due to the contribution of ^{222}Rn exhaled from soil. These concentrations decrease from the basement to the upper floors. Slightly increasing ^{222}Rn concentrations were found at the last levels in both types of blocks. These are probably, due to a reduced ventilation rate, and to the insulation of the roof, which is made of slag. The average indoor ^{222}Rn level for the blocks with red brick walls is not significantly higher than the level obtained for the blocks with prefabricated-concrete walls. In the block houses the main contribution to the increased ^{222}Rn level is coming from building materials while in detached houses the main part of indoor ^{222}Rn comes from underlying soil. ^{222}Rn activity concentration values are significantly smaller in block flats as compared to those found in the detached houses.

The difference between the mean ^{222}Rn activity concentration values determined by using scintillation cell and ionisation chamber, and long term samples was 29 %. The deviations of the daily means from the mean value of the previous day ranged between 10-24 %.

Table 4. The annual mean values of ^{222}Rn activity concentration in block-of-flats

BUILDING MATERIAL	STOREY of the rooms	^{222}Rn ACTIVITY CONCENTRATION ($Bq.m^{-3}$)
Brick	ground floor	71 + 34
	1st	68 + 35
	2nd	65 + 27
	3rd	49 + 28
	4th	56 + 14
Prefabricated concrete	ground floor	63 ± 25
	1st	59 ± 18
	2nd	64 ± 21
	4th	62 ± 32
	6th	66 ± 28
	7th	49 ± 24
	8th	68 ± 20
	9th	168 ± 26
OUTDOOR		29 ± 12

4. Conclusion

Measurements of the ^{222}Rn activity concentrations in the indoor air of the different type of buildings showed values with a great variation.

The ^{222}Rn activity concentrations in detached houses in Cluj have been found to be between 34 - 1850 $Bq.m^{-3}$. In same city, in block-of-flats these values were between 38 - 250 $Bq.m^{-3}$.

It has been shown that the concentration of ^{222}Rn inside a house varies according to the construction of its foundation. In the rooms with no cellar beneath the major contributor to the high indoor ^{222}Rn level is the ^{222}Rn entering the structure directly from the soil. In the buildings with stone and slag basement, the ^{222}Rn activity concentrations exceed the value of 200 Bq/m^3 recommended for action level for dwellings by ICRP 65/93 [6].

Comparisons were made between the mean values of ^{222}Rn activity concentration obtained by daily measurements and those determined by SSNTDs. The difference was 29 %.

The ^{222}Rn content in indoor air may be increased in the future by the utilisation of new building materials, energy conserving measures and by the change of living conditions. This may lead in the future to an increase of the indoor exposure of the population. The significance of ^{222}Rn for the human health is well known and thus the measurement of indoor radon is important for the evaluation of its impact in the public health field.

We propose to extend our measurements to other type of buildings, and to work out methods adapted to our economic resources, able to reduce the indoor ^{222}Rn concentration.

5. Acknowledgement

We acknowledge and thank to the inhabitants in whose flats these measurements were performed.

6. References

[1]. Wozniak S.J.: Handbook of radon, health, economic and building aspects, ISBN London, UK, 1992

[2]. Chang S.Y.,.Ha C.W.,Lee B.H.: Estimate of lifetime excess lung cancer risk due to radon-222 daughters exposure in Korea, Radiation Protection Dosimetry, 1992, 42, 2, pp.127-132

[3]. Poffijn A., Eggermont G., Hallez S., and Cohilis P.: Radon in Belgium: mapping and mitigation in the affected area of Vise, Radiation Protection Dosimetry, 1994, Vol.56, no. 1-4, pp. 77-80

[4]. Dudney C.S., Hawthorne A.R., Wallace R.G., and Reed R.P.: Radon-222, Radon-222 progeny and radon-220 progeny levels in 70 houses, Health Physics, 1990 Vol.58 pp.297-311

[5]. Mocsy I., Muntean E., Muntean N., and Vasarhelyi A.: Radon and its daughters in houses using methane gas for heating, In Proc. of the International Conference on Indoor Climate of Buildings, High Tatra, Slovakia, 1992, pp.233-241

[6]. International Commission on Radiological Protection ((ICRP): Protection against Radon-222 at home and at work, Annals of the ICRP Publication 65, 1993, Vol.23, No.2

MONITORING AND CONTROL OF THE RADON HAZARD IN POLISH COAL MINES

Jan Skowronek, Krystian Skubacz, Bogusław Michalik
Central Mining Institute, Katowice, Poland
Ryszard Kajdasz, Zbigniew Strześniewicz
Ziemowit Coal Mine, Lędziny, Poland

Earth's fumes, the bad air from centuries were known as a reason of a number of the different diseases of the human respiratory system. One of the oldest mention of the lung diseases, caused by „the bad air", is a part of „De Rerum Natura" written by Titus Lucretius Carus in first century BC [Titus Lucretius]. In this poem, he mentioned a very bad health and a very short live of a miners, searching for silver and gold, as a result of inhalation of a stinky earth fumes. Miners worked in the mines located in the northern Greece, near Aganion mountain, in the ancient times known as a centre of the gold mining as well as silver and other metals.

Probably the mentioned disease of the miners was radiogenic induced lung cancer rather than a silicosis. It can be proved by elevated radon concentrations measured in houses on that area and elevated gamma dose rates found there (the average concentration of ^{222}Rn in dwellings exceeds 100 Bq/m^3 while gamma dose rate outdoors is of about 100 nGy/h) [Lebecka, 1995].

In 1879 a first time a diagnosis of a lung cancer was proposed as an explanation of the disease of miners from Schneeberg and Jachimov, known since Avicenna's times as Schneebergkrankheit. 40 years later a possible connection between the lung cancer among miners and the enhanced radon concentration was mentioned first time. But for the full recognition of the sources of the disease miners had to wait till 50'ties. Nowadays augmented risk of a lung and larynx cancers as a result of the exposure on radon is not only known fact, but also the most important effect of the influence of the natural radioactivity on the general public. There is no doubt, that quite often miners are more endangered by enhanced radon exposures as other members of the society.

Upper Silesia is the region of the extended underground mining industry, mainly of a hard coal, but also the metal ore mining. Measurements of radon concentration in the mines were started in 70'ties and since late 80'ties are obligatory due to Polish mining law.

Radon and it's progeny

Radon is the main source of a radiation doses for any human being. As a gaseous isotope it emanates into the atmosphere from all substances, containing it's parent isotope - radium. Radium is widely spread in materials in the natural environment, therefore also radon can be found almost everywhere. The behaviour of radon and it's influence on the living creatures is due to the fact, that radon is a noble gas and it decays, producing radioactive isotopes of polonium, lead and bismuth. These radioisotopes are known as the short lived radon progeny, with relatively short half-lives. Moreover, these isotopes are chemically active, have an electric charge and decay with the emission of alpha or beta particles.

In the nature three isotopes of radon are present and all these isotopes are alpha emitters - ^{222}Rn ($T_{1/2}$ = 3,8 d), ^{220}Rn ($T_{1/2}$ = 55 s), ^{219}Rn ($T_{1/2}$ = 3,92s). They are a members, respectively, of the uranium, thorium and actinium chains. Only ^{222}Rn (known as radon), and ^{220}Rn (known as thoron), are present in the environment in the considerable amounts. The half-lives of ^{222}Rn progeny are shorter than 30 minutes. Due to this fact, a probability of its decay inside the respiratory tracts, when inhaled, is very high. Similar situation is observed in case of the thorium ^{232}Th chain. As a result of a decay of ^{220}Rn short lived progeny is produced, and the most important among them is ^{212}Pb with half-life 10,6 h.

Some of the radon daughter products are alpha emitters and this type of radiation can cause high doses in the tissues in the respiratory tracks. Therefore radon progeny plays so important role for the radiation hazard for the humans, more important as the role of the parent isotope - radon.

Relatively short half-lives of the isotopes known as radon and thoron progeny and the type of the emitted radiation (alpha and beta) cause severe problems with proper measurements of its concentration in air. Measurements are more difficult in underground mines, where environmental conditions are tough and very often instrumentation must fulfil additional requirements as intrinsic safety in mines with methane hazard. On the other hand, for the purposes of the radiological protection, it seems to be enough to measure the total amount of energy, released with alpha particles during decay of radon and thoron progeny, not a concentrations of particular isotopes. Therefore a new quantity was introduced - **potential alpha energy concentration of the short lived radon progeny** (in Polish standards [PN, 1988] it's symbol is C_α, in English the acronym PAEC - potential alpha energy concentration - is used). **C_α (PAEC) is defined as a total energy released in a form of an alpha radiation during decay of all nuclei of short lived radon progeny present in the unit volume of the air.** The unit of C_α is 1 J/m^3.

Radiation hazard caused by short lived radon progeny in coal mines

The guidelines of the monitoring and control of the radiation hazard in mines were issued in two acts, based on the Polish Geological and Mining Law [Geological Law, 1994].

In the Decree of the President of Chief Mining Authority [Mining Authority, 1994] criteria of the classification of the underground galleries or workplaces due to radiation hazard were set as well as requirements concerning a range and frequency of the monitoring. Underground galleries were classified into three groups: first one with no radiation hazard and two others called as galleries class A and galleries class B. The classification is based on assessment of the expected annual dose, which would be obtained by miners as a result of a work in such place during whole year.

In the Decree of Ministry of Industry and Trade [Ministry of Industry, 1995] the guidelines of the monitoring of the radiological hazard are described as well as prevention measures.

In this Decree two level are defined:

- the inspection level - when exceeded, more frequent and extended monitoring of the different factors of the radiation hazard is necessary on the particular workplace or gallery,
- the intervention level - the prevention measures leading to the diminishing of the radiation hazard must be undertaken by the management of a certain mine, when this level is exceeded.

Also the annual dose limit in units of the dose equivalent mSv (milisievert) is in the agreement with contemporary international guidelines: accordingly to the Decree the limit of the annual dose rate for miners is 50 mSv but not more than 100 mSv during 5 following years.

To ease the work of the different departments in coal mines all dose limits and certain levels were recalculated into respective values of risk factors, issued in the Decree of the President of the Chief Mining Authority. In case of the radon hazard the indicator of the risk is the potential alpha energy concentration of short lived radon progeny. The values of the dose limits and levels for radon progeny are shown in table 1.

Table 1. Values of potential alpha energy concentration for the levels and dose limits

The hazard factor	Inspection level (2 mSv)	Intervention level Class A (5 mSv)	Class B (20 mSv)	The average annual dose limit (20 mSv)	The annual dose limit (50 mSv)
Potential alpha energy concentration C_α, μJ/m^3	0,8	2	8	8	20

Analysis of the radiation hazard in underground coal mines, based on the obligatory law, shows that in 1996 [Skowronek et al., 1997] :

- the C_α was higher than the inspection level (0,8 μJ/m^3) in 106 underground galleries in 14 coal mines
- concentrations of potential alpha energy exceeded the intervention level (2 μJ/m^3) - were found in 20 galleries within 6 coal mines (28 measurements) and these galleries were treated as galleries with class A of the radiation hazard.

In 1996 no galleries with C_α above 8 μJ/m^3 were found, it means no galleries with class B of the radiation hazard.

The scale of the radiation hazard in the mining industry can be described not only by the value of the dose for a particular single miner, but mainly as the collective dose for the whole population of miners. Number of a new diseases among miners is correlated rather with the collective population dose due to relatively low doses, obtained by miners in non-uranium mines. Table 2 contains the values of the doses from particular sources in coal mines for the interval of the last 5 years. It can be clearly seen, that the biggest hazard is caused by the exposure on the short lived radon progeny.

Table 2. The effective annual collective dose equivalent from different sources of the radiation hazard in Polish coal mines, mansieverts/year.

The source of the hazard	1992	1993	1994	1995	1996
Short lived radon progeny	326	296	118	115	77
External γ radiation	5	3	2	0,5	0,5
radium-bearing waters and radioactive deposits	25	30	20	10	10
TOTAL:	356	329	140	125	87

Sources of radon and radon progeny in polish coal mines

Porous and fractured rocks with an extended surface forming the side walls and the abandoned workings, consist a source of continuos diffusion of radon into the mine atmospheres. Radon itself is not very harmful, the radiation hazard arises after radon daughter products are created [Ertle, 1981]. In other words, the radiation hazards begins some time after radon emission from the rocks into mine atmospheres. It also means, that the radiation hazard may occur on the pathway of the air stream in a place rather distant from the source of elevated radon emission.

In coal mines the radiation hazard caused by short lived radon progeny occurs mainly in the longwall area [Skowronek, 1992]. In this case, the gobs remaining after coal extraction with roof caving, present the source of radon. The small flow rate of air through the gobs causes a long residence time in gobs enabling the build up of radon daughter products resulting in a high equilibrium factor between radon and it's progeny. After the air flowing through the gobs is mixed with main ventilation stream, the PAEC may reach high values (over 10 $\mu J/m^3$). In these places high dust concentration is often observed as well.

The radiation hazard caused by radon progeny is highly dependent upon the system of ventilation of mine workings [Skowronek,1991]; ventilation enabling contact of air with abandoned mine workings increases the radiation hazard.

Other sources of radon may be waters with radium or radon concentration. Waters with high radon concentrations are rather rare in polish coal mines, they occur only in several mines in Lower Silesia, which are already closed. Radon emission from waters and deposits with high radium concentrations may cause elevated PAEC in mine workings like water galleries or settling tanks, with similar consequence as in case of long walls.

Recently, investigations of methods of control of concentrations of radon and its progeny have been performed in an active coal mine. The investigations have been concentrated mainly on following problems:

- influence of changes in the ventilation network on radon and radon daughters concentration,
- the efficiency of sealing of gobs in reduction of radon and its progeny concentration in mine atmospheres on work places,
- development of effective and efficient methods of filtration of air for removal of radioactive aerosols.

It has been proved that these methods may be effective in prevention as well as in reduction of the radiation hazard caused by radon progeny on work places.

Influence of changes in the ventilation net on the decrease of radon daughters concentration

The role of ventilation in protection against radon progeny is confirmed by the example of actions performed in the longwall 705 in the Ziemowit Colliery. The longwall 705 was mined as the last one in the SW part of the 207 coal seam, between the gobs remained after mining out the neighboring longwalls. before mining longwall 705 this section was connected with the levels II and III. After the longwall 705 was started, the connection with level III was sealed. the longwall was ventilated from level II, from the roadway E-2 in the „U" pattern, as shown. on. Fig.3.

Results of measurements of PAEC during the last months of longwall operation were between 2.10 and 2.89 $\mu J/m^3$. After closing the longwall 705, the air stream was reversed as shown. in Fig. 2 and Fig.3.

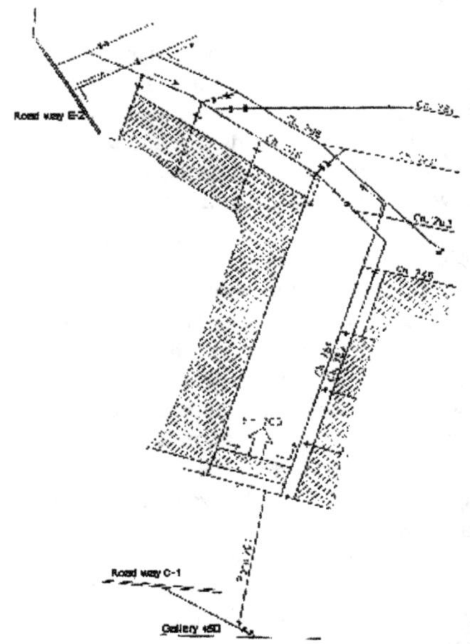

Fig.1 Ventilation scheme of 705 longwall in the last stage of mining

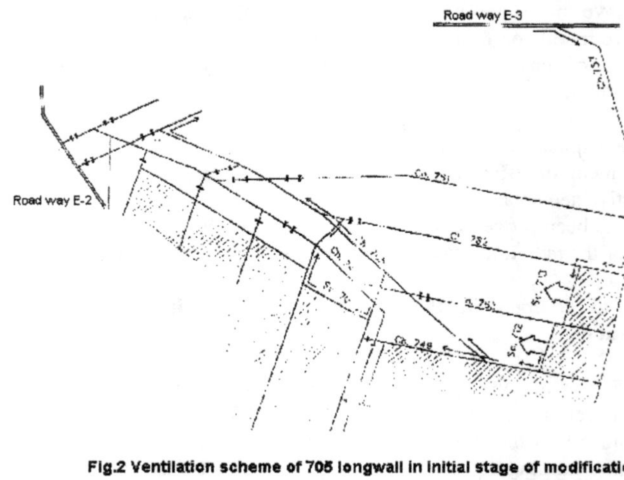

Fig.2 Ventilation scheme of 705 longwall in initial stage of modification

The changes in ventilation network caused diminishing of the air flow in the longwall being closed. The inflowing air passed before the new longwalls and passed along the gobs. As a results the first measurements performed after closing the longwall were 10.57 and 10.99

μJ/m^3. The estimated inflow of air from the gobs from the closed longwall 705 was of about 6 m^3/min.

The first stage of applied prevention measures included installing of auxiliary ventilation in the closed longwall and installing of fabric partition in the gallery 749 above the gallery 765 (Fig.3). These works were aimed to increase the amount of air in this area and to shorten the pathway of air along the gobs. At the same time measurements of aerodynamic potentials in gobs in the SW section. The results showed that the gobs are under depression of 40 Pa. This was caused by two stoppings situated in the cross-cut from the inclined drift 207 to the gallery 450, which were necessary for control of the air stream in another section.

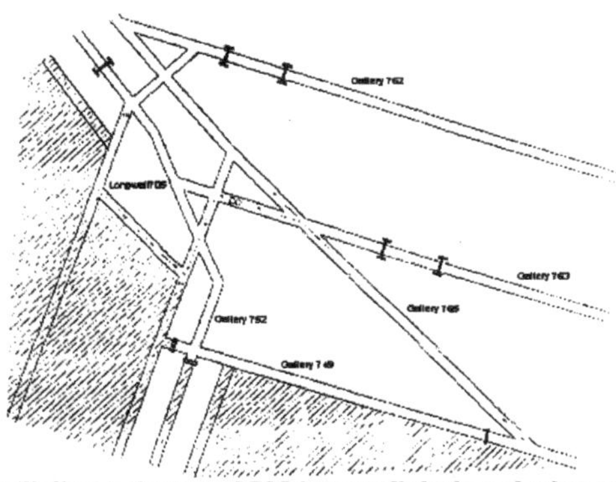

Fig.3 Ventilation scheme of 705 longwall during closing operation after modyfication

In order to reverse the disadvantageous relation of potentials urgently two new stoppings were build at the lower and of the drift 207, while the old stoppings, mentioned above were disassembled. The results were as anticipated, the depression 140 Pa in the opposite direction. Because there was necessary to minimize the amount of air inflowing from the 207 drift into the 450 gallery, which ventilated the G-IV flat, omitting other work places, it was impossible to even the potentials at the gobs of 705 longwall.

As a result, the situation was substantially improved. At the of closing of 705 longwall the average of three results of last measurements of PAEC was about 60% lower than at the beginning of closing of the longwall and was about 4.46 μJ/m^3. The mining crew was employed at the closing of the longwall during two months. The estimated doses were of about 6.2 μJ/m^3 (this value was estimated using the previously applied conversion factor - 2.94 mSv/μJ/m^3.

Restraining the radiation hazard caused by radon progeny, using sealing of gobs

Application of sealing of gobs in protection against radon progeny is discussed on the example of the longwall 920 of the Ziemowit Colliery. This longwall mined the upper layer of the seam 209. The mined before longwall 919 was ventilated as separately. When the mining of this longwall was almost finished, an operation of a new longwall 920 was started. It was necessary to use the same air stream for ventilation of both longwalls.. The return air from 919 longwall was directed to the longwall 920 through the gallery 145 along the 800 m of gobs of the section of 919 longwall. The side wall of the gallery was not insulated from the gobs of the longwall. As a result the PAEC in the longwall 920 was as high as 8 $\mu J/m^3$.

The first stage of prevention measures included increasing the air flow rate in the gallery 145 and insulation of the gobs adjacent to the longwall 919. At the beginning there was a slight increase of PAEC. Than, the PAEC returned to the previous value. Since these actions did not give the expected results, it was decided to use foam for suppression of outflow of air reach in radon and daughter products from the gobs.

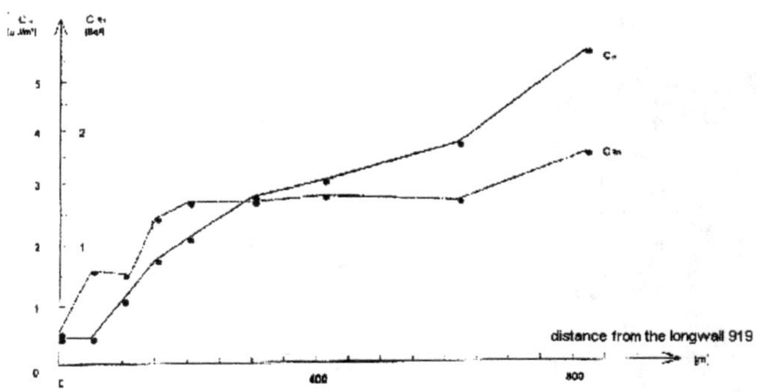

Fig.4 Changes of radon and radon progeny concentation in air along the gobs of the 919 longwall

To achieve improvement in a possibly short time, it was decided to start the insulation on the sections, with the highest gradient of PAEC. Measurements of radon concentrations and PAEC were performed along the 145 gallery, longwall 920 and gallery 146. The distance between the sampling points was about 50 m. (Fig.4). First the 150 m segment of the gallery, adjacent to the longwall was insulated. This insulation included also the boundary caving drift. of the 919 longwall. In order to limit the migration of air through the gobs, simultaneously stoppings of foam and fabric were placed in the 144 gallery in the back of 919 longwall.

The technique of insulation of the 145 gallery was adjusted to the local conditions. The thickness of the foam layer varied from 0.3 m to 0.5 m. In some sections of the gallery the output was left on the floor at the gobs side and covered with a layer of foam.

PAEC in the 920 longwall was measured during the insulation work. Decrease of PAEC followed the progress of insulation. As a result the PAEC became almost 4-times lower. It proves the effectiveness of such prevention measure in suppressing the hazard caused by short lived radon progeny on underground work places.

Investigations of air filtration

Removal of airborne radon daughter products from mine atmospheres consists serious technical problems. The main reasons are the high dust concentration causing fast clogging of filters and very small dimensions of radioactive aerosols. Most of the materials used in the filtration systems does not catch aerosols with diameters below 1 μm or does it with a very poor efficiency. After laboratory tests filter materials with high (98%) efficiency for radon progeny was chosen. The thickness of the material is 0.5 mm. It is made of super-fine glass fibers. It has good mechanical properties. Low content (<5%) of combustible enables application in mines.

The tests in underground mine workings were performed using a set of filters, which consisted of a filter for preliminary filtration (for removal of the large grains) and a fine filter. The set was adjusted for connecting with the ventilation duct of 600 mm of diameter. The casing was made of 1 mm passivated zinc coated sheet. The tightening were made of epoxide resin and polyurethane foam.

The filter cartridge was shaped as a symmetric bellows. To ensure the appropriate distance between the components of the filter separators made of aluminum foil were used. The preliminary filters, made of polypropylene fibers, were flat.

To connect the filter with the ventilation duct special joints, made of the fabric used in ventilation ducts, were developed. The entire set (containing the set of filters and the joints) could be clipped to the elements of lining.

This set was tested in two versions:

- with an auxiliary fan,
- with a wet dust collector.

During the test measurements of PAEC were performed. The sampling points were located 20 m before the position of the filter set and in the air stream just behind the filter. The results are presented in table 3.

Table 3. The results of tests of filtration efficiency

No	Position configuration	Mean PAEC [μJ/m^3]		Efficiency of filtration [%]
		inlet	behind the filter	
1	Position A filtration set with a fan	1.15	0.14	88
2	Position B filtration set with a fan	2.03	0.97	52
3	Position B filtration set with a wet dust collector	4.41	1.36	69

The efficiency of filtration of the set of filters with the fan in position A was high (88%). In position B only 52% efficiency was achieved. In case of using a wet dust collector with the set of filters, the efficiency was 69%. So far there is no explanation for the lower efficiency obtained in position B. The low efficiency was observed in this case for both type of arrangements: with a fan and with a wet dust collector. The possible cause could be smaller dust concentration in air in this place.

The results of tests performed with the sets of filters are encouraging. However, substantial decrease of air flow rate, caused by clogging of main filters. It is possible to avoid this problem using efficient preliminary filters or a dust collecting installation. The results of test showed, that one filter set could be used without changing the cartridges (preliminary and main filters) during at least three working shifts.

Summary

The system of the monitoring of the radiation hazard in Polish coal mines is based on the monitoring of the workplaces. This system works since 1986 in all coal mines. It gives a very good basis for further epidemiological investigation and assessment of the health detriment within the population of the mines as a result of the exposure for natural radiation. It is very important problem, due to the fact of the presence in the mines another factors, which probably have a synergetic effects on the respiratory tracts.

Several methods of suppression of radiation hazard caused by short lived radon progeny were tested in underground mine workings. the methods presented in the paper proved to be efficient. Insulation of gobs prevents the diffusion of radon and its progeny into mine atmospheres. This method enables prevention of radiation hazard caused by short lived radon progeny on workplaces. The method is ready for immediate application.

When there is radiation hazard on work places, filter sets can be used. The tested filter sets, after some development could be applied for removal of radon progeny from the ventilation air at the work place. This should enable to keep the PAEC below 2 $\mu J/m^3$ (which is the lower limit of class A of radiation hazard [Decree of Polish Ministry of Industry and Trade,1995, Regulation of the Chief Mining Authority,1994].

This work was sponsored by Polish Committee of Scientific Investigations and by the Ziemowit Colliery - research project No.4 1084 93 C/1536)

References:

Carus Titus; Carus Titus Lucretius: De Rerum Natura.

Lebecka, 1995; Lebecka J.: Radon in mines. Bezpieczeństwo jadrowe i ochrona radiologiczna, No23, 1995, pp. 21-40

PN, 1988; Polish standard PN-88/Z-70071 Radioprotection in underground minings. Limits of risks from the natural radiation sources and methods of monitoring.

Geological Law, 1994; Geological and Mining Law. Decree of the Polish Parliaments. Dz. U. No 27, 1994.

Mining Authority, 1994; Decree of the President of Chief Mining Authority. M.P. No 45, 1994

Ministry of Industry, 1995; Decree of the Ministry of Industry and Trade. Dz. U. No 67, 1995.

Skowronek et al., 1997; Skowronek J. et al.: Report on the natural radiation risk in mines in 1996. Report of the Central Mining Institute, Katowice, 1997. (in Polish)

Ertle, 1981; Ertle H.J., Reim K., Thome M., Haupt W.: Investigation into method for the protection of underground workers against radon and its daughter products. [in:] Gomey M. (ed.) Radiation Hazard in Mining, New York, 1981.

Skowronek, 1991; Skowronek J: Les descendants a vie courte du radon dans les mines de charbon en Pologne. Radioprotection, GEDIM 1991, vol. 26. No 1, pp. 65-74.

Skowronek, 1992; Skowronek J., Lebecka J., Skubacz K. et al.: Risque radiologique dans les mines de charbon en Pologne. Radioprotection, GEDIM 1992, vol. 27. No 4, pp. 451-456

RADON IN HOUSES ON AREA OF THE UPPER SILESIAN COAL BASIN

Małgorzata Wysocka, Antoni Mielnikow, Stanisław Chałupnik

Central Mining Institute, 40-166 Katowice, Pl. Gwarków 1, Poland

ABSTRACT

Upper Silesia, situated in southern part of Poland, is an industrial area, where over 60 coal mines are located. Geological structure of this area is very complicated. In the southern part of the coal basin, the carboniferous strata ate covered by thick layers of Miocene clays. In the north the carboniferous strata are covered by permeable limestones and Quaternary deposits. There are also numerous outcrops of coal seams. Such structure enables easy migration of gases. There are a lot of tectonic dislocations and faults. Additionally this area is strongly affected by geodynamic events caused by mining. Radioactive waters and waste rocks released from mines cause contamination of rivers, bottom sediments and adjacent areas.

Preliminary survey of radon in buildings in Upper Silesia was performed with use of charcoal detectors. The readout of the detectors was done by liquid scintillation counting.

The results show, that on areas with thick layer of impermeable clays an average radon concentration in dwelling is lower than on areas where such isolation does not exist. On the other hand we found no correlation between radon level in houses and the radioactive contamination of the environment in the vicinity.

We performed radon measurements in soils in these places where concentrations of radon in houses were highest. As predicted, we found highest radon levels in soil on areas with underlying permeable limestones and marls. This effect was additionally magnified by shallow underground mining. A lot of emptiness in coal beds near the surface are a source of radon exhalation.

1. INTRODUCTION

Preliminary monitoring of radon concentrations in buildings in the Upper Silesian Coal Basin have been performed by Central Mining Institute in years 1992-1993. These measurements were carried out with use of activated charcoal detectors followed by liquid scintillator counting. The applied method is described elsewhere [1].

Within the frame of the work it has been found that radon concentrations in buildings were within the range from 7 Bq/m^3 up to 360 Bq/m^3. It was also observed that radon concentrations in houses in northern part of Coal Basin were higher than in southern part. Such results of the preliminary monitoring of indoor radon concentrations in Upper Silesia pointed out that high levels of radon were correlated with geological factors such as lithology, structure and thickness of the underlying strata.

In order to prove mentioned above correlation, measurements of radon concentrations in soil gas have been undertaken. Measurements of radon in soil gas were performed in these sites where we found the highest concentrations of indoor radon.

2. INVESTIGATION SITES

Geological structure

The choice of the sites for a preliminary survey of radon indoor concentrations was done considering the following aspects:
- geological structure of the bedrock,
- environmental pollution caused by radium-bearing waters dumped out from coal mines.

Fig. 1. The sketch map of the Upper Silesian Coal Basin

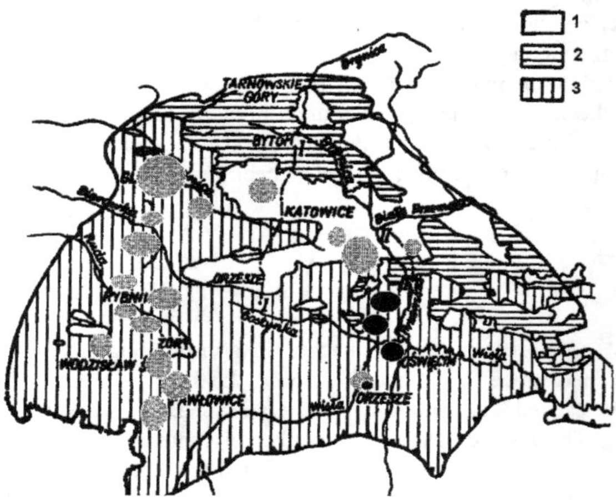

1. Quaternary sediments
2. Triassic sediments
3. Tertiary sediments

Upper Silesia (fig.1.), situated in the southern part of Poland is characterised by a very complicated and differentiated geological structure with numerous faults and other tectonic dislocations [2]. Moreover, Upper Silesia is an industrial region where over 60 coal mines and few zinc and lead ore mines are located. The area is strongly affected by geodynamic events caused by mining. Numerous mining excavations and underground workings are quite close to the ground surface. Accordingly to Mennin [3] and Bonetti [4], radon indoor concentration is correlated with geological structure of investigated area.

Within the frame of the work we wanted to find out, whether, despite of intense mining activity, the correlation between radon concentration and geological structure can be observed. Dwellings, located in different parts of Silesian Coal Basin have been chosen. First chosen area, located in southern and western Silesia is characterised by thick layer of sediments covering carboniferous formations. This overlay consists mainly of Miocene clays and silts. The thickness of these sediments is up to 700 m. Such strata makes the migration of water and gases almost impossible.

In northern and western part of Silesia Miocene sediments do not occur. Carboniferous strata are covered by Mesozoic and Quaternary slightly compacted sands, gravel and sandstones. The oldest part of the overburden on this area is built of an isolated and strongly fissured Permian or Triassic limestones. There are also numerous outcrops of coal seams. These formations enable very easy migration of water and gases.

Radioactive contamination of the environment

The second aspect which has been taken into consideration was the radioactive contamination caused by waste rocks and waste waters containing an enhanced amounts of radium isotopes. We chose for the further investigations areas in the vicinity of rivers and settling ponds in which waters and sediments with enhanced natural radioactivity have been found. The areas nearby two settling ponds and one small river were examined. Radioactive waters from underground workings are pumped out to the mentioned above ponds. We measured radium concentration in waters and natural radioisotopes in the bottom sediments.

Almost all measurements have been performed in single family houses, mainly on the ground floor level.

3. EXPERIMENTAL

The measurements of radon concentration in the soil gas were carried out with of the same method which had been applied for indoor radon measurements [1].

Boreholes of depth of 0.5 m were drilled in chosen sites (fig.2.). Glass scintillation vials with dried activated charcoal were placed in boreholes, encased by plastic pipes, closed by gum stopper and protected by polyethylene foil. After seven days exposition, detectors were collected, radon was washed out by scintillator and than measurements were performed by liquid scintillation spectrometer.

Fig.2 Method of measurements of Rn by activated charcoal detector

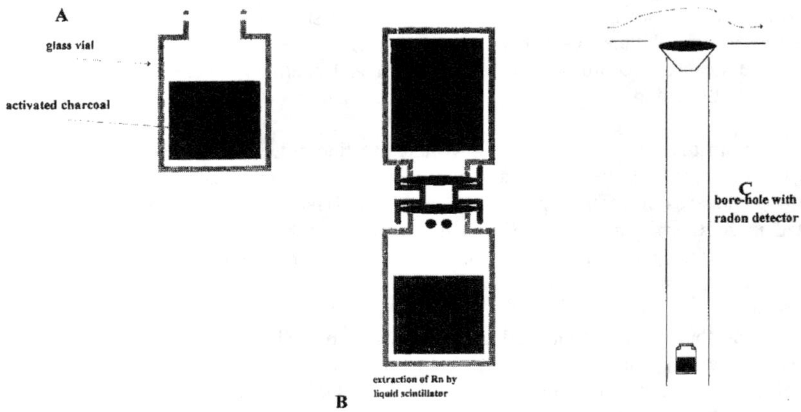

In aim to observe changes of radon concentrations series of measurements were performed in test boreholes drilled near our Laboratory building. Measurements were done in a period of a half of the year, during which detectors have been changed once a week. Fig. 3. shows results of measurements in one of the test borehole and changes of atmospheric pressure.

Fig. 3 Correlation between radon concentrations in soil and atmospheric pressure

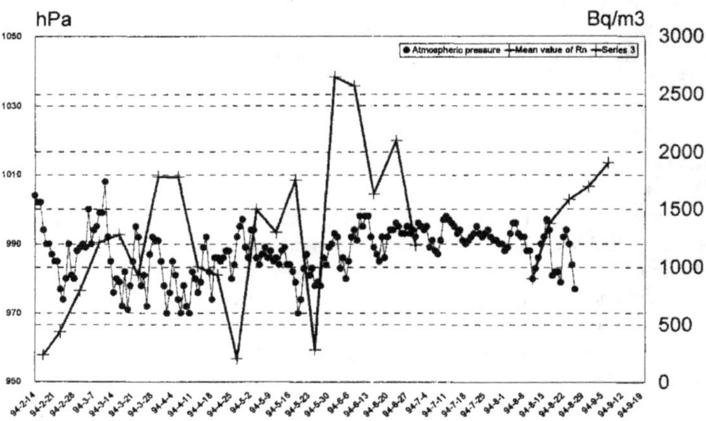

In order to determine the influence of a moisture on measurement's results one of the comparative borehole was dried with 25 grams of silicagel. Comparing results of measurements performed simultaneously in boreholes dried and not dried it have been found that influence of moisture was very high - calculated correction factor was 2,4 (fig.4.). Correction factor for the moisture have been taken into consideration.

Fig 4. Influence of moisture on radon concentration

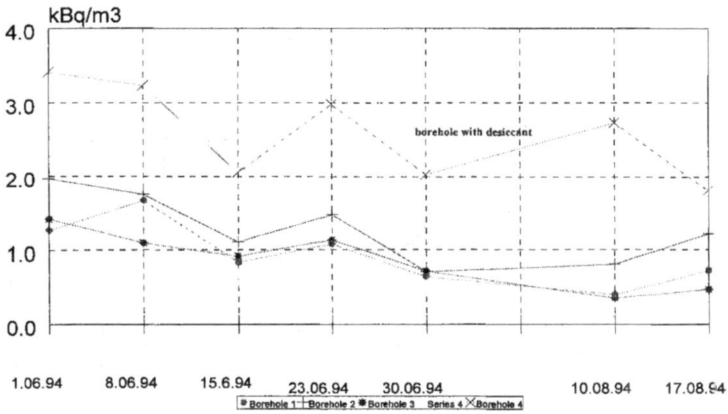

4. RESULTS OF MEASUREMENTS

Measurements of radon concentrations in the soil were performed near houses where relatively high concentrations of indoor radon were observed (up to 360 Bq/3 on first floor and 679 Bq/m^3 in cellars) during preliminary monitoring. Eleven family houses were chosen to perform measurements nearby. All investigation sites were situated in described above northern and north-eastern parts of Coal Basin.

In each investigated site three simultaneous measurements were performed in boreholes with distance 5 to 10 m between each other. At the same time charcoal detectors were placed in each house in rooms at ground-floor and in the cellar. Detectors were replaced once a week during the one month.

Radon concentrations measured in soil varied from 188 up to 22510 Bq/m^3. Radon concentration in buildings in all cases were much lower than concentrations observed during preliminary series of measurements and varied from 5 up to 311 Bq/m^3. Probably it is caused by the fact that in 1993 measurements were performed in cold days of autumn when the ventilation of buildings was rather poor. The next series of measurements were carried out very warm spring time, and intensive ventilation caused decrease of indoor radon concentrations.

5. ANALYSIS OF THE RESULTS

More detail analysis of geological structure of sites, where the highest concentrations of radon indoor and in soil gas were measured, have been done.

The highest concentration of radon in soil was measured on area with Triassic sediments filling synclinal forms of northern and eastern part of Coal Basin. The bedrock is built mainly of dolomites and limestones. Triassic formation is strongly weathered with many fissures and caves being good pathway for migration of water and gases (radon).

Fig.5. The sketch geological structure of the place with high radon concentration

On fig.5. a schematic plan of the geological profile of the strata beneath the house is presented. Directly under thin layer of the soil (0.3 - 1.0 m) strongly weathered Triassic rocks appear. Quaternary cover near investigated house is reduced, Tertiary sediments do not occur. There is no isolating shale-layer underneath dolomites and limestones. Under Triassic rocks with thickness up to 100 m carboniferous strata occur in which the exploitation of coal seams have been performed.

High concentrations of radon were also measured in places with different geological profile. In the area where such structural deformations as anticlines and dome-folds are present, carboniferous strata appear under thin layer of Quaternary sediments or directly at the surface in form of outcrops. These sites were intensively being exploited by small-scale mining till late

40's. In first years of this century and between first and second world war people living there were dug hundreds of small shafts and outcrops of coal seams have been exploited extensively. Additionally, mines applied a caving system for the excavation of a coal from beds on shallow depths. Situations very similar to described above were common in Upper Silesian Coal Basin.

Because of intensive exploitation the structure of bedrock were destroyed, rocks became cracked and fissured. Numerous of abandoned workings, excavations are present. The excavations are backfield with different fillings like sand, waste-rocks, wastes from power stations etc. Such transformations in bedrock caused by mining activity enables very easy migrations of gases and their penetration into houses.

5. SUMMARY

Within the frame of this work we performed 144 measurements in 87 dwellings. The maximum radon concentration, measured in houses was 362 Bq/m^3. The lowest concentration was 8 Bq/m^3 and the arithmetic mean value - 56 Bq/m^3.

The results obtained in cellars were higher than in the living areas in the same house. The highest result obtained in the cellar was 679 Bq/m^3, while the average radon concentration was calculated as 81 Bq/m^3.

Houses with highest indoor level concentrations are situated in northern part of Upper Silesia, where Miocene impermeable clay-silt cover do not occur near the surface. On this area dwellings are built on strongly fissured limestones or dolomites. Some of houses where high radon concentration have been measured are situated on the outcrops of coal seams covered with thin layer of permeable Quaternary deposits.

In southern part of Silesia, where thick strata of isolating Miocene deposits occur, radon concentration vary between 8 up to 128 Bq/m^3, with an average of 42 Bq/m^3.

The radon concentration on areas with enhanced natural radiation caused by radium-bearing waters released from coal mines, seldom exceeded the average value. Even the close vicinity of radium-bearing waters, sediments and waste rocks does not cause an increase of radon level in dwellings. It means, that geological structure is the main factor affecting the radon concentrations in buildings.

The influence of geology may be intensified by the activity of coal mines. The presence of abandoned workings, excavations and cavities enables easy migration of radon gas and it's penetration into houses. Moreover numerous faults and other tectonic dislocations occur in this area. Additionally, geodynamic events such as tremors, bursts and bumps are observed in carboniferous formation very often and may also be a cause of the variations in radon exhalation from soil.

The results of investigations allowed to characterise the geological structure of regions where high concentration of radon can be expected:

- under thin layer of soil strongly weathered limestone-dolomite formations with numerous fissures, caverns and debris are present,
- under thin layer of soil and reduced Quaternary sediments, outcrops of carboniferous strata with exploited coal beds appear.

Eleven houses with the highest radon indoor concentration were chosen to perform radon in soil measurements. Radon concentration measured in soil gas varied from 188 up to 22510 Bq/m^3.

The method of measurements of radon concentration with the use of activated charcoal detectors and liquid scintillator counter proved to be convenient not only for preliminary survey of radon indoor concentrations but also for screening of radon in soil gas as well. In this case the correction for moisture is needed.

Additional advantage of the method is the possibility of simultaneous screening of radon indoor and radon in soil gas concentrations.

References

1. Wysocka, M. *et al.* 1995. Survey of radon in houses in an area strongly affected by mining activity. *Gas geochemistry.*, 481-487.
2. Różkowski, A. and Wilk, A. 1982. Hydrology of Upper Silesia Coal Basin. Draft of Geological Structure. (in Polish). *Publication of Polish Geological Society*, vol.1, pp 72-89, Sosnowiec.
3. Mennin, A. 1993. Radon Monitoring in Dwellings and in Water in Localised Areas of Frulli Venecia-Giulia. *Proc.of the Second Workshop on Radon Monitoring in Radioprotection and/or Earth Sciences.*, 78-87.
4. Bonetti, R. *et al.* 1993. Measurement of Indoor Radon Concentration in High Natural Radiation Areas in Northern Italy. *Proc.of the Second Workshop on Radon Monitoring in Radioprotection and/or Earth Sciences.*,499-502.

E Radon in the water

Radon in natural water

Kimiko Horiuchi

School of Social Information Studies, Otsuma Women's University
Karakida Tama, 206 Japan

Abstract

There are three kinds of radioactive radon gases in nature, even though the word radon(Rn) often means ^{222}Rn. The two others are thoron(^{220}Rn) and actinon (^{219}Rn) and all of the three are decay products of the natural radioactive chains of Uranium, Thorium and Actinium. Since the half-life of ^{219}Rn is too short (4.0 seconds) and there has been little study done on it.

The study of radon in water started about 90 years ago since its existence in mineral springs was confirmed. Until the early 1950s radon research in water was measured with the fontactoscope. After that, liquid scintillation counter(LSC) was used in the measurement of radon in water: the toluene-extracted and LSC method since it was developed, and the method spread rapidly and made good progress. With the spread of the method more and more concern has been shown regarding the effects of environmental radiation in daily life. Radon research is presently being performed in environmental science, geochemical science, medical science, physics, biology and technology with many other applications.

The studied of radon in water resources, especially as hydrological tracer and in hot springs have been carried out and will be reported here.

1. Introduction

The presence of radon in natral water sample was recognized in 1903 at Kings Spring, Bath by H.S. Allen, who also suggested that the mineral spring water containing radon might have a therapeutic effect. Following his recognition, surveys and researches started on radon in mineral spring water at various places in the world, including the report by H. Mache on the measurement result of the Bad Gestein mineral springs. The Mache unit proposed by Mache for expressing the rate of the radium emanation concentration gradually gained general approval. This unit was used in literature on hot and cold mineral spring water primarily in Germany, and also was used in Japan until relative recently. The first report on the radioactivity measurement in Japan was made by means of Fontactoscopes by Denichiro Ishitani and Kaichiro Manabe from December 1909 to January 1910 at hot springs in Iizaka in Fukushima and Kinosaki in Hyogo prefectures.

It is no exaggeration to say that the research of the environmental radioactivity started with the measurement of the radon concentration in hot spring water.

2. Mechanism of radon release and transport in the ground

The physical process of radon release from the structure of the rocks is not yet fully understood. Only a part of radon atoms produced by α decay in a particle is released to the surrounding water or air. The fraction of the radon atoms formed in a solid which makes its way into the pores of the medium, and thereby becomes amenable to transport, is defined as the emanating power of the solid for radon[1].

The trasport and release of radon from a solid into air or water occurs through diffusion and flow of the air or water. High porosity increase the diffusion rate. In a very dry solid the release of radon is reduces by readsoption of radon atoms on surfaces in pores and fractures in the solid. If the solid is slightly moist, the radon release is enhanced up to a certain moisture content, above which it decreases again because of the lower diffusion rate in water-filled pores[2].

Once radon have entered the surrounding water or air phase, they are further transported by diffusion, mechanical and connective flow, and by percolation. The transport of radon through soil and bedrock by water probably depends mainly on the percolation of water through the pores and along fracture planes of the bedrock. In the main,however, the transport of radon through soil and bedrock by water probably depends on the percolation of water through the pores and along fracure planes of the bedrock.

The radon content in ground water and water in springs and wells is seen to depend on the radium concentration in the rock of the aquifer. Consequently, the highest radon concentrations are found in water in the vicinity of uranium ore bodies. High concentrations(>400 kBq/m^3) are also found in wells bored in granite, pegmatite and porphyry. Low radon concentrations occur in ground water from basic rocks and sedimentary rocks, such as limestone and sandstone[3].

3. Determination method of radon concentration in water

3.1 Toluene extraction-LSC method

Radon is easily dissolved in organic solvents including toluene as much as 45 timea in water(20℃).

The present method consisted firstly developed in Japan consisted of an extraction of radon into toluene containing scintillators, and a radioactivity measurement with the liquid scintillation counter (LSC) [4],[5],[6]. Scintillation cocktail(LS) was made from a 4 g of PPO and a 0.1 g of POPOP dissolved in a 1 l of toluene.

The radon extracted into LS decays in the counting vial and, after 5 nuclear transformations, reaches the long-lived daughter nuclide, ^{210}Pb. In a LSC 3 α + 2 β rays are counted after the radioactive equilibrium with short-lived daughters was established (aproximately 3.2 h after the extraction). The counting efficiency for these 5 radiations are estimated to be almost 100 % when the integral counting technique[7],[8],[9] is applied. The detection limit of this extraction method, if no other α or β contaminants are present, is about 0.02 Bq/l for a 1 liter water sample.

This method has advantages that 1)it is easily operated in the field, 2) the lower measureing limit is low and 3) many samples can be treated at a time. Because of this, it has rapidly spread. The method is now employed in the Guideline of Analytical Method of Mineral Spring of Japan. The method effective for the determination of small quantity of radon existing in the natural environment.

Toluene extraction-LSC method is the best method for determining radon concentration in water.

3.2 Direct LS counting method

Direct water system is commercially designed mainly determining for drinking water of the individual homes in the United States[10]. This is a direct LS counting method ; 10 ml of water sample and 10 ml of commercially ready-for-use LS cocktail placed in a 22 ml of counting vial and counted radioactivity in a LSC. The detection limit is about 4 Bq/l for a 10 ml water sample[11]. This direct method will be able to simple measurment of radon concentration in natural water samples applying integral counting technique.

4. Radon as hydrological tracer

Since, in general, radon concentration in ground water is lower than 10^{-10} ppm, and moreover, radon belongs to the rare gas elements which have lowest possibility of chemical reaction, the disturbance of the concentration caused by the chemical reaction with the rocks and soils is negligible. Therefore, from the variation of radon concentration due to ground water movement, the system and mechanism of ground water movement can be analyzed easily.

The change in the radon concentration of the ground water within the dam body or in the rock beds due to the impoundment of the dam is thought to have two characteristics tendencies.

The one is the increasing phenomenon of the radon concentration, owing the change of the retention water into the vadose water. The radon has such a short decay condition that is only 3.8 days half life, and the diffusion coefficient in the water is as small as 1.13 x 10^{-5} cm/s. Therefor, the radon concentration in the natural ground water is smaller in the moving vadose water which lies far from the surfaces of the rocks and soils than in the retention water which lies near the surfaces of the rocks and soils. In Japan, in most cases, this ratio is approximately 1 : 3. For this reason, the phenomenon forerunning the dam leakage, that is the phenomenon that the retention water under the natural conditions in the rock beds moves under the artificial condition of the impoundment, is shown by the increase in the radon concentration of the ground water in the rock beds.

4.1 An example of analysis of vertical hydrogeological structure

The analysis of the Nanao, in Ishikawa prefecture, depression zone on the coast of the Sea of Japan is shown in Fig. 1[12]. The distribution of the radon concentration of ground water coincides with the configuration of geological formations. The low concentration zone found at the upper left in the figure shows the old sand dunes that had not been confirmed by the geological techniques. However, the difference of the concentration between the Pliocene and the Diluvium, located at lower right in the figure, is small. Some geologists doubt the clear distinction of the two formations, and the ambiguity of the concentration seems to reflect the doubtful distinction.

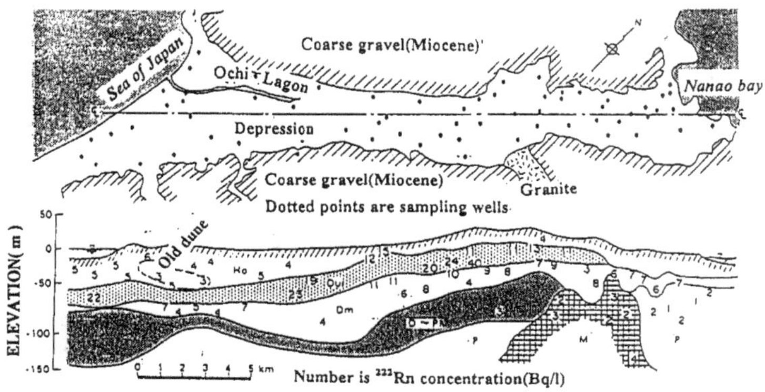

Fig. 1 Analysis of vertical hydrogeological stucture

4.2 Seepage ground water from the lake bottom

At Matunoura beach, on the west shore of the lake Biwa in Shiga prefecture, 14 seepage meters between 5 ~ 45 m were placed from the lakeshore to measure radon concentration, seepage flux and electric conductivity(EC) as a function of distance from lakeshore(Fig. 2). An increase in radon concentration was observed 30 ~ 35 m from the shoreline with little variation at 20 ~ 30 m. Seepage flux followed a broadly similar pattern with an increase over the same distance. EC showed a sharp dicrease between 25 and 35 m, in contrast to the increased radon concentrations observed over the same area. A repeat study carried out one year later at the same site gave the same result, with high radon concentrations beyond 30 m from the shoreline, again

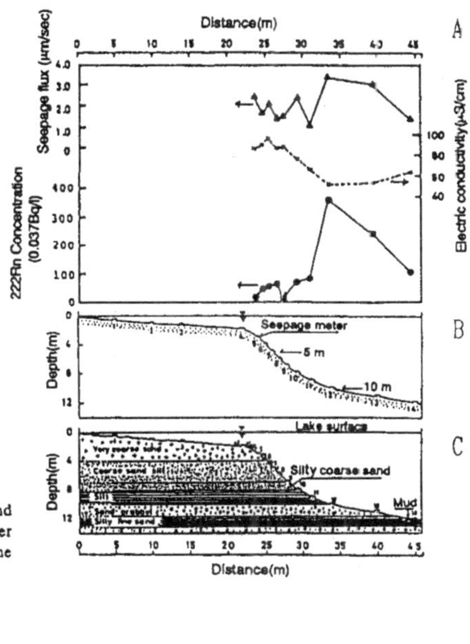

Fig. 2 A. Variation of Rn concentration, EC and seepage flux in seepage water. B. Seepage meter setting places tranced perpendicular to the shoreline of Station 5. C. Stratum profile of Station 5

correlated with high seepage flux and reduced EC. On the basis of stratigraphy of this area, it will be deduced that seepage water entered the lake through the permeable sand gravel from 10 to 12 m below the lake surface. Ground water of this stratum is known to have low EC, and thus lends weight to the hypothesis that high radon concentrations are associated with high seepage flux of ground water into the lake[13].

5. Radioactive hot springs

Radon in hot and mineral spring water is one of the analysis items of hot and mineral spring water, and the measurement has been conducted in two aspects, i.e. research and utilization. As radon is chemically inert, noble gas, it dose not make any complex with other chemical component present in hot and mineral spring water.

5.1. Definition of radioactive hot spring

The Japanese Hot Spring Law defines the radon concentration in hot and mineral spring water as follows[14].

Radon (Rn): 20 x 10^{-10} Ci/l or over : The concentration for distinguishing mineral spring water from common waters.
Radon (Rn): 30 x 10^{-10} Ci/l or over: Therapeutic spring waters containing specific minerals (radioactive hot spring)

When converted into the Bq indication, their concentration is 74 Bq/l and 111 Bq/l respectively. However, mineral springs for medical care are not covered by the Japanese Hot Spring Law, but are considered to be the springs that are expected to have a healing effect in view of experience acquired through the medical science on hot springs[15].

5.2 Features of radioactive springs in Japan as indicated in the hot and mineral springs distribution map

The hot and mineral springs distribution map of Japan[16] summarized recently indicates that the radioactive springs constitute 7. 73% of the total number of springs (springs with identified water types, total number 3,659). The numbers of the radioactive hot springs classified in accordance with prefectures indicate that the majority of the hot springs in Japan are located in the western part of Japan. It is known that radioactive hot springs are discharged from granite areas. This established theory can be substantiated by comparing the radioactive spring distribution map Fig.3 with the granite distribution map Fig. 4[17].

It is well known also that the hot spring phenomenon is closely related to the volcanic activity. Hot springs in the volcanic country of Japan are distributed all over the country, while so-called radioactive springs of high radon concentration are mainly distributed in the western half of the Honshu island. In Hiroshima prefecture where the largest number of springs are located, 89.1% of the springs are

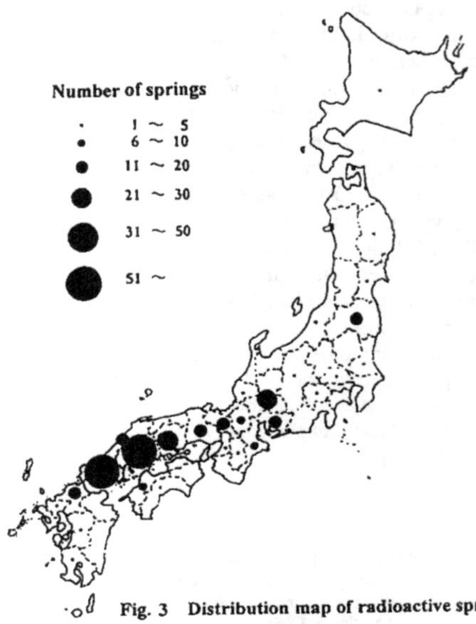

Fig. 3 Distribution map of radioactive springs in each prefecture

radioactive springs. Hiroshima is followed by Yamaguchi prefecture having radioactive springs, 63.2% of the total number of springs, and Fukuoka, Okayama and Gifu, each with radioactive springs of over 30% of the total springs in the respective prefectures. Most of the springs in Hiroshima prefecture are cold mineral springs flowing out from the granite areas, mainly from the matrixes consisting of Hiroshima granite (coarse Biotite granite).

Fig. 5 shows the frequency distribution map for the discharge volume of radioactive springs (l/min). The Environment Agency has reported that, as of March 1996, 25,129 springs (including un-utilized springs) are present in Japan, yielding the total volume of spring water of 2,504,700 l/min, one hundred l/min. per one spring. The type of the radioactive springs in terms of flowing is non-symmetric with the peak value at 20 ~ 50 l/min. Namely, the average volume of the yield of the radioactive springs in Japan seems to be below 50% of that of the other water types of spring.

The temperature of the majority of the spring water (86.5% of the total) is below 25℃, mostly between 16 and 20℃, which is close to the ground water temperature range.

The frequency distribution map (Fig. 6) classifying spring water by their properties indicates that strong acid springs(pH<3) constitute negligible part of the total, weak acid springs 4.3 %, neutral springs 49.4 % as many as a half of the total, weak alkaline springs 27.2 %, and alkaline springs 19.1%. When springs are classified by the chemical ingredients indication, simple weak radon springs constitute 56.1 %, simple radon springs 34.4 % and springs containing the chemical components other than radon (including radon plus other minerals) 9.4 %. The yield of the radioactive springs in Japan, therefore, is 1/2 ~ 1/3 of that of the remaining types of springs and are neutral or weakly alkaline, consisting mainly of

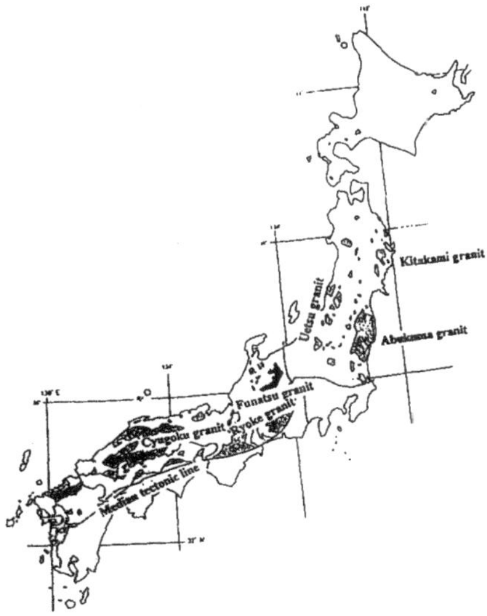

Fig. 4 Distribution map of Paleozoic granit zone cf Japan

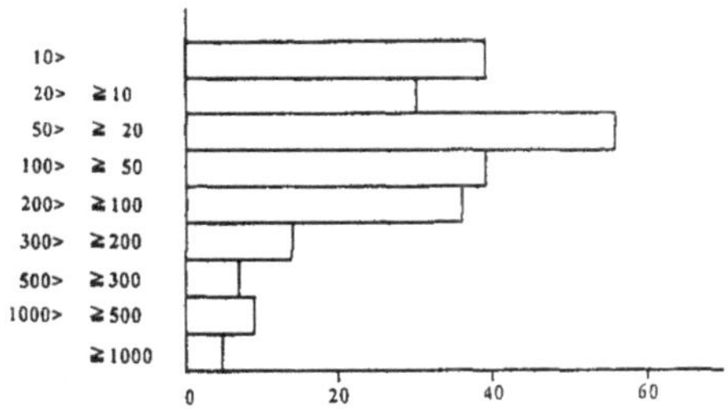

Fig.5 Frequency distribution of Japanese radioactive springs in terms of total discharge(l/mim)

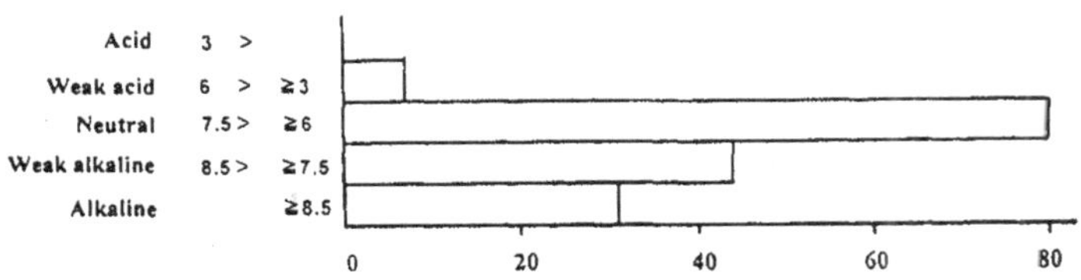

Fig. 6 Frequency distribution of Japanese radioactive springs in terms of pH value

simple cold mineral springs including small amount of soluble chemical components other than radon. Collected from the relevant reports Table 1 shows a typical example of mineral springs classified into three types[18], basing on the idea of radioactive equilibrium conditions of thoron, radon and radium .

1) Radium type: That contains radium only, including almost no radon which is the daughter nuclide of radium.
2) Radon type: That contains radon up to the amount far greater than the equilibrium amount of radium which is the parent nuclide of radon.
3) Thoron type: That contains thoron up to the amount far greater than that of radon.

Table 1 Three types of hot spring

Type	Name of Spring	Ra (Bq/l)	Rn (Bq/l)	Tn (Bq/l)
Radium type	Arima Karatokatakoshi	4.1	—	
	Tamatsukuri Matsunoyu	3.6	—	
	Masutomi Nakanoyu	1.2	—	
Radon type	Ikeda No.8	2.1	29.5	
	Sekigane Tohsenrou	0.06	0.2	
	Gifu Yunoshima No.1	—	5.3	
Thoron type	Misasa Ohhashireisen		1.0	309
	Masutomi Iwaburo		0.04	0.3
	Tamagawa No.4		0.1	3,262

Although the thoron type of the last item is rarely found presently, the number is expected to increase in the future with the advance of the research on thoron. However, in view of the determined examples, containing thoron only, no radon, is unlikely to occur except for extremely specific case.

5.3 Origin of radioactive springs

Although many reports have been made since many years ago on the origin of radioactive springs, none of the reports which show its origin have been established yet. The problem is still in the research stage. The fact that radioactive springs of the 2) and 3) types are abundant indicates that radon has not been created from radium in water, the parental nuclide. Normally, the radon concentration is 2 ~ 3 orders higher than the equilibrium amount of radium. In spite that no large scale mineral deposits of uranium or thorium containing minerals have been found in Japan, mineral springs at Masutomi and Ikeda in Yamanashi and Shimane prefectures are strongly radioactive as very well-known in the world, while the deposit acquired from the springs at Sarugajo, Misasa and Tamagawa in Kagoshima, Shimane and Akita prefecture respectively are reported to contain the large amount of radium which rank at the top or on the second place of the world(Table 2)[19],[20].

Table 2 ^{226}Ra Concentration in Radioactive Presipitation (maximum)

Name of Hot Spring and Main Element	^{226}Ra Concentration in the Sediment (Ci/l)
Sarugajo C (Mn, Fe)	$(64.2 \pm 3.4) \times 10^{-8}$(maximum)
Misasa OTR spring (Mn)	1.18×10^{-8}
Tamagawa (Ba, Pb, SO_4)	1.22×10^{-9}
Tamagawa (Ba, Pb, SO_4)	6.08×10^{-10}
Ikeda No.1 (Fe)	9.50×10^{-10}(maximum)
Ikeda No.2 (Ca,CO_3)	8.72×10^{-10}
Masutomi Tuganero Nakanoyu (Fe)	5.11×10^{-10}
Bad Gastein	5.66×10^{-8}
Ordinary Rock	$10^{-12} \sim 10^{-14}$

5.3.1 Mineral Spring Deposition Theory[19],[21],[22],

Although many of the radioactive springs occur in granite areas, the normal radium content in granite is not sufficient to account for the presence of the strongly radioactive springs. On the other hand, the presence of the condensed radium found in hot spring deposit is considered attributable to the cause of the radon's supply source. When radioactive substances are in contact with granite or spring deposit, only radon is dissolved normally.

Generally speaking, springs with high radon concentration discharge less amount of volume of spring water than that of other water types spring. Considering that the radon content is far greater than the amount of radium existing in the water in the radioactive equilibrium, it is possible that water of very high radon concentration may occur, if relatively small amount of mineral water comes in contact with radioactive deposit of high radium concentration in not-to-deep underground area.

5.3.2 "Radon-rich ground water" theory

Many researches have been made to explain the non-equilibrium condition of radon>radium in spring water mainly in strongly radioactive areas (Masutomi, Misasa and Arima). As the result of the study of the radon providing sources at Misasa, Kinosaki, Yumura and others, it has been clarified that spring water contains initially little amount of radon, but the concentration of the spring water is brought up by ground water of high radon concentration mixing-in[23].

The radon concentration of the public baths in Yamada district, Misasa, changes with seasons. It increases when water is introduced into the paddy field around. The spring deposit layer considered to have been evenly distributed in the area, would not be sufficient to cause to increase of the radon concentration. Ground water, from the depth, of high radon concentration, "radon-rich ground water" plus the ground water passing through the deposit layer of the spring near the ground surface is presumed to be mixed-in to cause the effect[24].

When compared with the amount of radium, far excessive amount of radon is present in the geothermal wells for the power generation and hot springs in Ohdake and Hatchobaru(Table 3). As most of the water in hot springs in Japan is heated ground water[25], it is more adequate to consider that the hot water is the mixture of the high concentration radon ground water with steam and hot water, instead of the magma-originated hot water and steam. The radon brought in from wide area by ground water has been dissolved in the hot water containing CO_2 and other substance at high temperature and high pressure. As the water shoots up from the depth at the speed of the sound, radon would not decay in the process. Instead, as most of radon is moved into the steam when released, the amount of radon remaining in water is minimal.

Table 3 Amount of Rn, Ra in vapour of geothermal power (Condenced water and uncondenced gas)

Geothermal Well	Depth (m)	Casing (m)	Bottom Temp. (℃)	Rn (x0.037Bq/l STP) Uncondenced Gas	Rn (x0.037Bq/l STP) Condenced Water	Ra (x0.037Bq/l) Hot Water	Gas Component (Vol.%) H_2S	CO_2	O_2	N_2
Ohdake No. 8	246	250	190			3.04				
Ohdake No. 9	550	250	150	4256	286.2*	2.71	0.7	87.4	0.9	11.0
Ohdake No.10	600	250	183			2.22				
Hatchobaru No.4				375	21.4**	3.03	4.9	59.4	-0.9	34.8

*Volum ratio of uncondenced gas : condenced water = 1 : 1.06 (27～28℃)
** 〃 〃 〃 = 1 : 1.3 (27～28℃)

6. Radioactive springs in the world

Various types of hot and mineral springs are present in many countries in the world, many of which are used for medical purposes. The same goes with radioactive springs. However, very few reports have been made about the radon concentration in the spring water recently.

Since the discovery of the radioactivity, however, radon concentration is determined in all over the world hot springs and number of radioactive springs have been found(Fig.7). Of presently known main radioactive springs in the world, low temperature and strongly radioactive springs are mainly distributed in the eastern part of Germany, Czecho, Slovakia and Japan, while strongly radioactive springs of the temperature over 37℃ are mainly distributed in Italy, Greece, Japan and Austria(Table 4)[26].

Fig. 7 Distribution map of the hot springs and radioactive springs in Europe

Table 4 Radioactive Spring in the world(Tw : >37℃, Rn concentration : >50nBq/l)

Name of Place	Origin of the Spring	Country	T℃	^{222}Rn(nCi /l)
Lacco Ameno	Sorgenta Sants Restituituta	Italy	59.7	1,019
Lacco Ameno	Fonte Greca	Italy	60	452
Therma	Artemis	Greece	50.4	274
Therma	Apollo	Greece	45.4	203
Therma	Moustapha Lintza	Greece	43.0	127
Kammena Vourla	Metaxa	Greece	38.1	127
Kammena Vourla	Papastoatoy	Greece	41.4	102
Lacco Ameno	Romana	Italy	60	99
Misasa	Gunze Saginoyu	Japan	42.6	92
Therma	Kratsa	Greece	46.8	87
Misasa	Yamadakukyohdoh	Japan	65.0	64
Gastein	Mesnil	Austria	37.0	59
Gastein	Sophien	Austria	38.1	59
Gastein	Reissacher	Austria	39.5	53
Gastein	Elisabeth	Austria	46.3	51

7. Conclusion

Radon in natural water has been utilized in various fields for its unique properties.

In the history of radioactivity since its discovery of 100 years ago, radon went through a variety of situations. Although radioactive springs have been used for the rehabilitation and pain-killing of rheumatism for its therapeutic effect, the term radioactivity tends to give bad impression after the world-war II.

In the 1980s, the radiation hormesis effect[27] based upon the abundant statistic data was introduced, allowing related research to be spread world-wide. The image of radioactive springs is changing along with this trend, and the popularity of the spring of the old days has come back. Of all the environmental radioactivities, we can never be able to keep away from radon which is present everywhere in the natural environment. Relevant researchers are being requested to investigate widely the effectiveness of radon in natural water so that it becomes more acceptable as handy and useful element.

8. References

[1] A.B.Tanner, Radon migration in the ground, A review, in The Natural Raddiation Environment (J.A.S.Adams and W.M.Lowder,eds), 161-190,Univ. Chicago Press, 1964

[2] Megumi, K. and T. Mamuro, Emanation and exhalation of radon and thoron gases from soil particles, J.Geophys.Res., 79, 3357-3360, 1974

[3] Knutsson, G.,Radon in various types of groungwater in south-eastern Sweden: a preliminary report. Striae 4: 55-59, 1977

[4] Noguchi, M., New method of radon activity mesurement with liquid scintillator, RADIOISOTOPES, 13, 362-367, 1964

[5] Horiuchi, K., and Y. Murakami, Basic studies on the new determination method of radon in mineral springs with a liquid scintillation counter, J. Balneol. Soc. Japan, 28, 39-52, 1977

[6] Horiuchi, K., Some remarks on the new determination method of Rn in hot springs by extraction liquid scintillation counting, J. Soc. Engineers for Mineral Spring, 13, 95-103, 1978

[7] Horrocks, D.L., and M.H. Studier, Determination of radioactive noble gases with a liquid scintillator, Anal. Chem., 36, 2077-2079, 1964

[8] Goldstein, G., Absolute liquid-scintillation counting of beta emitters, Nucleonics, 23, 67-69, 1965

[9] Homma, Y., and Y. Murakami, Study on the applicability of the integral counting method for the determination of ^{226}Ra in various sample forms using a liquid scintillation counter, J. Radioanal. Chem., 36, 173-184, 1977 R

[10] Kessler, M.J., Analysis of radon in air and water, The 2nd International Seminar for Liquid Scintillation Analysis Proceeding, 43-67, June 8-9 Tokyo, 1988

[11] Horiuchi, K., Method of determining radon concentration in natural water by liquid scintillation counting, Int. Conference on advances in LSC 1994, Proceeding, 127-135, 1994

[12] Kimura, S., and T. Komae, Applications of Environmental radon-222 to some cases of water circulation, Natural Radiation Environment III vol. 1 U.S.Department of energy 581-598, 1980

[13] Horiuchi, K., and Y. Murakami, Lake-shore groundwater seepage is characterized by its radon concentration, at lake Biwa, Envronmental Science, Report of Research B289-R12-2,91-106, 1986

[14] Environment Agency, Kosen bunsekihou sisin (kaitei), 1-75, 1978

[15] Murakami, Y., Thermalism and health resorts, Forum'89, 82-93, 1989

[16] Kimbara, K., Distribution map and catalogue of hot and mineral springs in Japan, Geological survey of Japan, 394, 1992

[17] Kimura, T., I. Hayami and S. Yoshida, Geology of Japan, University of Tokyo Press, 149, 1993

[18] Genser, C., Zeit. Dent. Geol. Gesell., 482, 1933

[19] Saito, N., Radioactive sinter deposit, J. Balneol. Soc. Japan, 27, 2/3, 31-35, 1976

[20] Sato, J., O., Yokozawa and N. Saito, Radioactive nuclides in the thermal spring water and its sinter deposits from Sarugajo, Tarumizu city, Kagoshima prececture, J. Soc. Engineers for Mineral Spring, 10, 47-60, 1975

[21] Iwasaki, I., J.Soc. Engineers for Mineral Spring,6, 18-28, 112-114, 165-168, 1968, 7, 16-24, 109-114, 1969

[22] Kikkawa, K., Study on radioactive springs, Jap. J. Geophys., 1, 1-25, 1954

[23] Okabe, S., On same relations between the hot spring and radioactivity, Mem. Coll. Sci., Univ. Kyoto, Ser. A, XXVIII, 39-71, 1956

[24] Horiuchi, K., and Y. Murakami, A new explanation on the presence of excessive amount of radon in mineral spring by the δ Rn term calculated from the amounts of radium and radon, Chikyukagaku(Geochemistry), 12, 59-70, 1978

[25] Fukutomi, T., a review, J.Balneol. Soc. Japan, 29, 88-102, 1978

[26] Mifune, M., Radioactive springs in Greece, J. Soc. Engineers for Mineral Spring, 19, 1-6, 1981

[27] Lucky, T.D., Hormesis with ionizing radiation, CRC Press,1980, Physiologicalbenefits from low levels of ionization radiation, Health Physics, 43, 771-789, 1982

Radon Concentrations in Neitaping Cold Spring

Cheng - Hsin Mao and Pao - Shan Weng*

Nuclear Science and Technology Development Center
and
*Department of Nuclear Science
National Tsing Hua University
Hsinchu, Taiwan 300, R.O.C.

Abstract

Neitaping cold spring is one of the recreation attractions in the northwestern part of Taiwan. It is located in Peipu Village of Hsinchu County about one-hour drive from the National Tsing Hua University. In summer time many tourists enjoy swimming in a small pool specially constructed for recreation purpose. There are several water sources situated in different places around the pool. The cold water flows into a creek called Shangping. All samples were taken from the sources, pool, and along the creek for a period starting in December 1996 and finished sampling in July 1997. During the eight-month monitoring period, electret ion chambers (EICs) have extensively been used.

1. Introduction

The island of Taiwan, is full of hot springs and also has a few cold ones. The hot-spring as well as cold-spring spa water is extensively used in public and private bathrooms and is distributed by pipelines to many hotels and houses. In the past, radioactivity, including radon concentration, has been investigated in some of these springs.[1-4]

Radon dissolves into groundwater from rocks or soils. When the water is exposed to the atmosphere, some of the dissolved radon is released. A similar process occurs in cold-spring spa water. Higher radon levels have been observed in individual rooms where water is heated or agitated, such as during shower use.

Several different methods of measuring radon in water have been developed and used; however, for routine waterborne radon monitoring, including use in field conditions, the technique based on electret ion chamber technology[5] seems to be the most suitable choice in many applications.

The goal of this investigation is to show how the electret ion chamber method can be used as a simple tool for waterborne radon monitoring, and to present the results of monitoring carried out in a cold spring called Neitaping which is located at Hsinchu county about an hour drive from National Tsing Hua University (NTHU) as shown in Fig.1. The spa water contains carbonate and is good for skin disease and gastrointentinal disorders either by bathing or drinking.

2. Experiment

The commercially available E-PERM (electret – passive environmental Rn monitor) is an integrated system for radon and radon-related measurements. It is based on the use of a digital electrostatic voltmeter to measure the difference in potential on the surface of an electret in response to the charge collected due to the ionization of air by radiation. An electret is a material specially manufactured to maintain a sufficiently suitable electrostatic charge. When the electret is housed in an electrically conducting chamber, it can efficiently collect the ion pairs that result from the interaction of radiation with air inside the chamber. The collection of ions opposite in polarity to the potential of the electret reduces its electrostatic voltage in direct proportion to the total ionization in the chamber.

Electret ion chambers were used for monitoring waterborne radon. This technique relies on the measurement of radon in air above a water sample enclosed in a sealed vessel. The concentration of airborne radon released from water is determined by means of the electret ion chamber method, i.e., using a short-term electret (STE) and a standard ion chamber (SIC), which is described elsewhere.[6]

The average radon concentration in air, C_a in Bq m^{-3}, based on a measurement using the monitoring arrangement with an STC and SIC, according to the manufacturer's recommendations (Rad Elec. Inc., U.S.A.), can be calculated as :

$$C_a = \frac{V_1 - V_2}{F_r T} - F_g X, \qquad (1)$$

where

V_1 = initial electret surface potential (V),

V_2 = final electret surface potential (V),

F_r = radon concentration factor (V day^{-1} per Bq m^{-3}),

T = period of measurement (day),

F_g = γ-correction factor (Bq m^{-3} per A kg^{-1}),

X = exposure rate of the ambient γ-field (A kg^{-1}).

The radon correction factor F_r is given by the experimentally derived formula :

$$F_r = A + BV_{AV}, \qquad (2)$$

where A and B are calibration constants depending on the type of electret and ion chamber used, and the voltage V_{AV} represents an average value of V_1 and V_2.

The manufacturer recommends using the following value of the relevant factors : $A=5.1 \times 10^{-2}$ Bq^{-1} m^3 V day^{-1}, $B=1.72 \times 10^{-5}$ Bq^{-1} m^3 V day^{-1} and $F_r=2.27 \times 10^{-13}$ Bq^{-1} m^{-3} A^{-1} kg. The original values mentioned above were expressed in conventional units, which were then converted into SI units. The electret, after their voltage drop below 150 V, are returned to the producer for recharging, or the user purchases new electret for further monitoring.

The water samples were collected in small standard-size bottles having a volume of 144 mL each. The sample was then carefully transferred into a radon leak-tight 4 L glass jar, where the average equilibrium ^{222}Rn concentration in the air phase using a standard electret ion chamber was determined. The analyses were performed immediately at the sampling site. From this airborne radon concentration the concentration of the dissolved ^{222}Rn in the original water sample was calculated.

The waterborne radon concentration C_w at the time of sample collection was determined using the following formula :

$$C_w = \frac{C_a \lambda T(V_a / V_w + O_s)}{[\exp(-\lambda D)][1-\exp(-\lambda T)]}, \qquad (3)$$

where

C_a = radon concentration in air in the jar (Bq m^{-3}),

λ = decay constant of ^{222}Rn (day^{-1}),

T = analysis time (day),

V_a = volume of air in the analysis jar (L),

V_w = volume of water in the analysis jar (L),

O_s = Ostwald coefficient,

D = decay time between the collection and the beginning of the analysis (day).

Twelve water samples were taken according to the terrain and the water eruption sources. Right in front of a small concrete pool served as a water reservoir, there is an eruption spot, so two samples were taken as I and J and are shown in Fig.2. There is one large eruption spot in the pool; four samples were taken in the pool as A,B,C and D. The overflow of pool water discharges to the Shangping Creek through pipe outlets. Four samples were taken along this creek since the creek water is used in private homes. They were labeled as E,F,G and H. In the downstream beyond a suspension bridge about 60 m from the pool, a small eruption spot is located yet with high radon concentration; two samples at K and L were taken.

3. Results and Discussion

The composition of the cold spring spa water was analyzed with an ICPMASS and the results are shown in Table 1. Rather high concentrations of Cl, Mg, S and Ca were found, where Mg and Ca are in the same family II as Ra, a precursor of Rn.

Table 1 Element content at sampling site K

Element	Content (ppm)
Li	2.48
Be	0.00
Na	5.40
Mg	174.30
Si	21.54
S	166.10
Cl	3438.00
K	4.86
Ca	50.40
Cr	0.24
Sc	0.01
Ti	0.05
Ba	1.79
Br	10.31
Sr	2.80
Au	0.10

Of all the samples taken, the highest radon concentration could reach 25.16 kBq m^{-3} as detected at site K in July 1997 as shown in Table 2. The lowest one was found at site I in December 1996 as a result of larger water flow in the Shangping Creek. Most radon concentrations were in the range of 0.4 - 3 kBq m^{-3}.

Table 2 Radon concentrations (k Bq m^{-3}) at sampling sites during 1996-97

	A	B	C	D	E	F	G	H	I	J	K	L
Dec 96	1.540	7.071	0.486	0.961	1.620	2.321	1.724	1.231	0.316	2.132	24.017	23.865
Jan 97	1.958	2.335	2.801	2.387	2.050	4.910	3.100	1.834	0.273	2.018	24.632	21.792
Feb 97	2.490	2.501	2.507	2.505	0.755	1.165	2.780	1.168	0.198	2.418	23.815	22.463
Mar 97	4.215	2.432	2.367	2.403	0.731	1.170	2.215	1.216	0.324	2.521	24.263	23.161
Apr 97	14.166	2.475	2.253	2.017	0.702	1.161	1.547	1.384	0.229	2.460	24.232	21.174
May 97	3.489	8.650	2.900	3.010	0.810	1.320	1.780	1.251	0.320	2.180	23.790	21.587
Jan 97	4.310	7.420	2.138	4.213	1.314	2.160	2.012	1.483	0.280	2.251	25.161	23.391
Jul 97	3.800	7.130	3.241	3.850	2.760	1.741	2.316	1.316	0.333	2.370	24.390	22.281

4. References

[1] Tsai, C.M., and P.S. Weng, Radium-226 concentrations in Taiwan hot springs and rivers, Hlth Phys. 24, 429-430, 1973

[2] Sabol, J., Weng, P.S. and C.H. Mao, Waterborne radon in some hot springs in Taiwan, Nucl. Sci. J. 30, 131-136, 1993

[3] Sabol, J., Weng, P.S., and C.H. Mao, Monitoring of ^{222}Rn in Taiwanese hot spring spa waters using a modified electret ion chamber method., Hlth Phys. 68, 100-104, 1995

[4] Weng, P.S., and C.L. Chen, Radon concentrations in spa water taken from hot and cold springs in Taiwan, Appl. Radiat. Isot. 46, 293-295, 1995

[5] Kotrappa, P., and W.A. Jester, Electret ion chamber radon monitors measure dissolved ^{222}Rn in water. Hlth Phys. 64, 397-405, 1993

[6] Kotrappa, P., Dempsey, J.C., Stieff, L.R., and R.W. Ramsey, A practical electret – passive environmental Rn monitor, Hlth Phys. 58, 461-467, 1990

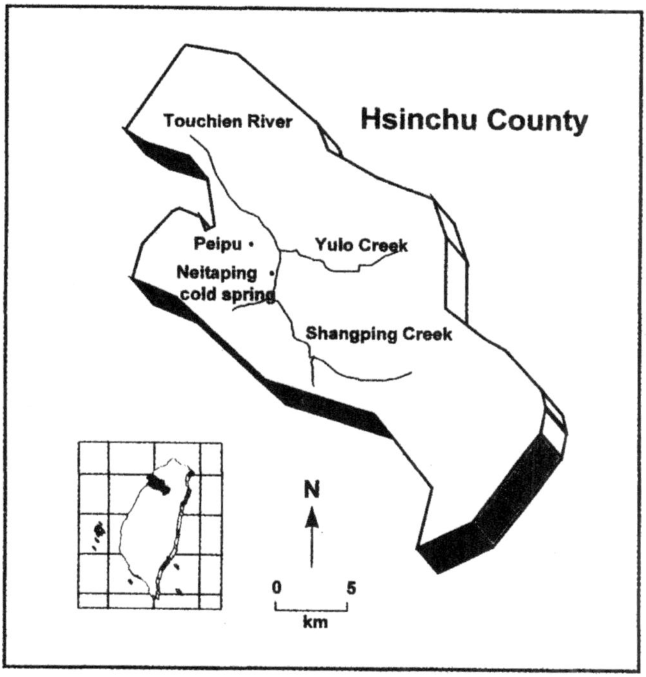

Fig.1 Map showing the location of Neitaping cold spring

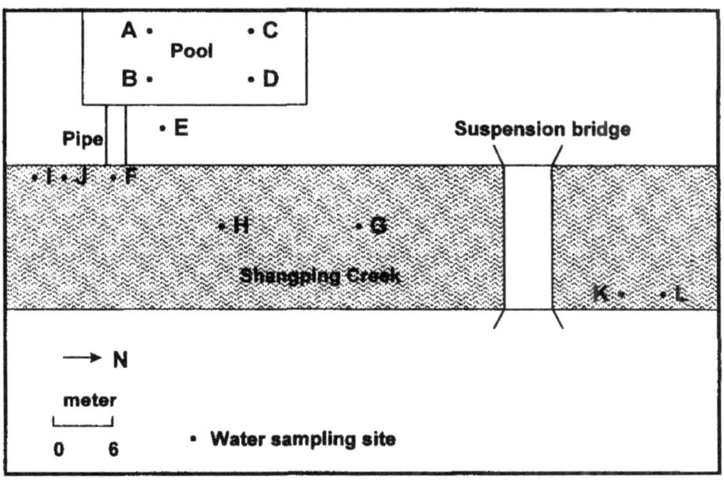

Fig.2 Water sampling sites at Neitaping cold spring

Radon concentrations in well water in Sichuan Province, China

Yibin Chen, Wu Qun, Zhang Bo, Daifu Chen

Sichuan Institute of Labor Hygiene, Chengdu,
Sichuan, 610041, China.

Abstract

There are 110 million people in Sichuan Province, China. Although most of the people in cities of Sichuan use river water, which contains low levels of radon, as potable water, people in countryside and in some communities of big cities still use well water as domestic consumption. This paper reports the radon concentrations in well water investigated in four cities, i. e. Chengdu, Chongqing, Leshan and Leijiang in Sichuan Province. Of the 80 wells investigated, the radon concentrations range from 3.5 to 181.6 $KBqm^{-3}$. Of the four cities, Chongqing has the highest well water radon concentration with the average 49.6 ± 54.1 $KBqm^{-3}$ and the greatest variation. The investigation in four cities showed that the radon concentrations in well water are much higher than that in tap-water. In Chongqing where there are complex geological structures, mainly granite stratum, for example, the average radon concentration in well water is 112 times higher than that in the tap-water, and even much higher than that in river water in Yangtse River, Jialing River, Jinsha River and Mingjiang River. The population in four cities is about one sixth of the total population in Sichuan Province. Because of the common use of well water and the high radon concentrations in well water in Sichuan Province, the health effect of radon in well water to the public should be stressed.

1. Introduction

Levels and the detriment of aerial radon have been widely investigated throughout the world and also there are some reports on radon concentrations in water in countries and districts where underground water was their water source[1-2]. A survey of radon contents in potable water in Shenyang, China has been conducted[3]. Another study showed that the estimated stomach dose equivalent of the residents of a countryside in Jiangxi Province, China, was about 20-42 times as much as the dose equivalent limit for the general public as result of the consumption of the water containing high levels of radon[4]. Researchers are more and more concerned over the health influence of water radon.

Sichuan Province, where there are 110 million people, is a large province of China. Although most of the people in cities of Sichuan use river water, which contains low levels of radon, as potable water, people in countryside and in some communities of big cities still use well water as domestic consumption. In order to know the radon levels in well water in Sichuan Province, we conducted the investigation of the water radon concentrations in 80 water wells in urban and suburban areas in four cities, i. e. Chengdu, Chongqing, Leshan and Leijiang.

2. Methods

(1) Each selected water well for sampling in urban areas in Chengdu, Chongqing, Leshan and Leijiang should supply water to at least 300 people. In suburban areas, each selected water well should supply water to at least 100 people in its village.

(2) In order to make the sampling representative, we take into account of the topographical features in dot design. We collected one third of samples in plains, mountaineous areas and hilly areas, respectively. Considering the effect of geological structure, we collected one third of samples in granite stratum, in quicksand stratum and in carbonatite stratum, respectively.

(3) Sampling method: 50ml water was sampled with a 100ml diffusion bottle, which have two branch tubes. The nitrogen gas was used to drive out the radon in water into the scintillation bottle, and the radon was measured three hours later. If the sample could not be sent to laboratory, the measured results were corrected according to the function $A = A_o e^{-\lambda t}$.

(4) Measure of Radon concentrations: A ZYW-8501 Radon Tester, which had a background of 0.1 cpm, and the K value is 19.24 Bq cpm^{-1}, was used to measure radon. The radon concentrations in water were calculated as follows.

$$C_{Rn} = \frac{K(I - I_b)}{V \cdot \eta}$$

Where C_{Rn} is the radon concentrations in water($Bq.L^{-1}$)

K is the calibration coefficient of the scintillation bottles

I is the counting rate of the samples(cpm)

I_b is the background counting rate(cpm)

V is the volume of the sampled water(L)

η is the transfer coefficient

The tester was calibrated with the liquid standard radium source provided by Radiochemical Center, UK.

3. Results and Discussions

(1)Radon concentrations in well water in Chengdu

Radon concentrations and the range in well water in Chengdu were shown in table 1.

Table 1. Radon concentrations in well water in Chengdu(KBqm^{-3})

Sampling place	Numbers of wells sampled	Range	Arithmetic mean
Eastern part	5	10.0±0.3~31.1±1.3	19.3±7.9
Western part	5	15.0±0.2~29.4±0.8	22.1±6.5
Southern part	10	5.9±0.2~59.1±3.3	27.4±17.6
Northern part	5	11.1±0.5~34.5±0.5	25.3±11.2
Mean in chengdu	25	5.9±0.2~59.1±3.3	24.3±12.8

It is shown in table 1 that the radon concentrations in well water in Chengdu ranged from 5.9 to 59.1 KBqm^{-3} with an arithmetic mean 24.3±12.8 KBqm^{-3}, which is a little lower than that in U.S.A., West Germany[6]. Chengdu is located in a silt plain which has little variation of geological structure, so there is low variation of radon concentrations in well water in different parts of Chengdu.

(2)Radon concentrations in well water in Chongqing

Radon concentrations in well water and the range were shown in Table 2.

Table 2. Radon concentrations in well water in Chongqing(KBqm^{-3})

Sampling place	Numbers of wells sampled	Range	Arithmetic mean
North bank part	3	4.1±0.1~85.1±7.0	33.1±45.1
South bank part	3	3.5±0.2~64.9±6.3	24.1±35.3
Central part	5	9.5±1.4~122.9±14.1	53.4±43.2
Shaping part	5	14.9±1.8~181.6±10.8	96.3±69.7
Baxian county	3	3.5±0.1~16.0±2.4	7.7±7.1
Mean in Chongqing	19	3.5±0.1~181.6±10.8	49.6±54.1

Table 2 shown that the radon concentrations in well water in Chongqing ranged from 3.5±0.1 to

181.6 ± 10.8 KBqm^{-3}.with a arithmetic mean 49.6 ± 54.1 KBqm^{-3}. The area of Chongqing is far larger than that of Chengdu, and the geological structure is complex in Chonging. Baxian county is a hilly land where the stratum is the carbonatite silt, so there is the lowest mean radon concentration(7.7 ± 7.2KBqm^{-3}) in well water in Baxian county. But the stratum in Shaping part and central part of Chongqing is granite stratum, so there are higher well water radon concentrations. The mean radon concentrations in well water in Shaping part is 96.3 ± 69.7KBqm^{-3} which is much higher than that(54.6KBqm^{-3})in Shenyang[3], where the water source is underground water.

(3)Well water radon concentrations in Leshan

Leshan city, which is located in Southern West side of Chengdu, 160km away, is a famous tourist city. The topography in Leshan is a mixture of hilly and mountainous land. The radon concentrations in each county of Leshan are listed in table 3.

Table 3. Radon concentrations in well water in Leshan(KBqm^{-3})

Sampling place	Numbers of wells sampled	Range	Arithmetic mean
Central part	10	3.8~46.8	18.1 ± 13.0
Shawan part	6	6.6~31.4	20.3 ± 8.1
Meishan county	3	7.8~24.3	14.0 ± 9.0
Jiajiang county	3	10.0~31.5	18.7 ± 11.3
Pengshan county	2	5.2~22.3	13.7 ± 12.1
Emei county	2	6.2~6.8	6.5 ± 0.4
Mean in Leshan	26	3.8~46.8	17.0 ± 10.5

(4)Radon Concentrations in well water in Leijiang.

The radon concentrations in well water in Leijiang, a hilly district, are shown in table 4.

Table 4. Radon concentrations in well water in Leijiang(KBqm^{-3})

Sampling place	Numbers of wells sampled	Range	Arithmetic mean
Shuguang village	3	22.9 ± 0.1~53.3 ± 0.4	37.2 ± 15.3
Xieli village	4	22.6 ± 0.3~60.3 ± 0.5	41.9 ± 18.8
Shujia village	3	31.2 ± 2.5~59.3 ± 1.2	40.9 ± 16.0
Mean in Leijiang	10	22.6 ± 0.3~60.3 ± 0.5	40.2 ± 15.2

4. Conclusions

(1) The arithmetic mean of radon concentrations in well water in Chengdu, Chongqing, Leshan and Leijiang is 24.3 ± 12.8, 49.6 ± 54.1, 17.0 ± 10.5 and 40.2 ± 15.2 KBq. m^{-3}, respectively. The provincial mean is 32.8 ± 23.2 KBq. m^{-3}. Of the districts investigated, the highest well water radon concentration, which is 112 times higher than that in river water, exists in the areas with granite stratum in Chongqing. The highest concentration is near the weighted mean 56 $KBqm^{-3}$ in Finland[6] . The well water radon concentrations in carbonatite stratum areas are much lower than that in granite stratum areas.

(2) Sichuan is a large agricultural province, where there are about 75% of farmers who use well water as a domestic consumption. According to the formula and factors recommended in reference[6] , and suppose that 0.5L water was consumpted each day per person and the well water radon concentration is the provincial mean. The estimated annual effective dose equivalent from consumption of well water is about 0. 015mSv. The estimated value has several influential factors which are not discussed here.

5. References

[1] Asikainen M, et al., Natural radioactivity of drinking water in Finland, Health Phys., 1980, 36(1):77.

[2] Dana P. Loomis, The relationship between water system size and ^{222}Rnconcentration in North Carolina Public water supplies, Health Phys. , 1987, 52(1):69.

[3] Shixian Zhou, et al., Survey of radon contents of potable water in Shenyang, Chinese Journal of Radiological Medicine and Protection, 1989, 9(3):196

[4] Qinqian Li, et al., Investigation on the health influence of consumption of the water containing high levels of radon, Radiation Protection, 1981, 1(4):56.

[5] Yibin Chen, Radon content in drinking water in 71 cities of China, Proceedings of the 6th International Conference on Indoor Air Quality and Climate, Vol 4: 453. Finland, 1993.

[6] UNSCEAR, Ionizing radiation sources and biological effects, 1982, 190 – 233.

Long-lived Radon Daughters, ^{210}Pb and ^{210}Po, in Volcanic Products and Hot and Mineral Spring Waters

M. Yamamoto*, K. Komura* and M. Sakanoue**

* Low Level Radioactivity Laboratory, Kanazawa University
Tatsunokuchi, Ishikawa 923-12, Japan
**Professor Emeritus, Kanazawa University
Miyano-cho 14-20, Takarazuka 995, Japan

Abstract

Lead-210 and ^{210}Po in volcanic gases and sulfur sublimates have been measured for several volcanoes in Hokkaido and kyushu districts. Furthermore, more than 40 water samples in mineral and hot springs were subjected to the analysis of ^{210}Pb and ^{210}Po, along with other natural radionuclides ^{238}U(^{234}U), ^{226}Ra (^{228}Ra) and ^{222}Rn. The sources of ^{210}Pb and ^{210}Po with regards to volcanic origin and behaviors of supply and removal of these nuclide were discussed based on the analytical date.

In volcanic samples, excess of ^{210}Po was observed and the results indicated that the concentrations of ^{210}Pb and ^{210}Po and the activity ratios of ^{210}Po/^{210}Pb in volcanic gases and also in sulfur sublimates depended highly on the temperature of fumarolic gases. Wide variations in ^{210}Po in ^{210}Pb and ^{210}Po concentrations have been observed in sulfur sublimates. On the other hand, even for hot springs in volcanic area the concentrations of ^{210}Po in the water samples were not so high. It was considered that ^{210}Po from volcanic gases was easily removed from water to the surface of rocks and soils, as suggested by the data of sulfur sublimates, The concentrations of ^{210}Pb and also ^{210}Po were, however, found to be higher in the springs of high ^{222}Rn concentrations, indicating that ^{210}Pb (and ^{210}Po) in springs water may be useful as a guide to elucidate the origin, ages of water sources and of the history of ^{222}Rn in the springs.

I. Introduction

As daughter nuclides of gaseous radioactive nuclide ^{222}Rn, ^{210}Pb ($T_{1/2}$=22.3 y, β^-) and ^{210}Po ($T_{1/2}$=138.38 d, α) exist widely in the environment. In recent studies[1-6], it has been observed that the significant quantities of these nuclides are released from active volcanoes as a volcanic gases. In Japan there are many active volcanoes, and hot and mineral springs closely relating to them. Information on ^{210}Pb and ^{210}Po both released from active volcanoes and in hot and mineral spring waters there is, however, very limited[7-9]. With regards to these volcanic origin ^{210}Pb and ^{210}Po, it is interesting to study the behavior of both radionuclides in the water samples of hot and mineral springs from geochemical point of view.

In this paper, ^{210}Pb and ^{210}Po contents in volcanic gas condensates and sulfur sublimates have been measured for several volcanoes in Hokkaido and Kyushu Districts where we can approach to the outlets of gases [10]. Furthermore, more than 40 water samples in hot and mineral springs, some of which are situated in the volcanic mountain areas and/or contain high ^{222}Rn concentrations, were subjected to the analysis of ^{210}Pb and ^{210}Po [11].

On the basis of these data, some geochemical considerations about ^{210}Pb and ^{210}Po in active volcanoes and spring waters are made in addition to the data on ^{238}U, ^{226}Ra and ^{222}Rn concentrations measured simultaneously.

2. Sample collection and analysis

2.1 Volcanic gases and sulfur sublimates

Sampling locations of volcanic gases and sulfur sublimates are shown in map of Japan (Fig.1). The samples were collected during 1989 at active volcanoes from two districts : Mt. Usu, Mt. Tarumae, Mt. Tokachi and Mt. Me'Akan in Hokkaido and Mt. Kujyu and Ebino Plateau (Mt. Kirishima) in Kyushu. For the volcanic gases, the condensate water (50~100 ml) was sampled in a polyethylene bottle by cooling volcanic gas from fumarole using cooling jacket by water or cooling by air using long titanium pipe. The temperature of each fumarole was measured by thermocouple. At the same time, the sulfur sublimate was also collected near the fumarole which condensate water was sampled.

Polonium-210 contents in these samples were determined by alpha-ray spectrometry after chemical separation, using ^{209}Po as a yield tracer. On the other hand, ^{210}Pb was indirectly evaluated by determining the ingrowth amount of its daughter, ^{210}Po. Details of the method are described elsewhere [10].

2.2 Hot and mineral spring waters

The major objective of the present paper is to assess the contents of ^{210}Po and ^{210}Pb in hot and mineral spring waters. However, other nuclides such as uranium isotopes (^{238}U and ^{234}U), ^{226}Ra (^{228}Ra) and ^{222}Rn were also determined by using the same sample to surve as an aid to resolve the mechanisms of supply and removal of ^{210}Po and ^{210}Pb nuclides.

2.2.1 Sequential determination of ^{210}Po, ^{210}Pb and U isotopes

The water samples in hot and mineral springs were collected from more than 40 springs as shown in Fig.1. A sample was collected of outlet at the springs or as near to the the well as possible. Water samples were collected in a 10~20 l (mostly 20 l) polyethylene bottle and about 200 ml of conc.HCl was added immediately after the sampling. After bring back to the laboratory, a 10~20 l aliquots were subjected to chemical analyses for Po, Pb and U, using ^{209}Po ($T_{1/2}$=102 y, α), stable Pb (50 mg per sample : this carrier was prepared from the old ^{210}Pb-free lead metal) and ^{232}U ($T_{1/2}$=71.7 y, α) as a yield tracer for each element . The chemical procedures used were developed previously by Yamamoto et al. and have been reported elsewhere, together with the precision and accuracy of the method [11].

Most of samples studied, except samples collected near our laboratory, were not measured immediately for ^{210}Pb and ^{210}Po after the sampling. Therefore, it is necessary to correct the measured ^{210}Pb and ^{210}Po concentrations for build-up from ^{222}Rn (for samples of high ^{222}Rn content) and for decay of ^{210}Po during storage.

In case of samples containing very low ^{222}Rn content or in which most of ^{222}Rn presented originally was removed in advance by bubbling Rn-free air through the sample , the activities of the ^{210}Pb and ^{210}Po at the time of collection, *Pb(0)* and *Po(0)*, as derived from the Beteman equation (Fig. 2) are calculated as follows:

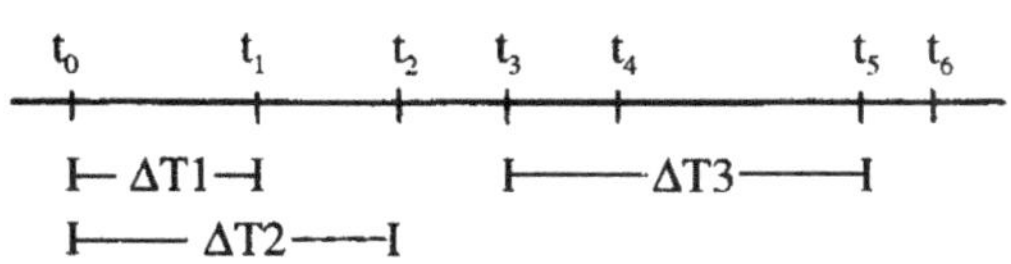

t_0 = Sampling date
t_1 = Starting date of analysis
t_2 = Date of preparation of counting source for Po-210
t_3 = Date of anion exchange treatment
t_4 = Date of measurement of Po-210
t_5 = Date of preparation of counting source for ingrowth Po-210
t_6 = Date of measurement of ingrowth Po-210

$$Pb(0) = Po(2)/G\{Po(\Delta T3)/Pb(0)\}/Y \quad \text{------------------------------}(1)$$

$$Po(0) = (Po(1)-G\{Po(\Delta T2)/Pb(0)\}Pb(0))/D\{Po(\Delta T2)\} \quad \text{--------}(2)$$

Po(1) = Concentration of Po-210 at t_2
Po(2) = Concentration of ingrowth Po-210 at t_5
Y = Chemical yield of Pb-210

In case of samples of high ^{222}Ra content (Fig. 3), the following equations are applied for calculating the content of ^{210}Pb and ^{210}Po, *Pb'(0)* and *Po'(0)* .

$$Pb'(0) = Pb(0) - G\{Pb(\Delta T1)/Rn(0)\}\ Pb(0) \quad \text{-----(3)}$$
$$Po'(0) = (Po(1)-G\{Po(\Delta T1)/Rn(0)\ Pn(0) - G\{Po(\Delta T2)/Pb(0)Pb'(0)\ /D(Po(\Delta T2) \quad \text{-----(4)}$$

Rn(0) = Concentration of Rn-222 at the date of sampling.

2.2.2 Determination of ^{226}Ra and ^{228}Ra

A 5 l of water samples were acidified to about pH 1 with conc. HCl before co-precipitation of radium nuclides by a $BaSO_4$ precipitation, with the addition of barium carrier (300 mg of Ba per sample) in which contamination levels of ^{226}Ra and ^{228}Ra were negligible. The precipitate was removed by filtering through a 0.45-μm milipore filter paper and washed sufficiently by HCl. After drying it and weighting to determine the chemical yield of the Ba carrier, the ^{226}Ra and ^{228}Ra concentrations were determined by γ-ray spectrometry of the short -lived radon progeny after a delay of about a month

2.2.3 Determination of ^{222}Rn

Following the techniques described by Horiuchi et al.[12], ^{222}Rn concentration in sample water was measured by low background liquid scintillation counter (Aloka LB-1) after preconcentration with a extraction of Rn in situ with toluene containing scintillator (PPO+POPOP) and/or a direct mixing of 40 ml of sample water and 60 ml of Instagel.

3. Results and discussion

3.1 ^{210}Pb and ^{210}Po in fumarolic gases and sulfur sublimates

The concentrations of ^{210}Pb and ^{210}Po observed in the gas condensates are summarized in Table 1, along with temperature of fumarolic gas, pH of condensate water, and concentrations of SO_4^{2-} and Cl^-. The fumarolic gases over 100℃ are discharged in every volcanoes, especially over 300 ℃ in Mt. Use and Mt. Tokachi, and their gases contain large quanties of SO_4^{2-} and Cl^-. The concentrations of ^{210}Pb and ^{210}Po ranged from 0.02 to 0.12 Bq/l and 0.02 to 0.73 Bq/l, respectively, and ^{210}Po/^{210}Pb activity ratios from 0.23 to as high as 42.9. In Mt. Usu whose temperature of fumarolic gas is the highest (557 ℃) among the samples measured, excess of ^{210}Po was evidently observed. Such phenomena on excess of ^{210}Po may closely related to the boiling points of several compounds of Po, especially Po chloride having lower boiling point, as shown in Table 2.

On the other hand, concentrations of ^{210}Pb and ^{210}Po in sulfur sublimates (Table 3) were from less than detection limit to 30.6 mBq/g-S and 0.07 to as high as 704.5 mBq/g-S, respectively, and ^{210}Po/^{210}Pb activity ratios from 0.29 to 24.7. Similarly to the volcanic gases, the highest value (24.7) of ^{210}Po/^{210}Pb activity ratios was observed in the sulfur sublimates from Mt. Usu. As the boiling point (390 ℃) of Po chloride is not largely different from that (444.6 ℃) of sulfur, Po seems to deposit near the fumarole togerther with sulfur. Although further study is needed for the quantitative evaluation of these data, the present results may indicate that the concentrations of ^{210}Pb and ^{210}Po and the activity ratios of ^{210}Po/^{210}Pb in volcanic gases and sulfur sublimates depend highly on the temperature of fumarolic gases (Figs. 4 and 5).

3.2 Hot and mineral spring waters

Before exact contents of ^{210}Pb and ^{210}Po at the date of sampling are evaluated, validity of the calculation by Eq.(3) and Eq.(4) was first checked using mineral spring water of high ^{222}Rn content in Masutomi Spa (Yamanashi Prefecture), which is one of the most strongly radioactive mineral springs in Japan. For two kinds of samples, each two samples were collected in a 20 l polyethylene bottles : for one sample ^{210}Pb and ^{210}Pb were pre-concentrated with $Fe(OH)_3$ immediately after sampling and the other sample was stored for 11 days until analysis with tight sealing. The results of these comparative data are listed in Table 4. As can be seen from this Table, after 11 days of storage the content of ^{210}Pb was found to increase remarkably, while that of ^{210}Po was not so evident as ^{210}Pb. The ^{210}Pb and ^{210}Pb contents at the date of sampling calculated by using Eq.(1)~Eq.(4) were confirmed to agree well each other within the errors, although the values found in the samples stored for 11 days were accompanied by relatively large errors. Thus, if ^{222}Rn concentration in spring water was measured before the storage and sample bottle was tightly sealed by a bottle caps, evaluation of the ^{210}Pb and ^{210}Pb contents at the date of sampling seems possible for the samples stored for long period of time.

In Table 5, the results of calculation are summarized. For the samples that the ^{222}Rn content was determined and which were not treated immediately after sampling , the ^{210}Pb and ^{210}Pb contents at the date of sampling were evaluated by using Eq.(3) and Eq.(4). The contents of ^{238}U, ^{226}Ra and ^{222}Rn are also given in the Appendix together with ^{234}U/^{238}U and ^{228}Ra/^{226}Ra activity ratios.

As shown in Table 5, high ^{210}Pb and ^{210}Po contents (13~565 mBq/l and 5.8~95 mBq/l, respectively) were found in Masutomi mineral spring waters of high ^{222}Rn contents (No.35~No.38 in Table 5). In Yunoshima mineral spring water (No.17), the highest ^{210}Po content (765 mBq/l) was detected, but ^{210}Pb was not detectable. For other spring waters, ^{210}Pb and ^{210}Po ranged in concentrations from less than detection limit to several mBq/l and 0.2 to several mBq/l, respectively. ^{210}Po contents was not so high even in the spring waters (No.2~4) which were situated in the volcanic mountain areas, indicating that ^{210}Po in spring waters related to active volcanoes is selectively removed even if high significant amounts of ^{210}Po are released from active volcanoes as a volcanic gases. The ^{210}Po/^{210}Pb activity ratios ranged from 0.1 to 2.9, mostly less than 1.0 .

The correlation between the concentrations of ^{210}Po (at the date of sampling) and those of ^{222}Rn and ^{226}Ra are shown in Figs. 6 and 7, respectively. The ^{210}Po contents appear to be correlated with ^{222}Rn contents rather than ^{226}Ra contents. The contents of ^{210}Po (and ^{210}Pb) are grossly deficient relative to their parent ^{222}Rn. The ^{222}Rn concentrations are about three orders of magnitude higher than the levels of its parent ^{226}Ra (see Appendix). The large excess of ^{222}Rn is well known to be due to its diffusion and transport through porous rock and soil grains into the aqueous phase. By taking into account such findings, the gross deficiency of ^{210}Po including ^{210}Pb may be attributed to the removal of these nuclides from the aqueous phase, probably due to the adsorption on to mineral grains. Assuming steady state between production and removal, the scavenging residence time for ^{210}Pb may be estimated simply using the following equation:

$$dN_{Pb}/dt = \lambda_{Rn}N_{Rn} - (\lambda_{Pb} + k_{Pb}) = 0 \quad \text{---------(5)}$$

$$\tau = 11753\ xR/(1-R) \quad \text{----------- (6)}$$

where k_{Pb} is the first order scavenging rate constant. Rearranging Eq.(5) and substituting 11753 days for the mean life of ^{210}Pb, the scavenging residence time τ is given by Eq.(6). The subscript R is the ^{210}Pb/^{222}Rn activity ratio in the spring waters. Using the ^{210}Pb/^{222}Rn activity ratios ranging from 10^{-4} to 10^{-5}, the value of τ is estimated to be about 1.0~0.1 days (Fig. 8).

The study of ^{210}Pb and ^{210}Po in hot and mineral spring waters (and/or in volcanic gases) may be useful in the determination of the origins and ages of water sources and of the history of ^{222}Rn in the water.

4. Summary

The concentrations of ^{210}Pb and ^{210}Po in fumarolic gas condensates and sulfur sublimates from several active volcanoes have been measured together with those in hot and mineral spring waters. The results of present study are summarized as follows:

1. By referring to the volcanic gas condensates, the excess of ^{210}Po was detectable and the concentrations of ^{210}Po ranged from 0.02 to 0.73 Bq/l, while those of ^{210}Pb from 0.02 to 0.12 Bq/l.
2. The concentrations of ^{210}Pb and ^{210}Po in sulfur sublimates ranged from 0.07 to as high as 705 mBq/g-S and from 0.23 to 30.6 mBq/g-S, respectively.
3. A general tendency was observed that the concentrations of both nuclides and $^{210}Po/^{210}Pb$ activity ratios depend highly on the temperature of fumarolic gases.
4. In Masutomi mineral spring waters of high ^{222}Rn contents, high ^{210}Pb and ^{210}Po contents (13~565 mBq/l and 5.8~95 mBq/l, respectively) were found. However, the contents of both nuclides were grossly deficient relative to their parent ^{222}Rn ($^{210}Pb/^{222}Rn$ activity ratio of 10^{-4}~10^{-5}). The ^{210}Pb and ^{210}Po in other spring waters ranged in concentrations from less than detection limit to several mBq/l and 0.2 to several mBq/l, respectively.
5. The contents of ^{210}Po was not so high even in the spring waters related to active volcanoes.
6. The scavenging residence time for ^{210}Pb in spring waters was estimated to be less than a day , on the basis of the $^{210}Pb/^{222}Rn$ activity ratio

5. References

[1] G. Lambert, P. Beisteau, G. Polian: Emission and enrichments of radon daughters from Etna volcano magma, Geochem. J., 3,724-726 (1976).

[2] G. Lambert, A. Busson, J. Sanak, B. Ardouin: Modification of the atmospheric Po-210 to Pb-210 ratio by volcanic emissions, J. Geophys. Res., 84, 6980-6986 (1979).

[3] G. Lambert, B. Ardouin, G. Polian: Volcanic output of long-lived radon daughters, J. Geophys. Res., 87, 11103-11108 (1982).

[4] P. K. Kuroda, J. C. H. Liou, A. D. Banavaali, J. D. Akridge, L. A. Burchfield: Po-210 fallout from the 1980 eruption of Mt. St. Helens and the mystery cloud of 1982, Geochem. J., 18, 55-60 (1984).

[5] T. Tada: Studies on airborne radionuclides in the atmosphere, Master Thesis of Kanazawa University (1988).

[6] K. Komura, K. Uchida, M. Yamamoto, K. Ueno: Pb-210 and Po-210 from active volcanoes in Japan, The 3rd Inter. Symp. on Advanced Nuclear Energy Research-Global Environment and Nuclear Energy, Mito,Japan (1991).

[7] P.K. Kuroda, Y. Yokoyama: On the equilibrium of the radioactive elements in the hydrosphere, II. Bull. Chem. Soc. Japan, 21, 58-63 (1948).

[8] I. Hatae: radiochemical determination of radium B and thorium B in radioactive springs, The memoirs of the Faculty of Science, Kyushu University, Series C, Chemistry, 3(2),71-81(1958).

[9] I. Hatae: Radiochemical studies on radioactive mineral springs (1), The memoirs of the Faculty of Science, Kyushu University, Series C, Chemistry, 3 (2), 83-91 (1958).

[10] K. Komura. M. Yamamoto, K. Ueno, K. Uchida, M. Sakanoue: Pb-210 and Po-210 in volcanic products and mineral springs (1); Pb-210 and Po-210 in volcanic gas and sulfur sublimates, J. Balneological Society of Japan, 44, 11-17 (1994) (in Japanese).

[11] M. Yamamoto, A. Toguchi, K. Sashimomo, K. Komura, K. Ueno, M. Sakanoue: Pb-210 and Po-210 in volcanic products and mineral springs(2); Measurement of Pb-210 and Po-210 contents in mineral springs, J. Balneological Society of Japan, 44, 18-27 (1994) (in Japanese).

[12] K. Horiuchi, Y. Murakami: Basic studies on the new determination method of radon in mineral springs with a liquid scintillation counter, J. Balneological Society of Japan, 28,39-52 (1977) (in Japanese).

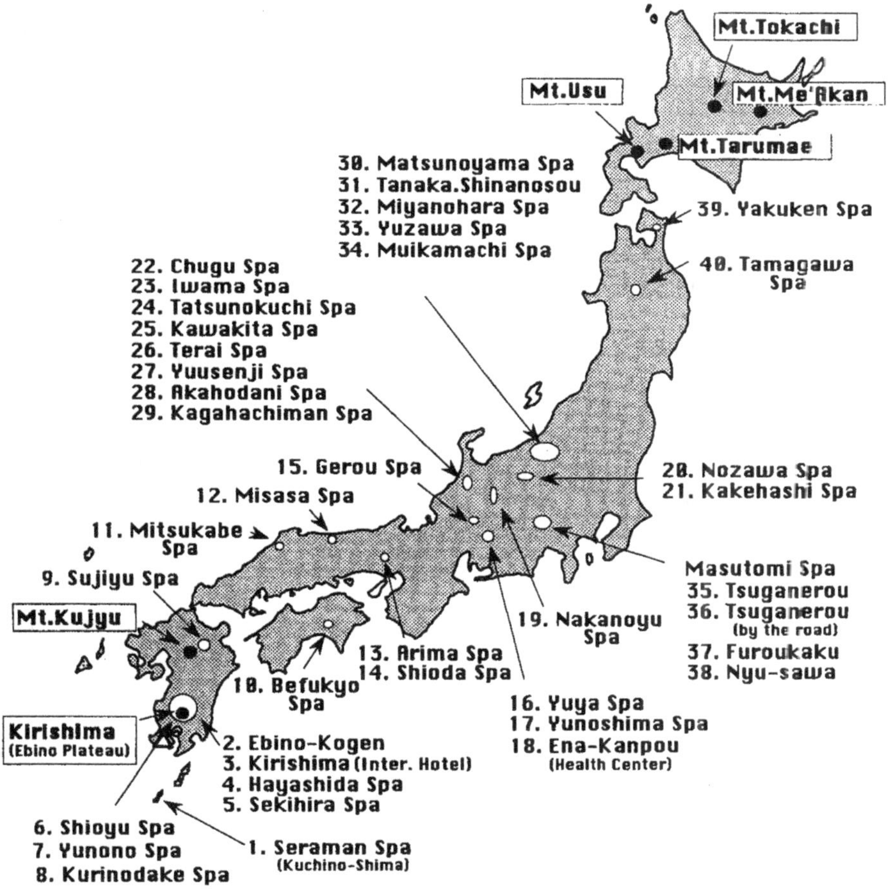

Fig. 1. Sampling location of volcanic gases, sulfur sublimates and spring waters

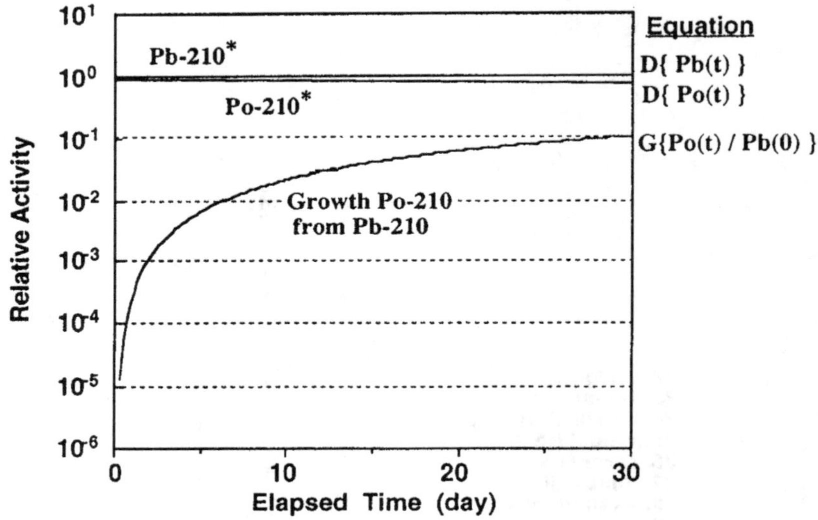

Fig. 2. Decay of Pb-210 and Po-210 and growth of Po-210 from Pb-210

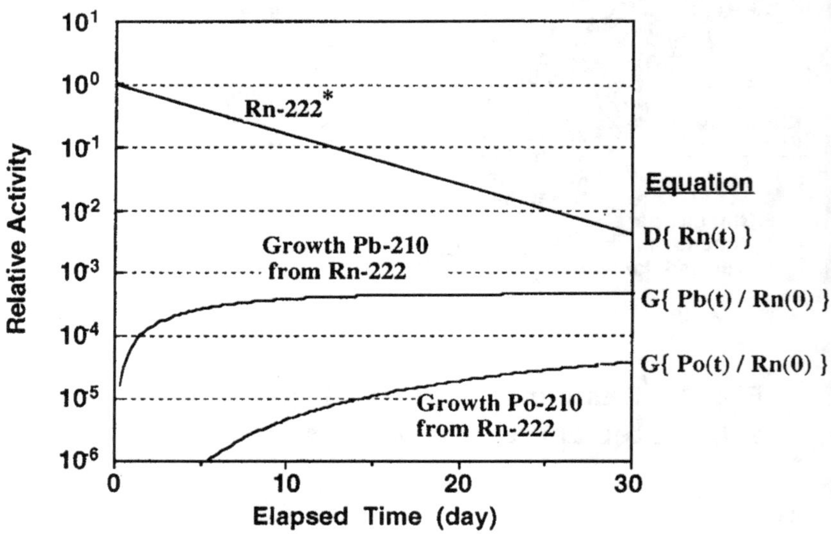

Fig. 3. Successive decay chain Rn-222 - Pb-210 - Po-210

Table 1. Contents of Pb-210 and Po-210 in the condensate water from fumarolic gases

Location	Fumarole Temp. (℃)	Concentration Pb-210 (Bq/l)	Concentration Po-210 (Bq/l)	Ratio Po-210/Pb-210	pH	SO_4^{2-} (mg/l)	Cl^- (mg/l)
Hokkaido							
Mt. Usu	557	0.017	0.73	42.9	1.45	76	1287
Mt. Tokachi	328	0.121	0.038	0.31	0.71	6246	2139
Mt. Me'Akan	256	0.066	0.157	2.38	0.92	2383	3330
Mt. Tarumae	210	0.098	0.214	2.18	0.93	2560	4030
Average		0.076	0.285	3.75			
Kyushu							
Mt. Kujyu-1	112	0.042	0.039	0.93	0.75	3200	1300
Mt. Kujyu-2		0.071	0.016	0.23			
Mt. Kujyu-3		0.032	0.066	2.06			
Average		0.048	0.041	0.85			

The chemical characteristics in the condensate waters from Hokkaido were measured by Dr. Abiko, Muroran Technical University.

Table 2. Melting and boiling points of Pb, Bi and Po compounds

Chemical form	Pb M.P.	Pb B.P.	Bi M.P.	Bi B.P.	Po M.P.	Po B.P.
Metal	328	1744	271	1560	254	962
Chloride	501	950	230	447	300	390
Oxide	888	1470	824	1900	885#	
Bromide	373	918	218	453	330	360
fluoride	855	1290	725			
Iodide	402	954	439	500*		
Sulfide	1114		685	685*		

M.P.: Melting point (℃) , B.P.: Boiling point (℃).
*: Decomposition, #: Sublimation

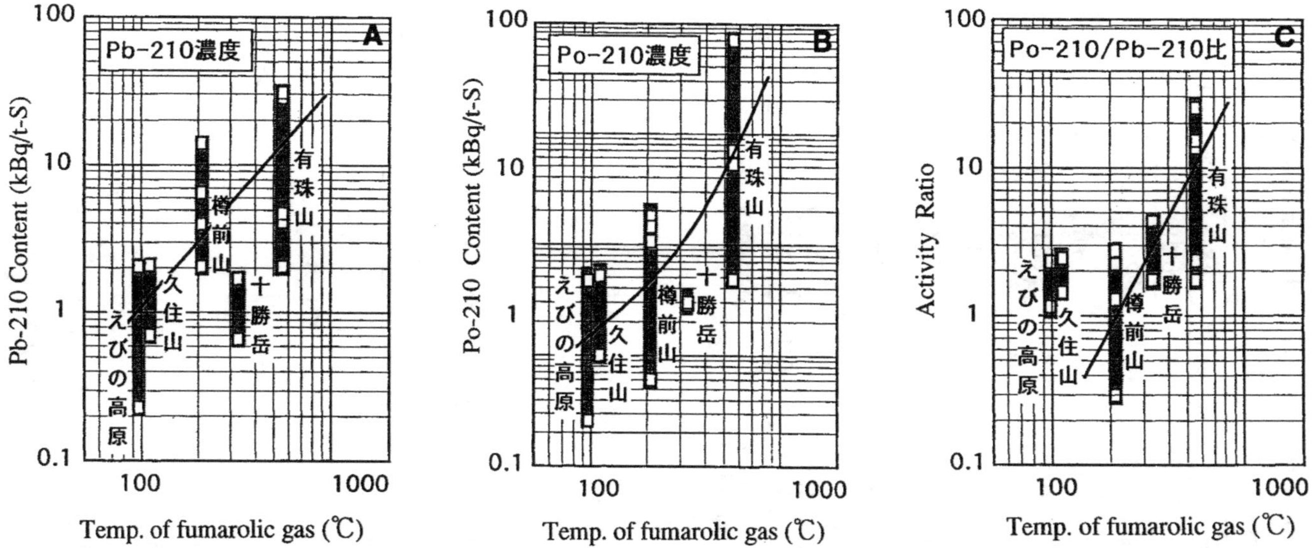

Fig. 4. Relationship between (a) Pb-210 contents, (b) Po-210 contents and (c) Po-210/Pb-210 ratios and temperature of fumarolic gases (℃). (有珠山: Mt. Usu, 十勝岳: Mt. Tokachi, 樽前山: Mt. Tarumae, 久住山: Mt. Kujyu, えびの高原:Ebino-Plateau (Mt. Kirishima)).

Table 3. Pb-210 and Po-210 contents of sulfur sublimates

Location (Temp.)	Sample No.	Concentration Pb-210 (mBq/g-S)	Concentration Po-210 (mBq/g-S)	Ratio Po-210/Pb-210
Hokkaido				
Mt. Usu (557 ℃)	1	4.08	54.6	13.4
	2	28.6	705	24.7
	3	30.6	53.2	1.74
	4	2.05	4.82	2.35
	5	4.67	70.1	15.0
Mt. Tokachi (328 ℃)	1	0.79	1.78	2.25
	2	13.4	4.31	0.32
	3	1.65	1.81	1.1
	4	0.67	2.83	4.22
	5	1.72	2.96	1.72
	6*	0.59	18.0	30.5
Mt. Tarumae (210 ℃)	1	3.97	10.7	2.7
	2	13.9	17.7	1.27
	3	—	0.07	—
	4	2.04	0.60	0.29
	5	6.44	14.3	2.22
	6	—	5.66	—
Kyushu				
Ebino Plateau (97 ℃)	1	0.23	0.26	1.13
	2	2.07	4.75	2.29
Mt. Kujyu (112 ℃)	1	2.11	5.13	2.43
	2	0.7	1.00	1.43

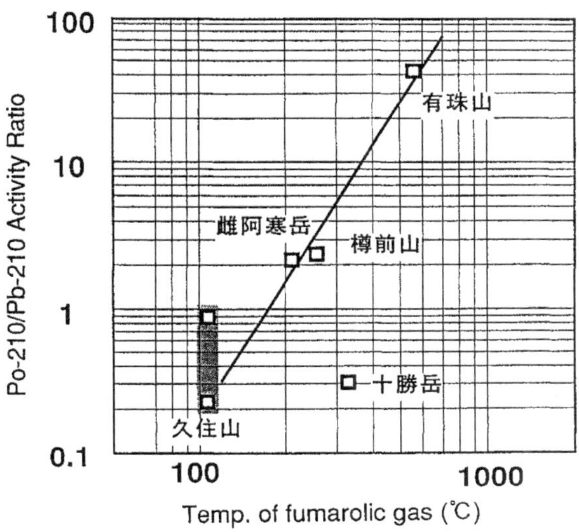

Fig. 5. Relationship between Po-210/Pb-210 ratio in sulfur sublimates and temperature of fumarolic gases (℃)

Table 4. Comparison of Pb-210 and Po-210 contents in two different pretreatments using Masutomi mineral spring waters (for one sample Po-210 and Pb-210 were immediately pre-concentrated with $Fe(OH)_3$ after sampling, and the other sample was stored for 11 days until analysis)

Location	ΔT1	ΔT2	Measured value (mBq/l)		Value at the sampling date(mBq/l)		
(Rn-222 Cont.)	(d)	(d)	Po-210*	Pb-210*	Cal. Eq.	Po-210	Pb-210
Tsuganerou (by the road)	0	8	40.7 ± 1.9	211 ± 16	(1), (2)	38.9 ± 2	211 ± 16
(1700 Bq/l)	11	15	50.3 ± 2.2	863 ± 41	(3), (4)	36.4 ± 3.4	170 ± 53
Furoukaku	0	8	65.2 ± 1.5	565 ± 45	(1), (2)	58.7 ± 1.7	565 ± 45
(5000 Bq/l)	11	15	92.1 ± 5.2	2490 ± 83	(3), (4)	48.8 ± 7.3	463 ± 107

ΔT1: Time from sampling date to the starting date of analysis.
ΔT2: Time from the sampling date to the date of preparation of counting source of Po-210.
*: Po-210; Values at the date of preparation of counting source, Pb-210; Values at the date of anion exchange treatment.

Table 5. Contents of Pb-210 and Po-210 in hot and mineral spring waters

No.	Sampling Location	Sampling Date	Elapsed time ΔT1 (d)	Elapsed time ΔT2 (d)	Contents at the measurement date Po-210* (mBq/l)	Contents at the measurement date Pb-210* (mBq/l)	Contents at the sampling date Po-210 (mBq/l)	Contents at the sampling date Pb-210 (mBq/l)	Activity ratio Po/Pb	Remarks Cal. Eq.
1	Seraman Spa (Kuchino island)	10/24/92	11	14	0.41 ± 0.03	2.54 ± 0.18				
2	Ebio Plateau	8/25/92	23	27	8.50 ± 0.47	25.8 ± 2.2				
3	Kirishima (Inter. Hotel)	8/27/92	21	25	0.20 ± 0.03	2.76 ± 0.19				
4	Hayashida Spa	8/24/92	24	28	0.28 ± 0.03	3.36 ± 0.25				
5	Sekihira Spa	8/27/92	21	25	0.17 ± 0.02	0.47 ± 0.04				
6	Shioyu Spa	8/27/92	21	25	0.16 ± 0.02	1.24 ± 0.11				
7	Yunoo Spa	8/28/92	20	24	0.21 ± 0.03	1.57 ± 0.11				
8	Kurinodake Spa	8/28/92	20	29	0.57 ± 0.04	2.34 ± 0.18				
9	Sujiyu Spa (Source)	8/29/92	19	28	0.51 ± 0.04	1.77 ± 0.12				
10	Befukyou Spa	10/21/92	14	17	0.68 ± 0.06	7.59 ± 0.61				
11	Mitsukabe Spa	8/30/92	18	31	0.36 ± 0.03	2.15 ± 0.15				
12	Misasa Spa	8/31/92	17	30	1.23 ± 0.08	21.5 ± 1.7				
13	Arima Spa (Tenjinnyu)	9/8/92	9	18	3.88 ± 0.31	10.6 ± 0.8				
14	Shioda Spa	10/19/92	19	23	312 ± 21	73.6 ± 5.8				
15	Gerou Spa	11/25/92	1	7	0.49 ± 0.04	2.24 ± 0.18	0.49 ± 0.04	1.4 ± 0.2	0.35	(3),(4)
16	Yuya Spa (No.6 spring)	3/3/93	7	10	0.29 ± 0.02	1.44 ± 0.11	0.30 ± 0.02	-0.7 ± 0.2		(3),(4)
17	Yunoshima Spa (Ra-No.1 spring)	3/3/93	7	10	733 ± 22	635 ± 32	765 ± 23	-532 ± 67		(3),(4)
18	Ena Kanpou-Health center	3/4/93	6	9	2.35 ± 0.13	17.3 ± 1.0	2.3 ± 0.1	3.3 ± 1.3	0.72	(3),(4)
19	Nakanoyu Spa	10/25/92	10	13	0.96 ± 0.06	2.63 ± 0.25	0.98 ± 0.06	0.6 ± 0.3	1.7	(3),(4)
20	Nozawa Spa	12/7/92	8	18	1.06 ± 0.09	2.44 ± 0.18	1.0 ± 0.1	2.3 ± 0.2	0.45	(3),(4)
21	Kakehashi Spa	12/8/92	7	17	3.57 ± 0.21	21.7 ± 1.8	3.2 ± 0.3	11 ± 2	0.29	(3),(4)
22	Chugu Spa	10/2/92	5	9	1.79 ± 0.07	3.89 ± 0.26				
23	Iwama Spa	10/2/92	5	9	0.24 ± 0.02	0.61 ± 0.05				
24	Tatsunokuchi Spa	5/26/93	0	7	0.23 ± 0.03	0.62 ± 0.05	0.23 ± 0.03	0.6 ± 0.1	0.38	(1),(2)
25	Kawakita Spa	12/10/92	0	15	1.44 ± 0.10	11.2 ± 0.9	1.0 ± 0.1	11 ± 1	0.09	(1),(2)
26	Terai Spa	12/10/92	0	15	1.24 ± 0.10	9.69 ± 0.08	0.90 ± 0.10	9.7 ± 0.1	0.09	(1),(2)
27	Yuusenji Spa	5/12/93	0	6	0.39 ± 0.04	0.64 ± 0.04	0.39 ± 0.04	0.6 ± 0.1	0.62	(1),(2)
28	Akahodani Sap	5/12/93	0	6	0.38 ± 0.04	0.56 ± 0.04	0.38 ± 0.04	0.6 ± 0.1	0.68	(1),(2)
29	Kagahachiman Spa	5/12/93	0	6	4.59 ± 0.18	1.63 ± 0.08	4.7 ± 0.2	1.6 ± 0.1	2.9	(1),(2)
30	Matsunoyama Spa	4/3/93	11	20	0.35 ± 0.03	0.56 ± 0.04	0.35 ± 0.04	-1.8 ± 0.2		(3),(4)
31	Tanaka. Shinano-sou	4/3/93	11	20	2.76 ± 0.15	4.51 ± 0.41	2.8 ± 0.2	2.5 ± 0.4	1.1	(3),(4)
32	Miyanohara Spa	4/3/93	11	20	1.02 ± 0.09	0.65 ± 0.07	1.1 ± 0.1	-0.3 ± 0.1		(3),(4)
33	Yuzawa Spa (Source)	4/3/93	11	20	0.66 ± 0.05	1.22 ± 0.12	0.71 ± 0.06	-0.1 ± 0.2		(3),(4)
34	Muikamachi Spa (Source)	4/3/93	11	20	0.38 ± 0.03	1.44 ± 0.12	0.30 ± 0.06	-6.7 ± 0.7		(3),(4)
35	Tsuganerou	10/27/92	11	15	6.36 ± 0.41	42.4 ± 2.6	5.8 ± 0.5	13 ± 3	0.44	(3),(4)
36	Tsuganerou (by the road)	10/27/92	0	8	40.7 ± 1.9	211 ± 16	39 ± 2	211 ± 16	0.18	(1),(2)
37	Furoukaku	10/27/92	0	8	65.2 ± 1.5	565 ± 45	59 ± 2	565 ± 45	0.10	(1),(2)
38	Nyu-sawa	10/27/92	11	15	123 ± 9	1860 ± 91	95 ± 11	340 ± 118	0.28	(3),(4)
39	Yakuken Spa	11/2/92	18	24	1.13 ± 0.08	1.84 ± 0.14	0.89 ± 0.11	-7.6 ± 0.7		(3),(4)
40	Tamagawa Spa	8/31/92	17	21	2.36 ± 0.17	73.6 ± 5.4				
41	Jachymov (Checo)	8/10/93	26	31	491 ± 23	2827 ± 207				
42	Lacco Ameno (Italy)	7/17/93	58	62	519 ± 26	1563 ± 154				

ΔT1: Time from sampling date to starting date of analysis. ΔT2: Time from sampling date to date of preparation of counting source of Po-210.

*: Po 210; Values at the date of preparation of counting source, Pb-210; Values at the date of anion exchange treatment.

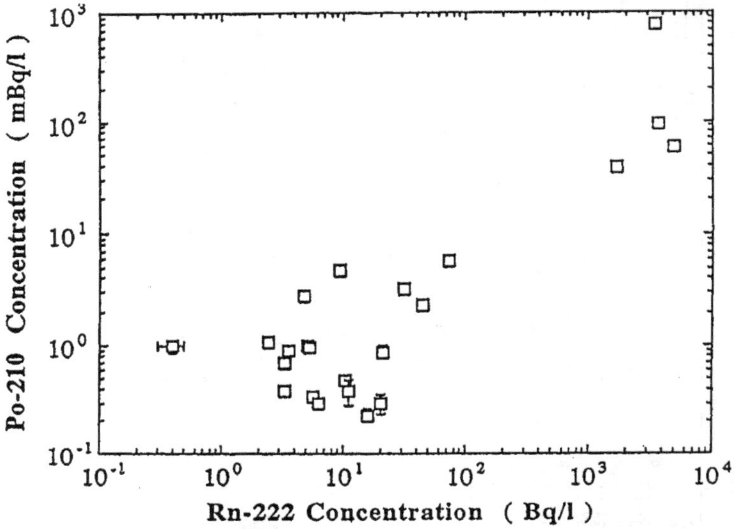

Fig. 6. Relationship between Rn-222 and Po-210 contents in hot and mineral spring waters

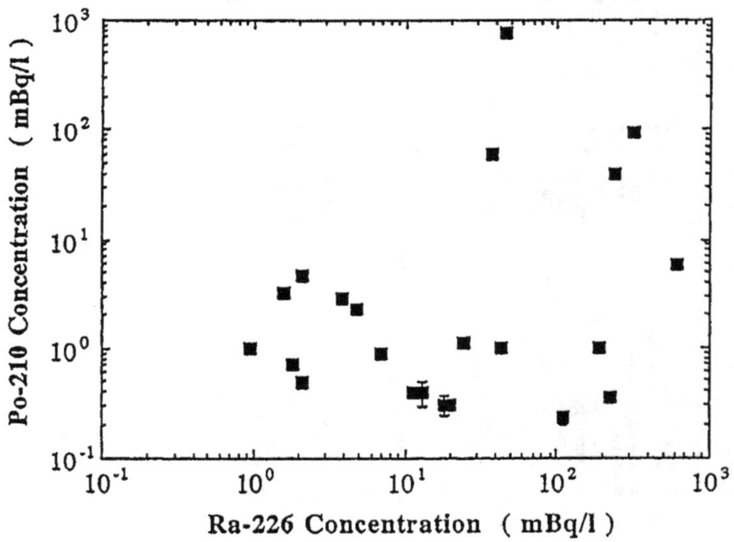

Fig. 7. Relationship between Ra-226 and Po-210 contents in hot and mineral spring waters

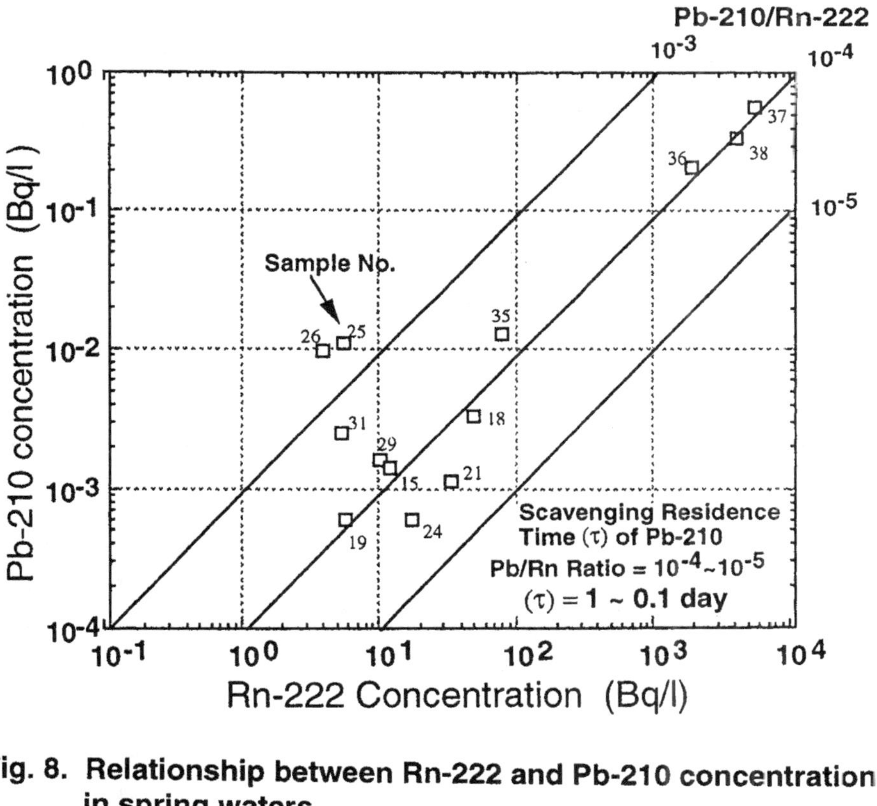

Fig. 8. Relationship between Rn-222 and Pb-210 concentrations in spring waters

Appendix Contents of uranium, radium and radon, and their activity ratios in spring waters

No.	Sampling location	Water Temp.	pH	Radioactivity concentration U-238 (mBq/l)	Ra-226 (mBq/l)	Rn-222 (Bq/l)	Activity ratio U-234/U-238	Ra-228/Ra-226
	(Kagoshima Pref.)							
1	Seraman Spa (Kuchino island)	54	7.0	1.68 ± 0.08	2.70 ± 0.41	n.d.	0.95 ± 0.07	0.31 ± 0.06
2	Ebio Plateau		1.5	27.0 ± 1.2			0.95 ± 0.06	
3	Kirishima (Inter. Hotel)		6.7	0.019 ± 0.002			1.20 ± 0.20	
4	Hayashida Spa		2.9	0.26 ± 0.03			1.21 ± 0.17	
5	Sekihira Spa		7.6	0.041 ± 0.007			1.18 ± 0.28	
6	Shioyu Spa	55	6.6	0.048 ± 0.007			1.31 ± 0.25	
7	Yunoo Spa		6.2	0.35 ± 0.03			0.96 ± 0.10	
8	Kurinodake Spa	92	2.5	3.00 ± 0.12			0.96 ± 0.06	
	(Oita Pref.)							
9	Sujiyu Spa (Source)		3.7	0.54 ± 0.03			1.85 ± 0.11	
	(Kochi Pref.)							
10	Befukyou Spa		8.6	0.49 ± 0.04			4.35 ± 0.40	
	(Tottori Pref.)							
11	Mitsukabe Spa		6.0	1.00 ± 0.09			1.41 ± 0.17	
12	Misasa Spa	58	7.3	18.1 ± 0.5			0.97 ± 0.04	
	(Hyogo Pref.)							
13	Arima Spa (Tenjinnyu)	98	4.6	0.10 ± 0.01			1.33 ± 0.21	
14	Shioda Spa	21	6.6	5.23 ± 0.24	185 ± 9		2.12 ± 0.12	2.11 ± 0.14
	(Gifu Pref.)							
15	Gerou Spa			0.30 ± 0.03	2.13 ± 0.41	11.0 ± 0.5	1.43 ± 0.16	2.21 ± 0.61
16	Yuya Spa (No.6 spring)	53		0.16 ± 0.01	20.2 ± 1.1	6.3 ± 0.1	11.4 ± 1.0	1.13 ± 0.08
17	Yunoshima Spa (Ra-No.1)	13		2.12 ± 0.11	46.3 ± 2.6	3460 ± 350	1.30 ± 0.08	0.042 ± 0.007
18	Ena Kanpou-Health center	15		0.43 ± 0.03	4.77 ± 0.33	45 ± 2	1.01 ± 0.08	1.28 ± 0.12
	(Nagano Pref.)							
19	Nakanoyu Spa	49	7.3	0.052 ± 0.007	4.50 ± 0.44	5.2 ± 0.2	1.71 ± 0.28	2.16 ± 0.29
20	Nozawa Spa	87	9.1	0.067 ± 0.007	0.95 ± 0.24	0.4 ± 0.02	1.06 ± 0.16	2.91 ± 0.31
21	Kakehashi Spa	12	5.0	0.30 ± 0.03	1.57 ± 0.12	31 ± 1	1.10 ± 0.13	0.58 ± 0.09
	(Ishikawa Pref.)							
22	Chugu Spa	43	7.4	0.10 ± 0.01	45.2 ± 0.8		2.96 ± 0.25	0.83 ± 0.02
23	Iwama Spa	59	7.7	0.11 ± 0.01	5.48 ± 0.52		1.69 ± 0.14	2.24 ± 0.26
24	Tatsunokuchi Spa	43	7.5	0.059 ± 0.007	112 ± 4	16 ± 1	45.9 ± 4.2	1.55 ± 0.08
25	Kawakita Spa	46	7.4	0.063 ± 0.007	191 ± 9	5.1 ± 0.3	1.47 ± 0.21	2.45 ± 0.19
26	Terai Spa	38	7.8	0.037 ± 0.004	7.07 ± 0.78	3.6 ± 0.1	1.00 ± 0.14	2.85 ± 0.26
27	Yuusenji Spa	50	8.0	0.014 ± 0.003	13.0 ± 0.9	11 ± 1	2.52 ± 0.81	3.54 ± 0.38
28	Akahodani Sap	25	8.0	0.007 ± 0.002	11.3 ± 0.8	3.3 ± 0.2	2.23 ± 1.03	3.74 ± 0.34
29	Kagahachiman Spa	19	8.1	0.052 ± 0.007	2.08 ± 0.27	9.3 ± 0.5	3.00 ± 0.49	1.49 ± 0.25
	(Niigata Pref.)							
30	Matsunoyama Spa	61		0.10 ± 0.01	224 ± 7	5.7 ± 0.3	1.29 ± 0.18	1.28 ± 0.06
31	Tanaka. Shinano-sou	31		0.27 ± 0.03	3.85 ± 0.48	4.8 ± 0.2	1.19 ± 0.15	1.21 ± 0.14
32	Miyanohara Spa	29		0.11 ± 0.01	24.9 ± 1.6	2.4 ± 0.1	1.02 ± 0.19	1.13 ± 0.12
33	Yuzawa Spa (Source)	41		0.085 ± 0.011	1.84 ± 0.41	3.3 ± 0.1	1.35 ± 0.25	1.61 ± 0.49
34	Muikamachi Spa (Source)	64		0.037 ± 0.007	18.7 ± 0.8	20 ± 1	1.12 ± 0.31	1.22 ± 0.08
	(Yamanashi Pref. Masutomi)							
35	Tsuganerou	26	6.3	0.32 ± 0.03	617 ± 9	72 ± 6	1.62 ± 0.19	2.00 ± 0.05
36	Tsuganerou (by the road)	21	6.0	0.53 ± 0.03	243 ± 7	1710 ± 50	1.09 ± 0.08	0.70 ± 0.05
37	Furoukaku	19	5.9	0.53 ± 0.05	37.4 ± 4.8	5000 ± 50	1.15 ± 0.14	0.69 ± 0.05
38	Nyu-sawa	20	6.4	4.45 ± 0.21	318 ± 6	3750 ± 40	2.21 ± 0.13	0.048 ± 0.008
	(Aomori Pref.)							
39	Yakuken Spa		7.1	n.d.	n.d.	21 ± 1		
	(Akita Pref.)							
40	Tamagawa Spa		1.3	14.0 ± 0.5			1.05 ± 0.05	
	Forigen							
41	Jachymov (Checo)			0.43 ± 0.03	718 ± 52		1.23 ± 0.04	0.29 ± 0.06
42	Lacco Ameno (Italy)			0.034 ± 0.001	n.d.		1.02 ± 0.06	

Development of a Drinking Water Standard for ^{222}Rn in the United States

Philip K. Hopke

Department of Chemistry
Clarkson University
Potsdam, NY 13699-5810

1. Abstract

The United States has been working to develop a regulatory standard for ^{222}Rn in public drinking water for about 12 years. There have been questions raised about the effectiveness of controlling radon in water without alternatively controlling the radon in indoor air arising from soil gas. This paper traces the development of the legislation that has established a new conceptual framework for the regulations in which a multimedia approach to risk reduction is now permitted. The actual regulated concentrations will be dependent on a new risk assessment that is currently being made by a National Academy of Sciences committee.

1. Introduction

The protection of public health through the regulation of the quality of potable water provided by public water supplies in the United States of America is provided for in the Safe Drinking Water Act. This legislation is updated periodically by Congress. The U.S. Environmental Protection Agency (EPA) then set regulations regarding allowable concentrations of contaminants in accordance with the provisions of the law.

After high concentrations of ^{222}Rn were found in homes in the United States in the mid-1980's, the revision to the Safe Drinking Water Act in 1986 specifically required EPA to set a standard for ^{222}Rn in drinking water [1]. Following a law suit and a resulting consent decree, EPA developed a criteria document in which the information defining the health effects of radon and its prevalence in drinking water are summarized [2]. On the basis of this document, the EPA then announced a tentative regulation which would set the Maximum Contaminant Limit for ^{222}Rn at 300 pCi/L (11 Bq/L) [3]. In the risk assessment, the risk of the increment of airborne radon arising from the use of drinking water in the home was estimated based on a transfer coefficient of radon from the water to the house air of 10,000 to 1 [4] (i.e. 10,000 Bq m^{-3} of ^{222}Rn in water would raise the concentration in house air by 1 Bq m^{-3}), and the risk per unit exposure developed by the National Academy of Sciences BEIR-IV Committee [5]. In addition the risk of stomach cancer due to the ingestion of the radon in the drinking water was taken from the work of Crawford-Brown [6]. At the proposed MCL, the EPA risk assessment suggested that the risks from the airborne radon and the ingested radon were approximately equal at about 10^{-4} for each.

Several significant problems were suggested to exist in the EPA analysis and proposed regulations. Groups representing environmental health suggested that the standard was insufficiently protective because there was known know threshold for radiation and thus the standard should be set at the minimum detectable value. Groups representing the water supply operators pointed to the general risk of radon in indoor air. The average concentration of ^{222}Rn in homes in the United States is 46.3 Bq m^{-3} [7]. Thus, the increase in airborne ^{222}Rn concentration in homes having water at the MCL would be 1.11 Bq m^{-3} or about a 2% relative to the existing risk from the soil derived radon in the home. There were also significant questions raised regarding the dosimetric model used to estimate the stomach cancer risk [8]. There was thus considerable political discussion of the issue and Congress intervened in 1992 and in the Appropriation Bill for the EPA (Public Law 102-389), the Administrator of the EPA was directed to prepare a multimedia risk and cost assessment for radon in drinking water and asked the Science Advisory Board (SAB) of the EPA to review these assessments. The SAB is a group of expects from outside of the EPA who provide advise to the Administrator on the variety of issues that EPA must confront. In this case the Radiation Advisory Committee created an *ad hoc* committee to review the risk assessment and again raised the issue of the small increment of risk from the water at concentrations close to the proposed MCL relative to the risk posed by soil gas radon. They felt that the risk estimate for the stomach was an upper limit estimate and was likely to be substantially lower than the inhalation risk [9]. In 1994, there was a change in the political control of Congress and at the same time, there was a need to undertake another reauthorization of the Safe Drinking Water Act. These issues then waited until the 1996 revisions to the Safe Drinking Water Act [10].

2. Current Activities

Under the revised law, the Administrator was required to withdraw all prior proposed rules related to radon in drinking water. A new risk assessment was mandated with the assessment to be performed by a committee established by the National Academy of Sciences. In addition this committee is to examine the "health risk reduction benefits associated with various mitigation to reduce radon levels in indoor air." Based on this risk analysis, the EPA will prepare a new prepare a health risk reduction and cost analysis for various possible MCL values and obtain public comment on these analyses. The Administrator will then propose and finalize regulations setting the maximum contaminant level for radon in public drinking water. However, the Administrator must "take into account the costs and benefits of control programs for radon from other sources." In other words the EPA is being required to look at the potential for controlling soil gas radon as a more cost effective way to reduce pubic health risks. In addition if the MCL for radon "is more stringent than necessary to reduce the contribution to radon in indoor air from drinking water to a concentration that is equivalent to the national average concentration of radon outdoor air," an alternative approach would be put into operation. An alternative maximum contaminant level (AMCL) will be set such that the "contribution of radon from drinking water to radon levels in indoor air" would be "equivalent to the national average concentration in outdoor air." States could then chose to establish multimedia radon mitigation strategies in which existing homes would be mitigated or "radon resistant homes would be constructed such that the overall health risk reductions expected to be achieved from the mitigation program are equal to or greater than would be achieved by having the water supply comply with the MCL. If a state does not chose to participate, there are provisions for local multimedia programs to be approved. Thus, it is possible in this approach to trade reductions in the air of specific houses in the community using the water supply for larger reductions of radon in water if the overall risk estimated reduction is at least that which would be achieved by reducing the water to the lower standard concentration.

3. Illustrative Example

To illustrate how this approach might work, it will be assumed that the only risk arises from the airborne radon. In addition it is assumed that the transfer coefficient is 10^{-4} and that the average outdoor ^{222}Rn concentration in the United States is 15 Bq m^{-3}. These values will also be the subject of the report by National Academy of Sciences Committee that is performing the risk analysis. The AMCL is then calculated by dividing the population-weighted mean outdoor ^{222}Rn concentration by the mean transfer coefficient. In this example, the AMCL would then be 150,000 Bq m^{-3} or 150 Bq L^{-1}. It will be assumed that the MCL will be 10,000 Bq m^{-3}. Suppose that a particular water supply contains 410,000 Bq m^{-3}. If the water was treated to reach the MCL, it would then contribute $[410{,}000 - 10{,}000]*10^{-4} = 40$ Bq m^{-3} to the home. For a water supply that provides water to 100 homes, the total reduction in radon resulting from the mitigation of the water to reach the MCL would be 4000 Bq m^{-3} houses. Thus, if 10 homes were mitigated so that the average ^{222}Rn reduction in each home was 400 Bq m^{-3}, then this mitigation of a limited number of homes would provide the same level of population risk reduction as would be achieved by reducing the radon in the drinking water to the MCL. Thus, if it is less expensive to reduce the radon in the water to the AMCL as well as mitigate the 10 homes, then the water supply utility might chose to adopt this multimedia approach to risk reduction rather than fully mitigate the water to the MCL. It is likely that only water supplies with radon concentrations between the MCL and the AMCL would seriously consider the multimedia remediation option since the equipment needed to reduce the ^{222}Rn to the AMCL would easily reduce the water to the MCL at essential no additional cost.

This alternative multimedia approach has several potential problems. It requires the water utility support activities to reduce indoor soil gas radon. The local water utility will generally have expertise in such activities. There are also important equity issues since the benefits of the risk reduction will go to the 10 homes that were mitigated rather than the 100 homes in the water supply district who are being exposed to the radon in the drinking water. It is not clear how these issues can be easily resolved.

4. Conclusions

It is not clear how many states or local water suppliers will actually adopt the alternative risk reduction approach. However, it is an important step that the Congress and the EPA have recognized that there is a multimedia risk and that addressing a portion of these risks without considering the entire problem may not lead to the best level of protection of public health. This multimedia approach is still evolving and the National Academy of Sciences report and the resulting actions of the EPA will be important in defining the directions where this changed view of risk management will go.

5. References

[1] Safe Drinking Water Act Amendments of 1986, Public Law 99-939, 100 STAT 642.

[2] U.S. Environmental Protection Agency, Final Draft for the Drinking Water Criteria Document on Radon, Report No. TR-1242-86, U.S. E.P.A., Washington, D.C. (1991).

[3] U.S. Environmental Protection Agency, *National Primary Drinking Water Regulations; Radionuclides; Proposed Rule 40 CFR Parts 141 and 142,* Federal Register, Vol. 56, No. 138, pages 33050 to 33127, July 18, 1991.

[4] Nazaroff, W.W., S.M. Doyle, A.V. Nero, and R.G. Sextro, Potable Water as a Source of Airborne ^{222}Rn in U.S. Dwellings: A Review and Assessment, *Health Phys.* 52:281-295 (1987).

[5] National Academy of Sciences, Health Risks of Radon and Other Internally Deposited Alpha-Emitters (BEIR-IV). National Academy Press, Washington, D.C. (1988).

[6] Crawford-Brown, D.J., The Biokinetics and Dosimetry of Radon-222 in the Human Body Following Ingestion of Groundwater, *Environ. Geochem. Health* 11:10-17 (1989).

[7] Marcinowski, F., R.M. Lucas, and W.M. Yeager, National and Regional Distributions of Airborne Radon Concentrations in U.S. Homes, *Health Phys.* 66:699-706 (1994).

[8] Harley, N.H. and E.S. Robbins, A Biokinetic Model for ^{222}Rn Gas Distribution and Alpha Dose in Humans Following Ingestion, *Environ. Int.* 20:605-610 (1994).

[9] U.S. Environmental Protection Agency, *Review of Uncertainty Analysis of Risks Associated with Exposure to Radon*, EPA Report No. EPA-SAB-RAC-93-014, July 1993.

[10] Safe Drinking Water Act Amendments of 1996, Public Law 104-182.

F Dosimetry and health physics

Medical experiences in Misasa Radon Spa

(Japan)

Masaaki Mifune [1) and Hiroshi Morinaga [2)]

Misasa Branch Hospital of Okayama University

[1)]888 Misasa-cho , Tottori-ken, 802-01 Japan

[2)]1-13-11 Soja, Soja-shi, Okayama-ken, 719-11 Japan

Abstract

Misasa Spa is one of the most highly radioactive hot springs in Japan, the waters of which contain mainly Rn-222(5 – 2544 Bq/1: average 437 Bq/1). Radon contents are 40 mBq/1 in the open air and 89 mBq/1 in indoor air. The environmental radiation dose rate was 12 μ R/h. The mean value of chromosome aberrations (dicentrics) in the peripheral blood lymphocytes of residents of Misasa were 0.21%. The 1952-88 cancer mortality records for inhabitants of Misasa were analyzed. The relative risk of lung cancer death was lower(0.55 times) in Misasa than the control area, although the difference was not significant.

1.Introduction

Misasa Spa was discovered during the 12th century. The town of Misasa is located in the mountainous region of Tottori Prefecture, Japan, covering an area of 9.9 sq. km with a population of 3034(1993). The mountains are composed of Cretaceous granite, Miocene tuffbreccias and various volcanic rocks of the Pleiocene age.

Thermal waters originate either at the surface spontaneously or are pumped fissures developed in granitic rocks. The major thermal area are confined within a narrow strip of about 1.5 km stretching along and down the Mitoku River. There are 90 wells and surface springs, and of which 70 are used for bathing. The total rate of hot water delivery by pumping was about 1690 liter/min in 1981 and that by spontaneous shallow underground springs, 390 liter/min. The temperature ranged from 36.5° C to 86.1° C, averaging 54° C. About 553,000 guests visited the spa in 1996.

The water is of the dilute neutral Natrium-Chloride-Bicarbonate type.

2.Radioactivity in the air

Misasa is often called "Radium spa", but most of the radioactivity in the Misasa hot spring waters is due to radon and its daughters. The radium concentration in hot waters is higher than in common springs but is still in a pronounced disequilibrium with radon. The shallow radioactive deposits are considered to be the most likely source of radon in hot waters.

Three million liters of thermal waters with a mean radon concentration of 437 Bq/1 are delivered daily by the 70 springs and cause a significant rise in the atmospheric radioactivity. A total of about 1.3 GBq.(35 mCi) per day of gaseous radon diffuses from the waters and soil into the air of this district. Table 1 shows mean values of the concentration of radon in outdoor and indoor air at various places.

Average radon concentrations are 40 mBq/1 in the open air, 89 mBq/1 in indoor air of ordinary houses and 807 mBq/1 in indoor air of bathrooms in Misasa Spa area, respectively, and 13 mBq/1 in open air in control areas without spa. Hot springs are commonly used by the residents in this district as a part of their regular life style, mainly for bathing twice a day.The authors calculated the mean annual exposure to radon daughters in background using the formula of ICRP(1987) and estimated the mean annual exposure level is 1.15 WLM.

3.Chromosome aberration in the peripheral blood cells of man

The frequencies of chromosome aberrations in peripheral blood lymphocytes are generally accepted as a suitable measure of the quantitative effects of ionising radiation. Chromosome analyses were carried out on 14 persons who had been living in central area of the spa for more than 10 years and 5 therapeutists for hydrotherapy of Misasa Branch Hospital of Okayama University who had been there for more than 5 years. Twenty persons resident in peripheral area of the spa were investigated to serve as a control group. The number of chromosomes and aberrations were scored in about 1000 cells for each person. The frequency of dicentrics, which is generally considered to be related to in vivo radiation exposure, ranged from 0 to 5 per 1000 cells and mean values of 0.21%, 0.10% and 0.12% were found for different groups. From the present results, a more detailed estimation of the individual doses of both α- and γ-radiations is needed to analyze relationship between chromosome aberration frequencies and accumulated radiation dose(Table 2).

4.Cancer Mortality survey in Misasa Spa area

Mortality records in Misasa Spa and control areas from 1952 to 1988 were those maintained at Health Centers of Tottori Prefecture. Lung cancer was the point of interest in this study in relation to radon inhalation. The standardized mortality ratio for lung cancer on female inhabitants in Misasa Spa area was significantly lower than that of in control area(Table 3). It seems a pertinent way to investigate for lung cancer mortalities on the scarcely smoking female inhabitants in the control area.

Therefore it appears that the increase in radon concentration in Misasa over that in the control areas probably has a negative correlation with a lung cancer incidence.

5.Conclusion

The authors could not recognize harmful effects of environmental radon on inhabitants of Misasa Spa.

6.References

[1] Morinaga, H., M.Mifune, and K.Furuno, Radioactivity of water and air in Misasa Spa, Japan. Radiation Protection Disimetry , 7(1-4), 295-297, 1984.

[2] Morinaga, H., Medical experiences in the Japanese radon spa Misasa, Z. Phys. Med. Baln. Med. Klim., 17(Sonderheft 1), 67-71, 1988.

[3] Mifune, M., T. Sobue, H. Arimoto, et al., Cancer mortality survey in a spa area Misasa, Japan) with a high radon background, Jpn. J. Cancer Res., 83, 1-5, 1992.

[4] ICRP, Lung cancer risk from indoor exposure to radon daughters, Annals of the ICRP 50, Pergamon Press, p.12, 1987.

Table 1. Mean air activities in the area of Misasa Spa

	^{222}Rn Bq.m^{-3}	Dose rate μ R. h^{-1}
Open air Misasa Spa	40±21	12±2
Room air hotels and private houses	89±67	14±2
Bathroom	807±78	13±1
Exercise pool	274±30	
Hubbard-tank bathroom	5306±2568	
Open air area without hot springs	13±4	9±1
Room air private houses without hot springs	14±5	

Table 2. Frequencies of dicentrics and rings in different population groups

Population group	No. of subjects	No. of cells observed	Dicentrics	Rings	WLM/a
Residents in central area	14	14,506	30 (0.21±0.08%)	2 (0.014±0.03%)	1.15
Residents in peripheral area	20	20,120	24 (0.12±0.06%)	3 (0.015±0.016%)	0.67
*Therapists for Hydrotherapy	5	4,926	5 (0.10±0.11%)	2 (0.041±0.132%)	1.41
	39	39,552			

*They work 3-4 hours daily in the room with facilities using radioactive hot spring waters(Okayama University).

Table 3. Relative Risks of Dying of Various Cancers for Inhabitants in Misasa versus Control Area Estimated by Poisson Regression Analysis[a)]

Site of cancer	Relative risk[b)]	95% confidence interval
All sites	0.67	0.53 – 0.85
Stomach	0.59	0.39 – 0.88
Lung	0.55	0.25 – 1.24
Colon, Rectum	0.32	0.10 – 1.06

a) Variables for sex, age and period were also included.

b) Relative risk for inhabitants in Misasa when mortality in the control area was treated as reference.

Biological and therapeutical effects of Radon

Peter Deetjen

Institute of Physiologie and Balneologie,

University of Innsbruck, Austria

Abstract

In spas with a somewhat elevated Radon^{222} (Rn) activity (between 300 and 3000 Bq/l), the empirical medicine ended - in all parts of the world - with the same list of indications. It mainly includes the more painful rheumatic diseases such as deformation or degeneration of the joints and non bacterial inflammation of muscles, tendons or joints; Morbus Bechterew and other diseases of the vertebral column like spondylosis, spondylarthrosis or osteochondrosis.

While in former times these effects were seldom documented in an objective manner, in recent years several prospective randomized double-blind studies proved the pain reducing efficacy of Radon therapy in patients with cervical pain syndromes, with chronic polyarthritis or with Morbus Bechterew.

Studies in experimental animal models have accumulated remarkable data in organs, tissue and cultured cells that provide a rationale to explain the observed effects of Radon therapy in patients.

Some of the oldest and most esteemed spas in Europe were found to have Radon as the effective ingredient of their mineral springs.
Interestingly enough, in all spas with comparable heigh Rn concentrations, in Europe as well as in other parts of the world, the empirical medicine ended in the same list of indications [1] (Table 1).

Tab. 1: List of indications for Radon therapy

Chronic rheumatic diseases
Degeneration or deforming diseases of joints
Spondylosis, Spondylathrosis
Osteochondrosis
Morbus Bechterew
Neuralgia
Chronic neuritis
Rehabilitation after injuries of locomotoric system
Ulcerations with poor healing tendency
Chronic skin diseases
Eczema, sclerodermia, psoriasis
Fertility problems and infecundity
Geriatic syndroms

It includes the more painful rheumatic diseases such as deformation or degeneration of the joints and non bacterial inflammation of muscles, tendons or joints; Morbus Bechterew and other diseases of the vertebral column like spondylosis, spondylarthrosis or osteochondrosis. Other

indications are neuralgia, chronic neuritis, restricted mobility of limbs after an accident or a sport injury; injuries with poor healing tendency (e.g. ulcer cruris, parodontosis), peripheral circulatory problems and some chronic skin diseases; also syndroms associated with old age and degeneration have all been treated in Radon spas with some degree of sucess. Finally, to this list should be added complications involving the endocrine system, preclimatic and climacteric difficulties and probably impotentia coeundi et generandi.
It may seem surprising that a single treatment is prescribed for such a wide range of conditions. Indeed, many of the effects are only known from experience, and have still to be properly explained on a medical basis. However, what we know already about the biological effects of Radon provides some rational basis for the scientific approach.

Radon is an inert gas and as such does not react with any chemical component of the body. Upon entry through lungs or through the skin, it reaches the bloodstream and is then distributed throughout the body. Being rather lipid soluble, Radon tends to accumulate in organs rich in fat, such as the endosecretory glands and also nerve fibres, which are surrounded and protected by a lipid-containing layer.

Retention time in the body is short, 50% having disappeared after only 15-30 minutes. It is during this short period, however, while Radon is in contact with the tissue that it launches the beneficial effects.
Radon-222 is a source of alpha radiation; it emits highly energized, positively charged Helium nuclei. Such particles have a relatively high atomic mass, and are immediately cushioned on contact with other matter. As a consequence, the radiation is only effective over very short distances and has little „penetrating power". It has a range of about 4cm in the air, it is completely stopped by a normal window pane and it can only travel a distance of about 20μm through body tissues. Thus, it is only with the help of Radon as an internal carrier substance that alpha radiation can be effective inside the body.
The relatively large transfer of energy which is associated with the absorption of alpha-particles causes a series of complicated reactions within the tissues. As yet the molecular processes involved are still poorly understood. It is safe, however, to assume that radiolytic radicals are released and these in turn stimulate detoxification processes and also may stimulate such processes as cell metabolism and energy conversion within mitochondria as well as biosynthesis of enzymes and other proteins or bioactive peptides.

From the many investigations we and others have performed, I can only mention a few examples. α-radiation increases superoxide dismutase (SOD) activity in cell cultures [2] as well as in rats [3].
SOD is a key encyme in the detoxification of superoxide radicals in the cytoplasma(Fig.1).

Rn exposure significantly increases the inner membranes of rat mitochondria in the zona fasciculata of adrenal glands which is the site of cortisol synthesis. The mitochndria of the zona glomerulosa at the same time remain uneffected [4] (Fig.2).

Radon stimulates the production of certain neuropeptides like substance P (Fig.3) or Calcitonin-Gen-Related-Peptide [5] (Fig. 4). Both are playing a role as co-transmitters in the perception and central processing of pain.

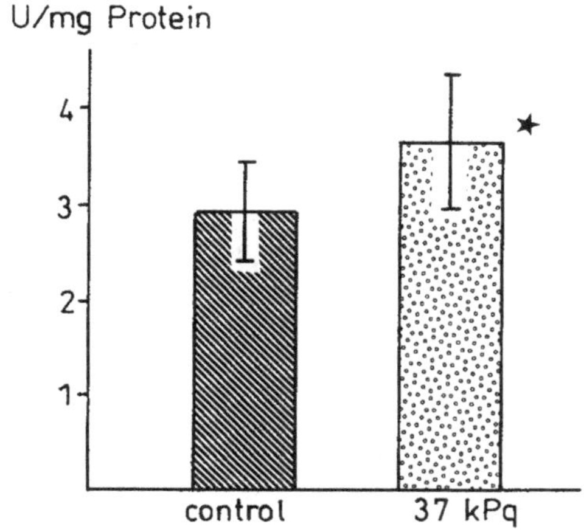

Fig.1: Superoxide dismutase activity in MDCK cell cultures in controls and after α-radiation of 3,7 kBq/24[h]. Mean values and SD, * $p < 0,05$ (Adapted from [2]).

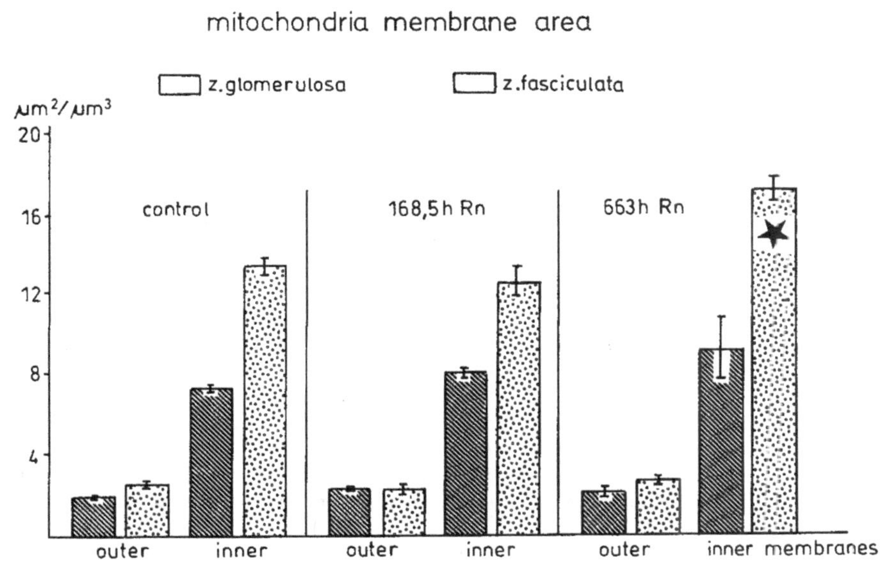

Fig.2: Membrane areas of mitochondria in rat adrenal glands. After exposure to 1 kBq Radon in inspiratory air a significant increase ($p < 0,05$) of mitochondria inner membranes is seen in the zona fasciculata (Adapted from [4]).

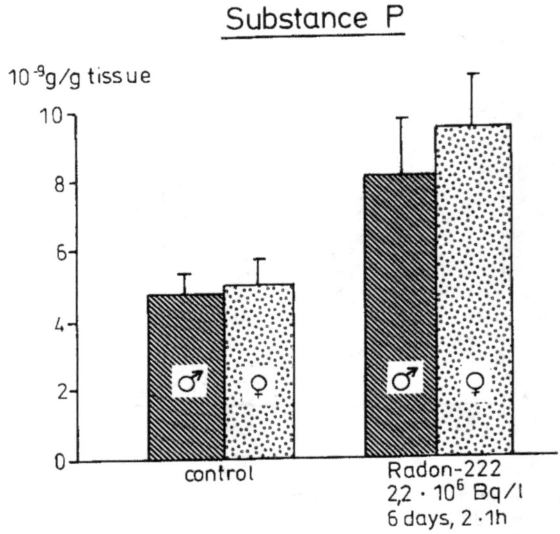

Fig.3: Substance P in rat bronchial tissue before and after exposure of 6 days to 2,2 x 10^3 k Bq/l Radon twice a day for one hour (Adapted from [5]).

In patients with spondilitis ancylans, also known as Morbus Bechterew, treatment with Rn increases the activity of natural killer cells [6]. Furthermore, in such patients the rate of DNA synthesis is increased while the DNA repair mechanism are partly inhibited, leading to prolonged errors in DNA structure. During curative treatment with Rn both parameters approach normal levels [7]

All these studies provide some rational hints to explain the analgesic and antiinflamatory effects of Rn in patients undergoing a sucessful treatment in Rn spas. However, in such spas, patients are exposed to several other factors different to their normal live. These also could influence their neurovegetative balance and could counterfeit a factual action of Rn.

To prove a specific efficacity of any agent in modern medicine, randomized double blind studies with placebo controls are essential for its acceptance. Recently 3 of such studies were performed to control the therapeutic effects of Rn.

In the first study [8] 46 patients suffering from cervical pains and degenerative alterations of the cervical spine were recruited and randomized into the verum or placebo group. Each group received 9 bath over a 3 week periode. The bath water had a temperature of 37° C and was prepared by addition of a canister with tap water or with Rn-containing water to make up a final Rn activity in the bath of 3 k Bq/l.

Each canister was only marked with a code number. Since Rn is colourless and odorless, neither the patients nor the supervising medical staff was aware of the actual treatment each patient was undergoing. Only after the termination of the study 4 month later the individual assignments were revealed to all the participants.

Subjective pain perception was determined with the aid of a visual analogue scale. As a more objective methode, however, the minimum pressure needed to cause pain at 16 typical myofacial pressure points was measured with a Pressure-Threshold-Meter.

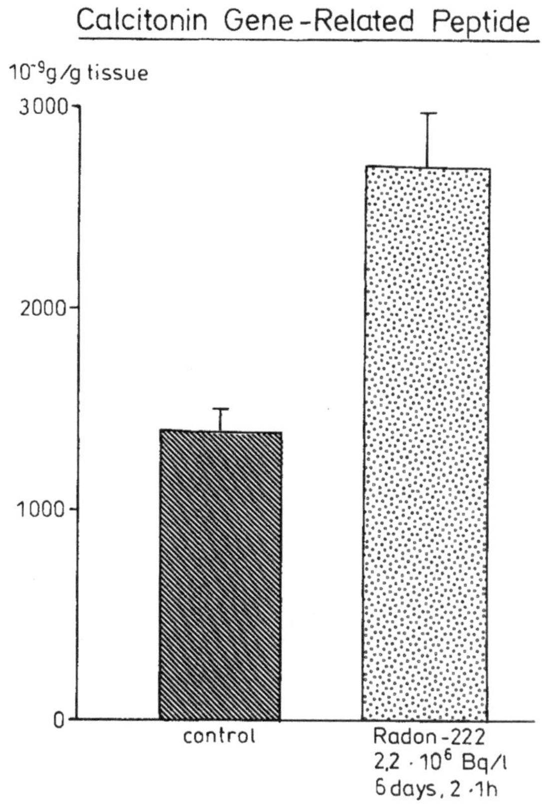

Fig.4: Calcitonin Gene Related Peptide in the spinal cord of male rats before and after exposure of 6 days to 2,2 x 10^3 k Bq/l Radon twice a day for one hour (Adapted rom [5]).

During treatment there was an almost equal improvement in both the verum and the placebo group resulting in reduced pains and an increase in well-being. After the end of the treatment, however, these paramenters gradually got worse again in the placebo group. In the Rn group, in contrast, all measured parameters continued to improve (Fig.5). This is in accordance with common experiences in Rn spas, that the maximal benefit is usually reached some time after the treatment periode.

Another study of almost identical design was performed in 52 patients with spondylosis or osteoarthrosis in the cervical or lumbal vertebral column. [9]
In spite of the fact that the Rn activity in the verum bath was only 800 Bq/l there also was a significant and transient improvement only in the Rn group.
In a third study 262 patients suffering from Morbus Bechterew received a comprehensive rehabilitation program [10]. 144 from these patients were randomly selected for an additional treatment in a Rn gallery with an average Rn activity in the air of 75 k Bq/m^3.

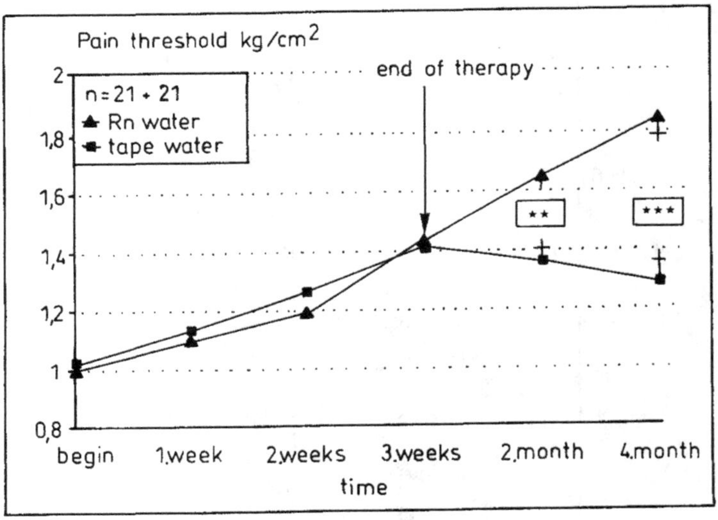

Fig.5: Pain threhold in patients with spondylarthrosis of the cervical spine undergoing balneotherapeutic treatment with 37° C warm bathes in tap water or in Radon water of 3 k Bq/l (Adapted from [8]).

At the end of the 4 weeks of treatment, in both groups a considerable reduction of the typical Bechterew pains were observed. 3 month later, however, only the radon patients on average still had a rather good pain relieve. This advantage was detectable up to 12 month.
Accordingly also the consumption of pain killers like nonsteroidal antirheumatics followed the same pattern. There was some reduction in both groups at the end of the rehabilitation procedures. But only in the Radon group this effect persisted during the following year (Fig.5).
These findings deserve special attention in regard of the sometimes severe side effects of pain killing drugs. Since on the other hand side effects of Rn therapy were never reported and therefore are obviously negligiable, especially the Rn therapy of severe rheumatic diseases appears to be a very valuable clinical measure.

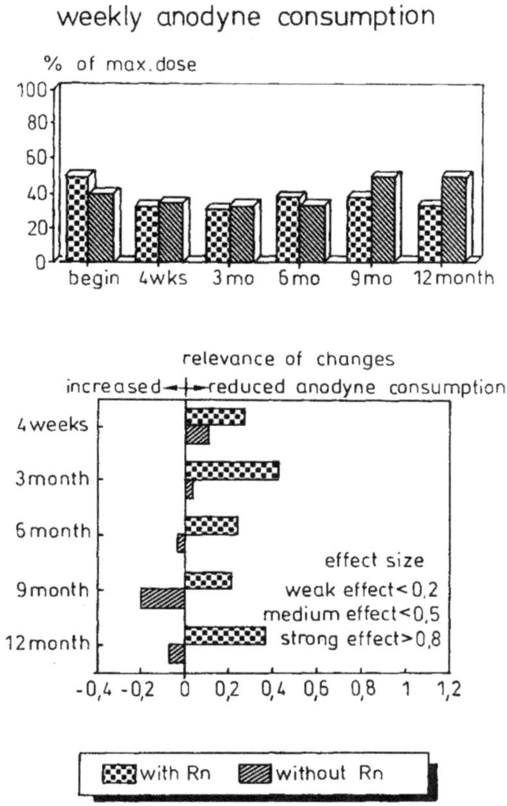

Fig.6: Weekly anodyne consumption in 262 patients with spondylitis ankylosans (Morbus Bechterew) undergoing a comprehensive rehabilitation program. The 144 patients receiving Radon of 75 k Bq/m^3 in a gallery for 9 days were randomly selected (Adapted from [10]).

References

[1] Deetjen, P. : Epidemiology and Biological Effects of Radon. In „Radon in der Kurmedizin“ Pratzel H.G. and P. Deetjen (Eds),
p. 32-38, I.S.M.H. Verlag Geretsried, 1997

[2] Frick H. und W. Pfaller: Die Auswirkung niedriger α-Strahlendosis auf epitheliale Zellkulturen. Z. Phys. Med. Baln. Med. Klin. 17, 23-3, (1988)

[3] Ma, J., H. Yonehara, M. Ikebuchi, and T. Aoyama: Effect of Radon Exposure on Superoxide Dismutase activity in rats.
J. Rad. Res. 37, 12-19 (1996)

[4] Pfaller, W.: Subzelluläre Veränderungen der Nebennierenrinde nach Inhalation von ^{222}Radon. Zsch. Bäder u. Klimahlkd. 26, 384-390, (1979)

[5] Bernatzky, G, A.Saria, H. Holzleithner, C. Kronberger, U. Wittauer, F. Blum, G.W. Hacker, W. Kullich, G. Leiner and H. Adam: Auswirkungen niedrig dosierter ionisierender Strahlung auf regulatorische Peptide im Blut und in Geweben.
Z. Phys. Med. Baln. Med. Klein. 19, 36-53, 1990

[6] Gastl, G., D. Egg, M. Herold, A.M. Födinger, Ch. Huber and R. Günther: Influence of Finnish Bath and Radon Balneotherapy on the Frequency and Activity of Neutral Killer Cells in peripheral Blood.
Z. Phys. Med. Baln. Med. Klin. 17, 47-53. 1988

[7] Günther, R., D. Egg and M. Herold: DNA-Repair und Cortisolproduktion unter Radontherapie bei Gesunden und Rheumakranken. Zsch. Bäder- Klimahlkd. 26, 336-355, 1979

[8] Pratzel H.G, B. Legler, K. Aurand, K. Baumann und Th. Franke: Wirksamkeitsnachweis von Radonbädern im Rahmen einer Kurortmedizinischen Behandlung des zervikalen Schmerzsyndroms
Phys. Rehab. Kur. Med 3, 76-82, 1993

[9] Heisig, S.: Zur analgetischen Wirkung von Radonbädern bei Patienten mit degenerativen Erkrankungen von Wirbelsäule und Gelenken.
Dissertation, Ludwigs-Maximilians-Universität München 1997

[10] Lind-Albrecht, G., and U. Droste: Zusatzeffekt der Radonstollentherapie im Rahmen der stationären Rehabilitation bei Spondylitis ankylosans (M. Bechterew)
Karl-Aschoff-Klinik, Bad Kreuznach 1996

Radon: characteristics in air and dose conversion factors

Justin Porstendörfer and August Reineking

Isotope Laboratory for Biological and Medical Research
Georg-August-University, Burckhardtweg 2, D-37077 Göttingen, Germany

Abstract

The dose conversion factor (DCF) which gives the relationship between effective dose and potential alpha energy concentration of inhaled short-lived radon decay products is calculated with a dosimetric approach. The calculations are based on a lung dose model with a structure that is related to the new recommended ICRP respiratory tract model (ICRP 66).

The characteristics of the radon decay products concerning the unattached fraction and the activity size distribution of the radon decay products are important input quantities for the calculation of DCF. The experimental data about these quantities obtained from measurements in homes, at work places, and in the free atmosphere near ground in the last past years are reported.

The DCF fraction of the unattached (DCF_u) and aerosol-attached (DCF_{ae}) radon decay products for different places are presented, taking into account the measured characteristics. The influence of the unattached radon daughters on the dose conversion factor DCF_u is reported and compared with the DCF_{ae} of the aerosol fraction.

1. Introduction

The inhalation of the airborne short-lived decay products constitutes the most important exposure at many working places and in homes. In almost all dosimetric models the maximum dose equivalent can be estimated for the epithelium of the bronchi. A dominant parameter is the activity size distribution of unattached and aerosol-attached decay products in the air, because place and amount of the inhaled activity depend on the particle size.

For the determination of dose, a relation between radon exposure in terms of potential alpha energy concentration (PAEC) and dose, the so-called dose conversion factor (DCF = Dose/PAEC) is needed.

At the time, in the radon field it is necessary to distinguish between two kinds of DCF's:

- DCF derived from lung model calculations, taking into account the characteristics of the radon decay products in air
- DCF derived from epidemiological studies.

There exist a discrepancy between the results of both approaches. Sensitivity analysis of model calculations show, that this discrepancy between the epidemiological and the dosimetric approaches can not be explained by the uncertainty of biological parameters like critical target cells for induction of lung cancer, retention, clearance rate and breathing mode [1] [2] [4].

This paper will show, how great is the variation of the DCF for different places with different characteristics of the radon decay products like unattached fraction (f_p) and the activity size distribution of the aerosol.

2. Quantities relevant for dose

The results of dose model calculations [1] [2] [3] [4] based on the new human respiratory tract model of ICRP 66 [5] show the strong influence of the particle size on the effective dose (Fig. 1). Especially the unattached radon daughter cluster of the size range between 0.5 – 5 nm have a great influence on the dose.

In our model calculations [1] [2] the nuclei of the secretory cells and basal cells are the targets. Differing from the ICRP 66 lung model, it was considered, that the density number distribution as function of the depth of epithelium are different for both target cells.

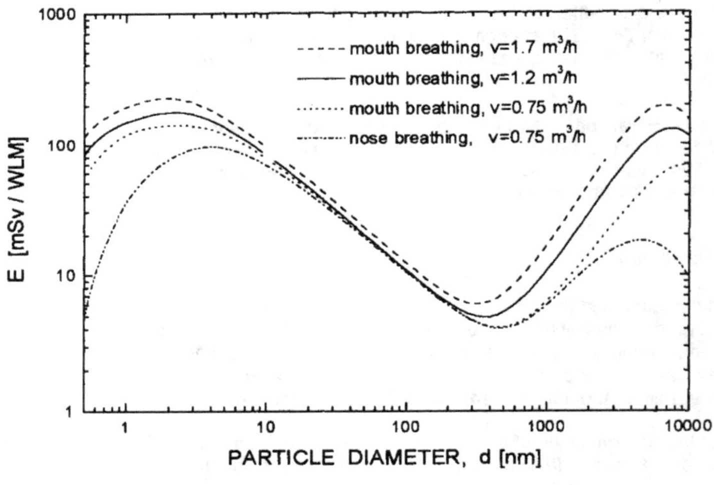

Fig. 1.: Dose conversion factor (in effective dose) as function of the particle diameter obtained from model calculations [1] based on the human respiratory tract model of ICRP 66 (target tissue: secretory and basal cells)

For dose estimation, we need the information of the activity size distribution in terms of the potential alpha energy concentration over the whole particle size range between 0.5 nm up to 10000 nm, as the decay products occur in air.

For practical reasons related to measurements it is reasonable to separate this size distribution in

- potential alpha energy concentration of the radon daughter aerosol in the size range 5 – 10000 nm: $PAEC_{ae}$
- potential alpha energy concentration of the radon daughter cluster with sizes between 0.5 – 5 nm: $PAEC_u$
- relative size distribution of the radon daughter aerosol
- relative size distribution of the radon daughter clusters

With the total potential alpha energy concentration in air

$$PAEC = PAEC_{ae} + PAEC_u,$$

and the unattached fraction of the clusters

$$f_p = PAEC_u/PAEC.$$

Especially, the measurements of the relative size distribution of the unattached decay product clusters under realistic environmental air conditions are difficult. Only some data of measurements in few indoor and working places with higher radon level exist [6]. Our actual data show, that the size distributions of the clusters are fairly narrow with geometric standard deviations (σ_{gu}) between 1.1 – 1.3. The activity median diameters (AMD_u) vary between 0.5 – 1.2 nm. Therefore up to now for dose estimations a constant average size distribution of the clusters with $AMD_u = 0.8$ nm and $\sigma_{gu} = 1.2$ for all air conditions is proposed. The inaccuracy of the dose estimation for the unattached fraction is about $\pm$ 30% or for the total dose (clusters and aerosols) is less than 15%.

With this assumption the dose conversion factor depends only on the following two quantities:

- Unattached fraction (f_p)
- Relative activity size distribution of the radon daughter aerosol ($1/C_p \cdot \delta C_p/\delta(\ln d)$)

The influence of f_p on the effective dose (mSv) per unit exposure (WLM) obtained from model calculations is shown in Fig. 2. Under normal aerosol conditions in homes and in working places the f_p-values are always smaller than 0.15. In this case, for nose-breathing the dose contribution of the clusters is smaller than the dose fraction of the aerosol attached radon daughters (E = 6.2 mSv/WLM). This effect is mainly caused by the high deposition efficiency of the nose for the small radon clusters with their high diffusivity. Only a small fraction reaches the bronchial and bronchiolar region of the lung.

For mouth-breathing and f_p-values >0.05 the dose contribution on DCF by the inhalation of radon daughter clusters is higher than the dose fraction by the radon daughter aerosol. The dose fraction by aerosols dominates only at places with higher number concentration of aerosol particles in air and therefore f_p-values <0.05.

The dose conversion factor as function of the activity median diameter (AMD) of the radon daughter aerosol for different geometric standard deviations σ_g is illustrated in Fig. 3. In the case of a multimodal size distribution the sum over all dose fractions of the modes yields the DCF.

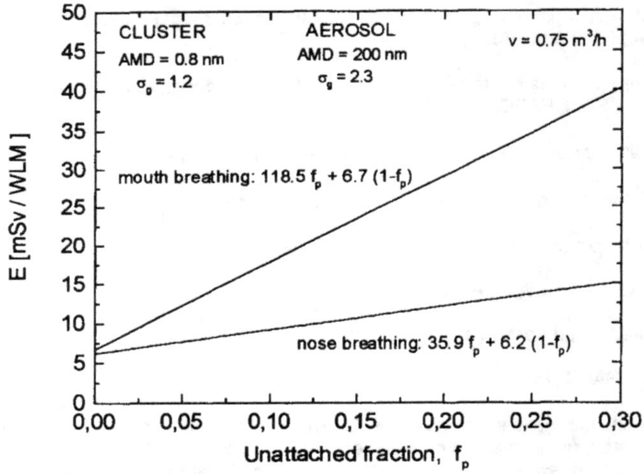

Fig.2: Dose conversion factor for mouth- and nose-breathing as function of the unattached fraction (f_p)

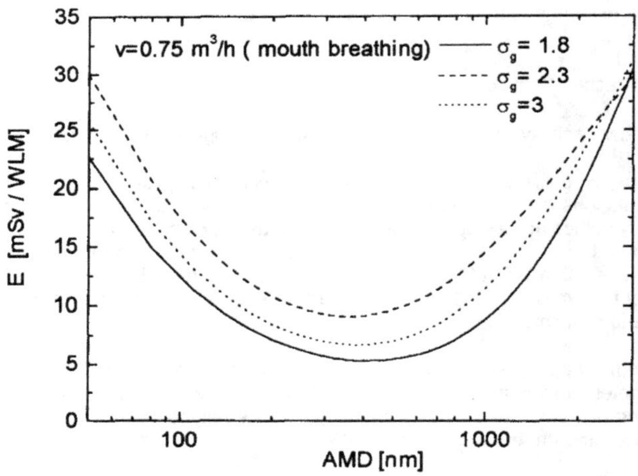

Fig.3: Dose conversion factor of the aerosol-attached radon decay products as function of the activity median diameter (inhalation rate: 0.75 m^3/h, geometric standard deviation σ_g)

3. Experimental data

Experimental data about the unattached fraction and the activity size distribution of the radon decay products obtained from our measurements at different places in the last years are summarized in Table 1.

Table 1: Log-normal parameters of the activity size distribution of the short-lived radon in the air of different places: Activity median diameter = AMD_i, geometric standard deviation = σ_{gi}, fraction of the mode = f_{pi}. The indices i = u, n, a and c represent the cluster, the accumulation, the nucleation and the coarse mode.

PLACE	ACTION/ AEROSOL SOURCE	CLUSTER f_p AMD_u=0.8 nm σ_{gu} = 1.2	AEROSOL			
			MODE i	f_{pi}	AMD_i [nm]	σ_{gi}
OPEN AIR		0.02 (0.01 – 0.04)	n a c	0.20 0.75 0.03	40 350 3000	1.5 2.3 1.5
DWELLING		0.1 (0.04 – 0.15)	n a	0.1 0.8	40 200	1.5 2.3
	+ cigarette smoke	0.005 (0.001-0.009)	a	0.995	300	2.0
INDOOR (AT WORK)		0.06	a	0.94	200	2.5
	+ cigarette smoke	0.005	a	0.995	300	2.0
MINE	working + diesel engine	0.006 (0.001-0.03)	a	0.994	200	2.0
	working (dry drill.), without diesel engine	0.005	a	o.995	550	4.0
	without working actions	0.02	a	0.98	350	2.3
WATER SUPPLY STATION		0.05 (0.03-0.09)	n a	0.15 0.80	50 300	1.5 1.8
TOURIST CAVE (POSTOJA)		0.10 (0.05 – 0.16)	a	0.9	330	2.5

The results of f_p-measurements show, that the dominant parameter for all investigated places is the aerosol concentration in air. The particle concentrations vary between 10^3 and 10^6 particle/cm^3 depending on aerosol sources and human activities indoors and outdoors. Fig. 4 shows the results of measurements in homes with different aerosols sources. The influence of the particle concentration in mines is illustrated in Fig. 5. In all places the dependence of the unattached fraction as function of the particle concentration Z (measured by means of a condensation nuclei counter (CNC)) can be approximated by a semi-empirical equation

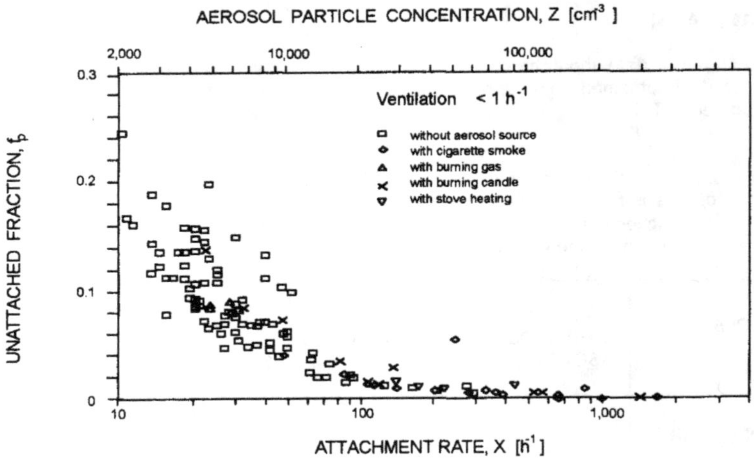

Fig. 4: Values of the unattached fraction (f_p) measured in rooms with different aerosol sources (aerosol measurement: condensation nuclei counter (CNC)) [7].

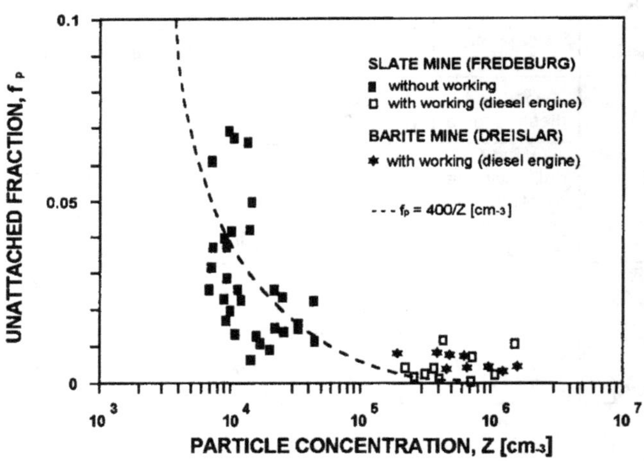

Fig. 5: Values of the unattached fraction f_p, of the radon decay products in two different mines as function of the aerosol particle concentration in mine air (measurement of the aerosol: concentration nuclei counter (CNC)) [8]

$$f_p = 400/Z[cm^{-3}]$$

Places with high particle concentration $Z > 4\ 10^4$ particles/cm^3 have f_p-values below 0.01. In rooms with low particle concentrations (< 10^3 particle/cm^3 with an operating air cleaner) the f_p-values are higher than 0.5.

In general, the relative activity size distribution of the radon daughter aerosol can be described by three modes, as illustrated by the example for outdoor air (Fig. 6). There is the nucleation mode (AMD_n values: 30 – 60 nm), the accumulation mode (AMD_a values: 100 – 1000 nm), and the coarse mode (AMD_c values: >1000 nm). However, in many places the activity size distribution can be approximated by one log-normal distribution (Table 1), because one aerosol source dominates.

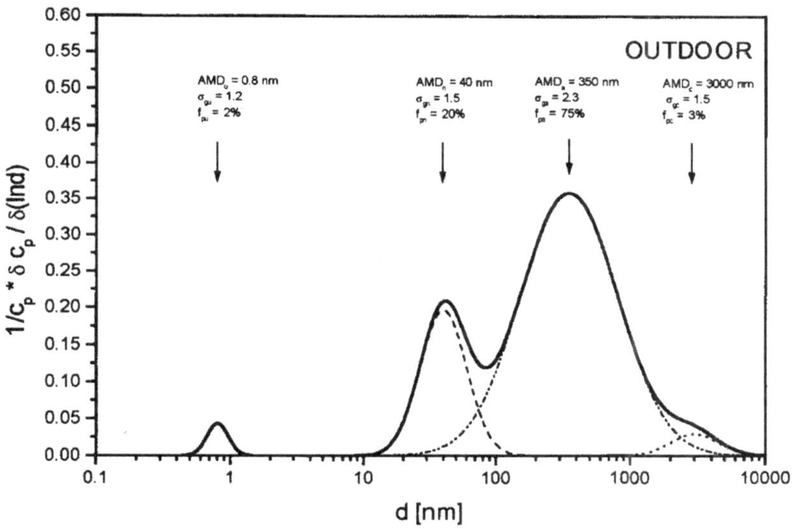

Fig. 6: A typical relative activity size distribution of the potentiel alpha energy concentration of the radon decay products in the open air (AMD_i = activity median diameter; σ_{gi} = geometric standard deviation; f_{pi} = fraction of the mode i) [7]

4. Dose conversion factors

In Table 2 the dose conversion factors for different exposure situations are summarised. These data are based on model calculation as reported before and illustrated in Fig. 2 and Fig. 3 and taking into account the different radon characteristics of Table 1.

The variation of the dose conversion factor for the radon daughter aerosol (DCF_{ae}) is small for all investigated places, if the result obtained of a mine with dry drilling working and without diesel engines is ignored. It ranges between 7 and 11 mSv/WLM.

The great variation of the unattached fraction in the air of different places caused a great variation of the dose conversion factor of the unattached clusters (DCF_u). In addition, a great difference of DCF_u is between nose and mouth breathing. The dose contribution by the unattached radon decay products varies between 0.1 – 11 mSv/WLM for different places. In some cases of air conditions the contribution of dose by the unattached clustêrs is about equal the dose fraction of the radon daughter aerosol.

The comparison with the dose conversion convention of ICRP 65 [9] based on the results of epidemiological studies shows, that the DCF values for working places (variation: 8 – 20 mSv/a) and homes (variation: 7 –18 mS/WLM) obtained by dosimetric model calculations are about 2 - 4 times higher than the values of the epidemiological approach.

Table 2: Dose conversion factors for the unattached (DCF_u) and aerosol attached (DCF_{ae}) radon daughters in the air of different places.
Assumed size distribution of the unattached fraction: AMD_u = 0.8 nm, σ_{gu} = 1.2

PLACE	ACTION/ AEROSOL SOURCE	INHALATION		DCF_u	DCF_{ae}	**DCF**
		NOSE/ MOUTH	v[$m^3 h^{-1}$]	[mSv/WLM] (effective dose)		
OPEN AIR		Mouth	1.2	2.1	11.1	13.2
DWELLING		Nose Mouth	0.75 0.75	1.7 9.4	9.0 9.0	10.7 18.4
	+ cigarette smoke	Nose Mouth	0.75 0.75	0.1 0.5	7.0 7.0	7.1 7.5
INDOOR (AT WORK)		Mouth	1.2	8.2	9.4	17.6
	+ cigarette smoke	Mouth	1.2	0.7	7.0	7.7
MINE	working + diesel engine	Mouth	1.7	1.0	10.0	11.0
	working (dry drill.), without diesel engines	Mouth	1.7	1.0	34	35
	without working actions	Nose	1.2	2.1	7.8	9.9
WATER SUPPLY STATION		Mouth	1.2	5.4	8.7	14.1
TOURIST CAVE (POSTOJA)	without diesel engine	Mouth	1.2	10.7	9.0	19.7
ICRP 65 - dose conversion convention					at home at work	4.0 5.0

5. References

[1] Zock, C.: Die Messung der Aktivitätsgrößenverteilung des radioaktiven Aerosols der Radonzerfallsprodukte und deren Einfluß auf die Strahlendosis beim Menschen, Dissertation, Georg-August-Universität Göttingen, 1996.

[2] Zock, C., J. Porstendörfer, and A. Reineking: The influence of biological and aerosol parameters of inhaled short-lived radon decay products on human lung dose, Radiation Protection Dosimetry, 63, 197-206, 1996.

[3] James, A.C.: Lung dosimetry. In: Radon and its decay products in indoor air, eds. W.M. Nazaroff, and A.V. Nero, (Chichester: John Wiley & Sons), 1988.

[4] Birchall, A. and A.C. James: Uncertainty analysis of the effective dose per unit exposure from radon progeny and implications for ICRP risk-weighting factors. Radiation Protection Dosimetry. 53, 133-144, 1994.

[5] International Commission on Radiological Protection: Human respiratory tract model for radiological protection. ICRP Publication 66, Pergamon Press, Oxford, 1994.

[6] Reineking, A.: unpublished results, 1979.

[7] Porstendörfer, J.: Radon: measurements related to dose, Environment International, 22, 563-583, 1996.

[8] Butterweck, G., J. Porstendörfer, A. Reineking, and J. Kesten: Unattached fraction and the aerosol size distribution of the radon progeny in a natural cave and mine atmospheres, Radiation Protection Dosimetry, 45, 167-170, 1992.

[9] International Commission on Radiological Protection: Protection against radon-222 at home and at work, ICRP Publication 65, Pergamon Press, Oxford, 1994.

Health effects and radiation dose from exposure to radon indoors

Gun Astri Swedjemark
Båtsmansvägen 11, S-192 48 Sollentuna, Sweden

Abstract

Radon exposure has been declared a health hazard by several organisations, for example the International Commission on Radiological Protection (ICRP) and the World Health Organisation (WHO). The basis for the risk estimate has been the results from epidemiological studies on miners exposed to radon, supported by the results of residential epidemiology. Only few of the many residential epidemiological studies carried out hitherto have a design applicable for a risk estimate. The largest is the Swedish national study but several large well designed studies are ongoing. An excess risk has also been found in animal research. The model describes smoking and radon exposure as between additive and multiplicative, found in both miners and residential studies. The relatively few non-smokers among the miners and also among the residents give a problem at estimating the radon risk for these groups. It would also be desirable to know more about the importance of the age and the time period at exposure. Lung dose calculations from radon exposure are not recommended by ICRP in their publication 66. For comparison with other radiation sources the ICRP recommends the concept "dose conversion convention" obtained as the risk estimate divided by the detriment. Other effects of radon exposure than lung cancer have not been shown epidemiologically, but dose calculations indicate an excess risk of about 5% of the excess lung cancer risk.

1. Introduction

The alpha dose to the critical bronchial cells from inhaled short-lived radon progeny, the clear evidence of an excess lung cancer risk, increasing with the exposure of the radon progeny which follows from epidemiological studies of radon exposed underground miners, as well as from animal experiments, is the basis for our concern about the lung cancer risk from indoor radon. Most of the residential studies with case control design support this. Radon with its progeny has been classified as a carcinogen to humans by the International Agency for Research on Cancer [1] and by other international organisations e.g. the International Commission on Radiological Protection, ICRP [2].

The problem is to quantify the possible lung cancer risk from indoor radon. Three types of approaches could be taken into consideration.

- Transfer of exposure-risk models from radon exposed miner's data, to the residential radon exposure to the general population.
- Well-designed epidemiological studies on residential radon.
- "A dosimetric approach" which proceeds from the observed excess risk of lung cancer among the atomic bomb survivors, exposed to a single short-term dose to the lung by external radiation.

2. Epidemiological studies

Epidemiological studies of radon exposed miners give strong evidence for an induction of lung cancer by inhaled radon progeny [3]. Miners with low exposure to radon and progeny have had little or no excess of lung cancer e.g. in coal [1] and potash mining [4] and also in iron mining [5]. Figure 1 shows the excess relative risk of lung cancer as function of the cumulative exposure to radon progeny for the combined data for 11 cohort studies of male underground miners [3]. The obtained relative risk coefficient is modified by various factors, such as age, time-period, and time since exposure, dose-rate etc.

For transfer of underground miner data to the general public several models have been published. They contain simplified assumptions particularly to females and children. Calculations by NEA 1983 [6] yield a factor of 0.65, by NCRP 1984 [7] a value of 1.4 for adult men, and by Harley 1984 [8] a range from 0.8 to 1.2. Using pooled data for the miners, analysis of lung cancer risk was carried out on data restricted to lower exposures, either < 50 WLM (≈3600 Bq y m^{-3}) or < 100 WLM (≈7300 Bq y m^{-3}) [9]. Risk models based on the unrestricted data and restricted data gave similar risk estimates.

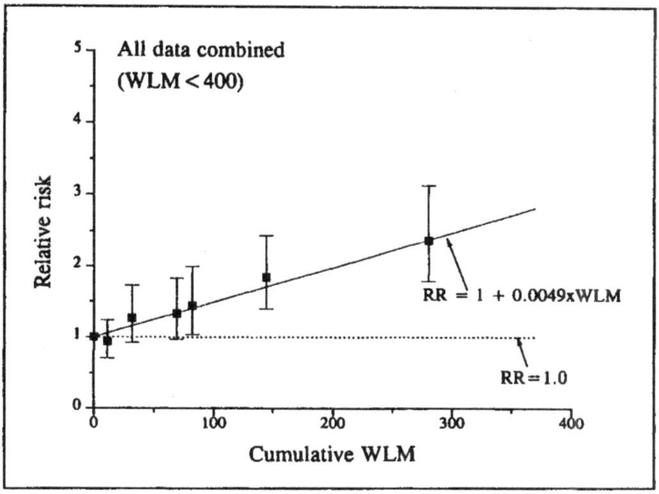

Figure 1. Relative risks (RR) of lung cancer by cumulative WLM and fitted linear excess RR model for each cohort and for all data combined [3].

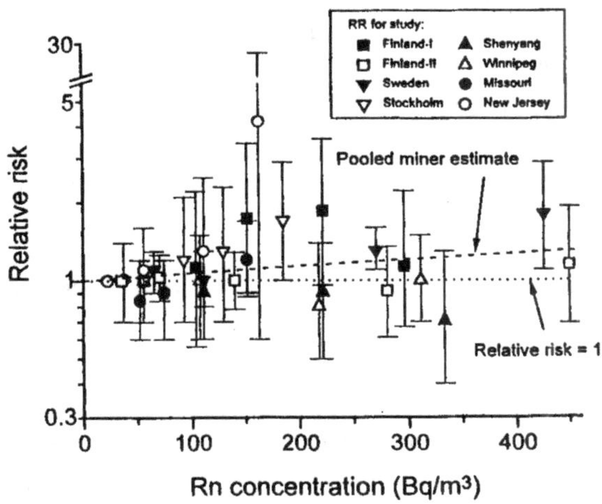

Figure 2. Relative risk (RR) of lung cancer by categories of Bq m^{-3} for eight residential radon case-control studies. Dashed line = extrapolation of RR from miners. The dotted line is RR = 1 [11].

The results of 12 published residential case-control studies are summarised and reviewed in UNSCEAR 1994 report [10] and in papers by Lubin et al [3, 11]. Most of these studies involve only a relatively small number of lung cancer cases, which leads to a large statistical uncertainty. This is illustrated in figure 2 where for 8 of these studies the observed relative lung cancer risk is plotted as function of the radon concentration indoors [11].

The largest study is the Swedish nation-wide study (1360 cases, 2847 controls) [13]. For the range 140-400 Bq/m^3 (mean 214 Bq/m^3) for an average exposure time period of about 30 years a mean relative risk of 1.3 (95% CI: 1.1-1.6) was found, figure 3. For concentrations of more than 400 Bq/m^3 (mean 734 Bq/m^3) the mean relative risk was found to be 1.8 (95% CI: 1.1-2.9). In addition to the random error in the confidence interval comes the systematic errors.

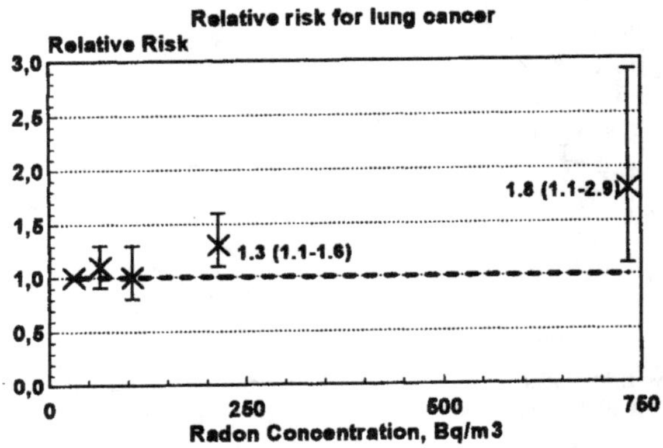

Figure 3. Relative risk for lung cancer related to estimated time weighted average for residential radon exposure in the nation-wide Swedish study [13]. Bars indicate 95% confidence intervals.

All residential case-control studies have limitations. This concerns particularly the assessment of the life-time exposure, taking into consideration residential mobility, measuring today for estimate of a lifetime exposure, different occupational habits etc. The radon exposure of most lung cancer cases in the studies were small compared with the corresponding lung cancer risk from smoking. To illustrate the influence of errors in exposure assessment, case-control cases have been simulated [12, 14]. Using Swedish data [14], the dominant source of error was found to be connected to the measuring of radon and the determination of the average air concentration, and that poor knowledge of living habits could contribute significantly to the resulting uncertainty. Alterations to the houses seemed on the other hand to be of little importance. Measurement of only the last home of the study persons was shown to be a strategy of very little value when assessing the exposure history of the study persons thereby jeopardising the analysis.

Using the simulation technique on the Swedish national residential study [13] adjustments for random errors in the radon exposure assessment were possible [15]. Without such adjustment the linear excess relative risk of about 0.10 per 100 Bq m^{-3} was found. Following adjustment for the random errors in the exposure estimate 0.15-0.20 per 100 Bq m^{-3} was suggested. The extrapolations from miner studies [3], and the meta-analysis of the combined residential studies (described below) give results within this range.

To provide more information on the risk of lung cancer from indoor radon Lubin et al [11] conducted a meta-analysis of all case-control studies that included at least 200 lung cancer cases each and that used long term indoor radon measurements. Eight studies were available including 4263 lung cancer cases and 6612 controls. From the published results from each study, confounder adjusted relative risk estimates and 95% confidence intervals for categories of radon concentrations were used for weighted linear regression analysis of the natural logarithm. The reasons for the departure from the linear excess relative risk modelling were: (i) the difference between linear and logarithmic linear models are minimal when the relative risks are small for the ingoing radon data, (ii) the variance estimate of the exposure response is minimised. For a linear model there is no symmetry of the errors around the averages of the exposures. The adaptation to the data points can only have been intended to investigate if there is a correlation between radon exposure and lung cancer, not to check the risk model.

The results are shown in figure 4. The combined results indicate a relative risk significantly (two sided P = 0.03) different from zero: 1.14 (95% CI: 1.0-1.3) at 150 Bq m^{-3}. No single study dominated the combined results. The trend was similar to model based extrapolations from miner results, and to relative risks computed directly from miners with low cumulative exposure. The meta-analysis should be evaluated with care when the single studies deal with low expected risks as for radon in dwellings, or when there is a risk for confounding or misclassification of the exposure, but is valuable to identify differences between studies and to summarise results.

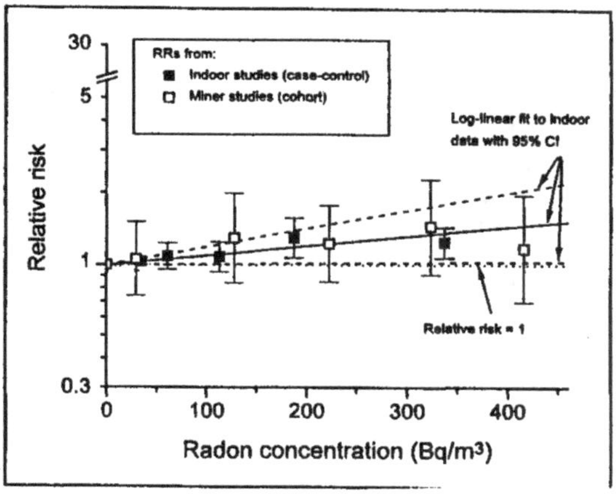

Figure 4. Summary relative risks (RR) from the meta-analysis of indoor radon studies, and RR from the pooled analysis of underground miner studies, restricted to exposures under 50 Working Level Months [11]. Also included are a RR of 1, the fitted exposure response and its 95% confidence interval (CI) from the indoor radon studies.

Analysis of miner data indicates that the relative risk for smoking and radon progeny exposure is most likely intermediate between a multiplicative and an additive model [3]. The Swedish national study [13] found a correlation between radon in dwellings and smoking of more than additive effect and near a multiplicative effect. Other residential studies have given various results due to the low number of non-smokers.

For residential exposure to radon progeny we do not know very much about the influence of age, or the **time period** of exposure. Dose calculations indicate that the effective dose would be 20-50 % higher for children than for adults [16, 17]. However, the sensitivity of the children for radon exposure is not known. According to the miner data the relative risk coefficient decreases with age and with the time since exposure. The miner data support the presence of an inverse **dose-rate effect** as well as a lower relative risk coefficient at low total dose [18]. Thus, a purely linear dose response model may be an oversimplification for radon exposure in dwellings.

The excess risks for **other organs** than the lungs from exposure to indoor radon have not been shown epidemiologically, neither in residential studies, nor for the miners. Dose calculations show an excess dose much lower than for the lung dose from inhalation [19]. Therefore, it will be more difficult to quantify the excess risk for other organs than the lungs by epidemiological studies. ICRP estimate an excess risk for other effects of radon inhalation than lung cancer of about 5% of the excess lung cancer risk [2].

3. Conversion models and radiation dose

Many physical, chemical and biological factors, now omitted, should be taken into account at calculation of the effective dose. Furthermore, the weighting factors at the transfer from the absorbed dose to the effective dose have large uncertainties. During the 1970s and 1980s similar risk estimates were obtained from studies of miners and from the dosimetric approach. At that time the basal cells were assumed to be the only critical cells [6] and the knowledge about the characteristics of the radon progeny in the air was unsatisfactory. Since then intensive research on the characteristics of the air and other factors has been performed. In the lung model of ICRP [20] inhalation of radon and progeny is exempted from the dose calculations.

For comparisons with the effective dose from other sources of ionising radiation ICRP has introduced the concept "dose conversion convention". This is the risk estimate based on epidemiological studies divided by the detriment. The detriment is a combination of the fatal cancer probability, the relative length of life lost, the relative non-fatal contribution, and genetical influence caused by irradiation of a population. The conversion convention is based on an equality of detriment, not on dosimetry. An exposure to radon progeny of 1 mJ h m^{-3} is equivalent to an effective dose of 1.4 mSv for workers and 1.1 mSv for members of the public [2] (for units see appendix).

4. Risk estimates for radiation protection

The organisations ICRP, WHO, BEIR etc. and many countries have made risk estimates for radiation protection purposes. The organisations all had the same basis but the studies have been judged in different ways. The large uncertainties in the risk estimates are emphasised. ICRP does not give any range for the estimate, SSI has decreased the uncertainty range compared with earlier risk estimates. The risk estimates have been given as relative risk (RR), attributable relative risk (ARR), and absolute risk. For calculation of the last two concepts, the lung cancer frequency in the actual population must be known.

The risk estimated for the recommendations of ICRP publication 65 [2] is based on the following principles:

- Relative risk models are based on studies of miners.
- ICRP's reference population is used.
- The same absolute risk coefficient is used for dwellings and workplaces, and for males, females and children.
- Additive effect of radon exposure and smoking is assumed.
- The uncertainty is emphasised. No uncertainty range has been given.

The absolute risk as basis for ICRP's recommendation was estimated to be 8×10^{-6} per mJ h m^{-3}, corresponding to 1.55×10^{-6} per Bq y m^{-3}.

WHO published their guidelines in 1987 [21]. These have now been revised but will not be published until 1998.

The Swedish Radiation Protection Institute (SSI) [22] has based its risk estimates on both miner studies and residential studies. This gives a somewhat lower risk estimate than ICRP's, 1.2×10^{-6} per Bq y m^{-3}.

Some of the residential studies, and the meta-analysis of the combined residential studies [11] were not published in time neither for the publication of the recommendations of ICRP, nor for the limitation system of SSI. The new results, however, do not alter the judgements made in 1994.

A somewhat different approach to the studying of the problem has been taken by Doi et al [23]. They based their calculations on the results from the Swedish national epidemiological study with two alternatives: (i) all study persons and (ii) for those not sleeping with open windows, and an average radon concentration of 100 Bq m^{-3} (13). The annual effective dose to the Swedish population was found to be 0.43 (CI: 0.04 - 1.08) - 0.99 (CI: 0.09 - 2.50) mSv y^{-1} and 0.85 (CI: 0.23-1.82) - 1.98 (CI: 0.53 - 4.20) mSv y^{-1} respectively. The official Swedish estimate of the dose to the population, 2 mSv y^{-1}, is calculated directly from the ICRP dose conversion convention coefficient. Another difference is the occupancy factor taken as 0.8 in the Swedish estimate (a conventional assumption) and as 0.6 by Doi et al as found for the Swedish population. The estimates are of the same magnitude.

Several well-designed epidemiological studies on indoor radon and lung cancer are ongoing and will be published within a couple of years. I hope that these results will decrease the uncertainty in the risk estimate for radon exposure to the general public. Particularly, the risk estimate for non-smokers is now very uncertain. A study in Sweden with about 500 non-smoking lung cancer cases is ongoing. Maybe the planned pooling of the ongoing studies can give a better estimate of the risk for the non-smokers.

5. Conclusions

There is no doubt that ionising radiation is a carcinogenous agent for humans. This statement is also valid for inhalation of radon and its progeny. The problem is how to quantify the magnitude of the risk for residential radon. The risk estimates for radiation protection purposes have been based on the data from studies of miners using models for transferring the conditions from miners to the general population. The case control studies in residents, hitherto published, do not contradict these estimates. The approach indicates that exposure to radon and its decay products are the second most important cause of lung cancer in the general population after smoking.

6. References

[1] International Agency for Research on Cancer, Evaluation of carcinogenic risks to humans: man-made fibres and radon, IARC Monograph 43, Lyon France, 1988.

[2] International Commission on Radiological Protection, ICRP, Protection against radon-222 at home and at work, Public. 65, Annals of the ICRP 23, 2, 1993

[3] Lubin J. H., J. D. Boice Jr., C. Edling, R. Hornung, G. Howe, E. Kunz, R.A. Kusiak, H.I. Morrison, E. P. Radford, J. M. Samet, M. Tirmarche, A. Woodward, Shu Xiang Yao, and D.A. Pierce, Radon lung cancer risk: a joint analysis of 11 underground miners studies, National Institutes of Health, National Cancer Institute, NIH Publ. No. 94-3644, 1994.

[4] Waxweiler R.J., J.K. Wagener, V.E. Archer, Mortality of potash workers. J. Occup. Med. 15, 486, 1973.

[5] Lawler A.B., J.S. Mandel, L.M. Schuman,and J.H. Lubin, A retrospective cohort mortality study of iron ore (hematite) miners in Minnesota. J. Occup. Med., 27, 507, 1985.

[6] Nuclear Energy Agency, OECD, Dosimetry aspects of exposure to radon and thoron daughter products, Report of a group of experts, 1983.

[7] National Council on Radiation Protection and Measurements, Evaluation of occupational and environmental exposures to radon and radon daughters in the United States, NCRP report No. 78, Bethesda, MD, 1984.

[8] Harley N. H., Comparing radon daughter dose: environmental versus underground mines, Radiat. Prot. Dosimetry, 7, 1-4, 371-375, 1984.

[9] Lubin J.H., L. Tomásek, C. Edling, R.W. Hornung, G. Howe, E. Kunz, R.A. Kusiak, H.I. Morrison, E.P. Radford, J.M. Samet, M. Timarche, A. Woodward, and Shu Xiang Yao, Estimating lung cancer mortality from residential radon using data for low exposures of miners, Radiation Research 147, 126-134, 1997.

[10] United Nations Scientific Committee on the Effects of Atomic Radiation. Sources and effects of ionising radiation. UNSCEAR 1994 Report to the General Assembly with scientific annexes. New York 1994

[11] Lubin J. H., and J. D. Boice, Jr, Lung cancer risk from residential radon: meta-analysis of eight epidemiological studies, J. Nat. Cancer Institute, 89, 49-57, 1997.

[12] Lubin J.H., J.D. Boice, Jr, and J.M. Samet, Errors in exposure assessment, statistical power and the interpretation of residential radon studies, Radiation Research, 144, 329-341, 1995.

[13] Pershagen G., G. Åkerblom, O. Axelson, B. Clavensjö, L. Damber, G. Desai, A. Enflo, F. Lagarde, H. Mellander, M. Svartengren, and G.A. Swedjemark, Residential radon exposure and lung cancer in Sweden. N. Engl. J. Med. 330, 159-164, 1994.

[14] Bäverstam U., and G. A. Swedjemark, Where are the errors when we estimate radon exposure in retrospect?, Radiat. Prot. Dosimetry. 36, 2/4, 107-112, 1991.

[15] Lagarde F., G. Pershagen, G. Åkerblom, O. Axelson, U. Bäverstam, L. Damber, A. Enflo, M. Svartengren, and G.A. Swedjemark, Residential radon and lung cancer in Sweden: risk analysis accounting for random error in the exposure assessment, Health Phys. 72, 269-276, 1997.

[16] Hofmann W. Dose calculations for the respiratory tract from inhaled natural radioactive nuclides as a function of age. Health Phys. 43, 1, 31-44, 1982.

[17] US National Research Council. Comparative dosimetry of radon in miners and homes. Washington, DC: National Academy Press 1991.

[18] Lubin J. H., J. D. Boice Jr., C. Edling, R. Hornung, G. Howe, E. Kunz, R.A. Kusiak, H.I. Morrison, E. P. Radford, J. M. Samet, M. Tirmarche, A. Woodward, and Shu Xiang Yao, Radon-exposed underground miners and inverse dose-rate (protraction enhancement) effects, Health Phys. 69, 494-500, 1995.

[19] Axelson O., and F. Forastiere, Radon as a risk factor for extra-pulmonary tumours, Med. Oncol. & Tumour Pharmacother, 10, 4, 167-172, 1993.

[20] International Commission on Radiological Protection, ICRP, Human respiratory tract model for radiological protection, Publ. 66, Annals of the ICRP, 24, 1-3, 1994.

[21] World Health Organisation regional Office for Europe, Air Quality Guidelines for Europe, WHO Regional Publications, European Series No. 23, 1987.

[22] Swedish Radiation Protection Institute, Radon - risk - stråldos, Bostäder och arbetsplatser, Information i 94-06, S-171 16 Stockholm Sweden, 1994.

[23] Doi M., F Lagarde, R. Falk, and G.A. Swedjemark, Annual effective dose due to residential radon progeny in Sweden: Evaluation based on current risk projection models and on risk estimates from a nation-wide Swedish epidemiological study. Report 96:16, Swedish Radiation Protection Institute, 1996.

Appendix: conversion coefficients

Conversion coefficients for the different concentration quantities and for the corresponding exposure quantities for radon-222 (ICRP-publication 65).

Quotient	Conversion coefficients
c_p / c_{eq}	5.56×10^{-9} (J m^{-3}) per (Bq m^{-3})
c_{eq} / c_p	1.80×10^{8} (Bq m^{-3}) per (J m^{-3})
P_p / P_{eq}	5.56×10^{-9} (J h m^{-3}) per (Bq h m^{-3})
	1.57×10^{-6} WLM per (Bq h m^{-3})
P_{eq} / P_p	1.80×10^{8} (Bq h m^{-3}) per (J h m^{-3})
	6.37×10^{5} (Bq h m^{-3}) per WLM

Quantities:

c_p - concentration of potential alpha energy,

c_{eq} - equilibrium equivalent concentration of radon,

P_p - time-integrated exposure to potential alpha energy concentration,

P_{eq} - time-integrated exposure to equilibrium equivalent concentration of radon.

Effective dose for radon progeny in Japan

Michikuni Shimo

Division of Radiotoxicology and Protection,
National Institute of Radiological Sciences
Anagawa, Inage-ku, Chiba, 263 Japan

Abstract

We estimated the effective dose for radon progeny in the Japanese population. The value was calculated with a modified UNSCEAR equation taking the following terms into consideration; radon concentration, radon equilibrium factor, occupancy factor or annual tidal volume, activity-energy conversion coefficient and dose conversion coefficient.

The following figures were used in the calculation; a radon concentration of 15.5 (7 ~ 27) Bq/m^3 for indoors and 5.0 (2.5~10) Bq/m^3 outdoors; an equilibrium factors of 0.45 (0.35~ 0.57) for indoors and 0.70 (0.50 ~ 0.90) outdoors; an occupancy factor of 0.87 for indoors and 0.13 outdoors for males, and 0.91 indoors and 0.09 outdoors for females; an annual tidal volume of 7,000 (6,350 ~ 7,900) m^3 for males and 6,600 (6,100 ~ 7,550) m^3 for females. The annual tidal volume was obtained the product of tidal volume and occupancy time, which were estimated in Japan and from surveys on Japanese life-style.

The effective annual doses for radon were estimated to be 0.55 mSv for males, 0.49 mSv for females and 0.52 mSv for the mean of both sexes. These were about 40 % of the value shown in the 1993 UNSCEAR Report. The dose range due to ranges of the factor was -75 ~ +175 % of the representative value.

1. Introduction

Much attention has been payed to exposure to indoor radon, ever since the annual dose was estimated at 1.0 mSv in the report published by UNSCEAR in 1982 [1]. Similar values have since been adopted throughout much of the world [2, 3, 4]. In Japan, however, the radon concentration is lower than the values adopted in the UNSCEAR Report and the life-style differs to some extent from that of Western societies. Therefore the effective dose in Japan may differ somewhat from the

UNSCEAR value [5].

The aim of this paper is to estimate the Japanese annual effective dose and range for radon progeny. The dose calculations were performed using a modified UNSCEAR equation with the radon concentration, various factors, the activity-energy conversion coefficient and the dose conversion coefficient. The factors used in the calculation were as follows; the equilibrium factor, the occupancy factor and the tidal volume, and were mainly evaluated using data obtained in Japan. For an exact measurement of effective dose, one must take into consideration differences of sex, location, whether one is indoors/outdoors, occupation and age.

2. Method for estimating doses of radon

2.1 Method

In a former study [6], the authors first distinguished workers from non-workers in the population, then took into account differences of sex, occupation (in the case of workers) and age (in the case of non-workers). Furthermore, differences of location, indoor/outdoor, occupancy factor and tidal volume were also considered as shown in Table 1. Numerous factors are used for estimating the effects on humans of radon,

Table 1 Taking account of several factors

	Worker	Non-worker
Sex	separated	separated
Location	considered	considered
Indoor/Outdoor	considered	considered
Occupation	considered	no
Age	no	considered
Occupancy factor	considered	considered
Tidal volume	ICRP Publ.23	considered

* The UNSCEAR equation is as follows;

$$E = Q \cdot F \cdot T \cdot K \qquad [\text{mSv/y}]$$

where Q is the radon concentration in Bq/m^3, F the radon equilibrium factor, T the occupancy factor, K the dose conversion coefficient 0.149 $mSv/(Bq \cdot m^{-3})$ (=17 $nSv/(Bq \cdot h \cdot m^{-3})$ x 8,760 h) outdoors, 0.076 $mSv/(Bq \cdot m^{-3})$ indoors for public, and 0.117 $mSv/(Bq \cdot m^{-3})$ for worker.

i.e. radon concentration, equilibrium factor, ratio of the unattached fraction, diameter of aerosol particle, the lung model, position of the target cell, thickness of the mucus layer, clearance from mucus layer, tidal volume, radiation weighting factor and tissue weighting factor as is shown in Fig.1. These factors are mutually correlated in many cases with often complex relations. For instance, the ratio of the unattached fraction and diameter of aerosol particle relate to the dose conversion coefficient [7]. The UNSCEAR Report showed that a simple equation* using four of the above factors was practical and useful for estimating effective dose of radon progeny. In this study, the effective dose was estimated with a modified UNSCEAR equation [1].

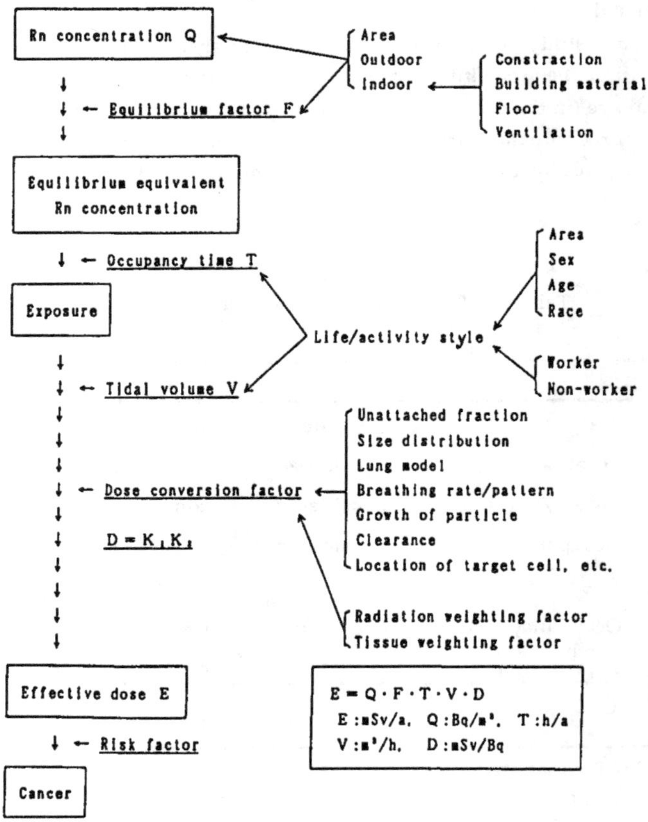

Fig.1 A flow chart for calculating the effective dose due to radon.

2.2 The modified equation

We modified the UNSCEAR equation for estimating the effective dose of radon progeny indoors or outdoors $E_{i,j}$ to:

$$E_{i,j} = Q_i \bullet F_i \bullet (T_{i,j} \times 8{,}760) \bullet V_{i,j} \bullet K_1 \bullet K_2 \tag{1}$$

where Q_i is the radon concentration in Bq/m^3 (sufix i means difference of district), F_i the radon equilibrium factor, $T_{i,j}$ the occupancy factor (sufix j means differences of occupation in workers or of age in non-workers), $V_{i,j}$ the tidal volume in m^3/h, K_1 the activity-energy conversion coefficient in J/Bq and K_2 the dose conversion coefficient in Sv/J [5], which are shown in Fig.1.

When mean values of radon concentration and equilibrium factor were adopted in the effective dose calculation, the mean value of the product between $T_{i,j}$ x 8,760 and $V_{i,j}$ was also used [6].

3. Dose estimation factors

3.1 The radon concentration

A representative radon concentration indoors of 15.5 Bq/m^3 and range of 7 ~ 27 Bq/m^3 was adopted from a nationwide survey from 1992 to 1996 (see Fig.2) [8] performed by the Japan Chemical Analysis Center with the support of the National Institute of Radiological Sciences. The outdoor radon concentration was 5.0 (2.5 ~ 10) Bq/m^3 which was reffered a semination wide survey performed by Abe *et al.* [9].

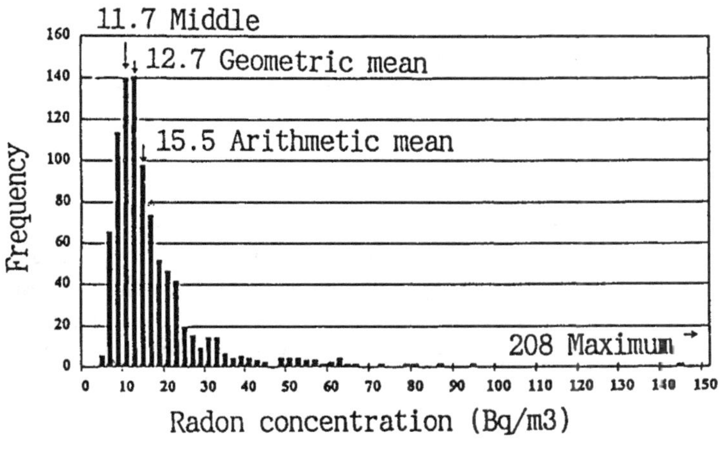

Fig.2 Indoor radon concentration in Japan (after ref.8).

3.2 The equilibrium factor

The equilibrium factors in Japan were estimated to be 0.45 (0.35 ~ 0.57) indoors and 0.70 (0.50 ~ 0.90) outdoors from the data of two studies [10, 11].

3.3 The occupancy factor

The occupancy factors were estimated at 0.87 indoors (including 0.6 in home) and 0.13 outdoors (including 0.04 in vehicles) for males, and 0.91 indoors (including 0.63 in home) and 0.09 outdoors (including 0.03 in vehicles) for females as shown in Fig.3. The values were obtained from surveys by a government office, broadcasting stations and publishing companies [6].

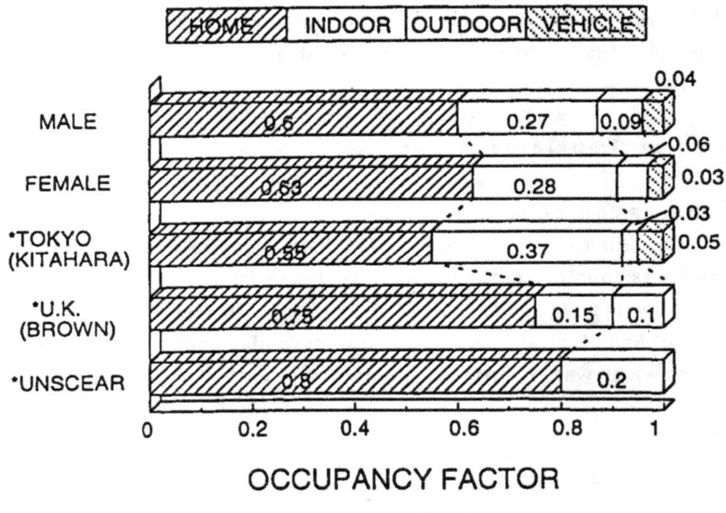

Fig.3 The occupancy factor obtained in Japan (after ref.6).

3.4 The annual tidal volume

In the worker's analysis, the value for tidal volume was referred from ICRP Publication 23, and for the non-workers, it was evaluated from Japanese physical health data obtained for each age. The annual tidal volume was obtained by multiplying tidal volume by occupancy time (occupancy factor x 8,760 hours). The mean annual tidal volume was estimated for males to be 6,000 (range: 3,850 ~ 6,550) m^3 indoors, 760 (400 ~ 3,800) m^3 outdoors and 6,760 (6,350 ~ 7,900) m^3 in total, and for females to be 6,000 (4,000 ~ 6,300) m^3 for indoors, 480 (300 ~ 3,300) m^3 for outdoors and 6,480 (6,100 ~ 7,550) m^3 in total whereas Fig.4 showed annual tidal volume in each occupation [6]. These values were also obtained with the analysis of occupancy factor.

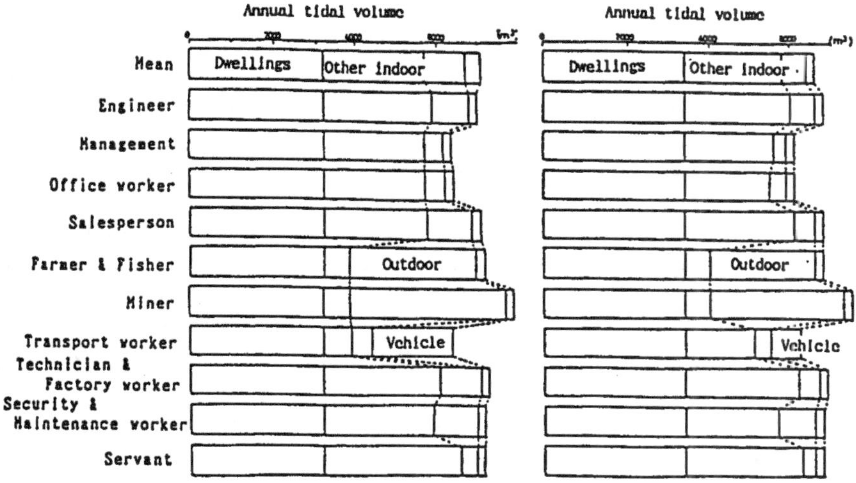

Fig.4 The annual tidal volumes in Japan (after ref.6)

3.5 The dose factor

The dose conversion factors are as follows: the coefficient of activity-energy conversion K_1 is 5.54 x 10^{-9} J/Bq, and K_2 was the dose conversion coefficient for public 2 Sv/J for indoors and 3 Sv/J for outdoors, although the factor was estimated to be 1.4 ~ 3.3 Sv/J using the data of aerosol size distribution and of unattached atom in a former study [12]. The authors have shown that particle sizes of radon progeny were 0.1 ~ 0.2 μm (σ_g: 2 ~ 4) indoors and 0.06 ~ 0.24 μm (σ_g: 1.6 ~ 4) outdoors, whereas in another study [13], the unattached atom ratios were 0.04 (0.02 ~ 0.06) for indoors and 0.037 (0.01 ~ 0.06) for outdoors. Jacobi & Eisfeld and James *et al.* showed that the unattached fraction and particle size distribution directly affected the dose conversion factor [7].

3.6 Other factors

In the calculation, biological and anatomical factors were not taken into account. Concerning the radiation weighting factor and the tissue weighting factor, only widely-adopted values were used.

4. Results and discussion

The mean annual effective dose for radon progeny in Japan was estimated to be 0.55 mSv for males and 0.49 mSv for females using the equation (1). The averaged value for both sexes was 0.52 mSv. This was only about 40 percent of the value in the 1993 UNSCEAR Report. The main reason for the discrepancy is the difference in the radon concentration: 15.5 Bq/m^3 as opposed to 40 Bq/m^3 in the report. The value for effective dose is only about 68 percent of that estimated from ICRP Publication 65 (radon concentration: 40 Bq/m^3), because of the dose conversion coefficient. If the ICRP calculation is applied to the estimation, the effective dose decreases from 0.52 mSv to 0.30 mSv.

The dose range was estimated to be -75 ~ +175 % of the mean value from the range of each factor. The dose range agreed well with that obtained from equation (1) taking into account difference of district, i.e. different radon concentration, equilibrium factor, occupancy time and tidal volume [6].

Table 2 Effective dose and variation due to radon progeny

	Effective dose	Range (%)
Adult	0.52 mSv/y	- 75 ~ + 175
Male	0.55 mSv/y	(0.14 ~ 1.5 mSv/y)
Female	0.49 mSv/y	(0.12 ~ 1.3 mSv/y)

5. Conclusion

The Japanese annual effective dose for radon progeny was calculated with a modified UNSCEAR equation. Figures for factors in the equation were mainly obtained in Japan and from surveys of the local life-style. The annual effective dose for radon progeny was estimated to be 0.55 mSv for males, 0.49 mSv for females and 0.52 mSv for the mean of both sexes. The dose range was -75 ~ +175 % of the mean value.

References

[1] UNSCEAR: "Ionizing Radiation: Sources and Biological Effects", United Nations, 1982.

[2] UNSCEAR: "Sources, Effects and Risks of Ionizing Radiation", United Nations, 1988.

[3] UNSCEAR: "Sources and Effects of Ionizing Radiation", United Nations, 1993.

[4] ICRP: "Protection Against Radon-222 at Home and at Work", (ICRP Publication 65), ICRP, Pergamon, 1993.

[5] Shimo,M.: Actuality of Exposure due to Radon Progeny, NIRS Seminor No.22, 49-60, 1996.

[6] Shimo,M., Seki, K. and Kikuchi, I.: Effective Dose to Radon Concerning People's Activities: On the Tidal Volume and Occupancy Factor, in Environmental Radon, p.579-592, 1991.

[7] Jacobi,W. and Eisfeld, K.: Dose to Tissues and Effective Dose Equivalent by Inhalation of Radon-222, Radon-220 and Their Short-lived Daughters, GSF-Report, S626, 1980.

[8] National Institute of Radiological Sciences: Final Report on Nationwide Survey of Indoor Radon Concentration, NIRS-R-32, 1997.

[9] Abe,S. and Abe, M.: A Nation-wide Survey of Radon Indoors and Outdoors in Japan, Radon in the Living Environment - Levels and Risks - , Ed. S.Kobayashi and T.Kankura, Nat.Instit.Radiol.Sci., p.79-88, 1989.

[10] Kojima,H.: private communication

[11] Hattori,T. and Ishida,K.: Equilibrium Factor and Unattached Fraction of Radon Progeny in Nuclear Power Plants, Rad.Prot.Dosim., **55**, 191-197, 1994.

[12] Shimo,M., Saito, H. and Ikebe, Y.: The Size Distribution of Radon Daughter Aerosol in Several Dwellings and Estimation of Human Lung Dose to these Particles, Proceedings of The Third International Aerosol Conference, Kyoto, p.1205-1208, 1990.

[13] Shimo,M. and Ikebe,Y.: Measurements of Radon and its Short-lived Decay Products and Unattached Fraction in Air, Rad.Prot.Dosim. **8**, 209-214, 1984.

TEST OF THE LINEAR - NO THRESHOLD THEORY OF RADIATION CARCINOGENESIS

Bernard L. Cohen
University of Pittsburgh
Pittsburgh, PA 15260, USA

1. ABSTRACT

It is shown that testing the linear-no threshold theory (L-NT) of radiation carcinogenesis is extremely important and that lung cancer resulting from exposure to radon in homes is the best tool for doing this. A study of lung cancer rates vs radon exposure in U.S. Counties, reported in 1975, is reviewed. It shows, with extremely powerful statistics, that lung cancer rates decrease with increasing radon exposure, in sharp contrast to the prediction of L-NT, with a discrepancy of over 20 standard deviations. Very extensive efforts were made to explain an appreciable part of this discrepancy consistently with L-NT, with no success; it was concluded that L-NT fails, grossly exaggerating the cancer risk of low level radiation. Two updating studies reported in 1996 are also reviewed. New updating studies utilizing more recent lung cancer statistics and considering 450 new potential confounding factors are reported. All updates reinforce the previous conclusion, and the discrepancy with L-NT is increased.

2. INTRODUCTION

One of the most important scientific questions of our era is the health impacts of low level radiation (LLR). Conventionally this has been addressed by use of the linear-no threshold theory (L-NT), assuming that the cancer risk is simply proportional to the dose; this allows one to use the well established effects of high level radiation to deduce the risk of LLR.

The original basis for the linear-no threshold theory (L-NT) of radiation carcinogenesis was the theoretical argument that a single particle of radiation can damage a DNA molecule so as to initiate a cancer; the number of such initiating events is then proportional to the number of particles of radiation, which is proportional to the dose. There has been no observational evidence for this theory. The problem with the theoretical reasoning is that it is now recognized that spontaneous thermal events and chemical processes cause this same type of damage to DNA molecules, creating many thousands of times as many cancer initiating events as low level radiation. The mechanism by which radiation causes cancer is, therefore, <u>not</u> through these initiating events. It must rather be through reducing the effectiveness of biological defense mechanisms (BDM).

But there is no reason to believe that this diminution of BDM is proportional to radiation dose down to the lowest doses. On the contrary, there is abundant indisputable evidence that low level radiation can <u>improve</u> the effectiveness of BDM. As a few examples, it has been shown that various type cells previously exposed to LLR suffer fewer chromosome aberrations when later exposed to large radiation doses, and this effect is ascribed to stimulated production of repair enzymes by the LLR; pre-exposure to LLR reduces induction of mutations and increases survival rates of various type cells later exposed to high level radiation; LLR has been found to stimulate immune systems in mice as measured by various indicators, and it is generally believed that the immune system is an important contributor to BDM. These types of evidence surely leave L-NT open to serious question in its applications to LLR. Such applications have cost the public hundreds of billions of dollars. It is therefore extremely important to test L-NT in the low dose region where there is no experimental evidence for its validity.

Since the expected effects are quite small, any epidemiological test requires very large numbers of subjects exposed at different levels. The only practical source of such exposures is natural radiation. It is very difficult to do such a study with natural background gamma radiation because: (1) according to L-NT, background gamma radiation is responsible for only 2% of all cancers; (2) variations in exposure levels are generally quite small; (3) populations exposed at different levels live in rather different geographical areas with many environmental and ethnic differences; and (4) there are large unexplained differences in cancer rates - e.g. 40% differences among U.S. States. The situation with lung cancer induced by radon in homes is much more favorable because: (1) according to L-NT, it is responsible for 10% of all lung cancers; (2) variations in exposure are much larger; (3) these variations are much less associated with different environmental conditions and different ethnic backgrounds; and (4) differences in lung cancer rates are largely explained by differences in cigarette smoking prevalence. It is therefore evident that lung cancer induced by radon in homes provides a much better test of L-NT and there have been many attempts to use it for that purpose. Testing L-NT may well be the most important application of radon in homes.

3. DATA AND ANALYSIS -- 1995 VERSION [1]

We developed a compilation of radon measurements from available sources which gives the average radon level, r, in homes for 1729 U.S. counties, well over half of all U.S. counties and comprising about 90% of the total U.S. population. Plots of age-adjusted lung cancer mortality rates, m, vs these r-values are shown in Fig. 1a, c where, rather than showing individual points for each county, we have grouped them into intervals of r (shown on the base-line along with the number of counties in each group) and we plot the mean value of m for each group, its standard deviation, and the first and third quartiles of the distribution. We see, in Fig. 1a,c, a clear tendency for m to decrease with increasing r, in sharp contrast to the increase expected from the fact that radon can cause lung cancer, shown by the line labelled "theory".

One obvious problem is migration: people do not spend their whole lives and receive all of their radon exposure in their county of residence at time of death where their cause of death is recorded. However, it is easy to correct the theoretical prediction for this, and the "theory" lines in Fig. 1 have been so corrected. As part of this correction, data for Florida, California, and Arizona, where many people move after retirement, have been deleted, reducing the number of counties to 1601. (This deletion does not affect results.)

A more serious problem is that this is an "ecological study", relating the average risk of groups (county populations) to their average exposure dose. In general, the average dose does <u>not</u> determine the average risk, and to assume otherwise is what epidemiologists call "the ecological fallacy". However, it is easily shown that the ecological fallacy does not apply in testing a L-NT theory. All other problems with ecological studies that have been discussed in the epidemiology literature have also been investigated and found not to be applicable here.

Epidemiologists normally study the mortality risk to individuals, m', from their exposure dose, r', so we start from that premise using the BEIR-IV version of L-NT (in simplified form; full treatment in Ref. 1),

$$m' = a_n (1 + b r') \qquad \text{non-smokers}$$

$$m' = a_s (1 + b r') \qquad \text{smokers}$$

Summing these over all people in the county and dividing by the population gives

$$m = [S a_s + (1 - S) a_n] (1 + b r) \qquad (1)$$

where m and r have the county average definitions given above in the presentation of Fig. 1, and S is the smoking prevalence--the fraction of the adult population that is smokers. Eq. (1) is the L-NT theory we are testing here (we also show that our test applies to all other L-NT theories); note that it is derived by rigorous mathematics from the risk to individuals, with no problem from the ecological fallacy.

The bracketed term in Eq. (1) which we call m_0, may be thought of as the correction for smoking. Fig. 1b,d show m/m_0 vs r. We fit the data (i.e. all 1601 points) to

$$m/m_0 = A + B r \qquad (2)$$

deriving values of B . The theory lines are from Eq. (1) with slight renormalization. We see from Fig. 1b,d that there is a huge discrepancy between measurements and theory, by about 20 standard deviations. The theory predicts B = +7.3% per pCi/L, whereas the data are fit by B = -7.3 (+/- 0.6) and -8.3 (+/- 0.8) % per pCi/L for males and females respectively.

All explanations for the discrepancy that we could develop or that have been suggested by others have been tested and found to be grossly inadequate. Three independent sources of radon data have been used, but all give the same result. Three different sources of data on smoking prevalence similarly fail, including one based on lung cancer rates. In fact, it is found that if our best estimate of the width of the distribution of S-values for U.S. counties is correct, even a perfect negative correlation between radon and smoking prevalence eliminates only half of the discrepancy. If the width of the S-value distribution were as wide as the distribution of lung cancer rates, which is the largest credible width since other factors surely contribute to lung cancer rates, an essentially perfect negative correlation between radon levels and smoking prevalence in U.S. Counties would be required to explain the discrepancy. It is shown that such a strong correlation is incredible.

It is shown that the strong correlation between radon exposure and lung cancer mortality, albeit negative rather than positive, is unique to lung cancer; no remotely comparable correlation was found for any of the other 32 cancer sites. We conclude that the observed behavior is not something that can easily occur by chance.

To investigate effects of a potential confounding variable, data are stratified into quintiles on the values of that variable, and a regression analysis is done separately for each stratum. Since the potential confounder has essentially the same value for all counties in a given stratum, its confounding effect is greatly reduced in these analyses. An average of the slopes, B, of the regression lines for the five quintiles then gives a value for B which is largely free of the confounding under investigation.

This test was carried out for 54 socioeconomic variables and none was found to be a significant confounder. In all 540 regression analyses (54 variables x 5 quintiles x 2 sexes), the slopes, B, were negative and the average B value for the five quintiles was always close to the value for the entire data set. Incidently, this means that the negative correlation between lung cancer rates and radon exposure is found if we consider only the very urban counties or only the very rural; if we consider only the richest counties or only the poorest; if we consider only the counties with the best medical care or only those with the poorest medical care; and so forth for all 54 socioeconomic variables. It is also found for all strata in between, as, for example, considering only counties of average urbanicity, only counties of average wealth, only counties of average medical care, etc.

The possibility of confounding by combinations of socioeconomic variables was studied by multiple regression analyses and found not to be an important potential explanation for the discrepancy.

The stratification method was also used to investigate the possibility of confounding by geography, and of confounding by physical features such as altitude, temperature, precipitation, wind, and cloudiness, but these factors were of little help in explaining the discrepancy. The negative slope and gross discrepancy with L-NT theory is found if we consider only the wettest areas, or only the driest; if we consider only the warmest areas, or only the coolest , etc.

The effects of the two principal recognized factors that correlate with both radon and smoking were calculated in detail: (1) urban people smoke 20% more but average 25% lower radon exposures than rural people; (2) houses of smokers have 10% lower average radon levels than houses of non-smokers. These were found to explain only 3% of the discrepancy. Since they are typical of the largest confounding effects one can plausibly expect, it is extremely difficult to imagine a confounding effect that can explain the discrepancy. Requirements on such an unrecognized confounder were listed, and they make its existence seem extremely implausible.

By far, the most plausible explanation I can find for this discrepancy is that the linear-no threshold theory fails, grossly over-estimating the cancer risk in the low dose, low dose rate region. There are no other data capable of testing the theory in that region.

3. ANALYSIS -- 1996 EXTENSIONS [2,3]

The effect of using non-linear data fits was studied [2]. It was found that a linear plus quadratic fit

$$m/m_0 = A + B\,r + C\,r^2 \qquad (3)$$

with A and C positive and B negative, gives a much improved fit to the entire data set and this fit is little improved by adding additional terms. Moreover, Eq. (3) provides a good fit to any subset of the data without changing the constants B and C. The subsets studied include counties stratified on the basis of geographic area of the U.S. and on the basis of various socioeconomic characteristics. In several cases, these characteristics correlate strongly with radon levels --for example, urban counties have much lower average radon levels than rural counties -- but Eq. (3) fits the data for each with the same values for B and C.

Also, m/m_0 is larger for the Northeast than for other sections of the country which requires a larger value of A, but B and C are unchanged, giving the curve the same shape. A similar situation arises in some cases with stratification on socioeconomic characteristics, requiring adjustment of the constant A in Eq. (3), but no adjustments of B and C are necessary. The slope of the curve given by Eq. (3) is (B + C r) which becomes less negative and eventually positive with increasing r; this explains why groups of counties with larger r values -- e.g. the Northcentral region --have a less negative value of B in Eq. (2) than groups of counties with smaller r-values -- e.g. the Southeastern region.

In a separate paper [3], it was pointed out that ignoring the discrepancy outlined here and continuing to use L-NT is a violation of "The Scientific Method". Various reasons for ignoring our discrepancy which have been passed around by word of mouth -- but never published or addressed to me personally -- were examined and shown to be invalid. It was shown that even if one ignores "The Ecological Fallacy" and interprets our data directly as risk to individuals vs their exposure to radon, there is no discrepancy with any other study. A simple procedure for settling the matter was proposed: All that is needed to show that our conclusion about failure of LNT is not valid is to find a not implausible explanation for our discrepancy. If people submit suggestions, I will carry out calculations to determine whether they resolve the discrepancy or how they must be modified to do so. (Alternatively, I have provided my data file to anyone requesting it so they can do the calculations themselves.) To show that this procedure is not unreasonable, I offer to provide at least three plausible potential explanations for any surprising finding in any other published ecological study; this alone should demonstrate how very different our study is from all other ecological studies.

4. ANALYSIS -- 1997 EXTENSIONS

The lung cancer rates, m, used up to this point were for whites only and for 1970-1979, the latest age-adjusted data then available. A new analysis was done for males using m values for 1979-1994, including all races. The results were essentially the same: for B in Eq. (2), the new analysis gives B = -6.9 (+/- 0.45) as compared with -7.3 (+/- 0.56) in the 1995 analysis and + 7.3 predicted by L-NT. Because of the better statistics, the discrepancy with L-NT is now by 30 standard deviations.

The 54 socioeconomic variables treated in the 1995 analysis were from the 1980 U.S. Census. Over 450 socioeconomic variables from the 1990 census have now been added to the analysis. These involve various measures of urban-rural differences, education, ethnicity, labor force characteristics, income and poverty, housing, population mobility, family and household structure, language, transportation and communication practices, business characteristics, age distribution, racial composition, etc. Examples of variables included are the fraction of the population in 31 different age ranges, in 10 different income ranges, and with 7 different educational attainments, with head of household in 7 different age ranges, with 32 different ancestries, with 13 different occupation types, and working in 26 different industries, houses in 8 different age ranges and with values in 20 different dollar ranges, etc.

Nearly all variables that had a non-negligible correlation with radon levels were easily recognized as due to the urban-rural effect, which had been thoroughly investigated previously. The few exceptions, including percent of German ancestry, percent working at home, percent with female householder with no husband present, percent married with no children, and percent black were investigated by stratification and found to be incapable of explaining any appreciable fraction of the discrepancy with L-NT.

Large monetary rewards, now totalling $10,000, have been offered for a "not implausible" explanation for the discrepancy, but no proposed explanations have been offered.

5. CONCLUSION

The studies reported here, plus the fact that no "not implausible" explanations have been offered for our discrepancy in the 2 1/2 years since its publication, lead to the conclusion that L-NT fails, grossly exaggerating the cancer risk of low level radiation.

6. REFERENCES

[1] Cohen, B.L. Test of the linear-no threshold theory of radiation carcinogenesis for inhaled radon decay products, Health Phys. 68: 157-174; 1995

[2] Cohen, B.L. Lung cancer rates vs mean radon levels in U.S. Counties of various characteristics, Health Phys. 72:114-119; 1996

[3] Cohen, B.L. Problems in the radon vs lung cancer test of the linear-no threshold theory and a procedure for resolving them, Health Phys. 72:623-628; 1997

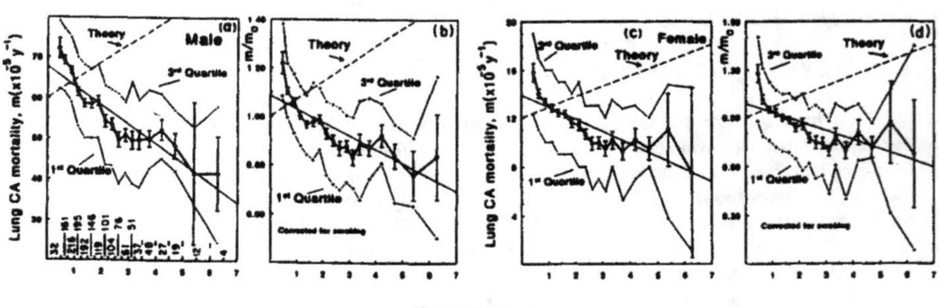

Fig. 1: Lung cancer mortality rates (age-adjusted) vs average radon level in homes for U.S. Counties. Fig. 1b,d are lung cancer rates corrected for smoking prevalence. See explanations in text.

Doses delivered to the lung by radon-decay products: Computations by a stochastic lung model

László Koblinger

KFKI Atomic Energy Research Institute,P.O. Box 49, H-1525 Budapest, Hungary

Werner Hofmann and Ralph Bergmann

University of Salzburg, A-5020 Salzburg, Austria

Abstract

Generation dependent deposition fractions are calculated by a stochastic lung model. In this model geometrical airway parameters are randomly selected from distributions and correlations derived from a statistical analysis of morphometric data.

Following the random walk of many particles individually, bronchial deposition patterns are determined. Deposited particles are subsequently cleared from the bronchial tree by mucociliary action. Residence times are computed for each particle at any generation. The number of radioactive decays per airway are calculated from the residence times and from the radioactive decays and build-ups. Doses normalized to unit surface areas are provided as final results.

1. Introduction

Because of the site specificity of radon-induced bronchial tumors, risk assessment for inhaled radon progeny must be based on the radiation burden in individual generations. The radiation burden can be computed in a three-step procedure:

- calculation of the deposition pattern,
- calculation of the times the radioactive particles spend in the individual generations during clearance,
- calculation of the number of alpha-particles emitted during these times.

To reflect the randomness of the lung stucture and the clearance processes involved, calculations are performed with a stochastic model.

2. The stochastic lung model

Morphometric measurements [1] have shown that the human bronchial tree is an asymmetrically dividing network (a sequence of bifurcations) with individual variations in dimensions of airways of the same generations.

To reflect these variations a stochastic lung model has been developed by the authors [2,3,4] in which an individual pathway is selected randomly for each particle. The selection rules are based on distributions of such parameters as diameters, lengths and branching angles of the parent and daughter tubes and on correlations found between them. The distributions and correlations applied [2] are derived from actual measurements [1]. Correlations are given either for different parameters of the same types of tubes (e.g. diameters and lengths of tubes of a given generation) or for the same type parameters of different airways (e.g. diameters of parent and daughter tubes).

The basic elements in our computations are the so-called bifurcation units. Each bifurcation unit is composed of half of the parent tube (except for the first unit, in which the total trachea is included) and the half length of the daughter selected for the airway chosen.

A bifurcation unit is considered as bronchial if both the parent and the daughter airways are bronchial.

In principle, there might be a theoretical maximum of 2^{n-1} tubes at generation n (n=1 for the trachea). In our stochastic model, termination of the bronchial tree is a random variable, alveolation of tubes starts at generation 12. At generation 15, only half of the tubes are still bronchial, and at generation 20 only 1.5 per cent of the theoretical maximum are found (the others are either pulmonary tubes, or do not exist due to shorter pathways selected).

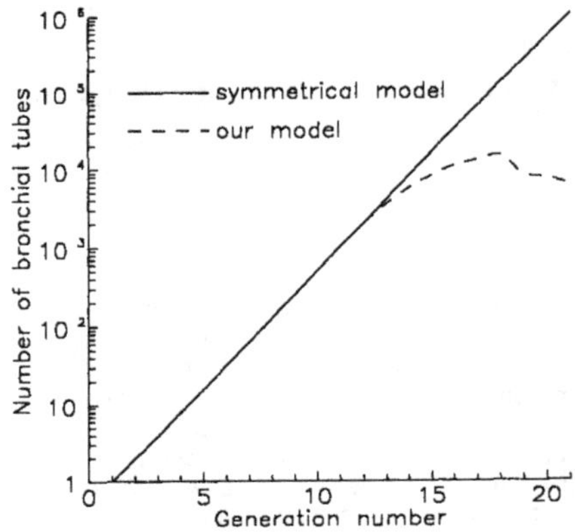

Figure 1. The number of brionchial tubes in the analytical, symmetrical models and in the stochastic model

3. The conditions of computations

In the present study no nasopharynx is considered, particles are assumed to be inhaled directly into the trachea. Deposition probabilities of the aerosol particles are computed by commonly used analytical formulae [3]. No axial mixing during the transport via the sequence of tubes is assumed, however an alveolar mixing factor of 0.25 is applied, i.e. we assume that in the alveoli one quarter of the freshly inhaled air parcels is totally mixed with the residual air.

The tidal volume of the lung is set to 1000 cm^3, the total lung volume is 3000 cm^3. The total cycle time of 4 s is divided into a 2 s inspiration and a 2 s expiration period, i.e. no breath-hold is assumed. Unit density aerosol particles are assumed to be inhaled uniformly over the inspiration period. The calculations refer to the attached decay products: a particle size of 0.2 μm is selected.

Secular equilibrium of Rn-daughters is assumed: 9, 6 and 4 Bq activities are selected for the ^{218}Po, ^{214}Pb and ^{214}Bi isotopes, respectively.

Aerosol particles deposited on bronchial tubes are assumed to be cleared by mucociliary clearance. Clearance is modeled by the following assumptions:

- the average mucus velocity in the trachea is set to 5.5 mm/min [5];
- at a given bifurcation, average mucus velocities in the asymmetrically branching daughters are proportional to their respective diameters;
- mucus transport is delayed at central bifurcation zones (the delay time is uniformly distributed between 0 and 30 min) [5];
- the slow bronchial clearance fraction (f) linearly decreases by the particle diameter: f = 0.77 - 0.12 d (μm) [6], i.e. at d = 0.2, f = 0.746;
- the slow-cleared particles are subject to an initial delay, with a half-life of 10 days;
- the upstream transport by clearance is simulated for the same pathway which was selected for the inhalation;
- radioactive decay is considered during clearance.

The validity of this model (with a comparison to experimental data) is given in an earlier paper [7].

4. Results

In Figure 2 the initial deposition efficiencies and the transit times in different bronchial airways (i.e. the times spent in the given tubes during clearance) are given. Due to the presumed relationship between mucus velocity and airway parameter, the decrease of diameters (by increasing generation numbers) results in increasing transit times. At the most distal regions, in contrast to symmetrical models, the number of bronchial tubes decreases by generation numbers, therefore, the transit times normalized to all inhaled particles also decrease in the stochastic model.

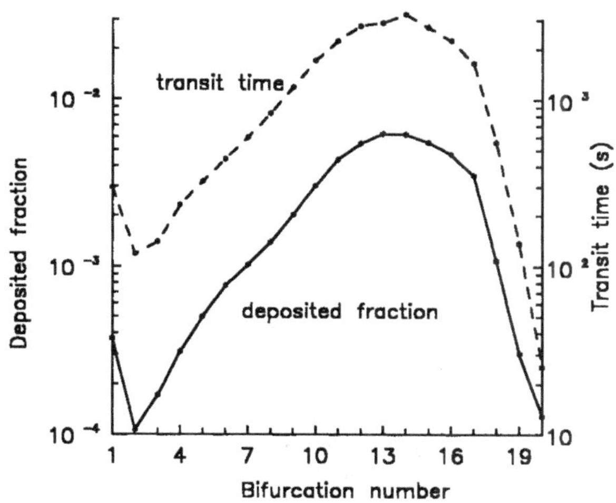

Figure 2. Initial deposition efficiencies for 0.2 μm unit density particles, and the transit times of these particles in individual bronchial airway bifurcation due to mucociliary clearance.

The total number of radioactive decay, illustrating the combined effect of deposition, clearance and radioactive decay, and the number of radioactive decays per unit area, representing the radiation dose distribution are given in Figure 3 for both alpha-emitting progenies.

Because of their relatively short half-lives, the distributions of both nuclide transformations reflect the transit time distributions.

In the upper regions of the bronchial tree, the number of nuclear transformations per unit surface area hardly changes due to the larger total surface area in more peripheral airway generations. Above about bifurcation 13, however, there is no further increase in the total surface area due to a decreasing fraction of bronchial airways and, as a consequence, the surface activities drop quite parallelly to the decay pattern.

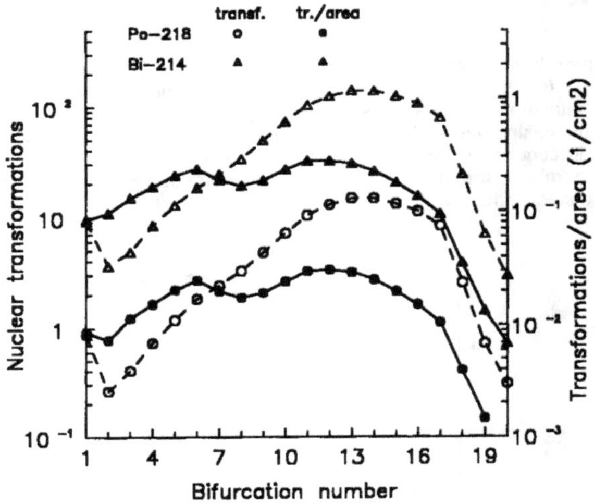

Figure 3. Number of nuclear transformations and number of decays per unit surface area in individual bronchial airway bifurcations for inhaled radon progeny with initial activities of 9 Bq ^{218}Po, 6 Bq ^{214}Pb and 4 Bq ^{214}Bi/^{214}Po.

5. Conclusions

Several factors (random geometry, random delay times) influencing the random walk of inhaled and, if deposited, cleared particles are studied by the stochastic model outlined. Further refinement of the model, needed to study the effect of individual variabilities in cellular doses, (with e.g. better description of the clearance mechanisms, inclusion of the random location of target cells) is limited by the lack of experimental data.

6. References

[1] O.G. Raabe, H.C. Yeh, G.M. Schum and R.F. Phalen: Report LF-53. Lovelace Foundation, Albuquerque (1976).

[2] L. Koblinger and W. Hofmann: *Phys. Med. Biol.*, **30**, 541 (1985).

[3] L. Koblinger and W. Hofmann: *J. Aerosol Sci.*, **21**, 661 (1990).

[4] W. Hofmann and L. Koblinger: *J. Aerosol Sci.*, **21**, 675 (1990).

[5] International Commission on Radiological Protection, ICRP Publication 66, *Ann. ICRP,* **24** (1994).

[6] W. Stahlhofen, G. Scheuch and M.R. Bailey: *Rad. Prot. Dosim.* **60**, 311 (1995).

[7] W. Hofmann and L. Koblinger: In: *1966 Int. Cong. Radiation Protection, Vienna,* **2**, 138 (1966).

Effects of Radon Inhalation on Physiology and Disorders

Kiyonori Yamaoka and Yoshiaki Komoto*

Central Research Institute of Electric Power Industry, 2-11-1 Iwado-kita, Komae-shi, Tokyo 201, Japan, *Takamatsu Hospital, 4-18 Tenzinmae, Takamatsu-shi, Kagawa 760 (formerly Okayama University Medical School), Japan

Abstract

In the first study, we administered Radon (Rn) to rabbits by inhalation and examined changes in the lipid peroxide (thiobalbituric acid reacting substances; TBARS) level, superoxide dismutase (SOD) activity and membrane fluidity in various organs to clarify the therapeutic effects of Rn. The TBARS level of the brain was significantly decreased immediately after Rn inhalation for 90 minutes in both the low concentration group (about 7-10 kBq/L) and the high concentration group (about 14-18 kBq/L) as compared with that in the control group. It further decreased in the low concentration group but slightly recovered in the high concentration group 2 hours after inhalation. The TBARS level of the lung showed no change immediately after inhalation but decreased significantly in both groups 2 hours after inhalation. With regard to SOD activity in the brain and lung, only that in the brain showed significant increase in the high concentration group immediately after inhalation; no other change was observed. Membrane fluidity, especially the fluidity of membrane protein, significantly increased in the brains of both groups immediately after inhalation as well as at the low concentration group, and that in the lung significantly increased 2 hours after inhalation in both groups. These findings suggest that the inhalation of Rn at Rn springs contributes to the prevention of brain disorders related to peroxidation reactions by promoting these physiologic changes. In the second study, we sprayed Rn spring water of various concentrations to rabbits to make the animals inhale them, and examined mainly the responses of biogenic amine neurotransmitters for clarifying the effects of Rn inhalation in the neuronal transmitter system. The following results suggested possible decrease in the activities of aromatic-L-amino acid decarboxylase, which are key enzymes of the metabolism of biogenic amines. i) In rabbit brain, noradrenaline (NA), serotonin (5HT) and 5-hydroxyindoleacetic acid (5HIAA) levels decreased significantly immediately after inhalation of Rn spring of 13 kBq/l or over, ii) changes in tyrosine, dopamine (DA) and homovanillic acid (HVA) levels did not depend on the concentrations of Rn inhaled, iii) the turnover ratios for these amines were evaluated. In the third study, indications for treatment at the Misasa Hot Spring, a Rn producing radioactive spring, include hypertension, diabetes mellitus and pain. To clarify its mechanisms of action on these conditions, we evaluated dynamic changes in blood components such as vasoactive substances after Rn inhalation. Vasodilation, alleviation of diabetic symptoms and morphine-like analgesic effects were observed, suggesting that these changes constitute part of the mechanisms of the Rn spring therapy on the above conditions.

1. Introduction

Rn is a radioactive gaseous element that emits α-particle. If Rn is inhaled, the lung will be subjected to the action of free radicals created by the radiation and may suffer inflammation [1]. Although Rn inhalation has been thought to be hazardous in general, Rn springs have been reported to have therapeutic effects on senile brain disorders and hypertension [2]. Another known effect of Rn spring is to promote the effects of such tissue perfusion agents as adrenaline in plasma, that is to say the level of plasma adrenaline is increased by Rn inhalation [3]. So far there are no epidemiologic data on the hazardous effects of Rn [4]. To clarify the mechanisms of therapeutic effects (e.g. clinical indications for Misasa Hot Spring, a Rn spring, include hypertension, diabetes mellitus and pain) of Rn, we administered spraied Rn (7-18 kBq/l) to rabbits by inhalation and examined the changes in TBARS level, SOD activity and membrane fluidity in various organs, biogenic amine neurotrsansmitters in brain, adrenal secretion of catecholamines and blood components such as vasoactive substances.

2. Materials and Methods

Animal model

Two kg rabbits were intubated by tracheotomy under intravenous anesthesia with 25 mg/kg sodium pentobarbital. Arterial pressure was monitored with a catheter introduced via the carotid artery into the thoracic aorta under 1 mg/kg general heparinization. Venous catheterization was performed via the external jugular vein into the right atrium. The chest of the rabbit was then shaved, and the rabbit was placed in a constant temperature-regulated 20 litre tub from which its neck and head protruded. Electrocardiogram (ECG) and arterial pressure were monitored using a polygraph, and anesthesia was maintained with a constant micro-drip infusion system of 15-20 guttae per minute with 0.3-0.4 % sodium pentobarbital. Respiration was maintained at 30 per minute (each, 20-30ml) and was assisted with 10 mmH_2O pressure, so that the minute ventilation was set at 250 ml/min/kg. Room temperature was maintained at 25 ±1 ℃ throughout the experiment.
Rn inhalation was maintained with an ultrasonic nebulizer during tap water bathing at 36-37 ℃. The Rn water (from Ikeda mineral springs; at 4 ℃) used in this study cotains mainly ^{222}Rn. The Rn concentration of the water was approximately 14-18 kBq/l (high concentration group: the original level) or 7-10 kBq/l (low concentration group). The controls group inhaled the same mineral spring water which had been left to stand for 40 days at 4 ℃; the radioactivity of this water was a thousandth or less of the original level. Rn concentration was measured with a liquid scintillation counter. Either immediately or 2 hours after Rn inhalation for 90 minutes, rabbits were sacrificed by exsanguination. The organs of interest were excised and rinsed in normal saline until free of blood.

Assay

TBARS was determined by the TBA method of Ohkawa et al. [5]. SOD activity was measured by the spin-trapping method using an electron spin resonance (ESR) spectrometer. Membrane fluidity was determined by the spin-label method using an ESR spectrometer. Catecholamines and indoleamines were analyzed by the high performance liquid chromatography (HPLC). Tissue perfusion rate (TPR) was evaluated 15 minutes after the beginning of inhalation by mass spectrometry. Levels of adrenaline and noradrenaline (NA) were analyzed by the HPLC. Vasoactive substances such as histamine, α atrial natriuretic polypeptide (α-ANP), vasopression, angiotension II and prostaglandin E_2 (PGE_2) were analyzed by radio immunoassay (RIA). As diabetes-associated substances, serum insulin (RIA), glucose 6-phosphate dehydrogenase (G6PDH, ultraviolet rays (UV) method), pancreatic glucagon (RIA) and blood glucose (glucose enzyme method) were analyzed by various methods indicated in parenthesis. As pain-associated substances, plasma β-endorphin and M-enkephalin were analyzed by RIA.

3. Results and Discussion

(1) Changes in TBARS level, SOD activity and membrane fluidity in various organs

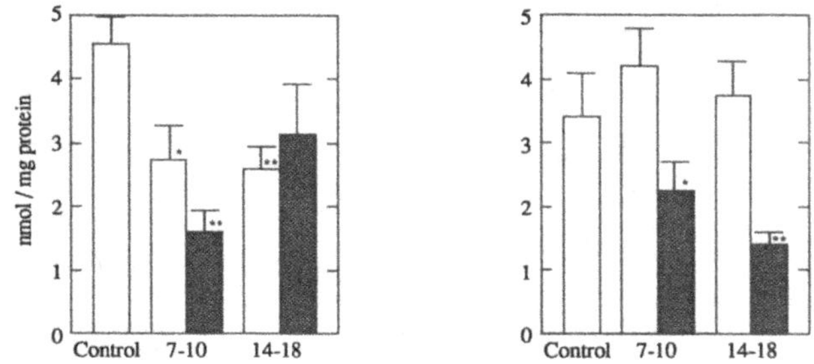

1) TBARS level (n=10-14)

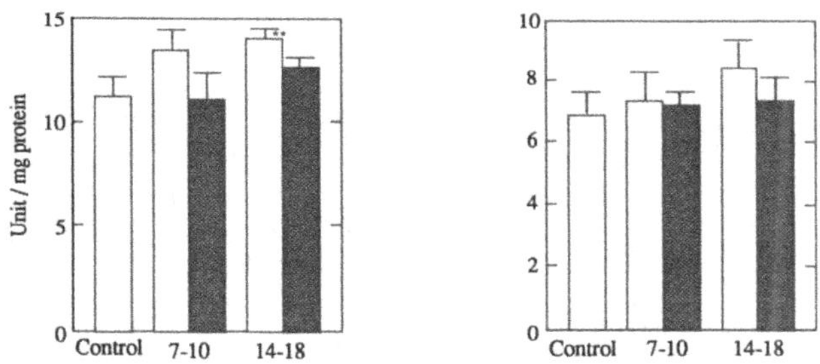

2) SOD activity (n=10-12)

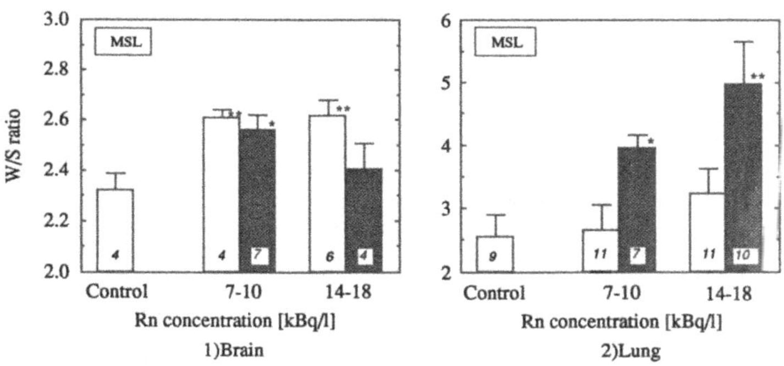

3) Membrane Fluidity

Fig. 1. Time-dependent Change in TBARS Level, SOD Activity and Membrane Fluidity in the Brain and Lungs of Rabbits after Rn Inhalation. Each value represents mean ±SEM, "□" immediately after inhalation; "■" 2 hours after inhalation ★P < 0.05 ;★★P < 0.01 vs control.

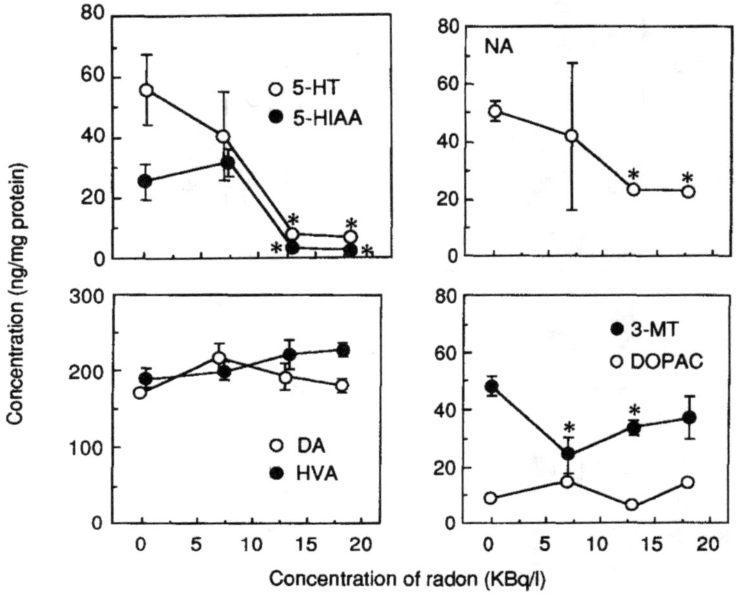

Fig. 2. Change in the Levels of 5-HT, 5-HIAA, Tyrosine, NA, DA, HVA, 3-MT and DOPAC in Rabbit Brain. Each value represnts mean±SEM. The number of rabbit per experimental point is 5 at 0 kBq/l, 4 at 7 kBq/l, 6 at 13 kBq/l and 5 at 18 kBq/l. *P<0.05 vs control.

Table 1 Effect of Rn Inhalation on Turn-over Ratios of Biogenic Amines in Rabbit Brains

Rn [kBq/l]	control	7	13	18
[DOPAC+3-MT+HVA]/[DA]	1.42±0.11	1.10±0.03*	1.36±0.06	1.54±0.13
[HVA]/[DOPAC]	21.5±3.08	13.3±2.2	40.7±19.8	16.0±2.5
[HVA]/[3-MT]	4.14±1.22	8.84±2.51*	6.73±1.20	6.72±2.11
[NA]/[DA]	0.297±0.047	0.183±0.141	0.126±0.026**	0.128±0.020**
[5-HIAA]/[5-HT]	0.443±0.085	0.857±0.284	0.429±0.088	0.499±0.036

Each value represents mean±SD. The number of rabbit per experiment was 5 at control, 2 at 7 kBq/l, 6 at 13 kBq/l and 3 at 18 kBq/l. *P<0.05, **P<0.01 vs control.

The TBARS level of the brain was significantly decreased immediately after Rn inhalation for 90 minutes in both the low concentration group and the high concentration group as compared with that in the control group. It further decreased in the low concentration group but slightly recovered in the high concentration group 2 hours after inhalation. The TBARS level of the lung showed no change immediately after inhalation but decreased significantly in both groups 2 hours after inhalation. With regard to SOD activity in the brain and lung, only that in the brain showed significant increase in the high concentration group immediately after inhalation; no other change was observed. Membrane fluidity, especially the fluidity of membrane protein, was significantly increased in the brains of both groups immediately after inhalation. In the lung, the membrane fluidity was significantly increased 2 hours after inhalation in both groups (see Fig. 1). These findings suggest that the inhalation of Rn at Rn springs contributes to the prevention of brain disorders related to peroxidation reactions by promoting these physiologic changes [6].

Table 2 Changes of Catecholamines and Tissue Perfusion Rate by Rn Inhalation

Rn [kBq/l]	control	14−18
plasma catecholamines [pg/mg protein]		
adrenaline	4.2±1.7	22.9±12.1**
noradrenaline	10.9±5.6	18.8± 8.2*
adrenal catecholamines [ng/mg protein]		
adrenaline	825±544	249± 129**
noradrenaline	134± 67	23.6±14.9**
tissue perfusion rate [ml/100g/min]		
no medication	15.8±1.8	21.4±2.4 **
phentolamine	18.4±1.7	22.6±2.2 **
propranolol	17.4±2.7	19.3±2.9
atenolol	16.6±2.6	18.4±3.0

Each value represents mean±SEM. The number of rabbits per experiment was 8-14 at control and 8-15 at 14-18 kBq/l. *P<0.05, **P<0.01 vs control.

Table 3 Dynamic Changes in Vasoactive, Diabetes-associated and Pain-associated Substances of Rabbit Blood by Rn Inhalation

Rn [kBq/l]	control	7−10	14−18
histamine [μg/dl]	67±12	146±31 **	52±62 **
α atrial natriuretic polypeptide [pg/ml]	1660±240	2170±170 *	3330±520 **
vasopression [pg/ml]	13.2±3.6	3.2±0.6 **	5.4±0.8 **
angiotensin II [pg/ml]	34±1	33±1	32±1
prostaglandin E_2 [pg/ml]	26±4	33±10	18±5
insulin [U/ml]	4.3±0.4	4.3±0.6	3.5±1.8 **
glucose-6-phosphate dehydrogenase [IU/37℃]	1.9±0.2	2.8±0.2 **	2.6±0.3 **
pancreatic glucagon [10^4 ×pg/ml]	1.6±0.1	1.9±0.1 *	2.4±0.3 **
blood glucose [mg/dl]	218±21	195±22	191±19
β endorphin [ng/wet・g]	16.2±2.5	19.0±1.9	22.4±3.5 *
M-enkephalin [ng/wet・g]	6.1±1.0	6.5±1.1	11.8±1.9 **

Each value represents mean±S.E.M.. The number of rabbits per experiment was 10 at control, 8 at 7-10 kBq/l and 9 at 14-18 kBq/l. *P<0.05, **P<0.01 vs control.

(2)Changes in biogenic amine neurotrsansmitters in brain

In rabbit brain immediately after Rn inhalation, NA, 5HT and 5HIAA levels decreased significantly by inhalation of Rn spring of 13 kBq/l or over. Changes in tyrosine, DA and HVA levels did not depend on the concentrations of Rn inhaled (see Fig. 2). The turnover ratios for these amines were evaluated (see Table 1). The results suggested possible decrease in the activities of aromatic-L-amino acid decarboxylase, which are key enzymes of the metabolism of biogenic amines [7].

(3)Changes in adrenal secretion of catecholamines in relation to increase of tissue perfusion rate

In the Rn group, plasma adrenaline and NA levels were significantly higher, while adrenaline and NA levels were significantly lower than those in the control group. In the no medication and phentolamine subgroups, tissue perfusion rates in the Rn group were significantly higher than those in the control group. It is suggested that catecholamines are secreted from the adrenal glands upon inhalation of Rn water and that the β_1-action of catecholamines contributes to the increase in tissue perfusion (see Table 2) [8].

(4)Changes in blood components indicated for Hypertension, Diabetes and Pain

Significant dose-dependent increases were observed in histamine and α-ANP and significant decrease in vasopression in both high and low concentration groups compared with the control group. Angiotensin II and PGE_2 showed no significant changes. Insulin significantly increased in the high concentration group, and G6PDH activity and glucagon significantly increased in both high and low concentration groups. The blood glucose level decreased slightly. β-endorphin and M-enkephalin dose-dependently increased with significant differences between the high concentration group and control group (see Table 3). Namely, vasodilation, alleviation of diabetic symptoms and morphine-like analgesic effects were observed, suggesting that these changes constitute part of the mechanisms of the Rn spring therapy include hypertension, diabetes mellitus and pain [9].

4. Conclusions

These findings suggest that an appropriate amount of active oxygen is produced in the body after Rn inhalation, and this contributes to the alleviation of the symptoms of active oxygen diseases such as hypertension after certain processes such as activation of the biological defense mechanism [10], or promoting these physiologic changes such as tissue perfusion, in contrast to the toxic effects of Rn inhalation at higher concentration.

5. References

[1]Hashimura, T., Kono, M. and Imajo, Y., NIPPON ACTA RADIOLOGICA 49, 335-343, 1989.

[2]Ooshima, Y., System of Current Internal Medicine II, Nakayama Shoten, Tokyo, pp. 122-144, 1966.

[3]Komoto, Y., Kohmoto, T., Nakao, T., Sunakawa, M. and Yorozu, H., Phys. Med. Baln. Med. Kli. 17, 72-78, 1988.

[4]Mifuna, M., Sobue, T., Arimoto, H., Komoto, Y., Kondo, S. and Tanaoka, H., Jpn. J. Cancer Res. 83, 1-5, 1992.

[5]Ookawa, H., Ooishi, S. and Yagi, K., Anal. Biochem. 95, 351-358, 1979.

[6]Yamaoka, K., Komoto, Y., Suzuka, I., Edamatsu, R. and Mori, A., Arch. Biochem. Biophys. 302, 37-41, 1993.

[7]Yamaoka, K., Ishii, K., Ito, T., Edamatsu, R., Komoto, Y., Suzuka, I. and Mori, A., Neurosciences 20, 17-22, 1994.

[8]Suzuka, I., Yamaoka, K. and Komoto, Y., J. Jpn. Coll. Angiol. 31, 1182, 1991.

[9]Yamaoka, K. and Komoto, Y, Physiol. Chem. Phys. & Med. NMR. 28, 1-5, 1996.

[10]Yamaoka, K., Edamatsu, R. and Mori, A., Free Radical Biol. Med. 11, 299-306, 1991.

Health Effects of Low Level Radiation

Sadao Hattori

Central Research Institute of Electric Power Industry(CRIEPI)
1-6-1 Otemachi, Chiyoda-ku, Tokyo 100, Japan

Abstract

In 1982, Prof. Thomas Don Luckey of Missouri Univ. asserted "Radiation Hormesis" on the Journal of Health Physics and he published two books [1,2]. CRIEPI initiated the research program on Radiation Hormesis following his assertion to confirm "is it true or not?".

After nearly ten year research activities on data surveys and animal tests with many Universities, we are realizing scientific truth of bio-positive effects by low level radiation exposures.

The interesting bio-positive effects we found could be categorized in following five groups.

1)Rejuvenation of cells such as increase of SOD and cell membrane permeability,

2)Moderation of psychological stress through response of key enzymes,

3)Suppression and therapy of adult-diseases such as diabetes and hypertension,

4)Suppression of cancer through enhancement of immune systems such as lymphocytes,

5)Suppression of cancer and ratio-adaptive response by activation of DNA repair and apoptosis.

In the responses of many specialists to our initiation of radiation hormesis research program following T.D.Luckey's claim about low level radiation, I have to pick up for the first, the great success of Prof. Sakamoto. Prof. Sakamoto had been already applying whole body low dose irradiation for ten years before our radiation hormesis research started on the therapy to suppress the cancer reappearing after treatment. He reported about his successful trial to real patients and showed an enhancement of immune system.

1.Introduction

In a review article entitled "Physical Benefits from Low Levels of Ionizing Radiation", published in Health Physics in December of 1982, Professor T.D. Luckey in the University of Missouri, asserted the "radiation hormesis" with 200 references[1].

This resulted in the first International Symposium on Radiation Hormesis in Oakland, California (August 1985). CRIEPI consulted many specialists about Luckey's paper and studied many other papers such as Lorenz, 1954 ; Luckey,1980, Liu et al.,1985[3].

Radiation hormesis research in Japan has been based on the rationale that if Luckey's claim were to be true, radiation management in Japan has been extremely erroneous.

CRIEPI organized a Hormesis Research Steering Committee composed of leading specialists in the field concerned, and began research in cooperation with a number of universities, as well as the National Cancer Research Institute, and the National Institute of Radiological Sciences. After obtaining interesting results in various experiments on the health effects of exposure to low doses of radiation, we have proceeded on an expanded program, which involves fourteen universities and two research institutes throughout Japan.

We are now carrying out experimental activities on the effects of low-dose radiation on mammals. After several years of research activities, we are recognizing Luckey's claim. Some basic surveys including Hiroshima, Nagasaki and animal experiments in Japan have brought us valuable information on the health effects of low-dose radiation. The followings are some topical research information obtained across Japan.

2.Topics of Low Dose Radiation Research

(1) Survey of A-bomb Survivors

Professor Okumura compiled the data from Hiroshima and Nagasaki, and raised new scientific information. The follow up data of the people who received radiation from the Atomic Bomb show us an interesting feature especially in the low dose range. A certain optimum dose for the suppression of leukemia was shown through the survey of the people of Hiroshima and Nagasaki exposed to the radiation of the Atomic Bomb[4].

(2) The Beneficial Effect of Misasa Spa

Emeritus Professor of Osaka University, Dr.Kondo and Dr.Tanooka, former Chairman of Japan Radiation Research Society, conducted statistical comparisons of cancer of the people in Misasa villages (i. e. high radon levels in drinking water), adjacent villages and Japan. The result was meaningful especially on the suppression of total cancer [5].

(3) Medical Application : Treatment of Cancer

Professor Sakamoto used low dose radiation to cure and to suppress the reappearance of cancer in the hospital of Tohoku University. For example, he applied 10 cGy three times a week for several weeks to liver cancer and lymphatic tumors. He applied whole body or half body low level dose combined with local high dose irradiation to treat non- hodgkin's lymphoma. The low survival rate of 36% in patients with non- hodgkin's lymphoma after five years of the therapy improved to a 90% survival rate with a low dose treatment schema.

Some analytical results demonstrate an increase of the ratio of the helper T cells to suppresser T cells.

(4) Cell Rejuvenation

Yamaoka of CRIEPI found cell rejuvenation on the properties of cell membranes and superoxide dismutase activities by 25 cGy to 50 cGy of irradiation.

After low dose of X-ray whole body irradiation, increase of SOD enzyme and decrease of Lipid-peroxide had been kept more than six weeks [6].

(5) Adaptive Response

Professor Ikushima of Kyoto Univ of Education examined the radio-adaptive response. Professor Yonezawa of University of Osaka prefecture confirmed two phases of radio-adaptive responses by using a priming dose and survival after a sublethal dose administration. He found that two windows of low level priming doses to bring enhanced resistances to sublethal X-radiation given two months after 5 to 10 cGy whole body dose, and two weeks after 30 to 50 cGy also [7].

(6) Response of p53

Professor Onishi of Nara Medical University discovered a marked increase of P53 protein production by p53 gene. Doses of 10 to 25 cGy were effective [8].

P53 protein induced by tumor suppression gene p53 brings important enhanced activities on DNA repair and Apoptosis.

(7) Importance of Low Dose Steady Irradiation

Prof. Nomura confirmed the importance of steady low dose irradiation for the enhancement of gene repairing activities, giving evidence that steady low dose administration is essential for obtaining beneficial health effects.

3. Research Program on the Adaptive Response to Low Dose Radiation in Japan

CRIEPI is running with fourteen Universities on twelve research subjects of adaptive responses, intending to examine the extensive range of biological functions related to low-dose radiation.

The subjects being interested in the research program are as follows :

(1) SOD response, Lipid-peroxide reduction, and membrane permeability

(2) Cell apoptosis and anti-cancer effects

(3) P53 response and its effects

(4) Response of the central nervous system

(5) Responses of the intercellular signal transduction

(6) Enhancement of immune system inhibitory effects on carcinogenesis

(7) Inhibitory effects of diseases of grown up people such as diabetes

(8) Effects on the energy metabolism regulation

(9) Radioadaptive response by DNA repair and apoptosis

(10) Enzyme response and stress relaxation

(11) Repair mechanism of DNA aberrations

(12) Epidemiological analysis of atomic bomb survivors

4. Closing Remarks

Formation of ions, free electrons and free radicals by ionizing radiation enhances and creates many comprehensive bio-chemical reactions, followed by significant biological responses.

Animal tests results give us a certain scientific synopsis on the adaptive response to low dose effect on carcinogenesis and malignant tumor; Suppression of cell aging. Activation of biological defense mechanisms.

We have such an impression that a certain low dose radiation raises some vitalizations of basic biological functions. The recent progress of analytical technique on the observation of DNA structural responses greatly contributes the unbelievable success of our research on the adaptive responses of low dose radiation.

Analysis of the history of the evolution of living materials through billions of years shows us all kinds of environmental conditions were put in the extremity of positive utilization without exception in our physical processes, air, water, sun light, temperature, salt, and radiation. This pursuit of the extremity of the positive utilization of the all kinds of environmental factor through the process of evolution is the life itself, in a sense. It is obvious the natural radiation background on the earth was fairly higher than present situation in the past days.

On this sense, the adaptive response is the fundamental characteristics of lives. All kinds of damages caused by environmental influence, diseases, loss of power are to be overcomed by enhancement of DNA repair and apoptosis activities. This is the way of adaptive response.

5.Acknowledgment

We appreciate the sincere advice and direction for research activities given by Dr.T.D.Luckey,Dr.S.Kondo,Dr.T.Sugawara,Dr.K.Sakamoto,Dr.T.Yamada,andDr.H.Tanooka.

Brain Cells

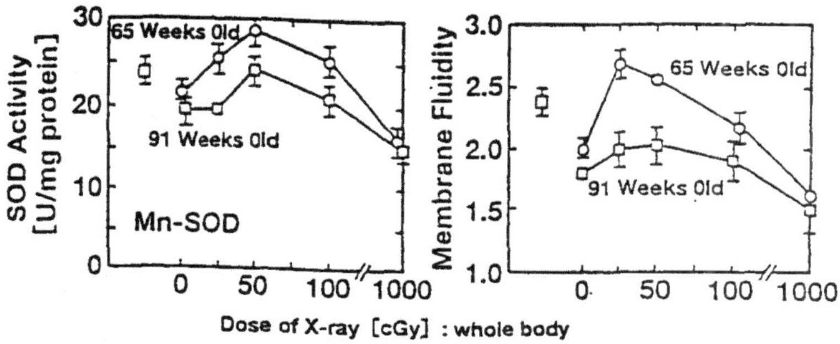

Dose and aging-dependent changes in SOD activity and membrane fluidity of rat's brain cortex by X-ray irradiation. (4 hrs after irradiation) Membrane fluidity was determined by spin-label method using ESR spectrometer. '□' shows the data from sham-irradiated 7 weeks old control. The number of rats per experimental point is 10-15 *p<0.05

K. Yamaoka (CRIEPI)

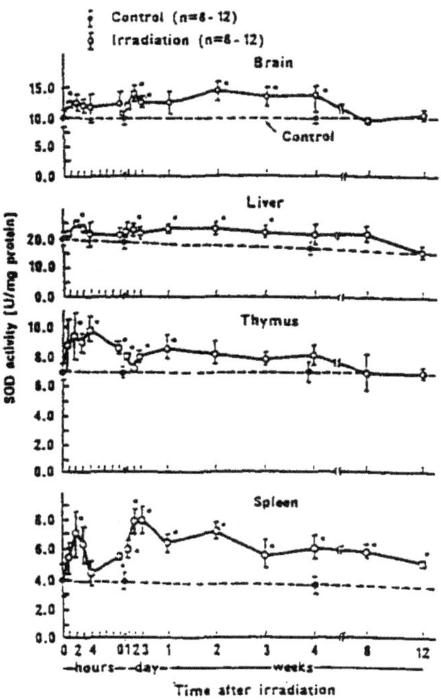

Time-dependent changes in SOD activity in organs of rats after X-irradiation at dose of 0.25 Gy.

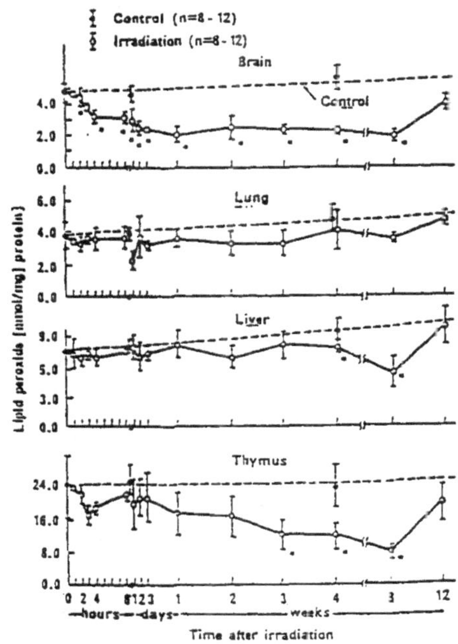

Time-dependent changes in Lipid peroxide level in organs of rats after X-irradiation at Dose of 0.25 Gy.

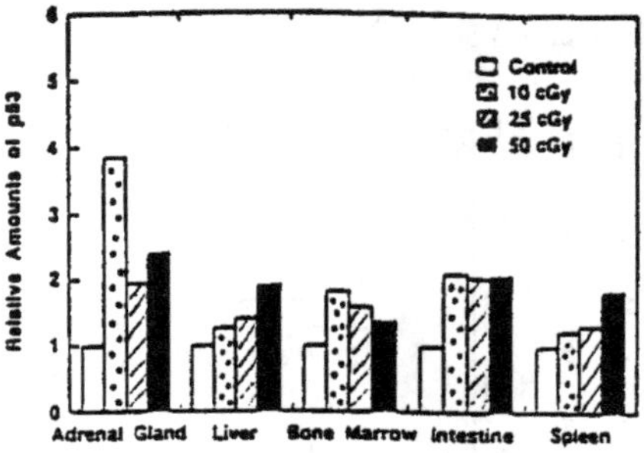

Relative Amounts of p53 at 6 h after X-ray Irradiation in Various Organs of Rats Prof. Ohnishi, Nara Medical Univ.

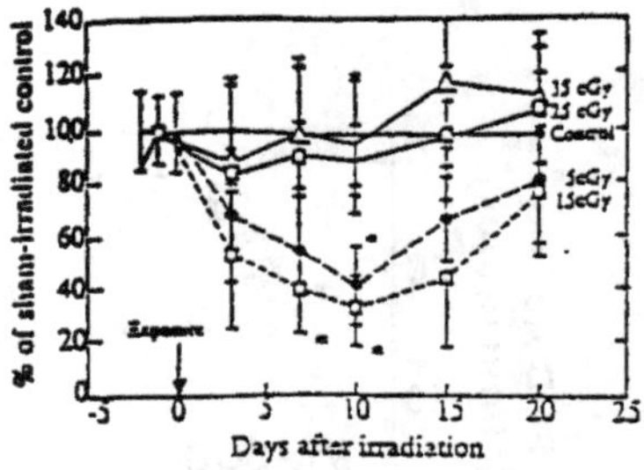

Effect of low-dose X-rays on aggression displayed by Isolated resident vs Isolated intruder.

Yamada, Miyachi, Tohoh Univ. 1993

6.References

[1] LUCKEY,T.D.(1982). "Physiological benefits from low levels of ionizing radiation." Health Phys. 43 : 771-789

[2] LUCKEY,T.D.(1991). Radiation Hormesis, CRC Press. Boca Raton, Florida,P.239

[3] LORENZ,E.(1954). Biological Effects of External Gamma Radiation, Part I (R.E. Zirkle, ed.), McGraw-Hill, New York, P.24.

[4] KONDO,S.(1993). Health Effects of Low-Level Radiation. Kinki University Press, Osaka, Japan and Medical Physics Publishing , Madison, Wisconsin.

[5] MIFUNE,M.,SOBUE,T.,ARIMOTO,H.,KOMOTO,Y.,KONDO,S.,andTANOOKA,H.(1992). "Cancer mortality survey in a spa area (Misasa, Japan) with a high radon background ." Jpn.J.Canser Res.83:1-5.

[6] YAMAOKA,K., EDAMATSU,R., and MORI,A.(1991). "Increased SOD activities and decreased lipid peroxide level in rat organs induced by low-dose x-irradiation. " Free Rad. Biol. Med. 11(3):299-306.

[7] YONEZAWA,M., TAKEDA,A., and MISONOH,J.(1990). "Acquired radioresistance after low-dose x-irradiation in mice." J.Radiat. Res.31:256-262

[8] OHNISHI,T., MATSUMOTO,H., OMATSU,T., and NOGAMI,M.(1993) "Increase of wpt53 pool size in specific organs of mice by low doses of Xrays." J.Radiat.Res.34:364.

[9] Cohen,A.F. and B.L.Cohen. Tests of the Linearity Assumption in Dose-Response Relationship for Radiation Induced Cancer. Health Phys. 38:53,1980

[10] ICLB(1992). Proceedings of the International Conference on Low-Dose Irradiation and Biological Defense Mechanisms, Kyoto, Japan,12-16 July, 1992(T.Sugahara,L.A.Sagan, and T.Aoyama,eds.). Elsevier Science Publishers B.V.,Amsterdam.

[11] SADAO HATTORI, State of Research and Perspective on Radiation Homesis in Japan International Journal of Occupational Medicine and Toxicology, Vol.3, No.2,1994.

Approach to publicly acceptable radiation dose unit based on annual dose equivalent dominated by radon exposure

Hidehiko Arima, Hiromi Etoh and Kenji Ishibashi

Department of Nuclear Engineering, Kyushu University
6-10-1 Hakozaki, Higashi-ku, Fukuoka, 812-81 Japan

Abstract

A new radiation dose unit is proposed which is based on the annual dose equivalent. The dose is designated as radiation environment dose (red). Red is the dose normalized by the annual effective dose equivalent. Various radiation exposures from natural origins and medical uses are expressed in terms of red. The new dose unit is suited to intuitive understanding of the degree of radiation exposure and is thought to be publicly acceptable. The comparison with actual quantities on the typical environmental toxic substance dioxin shows that the regulation standards for radiation exposure are as severe as those for the toxic substance.

1. Introduction

As is well known, people are exposed to a variety of radiation from such origins as cosmic rays, radioactive substances in the earth, radon and thoron gases, and natural radioisotopes in the body. In addition to natural irradiation, artificial irradiations in medical treatments are also well evaluated[1]. It is, however, still difficult for the magnitude of radiation exposure to be intuitively appreciated by the public. Even in advanced countries, most ordinary people do not have a basic knowledge of radiation, and few people recognize that radiation is related to the existence of background irradiation from natural origins. The radiation dose unit (RDU), which is currently used for these exposures, seems to be somewhat too complicated to be understood precisely by the public. This is because the units are hard to be intuitively imagined and the magnitude in RDU is unlike other common units such as meter, kg etc. As for the activity of radioactive materials, especially, the order of magnitude by Becquerel (Bq) becomes much larger than unity.

When radiation accidents take place, the magnitude of the RDU often confuses the public. Recently in Japan, radiation accidents occurred at some institutes and power plants, and they were widely reported. At most accidents, leakage of radioactive materials out of the institutes was far below the regulation standards and not severe. The amount seems to be negligible compared to natural radiation levels. The public are interested in whether there are human and/or environmental influences or not.

We propose a new RDU which is based on natural radiation exposure which is dominated by radon exposure. Since the unit is devised for easy understanding as with meter and kg, this is considered to be publicly acceptable. Exposures to the public in many fields are reduced to values in the new unit and summarized in the form of tables. Finally, the comparison with the typical toxic substance dioxin is simply discussed in the regulation standards in the latest issue.

2. New radiation dose unit

The International Commission on Radiological Protection (ICRP) began to apply the current international units system in 1982[2]. The current unit for dose equivalent was named as the Sievert (Sv). At the introduction of the present new unit, some requirements are considered. One of the important points is that the new unit should be intuitively understandable for ordinary people. Use of the current unit leads to the situation that few people precisely recognize how much dose we are exposed to from natural and artificial radiations. In the UNSCEAR 1988 Report[3], the mean annual effective dose equivalent (MAEDE) from natural sources was evaluated at 2.4 mSv for adults. For easy recognition of various exposures by the public, the use of a ratio of a dose equivalent to the MAEDE is proposed. This proposal would facilitate understanding of exposure on the basis of MAEDE. The new radiation dose unit was designated the radiation environment dose (red) instead of Sv. Thus, the unit becomes

$$1 \text{ red} = 2.4 \text{ mSv} \quad (1)$$

Radon exposure is the most important composition in MAEDE. The value of the mean annual radon exposure is 1.3 mSv and amounts to 54.2 % of MAEDE. In Table 1, exposures from natural sources are normalized by MAEDE as RED (red) and also by the mean annual radon exposure as relative exposure to radon, RER for comparison. As an interesting example, MAEDE for radon-222 inhalation is equivalent to 0.5 red. Medical exposures are summarized in Table 2. Most data in the tables were taken from UNSCEAR 1993. Four levels of health care have been defined by the committee, on the basis of the number of population per physician in the country. Most advanced countries are categorized into level I, because of lower populations per physician. Various examples of exposures from usual situation in daily life for the public in Japan are listed in Table 3. The expression in terms of red gives people intuitive recognition of degrees of radiation exposure at the incidents and/or accidents.

Table 1. Annual exposures from natural sources in units of Sv and red.

Source of irradiation	AEDE [mSv] [a]	RED [red] [b]	RER [c]	Reference
Cosmic rays				
Sea level	0.270	0.113	0.208	
La Paz (Bolivia)	2.020	0.842	1.554	
Denver (USA)	0.570	0.238	0.438	
Average	0.380	0.158	0.292	
Cosmogenic radionuclides	0.012	0.005	0.009	
Primordial radionuclides				
External irradiation (outdoor)				
Norway	0.195	0.081	0.150	[4]
China	0.146	0.061	0.112	[5]
Japan	0.148	0.062	0.114	[6]
USA	0.056	0.023	0.043	[7]
Average	0.070	0.029	0.054	
External irradiation (indoor)				
Norway	0.329	0.137	0.253	
China	0.376	0.157	0.289	
Japan	0.155	0.065	0.119	
USA	0.182	0.076	0.140	
Average	0.390	0.163	0.300	
Total				
Norway	0.524	0.218	0.403	
China	0.522	0.218	0.402	
Japan	0.303	0.126	0.233	
USA	0.238	0.099	0.183	
Average	0.460	0.192	0.354	
Internal irradiation [d]				
Arctic Area	0.300	0.125	0.231	[8]
Helsinki (Finland)	0.500	0.208	0.385	[9]
Average	0.230	0.096	0.177	
Internal irradiation				
Radon-222 inhalation	1.200	0.500	0.923	
Radon-220 inhalation	0.073	0.030	0.056	
Radon-222 oral-intake	0.005	0.002	0.004	
Total (rounded)	2.400	1.000	1.333	

Data from UNSCEAR 1993 and 1988. a : Annual effective dose equivalent. b : Radiation environment dose.
c : Relative exposure to radon. d : except irradiation by radon series.

Table 2. Annual exposures from medical uses of radiation.

	ADE per person			ADE per examination			Ref.
	AEDE[a] [mSv]	RED[b] [red]	RER[c]	AEDE[a] [mSv]	RED[b] [red]	RER[c]	
Diagnostic medical X-ray examinations							
Level I	0.93	0.39	0.72	0.11	0.46	0.85	[10] [11]
Level II	0.11	0.046	0.085	0.90	0.38	0.69	
Level III•IV	0.043	0.018	0.033	0.67	0.28	0.52	
Dental diagnostic medical X-ray examinations							
Level I	0.010	0.0042	0.0077				[12] [13]
Level II	0.0010	0.0004	0.0008				
Level III•IV	0.0003	0.0001	0.0002				
Diagnostic use of radiopharmaceuticals							
Level I	0.094	0.039	0.072	5.7	2.4	4.4	[14] [15]
Level II	0.0075	0.0031	0.0058	20.0	8.3	15	
Level III•IV	0.0076	0.0032	0.0058	27.0	11	21	
Total							
Level I	1.0	0.42	0.77				
Level II	0.12	0.050	0.092				
Level III•IV	0.051	0.021	0.039				
Therapeutic use of radiation							
Level I	0.73	0.30	0.56	310	129	238	[16] [17]
Level II	0.18	0.075	0.14	300	125	231	
Level III	0.031	0.013	0.024	310	129	238	
Level IV	0.015	0.0063	0.012	300	125	231	
Therapeutic use of radiopharmaceuticals							
Level I	0.0044	0.0018	0.0034	44	18	34	[18] [19]
Level II	0.00091	0.0004	0.0007	45	19	35	
Level III	0.00094	0.0004	0.0007	47	20	36	
Level IV	0.00043	0.0002	0.0003	40	17	31	
Total							
Level I	0.73	0.30	0.56				
Level II	0.18	0.075	0.14				
Level III	0.032	0.013	0.025				
Level IV	0.015	0.0063	0.0115				

Data from UNSCEAR 1993 and 1988. Most advanced countries are in health care level I.
a : Annual effective dose equivalent. b : Radiation environment dose. c : Relative exposure to radon.

Table 3. Examples of various exposures in daily life for the public in Japan.

	EDE [mSv]	RED [red]
Cancer therapy (only affected part)	60000	25000
Annual dose limit for occupational people	50	21
Average diagnostic x-ray examination of chest (by computer tomography)	6.9	2.9
Diagnostic x-ray examination of stomach per examination	2.8	1.17
Annual effective dose equivalent	2.4	1.0
Space trip between the earth and the moon (195 hours, total)	1.8	0.75
Annual dose limit for the public (except medical exposure)	1.0	0.42
Flight between Tokyo and New York (19 hours × 2)	0.19	0.079
Annual dose for the respiratory organs by smoking (20 cigarettes/day)	0.18	0.075
Collective diagnostic x-ray examination of chest per examnation	0.06	0.025
Annual dose objective on a boundary of nuclear reactors	0.05	0.021
Annual dose by watching TV at 10 cm from monitor surface (2 hours/day)	0.037	0.015

Most data are taken from UNSCEAR 1993.
EDE : Effective dose equivalent. RED : Radiation environment dose.

3. Regulation standards and lethal dose including toxic substance

In regulation standards for safe level exposure devised by MAEDE, the ratio is regarded as a measure of biological damage to the human body. The ratio coincides with the value in red, and is considered to give a simple guide line. The ratio R_1 can give 0.42 (1 mSv / 2.4 mSv). Another ratio of lethal dose to MAEDE also gives an interesting figure. This ratio R_2 is 1.7×10^3 (4 Sv / 2.4 mSv). Here, we adopted 4 Sv as a fifty percent lethal dose. As for lethal doses, radiation exposures are very restricted by severe limitation levels. Use of red automatically gives understanding of these situations.

For another completely different example, we selected toxicoids such as dioxin. Dioxin is one of the most toxic chemicals known in the world[20]. Dioxin is formed by burning chlorine-based chemical compounds with hydrocarbons. In general, the toxicity of other dioxins and chemicals such as PCBs that act like dioxin are measured in relation to toxic equivalents (TEQ) for representative toxic compound 2,3,7,8-tetrachlorodibenzo-p-dioxin (2,3,7,8-TCDD). According to the data of tolerable daily intake (TDI) of dioxins from the US Environmental Protection Agency (EPA) report, tolerable yearly intake of dioxins for reference human is 220 ng-TEQ/year. The quantity of human intake per year is scaled at 660 ng-TEQ/year [21]. The ratio for dioxin R_1 is 0.33 (220 ng / 660 ng). Since we estimated 4.2×10^6 ng as a fifty percent lethal dose of 2,3,7,8-TCDD for humans on the basis of the data for monkeys, the ratio for dioxin R_2 is 6.4×10^3 (4.2×10^6 ng / 660 ng). The comparison of radiation exposures and the toxic substance dioxin shows that the regulation for radiation exposure is as severe as that of toxic substances. Table 4 lists both radiation and dioxin. These approaches by the use of red give a simple and intuitive understanding for ordinary people and are applicable to other toxicoids.

Table 4. Comparison of exposures of radiation and dioxin.

	Radiation exposure	Dioxin exposure
Ratio for limitation R_1	0.42	0.33
	(1 mSv / 2.4 mSv)	(220 ng [a] / 660 ng [b])
Ratio for lethal dose R_2	1.7×10^3	6.4×10^3
	(4 Sv / 2.4 mSv)	(4.2×10^6 ng [c] / 660 ng [b])

a : Data is estimated from tolerable daily intake of dioxin (10 pg / kg-weight /day), assuming a person of 60 kg.

b : Concentration of dioxin in fat is 54.6 % (for Japannese), and fat is assumed to constitute 20% of human weight.

c : Data is estimated from 50% lethal dose for monkeys, taking human weight into consideration.

4. Summary

A new radiation dose unit was proposed which is based on the annual dose equivalent. The dose was designated as radiation environment dose (red). Red is considered to be suited for intuitively representing the degree of radiation exposure. Radiation exposures from natural sources and medical uses are tabled by means of red. The comparison of radiation exposures and the toxic substance dioxin shows that the regulation for radiation exposure is as severe as that of toxic substances. The use of red is thought to help ordinary people to simply and intuitively understand radiation exposures.

Acknowledgment

We would like to thank to Prof. J. Nagayama of Kyushu University, School of Health Sciences for useful comments and suggestions on the latest dioxin data.

References

[1] United Nations Scientific Committee on the Effects of Atomic Radiation, UNSCEAR 1993 Report, New York, United Nations, 1993.

[2] International Commission on Radiological Protection, ICRP Publication **38**, Pergamon Press, Oxford, 1983.

[3] UNSCEAR 1988 Report, United Nations Scientific Committee on the Effects of Atomic Radiation, New York, United Nations, 1988.

[4] Staranden, E., Population doses from environmental gamma radiation in Norway, Health Phys., **33**, 319-323, 1977.

[5] National Environmental Protection Agency, Natural survey of environmental radioactivity in China (1983-1990), 90-S315-206, The people's Republic of China, 1990.

[6] Abe, S., K. Fujitaka and K. Fujimoto, Natural radiation in Japan, 1034-1048, Natural Radiation Environment III, CONF-780422, 1980.

[7] Oakley, D.T., Natural radiation exposure in the United States, USEPA ORP/SID 72-1, 1972.

[8] Kauranen, P. and J.K. Miettinen, Po-210 and Pb-210 in the arctic food chain and the natural radiation exposure of Lapps, Health Phys., **16**, 287-295, 1969.

[9] Salonen, L., Natural radionuclides in ground water in Finland, Radiat. Prot. Dosim., **24**, 163-166, 1988.

[10] Maruyama, T., Y. Kumamoto, K. Nishikawa et al., Frequency of x-ray diagnostic examinations in Japan, 1986. Radioisotopes, **42**, 13-119, 1993.

[11] Nishikawa, K., T. Maruyama, M. Takayama et al., Determinations of organ doses and effective dose equivalents from computed tomographic examinatica, Br. J. Radiol., **64**, 20-28, 1991.

[12] Stenstrom, B., K. Bergman, P. G. Blomgren et al., Collective dose to the Swedish population from intraoral radiography, Swed. Dwet. J., **12**, 17-25, 1988.

[13] Benedittini, M., C. Maccia, C. Lefaure et al., Doses to patients from dental radiology in France, Health Phys., **56**, 903-910, 1989.

[14] Husak, V. and H. Rickova, Activities of administered radiopharmaceuticals and population dose from nuclear medicine in Czechoslovakia, Cesk. Radiol., **44**, 333-339, 1990.

[15] Zhang, L., J. Zhang, D. Jia et al., The frequencies and doses of medical exposure in China, Proc. CAMS and PUMC **5**, 200-206, 1990.

[16] International Commission on Radiological Protection, Protection of the patient in radiation therapy, ICRP publication **44**, Pergamon Press, Oxford, 1985.

[17] International Commission on Radiological Protection, Recommendations of the ICRP, ICRP publication **60**, Pergamon Press, Oxford, 1991.

[18] International Commission on Radiological Protection, Radiation dose to patients from radiopharmaceuticals, ICRP publication **53**, Pergamon Press, Oxford, 1987.

[19] Williams, L.E., J.Y.C. Wong, D.O. Findley et al., Measurement and estimation of organ bremsstrahlung radiation dose, J. Nucl. Med, **30**, 1373-1377, 1989.

[20] for example, the 17th international symposium on chlorinated dioxins and related compounds, Indianapolis, Indiana, USA, August 25 - 29, 1997.

[21] Hirakawa, H. T. Ida, J. Nagayama et al., Organohalogen Compounds, **10**, 93, 1992.

Size distributions of radon progeny and their influence on lung dose

Thomas Haninger

GSF-National Research Center for Environment and Health,
Institute of Radiation Protection, D-85764 Neuherberg, Germany.

Abstract

Measurements of the size distributions of short-lived radon progeny and the radon equilibrium factor F were carried out in the living room of a dwelling, under different aerosol conditions. With the PC-program RADEP of A. Birchall, the dose conversion coefficient DCC_{RADEP} (effective dose per unit exposure to the equilibrium equivalent radon concentration C_{eq}) was calculated. F and DCC_{RADEP} showed great variations, but they were inversely correlated (correlation coefficient: -0.83). Therefore the effective dose per unit of ^{222}Rn-concentration varied only within a factor of two. This result supports radiation protection measures which are based on long-term measurements of ^{222}Rn-concentrations with passive detectors.

1. Introduction

In its publication 65, "Protection against radon-222 at home and at work" [1], the International Commission on Radiological Protection (ICRP) has introduced a **dose conversion convention** to calculate the annual effective dose, which is caused by the inhalation of radon and its decay products:

$$E\,[\mathrm{nSv\ a^{-1}}] = C_{Rn} \cdot F \cdot t \cdot DCC_{ICRP} \tag{1}$$

C_{Rn} is the ^{222}Rn-concentration in Bq m^{-3}, F the equilibrium factor, t the occupancy time in h a^{-1} and DCC_{ICRP} the dose conversion coefficient (effective dose per unit exposure to the equilibrium equivalent radon concentration C_{eq}) in nSv Bq^{-1} h^{-1} m^{3}. Equation 1 includes the ^{222}Rn-concentration C_{Rn} as the measurement parameter. This is of great practical interest, because C_{Rn} can easily be measured with passive detectors over longer periods of time. If the reference values for homes, $F = 0.4$, $t = 6000$ h a^{-1} and $DCC_{ICRP} = 6.1$ nSv Bq^{-1} h^{-1} m^{3} are inserted in Equation 1, it follows that a radon concentration C_{Rn} of 600 Bq m^{-3} is equivalent to an annual effective dose of 10 mSv of the inhabitants.

The dose conversion coefficent DCC_{ICRP} was deduced from epidemiology studies on uranium miners. However, due to different environmental conditions, the equilibrium factor F and the size distributions of the short-lived radon decay products, and hence the DCC, can vary substantially. These variations cannot be taken into account, if the effective dose is calculated according to the dose conversion convention. It was the aim of this study to investigate the influence of different aerosol conditions on the effective dose per unit of ^{222}Rn-concentration. A series of measurements was carried out in the living room of a dwelling with a special aerosol spectrometer, under conditions of low ventilation. Room air was influenced systematically by several processes: air cleaning, smoking a cigarette or using an oil lamp.

2. Materials and Methods

The aerosol spectrometer was constructed to measure the size distributions of the short-lived radon decay products in the range of 0.5 to 10 000 nm. It consists of a wire screen diffusion battery, a low pressure impactor (type Berner LPI 25/0.018) and a detector system with 12 wide area proportional detectors. The diffusion battery can measure the size distribution of the unattached fraction and has three wire screens with different mesh sizes and a back-up filter. The 50%-cutoff diameters of the screens were calculated using the Cheng-Keating-Kanapilly equation [2] and are 1.5, 2.4 and 3.2 nm. For measurements of the aerosol-attached fraction, eight impactor stages with lower 50%-cutoff diameters between 18 and 2 000 nm [3] were used. After a sampling period of 10 minutes, the diffusion battery and the impactor were unmounted and the gross alpha activities on the sampling media were recorded in 40 one-minute intervals. These data were analysed by a least-squares procedure to determine the equilibrium equivalent radon concentration C_{eq}. For the reconstruction of the size distribution a sum of up to five lognormal distributions was fitted to the data with a non-linear algorithm. Additionally, the ^{222}Rn-concentration was measured with an ionisation chamber (AlphaGUARD, Genitron Instruments GmbH) to determine the equilibrium factor F.

The measured size distributions of C_{eq} were used as input parameters for the PC-program RADEP, which was developed by A. Birchall at NRPB [4] to calculate the dose conversion coefficient and is based on the respiratory tract model of the ICRP [5]. A breathing rate of 0.78 m^3 h^{-1} was regarded as representative for an exposure at home. If RADEP is applied to the exposure conditions of a uranium miner, the calculations result in a DCC_{RADEP} of 21 nSv Bq^{-1} h^{-1} m^3, which is approx. three times higher than the DCC_{ICRP}, which is recommended by the ICRP [4]. To overcome this discrepancy a normalization factor of 6.1/21 = 0.29 was used to adjust the lung model to the findings of the epidemiology studies. For a comparison of different environmental conditions a factor K_{Rn} was defined:

$$K_{Rn} = \frac{DCC_{RADEP} \cdot 0.29}{6.1} \cdot \frac{F}{0.4} \quad (2)$$

If K_{Rn} is greater than one, this is an indicator that the effective dose may be underestimated by the dose conversion convention of the ICRP [1].

3. Results and Discussion

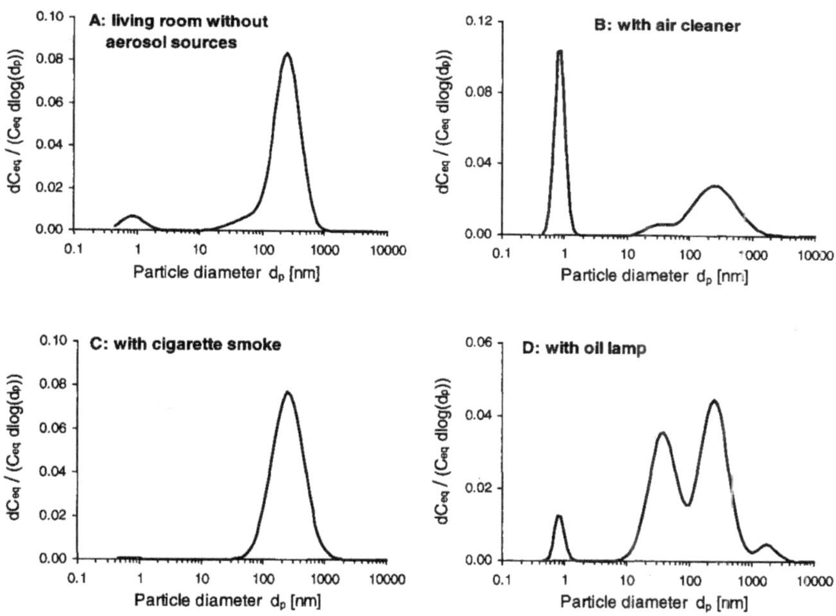

Fig. 1: Size distributions of the short-lived radon decay products under different aerosol conditions.

Some results of the measurements are shown in Fig. 1 and summarized in Table 1. The size distributions were normalized to an area of one beneath the curves. Fig. 1A shows the size distribution of an aged indoor aerosol: 6% percent of the radon progeny were unattached and had a median diameter of 0.85 nm. The main part of the radon decay products were attached to aerosol particles in the accumulation mode with a median diameter of 250 nm. Air cleaning reduced the activity concentrations of the radon decay products, because aerosol particles were removed from the air. This process however, did not influence the ^{222}Rn-concentration C_{Rn}. Hence, the equilibrium factor F decreased. Simultaneously the unattached fraction f_p increased due to the low aerosol particle concentration (Fig. 1B). Smoking a cigarette produced a high aerosol concentration and this resulted in an unattached fraction f_p of nearly zero (Fig. 1C). The equilibrium factor F increased, because aerosol-attached radon progeny have a lower plate-out rate on room surfaces than unattached radon decay products. A burning oil lamp produced ultrafine particles. In Fig. 1D an additional mode with a median diameter of 40 nm can be seen.

Table 1: Summary of the measurement parameters.

Meas.-no.	f_p (%) (d_p < 10 nm)	F	DCC_{RADEP} (nSv Bq^{-1} h^{-1} m^3)	K_{Rn}
1A	6	0.54	18.1	1.16
1B	47	0.11	48.3	0.61
1C	<1	0.71	12.8	1.09
1D	5	0.36	29.4	1.26

The dose conversion factors DCC_{RADEP} and the equilibrium factors F varied substantially, whereas the factors K_{Rn} always had values close to unity. In 40 measurements K_{Rn}-factors in the range of 0.6 to 1.9 were found. This result can be explained by the correlation between F and DCC_{RADEP}, which is shown in Fig. 2. From the data, a correlation coefficient of -0.83 was calculated. It can be deduced from Equation 2, that this negative correlation prevented great variations in the K_{Rn}-factors, because a deviation of F from its reference value was partly counterbalanced by the simultaneous deviation of DCC_{RADEP}.

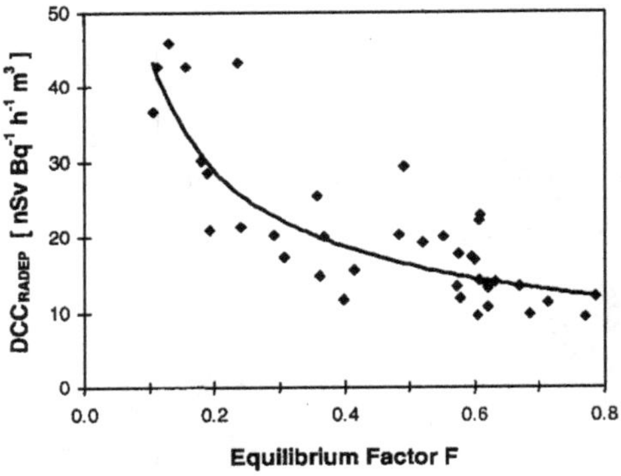

Fig. 2: Correlation between the equilibrium factor F and the dose conversion coefficient DCC_{RADEP}.

4. Conclusions

The findings of this study indicate that the dose conversion convention of the ICRP [1] can be used to estimate the effective dose per unit of ^{222}Rn-concentration within a factor of two. This result is of great importance, because radiation protection measures against radon at home are often based on long-term measurements of ^{222}Rn-concentrations with passive detectors.

5. References

[1] The International Commission on Radiological Protection, Protection against radon-222 at home and at work, ICRP Publication 65, Pergamon Press, Oxford, 1993.

[2] Cheng, Y. S., J. A. Keating and G. M. Kanapilly, Theory and calibration of a screen type diffusion battery, J. Aer. Sci. 11, 549-556, 1980.

[3] Hauke GmbH & Co. KG, impactor data sheet, model LPI 25/0.018/2.0, Gmunden, Austria, 1994.

[4] Birchall, A. and A. C. James, Uncertainty analysis of the effective dose per unit exposure from radon progeny and implications for ICRP risk weighting factors, Rad. Prot. Dos. 53, 133-140, 1994.

[5] The International Commission on Radiological Protection, Human respiratory tract model for radiological protection, ICRP Publication 66, Pergamon Press, Oxford, 1994.

Author Index

www.ingramcontent.com/pod-product-compliance
Lightning Source LLC
LaVergne TN
LVHW010536300926
849162LV00002B/157